Composition and Properties of Drilling and Completion Fluids

Composition and Properties of Drilling and Completion Fluids

Seventh Edition

Ryen Caenn
Consultant, Laguna Beach, CA, USA

H.C.H. Darley[†]
Former researcher for Shell Development Company and a consultant, Houston, TX, USA

George R. Gray[†]
Former drilling fluids technologist and sales Manager for NL Baroid, Houston, TX, USA

ELSEVIER

AMSTERDAM • BOSTON • HEIDELBERG • LONDON
NEW YORK • OXFORD • PARIS • SAN DIEGO
SAN FRANCISCO • SINGAPORE • SYDNEY • TOKYO
Gulf Professional Publishing is an imprint of Elsevier

Gulf Professional Publishing is an imprint of Elsevier
50 Hampshire Street, 5th Floor, Cambridge, MA 02139, United States
The Boulevard, Langford Lane, Kidlington, Oxford, OX5 1GB, United Kingdom

Copyright © 2017 Elsevier Inc. All rights reserved.

No part of this publication may be reproduced or transmitted in any form or by any means, electronic or mechanical, including photocopying, recording, or any information storage and retrieval system, without permission in writing from the publisher. Details on how to seek permission, further information about the Publisher's permissions policies and our arrangements with organizations such as the Copyright Clearance Center and the Copyright Licensing Agency, can be found at our website: www.elsevier.com/permissions.

This book and the individual contributions contained in it are protected under copyright by the Publisher (other than as may be noted herein).

Notices
Knowledge and best practice in this field are constantly changing. As new research and experience broaden our understanding, changes in research methods, professional practices, or medical treatment may become necessary.

Practitioners and researchers must always rely on their own experience and knowledge in evaluating and using any information, methods, compounds, or experiments described herein. In using such information or methods they should be mindful of their own safety and the safety of others, including parties for whom they have a professional responsibility.

To the fullest extent of the law, neither the Publisher nor the authors, contributors, or editors, assume any liability for any injury and/or damage to persons or property as a matter of products liability, negligence or otherwise, or from any use or operation of any methods, products, instructions, or ideas contained in the material herein.

British Library Cataloguing-in-Publication Data
A catalogue record for this book is available from the British Library

Library of Congress Cataloging-in-Publication Data
A catalog record for this book is available from the Library of Congress

ISBN: 978-0-12-804751-4

For Information on all Gulf Professional Publishing
visit our website at https://www.elsevier.com

Working together
to grow libraries in
developing countries

www.elsevier.com • www.bookaid.org

Publisher: Joe Hayton
Senior Acquisition Editor: Katie Hammon
Senior Editorial Project Manager: Kattie Washington
Production Project Manager: Kiruthika Govindaraju
Cover Designer: Mark Rogers

Typeset by MPS Limited, Chennai, India

Transferred to Digital Printing in 2017

Dedication

The dedication to the 6th edition of this book was to the men who wrote the first five editions—Walter Rogers, Doc Gray, and Slim Darley. They were among the early inductees to the American Association of Drilling Engineers Fluids Hall of Fame (http://www.aade.org/fluids-hall-of-fame/). The purpose of the Fluids Hall of Fame as stated by the AADE is:

> *To recognize key individuals who have contributed to the advancement of our understanding and knowledge of drilling and completion fluids — To share their accomplishments with the rest of the drilling industry in order that they are not forgotten.*

This book is dedicated to the remarkable men and women who have advanced our knowledge and use of drilling fluids, completion fluids, fracturing fluids, and cements. They are represented by the inductees into the Fluids Hall of Fame. Some of the inductees gathered for this photo in 2010.

Back row, left to right: Leon Robinson, Jay Simpson, Max Annis, Don Weintritt, Martin Chenevert, Bill Rehm, Jack Estes, Bob Garrett, George Savins, Charles Perricone, Dorothy Carney (For Leroy Carney), Tommy Mondshine, Billy Chesser.
In front: George Ormsby

Contents

Preface	xvii

1. Introduction to Drilling Fluids

Functions of Drilling Fluids	2
Principle Functions	2
Associated Functions	3
Limitations	3
Composition of Drilling Fluids	3
Weight Materials	4
Drilling Fluid Systems	5
Types of Drilling Fluid Systems	9
Properties of Drilling Fluids	13
Density	13
Flow Properties	15
Filtration Properties	19
Drilling Fluid Selection	22
Location	22
Mud-Making Shales	25
Geopressured Formations	26
High Temperature	26
Hole Instability	27
Fast Drilling Fluids	28
Rock Salt	28
High Angle Holes	29
Formation Evaluation	29
Productivity Impairment	29
Mud Handling Equipment	30
References	32

2. Introduction to Completion Fluids

Completion Fluid Selection	36
Data Collection and Planning	36
Corrosion Resistant Alloys	37
Safety and the Environment	37
Wellbore Cleanout	38
Composition of Completion Fluids	39
Clear Brine Fluids	39

	Monovalent Inorganic Brines	39
	Divalent Inorganic Brines	42
	Monovalent Organic Brines	44
	Brine Test Procedures	44
	Density	44
	Thermodynamic Crystallization Temperature	48
	Brine Clarity	49
	Contaminants	51
	References	52
3.	**Evaluating Drilling Fluid Performance**	
	Introduction	55
	LAB Sample Preparation	56
	Properties Measured	57
	Density	57
	Viscosity	60
	Gel Strength	64
	API Fluid Loss	65
	Determination of Gas, Oil, and Solids Content	71
	Gas Content	71
	Oil and Solids Content	71
	Do Not Use the API Retort with Formate Fluids	72
	The API Sand Test	72
	Sieve Tests	73
	Methylene Blue Adsorption	73
	Electrical Properties	74
	Stability of Water-in-Oil Emulsions	74
	Resistivity of Water Muds	75
	Hydrogen Ion Concentration (pH)	76
	Lubricity	76
	Differential-Pressure Sticking Test	77
	Corrosion Tests	79
	Flocculants	82
	Foams and Foaming Agents	82
	Rheological Properties of Foams	82
	Evaluation of Foaming Agents	82
	Aniline Point	84
	Chemical Analysis	84
	Chlorides	84
	Alkalinity and Lime Content	85
	Total Hardness: Calcium and Magnesium, Calcium, and Calcium Sulfate	85
	Sulfides	86
	Potassium	86
	Evaluation of Drilling Fluid Materials	86
	General Principles	86
	Aging at High Temperature	87
	References	89

4. Clay Mineralogy and the Colloid Chemistry of Drilling Fluids

Introduction	93
Characteristics of Colloidal Systems	93
Clay Mineralogy	96
The Smectites	100
Illites	102
Kaolinite	105
Chlorites	105
Mixed-Layer Clays	106
Attapulgite	106
Origin and Occurrence of Clay Minerals	107
Ion Exchange	108
Clay Swelling Mechanisms	109
The Electrostatic Double Layer	115
Particle Association	117
Flocculation and Deflocculation	117
Aggregation and Dispersion	121
The Mechanism of Gelation	125
Polymers	127
References	132

5. Water-Dispersible Polymers

Introduction	135
Polymer Classification	135
Polysaccharides	136
Types of Polymers	137
Guar Gum	139
Starch	141
Fermentation Biopolymers	142
Cellulosics	147
Synthetics	147
References	150

6. The Rheology of Drilling Fluids

Laminar Flow Regime	152
Laminar Flow of Newtonian Fluids	152
The Bingham Plastic Flow Model	155
The Concentric Cylinder Rotary Viscometer	159
Behavior of Drilling Fluids at Low Shear Rates	163
The Effect of Thixotropy on Drilling Muds	166
Pseudoplastic Fluids	175
The Generalized Power Law	179
Turbulent Flow Regime	184
Turbulent Flow of Newtonian Liquids	184
Turbulent Flow of Non-Newtonian Fluids	185

	The Onset of Turbulence	191
	Friction Reducers	191
	Influence of Temperature and Pressure on the Rheology of Drilling Fluids	193
	Application of Flow Equations to Conditions in the Drilling Well	203
	Flow Conditions in the Well	204
	Hydraulic Calculations Made at the Well	208
	Determining Pressure Gradient in the Drill Pipe	210
	Determining Pressure Gradient in the Annulus	213
	Rheological Properties Required for Optimum Performance	215
	Pumping Capacities	216
	Effect of Mud Properties on Bit Penetration Rate	216
	Hole Cleaning	216
	Optimum Annular Velocity	222
	Optimum Rheological Properties for Hole Cleaning	224
	Inclined Holes	226
	Release of Cuttings and Entrained Gas at the Surface	228
	Transient Borehole Pressures	228
	Practical Drilling Fluid Rheology	236
	Water-Based Muds	236
	Nonaqueous Drilling Fluids	236
	Annular Effective Viscosity	237
	Notation	239
	References	240
7.	**The Filtration Properties of Drilling Fluids**	
	Introduction	245
	Static Filtration	246
	The Theory of Static Filtration	246
	Relationship Between Filtrate Volume and Time	246
	Relationship Between Pressure and Filtrate Volume	248
	Relationship Between Temperature and Filtrate Volume	249
	The Filter Cake	251
	Cake Thickness	251
	Cake Thickness Test	253
	The Permeability of the Filter Cake	255
	The Effect of Particle Size and Shape on Cake Permeability	256
	Polymer Fluid Loss Agents	258
	Nonaqueous Fluid Loss	258
	Effect of Flocculation and Aggregation on Cake Permeability	260
	The Bridging Process	260
	Dynamic Filtration	264
	Filtration in the Borehole	270
	The Filtration Cycle in a Drilling Well	270
	Filtration Beneath the Bit	271
	Evaluation of Downhole Filtration Rates	275
	Notation	281
	References	281

Contents xi

8. The Surface Chemistry of Drilling Fluids

Introduction	285
Surface Tension	285
Measurement of Surface Tension	285
Wettability	287
Capillarity and Wettability	290
Surface Free Energy	290
Adhesion	290
Surfactants	291
Emulsions	293
Oil-Wetting Agents	297
Foams	300
Foams and Mists	300
Defoamers	303
Notation	306
References	306

9. Wellbore Stability

Introduction	309
The Mechanics of Borehole Stability	309
Brief Review of the Geology and Geophysics of Sedimentary Basins	309
The Geostatic Gradient	310
Hydrostatic Pore Pressure Gradients	311
Abnormal or Geopressured Gradients	312
The Behavior of Rocks Under Stress	315
The Subsurface Stress Field	317
Stresses Around a Borehole	319
The Influence of Hydraulic Pressure Gradient on Hole Stability	325
Occurrence of Plastic Yielding in the Field	330
Brittle-Plastic Yielding	334
Hole Enlargement	337
Formations with No Cohesive Strength	337
Coal Seams	338
Hole Instability Caused by Interaction Between the Drilling Fluid and Shale Formations	338
Adsorption and Desorption of Clays and Shales	338
Hydration of the Borehole with Water-Based Drilling Fluids	342
Brittle Shales	343
Control of Borehole Hydration	344
Nonaqueous Fluids and Wellbore Stability	345
Cation Exchange Reactions	347
Diffusion Osmosis and Methyl Glucoside	352
Selection of Mud Type for Maintaining Borehole Stability	353
Laboratory Tests	354
Properties for Wellbore Stability by Water-Based Fluids	361
Notation	362
References	363

10. Drilling Problems Related to Drilling Fluids

Introduction	367
Drill String Torque and Drag	368
Lubricants	369
Differential Sticking of the Drill String	372
Mechanism of Differential Sticking	372
Prevention of Differential Sticking	379
Freeing Stuck Pipe	381
Slow Drilling Rate	383
Laboratory Drilling Tests	383
Static Chip Hold-Down Pressure	383
Dynamic Chip Hold-Down Pressure	389
Bottomhole Balling	391
Bit Balling	392
The Effect of Mud Properties on Drilling Rate	394
Loss of Circulation	401
Induced Fractures	402
Natural Open Fractures	411
Openings With Structural Strength	411
Lost Circulation Materials	413
Regaining Circulation	416
Loss Into Structural Voids	417
Losses Induced by Marginal Pressures	418
Losses Caused by the Hydrostatic Pressure of the Mud Column Exceeding the Fracture Pressure	419
Natural Open Fractures	421
Wellbore Strengthening	421
High Temperatures	423
Geothermal Gradients	423
High-Temperature Drilling Fluids	427
Corrosion of Drill Pipe	437
Electrochemical Reactions	437
Stress Cracking	440
Control of Corrosion	443
Notation	454
References	455

11. Completion, Workover, Packer, and Reservoir Drilling Fluids

Introduction	461
Expense Versus Value	462
The Skin Effect	462
Capillary Phenomena	464
Permeability Impairment by Indigenous Clays	469
Mechanism of Impairment by Indigenous Clays	470
Effect of Base Exchange Reactions on Clay Blocking	475

The Effect of pH	478
Permeability Impairment by Particles From the Drilling Mud	481
Occurrence of Mud Particle Damage in the Field	484
Workover Wells	487
Gravel Pack Operations	488
Completion of Water Injection Wells	489
Prevention of Formation Damage	489
Completion Workover Procedures	490
Selection of Completion and Workover Fluids	493
Deepwater Completion Fluid Selection	493
Solids-Free Brines	494
Viscous Brines	495
Water-Base Fluids Containing Oil-Soluble Organic Particles	497
Acid-Soluble and Biodegradable Systems	498
Fluids With Water-Soluble Solids	499
An Oil-in-Water Emulsion for Gun Perforating	499
Nonaqueous Fluids	499
Tests for Potential Formation Damage by Completion Fluids	500
Packer Fluids and Casing Packs	503
Functions and Requirements	503
Aqueous Packer Fluids	503
Low-Solid Packer Fluids	504
Clear Brines	504
Brine Properties	505
Oil-Base Packer Fluids and Casing Packs	505
Reservoir Drilling Fluids	506
Why a Special Reservoir Drilling Fluid?	506
Cost Versus Value	507
Reservoir Drilling Fluid Properties	507
Fluid Types	508
Fluid Design	508
Formate Brines	511
Formate Test Procedures	513
Calcium Carbonate	513
Drill Solids	515
References	515

12. Introduction to Fracturing Fluids

Introduction	521
Fracturing Fluid Types	522
Purpose of a Fracturing Fluid	522
Fracturing Fluids Composition	523
Basic Components and Additives	523
Selecting a Fracturing Fluid System	525
Water-Based	525
Additives	532
References	534

13. Drilling Fluid Components

Introduction	537
Summary of Additives	537
Weighting Agents	538
Barite (BaSO$_4$)	540
Iron Oxides—Fe$_2$O$_3$	541
Ilmenite (Fe-TiO$_2$)	541
Galena (PbO)	542
Calcium Carbonate—CaCO$_3$	542
Manganese Tetroxide	542
Hollow Glass Microspheres	543
Viscosity Modifiers	543
pH Responsive Thickeners	543
Mixed Metal Hydroxides/Silicates (MMH/MMS)	543
Nonaqueous Gelling Agent	544
Thinners	545
Lost Circulation Additives	546
Water Swellable Polymers	546
Anionic Association Polymer	547
Permanent Grouting	547
Lubricants	547
Polarized Graphite	548
Ellipsodial Glass Granules	549
Paraffins	549
Olefins	549
Phospholipids	549
Alcohols	549
Ethers and Esters	550
Ethers and Esters	550
Biodegradable Lubricants	551
Clay and Shale Stabilizers	551
Salts	552
Quaternary Ammonium Salts	552
Amine Salts of Maleic Imide	554
Potassium Formate	554
Saccharide Derivatives	555
Sulfonated Asphalt	555
Grafted Copolymers	556
Poly(oxyalkylene Amine)s	557
Anionic Polymers	557
Shale Encapsulator	558
Membrane Formation	558
Formation Damage Prevention	559
Surfactants	560
Emulsifiers	561
Invert Emulsions	561
Breakers	562

Drilling Fluid Systems	562
Aphrons	562
Low-Fluorescent Emulsifiers	565
Bacteria Control	565
Corrosion Inhibitors	566
Oxygen Scavenger	571
Hydrogen Sulfide Removal	571
Special Additives for Water-Based Drilling Muds	572
Improving the Thermal Stability	572
Dispersants	573
Synthetic Polymers	573
Maleic Anhydride Copolymers	573
Acrylics	574
Amine Sulfide Terminal Moieties	575
Polycarboxylates	575
Allyloxybenzenesulfonate	575
Sulfonated Isobutylene Maleic Anhydride Copolymer	575
Modified Natural Polymers	576
Special Additives for Nonaqueous Drilling Muds	577
Poly(ether)cyclicpolyols	577
Emulsifier for Deep Drilling	577
Esters and Acetals	577
Antisettling Properties	578
Glycosides	579
Wettability	579
Surfactants	580
Mud to Cement Conversion	580
References	580

14. Drilling and Drilling Fluids Waste Management

Introduction	597
Drilling Wastes	598
Minimizing Waste Problems	599
Waste Disposal Options	600
Slurry Fracture Injection	600
Offshore Waste Disposal for NADFs	603
Group I NABFs (High Aromatic Content)	603
Group II NABFs (Medium Aromatic Content)	604
Group III NABFs (Low to Negligible Aromatic Content)	604
Offshore Discharge	608
Onshore Disposal	608
Evaluation of Fate and Effects of Drill Cuttings Discharge	612
Initial Seabed Deposition	612
Physical Persistence	614
Benthic Impacts and Recovery	614
Biodegradation and Organic Enrichment	615
Chemical Toxicity and Bioaccumulation	616

Recovery	616
Laboratory Studies	617
Characterization of NADF Biodegradability	618
Standard Laboratory Biodegradation Tests	619
ISO 11734 Modified for NADF Biodegradation	619
The SOAEFD Solid Phase Test	621
Simulated Seabed Studies	623
Aquatic and Sediment Toxicity of Drilling Fluids	624
Aquatic Toxicity and Regulations	625
Characterization of NABF Bioaccumulation	627
OGP Document Conclusions	629
Waste Reduction and Recycling	630
Prototype Small Footprint Drilling Rig	631
Disappearing Roads	631
NO_x Air Emissions Studies	632
Drilling Waste Management Website	632
New Product R&D	633
References	633
US Waste Regulation Bibliography	635
Appendix A: Conversion Factors	637
Appendix B: Abbreviations	641
Appendix C: The Development of Drilling Fluids Technology	643
Author Index	695
Subject Index	707

Preface

Readers of the 7th edition will find much that is in common with the 4th through the 6th editions, but updated with modern technology and terminology. Chapter 4, Clay Mineralogy and the Colloid Chemistry of Drilling Fluids, Chapter 6, The Rheology of Drilling Fluids, Chapter 7, The Filtration Properties of Drilling Fluids, Chapter 8, The Surface Chemistry of Drilling Fluids, and Chapter 14, Drilling and Drilling Fluids Waste Management remain mostly as written in the 6th edition. All other Chapters have been completely rewritten and two new chapters are included. The new chapters are on Completion Fluids and Fracturing Fluids.

Since the title of this book is about drilling and completion fluids, it was felt that completion brines needed a complete chapter to put them on the same level as drilling fluids. The first four editions merely titled "Composition and Properties of Oilwell Drilling Fluids." Even though the title was changed in the 5th edition (1988) to drilling and completion fluids, there was only one chapter, Completion, Workover, and Packer Fluids, that covered completions. That chapter was updated somewhat in the 6th edition (2011) and was called Completion, Reservoir Drilling, Workover, and Packer Fluids.

This edition has two complete chapters on completion fluids: Introduction to Completion Fluids and a chapter called Fracturing Fluids. It's obvious that the technology surrounding those two topics has expanded greatly since the 5th edition and should have been included in the 6th edition. However, catching up on technology changes between 1988 and 2011 meant more effort spent on updating each of the chapters.

The reader will find less emphasis on bentonite water-based muds. To be sure, clay-based systems are still prominent around the world, but brine-based systems and synthetic base fluid systems are rapidly becoming the drilling fluids of choice. Solids in general are being replaced to make more environmentally and operationally sound fluids. First, by improved solids control equipment and closed loop systems to remove drilled solids. Second, by eliminating solid weight material by the use of high-density brines.

A new chapter included in this addition is called Polymers. Colloidal polymers have been used in muds since the late 1930s, but the designer fluids of today are extending the usefulness of water/brine based fluids in drilling operations. These fluids are replacing nonaqueous fluids in many operations.

This book would not be possible without the contributions through the years of innumerable drilling and completion fluid specialists. Some of those individuals are mentioned in the dedication.

Walter Rogers said in the preface to the 1st edition, "...it is the author's earnest hope that the contents will be of value as a training and reference text to those interested in the subject. Above all it is hoped the data will aid in eliminating some of the haphazardness now followed in mud practices and in removing some of the mystery now surrounding the art."

That is my hope also. Given the internet's World Wide Web of haphazardness, perhaps this book can untangle some of those webs.

Ryen Caenn

Chapter 1

Introduction to Drilling Fluids

The successful completion of an oil well depends to a considerable extent on the properties of the drilling fluid. The cost of the drilling fluid itself is relatively small in comparison to the overall cost of drilling a well, but the choice of the right fluid and maintenance of its properties while drilling profoundly influence the total well costs. For example, the number of rig days required to drill to total depth depends on the rate of penetration of the bit, and on the avoidance of delays caused by caving shales, stuck drill pipe, loss of circulation, etc., all of which are influenced by the properties of the drilling fluid. In the case of some critical wells, such as deepwater operations, these excess costs can run into the millions of dollars. In addition, the drilling fluid affects formation evaluation and the subsequent productivity of the well. The fluid also needs to be environmentally benign and generate minimal waste.

It follows that the selection of a suitable drilling fluid and the day-to-day control of its properties are the concern not only of the mud specialist, but also of the drilling supervisor, the drilling foreman, and drilling, logging, and production engineers, as well as other service company specialists. Drilling and production personnel do not need a detailed knowledge of drilling fluids, but they should understand the basic principles governing their behavior, and the relation of these principles to drilling and production performance. The object of this chapter is to provide this knowledge as simply and briefly as possible, and to explain the technical terms so that the information provided by the mud specialist may be comprehensible. Aspiring drilling fluid specialists who have no previous knowledge of drilling fluids should also read this chapter before going on to the more detailed coverage in the subsequent chapters.

Traditionally the common term for a field drilling fluids specialist has been "mud engineer." In practice, the drilling fluids specialist will likely not have an engineering degree. Some drilling fluids specialists have risen through the ranks, starting as a laborer on the rig working up to being a derrick man or driller and is then hired by a drilling fluid service company.

These individuals have unique experience in the practical application of drilling fluids and are very successful in their specific operational area.

The American Petroleum Institute (API) publishes documents relating to oilfield standards, including drilling fluids testing procedures. One document they publish, "Recommended Practice for Training and Qualification of Rig Site drilling Fluids Technologists and Drilling Fluids engineers, API RP 13L (2015)," sets the structure needed to train a drilling fluids specialist. This document gives the following definitions for drilling fluids specialists:

Drilling fluids engineer: A drilling fluids technologist with an engineering or sub-discipline appropriate degree from an accredited university.
Drilling fluids technologist: Individual with specialized knowledge of the application of drilling fluids during the drilling operation.
Drilling fluids technician: An individual skilled in the art of testing drilling fluids in the field or the laboratory
Senior drilling fluids technologist/engineer: An individual who by training and experience has advanced knowledge of drilling fluids, drilling fluids chemistry, and their varied applications.

FUNCTIONS OF DRILLING FLUIDS

Many requirements are placed on the drilling fluid. Historically, the first use of a drilling fluid, was to serve as a vehicle for the removal of cuttings from the bore hole and controlling downhole pressures to prevent blowouts. The fluid was primarily dirt and water and was called "mud." Today, however, the diverse drilling fluid applications and their complex chemistry make the assignment of specific functions difficult. At any point in the drilling operations, one or more of these functions may take precedence over the others.

Principle Functions

The principle functions of a drilling fluid are those that require continued observation and intervention by the drilling fluid specialist. Usually every day that a rig is drilling, a mud report form is issued. This document lists the current properties as tested by the fluids specialist. These numbers and vigilant observation of the drilling operations, allow the specialists to adjust the properties to optimize its functionality.

In rotary drilling, the principal functions performed by the drilling fluid are:

1. Prevent the inflow of fluids—oil, gas, or water—from permeable rocks penetrated and minimize causing fractures in the wellbore. These functions are controlled by monitoring the fluid's density (mud weight) and the equivalent circulating density (ECD). The ECD is a combination of the hydrostatic pressure and the added pressure needed to pump the fluid up the annulus of the wellbore.

2. Carry cuttings from beneath the bit, transport them up the annulus, and permit their separation at the surface. The fluids specialist must manipulate the viscosity profile to ensure good transport efficiency in the wellbore annulus and to help increase the efficiency of solids control equipment.
3. Suspend solids, particularly high specific gravity weight materials. The effective viscosity and gel strengths of the fluid are controlled to minimize settling under either static or dynamic flow conditions.
4. Form a thin, low-permeability filter cake that seals pores and other openings in permeable formations penetrated by the bit. This is done by monitoring the particle size distribution of the solids and maintaining the proper wellbore strengthening materials.
5. Maintain the stability of uncased sections of the borehole. The fluid specialist monitors the mud weight and mud/wellbore chemical reactivity to maintain the integrity of the wellbore until the next casing setting point is reached.

Associated Functions

These functions intrinsically arise from the use of a drilling fluid. The mud specialist does not necessarily routinely monitor the fluid for properties affecting these functions, or has no control over them.

1. Reduce friction between the drilling string and the sides of the hole.
2. Cool and clean the bit.
3. Assist in the collection and interpretation of information available from drill cuttings, cores, and electrical logs.

Limitations

In conjunction with the above functions, certain limitations—or negative requirements—are placed on the drill fluid. The fluid should:

1. Not injure drilling personnel nor be damaging or offensive to the environment.
2. Not require unusual or expensive methods to complete and produce the drilled hole.
3. Not interfere with the normal productivity of the fluid-bearing formation.
4. Not corrode or cause excessive wear of drilling equipment.

COMPOSITION OF DRILLING FLUIDS

All drilling fluids systems are composed of:

- Base fluids—Water, Nonaqueous, Pneumatic
- Solids—Active and Inactive (inert)
- Additives to maintain the properties of the system.

Additives in a drilling fluid are used to control one or more of the properties measured by the drilling fluids specialist. These properties can be classified as controlling:

- Mud weight—Specific Gravity. Density
- Viscosity—Thickening, Thinning, Rheology Modifiction
- Fluid loss—API Filtrate, Seepage, Lost Circulation, Wellbore Strengthening
- Chemical reactivity—Alkalinity, pH, Lubrication, Shale Stability, Clay Inhibition, Flocculation, Contamination Control, Interfacial/Surface Activity, Emulsification

Weight Materials

Weight materials are inert high-density minerals. The most common are barite and hematite. The specifications for the API weight materials are shown in Tables 1.1 and 1.2.

Many other high-density minerals have been used over the years. These are shown in Table 1.3.

Commercially available weighting agents are also available as micronized ($\sim 5 \mu m$) minerals—micronized barite and micronized manganese tetroxide. These were developed to combat dynamic weight material SAG and to minimize formation damage that sometimes occurs with regular grind barite (Al-Yami and Nasr-El-Din, 2007).

The quantity of weight material in the fluid is set by the required mud weight. Barite is the primary weight material used in drilling fluids. It is a relatively soft material that over time will grind down to very small sizes and can give fluctuating viscosities and gel strengths and can possibly damage the producing zone. Hematite is an abrasive material and is not normally used in water-based fluids.

TABLE 1.1 Barite Specifications—API Specification 13A (2010)

Requirement	Standard
Density	4.10, minimum
Water soluble alkaline earth metals, as calcium	250 mg/kg, maximum
Residue greater than 75 μm	Maximum mass fraction, 3.0%
Particles less than 6 μm in equivalent spherical diameter	Maximum mass fraction, 30.0%

TABLE 1.2 Hematite Specifications—API Specification 13A (2010)

Requirement	Standard
Density	5.05, minimum
Water soluble alkaline earth metals, as calcium	100 mg/kg
Residue greater than 75 μm	Maximum mass fraction, 1.5%
Residue greater than 45 μm	Maximum mass fraction, 15%
Particles less than 6 μm in equivalent spherical diameter	Maximum mass fraction, 15.0%

TABLE 1.3 Densities of Common Drilling Fluids Weighting Agents

Material	Principal Component	Specific Gravity	Hardness (Moh's Scale)
Galena	PbS	7.4–7.7	2.5–2.7
Hematite	Fe_2O_3	4.9–5.3	5.5–6.5
Magnetite	Fe_3O_4	5.0–5.2	5.5–6.5
Iron Oxide (manufactured)	Fe_2O_3	4.7	–
Iilmenite	$FeO.TiO_2$	4.5–5.1	5.0–6.0
Barite	$BaSO_4$	4.2–4.5	2.5–3.5
Siderite	$FeCO_3$	3.7–3.9	3.5–4.0
Celesite	$SrSO_4$	3.7–3.9	3.0–3.5
Dolomite	$CaCO_3.MgCO_3$	2.8–2.9	3.5–4.0
Calcite	$CaCO_3$	2.6–2.8	3.0

Drilling Fluid Systems

Fluid systems are classified according to their base fluid (Fig. 1.1):

- *Water-Based Muds (WBM)*. Solid particles are suspended in water or brine. Oil may be emulsified in the water, in which case water is termed the *continuous phase*.
- *Nonaqueous-Based Drilling Fluids (NADF)*. Solid particles are suspended in oil. Brine water or another low-activity liquid is emulsified in the oil, i.e., oil is the *continuous phase*.

6 Composition and Properties of Drilling and Completion Fluids

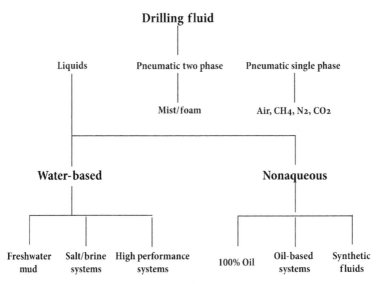

FIGURE 1.1 Drilling fluid systems by Base Fluid.

- *Pneumatic Systems.* Drill cuttings are removed by a high-velocity stream of air, natural gas, nitrogen, carbon dioxide, or some other fluid that is injected in a gaseous phase. When minor inflows of water are encountered these systems are converted to gas/liquid systems—mist or foam.

Over the years a considerable number of drilling fluid formulations have been developed to suit various subsurface conditions. Selection of the best fluid to meet anticipated conditions will minimize well costs and reduce the risk of catastrophes, such as wellbore instability, stuck drill pipe, loss of circulation, and gas kicks. Consideration must also be given to obtaining adequate formation evaluation and maximum productivity.

Inorganic Colloids

Inorganic colloids are mainly composed of active clay material that has completely hydrated and dispersed. The activity of the inorganic colloidal fraction is derived from the very small size of the particle (and consequent high surface area) relative to its weight and the electrostatic forces on its surfaces and edges. Because of the high *specific surface*, the behavior of the particles is governed primarily by the electrostatic charges on their surfaces, which give rise to attractive or repulsive interparticle forces. Clay minerals are particularly active colloids, partly because of their shape—tiny crystalline platelets or packets of platelets—and partly because of their molecular structure, which results in high negative charges on their basal surfaces, and positive charges on their edges. Interaction between these opposite charges profoundly influences the viscosity of clay muds at low flow velocities, and is responsible for the formation of a reversible gel structure when the mud is at rest.

Organic Colloids

Organic colloids, usually called polymers, are high molecular weight, water-dispersible organic polymers that control either the rheological or fluid loss properties of the fluid. They are essential for maintaining those fluid properties in brine-based fluids. Detailed information on polymers can be found in Chapter 5, Water Dispersible Polymers.

Drilled Solids

Drilled solids are composed of both active clays and inert mixtures of various minerals. Most water-based fluids problems are related in some way to the solids in the system, primarily the buildup of small sizes that cannot be removed by solids control equipment. To maintain a properly functioning drilling fluid, the drilled solids are best kept below 4 vol%, but ideally below 2 vol%.

Water-Based Muds

These systems may be either fresh water or brines of various types, from lightly treated to fully saturated. The solids consist of clays and organic colloids added to provide the necessary viscous and filtration properties. Heavy minerals (usually barite) are added to increase the mud weight. In addition, solids from the formation (drill solids) become dispersed in the mud during drilling. The water can contain dissolved salts, either derived from contamination with formation water or added for various purposes.

Many WBMs are formulated with a saturated brine to replace all or part of the mineral weight material. By eliminating these minerals, the well can usually be drilled faster. Also, the likelihood of formation damage by particle plugging is lessened.

Two types of brines are used:

1. Inorganic—usually containing a chloride anion
2. Organic—usually derived from formic acid.

Figs. 1.2 and 1.3 show the densities obtained with these brines at saturation.

Most of the time, these brines are used with either the sodium or potassium cation—Na^+ or K^+. The sodium form is usually less expensive, but the potassium form is preferred to minimize the swelling of clays. This is discussed further in Chapter 4, Clay Mineralogy and the Colloid Chemistry.

The main disadvantage of using chloride brines is that waste disposal costs can increase dramatically. Waste costs in drilling operations are primarily driven by the amount of drill cuttings generated. All cuttings will be covered with mud and if the base fluid contains chlorides, that increases disposal costs.

Formates, on the other hand are easily disposed of in a regular landfill or land farming operation. The formate degrades into CO_2 and water. Formates

8 Composition and Properties of Drilling and Completion Fluids

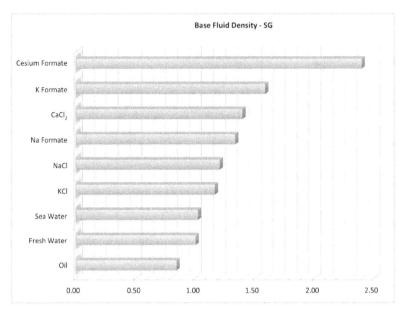

FIGURE 1.2 Base Fluid Specific Gravities.

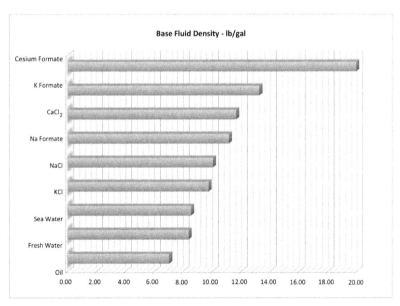

FIGURE 1.3 Base Fluid Densities—lb/gal.

have other advantages that are discussed in Chapter 2, Introduction to Completion Fluids.

Solid particles in the mud may conveniently be divided into four groups according to size:

1. Inorganic and organic colloids, from about 0.005−2 μm
2. Drill solids, 2−74 μm
3. Weight material, 5−74 μm
4. Sand, 74−1000 μm.

Types of Drilling Fluid Systems

The different types of drilling fluids that are available are summarized below. The following classifications are modified from the World Oil magazine's annual Drilling, Completion, and Workover Fluids Supplement (World Oil, 2015) published in June every year.

Water-Based Muds

Non-Dispersed—This term is used to describe drilling fluids which have not been chemically treated to disperse any clays present. These are normally fresh water muds with bentonite clay and no barite. Examples are spud muds or native/natural muds. Also in this category are seawater pump and dump riserless muds in offshore drilling.

Dispersed Bentonitic—These are fluids used as the well is drilled deeper requiring greater control of its properties, i.e., weight, rheology and fluid loss. They will usually contain chemical thinners, such as lignosulfonate, lignite, or tannins. They require alkalinity control, so significant quantities of caustic soda is used to maintain the proper pH level.

Calcium Treated—Calcium-based fluids use $CaCl_2$ brine as the base fluid or additions of lime or gypsum, to make a lime mud or gyp mud. These fluids were designed to assist in wellbore stability as they give high levels of soluble calcium to minimize clay swelling. They are contaminant resistant, but can experience high gel strengths, especially at high temperatures.

Polymer Systems—A polymer system is one that uses an organic, usually long chain and high molecular weight, natural or synthetic material used to control one or more properties of the fluid. The polymer replaces the traditional water-based additives such as bentonite and chemical thinners. Bentonite does not yield when the chloride content is above about 5000 mg/L. It becomes inert and acts like a drill solid. More information is given in Chapter 5, Water Dispersible Polymers.

High Performance or Enhanced Performance Water-Based Muds (HPWBM)—These fluids are formulated to solve drilling problems such as wellbore stability, solids inhibition, wellbore strengthening, drilling rate enhancement, HTHP stability, and reservoir drill-in fluids (RDIF).

Most high performance fluids are formulated with a brine base fluid and polymers for rheology and fluid loss control. Depending on the end use of the fluid additional surfactants, wellbore strengthening materials, antiaccretion agents, film formers, etc. Specific formulations are discussed in other chapters.

Saltwater Systems—These fluids are used to drill salt formations, when fresh water is not available or selected for one of the systems described above. The most common workover system is saturated sodium chloride, using attapulgite or sepiolite clay augmented by polymers to control the properties. Offshore drilling will normally use seawater for the top hole water-based system, before displacing to a nonaqueous fluid.

Nonaqueous-Based Systems

The original nonaqueous fluid was the use of crude oil or refined oils, such as stove oil, as the base fluid and was called an oil-based mud (OBM) (Swan, 1923). Crude oil muds were difficult to maintain so the industry settled on diesel oil as the base fluid for use in an OBM Alexander, 1944). One of the disadvantages of an OBM is that could not handle concentrations of water above 5–10%. In addition, fluid loss was difficult to control. The additives used for fluid loss were asphaltic materials and blends of different softening point gilsonites. Viscosity and suspension properties of the OBMs were modified by sodium soap emulsions (Miller, 1944).

In the 1950s, a new type of oil mud was developed called an invert emulsion mud (Wright, 1954). Since the industry already had fluids called "emulsion muds," oil in water emulsions, it was deemed advisable to call this new system "invert emulsions," water emulsified into oil. While the traditional OBM could only hold its properties by limiting the amount of water, the invert could routinely handle 40 or 50% water. The development of strong emulsifiers and wetting agents, as well as organophyllic clays, made invert emulsions acceptable as a replacement to using water-based muds (Jordon et al., 1965; Simpson et al., 1961).

Diesel oil contains aromatic compounds. These compounds are toxins. In the 1980s new environmental rules and regulations were put in place that banned the use of oil-based drilling fluids in offshore operations, specifically diesel and mineral oils (Ayers et al., 1985; Bennett, 1984). The industry then developed new base fluids to formulate what was called a nonaqueous drilling fluid (NADF). These were synthesized fluids that were nontoxic to marine organisms and were accepted for use by environmental governing agencies around the world (Boyd et al., 1985).

The base fluids (NAF) used to formulate an NADF include:

Oils—diesel, mineral oil and low toxicity mineral oil
Synthetics—alkanes, olephins and esters.

All of the above NAFs are currently being used. Most of the diesel and mineral oil use is for land based drilling, while the synthetics are primarily used in offshore drilling.

Base Oils

With the advent of the Clean Water Act environmental rules and regulations, the use of NADFs in offshore drilling has been strictly regulated. The aromatic chemical content in diesel and mineral oils have been shown to be toxic to some marine organisms as well as hazardous to oilfield workers. The most serious problem with NADF use offshore is with cuttings disposal. Cuttings generated with oil muds will always be coated with oil, which has been deemed unacceptable.

As a result a number of environmental guidelines have been developed concerning drilling fluid use and disposal, especially the use of NADFs. In 2001 the United States Environmental Protection Agency published Effluent Limit Guidelines for offshore drilling operations in the Federal Register (EPA, 1985). These guidelines were incorporated into the EPA's NPDES general permit for offshore drilling. The permit specifies that, among other restrictions, there would be no discharge of NADF into the water and that to dispose of cuttings into the water, the oil on cuttings should be less than 8%. In addition, the use of diesel oil and mineral oil were banned.

The industry responded to this by developing "synthetic oils" as the base for a new class of NADF called synthetic-based muds (SBM). Diesel and mineral oil muds are still used onshore if disposed of properly and can be used offshore if all waste muds and cuttings are transported to shore for proper disposal.

The current base oils in use are:

- *Paraffin (alkanes)*. These oils have carbon-hydrogen single bonds and are either linear (normal), branched (iso), or cyclic (ring) with carbon chain link of C10 to C22.
- *Olephins (alkenes)*. These oils have carbon-hydrogen double bonds, either alpha olephins with the double bond between the number 1 and number 2 carbon or internal olephins with the double bond moved internally and carbon chain lengths of C15 to C18.
- *Esters*. These oils are modified vegetable oils, primarily from palm oil.

The characteristics of these oils are

- Traditional hydrocarbons
  - Diesel
    - Low viscosity
    - Inexpensive
    - Easy to maintain

- Toxic
- Good biodegradation
- Mineral Paraffin (Alkane)
  - Lower toxicity
  - Higher cost
  - Higher viscosity
- Modified vegetable oils (Ester)
  - High viscosity
  - Good biodegradation
  - Low toxicity
  - High cost
  - Low temperature stability
  - Difficult to maintain
- Olephins—Linear, iso, blends
  - Moderate viscosity
  - Good biodegradation
  - Low toxicity
  - Moderate cost
  - Easier to maintain
  - Less sheen forming

The NADF also contains a solids phase, as described above, and a discontinuous or internal phase emulsified into the base fluid. This emulsified phase can be a brine water or other fluid that has a relatively low activity vis-à-vis the formation being drilled. Water activity (balanced activity) is covered in detail in Chapter 9, Wellbore Stability.

In addition to the properties discussed above, other additives in a NADF are used to control the emulsion stability and to maintain any solids in the fluid in an oil-wet condition. These additives are covered in Chapter 8, The Surface Chemistry of Drilling Fluids and Appendix C, The Development of Drilling Fluids Technology.

Pneumatic Drilling Fluids

Also called air or gas drilling, one or more compressors are used to clean cuttings from the hole. In its simplest form, a pneumatic drilling fluid is a dry gas, either air, natural gas, nitrogen, carbon dioxide, etc. The dry gas drilling is best suited for formations that have a minimal amount of water. Once water is encountered while drilling, the dry gas system is converted by the addition of a surfactant and called a "mist" system. Increasing amounts of water in the formation require additional amounts of foaming surfactants and polymer foam stabilizers. More information on these additives are presented in Chapter 8, The Surface Chemistry of Drilling Fluids.

A subset of pneumatic drilling is by using aerated drilling fluids. Air is injected into the mud stream in the upper part of the annulus, thereby reducing the hydrostatic pressure. This results in faster drilling and little or no formation damage. To use this technique a rotating blowout preventer must be used. Several service companies have this service called managed pressure drilling. Appendix C, The Development of Drilling Fluids Technology covers the development of gas drilling and the process of managed pressure drilling (MPD).

PROPERTIES OF DRILLING FLUIDS

Density

Density is defined as weight per unit volume. It is expressed either in *pounds per gallon* (lb/gal), *pounds per cubic foot* (lb/ft^3), or in *kilograms per cubic meter* (kg/m^3), or compared to the weight of an equal volume of water, as *specific gravity* (SG). The pressure exerted by a static mud column depends on both the density and the depth; therefore, it is convenient to express density in terms of *pounds per square inch per foot* (psi/ft), or *kilograms per square centimeter per meter* (kg/cm^2 per m). The densities of some drilling fluid components are given in Table 1.4

To prevent the inflow of formation fluids the pressure of the mud column must exceed the *pore pressure*, the pressure exerted by the fluids in the pores of the formation. The pore pressure depends on the depth of the porous formation, the density of the formation fluids, and the geological conditions. Two types of geological conditions affect pore pressure: *normally pressured formations*, which have a self-supporting structure of solid particles (so the pore pressure depends only on the weight of the overlying pore fluids), and *abnormally pressured* or *geopressured formations*, which are not fully compacted into a self-supporting structure (so the pore fluids must bear the

TABLE 1.4 Densities of Common Mud Components

Material	g/cm^3	lb/gal	lb/ft^3	lb/bbl	kg/m^3
Water	1.0	8.33	62.4	350	1000
Oil	0.8	6.66	50	280	800
Barite	4.1	34.2	256	1436	4100
Clay	2.5	20.8	156	874	2500
NaCl	2.2	18.3	137	770	2200

weight of some or all of the overlying sediments as well as the weight of the overlying fluids). The hydrostatic pressure gradient of formation fluids varies from 0.43 psi/ft to over 0.52 psi/ft (0.1–0.12 kg/cm^2 per m), depending on the salinity of the water.

The bulk density of partially compacted sediments increases with depth, but an average SG of 2.3 is usually accepted, so that the *overburden* (or *geostatic* or *litholostatic*) *pressure gradient* is about 1 psi/ft (0.23 kg/cm^2 per m), and the pore pressure of geopressured formations is somewhere between the normal and the overburden pressure gradients, depending on the degree of compaction. Besides controlling pore fluids, the pressure of the mud column on the walls of the hole helps maintain borehole stability. In the case of plastic formations, such as rock salt and unconsolidated clays, the pressure of the mud is crucial. The buoyant effect of the mud on the drill cuttings increases with its density, helping transport them in the annulus, but retarding settling at the surface. Very rarely is an increase in mud density justified as a means of improving cutting carrying capacity.

In the interest of well safety, there is a natural tendency to carry a mud density well above that actually needed to control the formation fluids, but this policy has several major disadvantages. In the first place, excessive mud density may increase the pressure on the borehole walls so much that the hole fails in tension. This failure is known as *induced fracturing*.

In induced fracturing, mud is lost into the fracture so formed, and the level in the annulus falls until equilibrium conditions are reached. The problem of maintaining mud density high enough to control formation fluids, but not so high as to induce a fracture, becomes acute when normally pressured and geopressured formations are exposed at the same time. Under these circumstances, it is generally necessary to set a string of casing to separate the two zones. Several methods have been developed for predicting the occurrence of geopressures (Fertl and Chilingar, 1977). Knowledge of the expected pore pressure and fracture gradients (Breckels and Van Eekelen, 1982; Daines, 1982; Stuart, 1970) usually enables casing to be set at exactly the right depth, thereby greatly reducing the number of disaster wells. Recent advances in prediction uses seismic while drilling techniques (Poletto and Miranda, 2004; Dethloff and Petersen, S., 2007).

Another disadvantage of excessive mud densities is their influence on drilling rate (*rate of penetration—ROP*). Laboratory experiments and field experience have shown that the rate of penetration is reduced by *mud overbalance pressure* (the differential between the mud pressure and the pore pressure when drilling in permeable rocks) (Murray and Cunningham, 1955; Eckel, 1958; Cunningham and Eenink, 1959; Garnier and van Lingen, 1959; Vidrine and Benit, 1968) and by the absolute pressure of the mud column when drilling rocks of very low permeability. A high overbalance pressure also increases the risk of sticking the drill pipe.

Lastly, excessive mud densities are a disadvantage because they unnecessarily increase mud costs. Mud costs are not a very important consideration when drilling in normally pressured formations, because adequate densities are automatically obtained from the formation solids that are dispersed into the mud by the action of the bit. Mud densities greater than about 11 lb/gal (1.32 SG) cannot be obtained with formation solids because the increase in viscosity is too great. Higher densities are obtained with barite, which has a specific gravity of about 4.1, as compared to about 2.6 SG for formation solids, so that far fewer solids by volume are required to obtain a given density. Mud costs are increased not only by the initial cost of the barite, but also, and to a greater extent, by the increased cost of maintaining suitable properties, particularly flow properties. Because of the incorporation of drilled solids, the viscosity continuously increases in WBMs as drilling proceeds, and must be reduced from time to time by the addition of water and more barite to restore the density. In NADFs, drilled solids have little effect on its properties as long as the solids are kept in an oil wet state.

An option to the use of mineral weight materials is to use a high-density brine to increase the specific gravity of the drilling fluid, thereby minimizing or eliminating the use of those solids. The specific gravity of brines at saturation are shown in Figs. 1.2 and 1.3.

Flow Properties

The flow properties of the drilling fluid play a vital role in the success of the drilling operation. These properties are primarily responsible for removal of the drill cuttings, but also influence drilling progress in many other ways. Unsatisfactory performance can lead to such serious problems as bridging across the wellbore, filling the bottom of the hole with drill cuttings, reduced penetration rate, hole enlargement, stuck pipe, loss of circulation, and even a blowout.

The flow behavior of fluids is governed by *flow regimes*, the relationships between pressure and velocity. There are two such flow regimes, namely *laminar flow*, which prevails at low flow velocities and is a function of the viscous properties of the fluid, and *turbulent flow*, which is governed by the inertial properties (solids and mud weight) of the fluid and is only indirectly influenced by the viscosity. As shown in Fig. 1.4, pressure increases with velocity increase much more rapidly when flow is turbulent than when it is laminar.

Laminar Flow

Laminar flow in a round pipe may be visualized as infinitely thin cylinders sliding over each other. The velocity of the cylinders increases from zero at the pipe wall to a maximum at the axis of the pipe. The

16 Composition and Properties of Drilling and Completion Fluids

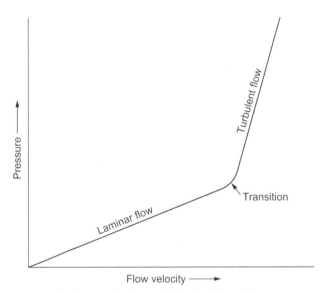

FIGURE 1.4 Schematic diagram of laminar and turbulent flow regimes.

difference in velocity between any two such cylinders, divided by the distance between them, defines the *shear rate*. The axial force divided by the surface area of a cylinder defines the shear stress. The ratio of shear stress to shear rate is called the *viscosity*, and is a measure of the resistance to flow of the fluid. The unit of viscosity is the *Poise*; the shear stress in dynes/cm^2 divided by the shear rate in reciprocal seconds gives the viscosity in Poises. The unit employed in mud viscometry is the *centipoise* (cP), which is one hundredth of a Poise. (See Table A.1 for the SI unit.)

Newtonian Fluids

A plot of shear stress versus shear rate is known as a *consistency curve*, or *rheogram* (Fig. 1.4), the shape of which depends on the nature of the fluid being tested; with fluids that contain no particles larger than a molecule (e.g., water, brine solutions, sugar solutions, oil, glycerine), the consistency curves are straight lines passing through the origin. Such fluids are called *Newtonian* because their behavior follows the laws first laid down by Sir Isaac Newton. The viscosity of a Newtonian fluid is defined by the slope of its consistency curve (see Fig. 1.4). Since the viscosity of a Newtonian fluid does not change with shear rate, a viscosity determined at a single shear rate may be used in hydraulic calculations involving flow at any other shear rate.

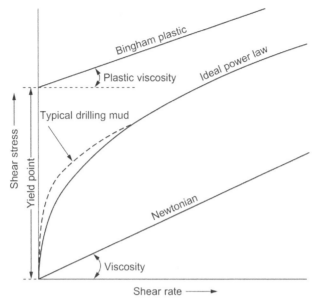

FIGURE 1.5 Ideal consistency curves for common flow models.

Bingham Plastic Fluids

Suspensions such as drilling muds that contain particles larger than molecules in significant quantities do not conform to Newton's laws, and thus are classified under the general title of *non-Newtonian* fluids. The shear stress/shear rate relationship of non-Newtonian fluids depends on the composition of the fluid. Clay muds having a high solids content behave approximately in accordance with the Bingham theory of plastic flow, which postulates that a finite stress must be applied to initiate flow, and that at greater stresses the flow will be Newtonian. The consistency curve of a *Bingham plastic* fluid must therefore be described by two parameters, the *yield point* and the *plastic viscosity*, as shown in Fig. 1.4. The shear stress divided by the shear rate (at any given rate of shear) is known as the *apparent viscosity*. Fig. 1.4 shows that apparent viscosity decreases with increase in shear rate, and is therefore a valid parameter for hydraulic calculations only at the shear rate at which it was measured. Indeed, as shown by Fig. 1.5, effective viscosity is not even a reliable parameter for comparing the behavior of two different muds.

Power Law Fluids

The decrease in apparent viscosity with increase in shear rate is known as *shear thinning*, and is a desirable property, because the effective viscosity at the shear rate of interest for hydraulics calculations will be relatively low

18 Composition and Properties of Drilling and Completion Fluids

at the high shear rates prevailing in the drill pipe, thereby reducing pumping pressures, and relatively high at the low shear rates prevailing in the annulus, thereby increasing cutting carrying capacity. In addition highly shear thinning fluids enhance the efficiency of solids control equipment.

Many water-based drilling muds consist of polymers and little or no particulate solids, especially if a high-density brine is used for weight control, are highly shear thinning and cannot be described by the Bingham Plastic Viscosity and Yield Point. The behavior of these *pseudo-plastic* fluids is better described by the power law rheology model, also known as the Ostwald-deWaele equation, which states that

$$\text{shear stress} = K(\text{shear rate})^n$$

The parameter K, *(consistency factor)* is the shear strength at a shear rate of 1.0 s^{-1} and is usually expressed in lb/100 ft^2 or dynes/cm^2, and corresponds approximately to a relative viscosity at 1.0 s^{-1}. K can be converted to viscosity units of cP or Pascal.sec by using the appropriate constants. The n factor (degree of deviation from the Newtonian) is a measure of the rate of change of viscosity with rate of shear, and thus is an alternative measure of shear thinning: the lower the value of n, the greater the shear thinning. The power law may be used to describe the behavior of all flow models by inserting the proper value of n; thus n is 1.0 for a Newtonian fluid, and less than 1.0 for shear-thinning muds.

Most drilling muds exhibit behavior intermediate between ideal Bingham plastics and ideal power law fluids. High solids laden fluids can be described by the Bingham Plastic Model (as specified by the API 13D (2010) recommended practice). Low solids, polymer water-based fluids are better described by the power law model, since n and K are not constant in laminar flow. Muds have a rather indefinite yield point that is less than would be predicted by extrapolation of shear stresses measured at high shear rates. Fig. 1.5 compares the consistency curves of the three flow models just discussed.

Modified Power Law

The modified power law fluid flow model was developed by Winslow Hershel and Ronald Bulkley in 1926. The dashed line in Fig. 1.5 represents a typical drilling fluid showing a true yield point. The Hershel-Bulkley model states:

$$\text{shear stress} = \text{true yield point} + K(\text{shear rate})^n$$

The true yield point is calculated by extrapolation to 0 shear rate. Whereas the Bingham and power law parameters are relatively easy to calculate the modified power law yield point parameter (called tau zero—T_0) requires a computer and suitable software to approximate tau zero. Further details on calculating and using these parameters are covered in Chapter 6, The Rheology of Drilling Fluids.

The fact that the consistency curve of clay muds intercepts the stress axis at a value greater than zero can indicate the development of a gel structure. In water-based fluids this structure results from the tendency of the clay platelets to align themselves so as to bring their positively charged edges towards their negatively charged basal surfaces. This interaction between the charges on the platelets also increases the effective viscosity at low rates of shear, thereby influencing the values of the Bingham Yield Point, n, and K, and is responsible for the formation of a gel when agitation ceases.

The gel strength of some muds, notably fresh-water clay muds, increases with time after agitation has ceased, a phenomenon that is known as *thixotropy*. Furthermore, if after standing quiescent the mud is subjected to a constant rate of shear, its viscosity decreases with time as its gel structure is broken up until an equilibrium viscosity is reached. Thus the effective viscosity of a thixotropic mud is time-dependent as well as shear-dependent.

Turbulent Flow

Flow in a pipe changes from laminar to turbulent when the flow velocity exceeds a certain critical value. Instead of layers of water sliding smoothly over each other, flow changes locally in velocity and direction, while maintaining an overall direction parallel to the axis of the pipe. Laminar flow may be compared to a river flowing smoothly over a plain, and turbulent flow to flow over rapids where interaction with irregularities on the bottom causes vortices and eddies.

The critical velocity for the onset of turbulence decreases with increase in pipe diameter, with increase in density, and with decrease in viscosity, and is expressed by a dimensionless number known as the *Reynolds number*. With most drilling muds the critical value of the Reynolds number lies between 2000 and 3000.

The pressure loss of a fluid in turbulent flow through a given length of pipe depends on inertial factors, and is little influenced by the viscosity of the fluid. The pressure loss increases with the square of the velocity, with the density, and with a dimensionless number known as the *Fanning friction factor*, which is a function of the Reynolds number and the roughness of the pipe wall.

Filtration Properties

The ability of the mud to seal permeable formations exposed by the bit with a thin, low-permeability filter cake is another major requirement for successful completion of the hole. Because the pressure of the mud column must be greater than the formation pore pressure in order to prevent the inflow of formation fluids, the mud would continuously invade permeable formations if a filter cake were not formed.

For a filter cake to form, it is essential that the mud contains some particles of a size only slightly smaller than that of the pore openings of the formation. These particles, which are known as bridging particles, are trapped in the surface pores, while the finer particles are, at first, carried deeper into the formation. The bridged zone in the surface pores begins to trap successively smaller particles, and, in a few seconds, only liquid invades the formation. The suspension of fine particles that enters the formation while the cake is being established is known as the *mud spurt*. The liquid that enters subsequently is known as the *filtrate* which, in a water-based fluid, is analyzed to determine the soluble chemical entities in the fluid.

The rate of filtration and the increase in cake thickness depend on whether or not the surface of the cake is being subjected to fluid or mechanical erosion during the filtration process. When the mud is static, the filtrate volume and the cake thickness increase in proportion to the square root of time (hence, at a decreasing rate). Under dynamic conditions, the surface of the cake is subjected to erosion at a constant rate, and when the rate of growth of the filter cake becomes equal to the rate of erosion, the thickness of the cake and the rate of filtration remain constant. In the well, because of erosion by the mud and because of mechanical wear by the drill string, filtration is dynamic while drilling is proceeding; however, it is static during round trips. All routine testing of the filtration property is made under static conditions because dynamic tests are time-consuming and require elaborate equipment. Thus, filtration rates and cake thicknesses measured in surface tests correlate only approximately to those prevailing downhole and can be grossly misleading. The permeability of the filter cake—which may readily be calculated from static test data—is a better criterion because it is the fundamental factor controlling both static and dynamic filtration.

The permeability of the filter cake depends on the particle size distribution in the mud and on the electrochemical conditions. In general, the more particles there are in the colloidal size range, the lower the cake permeability. The presence of soluble salts in clay muds increases the permeability of the filter cake sharply, but certain organic colloids enable low cake permeability to be obtained even in saturated salt solutions. Chemical thinners usually decrease cake permeability because they disperse clay aggregates to smaller particles.

The filtration properties required for the successful completion of a well depend largely on the nature of the formations to be drilled. Stable formations with low permeability, such as dense carbonates, sandstones, and lithified shales, can usually be drilled with little or no control of filtration properties. But many shales are *water sensitive*, i.e., on contact with water, they develop swelling pressures which cause caving and hole enlargement. Sealing of incipient fractures by mud filter cake will help control the caving, but the type of mud used and the chemical composition of its filtrate are more important factors. Superior hole stabilization is obtained with oil-base mud when the salinity of the filtrate is adjusted to prevent swelling pressures from developing in the shales.

In permeable formations, filtration properties must be controlled in order to prevent thick filter cakes from excessively reducing the gauge of the borehole. Furthermore, thick filter cakes may cause the drill pipe to become stuck by a mechanism known as *differential sticking* (Helmick and Longley, 1957; Outmans, 1958). See Chapter 10, Driling Problems Related to Drilling Fluids for a detailed description of drillpipe sticking and remediation of stuck pipe.

Filtration performance in the well is routinely judged by means of the standard API filtration test. In this test, the mud is subjected to static filtration through filter paper for 30 min, and the volume of filtrate and the cake thickness are then measured. In planning a mud program, a certain maximum API filter loss is often specified, with the thought that as long as the filter loss is kept below this figure, adequate control of downhole filtration properties will be maintained. From what has already been said in this section, it will be appreciated that reliance on API filter loss alone for control of downhole filtration performance is a highly dubious procedure, which can lead to poor drilling and production performance and greatly increased well costs. A major problem is that downhole cake thickness depends to a considerable extent on cake erodability, which does not affect the static filter loss. For example, laboratory tests have shown that emulsification of diesel oil in water-base muds decreases the API filter loss, but sharply increases the dynamic rate, because of cake erodability. Tests have also shown that a commercial additive that decreases the API filter loss might have little or no effect on the dynamic rate, while with another additive the opposite might be true (see Chapter 7, The Filtration Properties of Drilling Fluids). It is essential, therefore, that filtration control agents be evaluated at least once in local muds and under local downhole conditions in a dynamic tester.

For practical reasons, the API filter test must be used at the well site, but the results should be interpreted in the light of laboratory data. Also, the particular reason for filtrate control should be kept in mind. For example, if differential sticking is the problem, the filter cake thickness is more important than the filter loss, or if productivity impairment is the reason for filter loss control, the salinity of the filtrate or having sufficient bridging particles may be the critical property.

Physical-Chemical Properties

In a water-based fluid, the tests to determine the chemical attributes of the system are:

- Alkalinity
- Hardness
- Potassium
- pH
- Cation Exchange Capacity
- Electrical conductivity.

In a nonaqueous fluid additional tests are made to determine:

- Emulsion stability
- Aniline point
- Emulsified water activity
- Oil/water ratio.

The following can be performed on any drilling fluids system:

- Corrosion
- Lubricity
- Solids content
- Salts
- H_2S.

Details of these tests are in Chapter 3, Evaluating Drilling Fluid Performance.

DRILLING FLUID SELECTION

Location

Table 1.5 summarizes the basic characteristics of the majority of mud systems available to the operator. In remote locations, the availability of supplies must be considered; e.g., in an offshore well there is an obvious advantage in selecting a mud that can tolerate seawater for its aqueous phase. Other remote locations, such as desert, arctic, jungle, and deepwater drilling, also require attention to supply change management. Lack of ready access to Nonaqueous base fluids and products in remote locations may force the driller to only use water-based muds. Available makeup waters, such as alkaline fluids, and high hardness solutions may require special additives to formulate an acceptable system.

Government regulations designed to protect the environment restrict the choice of muds in some locations (Environmental Aspects of Chemical Use in Well-Drilling Operations, 1975; McAuliffe and Palmer, 1976; Monaghan et al., 1976; Ayers et al., 1985; Candler and Friedheim, 2006). Such regulations have made the handling of oil-based muds difficult and expensive, particularly in offshore wells. The substitution of low-toxicity mineral oils

TABLE 1.5 Selection of the Drilling Fluid

Classification	Principal Ingredients	Characteristics
Pneumatic		
Dry air	Dry air	Fast drilling in dry, hard rock
		No water influx
		Dust
Mist	Air, water and drying agents	Wet formations but little water influx
		High annular velocity
Foam	Air, water, foaming agent	Stable rock
		Minimal water flow tolerated
Stable foam	Air, water containing polymers, foaming agent, K^+ stabilizer	All "reduced-pressure" conditions: Large volumes of water, big cuttings removed at low annular velocity
		Select polymer and foaming agent to afford hole stability and tolerate salts
		Foam can be formed at surface
Water-Based		
Fresh	Fresh water	Fast drilling in stable formations
	Brackish water <5000 mg/L	Need large settling area, flocculants, or ample water supply and easy disposal
Salt additive or saturated brine	Seawater	Mud weight
	NaCl	Fresh water unavailable
		Wellbore stability
	KCl	Contaminate resistance
	$CaCl_2$	Massive salt formations
		Corrosive
		Waste management expenses
		Workover fluid
		Polymers for rheology and filtrate control

(Continued)

TABLE 1.5 (Continued)

Classification	Principal Ingredients	Characteristics
Inorganic partial or fully saturated brine	Na-Formate	Mud weight
	K-Formate	Wellbore stability
	Cs-Formate	Contaminate resistance
		Biodegradable
		Lubricious
		Noncorrosive
		Polymers for rheology and filtrate control
		Polymer temperature stabilization
Low-solids muds	Fresh water, polymer, bentonite	Fast drilling in competent rocks
		Mechanical solids removal equipment needed
		Contaminated by cement, soluble salts
Spud mud	Bentonite and water	Inexpensive—dirt and water
Salt water muds[b]	Seawater, brine, saturated salt water, salt-water clays, starch, cellulosic polymers	Drilling rock salt (drilling salts other than halite may require special treatment), formulating workover fluids, wellbore stability mitigation
Chemically thinned muds	Fresh or brackish water, bentonite, caustic soda, lignite, lignosulfonate, tannin, sulphonated tannin	Shale drilling
		Simple maintenance
		Max. temp. 350°F (180°C)
	Surfactant added to coat solids and for high temperature	Same tolerance for contaminants
		pH 9–10
		Waste management issues
Nonaqueous Based		
Oil-Based	Weathered crude oil Asphaltic crude + soap	Low-pressure well completion and workover
	Emulsifiers water 2–10%	Drill shallow, low-pressure productive zone
		Waste management issues
		HT stability to 600°F (315°C)

(*Continued*)

TABLE 1.5 (Continued)

Classification	Principal Ingredients	Characteristics
Invert emulsions	Diesel or Mineral oil, emulsifiers, organophilic clay, modified resins and calcium soaps, lime 5–40% water	High initial cost and environmental restrictions, but low maintenance cost
		Wellbore stability
		HT stability
		Contaminate resistance
Synthetic invert emulsions	Synthesized alkane and alkenes, esterified vegetable oils emulsifiers, organophilic clay, modified resins and calcium soaps, lime 5–40% water	Highest initial cost
		Reduced environmental restrictions
		Wellbore stability
		HT stability
		Contaminate resistance

Notes:
Detergents, lubricants, and/or corrosion inhibitors may be added to any water composition.
Density of oil muds can be raised by addition of calcium carbonate, barite, hematite, illmenite, or manganese tetroxide.
Calcium chloride or other low activity emulsified liquids are added to the emulsified water phase in Nonaqueous drilling fluids to increase shale stability.

(Hinds and Clements, 1982; Boyd et al., 1985; Jackson and Kwan, 1984; Bennett, 1984) for diesel oil helped. But the use of synthetic oils has greatly alleviated this problem.

Methods and equipment for preventing environmental damage by drilling fluids and drill cuttings have been described by several authors (Kelley et al., 1980; Carter, 1985; Johancsik and Grieve, 1987; Candler and Friedheim, 2006). Tests for evaluating the toxicity and other relevant properties of mineral oils have been described by Burton and Ford (1985). Thawing of the permafrost by the drilling mud is a problem in the Arctic (Goodman, 1977). The hole may enlarge up to 8 ft, and special foams are necessary to clean the hole (Fraser and Moore, 1987).

Mud-Making Shales

Thick shale sections containing dispersible clays, such as montmorillonite, cause rapid rises in viscosity as drilled solids become incorporated in a water-based mud. High viscosities are not a serious problem with unweighted muds because they can easily be reduced by dilution and light chemical treatment, but with weighted muds dilution is costly because of the barite and chemicals required to restore the mud properties. Drillers are now more and more using organic or inorganic brine water systems (formate or chloride) to minimize the breakdown of drill solids as well as to provide shale wellbore stability. Nonaqueous fluids can tolerate higher solids buildup, but if not controlled can cause eventually cause problems.

The drilled solids can, of course, be removed mechanically, but the presence of barite greatly complicates the process. Therefore, a mud which inhibits the dispersion of clays, such as a lime, gyp, or other brine water systems, should be used when drilling thick sections of mud-making shales.

Geopressured Formations

Shallow formations are usually normally pressured and can be drilled with unweighted muds. When geopressured formations are encountered, the density of the mud must be increased so that the pressure of the mud column exceeds the formation pore fluid pressure by a safe margin. Exactly what is a safe margin is a matter of opinion, but remember that excessive density adds greatly to drilling costs and increases the risk of stuck pipe and loss of circulation, and possible formation damage. Lower margins can be carried if the viscosity and gel strengths are kept to a minimum so as to avoid swabbing the well when pulling pipe, and to facilitate the removal of entrained gas.

In a water-based system the presence of weight material of the solids content of the mud is high, and the incorporation of drilled solids soon increases the total solids content to the point where the viscosity rises rapidly. Therefore the solids content must be kept to a minimum *before* adding the barite, or the old mud should be discarded and a fresh mud prepared containing only barite and just enough bentonite to suspend it. When densities over 14 lb/gal (1.68 SG) are required, a mud that tolerates a high solids content, such as one of the inhibitive muds discussed above, or an nonaqueous mud, should be used.

An alternative to high-density minerals for increasing the mud weight is to use a high-density brine or brine blend as the base fluid.

High Temperature

The constituents of drilling muds degrade with time at elevated temperatures: the higher the temperature the greater the rate of degradation. Water-based fluids are more susceptible to high temperature (HT) degradation than NADFs, but even NADF additives and base fluids can degrade or rendered ineffective at extreme HT drilling. Both the temperature and the rate of degradation at that temperature must be taken into account when specifying the temperature stability of a fluid system or additive. The critical temperature is that at which the cost of replacing the degraded material becomes uneconomical, which is generally established by experience, but may be calculated (St Pierre and Welch, 2001).

Bentonite, a principle ingredient in most WBMs will start changing form, to a less active clay, in the 350–400°F range. The critical temperature for common WBM additives ranges from about 225°F (107°C) to 350°F (107°C) for some organic polymers and 275°F (135°C) for starch and cellulosic polymers. Chemically thinned muds can be used at temperatures up to 400°F (204°C), and invert emulsions up to about 500°F (204°C). Synthetic polymers, such as polyacrylates and copolymers with polyacylamides, are stable up to 500°F (204°C). Certain other acrylic copolymers are stable to 600°F (316°C) (see Chapter 10, Drilling Problems Related to Drilling Fluids).

Hole Instability

There are two basic forms of hole instability encountered while drilling—hole contraction and hole enlargement.

Hole Contraction

If the lateral earth stresses bearing on the walls of the hole exceed the yield strength of the formation, the hole slowly contracts. In soft plastic formations, such as rock salt, gumbo shales, and geopressured shales, large quantities of plastic spallings are generated, but the pipe becomes stuck only in severe cases. Hard shales and rocks are generally strong enough to withstand the earth stresses, unless there are stress concentrations at specific points on the circumference of the hole (e.g., at keyholes), in which case spalling occurs at the point of maximum stress. Although spalling and tight hole can be alleviated by shale stabilizing muds, only raising the density of the mud to balance the earth stresses will prevent the problem.

Hole Enlargement

This problem occurs in hard shales that can withstand the earth stresses unless destabilized by interaction with the WBM filtrate or lack of proper osmotic control in a NADF (see Chapter 9, Wellbore Stability), in which case the contaminated zone caves in hard fragments and the hole gradually enlarges. Such shales are called *water sensitive*. Hole enlargement can be prevented only by the use of shale stabilizing muds. NADF muds are best for shale stabilization, provided the salinity of the aqueous phase is high enough to balance the swelling pressure of the shale. Potassium chloride polymer muds, along with other additives, are the best water-based muds for stabilizing hard shales. Potassium chloride muds are less suitable for stabilizing soft, dispersible shales because high concentrations are required and initial and maintenance costs are high. Gyp, lime, or modifications thereof (Walker et al., 1984; de Boisblanc, 1985) are suitable alternatives as well as the so-called high-performance (HPWBM) or enhanced-performance (EPWBM) fluids (Riley et al., 2012; Al-Ansari et al., 2014).

Both nonaqueous and, to a lesser extent, potassium-polymer WBMs are expensive, but their cost can easily be justified if they prevent caving and hole enlargement, which greatly increase drilling time and costs. Besides the obvious costs of cleaning out cavings and bridges, freeing stuck pipe, bad cement jobs, and poor formation evaluation, there are the less obvious costs resulting from the necessity of carrying high viscosities and gel strengths, which reduce rate of penetration, cause swabbing and surge pressures, gas cutting, etc. Reduction of rate of penetration is especially damaging because caving is time-dependent; the longer the shale is exposed to the mud, the greater the hole enlargement.

Fast Drilling Fluids

The characteristics of fast drilling fluids are low density, low viscosity, and low solids content. Air is by far the fastest fluid, but can only be used in stable, nonpermeable formations that do not permit any significant inflow of water (Sheffield and Sitzman, 1985). Clear brines can be used to drill hard rocks, but only if a gauge or near-gauge hole can be maintained, because of their poor cutting carrying capacity (Zeidler, 1981). Low-solid nondispersed muds and potassium muds using soluble salts as weighting agents drill fast in hard rocks and nondispersible shales, but the solids content must be kept below 10% by mechanical means. Rates approaching clear brines can be obtained if the solids can be kept below about 2%. Antiaccretion agents to agglomerate cuttings and prevent bit balling are also used to enhance rate of penetration (Cliffe and Young, 2008).

Because of their high viscosity, standard NADFs do not permit fast drilling rates in comparison to most WBMs. But a special low-viscosity NADF, which has proved particularly successful when drilling with polycrystalline bits, is available (Simpson, 1979). This mud is less stable and has somewhat poorer filtration characteristics than standard NADFs, and therefore should not be used when conditions are severe, as with highly deviated holes (because of the danger of sticking the pipe), deep hot holes (break the emulsion), etc. (Golis, 1984).

Rock Salt

To prevent the salt from dissolving and consequently enlarging the hole, either a NADF or a saturated brine must be used. The chemical composition of the brine should be approximately the same as that of the salt bed. A slight tendency for the salt to creep into the hole may be offset by maintaining the salinity slightly below saturation. As previously mentioned, high densities are essential for deep salt beds.

High Angle Holes

In highly deviated holes, such as are drilled from offshore platforms, torque and drag are a problem because the pipe lies against the low side of the hole, and the risk of the pipe sticking is high. Because of its thin, slick filter cake, the cost of a NADF can often be justified in these wells. If a water-based mud is used, lubricants must be added, and good filtration properties maintained. Hole cleaning is also a problem because the cuttings fall to the low side of the hole. Muds must, therefore, be formulated to have a high low-shear rate viscosity.

A significant problem with NADFs is called dynamic weight material sag. Weight material settling is always a problem in any drilling fluid, but is particularly so in nonaqueous fluids. Water-based muds can be formulated to be highly shear thinning with good low-rate viscosities. NADFs, on the other hand, are usually less shear thinning and have much lower low shear rate viscosities. Dynamic sag is particularly troubling since it occurs while pumping the mud, and can significantly lower the hydrostatic pressure in the annulus possibly initiating a blowout.

Formation Evaluation

Sometimes the mud that is best for drilling the well is not suitable for logging it, and alternatives must be considered. For example, NADFs are excellent for maintaining hole stability, but, being nonconductive, require special logging techniques. These present no problem in development wells, but are a problem in wildcat wells where there are no previously drilled wells available for correlation. In such cases, a brine mud can be used, but if its low resistivity interferes with resistivity interpretations, additional lab work must be done for correlation. A logging engineer should always be consulted when selecting a mud for a wildcat well.

Productivity Impairment

The various mechanisms by which drilling fluids reduce well productivity were discussed earlier in this chapter in the section on filtration properties and additional discussions are found in the completion fluid and drill-in fluid sections (see Chapter 11). Preventive measures are as follows:

1. Dispersion of indigenous clays will be prevented if the mud filtrate contains at least 3% sodium chloride, or 1% of either potassium or calcium chloride. Treatment with thinners must be discontinued a day or so before drilling into the reservoir.
2. Impairment by mud solids is insignificant when drilling through the reservoir is done with the mud that has been used to drill the upper part of the hole, because it will contain enough particles to bridge the pores

of the formation and establish an external filter cake quickly. However, special degradable fluids should be used when drilling into reservoirs with open fractures, when drilling water injection wells, and when gun perforating or gravel packing.
3. Impairment by waterblock or emulsion blocks and other capillary mechanisms is confined to the zone immediately surrounding the producing zone, and may be eliminated by gun perforating.
4. As mentioned earlier, laboratory tests should always be done on cores from a newly discovered reservoir to determine its characteristics and the best completion fluid for preventing impairment. The cost will be offset many times by improved productivity.

Mud Handling Equipment

Various pieces of drilling rig equipment capability may affect the drilling fluids operation. Inadequacies in pumps, pipe configuration, mixing equipment, or solids removal equipment will likely increase consumption of materials, and sometimes the preferred mud program must be modified to compensate for deficiencies in the equipment (Omland et al., 2012). The importance of the drilling equipment to the success of the mud program deserves a more extensive treatment than can be given here.

The equipment employed in the mechanical separation of cuttings from the active mud system vitally affects mud costs. The mud specialist is usually not assigned the responsibility for this equipment. A separate solids control equipment company is hired to set up and maintain the equipment.

Solids Removal Equipment

The importance of removing drilled solids has been emphasized several times in the preceding sections. The advantages of doing so may be summarized as follows (Robinson and Heilhecker, 1975):

1. Increase in drilling rates
2. Longer bit life
3. Better filter cakes
   a. Less differential pressure sticking
   b. Less torque and drag
   c. Minimize filtrate entry into producing zone
   d. Reduced surge and swab pressures
4. Lower drilling fluids costs
5. Reduced flow pressure losses due to high gels and viscosity
6. Easier to drill a gauge hole
   a. Better hole cleaning
   b. Better formation evaluation
   c. Better cement jobs.

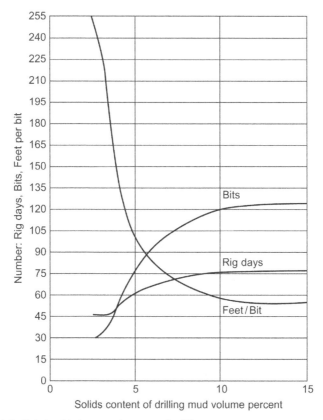

FIGURE 1.6 Relationship between rig costs and solids content.

The results of a field study, shown in Fig. 1.6, illustrate the relationship between rig costs and solids content. Such savings are generally much greater than the cost of renting and running solids removal equipment.

Planning for Successful Solids Control

The common types of solids control equipment are

- Gumbo, highly active, swelling clays, removal (optional)
- Scalping shaker (optional)
- Main shale shaker—Use as small a mesh size as possible
- Degasser
- Desander
- Desilter
- Centrifuge—discard all solids (optional)
- 2 Centrifuges—reclaim barite, recapture base fluid
- Dewatering (flocculation) (optional)

For effective removal of drilled solids, it is essential that the shakers and hydrocyclones have enough capacity to handle the whole mud stream. Centrifuges are designed to handle from 5 to 15% of mud pump output. More detail on solids control and equipment can be found in the ASME publication "Drilling Fluids Processing Handbook."

REFERENCES

Al-Ansari, A., Musa, I., Abahusain, A., Olivares, T., El-Bialy, M.E., Maghrabi, S., 2014. Optimized High-Performance Water-Based Mud Successfully Drilled Challenging Sticky Shales in a Stratgraphic Well in Saudia Arabia. AADE Fluids Technology Conference. Houston, TX. Paper 14-FTCE-51.

Al-Yami, A.S., Nasr-El-Din, H.A., 2007. An innovative manganese tetroxide/KCl water-based drill-in fluid for HT/HP Wells. Soc. Petrol. Eng. Available from: http://dx.doi.org/10.2118/110638-MS.

Alexander, W., 1944. Oil base drilling fluids often boost production. Oil Weekly, pp. 36−40.

Annis, M.R., Monaghan, P.H., 1962. Differential pressure sticking—laboratory studies of friction between steel and mud filter cake. J. Petrol. Technol. vol. 14, 537−543, http://dx.doi.org/10.2118/151-PA.

American Petroleum Institute, <www.API.org>.

ASME, 2005. Shale Shakers and Drilling Fluids Systems. Gulf Professional Publishing, Houston, TX.

Ayers, R.C., Sauer, T.C., Anderson, R.W., 1985. The generic mud concept for NPDES permitting of offshore drilling discharges. J. Petrol. Technol. 37, 475−480.

Bennett, R.B., 1984. New drilling fluid technology—mineral oil mud. J. Petrol. Technol. 975−981.

Boyd, P.A., Whitfill, D.L., Carter, D.S., Allamon, J.P., 1985. New base oil used in low-toxicity oil muds. J. Petrol. Technol. 137−142. In: SPE Paper 12119, Annual Meeting, San Francisco, CA, October 1983.

Breckels, I.M., Van Eekelen, H.A.M., 1982. Relationship between horizontal stress and depth in sedimentary basins. J. Petrol. Technol. 34, 2191−2199.

Burton, J., Ford, T., 1985. Evaluating mineral oils for low-toxicity muds. Oil Gas J. vol. 65, 129−131.

Candler, J., Friedheim, J., 2006. Designing environmental performance into new drilling fluids and waste management technology. In: 13th International Petroleum Environmental Conference, San Antonio, TX, October 17−20.

Carter, T.S., 1985. Rig preparation for drilling with oil-based muds. In: IADC/SPE Paper 13436, Drilling Conference, New Orleans, LA.

Cliffe, S., and Young, S., 2008. Agglomeration and accretion of drill cuttings in water-based fluids. In: AADE Fluids Conference and Exhibition. Houston, TX, April 8−9.

Cunningham, R.A., Eenink, J.G., 1959. Laboratory study of effect of overburden, formation and mud column pressures on drilling rate of permeable formations. Trans. AIME 216, 9−17.

Daines, S.R., 1982. Predictions of fracture pressures in wildcat wells. J. Petrol. Technol. 34, 863−872.

de Boisblanc, C.W., 1985. Water mud gives advantages with PCD bits. Oil Gas J. 83, 134−137.

Dethloff, M. 2007. Seismic-While-Drilling Operation and Application. SPE Annual Technology Conference, Anaheim, CA, November 11−14.

Eckel, J.R., 1958. Effect of pressure on rock drillability. Trans. AIME 213, 1−6.
Environmental Aspects of Chemical Use in Well-Drilling Operations, 1975. Conference Proceedings, Houston, May 1975. Office of Toxic Substances, Environmental Protection Agency, Washington, DC.
EPA 1985. Federal Register 40 CFR, Part 425, Subpart A, Appendix 8.
Fertl, W.H., Chilingar, G.V., 1977. Importance of abnormal formation pressures. J. Petrol. Technol 29, 347−354.
Fraser, I.M., Moore, R.H., 1987. Guidelines for stable foam drilling through permafrost. In: SPE/IADC Paper 16055, Drilling Conference, New Orleans, LA, March 15−18, 1987.
Garnier, A.J., van Lingen, N.H., 1959. Phenomena affecting drilling rates at depth. J. Petrol. Technol. 216, 232−239.
Golis, S.W., 1984. Oil mud techniques improve performance in deep, hostile environment wells. In: SPE Paper 13156, Annual Meeting, Houston, TX.
Goodman, M.A., 1977. Arctic drilling operations present unique problems. World Oil 85, 95−110.
Helmick, W.E., Longley, A.J., 1957. Pressure-differential sticking of drill pipe and how it can be avoided or relieved. API Drill. Prod. Prac. 1, 55−60.
Hinds, A.A., Clements, W.R., 1982. New mud passes environmental tests. SPE Paper 11113, Annual Meeting, New Orleans, LA.
Jackson, S.A., Kwan, J.T., 1984. Evaluation of a centrifuge drill-cuttings disposal system with a mineral oil-based drilling fluid on a Gulf Coast offshore drilling vessel. SPE Paper 13157, Annual Meeting, Houston, TX.
Johancsik, C.A., Grieve, W.A., 1987. Oil-based mud reduces borehole problems. Oil Gas J. 42−45 (April 27), 46−58. (May 4).
Jordon, J.W., Nevins, M.J., Stearns, R.C., Cowan, J.C., Beasley, A.E. Jr. 1965. Well Working Fluids US Patent 3168471 February 2, 1965.
Kelley, J., Wells, P., Perry, G.W., Wilkie, S.K., 1980. How using oil mud solved North Sea drilling problems. J. Petrol. Technol. 931−940.
McAuliffe, C.D., Palmer, L.L., 1976. Environmental aspects of offshore disposal of drilling fluids and cuttings. SPE Paper 5864, Regional Meeting, Long Beach, CA.
Miller, G. 1944. Composition and Properties of Oil Base Drilling Fluids. US Patent 2356776 August 24, 1944.
Monaghan, P.H., McAuliffe, C.D., Weiss, F.T., 1976. Environmental Aspects of Drilling Muds and Cuttings from Oil and Gas Extraction Operations in Offshore and Coastal Waters. Offshore Operators Committee, New Orleans, LA.
Murray, A.S., Cunningham, R.A., 1955. Effect of mud column pressure on drilling rates. Trans. AIME 204, 196−204.
Omland, T., Vestbakke, A., Aase, B., Jensen, E., Steinnes, I., 2012. Criticality Testing of Drilling Fluid Solids Control Equipment. Paper 159894-MS presented at SPE Annual Technical Conference and Exhibition, 8-10 October, San Antonio, Texas, USA. http://dx.doi.org/10.2118/159894-MS.
Outmans, H.D., 1958. Mechanics of differential-pressure sticking of drill collars. Trans. AIME 213, 265−274.
Poletto, F., Miranda, F., 2004. Seismic While Drilling, 1st Edition Elsevier, Amsterdam.
Riley, M., Young, S., Stamatakis, E., Guo, Q., Ji, L., De Stefano, G., et al., 2012. Wellbore Stability in Unconventional Shales-The Design of a Nano-Particle Fluid. SPE paper 153729-MS, Oil and Gas India Conference and Exhibition, March 28-30, Mumbai, India. http://dx.doi.org/10.2118/153729-MS.

Robinson, L.H., Heilhecker, J.K., 1975. Solids control in weighted drilling fluids. J. Petrol. Technol. 1141–1144.

Sheffield, J.S., Sitzman, J.J., 1985. Air drilling practices in the Midcontinent and Rocky Mountain areas. In: IADC/SPE Paper 13490, Drilling Conference, New Orleans, LA.

Simpson, J.P., Cowen, J.C., Beasley Jr., A.E., 1961. The New Look in Oil-Mud Technology. J. Petroleum Technol. 13, 1177–1183.

Simpson, J., 1979. A new approach to OBM low cost drilling. J. Pet. Tech643–650, February 6.

St Pierre, R., Welch, D., 2001. High performance/High temperature water based drills Wilcox test. In: AADE National Drilling Conference Paper 01-NC-HO-54.

Stuart, C.A., 1970. Geopressures. Supplement to Proc. Abnormal Subsurface Pressure, Louisiana State University.

Swan, J., 1923. Method of drilling wells. US Patent 1455010, May.

Vidrine, D.J., Benit, E.J., 1968. Field verification of the effect of differential pressure on drilling rate. J. Petrol. Technol. 676–681.

Walker, T.O., Dearing, H.L., Simpson, J.P., 1984. The role of potassium in lime muds. SPE Paper 13161, Annual Meeting, Houston TX, September 16–19, 1984.

World Oil Magazine, 2015. Supplement, Guide to Drilling, Completion and Workover Fluids. June 2015.

Wright, C., 1954. Oil base emulsion drilling fluids. Oil Gas J88–90.

Zeidler, H.U., 1981. Better understanding permits deeper clear water drilling. World Oil, 167–178.

Chapter 2

Introduction to Completion Fluids

The producing reservoir is an operators most important asset. A completion fluid is placed in the well to facilitate final operations prior to producing the well and must be designed to do no harm. There are many different types of completion operations, such as casing cleanout, running production tubing with packers, open hole completions, setting screens and production liners, gravel packs, downhole valves, perforating into the producing zone, fracturing and downhole pumps. The completion fluid is meant to control the pressure in a well should downhole hardware fail. However, of prime importance is that the completion fluid be designed to be nondamaging to the producing formation as well as to the completion equipment. Traditionally, completion fluids are typically clear brines (chlorides, bromides, phosphates, formates, or acetates). Attention was placed on cleaning any residual solids in the fluid, primarily by ultrafiltration. The fluid also needed to be chemically compatible with the reservoir formation and interstitial fluids. Clear brines were filtered with small micron cartridge filters to obtain a less than 100 mg/L total solids content.

In the past, the drilling fluid was used to set the production casing and bring the well onto production. Today, a regular drilling fluid is seldom used. It is unsuitable for completion operations due to its solids content, pH, and ionic composition. A special drilling fluid called a reservoir drill-in fluid has been developed to be used as both a drilling fluid and completion fluid. These are designer fluids based on extensive laboratory testing. Reservoir drill-in fluids are discussed in Chapter 11, Completion, Workover, Packer, and Reservoir Drilling Fluids.

Chapter 1, Introduction to Completion Fluids, discussed the blending of personnel for both the drilling operations and completion operations groups to design a total drilling fluids and completion fluids program. With the trend toward more open hole completions, that concept has encompassed reservoir engineering input to develop complete drilling, completion, and production methodologies. These programs are based on intensely gathering data on the geology of the formations and complete characterization of the reservoir. In addition to field data, laboratory work is conducted for analysis.

COMPLETION FLUID SELECTION

Data Collection and Planning

Fig. 2.1 shows the process of gathering data, determining operating parameters, selecting equipment, and then deciding on the best completion fluid to use. The planning process steps (Jeu et al. 2002) are arranged to enable you to:

- know the bottomhole pressure (BHP)
- know bottomhole temperature (BHT)
- determine appropriate fluid density using true vertical depth (TVD)
- select the correct true crystallization temperature (TCT) of the brine
- estimate the volume of clear brine fluid (CBF) for the job
- determine compatibility issues, corrosion concerns, or formations sensitivity

Then the proper CBF family (single, two, or three-salt; inorganic or organic) can be selected.

Bottomhole conditions are the basic criteria that influence the selection of a clear brine completion fluid. When a brine is put into service, the downhole temperature profile will cause the brine to expand, lowering the average density of the fluid column. Pressure has the opposite effect and increases the density. Adjustments must be made to the fluid density to compensate for the combination of BHP and BHT.

The planning process includes:

Data acquisition—Depending on the importance of the completion project, a great deal of time and money can be spent to make certain the reservoir comes in contact with the appropriate completion fluid, as well as the best drilling fluid. Extensive laboratory core work may be necessary.

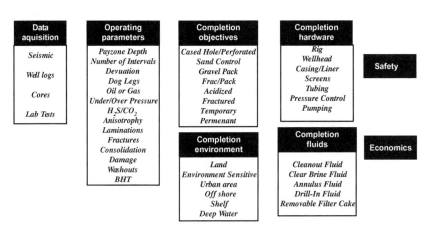

FIGURE 2.1 A typical process for planning a completion operation.

Operating parameters—This phase is critical for determining the details of subsequent decisions

Completion objectives and environment—The planning of the final objectives for the completion is greatly effected by the condition of the wellbore handed over by the drilling department. All local, regional, and national government rules and regulations are involved at this decision point. The various completion fluid options are also decided at this point, i.e., clear brine versus suspended solids, perhaps followed by acidizing or fracturing.

Completion hardware—This point in the planning process is when the cost factors are analyzed and acted upon.

Completion fluids—Throughout the planning process to develop the completion program, fluid decisions have been address and finalized. High-density clear brines are usually very expensive and are normally highly regulated in environmentally sensitive areas. Fluid densities above about 17.0 lb/gal (2.05 sg), usually either zinc and calcium chloride/bromide blends or cesium formate, are cost prohibitive in some situations. This has led to operators leasing the fluid from a service company. The operator pays for the fluid lost or left in the well and the rest is returned to the service company, cleaned up, and reused. Also adding to the cost of clear brines is the necessity of filtering the fluid to a very low total suspended solids content (SS). This involves different types of filtration equipment, such as diatomaceous earth plate filters to low micron size cartridge filters. The SS standard, as well as particle size distribution standard is dependent on lab analysis of cores, and is different for each producing formation.

Corrosion Resistant Alloys

In HPHT wells when corrosion resistant alloy (CRA) tubing is used environmentally assisted cracking (EAC) may occur. Where CRA tubulars are used the fluid selection process is different than for traditional well completions. Instead of selecting the fluid at the end of the process, metallurgy and fluids are selected concurrently for wells where a CRA will be used with a packer fluid. In these wells, it is important to take steps to decrease the probability of EAC by selecting the best combination of metallurgy and CBF for the specific well conditions (Leth-Olsen, 2004; Scoppio et al., 2004; TetraTec, 2016).

Safety and the Environment

Inorganic CBFs and are not highly reactive. Chemical constituents of CBFs include the positively charged cations of sodium (Na^+), potassium (K^+), ammonium (NH^{4+}), calcium (Ca^{2+}), and zinc (Zn^{2+}). Of these, only zinc and ammonium are regulated environmentally. None of them are considered highly toxic. The negatively charged anions are chloride (Cl^-) and bromide (Br^-), both of which are found in seawater, but at a much lower concentration than is found in a CBF (TetraTec, 2016). Organic CBFs are based on the formate or acetate ion. These CBFs are used with sodium, potassium, and cesium. They are low molecular weight materials and the anion is biodegradable.

Under EPA regulations, spills of completion fluids containing zinc bromide or ammonium chloride must be immediately reported to the National Response Center at 1.800.424.8802 if:

- the quantity of zinc bromide in the spill exceeds the 1000 lb RQ for zinc bromide; or
- the quantity of ammonium chloride in the spill exceeds the 5000 lb RQ for ammonium chloride.

WELLBORE CLEANOUT

Most service companies offer special cleanout services after the casing is set, prior to perforating. Cleanout operations involve the use of large amounts of water, either fresh or brine water, and surfactants to change oil-wet surfaces to water-wet, and mechanical scrapers, brushes, and water jets to remove scale and debris from the casing. Large volumes of fluids and lengthy time periods may be required to achieve the proper SS level. The cleanout process may also be required after perforating to help remove debris from the open perforations in overbalanced operations.

High-density clear brines have been and are used primarily for cased perforated holes. The goal in cased holes is to control the well with the high-density fluid instead of using mechanical means such as snubbing. For workover operations clear brines are used as kill fluids. The trend in recent years has been toward more open hole completions and a reduced use of clear brines. Techniques for sand control, removal of mud filter cake damage from the open hole and screens are taking more prominence in completion planning. Many times this requires adding solid bridging materials, such as sized salt or carbonates. The resulting filter cakes must be designed to be easily removed either prior to production or during the initial flow stages. See Chapters 7, The Filtration Properties of Drilling Fluids and Chapter 11, Completion, Workover, Packer, and Reservoir Drilling Fluids, for more details on bridging agents and reservoir drill-in fluids. In addition, the advent of hydraulic fracturing of shales in horizontal wells has resulted in a different class of fluid additives. In fracturing operations, a high-density base fluid is not needed as pressure control is done with applied pressure. See Chapter 12, Introduction to Fracturing Fluids, for more details on fracturing fluids.

COMPOSITION OF COMPLETION FLUIDS

Clear Brine Fluids

Table 2.1 lists the various brines used for solids free completion fluids. The primary use for these brines is to control pressures during the completion process until the production tubing is in place and the packer set. The brine may also be used as a packer fluid in the annular space above the packer, to control pressure if the packer fails or is released to workover the well. Fig. 2.2 shows the process of finding the final specification for the fluid to be used.

Inorganic brines include both monovalent and divalent cations. In general, monovalent cations are less hazardous and more environmentally acceptable for disposal than the divalent cation brines. The anion is either a chloride or bromide ion. These anions are restricted for waste disposal as they are not biodegradable and, in excess, will inhibit vegetation growth. Figs. 2.3 and 2.4 show the maximum specific gravities attainable for each brine.

NOTE: The following are inorganic brine descriptions from the Tetra Technologies, Inc., (2016) users manual.

Monovalent Inorganic Brines

Sodium Chloride (Dry)—a high-purity salt used in brines with a density range between 8.4 and 10.0 lb/gal (1.008−1.200 sg). When mixed with NaBr, the densities can reach up to 12.5 lb/gal (1.501 sg). It is normally packaged in 100-lb (45.4-kg), 80-lb (36.3-kg), and 50-kg (110-lb) sacks.

TABLE 2.1 Base Fluid Types

Clear Brine Base Fluids	
Inorganic Brines	**Organic Brines**
Monovalent	**Monovalent**
Sodium Chloride, NaCl	Sodium Formate, $NaCHO_2$
Sodium Bromide, NaBr	Potassium Formate, $KCH\,O_2$
Potassium Chloride, KCl	Cesium Formate, $CeCHO_2$
Potassium Bromide, KBr	Sodium Acetate, NaC_2HO_2
Ammonium Chloride, NH_4Cl	Potassium Acetate, KC_2HO_2
Divalent	Cesium Acetate, CeC_2HO_2
Calcium Chloride, $CaCl_2$	
Calcium Bromide, $CaBr_2$	
Zinc Bromide, $ZnBr_2$	

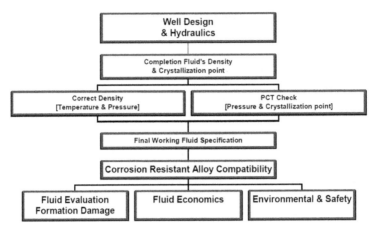

FIGURE 2.2 Flow chart for completion fluid selection (Tetra Technologies, Inc., 2016).

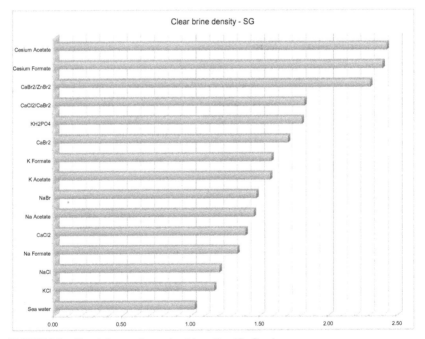

FIGURE 2.3 Clear brine maximum densities—Specific Gravity.

Potassium Chloride (Dry)—a high-purity salt that can reach brine densities from 8.4 lb/gal to 9.7 lb/gal (1.008–1.164 sg). It is used at 2–4% as a clay and shale stabilizer in drilling fluids, workover fluids, and reservoir drill-in fluids. It is packaged in 50-lb (22.7-kg) and 100-lb (45.4-kg) sacks.

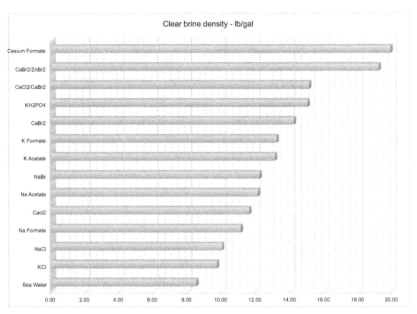

FIGURE 2.4 Clear brine maximum densities—Pounds/gallon.

Ammonium Chloride (Dry)—a high-purity salt that can produce brine densities from 8.4 to 9.7 lb/gal (1.008−1.164 sg). It is also used at 2−4% as a clay and shale stabilizer. It may release ammonia gas with pH above 9.0. Ammonium chloride (dry) is packaged in 50-lb (22.7-kg) and 25-kg sacks.

Sodium Bromide (Liquid)—a single-salt CBF. Pure sodium bromide solutions can be prepared with densities between 8.4 and 12.8 lb/gal (1.008−1.537 sg). Normally, it can be mixed with NaCl to prepare brines with densities between 10.0 and 12.5 lb/gal (1.200 and 1.501 sg). It is used where formation waters contain high concentrations of bicarbonate or sulfate ions. It can be formulated for various crystallization temperatures for summer or winter blends. It is packaged in bulk liquid quantities.

Sodium Bromide (Dry)—a high-purity salt. Pure sodium bromide solutions can be prepared with densities between 8.4 and 12.8 lb/gal (1.008−1.537 sg). Normally, it can be mixed with NaCl to prepare brines with densities between 8.4 and 12.5 lb/gal (1.008 and 1.501 sg). It is used where formation waters contain high concentrations of bicarbonate or sulfate ions and is packaged in 55-lb (25-kg) sacks.

Phosphate—inorganic salt of phosphoric acid. Potassium hydrogen phosphate brine, comprising of a blend of dipotassium hydrogen phosphate (K_2HPO_4) and potassium dihydrogen phosphate (KH_2PO_4), has been used as completion and workover fluid in China and Indonesia (Sangka and Budiman, 2010). The maximum density achievable with potassium hydrogen phosphate brine is thought to be about 1.78 g/cm^3/14.9 lb/gal, although lower density brines have been found to crystallize at ambient conditions. They are low cost and environmentally friendly.

However, Downs reported (Downs, 2012) that when these brines invade a reservoir they can cause two types of formation damage:

- Formation of insoluble scales on contact with multivalent cations in formation water. For example, in a very typical formation water containing soluble calcium and iron, the trivalent phosphate anion reacts to form tricalcium phosphate [$Ca_3(PO_4)_2$], hydroxylapatite [$Ca_5(PO_4)_3 \cdot OH$], ferric hydroxide [$Fe(OH)_3$], and strengite [$FePO_4 \cdot 2H_2O$]. The phosphate scales form hard bone-like deposits, whereas the hydroxide forms gels.
- Phosphates absorb strongly onto mineral surfaces where they can form precipitates and complex insoluble salts from exposure to multivalent cations (Nowack and Stone, 2006, Goldberg and Sposito, 1984). These absorption reaction products block pore throats and reduce formation permeability. A laboratory study that showed potassium hydrogen phosphate brine invasions of low-permeability sandstone cores under HPHT conditions was reported by Downs (2012).

Divalent Inorganic Brines

Calcium Chloride (Dry or Liquid)—available in sacks or as a concentrated solution. It is manufactured at two different densities depending on the source, i.e., 11.6 lb/gal (1.392 sg) and 11.3 lb/gal (1.356 sg). Liquid calcium chloride is the most inexpensive form. It can be used to prepare solids free of brine with a density from the concentrate to 8.4 lb/gal (1.008 sg). A pelletized solid form of $CaCl_2$ is used at the rig site to alter the fluid density if needed.

The solid form of calcium chloride contains trace amounts of insoluble impurities which cause brines mixed on location to be more turbid than premixed brines. These impurities are not present in plant-manufactured brines for two reasons:

- The solids in plant-manufactured brines settle out over time.
- Plant-manufactured brines are filtered to two microns prior to shipment to eliminate insoluble contaminants.

When solid calcium chloride is added to water, a great amount of heat is produced. Adding the solid calcium chloride too fast could result in enough heat to bring the temperature of the solution to above 200°F. Safe handling must be used at all times to avoid individuals from being burned by the hot fluids or equipment. Less heat is produced when the concentrated solution is diluted to prepare the desired density. As a result, problems related to heat are generally not encountered. Produced brines or seawater should not be used to prepare calcium chloride completion fluids because sodium chloride and/or insoluble calcium salts may precipitate.

Calcium Chloride/Bromide blends—clear brines having a density range of 11.7 lb/gal (1.404 sg) and 15.1 lb/gal (1.813 sg) are prepared using a combination of calcium chloride and calcium bromide. Liquid $CaCl_2$, pelletized calcium chloride, concentrated liquid $CaBr_2$, or solid calcium bromide powder is used in combination to prepare these brines. $CaBr_2$ concentrate is produced at a density of 14.2 lb/gal (1.705 sg). Calcium bromide costs about five times as much as calcium chloride. When TCT and density requirements allow, field-prepared brines should contain as much calcium chloride as is practical.

Increasing the density of $CaCl_2-CaBr_2$ blended brine by adding dry salts can cause problems in the wells if proper blending techniques are not employed. For example, the addition of calcium bromide powder to a saturated blend can result in the precipitation of calcium chloride. Under these conditions, both water and calcium bromide must be added to avoid precipitation.

High-density, solids free brines ranging up to 15.3 lb/gal (1.837 sg) can be prepared using either calcium bromide or the combination of calcium bromide and calcium chloride. The ratio of bromide-to-chloride in any particular density determines the true crystallization temperature (TCT), or "freezing point." Crystallization temperature must always be considered when blending brines of any type, however, the chloride–bromide brines are particularly sensitive because small changes in the ratio of the two salts can result in significant changes in TCT. Environmental factors such as surface temperature, water depth, and water temperature and the influence of pressure on the crystallization point are important considerations and must be taken into account when formulating the proper blend.

NOTE: High-density slugs are used to insure that a dry string is pulled when coming out of the hole. This is an important safety consideration since calcium bromide brines may cause irritation to the skin and eyes if they come into contact.

NOTE: When solid calcium bromide is added to freshwater, a significant amount of heat is released. Precautions must be taken to avoid getting splashed by the hot liquid or burned by hot equipment. Unlike calcium chloride, this is not a problem when liquid calcium bromide is added to water because very little heat is generated.

Calcium chloride, Calcium bromide, and zinc bromide—concentrated zinc bromide–calcium bromide solutions are manufactured to a density of 19.2 lb/gal (2.305 sg). Solution densities between $\pm$ 14.0 and 19.2 lb/gal (1.681–2.305 sg) are prepared by blending this 19.2 lb/gal "stock" fluid with lower density calcium bromide or calcium bromide–calcium chloride brines. The three-salt formulations are less expensive due to the presence of calcium chloride. As with the lower density chloride–bromide brines, special blend formulations are used to attain a specific density and TCT.

NOTE: Zinc bromide or zinc bromide–calcium bromide solutions of up to 20.5 lb/gal are also offered in smaller quantities for slugging or spiking purposes. When agitated in pits which are exposed to the atmosphere for as little as four hours, the density of these concentrated liquids can drop by as much as 0.02 lb/gal. A calm solution does not pick up moisture as readily and will not lose density as quickly. To avoid absorption of moisture from the atmosphere, these high-density brines should be mixed and kept in covered tanks.

Monovalent Organic Brines

Sodium Formate (Dry)—a high-purity, organic salt that can deliver brine fluid densities ranging from 8.4 lb/gal (1.008 sg) to 11.1 lb/gal (1.330 sg). It is packaged in 25-kg (55-lb) sacks and 1000-kg "big" bags.

Potassium Formate (Liquid)—a single-salt CBF. Pure potassium formate solutions can be prepared with densities between 8.4 lb/gal (1.08 sg) and 13.1 lb/gal (1.571 sg). Potassium formate provides excellent thermal stabilization effects on natural polymers. The potassium ion provides exceptional clay stabilization and swelling inhibition of shales.

Potassium Formate (Dry)—a high-purity, organic salt with eventual densities between 8.4 lb/gal (1.008 sg) and 13.1 lb/gal (1.573 sg). It is packaged in 55-lb (25-kg) sacks or in 1000-kg "big" bags.

Cesium Formate (Liquid)—a single-salt CBF. Pure cesium formate systems can be prepared with densities between 8.7 lb/gal (1.05 sg) and 20.0 lb/gal (2.40 sg), but cesium formate is most often commercially available at 17.5 lb/gal (2.10 sg) and 18.3 lb/gal (2.20 sg). Like potassium formate, cesium formate provides excellent thermal stability on natural polymers, clay stabilization, and shale swelling inhibition.

BRINE TEST PROCEDURES

Density

Brine densities are set by the salt type and its concentration. The brine density decreases with temperature increases and increases with pressure. As such, the brine density at ambient atmospheric conditions is not a reliable indicator of brine density downhole.

The reference temperature to report densities of heavy brines is 70°F (20°C). Hydrometer readings at the surface, corrected to the reference temperatures, are not used to calculate hydrostatic pressures. Accurate hydrostatic pressures in the wellbore must be calculated by integrating the density changes due to changing temperatures and pressures in the wellbore. See the following section—*Temperature and Pressure Effect on Density.*

Glass hydrometers are normally used to measure the density of a brine. Corrections need to be applied to the readings to allow for glass thermal expansion and contraction. Sometimes gases are entrained in the brine. The API pressurized mud balance can be used to more accurately measure the density. See API RP 13J (2014) for details on density measurements.

Temperature and Pressure Effect on Density

In the wellbore, brine densities are greatly affected by the changing temperatures and pressures encountered. Fig. 2.5 (King, 2016) shows the changes in equivalent static density (ESD) of a brine column from the surface to TVD. The highest density is at the mudline. Mudline temperatures around the world are approximately 40°F (4°C).

NOTE: The following is adapted from the Tetra Technologies, Inc. completion fluids users manual (2016).

Completion fluids exhibit the typical volumetric response to temperature and pressure, i.e., expanding with increasing temperature and compressing with increasing pressure. In shallow waters or land-based wellbores,

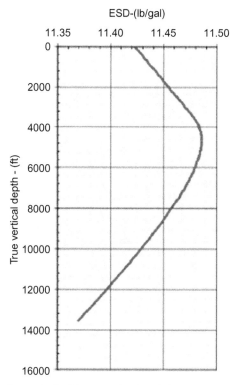

FIGURE 2.5 Equivalent static densities at varying downhole temperatures (King, 2016).

the expansion of completion fluids with temperatures produce a more pronounced effect on volume than pressure. This overall increase in volume results in a fluid of lower density at the bottom of the well rather than at the surface. In deep-water environments, the depth of cold water will impact the expansion/compression relationship such that the fluid at the mudline is heavier than that at the surface. The combination of hydrostatic pressure and cold temperature can have catastrophic effects unless the fluid is properly formulated to account for its environment.

For fluids with densities less than approximately 12.0 lb/gal (1.45 sg), thermal expansion will typically be in the range of 0.26 lb/gal (0.03 sg) to 0.38 lb/gal (0.046) per 100°F (38°C) increase in temperature. From 12.0 lb/gal (1.45 sg) to 19.0 lb/gal (2.28 sg), the expansion ranges from 0.33 lb/gal (0.04 sg) to 0.53 lb/gal (0.06 sg) per 100°F (38°C) increase. Typically, the density correction is made for the average temperature of the fluid column. Pressure effects are much smaller and range from 0.024 lb/gal (0.003 sg) per thousand psi (6.9 kPa) (TetraTec, 2016)

Density Prediction

The ability to calculate the hydrostatic pressure at any point in a wellbore containing a column of completion fluid is necessary for proper use—see Fig. 2.6 (King, 2016). Because hydrostatic pressure increases with depth and is directly related to density, which may be increasing with depth in

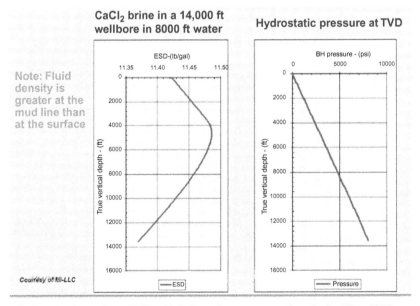

FIGURE 2.6 Comparison of ESD with hydrostatic pressure at various depths (King, 2016).

deep-water or decreasing with depth as the temperature increases, it is necessary to mathematically predict the density of the completion fluid under the combined influence of compression and temperature.

The bottomhole density can be calculated with detailed PVT data of the fluid in question. If such data is nonexistent, downhole density and total hydrostatic pressure at depth can be approximated by the following equations. Thermal expansion and compressibility factors are provided in Tables 2.3.

Total Hydrostatic Pressure in the Wellbore.

$$Psi_h = 0.052 \times D_{avg} \times TVD$$

where,

Average Brine Density in a Wellbore.

$$D_{avg} = \frac{(2000 - 0.052 \times C_f \times TVD) \times DD_{surf} - 10 \times V_e \times (BHT - T_s)}{2000 - 0.104 \times C_f \times TVD}$$

V_e = Temperature expansion factor, lbm/gal/100°F (Table 2.2)
C_f = Pressure compressibility factor, lbm/gal/1000 psi (Table 2.3)
TVD = Total vertical depth, ft
D_{surf} = Density at surface, lbm/gal
BHT = Bottomhole temperature, °F
T_s = Temperature at surface, °F
Psi_h = Hydrostatic pressure, psi

Formate Brines Densities and PVT Data

Cabot Special Fluids has published PVT data for formates and formate blends in their Formate Manual (Cabot, 2016). Cabot also has a free software package that calculates densities from actual field data using their published PVT data.

TABLE 2.2 Expansibility of Brine at 12,000 psi from 76°F to 198°F (Tetra Technologies, Inc., 2016)

Brine Type	Density (lbm/gal)	V_e (lbm/gal/100°F)
NaCl	9.42	0.24
CaCl$_2$	11.45	0.27
NaBr	12.48	0.33
CaBr$_2$	14.13	0.33
ZnBr$_2$/CaBr$_2$/CaCl$_2$	16.01	0.36
ZnBr$_2$/CaBr$_2$	19.27	0.48

TABLE 2.3 Compressibility of Brines at 198°F from 2000 to 12,000 psi (Tetra Technologies, Inc., 2016)

Brine Type	Density (lbm/gal)	C_f (lbm/gal/1000 psi)
NaCl	9.49	0.019
$CaCl_2$	11.45	0.017
NaBr	12.48	0.021
$CaBr_2$	14.30	0.022
$ZnBr_2/CaBr_2/CaCl_2$	16.01	0.022
$ZnBr_2/CaBr_2$	19.27	0.031

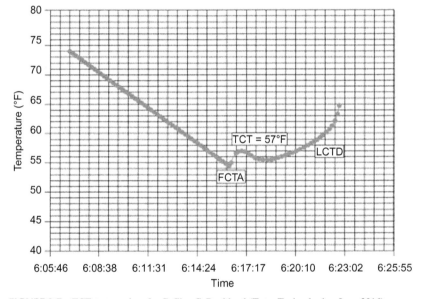

FIGURE 2.7 TCT test results of a $CaCl_2-CaBr_2$ blend (Tetra Technologies, Inc., 2016).

Thermodynamic Crystallization Temperature

Thermodynamic crystallization temperature, also called the true crystallization temperature (TCT), is that temperature at which the brine solution is fully saturated with respect to the least soluble salt. Fig. 2.7 represents the TCT test results of an example $CaCl_2-CaBr_2$ completion brine. Included in the diagram is the first crystal to appear (FCTA) and the last crystal to dissolve (LCTD). Fig. 2.8 presents the phase diagram (TCT vs. Temperature) for various common completion fluids. Crystallization of the fluid is a result

Introduction to Completion Fluids **Chapter | 2** 49

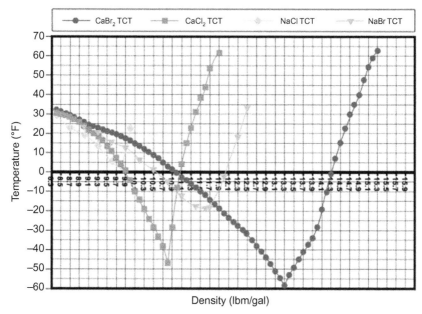

FIGURE 2.8 Phase diagrams, TCT versus temperature, for common clear brine fluids (Tetra Technologies, Inc., 2016).

of hydrostatic pressure and is referred to as pressurized crystallization temperature (PCT). Fig. 2.9 shows the impact of pressure on the TCT of a $CaCl_2-CaBr_2$ completion brine with a TCT of 40°F (4.4°C).

TCT and Supercooling of Formate Brines

The standard API TCT procedure for halide brines does not work for some formate brines. Potassium and cesium formate brines and their blends behave differently due to strong kinetic effects. Potassium formate can form a metastable crystal. The crystallization of metastable crystals is lower than the TCT standard measurement. Figs. 2.10 and 2.11 show the TCT and metastable temperatures for K-formate. The supercooling points indicate the temperature where the brine has been successfully kept for at least two weeks with and without seeding material present. The formate technical manual gives an alternate to the standard API TCT test (Cabot, 2016).

Brine Clarity

Solids contamination can cause substantial formation damage to a producing zone. The suspended solids concentration is estimated by the clarity (turbidity) of the brine as measured by a nephelometer. The nephelometer measures light transmission through the sample and provides data expressed as

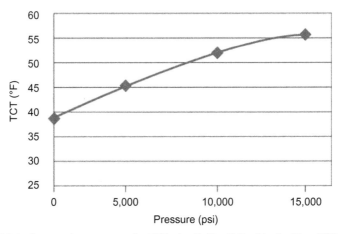

FIGURE 2.9 Impact of pressure on the TCT of a $CaCl_2-CaBr_2$ blend with a TCT of 40°F (4.4°C) (Tetra Technologies, Inc., 2016).

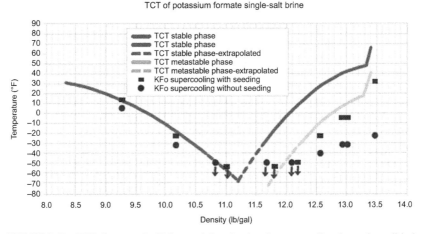

FIGURE 2.10 TCT diagram of a K-formate brine showing the supercooling data points—lb/gal density (Cabot, 2016).

nephelometric turbidity units (NTU). The NTU of the test sample is compared to a standard sample of known NTU (API RP13J 2014).

Turbidity is generally used by most operators. The nephelometer measurement however, is an indirect measurement of solids in the fluid. Standards used by operators can range from 10 NTU or higher. Solids can also be measured gravimetrically, which is used by operators requiring tighter solids control. The fluid is filtered on predetermined sized filters and the absolute weight of solids is determined. Standards used by operators can range from 10 mg/L to higher.

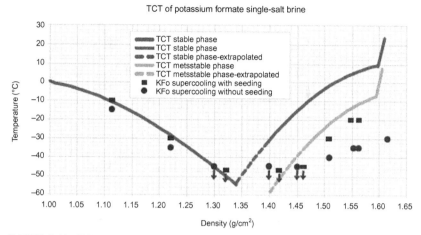

FIGURE 2.11 TCT diagram of a K-formate brine showing the supercooling data points—SG density (Cabot, 2016).

The most stringent method for solids content used by operators with highly sensitive formations is by using a particle counter, which is typically a laser particle counter. This device monitors particle size and concentration. Standards used by operators range from specified maximum particle size, minimum particle count reduction, and maximum particle concentrations.

Contaminants

Contaminants that can affect completion and workover fluids include:

- Iron
- Solids
- Oil, distillate, grease, and pipe dope
- Polymers
- Surfactants

Iron—Iron can be a contaminant in either soluble or insoluble form. Soluble iron is a product of corrosion and is common in zinc fluids. Soluble iron can form a precipitate, which can cause formation damage. Iron should be removed at the brine plant by adding hydrogen peroxide, flocculation, and then filtering the fluid. On location, treating a fluid for iron is very difficult and is usually successful only in low-density brines. The treatment consists of increasing pH with caustic or lime and removing the precipitated iron by filtration.

Solids—Solids that are not added to the system to enhance its performance are considered contaminants. These include formation clays, precipitates, polymer residues, corrosion by-products, and scales, among other things.

Contaminants can be filtered at the brine plant or wellsite using diatomaceous earth, a plate and frame press, and/or two-micron absolute cartridges. It is recommended that a clear completion fluid not be sent to the wellsite with an NTU greater than 40 or a suspended solids concentration greater than 50 ppm.

Oil, distillate, grease, and pipe dope—Produced oils and other hydrocarbons affect brine density and can also blind filtration units. Hydrocarbons will form a separate layer above heavy brine and can be pumped off the surface.

Polymers—Brines contaminated with polymers usually cannot be filtered without chemical and/or special mechanical treatment at the plant site where hydrogen peroxide can be used to oxidize polymers and permit filtration. At the wellsite, polymer pills used in displacement should be caught and isolated from the active brine system.

Surfactants—Any surfactant package used in wellbore cleanout or as an additive to the brine can cause compatibility problems and possible formation damage. Lab tests, including return permeability tests, should be done prior to displacing the wellbore to the clear brine.

REFERENCES

API 2014. Testing of Heavy Brines. Recommended Practice RP 13J, American Petroleum Institute, Washington, DC. www.API.org.

Brangetto, M., Pasturel, C., Gregoire, M., Ligertwood, J., Downs, J., Harris, M. et al., 2007. Caesium formate brines used as workover, suspension fluids in HPHT field development IADC Drilling Contractor, May/June.

Cabot 2016. Formate Technical Manual. Cabot Specialty Fluids, Aberdeen, Scotland. www.formatebrines.com.

Carpenter, J., 2004. A new field method for determining the levels of iron contamination in oilfield completion brine. In: SPE Paper 86551 Presented at the SPE Formation Damage Control Symposium, Lafayette, LA, 18–20 February.

Downs, J., 2012. Exposure to phosphate-based completion brine under HPHT laboratory conditions causes significant gas permeability reduction in sandstone cores. In: IPTC 14285 Prepared for Presentation at the International Petroleum Technology Conference, Bangkok, 7–9 February.

Goldberg, S., Sposito, G.,, 1984. A chemical model of phosphate adsorption by soils: I. Reference oxide minerals.. Soil Sci. Soc. Am. J. 48, 772–778.

Jeu, S., Foreman, D., Fisher, B., 2002. Systematic approach to selecting completion fluids for deep-water subsea wells reduces completion problems. In: AADE -02-DFWM-HO-02 AADE 2002 Technology Conference "Drilling & Completion Fluids and Waste Management", held at the Radisson Astrodome Houston, TX, 2–3 April.

King, G., 2016. www.GEKINGConsulting.com.

Leth-Olsen, H., 2004. CO2 Corrosion of Steel in Formate Brines for Well Applications. NACE 04357.

MISwaco, Completion Fluids Manual 2016.

Morgenthaler, L., 1986. Formation damage tests of high-density brine completion fluids. In: SPE paper 14831, SPE Symposium on Formation Damage Control, Lafayette, LA, 26–27 February.

Messler, D., Kippie, D., Broach, M., Benson, D., 2004. A potassium formate milling fluid breaks the 400°fahrenheit barrier in the deep tuscaloosa coiled tubing clean-out. In: SPE Paper 86503, SPE International Symposium on Formation Damage Control, Lafayette, LA, 18–20 February.

Nowack, B., Stone, A., 2006. Competitive adsorption of phosphate and phosphonate onto goethite. Water Res. 40, 2201–2209.

Rodger, P., Wilson, M., 2002. Crystallization Suppression of Cesium Formate. Department of Chemistry Report, University of Warwick, June.

Sangka, N.B., Budiman, H., 2010. New high density phosphate completion fluid: a case history of exploration wells: KRE-1, BOP-1, TBR-1 and KRT-1 in Indonesia. In: SPE Paper 139169 Presented at the SPE Latin American and Caribbean Pteroleum Engineering Conference, Lima ,Peru 1–3 December.

Scoppio, L., Nice, P., Nødland, S., LoPiccolo, E., 2004. Corrosion and Environmental Cracking Testing of a High-Density Brine for HPHT Field Application. NACE 04113.

Tetra Technologies Inc, 2016. Completion Fluids Technical Manual. Tetra Technologies, Inc, The Woodlands, TX, www.tetratec.com.

Chapter 3

Evaluating Drilling Fluid Performance

INTRODUCTION

Devising tests that will accurately describe how drilling fluids behave downhole is virtually an impossible task. Most drilling fluids, especially water-based muds (WBMs), are complex mixtures of interacting components, and the properties can change markedly with changes in temperature, pressure, mixing shear, and time. As they circulate through the wellbore, drilling fluids are subjected to ever-changing conditions—turbulent flow in the drill pipe, intense shearing at the bit and laminar flow in the annulus at frequently changing flowing shear rates, and a constantly changing geothermal temperature gradient. The viscous and elastic properties of most muds are time, temperature, and pressure sensitive and they seldom have time to adjust to any one set of conditions while circulating. In addition, there is a continuous composition change as solids and liquids from the formation are incorporated into the mud by the drilling process. Another problem is that tests at the wellsite must be performed quickly and with simple, rugged apparatus. To a lesser extent, this limitation also applies to laboratory tests made in support of field operations.

It is not surprising, therefore, that the standard field tests that have been accepted by industry are quick and practical, but only approximately reflect downhole behavior. Nevertheless, these tests serve their purpose very well if their limitations are understood and if the data obtained from them are correlated with experience.

Over the years, various tests that more closely simulate downhole conditions have been devised by individual investigators. These tests require more elaborate and expensive equipment, are time-consuming, and are therefore more suited to laboratory use and for research and development. In many cases, the results from these tests are discussed in subsequent chapters; in this chapter, the standard API equipment and procedures will be briefly described and the appropriate references given.

The future of the drilling industry, and the petroleum industry in general, will be the use of automation in all operational aspects. Data will be electronically captured at the wellsite and transmitted to a central location for analysis. This will occur in drilling fluids testing also. Feedback loops will be in place so that the changing data will automatically add the proper additives to maintain optimum functioning of the drilling fluids. Some automated mud tests are already in place, but more work needs to be done to truly automate drilling fluids operations (Magalhaes et al., 2014; MacPherson et al., 2013; Oort et al., 2016; Vajargah and van Oort, 2015).

LAB SAMPLE PREPARATION

Since the properties of muds depend so much on shear history and on temperature, it is of the utmost importance that muds tested in the laboratory first be subjected to conditions similar to those prevailing in the drilling well. Samples from the wellsite will have had time to cool, and, if thixotropic, will have set up a gel structure. Such muds must be sheared at the temperature observed in the flow line until the viscosity corresponds to that measured at the rig.

Muds made in the lab from dry materials must be given a preliminary mixing, and then be aged for a day or so to allow the colloids time to hydrate and any chemical reactions to proceed to completion. Then, the mud is subjected to a high rate of shear until a constant viscosity is obtained, and then a full set of properties are tested at ambient temperature. If the mud is intended for use in a well with a bottomhole temperature greater than 212°F (100°C), it must be aged at the temperature of interest, as outlined later in this chapter.

Mixers such as the Hamilton Beach (Fig. 3.1) and the multispindle mixer (Fig. 3.2) are used in laboratory tests of mud materials. They do not, however, produce the high rates of shear that exist in circulation of the drilling fluid in wells. High rates of shear are only obtained when there is little clearance between the stator and the rotor, or when the mud is pumped through a small orifice or opening. Food blenders in which the blades rotate in a recessed section at the bottom of container (Fig. 3.3) provide high shearing rates, and are suitable for shearing small quantities (about a liter) of mud. They can only be used for short periods of time because the temperature rises rapidly, with consequent loss of water by evaporation and possible additive degradation.

For larger amounts of mud and to ensure complete emulsification of nonaqueous fluids, it is best to use a high-shear mixer such as the Silverson dispersater (Fig. 3.4). This instrument consists essentially of a circulating unit that is mounted on rods extending from the base of the driving motor. This arrangement enables the unit to be lowered into a large vessel (about 8 L) of mud, and to circulate the mud throughout the vessel. The clearance between the rotor blades and the baffles on the housing is close so that a high rate of shear is maintained in the circulating unit.

FIGURE 3.1 Hamilton Beach mixer and cup. *Courtesy of Fann Instrument Company.*

PROPERTIES MEASURED

Density

Density, or mud weight, is determined by weighing a precise volume of mud and then dividing the weight by the volume. The mud balance provides the most convenient way of obtaining a precise volume. The procedure is to fill the cup with mud, put on the lid, wipe off excess of mud from the lid, move the rider along the arm till a balance is obtained, and read the density at the side of the rider towards the knife edge. Fig. 3.5 shows a standard mud balance. Fig. 3.6 is of the pressurized balance, developed to minimize the effect of entrained air in the sample.

FIGURE 3.2 Hamilton Beach three spindle mixer. *Courtesy of Fann Instrument Company.*

Density is expressed in pounds per gallon (lb/gal), pounds per cubic foot (lb/ft^3), grams per cubic centimeter (g/cm^3), specific gravity (SG), or as a gradient of pressure exerted per unit of depth (psi/ft). Conversion factors are as follows:

$$\text{Specific gravity (SG)} = \text{g/cm}^3 = \text{lb/gal}/8.33 = \text{lb/ft}^3/62.3 \quad (3.1)$$

$$\text{Mud gradient in psi/ft} = \text{lb/ft}^3/144 = \text{lb/gal}/19.24 = \text{SG} \times 0.433 \quad (3.2)$$

$$\text{Mud gradient in kg/cm}^2/\text{m} = \text{SG} \times 0.1 \quad (3.3)$$

The mud balance can be calibrated with fresh water. At 70°F (21°C) the reading should be 8.33 lb/gal, 62.3 lb/ft^3, or 1.0 SG. Further calibration can

FIGURE 3.3 High-speed blenders. Narrow clearance between blades and baffles results in high shear rates. *Courtesy of Fann Instrument Company.*

FIGURE 3.4 Silverson dispersator, Model L5M-A.

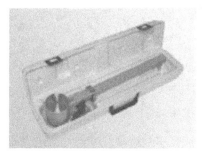

FIGURE 3.5 Standard mud balance. *Courtesy of OFI Testing Equipment.*

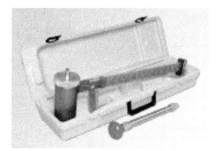

FIGURE 3.6 Pressurized mud balance. *Courtesy of OFI Testing Equipment.*

be done by weighing a saturated salt solution of known SG. Instructions for adjusting the calibration are given in the manufacturer's manual or website.

Viscosity

The Marsh funnel. Harlan Marsh, General Petroleum Company, developed a funnel device to measure a relative flowing viscosity of a drilling fluid. He donated this concept to the drilling industry (Marsh, 1931). This instrument is useful on the drilling rig, where it enables the crew to periodically report the consistency of the mud so that the mud specialist can examine any significant changes by detailed analysis.

The Marsh funnel consists of a funnel and a measuring cup (Fig. 3.7), and gives an empirical value for the consistency of the mud. The procedure is to fill the funnel to the level of the screen and to then observe the time (in seconds) of efflux of one quart (946 cc). The number obtained depends partly on the effective viscosity at the rate of shear prevailing in the orifice, and partly on the rate of gelation. The time of efflux of fresh water at $70 \pm 5°F$ ($21 \pm 3°C$) is 26 ± 0.5 s. In some areas of the world, Marsh funnel viscosity is reported as seconds per liter.

FIGURE 3.7 Marsh funnel and measuring cups. *Courtesy of Fann Instrument Company.*

Direct-indicating viscometers. These instruments are a form of concentric cylinder viscometer that enable the variation of shearing stress with shear rate to be observed. The essential elements are shown in Fig. 3.8. A bob suspended from a spring hangs concentrically in an outer cylinder. The assembly is lowered to a prescribed mark in a cup of mud, and the outer cylinder rotated at a constant speed. The viscous drag of the mud turns the bob until balanced by the torque in the spring. The deflection of the bob is read from a calibrated dial on the top of the instrument. Multiplying the dial reading by 1.07 gives the shear stress in lb/100 ft^2 at the surface of the bob.

The direct-indicating viscometer is available in several forms. The two-speed viscometer was designed to enable the Bingham Plastic rheological parameters to be easily calculated (Savins and Roper, 1954). The plastic viscosity (PV), in centipoises, is calculated by subtracting the 300 rpm dial reading from the 600 rpm reading, and the yield point (YP), in lb/100 ft^2 (multiply by 0.05 to obtain kg/m^2), is obtained by subtracting the PV from the 300 rpm reading. The API apparent viscosity (AV), in centipoises, is obtained by dividing the 600 rpm reading by 2.

The Power Law constants (n and K) are calculated from any two dial readings from a multispeed viscometer as follows:

$$n = \log(DR_2) - \log(DR_1)/||\log(RPM_2) - \log(RPM_1)|| \qquad (3.4)$$

$$K, \text{ dyn sec}/\text{cm}^2 = 5.11(\text{Dial Reading})^n/\text{Shear Rate} \qquad (3.5)$$

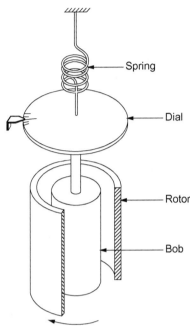

FIGURE 3.8 Schematic diagram of the direct indicating viscometer. The deflection in degrees of the bob is read from the graduated scale on the dial. *Courtesy of J.D. Fann.*

The theory underlying the above calculations is discussed in Chapter 6, Rheology and Hydraulics.

The shear rates prevailing in the two-speed direct-indicating viscometer are much higher than those usually prevailing in the annulus. At lower shear rates, such as those in the annulus, the effective viscosity (EV) and K value of many muds increase; the true YP is less than predicted from the 600 and 300 rpm readings; and n decreases. Therefore, when flow parameters are being determined for the purpose of calculating pressures in the annulus, it is advisable to use a multispeed viscometer. The common rigsite multispeed, standalone viscometer is available from drilling fluid equipment suppliers in rotary speeds of 6, 8, 12, and 16 speeds. The most common field units are the six- and eight-speed devices (Fig. 3.9). All these viscometers are manufactured to API rotor and bob specifications. Table 3.1 shows the approximate shear rates from the six- and eight-speed viscometers.

These shear rates cover most of the fluid flow situations while pumping a drilling fluid. Table 3.2 shows the ranges of shear rates encountered while pumping a drilling fluid.

It is important to realize that the viscometer dial reading is not viscosity. Table 3.3 shows the conversion constants to calculate the EV at each speed on the standard field viscometer. The dial reading for each speed is

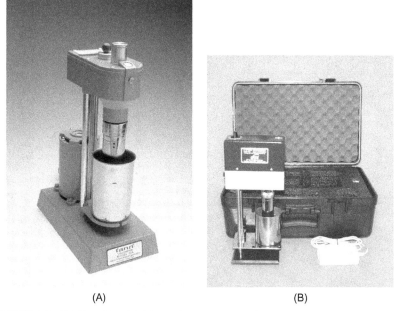

(A)   (B)

FIGURE 3.9 (A) Fann Model 35 6-speed viscometer. (B) OFI Testing Equipment Model 800 8-speed viscometer.

TABLE 3.1 Approximate Shear Rates for API Standard Rotor/Bob Configuration

RPM	Shear Rate, s^{-1}
600	1022
300	511
200	370
100	170
60	102
30	51
6	10.2
3	5.1

multiplied by the appropriate factor to obtain an approximation of the effective vicscosity in cP for that speed.

Continuous speed viscometers are also available for drilling fluid use. These can either be manually operated or computer controlled. Fig. 3.10 shows two

TABLE 3.2 Ranges of Shear Rates Encountered in a Circulating Drilling Fluid System

Standpipe and inside the drill pipe and collars	500–10,000 s^{-1}
Solids removal equipment	400–4000 s^{-1}
Annulus flow rates—for hole cleaning	1–170 s^{-1}
Flow through surface tanks	1–10 s^{-1}
Particle suspension	<1.0 s^{-1}

TABLE 3.3 Multiplication Factors to Convert Viscometer Dial Readings to Effective Viscosities in cP

600	0.5
300	1.0
200	1.5
100	3.0
60	5.0
30	10
6	50
3	100

types of unpressurized, continuous speed viscometers. In continuous mode, the shear stress versus time is either saved manually, saved in digital format, or plotted on a computer screen and saved. It is useful for observing hysterisis loops (See Chapter 6) and for observing changes in shear stress with time at constant speed and temperature. These devices operate at atmospheric pressure and can be heated to about 180°F (80°C).

Several viscometers have been developed to automatically measure shear stress, viscosities, and rheological properties of drilling and completion fluids at any shear rate of interest and at elevated temperature and pressure. Many of these devices can operate at shear rates from 0.001 to above 1000 s^{-1}. These viscometers can test fluids to above 600°F (315°C) and up to 30,000 or 40,000 psi (>275,000 kPa). It is also possible to attach a chiller to these viscometers to measure low-temperature viscosities, such as are encountered in deep-water drilling.

Gel Strength

Gel strengths are determined in the two-speed direct-indicating viscometer by slowly turning by hand the driving wheel on the top or side of the

FIGURE 3.10 Continuous speed viscometers. (A) OFI Testing Equipment Model 900 viscometer and (B) Fann Instrument Company RheoVADR rheometer.

instrument and observing the maximum deflection before the gel breaks. The same procedure is followed in the multispeed viscometer, except that the cylinder is rotated at 3 rpm with the motor. The maximum deflection is the gel strength. Gel strengths are measured after allowing the mud to stand quiescent for any time interval of interest, but they are routinely measured after 10 s (initial gel strength) and 10 min. The dial reading gives the gel strength in approximately pounds per hundred square feet.

API Fluid Loss

Static Filtration

The low-pressure static filtration press in use today is based on an original design by Jones (1937). The essential components are shown in Fig. 3.11. Several modifications of this cell are commercially available. The standard dimensions are: filtration area, 7.1 in^2 (45.8 cm^2); minimum height, 2.5 in (6.4 cm); and standard filter paper, Whatman 50, S & S No. 576, or equivalent. Pressure of 100 psi (7.0 kg/cm^2), from either a nitrogen cylinder or a carbon dioxide cartridge, is applied at the top of the cell. The amount of filtrate discharged in 30 min is measured in mL, and the thickness of the filter

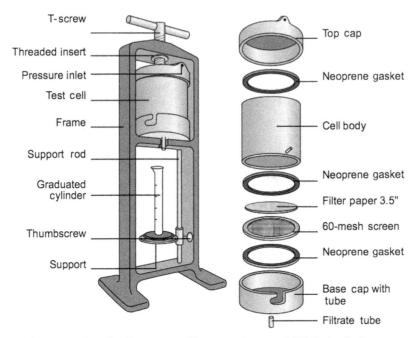

FIGURE 3.11 Schematic of low pressure filter press. *Courtesy of OFI Testing Equipment.*

cake to the nearest 1/32 inch (1 mm) after removing excess mud on the cake with a gentle stream of water.

Filtration properties at high temperatures and pressures (HTHP) are usually measured in cells similar to those shown in Figs. 3.12 and 3.13. The 175 mL cell has a maximum pressure 1500 psi (10,343 kP) and the maximum temperature of 450°F (232°C). The 500 mL cell has a maximum pressure 5000 psi (13,880 kP) and the maximum temperature of 500°F (260°C). To avoid flashing or evaporation of the filtrate at high temperatures, a back pressure of 100 psi (7.0 kg/cm^2) is held on the filtrate discharge when the test temperature is less than 300°F (149°C), and 450 psi (31.6 kg/cm^2) when the temperature is between 300°F and 450°F (149°C and 232°C). For temperatures up to 400°F (204°C), a Whatman 50 filter paper can be used, and for temperatures above 400°F, a new ceramic disc is used. *It is recommended that ceramic cores be used for all HTHP tests.* Ceramic disks are available in sizes from 10 to 120 μm. Filtration time is 30 min at the temperature of interest, but the volume of filtrate collected is doubled to allow for the difference in filtration area between the high- and low-pressure filtration cells, 22.9 cm^2 versus 45.8 cm^2.

Strict safety precautions must be followed in making filtration tests at HTHP. The procedure recommended in API RP 13B-1 and 13B-2 should be closely followed. In particular, the cell must not be filled above the

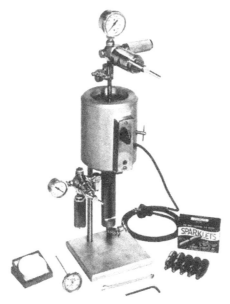

FIGURE 3.12 HPHT filter tester with pressure receiver—175 mL volume. *Courtesy of Fann Instrument Company.*

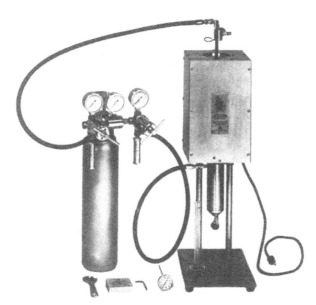

FIGURE 3.13 HPHT filter tester—500 mL volume. *Courtesy of Fann Instrument Company.*

manufacturer's recommendations and, at the conclusion of the test, the cell must be allowed to cool to room temperature before disassembly. Both size cells are available with threaded end caps. These are recommended from a safety standpoint.

Dynamic Filtration

To more closely simulate filtration in the drilling well, it is necessary to dynamically erode the filter cake. Over the years, a number of investigators have studied dynamic filtration in specially designed apparatus (Bezemer and Havenaar, 1966; Ferguson and Klotz, 1954; Horner et al., 1957; Prokop, 1952; Williams, 1940). The most meaningful results were obtained in systems that either closely simulated conditions in a drilling well, or that permitted the rate of shear at the surface of the cake—which is the critical factor limiting growth—to be calculated. Ferguson and Klotz (1954) came close to simulating well conditions by measuring filtration rates through permeable lumnite cement and sand cylinders in a model well using full-size drilling tools. Horner et al. (1957) used a microbit drilling machine and rock cores. Nowak and Krueger (1951) observed filtration rates through cores exposed on the side of an annulus through which mud was being circulated. A mechanical scraper enabled filtration conditions under the bit to be simulated when desired.

The rate of shear at the cake surface can be calculated in systems, such as that of Prokop (1952), in which mud is circulated under pressure through a permeable cylinder. The internal diameter of the cylinder should be large relative to the thickness of the filter cake so that the growth of the cake does not change the internal diameter significantly, and thereby change the rate of shear. Bezemer and Havenaar (1966) developed a compact and very convenient dynamic filtration apparatus, in which mud was filtered into a central core or sleeve of filter paper, while being sheared by an outer concentric cylinder rotating at constant speed. The equilibrium filtration rate and cake thickness were related to the rate of shear prevailing at the conclusion of the test.

One version of a commercial version of dynamic fluid loss tester, the Fann dynamic HPHT filtration tester, is shown in Fig. 3.14. This tester uses a rotating rod inside a ceramic cylinder through which the fluid flows, thus simulating radial flow. Test results include two numbers: the dynamic filtration rate and the cake deposition index (CDI). The dynamic filtration rate is calculated from the slope of the curve of volume versus time. The CDI is calculated from the slope of the curve of volume/time versus time. A different radial flow filtration device is available from Grace Instruments (Fig. 3.15).

Another commercial filtration tester based on a modified HPHT filter press is shown in Fig. 3.16. This device is manufactured by OFI Testing

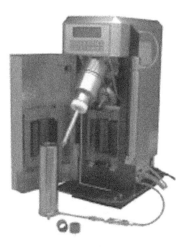

FIGURE 3.14 Dynamic HTHP fluid loss tester Model 90. *Courtesy of Fann Instrument Company.*

FIGURE 3.15 M2200 HPHT lubricity, dynamic filtration, and drilling simulator. *Courtesy of Grace Instruments, Inc.*

FIGURE 3.16 Dynamic HTHP filter tester. *Courtesy of OFI Testing Equipment.*

Equipment (OFITE). A mixing spindle is suspended above the filter media thereby causing the filter cake to erode. The OFITE tester can use either standard API filter paper or ceramic disks.

These devices will give different filtration volumes over the 30 min test, but relative changes will be similar when analyzing the effects of additives to a base mud.

DETERMINATION OF GAS, OIL, AND SOLIDS CONTENT

Gas Content

A measure of the amount of gas or air entrained in a mud may be obtained by diluting the mud substantially, stirring to release the gas, weighing the gas-free mud, and then back-calculating the density of the gas-free mud without dilution. For example, if ρ_1 is the density of the gas-cut mud, ρ_2 the density after diluting 1 volume of mud with 1 volume of water and removing the gas, ρ_3 the density of the gas-free, undiluted mud, and ρ_w the density of water, and x is the volume fraction of gas in the original mud, then:

$$\rho_1 = \frac{(1-x)\rho_3}{1} \tag{3.6}$$

$$\rho_2 = \frac{(1-x)\rho_3 + 1 \times \rho_w}{2-x} \tag{3.7}$$

Solving for x from the above equations:

$$x = \frac{2\rho_2 - \rho_1 - \rho_w}{\rho_2} \tag{3.8}$$

ρ_3 may then be calculated from either of the first two equations.

Oil and Solids Content

The volume fractions of oil, water, and solids in a mud are determined in a retort such as that shown in Fig. 3.17. It is important that any air or gas entrained on the mud be removed before retorting; otherwise, the solids content will be in considerable error. Removal of gas by substantial dilution is undesirable because of the loss of accuracy involved, especially with low-solids muds. Gas may often be removed by adding a defoamer, plus a thinner if necessary to break the gel.

Retorting involves placement of a precise volume of mud in a steel container, and heating it in a retort, in which the temperature is about 1000°F (540°C) until no more distillate collects in the graduated cylinder. The volume of oil and water are read in the graduated cylinder, and their sum subtracted from the volume of the mud sample to obtain the volume of solids. The method is rather inaccurate with low-solids muds because the result depends on the difference between two large numbers.

If the mud contains substantial amounts of salt, the volume occupied by the salt must be subtracted from the volume of solids. The API Recommended Practices have equations to calculate the corrected water content based on the chloride titration. The API equations, however, are based on sodium chloride salt, so if a different cation salt is added, further correction is necessary.

FIGURE 3.17 50 mL drilling fluid retort kit—10 mL. *Courtesy of OFI Testing Equipment.*

Do Not Use the API Retort with Formate Fluids

The standard API retort test should never be used with formate fluids because the condensation chamber of the standard retort could get plugged with salt crystals, causing the retort to burst (Cabot, 2016). Even if the retort test could be performed safely, the results are invalid since most solids are formed from formate salts crystallizing out of the highly concentrated brines. See the Formate Brines section in Chapter 2, Introduction to Completion Fluids, for an alternate solids content procedure with formate fluids.

The API Sand Test

The sand content is a measure of the amount of particles larger than 200 mesh present in a mud. Even though it is called a sand test, the test defines the size, not the composition, of the particles. The test is conveniently made in the apparatus shown in Fig. 3.18. The mud is first diluted by adding mud and water to the respective marks inscribed on the glass tube. The mixture is then shaken and poured through the screen in the upper cylinder, and then washed with water until clean. The material remaining on the screen is then backwashed through the funnel into the glass tube and allowed to settle, and, finally, the gross volume is read from the graduations on the bottom of the tube.

FIGURE 3.18 Standard API sand sieve. *Courtesy of Fann Instrument Company.*

Sieve Tests

Sieves are used to determine the size distribution of the coarser particles in commercial clays, bridging materials, and barite. The procedure is to shake the material through a nest of sieves, preferably by means of a vibrating shaker, and weigh both the residue left on each sieve and the material in the pan. Sieve sizes to suit the particular material being tested may be chosen, but the mesh sizes should correspond to the American Society for Testing Materials recommendations or API shaker screen specifications. The ASTM has recommended procedures for sieve tests (ASTM, 2014) and the API Committee on Standardization of Drilling Fluid Materials procedure to determine the particle size distribution of drilling fluid solids is in RP13C (API 2010). Commonly used oilfield particle size definitions are shown in (Table 3.4).

Methylene Blue Adsorption

A rapid estimate of the amount of montmorillonite present in a mud or clay can be obtained by means of the methylene blue test (Jones, 1964; Nevins and Weintritt, 1967). This test measures the amount of methylene blue dye adsorbed by clays, which is a function of their base exchange capacity. Since montmorillonite has a much larger base exchange capacity than other clay minerals, the test has come to be regarded as a measure of the amount of montmorillonite present, reported as estimated bentonite content.

TABLE 3.4 Definition of Particle Sizes[a]

Particle Size	Particle Classification	Sieve Size
Greater than 2000 μm	Coarse	10
2000–250 μm	Intermediate	60
250–74 μm	Medium	200
74–44 μm	Fine	325
44–2 μm	Ultra fine	
2–0 μm	Colloidal	

[a]*From API Bul. 13C (June 1974). American Petroleum Institute, Dallas.*

The test is made by diluting a sample of the mud, adding hydrogen peroxide to remove organic matter, such as polymers and thinning agents, and adding methylene blue solution until a drop of the suspension when placed on a filter paper appears as a blue ring surrounding the dyed solids (see the color chart in API RP 13B-1a). The methylene blue capacity is defined as the number of cm^3 of methylene blue solution (0.01 meq/cm^3) added per cm^3 of mud. The estimated bentonite is obtained in pounds per barrel by multiplying the methylene blue capacity by 5, and in kg/m^3 by multiplying the methylene blue capacity by 14.25.

ELECTRICAL PROPERTIES

Stability of Water-in-Oil Emulsions

The stability of water-in-oil emulsions is tested in an emulsion tester (Fig. 3.19), which permits a variable voltage to be applied across two electrodes immersed in the emulsion (Nelson et al., 1955). The voltage is increased until the emulsion breaks and a surge of current flows between the

FIGURE 3.19 Electrical stability meter for NADF. *Courtesy of OFI Testing Equipment.*

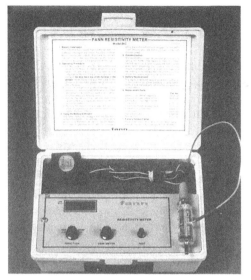

FIGURE 3.20 Resistivity meter. *Courtesy of Fann Instrument Company.*

electrodes. The voltage required for breakdown is regarded as a measure of the stability of the emulsion, the higher the voltage the greater the stability.

Research by Ali et al. (1987) has shown that results depend on a number of variables: emulsifier concentration, hot roll, oil/water ratio, mud density, and composition variables. They recommend that only trends in electrical stability be used for making treatment decisions.

Resistivity of Water Muds

Measurements of the resistivity of water muds, filtrates, and filter cakes are routinely applied in electrical logging. Under some conditions, better evaluation of formation characteristics may be had by controlling resistivity while drilling is in progress. Resistivity measurement provides a rapid means of detecting soluble salts in barite and in waters, such as makeup or produced waters.

Resistivity is measured by placing the sample in a resistive container having two electrodes spaced so that electrical current can flow through the sample. The resistance is measured with a suitable meter (see Fig. 3.20). If the instrument indicates the sample resistance in ohms, the cell constant must be determined by calibration with standard solutions of known resistivity to convert the measured value to ohm-meters. Most instruments, however, read directly in ohm-meters because the cell constant has been accounted for in the circuitry of the electrical meter. Details of operation of the resistivity meter are supplied by the manufacturer. The conductivity of the sample is the reciprocal of the measured resistivity.

Hydrogen Ion Concentration (pH)

The significant influence of the hydrogen ion concentration on the properties of water-based drilling fluids has long been recognized and has been the subject of numerous studies. Hydrogen ion concentration is more conveniently expressed as pH, which is the logarithm of the reciprocal of the hydrogen ion concentration in gram moles per liter. Thus, in a neutral solution the hydrogen ion (H^+) and the hydroxyl ion (OH^-) concentrations are equal, and each is equal to 10^{-7}. A pH of 7 is neutral. A decrease in pH below 7 shows an increase in acidity (hydrogen ions), while an increase in pH above 7 shows an increase in alkalinity (hydroxyl ions). Each pH unit represents a 10-fold change in concentration.

Two methods for the measurement of pH are in common use: (1) a colorimetric method using paper test strips impregnated with indicators; and (2) an electrometric method using a glass electrode instrument.

Colorimetric method. Paper test strips impregnated with organic dyes, which develop colors characteristic of the pH of the liquid with which they come in contact, afford a simple and convenient method of pH measurement. The rolls of indicator paper are taken from a dispenser that has the reference comparison colors mounted on its sides. Test papers are available in both a wide-range type, which permits estimation of pH to 0.5 units, and a narrow-range type, which permits estimation to 0.2 units of pH. The test is made by placing a strip of the paper on the surface of the mud (or filtrate), allowing it to remain until the color has stabilized (usually <30 s), and comparing the color of the paper with the color standards. High concentrations of salt in the sample may alter the color developed by the dyes and cause the estimate of pH to be unreliable.

Glass electrode pH meter. When a thin membrane of glass separates two solutions of differing hydrogen ion concentrations, an electrical potential difference develops that can be amplified and measured. The pH meter consists of (1) a glass electrode made of a thin-walled bulb of special glass within which is sealed a suitable electrolyte and electrode; (2) the reference electrode, a saturated calomel cell; (3) a means of amplifying the potential difference between the external liquid (mud) and the glass electrode; and (4) a meter reading directly in pH units. Provision is made for calibrating with standard buffer solutions and for compensating for variations in temperature. A special glass electrode (less affected by sodium ions) should be used in measuring the pH of solutions containing high concentrations of sodium ions (high salinity or very high pH).

LUBRICITY

Extreme pressure lubricants were originally added to drilling muds as a means of increasing the life of bit bearings (Rosenberg and Tailleur, 1959). A Timken lubrication tester was modified to permit the mud under test to circulate between the rotating ring and the block upon which the ring bears, as shown in the diagram in Fig. 3.21. The load-carrying capacity is indicated

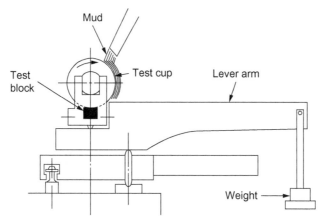

FIGURE 3.21 Timken lubrication tester. Diagram of loading lever system showing test block and cup in place.

by the maximum weight that can be used without seizure. Film strength is calculated from the area of the scar and from the load that is used on the block that passes the highest load test.

With the advent of sealed bit bearings, extreme pressure lubricants are no longer added to drilling muds to reduce bearing wear, but they, and other surfactants, are added to reduce drill pipe torque, as discussed in Chapter 10. For this purpose, the Timken tester has been further modified (Mondshine, 1970) as shown in Fig. 3.22. The recommended procedure is to apply a 150-lb load with the torque arm, adjust the shaft speed to 60 rpm, and read the amperes on the meter. Amperes are converted to the lubricity coefficient by means of a calibration chart.

Fig. 3.23 shows a more versatile lubricity tester (Alford, 1976). Mud is continuously circulated through the hole in the sandstone core; the stainless steel shaft rotates under load against the side of the hole; and torque in the shaft is monitored by a torque transducer and automatically plotted against time. Tests may be made with the shaft bearing against either the bare sandstone, a filter cake, or the inside of a steel pipe (to simulate torque conditions in casing). Each set of conditions is tested at several applied loads, and torque is plotted against load.

DIFFERENTIAL-PRESSURE STICKING TEST

There is no standard test for evaluating muds or mud additives with respect to their influence on the differential-pressure sticking of drill pipe, but various procedures and types of apparatus have been described in the literature. Several investigators (Albers and Willard, 1962; Haden and Welch, 1961; Helmick and Longley, 1957) have measured the pull-out force

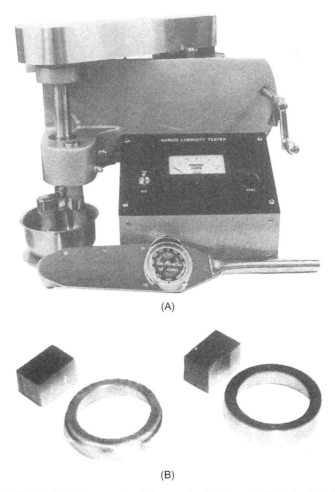

FIGURE 3.22 (A) Lubricity tester for drilling muds. (B) Ring and block for lubricity tester. *Courtesy of Fann Instrument Company.*

required to free a rod stuck in the filter cake on a round hole in a permeable medium, while others (Annis and Monaghan, 1962) have determined the coefficient of friction between a flat steel plate and a filter cake. A common and convenient form of apparatus, available from drilling fluid equipment manufacturers, consists of a modified filter cell, such as that shown in Fig. 3.24, and a disc or rod to simulate the drill pipe (Haden and Welch, 1961; Annis and Monaghan, 1962; Park and Lummus, 1962). Another sticking device is described by Simpson (1962). In general, the procedure is to lay down a filter cake, place the disc or rod in contact with the cake, continue filtration for a specified time, and then measure the torque or pull required to free the disc or rod (Fig. 3.25).

Evaluating Drilling Fluid Performance Chapter | 3 | 79

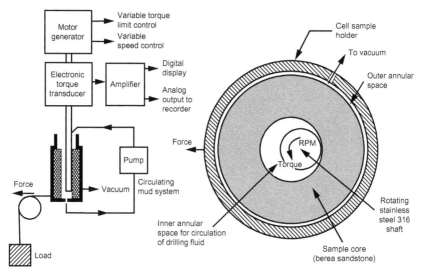

FIGURE 3.23 Simplified schematic of lubricity evaluation monitor, left. Sample cell, right, has an inner annular space for circulation of drilling fluid while a metal shaft is rotated against a sample core. A vacuum pump connected to the sample cell allows deposition of filter cake on the sample core. *From Alford, S.E., 1976. New technique evaluates drilling mud lubricants. World Oil Vol. 197, (July), 105–110.*

In the section on stuck drill pipe in Chapter 10, Drilling Problems Related to Drilling Fluids, it is shown that the force required to free the pipe is influenced by the initial thickness of the filter cake. Therefore, when comparing the effects of different mud compositions or mud additives on pipe sticking, it is essential that the thickness of the preliminary filter cake be the same in all cases.

CORROSION TESTS

Tests for corrosivity may be made in the laboratory by putting steel coupons and the mud to be tested in a container, tumbling the container end over end or rotating it on a wheel for a prolonged period, and then determining the weight lost by the coupon. If the test is made at temperatures or pressures that require the use of a steel cell, the coupon must not be in electrical contact with the cell (see Fig. 3.26). Results are reported as loss of weight per unit area per year, or as mils per year (mpy). With steel coupons of SG 7.86, the formula is

$$\text{mpy} = \frac{\text{weight loss, mg} \times 68.33}{\text{area, in}^2 \times \text{hours exposed}} \quad (3.9)$$

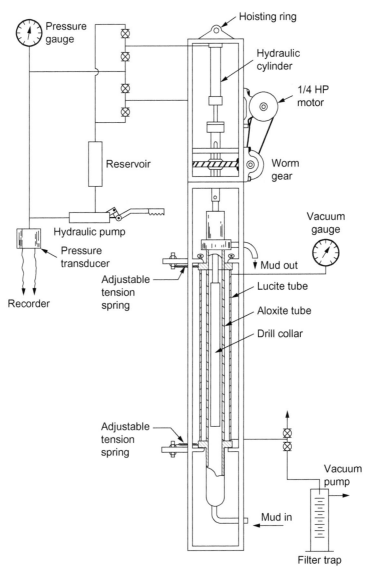

FIGURE 3.24 Schematic diagram of low differential-pressure test apparatus. *From Haden, E.L., Welch, G.R., 1961. Techniques for preventing differential pressure sticking of drill pipe. API Drill. Prod. Prac. 36–41. Copyright 1961 by API.*

Steel rings machined to fit into a tool joint box recess are commonly used to measure the corrosion that occurs in a drilling well (API Recommended Practice 13B-1, 2009a; Behrens et al., 1962). Recommended exposure times in the well vary from 40 h to 7 days. The rings are then

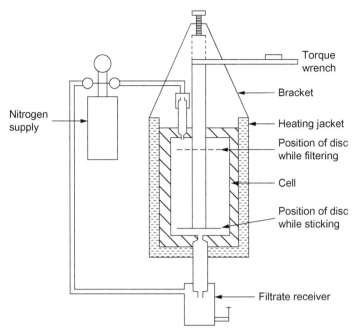

FIGURE 3.25 Apparatus for static testing of differential-pressure sticking. *From Simpson, J.P., 1962. The role of mud in controlling differential-pressure sticking of drill pipe. SPE Upper Gulf Coast Drill. and Prod. Conf. April 5–6, Beaumont. Copyright 1962 by SPE-AIME.*

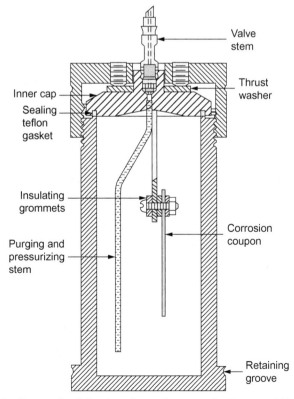

FIGURE 3.26 Cross-sectional drawing of corrosion test cell. *Courtesy of Fann Instrument Company.*

removed, cleaned, and examined for type of corrosion, and the loss of weight determined.

A sensitive test for hydrogen embrittlement is obtained by using solar steel roller bearings that have been permanently stressed by a 40,000 psi (3000 kg/cm^2) load, instead of the steel coupons (Bush et al., 1966). Rubber or Teflon O rings may be placed around the bearings to simulate the corrosive conditions that prevail under scale or mud cake on the drill pipe.

FLOCCULANTS

The standard method of determining flocculation value is described in Chapter 4, Clay Mineralogy and the Colloid Chemistry of Drilling Fluids. A variation suggested by Lummus (1965) for evaluating polymer flocculants is to add 0.01 lb/bbl of the flocculant to a 4% clay suspension, allow it to stand quiescent, and note the level of clear supernatant liquid after various intervals of time. Empirical tests for determining the degree to which treating agents will disperse clay mineral aggregates are also described in Chapter 4, Clay Mineralogy and the Colloid Chemistry of Drilling Fluids.

FOAMS AND FOAMING AGENTS

Rheological Properties of Foams

The rheological behavior of foams depends on foam quality (ratio of gas volume to total volume). In the quality range of interest in oilfield operations the flow model of foams is that of Bingham plastic and the parameters yield stress, PV, and effective viscosity are used to describe foam consistency curves. A multispeed viscometer, and capillary viscometers have been used to determine these parameters (Marsden and Khan, 1966). Mitchell (1971) used a capillary viscometer in which the flow velocity was measured by the transit time of a dye between two photoelectric cells, and the pressure drop across the capillary was measured by means of a differential transducer or a high-pressure manometer. Further information on foams can be found in Air and Gas Drilling Manual (Lyons, 2009).

Evaluation of Foaming Agents

The apparatus recommended in API publication RP46 (1966) for evaluating foaming agents is shown in Fig. 3.27. The foams are tested in four standard solutions whose composition is shown in Table 3.5. To test the effect of solids on the stability of the foam, 10 g of silica flour are placed in the bottom of the

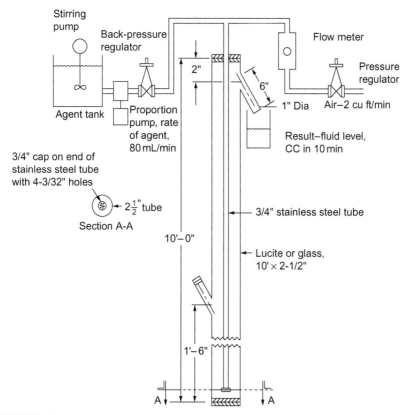

FIGURE 3.27 Laboratory apparatus for testing foam agents. *From API RP 46 (1966).*

TABLE 3.5 Evaluation of Foaming Agents[a]

Component	Fresh Water	Fresh Water Plus 15% Kerosene	10% Brine	10% Brine Plus 15% Kerosene
Distilled water (cm^3)	4000	3400	3800	3230
Kerosene (cm^3)	—	600	—	600
Sodium chloride (g)	—	—	400	340
Foamer (% by volume)	0.15	1.0	0.45	1.5

[a]*From API RP46. 1966. First Edition (Nov. 1966). American Petroleum Institute, Dallas.*

tube. One liter of test solution is poured into the tube and the remainder put in the reservoir. Air and solution are then flowed down the tube at the rates shown in Fig. 3.27. The volume of liquid carried out of the top of the tube in 10 min is taken as a measure of the effectiveness of the foam.

For specific field applications, samples of the liquids and solids from the well of interest should be used instead of the solutions and solids recommended above. Since the results of the tests are comparative, any apparatus embodying the essential features shown in Fig. 3.27 will give satisfactory results.

ANILINE POINT

The aniline point determination indicates the relative aromatic content of an oil. The aniline point is the lowest temperature at which equal volumes of aniline and oil are completely miscible. Oils having a high aromatic content have a low aniline point. An aniline point of 150°F (65°C) or higher indicates a low aromatic content and consequently the oil is less likely to damage rubber with which it comes into contact. The aromatic content of the oil (measured as the aniline point) used in oil muds is an important property if filtration control is dependent on asphalt and its degree of dispersion in the oil. High aromatic contents are also hazardous to marine organisms and, as a result, these fluids are banned in most offshore operations.

CHEMICAL ANALYSIS

Experience has shown that certain chemical analyses are useful in the control of mud performance, e.g., an increase in chloride content may adversely affect WBM properties unless the mud has been designed to withstand contamination by salt. Those analyses that have been found to be adaptable to use in the field have been included in API RP 13B-1 (2009). The detailed procedures recorded there will not be repeated here. Some literature references will be cited on tests developed specifically for use with muds and mud filtrates.

Chlorides

A sample of mud filtrate (neutralized, if alkaline) is titrated with standard silver nitrate solution, using potassium chromate as indicator. The results are reported as parts per million chloride ion, although actually measured in terms of mg Cl ion per 1000 cm^3 of filtrate. To determine the chloride content of an oil mud, the sample is diluted with the solvent propylene glycol normal-propyl ether neutralized to the phenolphthalein end point, and then titrated in the usual way. The end point is difficult to detect using the silver nitrate method, and the results depend to some extent on the operator.

Alkalinity and Lime Content

Measurements of the alkalinity and the lime content of lime-treated muds are useful in the control of their properties (Battle and Chaney, 1950). The procedure involves titration of the filtrate with standard sulfuric acid to the phenolphthalein end point (P_f), and to the methyl orange end point (M_f), and titration of the mud to the phenolphthalein end point (P_m). The lime content is calculated as

$$\text{lime, lb/bbl} = 0.26(P_m - F_w P_f) \text{ or} \tag{3.10}$$

$$\text{lime, kg/m}^3 = 0.74(P_m - F_w P_f) \tag{3.11}$$

where F_w is the volume fraction of water in the mud (Nelson and Watkins, 1950).

Routine methods of water analysis permit the calculation of hydroxide, carbonate, and bicarbonate concentrations from the simple titration to the phenolphthalein and methyl orange end points (API RP 13B-1, 2009). The composition of mud filtrates, however, usually is so complex that such an interpretation cannot be justified. The methyl orange end point, in particular, is of dubious significance because of the presence in mud filtrates of the reaction products of various organic additives, and of silicates resulting from the action of sodium hydroxide on clays.

Consequently, methods have been sought that would give a more reliable estimation of carbonate and bicarbonate contents. An alternate filtrate alkalinity procedure involves three titrations and careful attention to technique, but does avoid the methyl orange indicator (API 13B-1, 2009; Green, 1972). This method has been used mainly on filtrates of muds heavily treated with lignites. Another method (Garrett, 1978) uses the Garrett gas train employed in the analysis for sulfides, and a Dräger tube designed for the detection of carbon dioxide, for the direct measurement of the CO_2 liberated on acidification of a sample of the filtrate.

Total Hardness: Calcium and Magnesium, Calcium, and Calcium Sulfate

Estimations of total hardness and of calcium ion are based on titration with standard versenate solution. The use of different buffer solutions and different indicators makes possible the separate estimation of calcium, and thus estimation of magnesium by difference from the total hardness value. Calcium sulfate (undissolved) is calculated from total hardness titrations of the filtered diluted mud and the original mud filtrate.

Sulfides

The reliable estimation of sulfides in muds is of concern in the avoidance of corrosion and injury to personnel, and in the detection while drilling of formations containing hydrogen sulfide.

In alkaline muds, hydrogen sulfide is neutralized, and can be detected as soluble sulfides in the filtrates. The Hach sulfide test, based on the darkening of paper impregnated with lead acetate, appeared in the fourth through sixth editions of API RP 13B, but was omitted from the seventh edition. Recognized limitations in the accuracy of the Hach test led to the development of more reliable methods involving the use of the Garrett gas train (Garrett and Carlton, 1975) and subsequent adoption of this device with the Dräger H_2S detector tubes as the API recommended practice. Ion-selective electrodes have been used with special electronic circuitry to measure sulfides in the drilling mud (Hadden, 1977). The device has been used commercially in mud logging operations.

Potassium

The application of potassium chloride-polymer muds for hole stabilization led to the development of a field procedure for the determination of potassium (API 13B-1, 2009; Steiger, 1976). In this method, the concentration of potassium in a sample of the mud filtrate is estimated by comparison of the volume of potassium perchlorate precipitate formed in the sample with the volume produced by a standard potassium chloride solution subjected to the same conditions of treatment.

A method involving the use of sodium tetraphenylboron (NaTPB) has been proposed as a replacement for API RP 13B (the perchlorate method) because it is more accurate, faster, and more readily performed at the well (Zilch, 1984). The proposed method is a volumetric analysis of potassium in the mud filtrate using NaTPB to precipitate the potassium, then back-titrating the unused portion of the NaTPB with cetyltrimethyl ammonium bromide, using Titan yellow (Clayton yellow) as an indicator.

EVALUATION OF DRILLING FLUID MATERIALS

General Principles

There is general agreement that the components of drilling fluids should be evaluated in terms of performance under conditions as similar as possible to the conditions of use. Because of the wide variety of conditions, and because the same product may serve different functions, there has been little success in efforts to standardize methods of product evaluation. Such physical properties as density, moisture content, and sieve size analysis can be measured by generally accepted methods. But when a component such as bentonite requires testing, the question arises as to the purpose of adding the component, whether for use as a thickening agent or to reduce filtration.

In evaluating barite, for example, the presence of a small amount of calcium sulfate in the product would not affect its performance in a mud already saturated with gypsum, although the same barite might be unsatisfactory in fresh-water mud.

Laboratory tests to evaluate performance of products are necessarily limited in scope, and must be directed to the examination of specific properties under controlled conditions. For routine measurement of product quality, the methods must not be too involved or too time-consuming. The procedures usually followed are the simplest of those that can establish the quality level of a material. In routine laboratory tests, for example, muds are prepared by stirring for specified times with the Hamilton Beach mixer (Fig. 3.1) or the multispindle mixer (Fig. 3.2) at room temperature.

Some consideration always must be given, however, to possible interactions between components under conditions of actual use. Special tests may be required to show the effect of prolonged or intense agitation, elevated temperatures, or unusual contaminants.

The cost of the mud additive under investigation is of vital concern in any evaluation of performance. Tests should be designed to afford a comparison of costs between the product being examined and products that have shown satisfactory performance in field use.

In summary, laboratory testing of a submitted product should either provide an economic evaluation of its principal functions in comparison with the performance of an acceptable existing product, or establish that the product has unusual qualities justifying field trials.

AGING AT HIGH TEMPERATURE

Many water-based fluid constituents degrade slowly at high temperatures. Such degradation occurs while circulating, but is more severe with the mud left in the lower part of the hole when making a trip, because of the higher temperature involved. Consequently, the effect of aging at elevated temperatures should be observed on all WBM compositions and additives.

Such tests are usually made in stainless steel or aluminum bronze pressure cells (Gray et al., 1951; Fann, 2016; OFITE, 2016), which are commercially available in 260 or 500 cm^3 sizes (Fig. 3.28). To prevent boiling of the liquid phase, the cells are pressurized with nitrogen or carbon dioxide through connections provided for the purpose (Cowan, 1959). The applied pressure must be at least equal to the vapor pressure of the liquid at the test temperature.

Nonaqueous fluids are less likely to be dramatically effected by bottom-hole temperatures less than about 500°F. An extreme buildup of fine drilled solids or water-wet solids, however, may result in erratic viscosities effecting

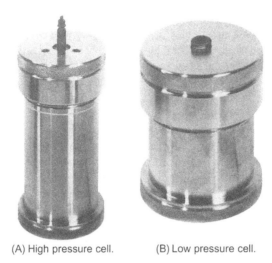

(A) High pressure cell.   (B) Low pressure cell.

FIGURE 3.28 Cells for aging muds at elevated temperatures in roller oven. *Courtesy of Fann Instrument Company.*

FIGURE 3.29 Roller oven. *Courtesy of Fann Instrument Company.*

the drilling operations. Long-term aging of NADF may be necessary if weight material settling is a problem.

To simulate aging of the mud while it is circulating in the well, the cells are rolled in an oven, such as that shown in Fig. 3.29, for at least 16 h at the average well circulating temperature. The cells are then cooled to room temperature, and the rheological and filtration properties are measured and compared to the same properties before aging.

When a mud is left in a high-temperature hole during a round trip, the crucial factor is the undisturbed gel strength, which determines the pressure required to break circulation. When testing, therefore, the cells are aged statically in an oven heated to the temperature of interest for the required length of time; cooled to room temperature; and the undisturbed gel strength measured in the cell with a shearometer tube (Watkins and Nelson, 1953).

REFERENCES

Albers, D.C., Willard, D.R., 1962. The evaluation of surface-active agents for use in the prevention of differential pressure sticking of drill pipe. SPE Paper 298, Prod. Res. Symp. April 12–13, Tulsa.
Alford, S.E., 1976. New technique evaluates drilling mud lubricants. World Oil Vol. 197, July, 105–110.
Ali, A., Schmidt, D.D., Harvey, J., 1987. Investigation of the electrical stability test for oil muds. SPE/IADC Paper 16077, Drill. Conf. March 15–18, New Orleans, LA.
Annis, M.R., Monaghan, P.H., 1962. Differential pressure sticking laboratory studies of friction between steel and mud filter cake. J. Petrol. Technol. May, 537–542, Trans. AIME 225.
American Petroleum Institute, 1966. RP46. Recommended practice testing foaming agents for mist drilling.
American Petroleum Institute, 2009a. RP 13B-1. Recommended practice for field testing water--based drilling fluids.
American Petroleum Institute, 2009b. RP 13I. Recommended practice for standard procedure for laboratory testing drilling fluids.
American Petroleum Institute, 2014a. Bulletin 13C. Drilling fluids processing equipment.
American Petroleum Institute, 2014b. RP 13B-2. Recommended practice for field testing oil--based drilling fluids.
American Petroleum Institute, 2015. Specification 13A. Specification for oil-well drilling fluid materials.
American Society for Testing Materials (ASTM), 2014. Test sieving methods: guidelines for establishing sieve analysis procedures.
Battle, J.L., Chaney, P.E., 1950. Lime base muds. API Drill. Prod. Prac. 99–109.
Behrens, R.W., Holman, W.E., Cizek, A., 1962. Technique for evaluation of corrosion of drilling fluids. API Paper 906–7-G, Southwestern District Meeting. March 21–23, Odessa.
Bezemer, C., Havenaar, I., 1966. Filtration behavior of circulating drilling fluids. Soc. Petrol. Eng. J.292–298, Trans. AIME 237.
Bush, H.E., Barbee, R., Simpson, J.P., 1966. Current techniques for combating drill-pipe corrosion. API Drill. Prod. Prac. 59–69.
Cabot Specialty Fluids, 2016. Formate Technical Manual. www.FormateBrines.com.
Cowan, J.C., 1959. Low filtrate loss and good rheology retention at high temperatures are practical features of this new drilling mud. Oil Gas J. November 2, 83–87.
Fann Instruments, 2016. 2016 Catalog. Houston. www.OFITE.com.
Ferguson, C.K., Klotz, J.A., 1954. Filtration from mud during drilling. Trans. AIME 201, 30–43.

Garrett, R.L., 1977. A new field method for the quantitative determination of sulfides in water-base drilling fluids. J. Petrol. Technol. September, 1195–1202, Trans. AIME 263.

Garrett, R.L., 1978. A new method for the quantitative determination of soluble carbonates in water- base drilling fluids. J. Petrol. Technol. June, 860–868.

Garrett, R.L., Carlton, L.A., 1975. Iodometric method shows mud H_2S. Oil Gas J. January 6, 74–77.

Gray, G.R., Cramer, A.C., Litman, K.K., 1951. Stability of mud in high-temperature holes shown by surface tests. World Oil August, 149, 150.

Green, B.Q., 1972. Carbonate/bicarbonate influence in water-base drilling fluids. Petrol. Eng. May, 74–76.

Hadden, D.M., 1977. Continuous on-site measurement of sulfides in water-base drilling muds. SPE Paper. 6664, Symp. Sour Gas and Crude. November 14–15, Tyler.

Haden, E.L., Welch, G.R., 1961. Techniques for preventing differential pressure sticking of drill pipe. API Drill. Prod. Prac. 36–41.

Helmick, W., Longley, A., 1957. Pressure-differential sticking of drill pipe and how it can be avoided or relieved. API Drill. Prod. Prac.55–60.

Horner, V., White, M.M., Cochran, C.D., Deily, F.H., 1957. Microbit dynamic filtration studies. Trans. AIME 210, 183–189.

Jones, F.O., 1964. New fast, accurate test measures bentonite in drilling mud. Oil Gas J. June 1, 76–78.

Jones, P.H., 1937. Field control of drilling mud. API Drill. Prod. Prac. 24–29.

Lummus, J.L., 1965. Chemical removal of drilled solids. Drill. Contract. March/April, 50–54, 67.

Lyons, W.C., 2009. Air and Gas Drilling Manual. Gulf Professional Publishing, Elsevier, Inc.

Macpherson, J.D., de Wardt J.P., Florence, F., Chapman, C.D., Zamora, M., 2013. Drilling-systems automation: current state, initiatives, and potential impact. SPE ATCE New Orleans 2013. Paper SPE 66263.

Magalhaes, S., Scheid, C. M., Calcada, L. A., Folsta, M., Martins, A. L., Marques deSa, C. H., 2014. Development of on-line sensors for automated measurement of drilling fluid properties. SPE ATCE 2014. Paper SPE 167978-MS.

Marsden, S., Kahn, S., 1966. The flow of foam through short porous media and apparent viscosity measurements. Soc. Petrol. Eng. March, 17–25, Trans AIME 237.

Marsh, H., 1931. Properties and treatment of rotary mud. Petroleum Development and Technology Transactions of the AIME. pp. 234–251.

Mitchell, B., 1971. Test data fill theory gap on using foam as a drilling fluid. Oil Gas J. September 6, 96–100.

Mondshine, T.C., 1970. Drilling mud lubricity: guide to reduced torque and drag. Oil Gas J. December 7, 70–77.

Nelson, M.D., Crittenden, B.C., Trimble, G.A., 1955. Development and application of a water-in-oil emulsion drilling mud. API Drill. Prod. Prac.238.

Nelson, M., Watkins, T., 1950. Lime content of drilling mud-Calculation method. Trans. AIME 189, 366–367.

Nevins, M.J., Weintritt, D.J., 1967. Determination of cation exchange capacity by methylene blue adsorption. Amer. Ceramic. Soc. Bull 46, 587–592.

Nowak, T., Krueger, R., 1951. The effect of mud filtrates and mud particles on the permeability of cores. API Drill. Prod. Prac.164–181.

OFITE, 2016. OFI Testing Equipment Catalog. Houston.

Oort, E. van, Hoxha, B. B., Yang, L., Hale, A., 2016. Automated drilling fluid analysis using advanced particle size analyzers. Society of Petroleum Engineers. Paper SPE178877-MS.

Park, A., Lummus, J.L., 1962. New surfactant mixture eases differential sticking, stablizes hole. Oil Gas J. November 26, 62–66.

Prokop, C.L., 1952. Radial filtration of drilling mud. Trans. AIME 195, 5–10.

Rosenberg, M., Tailleur, R.J., 1959. Increased drill bit life through use of extreme pressure lubricant drilling fluids. Trans. AIME 216, 195–202.

Savins, J.G., Roper, W.F., 1954. A direct indicating viscometer for drilling fluids. API Drill. Prod. Prac. 7–22.

Simpson, J.P., 1962. The role of mud in controlling differential-pressure sticking of drill pipe. SPE Upper Gulf Coast Drill. and Prod. Conf. April 5–6, Beaumont.

Steiger, R.P., 1976. A new field procedure for determining potassium in drilling fluids. J. Petrol. Technol. August, 868–869.

Vajargah, A.K., van Oort, E., 2015. Automated drilling fluid rheology characterization with downhole pressure sensor data. Society of Petroleum Engineers. Available from: http://dx.doi.org/10.2118/173085-MS.

Watkins, T.E., Nelson, M.D., 1953. Measuring and interpreting high-temperature shear strengths of drilling fluids. Trans. AIME 198, 213–218.

Williams, M., 1940. Radial filtration of drilling muds. Trans. AIME 136, 57–69.

Zilch, H.E., 1984. New chemical titration proposed for potassium testing. Oil Gas J. January 16, 106–108.

Chapter 4

Clay Mineralogy and the Colloid Chemistry of Drilling Fluids

INTRODUCTION

Anyone concerned with drilling fluids technology should have a good basic knowledge of clay mineralogy, as clay provides the colloidal base of nearly all aqueous muds, and is also modified for use in oil-based drilling fluids. Drill cuttings from argillaceous formations become incorporated in any drilling fluid, and can profoundly change its properties. The stability of the borehole depends to a large extent on interactions between the drilling fluid and exposed shale formations. The development of "inhibitive" aqueous drilling fluids was initiated to control clay hydration and swelling. In nonaqueous fluids, the introduction of water-wet solids can upset the properties of the fluid. Interactions between the mud filtrate, whether water or oil, and the clays present in producing horizons may restrict productivity of the well. All of these point out the need for the drilling fluids technologist to have knowledge of clay mineralogy.

The technologist should also have a basic knowledge of colloid chemistry as well as clay mineralogy, because clays form colloidal suspensions in water, and also because a number of organic colloids are used in water-based drilling muds.

As well as colloid chemistry, nanotechnology is entering the drilling and completion fluids industry. Nanoparticles are new to the industry and there has not been very much data on their effectiveness. The current uses of these nanomaterials are covered in the appropriate chapter in which they are used.

Both clay mineralogy and colloid chemistry are extensive subjects. In this chapter it will only be possible to summarize briefly those aspects that affect drilling and completion fluids technology.

CHARACTERISTICS OF COLLOIDAL SYSTEMS

Colloids are not, as is sometimes supposed, a specific kind of matter. They are particles whose size falls roughly between that of the smallest particles that can be seen with an optical microscope and that of true molecules, but they may be of any substance. Actually, it is more correct to speak of colloidal systems, since the interactions between two phases of matter are an essential part of colloidal behavior. *Colloidal systems* may consist of solids dispersed in liquids (e.g., clay suspensions, polymers), liquid droplets dispersed in liquids (e.g., emulsions), or solids dispersed in gases (e.g., smoke). In this chapter, we shall only be concerned with solids and polymers dispersed in water-based drilling fluids (WBM).

One characteristic of aqueous colloidal systems is that the particles are so small that they are kept in suspension indefinitely by bombardment of water molecules, a phenomenon known as the Brownian movement. The erratic movements of the particles can be seen by light reflected off them when they are viewed against a dark background in the ultramicroscope.

Another characteristic of clay colloidal systems is that the particles are so small that properties like viscosity and sedimentation velocity are controlled by surface phenomena. Surface phenomena occur because molecules in the surface layer are not in electrostatic balance, i.e., they have similar molecules on one side and dissimilar molecules on the other, whereas molecules in the interior of a phase have similar molecules on all sides. Therefore, the surface carries an electrostatic charge, the size and sign of which depends on the coordination of the atoms on both sides of the interface. Some substances, notably clay minerals, carry an unusually high surface potential because of certain deficiencies in their atomic structure, which is explained later.

The greater the degree of subdivision of a solid, the greater will be its surface area per unit weight, and therefore the greater will be the influence of the surface phenomena. For example, a cube with sides 1 mm long would have a total surface area of 6 mm^2. If it were subdivided into cubes with 1 μm sides (1 μm = 1×10^{-3} mm) there would be 10^9 cubes, each with a surface area of 6×10^{-6} mm^2 and the total surface area would be 6×10^3 mm^2. Subdivided again into millimicron cubes, the total surface area would be 6×10^6 mm^2 or 6 m^2.

The ratio of surface area per unit weight of particles is called the *specific surface*. Thus if a 1 cm^3 cube were divided into micron-sized cubes, the specific surface would be $6 \times 10^6/2.7 = 2.2 \times 10^6$ mm^2/g = 2.2 m^2/g assuming the specific gravity of the cube to be 2.7.

Fig. 4.1 shows specific surface versus cube size. To put the values in perspective, the size of various particles, expressed in equivalent spherical radii (esr), are shown at the top. The esr of a particle is the radius of a sphere that would have the same sedimentation rate as the particle. The esr may be determined by applying Stokes' law (Caenn et al., 2011) to the measured sedimentation rate.

The division between colloids and silt, shown in Fig. 4.1, is arbitrary and indefinite, because colloidal activity depends (1) on specific surface, which varies with particle shape, and (2) on surface potential, which varies with atomic structure. Another solids classification method is shown in Table 4.1.

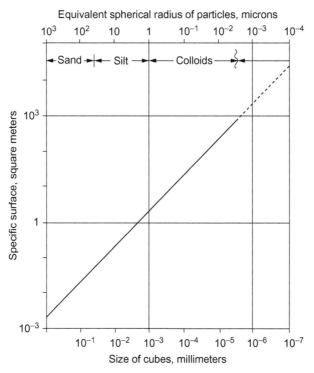

FIGURE 4.1 Specific surface of cubes, assuming specific gravity of 2.7.

TABLE 4.1 The Suggested Particle Size Classification Published in the API Recommended Practice 13C 2009

Particle Size, μm	Particle Classification
Coarser than 2000	Coarse
2000 – 250	Intermediate
250 – 74	Medium
74 – 44	Fine
44 – 2	Ultrafine
<2	Colloidal

A large proportion of the solids in drilling muds falls in the silt size range. These particles are derived either from natural silts picked up from the formation, from larger particles manufactured by the action of the bit, or from barite added to raise the density. Particles in this size fraction are commonly called the *inert solids*, but the term is relative, and when present in high enough concentrations, inert solids exercise a considerable influence on the viscous properties of a mud. Even NADFs can experience erratic viscous properties from inert solids if they are allowed to build up.

Colloids, on the other hand, usually constitute a small proportion of the total solids, but exercise a relatively high influence on mud properties because of their high degree of activity. They may be divided into two classes: (1) clay minerals, and (2) organic colloids, such as starches, the carboxycelluloses, biopolymers, and the polyacrylamide derivatives. These substances have macromolecules, or are long-chain polymers, whose size gives them colloidal properties. We will consider the clay minerals first.

CLAY MINERALOGY

The upper limit of the particle size of clays is defined by geologists as 2 μm, so that virtually all bentonitic particles fall within the colloidal size range. As they occur in nature, clays consist of a heterogeneous mixture of finely divided minerals, such as quartz, feldspars, calcite, pyrites, etc., but the most colloidally active components are one or more species of clay minerals.

Ordinary chemical analysis plays only a minor part in identifying and classifying clay minerals. Clay minerals are of a crystalline nature, and the atomic structure of their crystals is the prime factor that determines their properties. Identification and classification is carried out mainly by analysis of X-ray diffraction patterns, adsorption spectra, and differential thermal analysis. These methods have been summarized in the abundant literature on the subject (Grim, 1953, 1962; Marshall, 1949; Weaver and Pollard, 1973).

Most clays have a mica-type structure. Their flakes are composed of tiny crystal platelets, normally stacked together face-to-face. A single platelet is called a *unit layer*, and consists of:

1. An *octahedral* sheet, made up of either aluminum or magnesium atoms in octahedral coordination with oxygen atoms as shown in Fig. 4.2. If the metal atoms are aluminum, the structure is the same as the mineral *gibbsite*, $Al_2(OH)_6$. In this case, only two out of three possible sites in the structure can be filled with the metal atom, so the sheet is termed *dioctahedral*. If, on the other hand, the metal atoms are magnesium, the structure is that of *brucite*, $Mg_3(OH)_6$. In this case all three sites are filled with the metal atom, and the structure is termed *trioctahedral*.
2. One or two sheets of silica tetrahedra, each silicon atom being coordinated with four oxygen atoms, as shown in Fig. 4.3. The base of the tetrahedra form a hexagonal network of oxygen atoms of indefinite areal extent.

The sheets are tied together by sharing common oxygen atoms. When there are two tetrahedral sheets, the octahedral sheet is sandwiched between them, as shown in Fig. 4.3. The tetrahedra face inwards and share the

Clay Mineralogy and the Colloid Chemistry of Drilling Fluids Chapter | 4 97

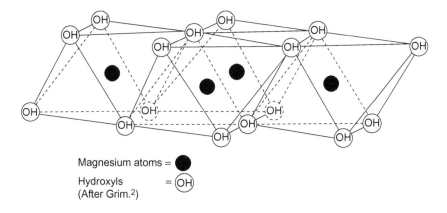

FIGURE 4.2 Octahedral sheet; structure shown is that of brucite.

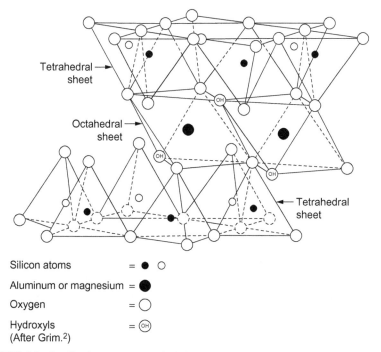

FIGURE 4.3 Bonding between one octahedral sheet and two tetrahedral sheets through shared oxygen atoms.

oxygen atom at their apexes with the octahedral sheet, which displaces two out of three of the hydroxyls originally present. This structure is known as the *Hoffmann structure* (Hofmann et al., 1933), the dimensions of which are shown in Fig. 4.4.

98 Composition and Properties of Drilling and Completion Fluids

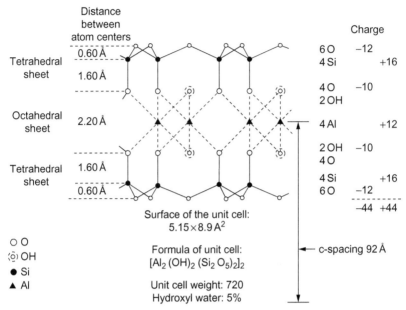

FIGURE 4.4 Atom arrangement in the unit cell of a three-layer mineral (schematic). *Courtesy of John Wiley & Sons.*

Note that the oxygen network is exposed on both basal surfaces. When there is only one tetrahedral sheet, it is bonded to the octahedral sheet in the same manner, so that, in this case, the oxygen network is exposed on one basal surface, and hydroxyls are exposed on the other, as shown in Fig. 4.5.

The unit layers are stacked together face-to-face to form what is known as the *crystal lattice*. The distance between a plane in one layer and the corresponding plane in the next layer (see Fig. 4.6) is called either the *c-spacing*, the *001*, or the *basal spacing*. This spacing is 9.2 Å for the standard three-layer mineral (Grim, 1953) and 7.2 Å (Angstrom (A) = 10^{-7} mm) for a two-layer mineral. The crystal extends indefinitely along the lateral axes, a and b, to a maximum of about 1 μm.

The sheets in the unit layer are tied together by covalent bonds, so that the unit layer is stable. On the other hand, the layers in the crystal lattice are held together only by van der Waals forces (Grim, 1953) and secondary valencies between juxtaposed atoms. Consequently, the lattice cleaves readily along the basal surfaces, forming tiny mica-like flakes.

The chemical composition of the dioctahedral structure shown in Fig. 4.4 is that of the mineral pyrophyllite. The analogous trioctahedral mineral is talc, which is similar, except for the presence of magnesium instead of aluminum. Pyrophyllite and talc are prototypes for clay minerals in the *smectite*

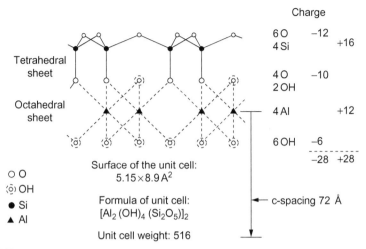

FIGURE 4.5 Atom arrangement in the unit cell of a two-layer mineral (schematic). *Courtesy of John Wiley & Sons.*

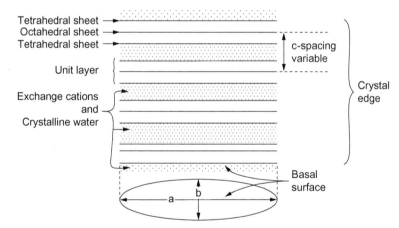

FIGURE 4.6 Diagrammatic representation of a three-layer expanding clay lattice.

group, but are not themselves true clay minerals. They cleave to, but do not break down to, the very small platelets that are characteristic of clay minerals. The fundamental difference between the two types of minerals is that the structures of the prototypes are balanced and electrostatically neutral, whereas the clay mineral crystals carry a charge arising from isomorphous substitutions of certain atoms in their structure for other atoms of a different valence (Marshall, 1935). For example, if an atom of Al^{+3} is replaced by an atom of Mg^{+2} a charge deficiency of one results. This creates a negative potential at the surface of the crystal, which is compensated for by the adsorption of a cation. In the presence of water, the adsorbed cations can exchange with cations of another species in the water, and they are therefore known as the *exchangeable cations*. Substitutions may occur in either the octahedral or tetrahedral sheets, and diverse species may be exchanged, giving rise to innumerable groupings and subgroupings of clay minerals.

The degree of substitution (DS), the atoms involved, and the species of exchangeable cations are of enormous importance in drilling fluids technology because of the influence they exert on such properties as swelling, dispersion, and rheological and filtration characteristics. The clay mineral groups of interest, and their characteristics, are discussed next.

The Smectites

As mentioned earlier, pyrophyllite and talc are the prototype minerals for the smectite group. In their crystal lattice, the tetrahedral sheet of one layer is adjacent to the tetrahedral sheet of the next, so that oxygen atoms are opposite oxygen atoms. Consequently, bonding between the layers is weak and cleavage is easy (Grim, 1953, p. 16). Partly because of the weak bonding, and partly because of high repulsive potentials on the surface of the layers arising from isomorphous substitutions, water can enter between the layers, thereby causing an increase in the c-spacing. Thus, smectites have an expanding lattice, which greatly increases their colloidal activity, because it has the effect of increasing their specific surface many times over. All the layer surfaces, instead of just the exterior surfaces, are now available for hydration and cation exchange, as shown in Fig. 4.6.

Members of the smectite group are differentiated on the basis of the prototype mineral, the relative amounts of substitutions in the octahedral or tetrahedral layer, and on the species of atoms substituted. Table 4.2 lists the principal members of the group (Brindley and Roy, 1957).

Note that the convention for writing the formulas of clay minerals is as follows:

Suppose the prototype mineral is pyrophyllite, which has the formula:

$$2[Al_2Si_4O_{10}(OH)_2].$$

If one aluminum atom in six in the octahedral sheet is replaced by one atom of magnesium, and one atom of silicon in eight in the tetrahedral sheet is replaced by one atom of aluminum, then the formula would be written:

$$2[(Al_{1.67}Mg_{0.33})(Si_{3.5}Al_{0.5})O_{10}(OH)_2].$$

Montmorillonite is by far the best known member of the smectite group, and has been extensively studied because of its common occurrence and economic importance. It is the principal constituent of Wyoming bentonite,

TABLE 4.2 Smectites

Principal Substitutions	Trioctahedral Minerals	Dioctahedral Minerals
Prototype (no substitutions)	Talc $(Mg_3Si_4{}^a)$	Pyrophyllite (Al_2Si_4)
Practically all octahedral	Hectorite $(Mg_{3-x}Li_x)(Si_4)$	Montmorillonite $(Al_{2-x}Mg_x)(Si_4)$
Predominantly octahedral	Saponite $(Mg_{3-x}Al_x)(Si_{4-y}Al_y)$ Sauconite $(Zn_{3-x}Al_x)(Si_{4-y}Al_y)$	Volchonskoite $(Al,Cr)_2(Si_{4-y}Al_y)$
Predominantly tetrahedral	Vermiculite $(Mg_{3-x}Fe_x)(Si_3Al)$	Nontronite $(Al,Fe)_2(Si_{4-y}Al_y)$

^{a}To each formula the group $O_{10}(OH)_2$ should be added as well as the exchangeable cation.
Source: From Brindley, G.W., Roy, R., 1957. Fourth Progress Report and First Annual Report. API Project 55. Copyright 1957 by API.

and of many other clays added to drilling fluids. It is the active component in the younger argillaceous formations that cause problems of swelling and heaving when drilled.

The predominant substitutions are Mg^{+2} and Fe^{+3} for Al^{+3} in the octahedral sheet, but Al^{+3} may be substituted for Si^{+4} in the tetrahedral sheet. If the substitutions in the tetrahedral sheet exceed those in the octahedral, the mineral is termed a *beidellite* (Weaver and Pollard, 1973, p. 63), so that montmorillonite and beidellite may be regarded as end members of a series.

The charge deficiency varies over a wide range, depending on the DS. The maximum is approximately 0.60, and the average 0.41 (Weaver and Pollard, 1973, p. 73). The specific surface may be as much as $800 \, m^2/g$ (Dyal and Hendricks, 1950).

Like other smectites, montmorillonite swells greatly because of its expanding lattice. The increase in c-spacing depends on the exchangeable cations. With certain cations (notably sodium), the swelling pressure is so strong that the layers separate into smaller aggregates and even into individual unit layers (see Fig. 4.7). A number of attempts have been made to determine the particle size of sodium montmorillonite, but the determination is difficult because of the flat, thin, irregular shape of the platelets, and because of the wide range of sizes. In a comprehensive study, Kahn (1957) separated sodium montmorillonite into five size fractions in an ultracentrifuge. Using a combination of methods, he then determined the maximum width and the thickness of the platelets in each fraction. The results, summarized in Table 4.3, show that both the width and thickness decrease with decrease in equivalent spherical radius. If the c-spacing in the aggregates is assumed to be 19 Å (see Fig. 4.14), then the number of layers in the coarsest fraction

FIGURE 4.7 Electron micrograph of montmorillonite. Magnification ×87,500. *Courtesy of J.L. McAtee, Baylor University.*

was eight, and the average was a little over one in the three finest fractions, which represented 57% by weight of the sample.

Small angle X-ray diffraction studies (Hight et al., 1962) of the same three fine fractions also indicated monolayers. Light scattering studies (Melrose, 1956) indicated one to two layers per aggregate in fractions with less than 60 Å esr, and somewhat smaller maximum widths than found by Kahn. An electron micrograph (Barclay and Thompson, 1969) of the edge of a flake of sodium montmorillonite, which was taken from the coarse fraction in an ultracentrifuge, showed aggregates of three to four layers each, stacked together to form the flake (see Fig. 4.8).

Illites

Illites are hydrous micas, the prototypes for which are muscovite (dioctahedral mica) and biotite (trioctahedral mica). They are three-layer clays, with a structure similar to montmorillonite, except that the substitutions are

TABLE 4.3 Dimensions of Sodium Montmorillonite Particles in Aqueous Suspension[a]

Fraction Number	Percent by Weight	Equivalent Spherical Radius (μm)	Maximum Width By Electrooptical Birefringence (μm)	Maximum Width By Electron Microscope (μm)	Thickness, Å[b]	Average[c] Number per Particle
1	27.3	>0.14	2.5	1.4	146	7.7
2	15.4	0.14–0.08	2.1	1.1	88	4.6
3	17.0	0.08–0.04	0.76	0.68	28	1.5
4	17.9	0.04–0.023	0.51	0.32	22	1.1
5	22.4	0.023–0.007	0.49	0.28	18	1

[a] Kahn, 1957.
[b] 10,000 Å = 1 μm.
[c] Assuming a c-spacing of 19 Å.

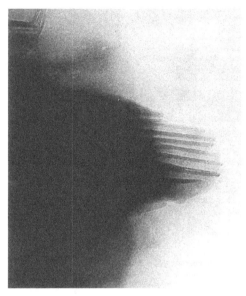

FIGURE 4.8 Edge view of a flake of sodium montmorillonite. The dark parallel lines, which are 10 Å ± thick, are the unit layers. *Electron micrograph by Barclay, L.M. and Thompson, D.W., 1969. Electron microscopy of montmorillonite. Nature 222, 263. Courtesy of Nature.*

predominately aluminum for silicon in the tetrahedral sheet. In many cases, as much as one silicon in four may be so replaced. Substitutions may also take place in the octahedral sheet, typically magnesium and iron for aluminum. The average charge deficiency is higher than that of montmorillonite (0.69 vs 0.41) (Weaver and Pollard, 1973, p. 63), and the balancing cation is always potassium.

Illites differ sharply from montmorillonite in that they do not have an expanding lattice and in that no water can penetrate between the layers. The strong interlayer bonding is probably because of the higher layer charge, because the site of the charge is nearer the surface in the tetrahedral sheet, and because the size of the potassium ion is such that it just fits into the holes in the oxygen network and forms secondary valance links between adjacent layers. Thus, the potassium normally is fixed, and cannot be exchanged. Ion exchange can, however, take place at the exterior surfaces of each aggregate. Since hydration is also confined to the exterior surfaces, the increase in volume is much less than that caused by the hydration of montmorillonite. Illites disperse in water to particles having an equivalent spherical radius of about 0.15 μm, widths of about 0.7 μm, and thickness of about 720 Å.

Some illites occur in degraded form, brought about by leaching of potassium from between the layers. This alteration permits some interlayer hydration and lattice expansion, but never to the degree attained by montmorillonite.

Kaolinite

Kaolinite is a two-layer clay with a structure similar to that shown in Fig. 4.5. One tetrahedral sheet is tied to one octahedral in the usual manner, so that the octahedral hydroxyls on the face of one layer are juxtaposed to tetrahedral oxygens on the face of the next layer. In consequence, there is strong hydrogen bonding between the layers, which prevent lattice expansion. There is little, if any, isomorphous substitutions, and very few, if any, cations are adsorbed on the basal surfaces.

Not surprisingly, therefore, most kaolinites occur in large, well-ordered crystals that do not readily disperse to smaller units in water. The width of the crystals ranges from 0.3 to 4 μm and a thickness from 0.05 to 2 μm. *Dickite* and *nacrite* are two other members of the kaolinite group. They differ from kaolinite in their stacking sequences.

Chlorites

Chlorites are a group of clay minerals whose characteristic structure consists of a layer of brucite alternating with a three-sheet pyrophyllite-type layer, as shown in Fig. 4.9. There is some substitution of Al^{+3} for Mg^{+2} in the brucite layer, giving it a positive charge which is balanced by a negative charge on the three-sheet layer, so that the net charge is very low. The negative charge is derived from the substitution of Al^{+3} for Si^{+4} in the tetrahedral sheet. The general formula is

$$2[(Si, Al)_4(Mg, Fe)_3O_{10}(OH)_2] + (Mg, Al)_6(OH)_{12}.$$

The members of the chlorite group differ in the amount and species of atoms substituted in the two layers, and in the orientation and stacking of the layers. Normally, there is no interlayer water, but in certain degraded chlorites, part of the brucite layer has been removed, which permits some degree of interlayer hydration and lattice expansion.

Chlorites occur both in macroscopic and in microscopic crystals. In the latter case, they always occur in mixtures with other minerals, which makes determination of their particle size and shape very difficult. The c-spacing, as determined from macroscopic crystals, is 14 Å, reflecting the presence of the brucite layer.

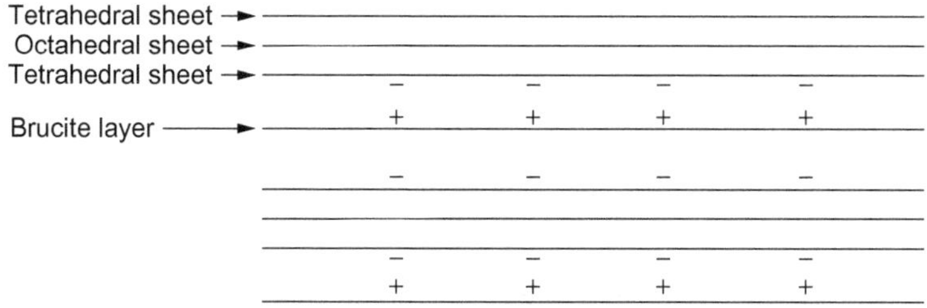

FIGURE 4.9 Diagrammatic representation of chlorite.

Mixed-Layer Clays

Layers of different clay minerals are sometimes found stacked in the same lattice. Interstratified layers of illite and montmorillonite, and of chlorite and vermiculite, are the most common combinations. Generally, the layer sequence is random, but sometimes the same sequence is repeated regularly. Usually, mixed-layer clays disperse in water to smaller units more easily than do single mineral lattices, particularly when one component is of the expanding type.

Attapulgite

Attapulgite particles are completely different in structure and shape from the mica-type minerals discussed so far. They consist of bundles of laths, which separate to individual laths when mixed vigorously with water (see Fig. 4.10). The structure of these laths has been described by Bradley (1940).

FIGURE 4.10 Electron micrograph of attapulgite clay showing open mesh structure magnified 45,000 times. *Courtesy of Attapulgus Minerals and Chemical Corporation.*

There are very few atomic substitutions in the structure, so the surface charge on the particles is low. Also, their specific surface is low. Consequently, the rheological properties of attapulgite suspensions are dependent on mechanical interference between the long laths, rather than on electrostatic interparticle forces. For this reason, attapulgite makes an excellent suspending agent in salt water.

Sepiolite is an analagous clay mineral, with different substitutions in the structure, and wider laths than attapulgite. Sepiolite-based muds are recommended for use in deep, hot wells because their rheological properties are not affected by high temperatures (Carney and Meyer, 1976).

ORIGIN AND OCCURRENCE OF CLAY MINERALS

Clay minerals originate from the degradation of igneous rocks in situ. The parent minerals are the micas, which have already been discussed, the feldspars, $[(CaO)(K_2O)Al_2O_36SiO_2]$ and ferromagnesium minerals, such as horneblende $[(Ca, Na_2)_2 (Mg, Fe, Al)_5 (Al, Si)_8O_{22} (OH, F)_2]$. Bentonite is formed by the weathering of volcanic ash.

The weathering process, by which the clay minerals are formed from the parent minerals, is complex and beyond the scope of this chapter. Suffice it to say that the main factors are climate, topography, vegetation, and time of exposure (Jackson, 1957). Of major importance are the amount of rainfall percolating downwards through the soil and the soil's pH. The pH is determined by the parent rock, the amount of carbon dioxide in the atmosphere, and the vegetation. Silica is leached out under alkaline conditions, and alumina and the ferric oxides under acid conditions. Leaching and deposition lead to the various isomorphous substitutions discussed previously.

Clays formed in situ are termed *primary clays. Secondary clays* are formed from primary clays carried down by streams and rivers, and deposited as sediments in freshwater or marine environments. Their subsequent burial and transformation by diagenesis is discussed in Chapter 8, The Surface Chemistry of Drilling Fluids.

The various species of clay minerals are not distributed evenly throughout the sedimentary sequence. Montmorillonite is abundant in Tertiary sediments, less common in Mesozoic, and rare below that. Chlorite and illite are the most abundant clay minerals; they are found in sediments of all ages and predominate in ancient sediments. Kaolinite is present in both young and old sediments, but in small amounts.

Montmorillonite occurs in its purest form in primary deposits of bentonite. Wyoming bentonite is about 85% montmorillonite. Sodium, calcium, and magnesium are the most common base exchange ions. The ratio of monovalent to divalent cations varies over a range of approximately $0.5-1.7$ (McAtee, 1956), even within the same deposit. Other montmorillonites, of various degrees of purity, have been found in many places all over the world. They seem to be particularly abundant in formations of Middle Tertiary and Upper Cretaceous (Grim, 1962, pp. 43–44).

Note that the term *bentonite* was originally defined as a clay produced by in situ alteration of volcanic ash to montmorillonite, but the term is now used for any clay whose physical properties are dominated by the presence of a smectite.

ION EXCHANGE

As already mentioned, cations are adsorbed on the basal surfaces of clay crystals to compensate for atomic substitutions in the crystal structure. Cations and anions are also held at the crystal edges, because the interruption of the crystal structure along the c axis results in broken valence bonds. In aqueous suspension, both sets of ions may exchange with ions in the bulk solution.

The exchange reaction is governed primarily by the relative concentration of the different species of ions in each phase, as expressed by the law of mass action. For example, for two species of monovalent ions, the equation may be written

$$[A]_c/[B]_c = K[A]_s/[B]_s$$

where $[A]_s$ and $[B]_s$ are the molecular concentrations of the two species of ions in the solution, and $[A]_c$ and $[B]_c$ are those on the clay. K is the ion exchange equilibrium constant, e.g., when K is greater than unity, A is preferentially adsorbed.

When two ions of different valencies are present, the one with the higher valence is generally adsorbed preferentially. The order of preference usually is (Hendricks et al., 1940)

$$H^+ > Ba^{++} > Sr^{++} > Ca^{++} > Cs^+ > Rb^+ > K^+ > Na^+ > Li^+$$

but this series does not strictly apply to all clay minerals; there may be variations. Note that hydrogen is strongly adsorbed, and therefore pH has a strong influence on the base exchange reaction.

The total amount of cations adsorbed, expressed in milliequivalents per hundred grams of dry clay, is called the *base exchange capacity* (BEC), or the *cation exchange capacity* (CEC). The value of the BEC varies considerably, even within each clay mineral group, as shown in Table 4.4. With montmorillonite and illite, the basal surfaces account for some 80% of the BEC. With kaolinite, the broken bonds at the crystal edges account for most of the BEC.

The BEC of a clay and the species of cations in the exchange positions are a good indication of the colloidal activity of the clay. A clay such as montmorillonite that has a high BEC swells greatly and forms viscous

TABLE 4.4 Base Exchange Capacities of Clay Minerals

Meq/100 g of Dry Clay	
Montmorillonite	70–130
Vermiculite	100–200
Illite	10–40
Kaolinite	3–15
Chlorite	10–40
Attapulgite-sepiolite	10–35

Source: From Grim, R.E., 1953. Clay Mineralogy. McGraw Hill Book Co., New York and Weaver, C.E. and Pollard, L.D., 1973. The Chemistry of Clay Minerals. Elsevier Scientific Publ. Co., New York.

suspensions at low concentrations of clay, particularly when sodium is in the exchange positions. In contrast, kaolinite is relatively inert, regardless of the species of exchange cations.

The BEC and the species of exchange cations may be determined in the laboratory by leaching the clay with excess of a suitable salt, such as ammonium acetate, which displaces both the adsorbed cations and those in the interstitial water. Then, another sample is leached with distilled water, which displaces only the ions in the interstitial water. Both filtrates are analyzed for the common exchange cations: the difference between the ionic content of the acetate and water leachates gives the meq of each species adsorbed on the clay, and the total meq of all species of cations gives the BEC. A field test for the approximate determination of the BEC (but not the species of cations) based on the adsorption of methylene blue is given in Chapter 3, Evaluating Drilling Fluid Performance.

Clays with a single species of exchange cation may be prepared by leaching with an appropriate salt and washing to remove excess ions. Alternatively, they may be prepared by passing a dilute clay suspension through an exchange resin which has been saturated with the desired cation. Since clays are analogous to large multivalent anions, it is customary to call monoionic clays by the name of the adsorbed cation; thus we speak of sodium montmorillonite, calcium montmorillonite, etc.

Anion exchange capacities are much less than base exchange capacities, about 10–20 meq/100 g for minerals in the smectite group. With some clay minerals, anion exchange capacities are difficult to determine because of the small amounts involved.

CLAY SWELLING MECHANISMS

All classes of clay minerals adsorb water, but smectites take up much larger volumes than do other classes because of their expanding lattice. For this reason, most of the studies on clay swelling have been made with smectites, particularly with montmorillonite.

Two swelling mechanisms are recognized: crystalline and osmotic. *Crystalline swelling* (sometimes called *surface hydration*) results from the adsorption of monomolecular layers of water on the basal crystal surfaces—on both the external, and, in the case of expanding lattice clays, the interlayer surfaces (see Fig. 4.6). The first layer of water is held on the surface by hydrogen bonding to the hexagonal network of oxygen atoms (Hendricks and Jefferson, 1938), as shown in Fig. 4.11. Consequently, the water molecules are also in hexagonal coordination, as shown in Fig. 4.12. The next layer is similarly coordinated and bonded to the first, and so on with succeeding layers. The strength of the bonds decreases with distance from the surface, but structured water is believed to persist to distances of 75–100 Å from an external surface (Low, 1961).

The structured nature of the water gives it quasicrystalline properties. Thus, water within 10 Å of the surface has a specific volume about 3% less than that of free water (Low, 1961, p. 291) (compared with the specific volume of ice, which is 8% greater than that of free water). The structured water also has a viscosity greater than that of free water.

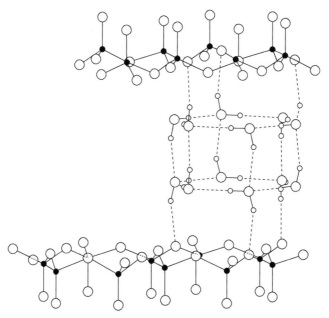

FIGURE 4.11 Combined water layers between layers of partially dehydrated vermiculite. *From Hendricks, S.B. and Jefferson, M.E., 1938. Structure of kaolin and talc-pyrophyllite hydrates and their bearing on water sorption of the clays. Am. Mineral. 23, 863–875. Courtesy of American Mineralogist.*

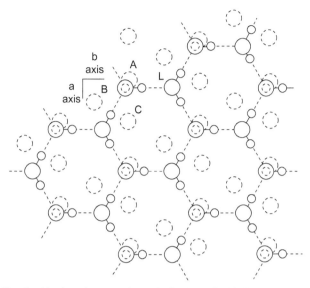

FIGURE 4.12 Combination of water and vermiculite layers by binding through hydrogen. The oxygen atoms represented by large dotted circles are 2.73 Å below the plane of the water molecules. *From Hendricks, S.B. and Jefferson, M.E., 1938. Structure of kaolin and talc-pyrophyllite hydrates and their bearing on water sorption of the clays. Am. Mineral. 23, 863−875. Courtesy of American Mineralogist.*

The exchangeable cations influence the crystalline water in two ways. First, many of the cations are themselves hydrated, i.e., they have shells of water molecules (exceptions are NH_4^+, K^+, and Na^+). Second, they bond to the crystal surface in competition with the water molecules, and thus tend to disrupt the water structure. Exceptions are Na^+ and Li^+, which are lightly bonded and tend to diffuse away.

When dry montmorillonite is exposed to water vapor, water condenses between the layers, and the lattice expands. Fig. 4.13 shows the relationships between the water vapor pressure, the amount of water adsorbed, and the increase in c-spacing (Ross and Hendricks, 1945). It is evident that the energy of adsorption of the first layer is extremely high, but that it decreases rapidly with succeeding layers. The relation between the vapor pressure and the potential swelling pressure is given in Eq. 9.8 Chapter 9, Wellbore Stability.

Norrish (1954) used an X-ray diffraction technique to measure the c-spacing of flakes of monoionic montmorillonites while they were immersed in saturated solutions of a salt of the cation on the clay. Then, the spacing was observed in progressively more dilute solutions and, finally, in pure water.

In all cases, the spacing at first increased in discrete steps with decrease in concentration, each step corresponding to the adsorption of a monolayer

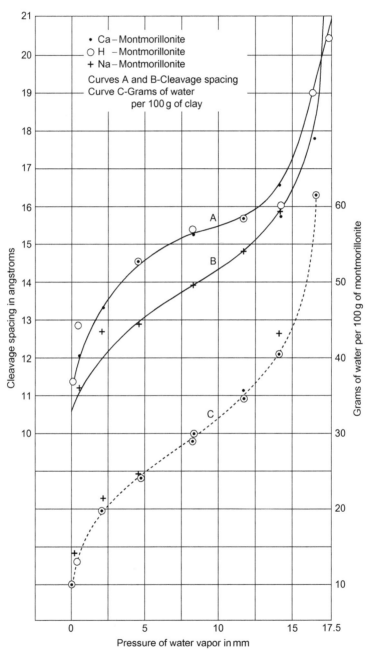

FIGURE 4.13 Curves showing relation of vapor pressure and cleavage in montmorillonite. Experiment conducted as 21°C. *From Ross, C.S. and Hendricks, S.B., 1945. Minerals of the montmorillonite group. Professional Paper 205B, U.S. Dept. Interior, p. 53.*

TABLE 4.5 c-Spacing of Monoionic Montmorillonite Flakes in Pure Water

Cation on the Clay	Maximum c-Spacing, Å
Cs^{+1}	13.8
NH_4^{+1}, K^{+1}	15.0
Ca^{+2}, Ba^{+2}	18.9
Mg^{+2}	19.2
Al^{+3}	19.4

Source: After Norrish, K., 1954. The swelling of montmorillonite. Discuss. Faraday Soc. 18, 120−134.

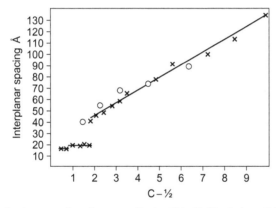

FIGURE 4.14 Lattice expansion of montmorillonite. X in NaCl solutions; O in Na_2SO_4 solutions. *From Norrish, K., 1954. The swelling of montmorillonite. Discuss. Faraday Soc. 18, 120−134. Courtesy of Discussions of the Faraday Society.*

of water molecules. Table 4.5 shows the maximum spacing observed with most monoionic clays. The values obtained indicate that no more than four layers of water were adsorbed.

With monoionic sodium montmorillonite, however, a jump in spacing from 19 to 40 Å was observed at a concentration of 0.3 N, and the X-ray patterns changed from sharp to diffuse. At still lower concentrations, the spacings increased linearly with the reciprocal of the square root of the concentration as shown in Fig. 4.14. The patterns became more diffuse as the spacing increased, so that spacings above the maximum of 130 Å shown in Fig. 4.14 may have occurred, but could not be detected. Similar behavior was observed with lithium chloride and hydrogen chloride, except that the stepwise expansion persisted until the spacing was 22.5 Å, which occurred at

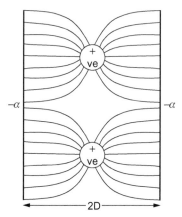

FIGURE 4.15 Cations between the sheets of montmorillonite. *From Norrish, K., 1954. The swelling of montmorillonite. Discuss. Faraday Soc. 18, 120–134. Courtesy of Discussions of the Faraday Society.*

a concentration of 0.66 N. However, the diffuse spacings observed in dilute hydrogen chloride solutions collapsed on aging, probably because of acid attack on the crystal structure, and because of the consequent release of Al^{+3} ions and conversion of the clay to the aluminum form.

Norrish explained the c-spacing changes in terms of repulsive swelling forces arising from hydration of the interlayer cations, and opposing attractive forces arising from electrostatic links between the negatively charged layer surface and the interlayer cations, as shown in Fig. 4.15. In the case of the salts shown in Table 4.5, the swelling forces were not strong enough to break the electrostatic links, and only crystalline swelling was observed. On the other hand, the repulsive forces developed in dilute solutions of sodium, lithium, and hydrogen chlorides were strong enough to break the links, thereby permitting osmotic swelling to take place.

Osmotic swelling occurs because the concentration of cations between the layers is greater than that in the bulk solution. Consequently, water is drawn between the layers, thereby increasing the c-spacing and permitting the development of the diffuse double layers that are discussed in the next section. Although no semipermeable membrane is involved, the mechanism is essentially osmotic, because it is governed by a difference in electrolyte concentration.

Osmotic swelling causes much larger increases in bulk volume than does crystalline swelling. For example, sodium montmorillonite adsorbs about 0.5 g water per gram of dry clay, doubling the volume, in the crystalline swelling region, but about 10 g water per gram dry clay, increasing the volume 20-fold, in the osmotic region. On the other hand, the repulsive forces between the layers are much less in the osmotic region than in the crystalline region.

If hydrated aggregates of sodium montmorillonite are left on the bottom of a beaker full of distilled water, they will, in time, subdivide, and autodisperse throughout the body of the liquid. The size of dispersed montmorillonite particles, and the number of unit layers per particle, are discussed in the section on smectites earlier in this chapter.

THE ELECTROSTATIC DOUBLE LAYER

At the beginning of this chapter, we said that particles in colloidal suspension carried a surface charge. This charge attracts ions of the opposite sign, which are called counter ions, and the combination is called the *electrostatic double layer*. Some counter ions are not tightly held to the surface and tend to drift away, forming a diffuse ionic atmosphere around the particle. In addition to attracting ions of the opposite sign, the surface charge repels those of the same sign. The net result is a distribution of positive and negative ions, as shown schematically in Fig. 4.16 (van Olphen, 1977, p. 30). In the case of clays, the surface charge is negative, as we have seen, and the exchangeable cations act as counter ions.

The distribution of ions in the double layer results in a potential grading from a maximum at the clay surface to zero in the bulk solution, as shown in Fig. 4.16 (Engelmann et al., 1967).

The layer of cations next to the surface of the particle, known as the *Stern layer*, is bound to and moves with the particle whereas the diffuse ions are independently mobile. Thus, if a clay suspension is placed in a cataphoretic cell, the particle, the Stern layer, and some of the diffuse ions move toward the cathode. The potential difference from the plane of shear to the bulk of the solution is known as the *zeta potential*, and is a major factor controlling the behavior of the particle. Again, water flowing past stationary particles, as in the case of water flowing through the pores of a shale, removes the mobile ions, thereby generating a potential, which is known as the *streaming potential*.

The zeta potential is maximum, and the mobile layer is most diffuse when the bulk solution is pure water. Addition of electrolytes to the suspension compresses the diffuse layer, and reduces the zeta potential. The zeta potential decreases greatly with increase in valence of the added cations, especially if low valence ions are replaced by higher valence ones through base exchange, the ratio being approximately 1 to 10 to 500 for monovalent, divalent, and trivalent cations, respectively (Engelmann et al., 1967). The zeta potential is also reduced by the adsorption of certain long-chain organic cations. In some cases, it is possible to neutralize and reverse the zeta potential.

The potential difference between the surface of the particle and the bulk solution is known as the *Nernst potential*. In a clay suspension this potential is independent of the electrolytes in solution.

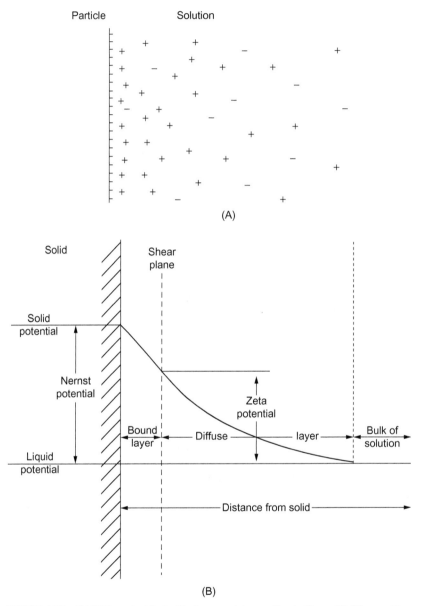

FIGURE 4.16 (A) Diffuse electric double layer model according to Gouy. (B) Diagram illustrating the zeta potential. *From (A) van Olphen, H., 1977. An Introduction to Clay Colloid Chemistry, second ed. John Wiley & Sons, New York. Courtesy of John Wiley & Sons; (B) Engelmann, W.H., Terichow, O., Selim, A.A., 1967. Zeta potential and pendulum sclerometer studies of granite in a solution environment. U.S. Bur. Mines. Report of Investigations 7048.*

Although ions are adsorbed mostly on the basal surfaces, they are also adsorbed at the crystal edge, and consequently a double layer also develops there. It must be remembered, however, that the crystal structure is interrupted at the edge, so that, in addition to physical (electrostatic) adsorption, there may be specific chemical reactions with broken valences. *Chemisorption*, as it is called, is similar to ordinary chemical reactions, and only occurs under the appropriate electrochemical conditions. The edge charge is less than the basal surface charge, and may be positive or negative, largely depending on pH. For example, if kaolinite is treated with HCl, it has a positive charge, and if treated with NaOH, it has a negative charge (Schofield and Samson, 1954). The reason for this behavior is that aluminum atoms at the edge react with HCl to form $AlCl_3$, a strong electrolyte which dissociates to $Al^{+3} + 3Cl^-$, whereas with NaOH, aluminum forms aluminum hydroxide, which is insoluble. (Remember that ion adsorption in kaolinite takes place almost entirely at the edge, so that the charge on the particle is determined by the charge on the edge.)

The existence of positive sites on the edges of kaolinite has also been demonstrated by an experiment in which a negative gold sol was added to a kaolinite suspension (van Olphen, 1977, p. 95). An electron micrograph showed the gold particles adsorbed only at the crystal edges (see Fig. 4.17).

PARTICLE ASSOCIATION

Flocculation and Deflocculation

As mentioned in the beginning of this chapter, colloid particles remain indefinitely in suspension because of their extremely small size. Only if they agglomerate to larger units do they have finite sedimentation rates. When suspended in pure water, they cannot agglomerate, because of interference between the highly diffuse double layers. But if an electrolyte is added, the double layers are compressed, and if enough electrolyte is added, the particles can approach each other so closely that the attractive forces predominate, and the particles agglomerate. This phenomenon is known as *flocculation*, and the critical concentration of electrolyte at which it occurs is known as the *flocculation value*.

The flocculation value of clays may be readily determined by adding increasing amounts of electrolyte to a series of dilute suspensions. The change from a deflocculated suspension to a flocculated one is very marked. Before flocculation, the coarser particles may sediment out, but the supernatant fluid always remains cloudy. Upon flocculation, clumps of particles big enough to be seen by the naked eye are formed; these sediment, leaving a clear supernatant liquid. The particles are very loosely associated in the flocs, which enclose large amounts of water (see Fig. 4.18), and consequently form voluminous sediments.

FIGURE 4.17 Electron micrograph of a mixture of kaolinite and a gold sol. *From van Olphen, H., 1977. An Introduction to Clay Colloid Chemistry, second ed. John Wiley & Sons, New York, photographed by H.P. Studer. Courtesy of John Wiley & Sons.*

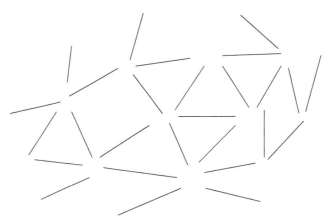

FIGURE 4.18 Schematic representation of flocculated clay platelets (assuming negative edge potential).

The flocculation value depends on the species of clay mineral, the exchange cations thereon, and on the kind of salt added. The higher the valence of the cations (either on the clay or in the salt) the lower the flocculation value. Thus, sodium montmorillonite is flocculated by about 15 meq/L of sodium chloride, and calcium montmorillonite by about 0.2 meq/L of calcium chloride. The situation is more complicated when the cation of the salt is different from the cation on the clay, because then base exchange occurs, but the flocculation value is always much lower whenever polyvalent cations are involved. For instance, the flocculation value of sodium montmorillonite by calcium chloride is about 5 meq/L, and of calcium montmorillonite by sodium chloride about 1.5 meq/L.

There is a slight difference in the flocculating power of monovalent salts, as follows: $Cs > Rb > NH_4 > K > Na > Li$. This series is known as the *Hoffmeister series*, or as the *lyotropic series*.

If the concentration of clay in a suspension is high enough, flocculation will cause the formation of a continuous gel structure instead of individual flocs. The gels commonly observed in aqueous drilling fluids are the result of flocculation by soluble salts, which are always present in sufficient concentrations to cause at least a mild degree of flocculation.

Gel structures build up slowly with time, as the particles orient themselves into positions of minimum free energy under the influence of Brownian motion of the water molecules (a position of minimum free energy would be obtained, for instance, by a positive edge on one particle moving towards the negative surface on another). The length of time required for a gel to attain maximum strength depends on the flocculation value for the system, and the concentration of the clay and of the salt. At a very low concentration of both, it may take days for gelation even to be observable (Hauser and Reed, 1937), whereas at high concentrations of salt, gelation may be almost instantaneous.

Flocculation may be prevented, or reversed, by the addition of the sodium salts of certain complex anions, notably polyphosphates, tannates, and lignosulfonates. For instance, if about 0.5% of sodium hexametaphosphate is added to a dilute suspension of sodium montmorillonite, the flocculation value is raised from 15 to about 400 meq/L of sodium chloride. A similar amount of a polyphosphate will liquify a thick gelatinous mud. This action is known as *peptization*, or *deflocculation*, and the relevant salts are called *deflocculants* or *thinners* in the drilling mud business.

There is little doubt that thinners are adsorbed at the crystal edges. The small amounts involved are comparable to the anion exchange capacity, and there is no increase in the c-spacing, such as one would expect if they were adsorbed on the basal surfaces. The mechanism is almost certainly chemisorption, because all the common thinners are known to form insoluble salts, or complexes, with the metals such as aluminum, magnesium, and iron, whose atoms are likely to be exposed at the crystal edges (van Olphen, 1977, p. 114).

Furthermore, Loomis et al. (1941) obtained experimental evidence indicating chemisorption. They treated clay suspensions with sodium tetraphosphate, centrifuged the suspensions, and analyzed the supernatant liquor. They found that the phosphate was adsorbed, and that the amount required to effect maximum reduction in viscosity depended on the amount of phosphate per unit weight of clay, not on the phosphate concentration in the water phase, indicating chemisorption. In contrast, when suspensions were treated with sodium chloride, it was not adsorbed, and the amount required to cause maximum gelation depended on the concentration of the salt in the water. In addition, they found that when suspensions were treated with a series of complex phosphates of increasing molecular weight or with a tannate, the amount of each required to produce maximum reduction in viscosity was in proportion to the area that would be covered if the molecules were adsorbed on the clay.

van Olphen (1977, p. 114) has postulated that the complex phosphate molecules are oriented on the clay edge surfaces by bonding with the exposed positive-charged aluminum atoms, as shown in Fig. 4.19. Dissociation of the sodium ions then produces a negative edge surface, thus preventing the buildup of gel structures by positive edge to negative basal surface linkages. The creation of a negative edge is supported by the observation that the cataphoretic mobility increases after treatment with phosphates.

Although chemisorption appears to be the mechanism involved with small amounts of thinner, another mechanism must be responsible for the decrease in gel strength that is observed when larger quantities are added. In this case, the reduction in gel strength and the amount of thinner adsorbed, relative to the amounts added, are smaller. Probably the mechanism in this case is the exchange of simple anions in the double layer at the crystal edge with the large, multivalent anions of the thinner.

Some doubts have been raised about the action of ferrochrome lignosulfonate because it has been observed that base exchange occurs between Fe^{+2} and Cr^{+3} from the lignosulfonate, and Na^+ and Ca^{+2} on the clay (Jessen and Johnson, 1963), and this exchange would suggest that lignosulfonates were adsorbed on the basal surfaces. On the other hand, X-ray diffraction studies

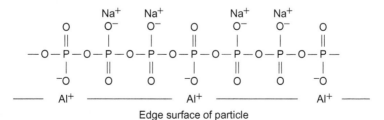

FIGURE 4.19 Schematic representation of a polyphosphate molecule adsorbed on clay crystal edge by bonding with exposed aluminum atoms. *From van Olphen, H., 1977. An Introduction to Clay Colloid Chemistry, second ed. John Wiley & Sons, New York. Courtesy of John Wiley & Sons.*

have shown no significant change in c-spacing. A possible explanation is that the lignosulfonates react with the aluminum at the crystal edges, but in doing so, release chromium and ferrous ions, which subsequently exchange with sodium and calcium ions from the basal surfaces (McAtee and Smith, 1969).

Aggregation and Dispersion

Although all forms of particle association are termed flocculation in classical colloid chemistry, in drilling fluid technology it is necessary to distinguish between two forms of association, because they have a profoundly different effect on the rheology of suspensions. The term *flocculation* is limited to the loose association of clay platelets that forms flocs or gel structures, as discussed in the preceding section. The term *aggregation*, as used here, refers to the collapse of the diffuse double layers and the formation of aggregates of parallel platelets spaced 20 Å or less apart (Mering, 1946), Aggregation is the reverse of the sudden increase in c-spacing that Norrish (1954) observed when the layers in a flake of sodium montmorillonite overcame the attractive forces between them, and expanded to virtually individual units (see the section on clay swelling mechanisms, earlier in this chapter). Thus, whereas flocculation causes an increase in gel strength, aggregation causes a decrease because it reduces (1) the number of units available to build gel structures and (2) the surface area available for particle interaction.

The term *dispersion* is commonly used to describe the subdivision of particle aggregates in a suspension, usually by mechanical means. Garrison and ten Brink (1940) proposed extending the term to the subdivision of clay platelet stacks, which is usually the result of electrochemical effects, and thus to distinguish between the dispersion−aggregation process and the deflocculation−flocculation process. In the technical literature, the term *dispersion* is unfortunately still sometimes applied to the deflocculation process. The difference between the two processes (the flocculation−deflocculation process on the one hand, and the aggregation−dispersion process on the other) is illustrated schematically in Fig. 4.20. The two left-hand pictures show 1% suspensions of calcium bentonite and of sodium bentonite in distilled water. The calcium bentonite is aggregated and the sodium bentonite is dispersed, but both are deflocculated, as shown by the misty supernatant liquid after centrifuging. The picture on the lower right shows the calcium bentonite suspension after the addition of 0.01 N calcium chloride; the upper right-hand picture shows the sodium suspension after the addition of 0.1 N sodium chloride. Both are flocculated, as shown by the clear supernatant, but the calcium bentonite suspension is aggregated and the sodium bentonite suspension is dispersed, as shown by the much greater volume of sediment.

Although low concentrations of sodium chloride cause only flocculation, high concentrations cause aggregation as well. This was shown by some experiments in which increasing amounts of sodium chloride were added to

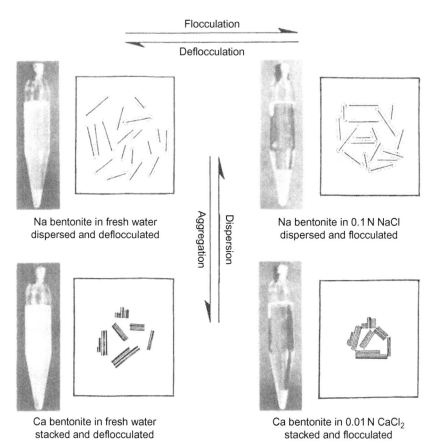

FIGURE 4.20 Schematic representation of the flocculation–deflocculation mechanism and the aggregation-dispersion mechanism. *Courtesy of Shell Development Co.*

a 2.7% suspension of sodium bentonite (Darley, 1957). The onset of flocculation is shown in Fig. 4.21 by the rise in gel strengths from 10 meq/L or greater. This value correlates well with the flocculation value given in the section "Flocculation and Deflocculation" earlier in this chapter. The gel strength continues to rise with concentration of sodium chloride up to 400 meq/L, but the particles reach equilibrium positions slowly, as indicated by the difference between the initial and 10-min gel strengths. Evidently, the attractive and repulsive forces are nearly in balance. At concentrations above 400 meq/L the gel strengths decline, and at 1000 meq/L, the initial and 10-min gel strengths coincide, suggesting that the attractive forces are now overwhelming.

Data from two tests for aggregation are plotted at the top of Fig. 4.21. An increase in optical density indicates an increase in particle size, and a

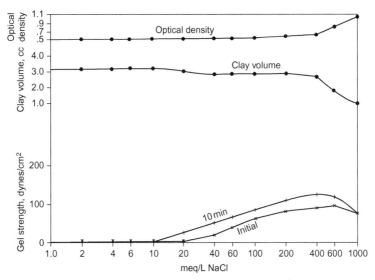

FIGURE 4.21 Flocculation and aggregation of sodium bentonite by sodium chloride. *From Darley, H.C.H., 1957. A test for degree of dispersion in drilling muds. Trans. AIME 210, 93–96. Copyright 1957 by SPE-AIME.*

decrease in "clay volume" indicates a decrease in the volume of centrifuged sediment. Both tests indicate that aggregation starts around 400 meq/L. This conclusion is in accordance with the X-ray diffraction studies of sodium bentonite discussed in the section on clay swelling mechanisms earlier in this chapter, which showed that the critical change in c-spacing occurred at a concentration of 300 meq/L of sodium chloride. In addition, X-ray scattering studies have shown independent clay platelets in 0.1 N sodium chloride, and aggregates of six to eight layers in 1 N solutions (Hight et al., 1962).

The addition of polyvalent salts to sodium bentonite suspensions show flocculation at first, and then aggregation as the concentration increases (see Figs. 4.22 and 4.23). Note that the critical concentrations decrease with increase in valency of the cation. The mechanism is complicated by base exchange reactions. Other studies have shown that the maximum gel strength occurs when the amount of calcium added is 60% of the BEC, and the minimum is reached when 85% has been added (Williams et al., 1953).

Many clays encountered in drilling are predominantly calcium and magnesium clays, and hence are aggregated. When treated with thinner, both deflocculation and dispersion occur simultaneously—deflocculation because of the action of the anion, and dispersion because of the conversion of the clay to the sodium form. Dispersion is undesirable because it increases the plastic viscosity. Dispersion may be avoided by the simultaneous addition of a polyvalent salt or hydroxide with the thinner.

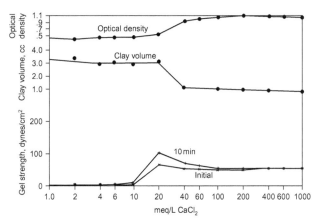

FIGURE 4.22 Flocculation and aggregation of sodium bentonite by calcium chloride. *From Darley, H.C.H., 1957. A test for degree of dispersion in drilling muds. Trans. AIME 210, 93–96. Copyright 1957 by SPE-AIME.*

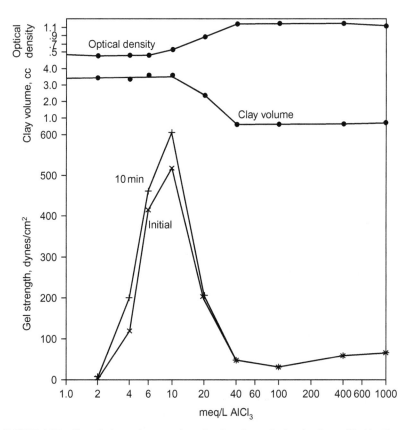

FIGURE 4.23 Flocculation and aggregation of sodium bentonite by aluminum chloride. *From Darley, H.C.H., 1957. A test for degree of dispersion in drilling muds. Trans. AIME 210, 93–96. Copyright 1957 by SPE-AIME.*

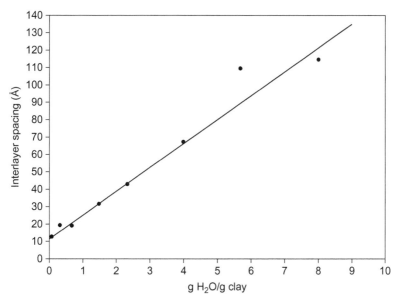

FIGURE 4.24 Lattice expansion of Na Wyoming bentonite with varying water contents. The dashed line represents the relationship that holds if the lattice expansion is continuous throughout the range of water contents. *From Hight, R., Higdon, W.T., Darley, H.C.H., Schmidt, P.W., 1962. Small angle scattering from montmorillonite clay suspension. J. Chem. Phys. 37, 502–510. Courtesy of Journal of Chemical Physics.*

Up to this point, we have been concerned with the gelation of comparatively dilute (3%) bentonite suspensions, in which gel structures are not apparent unless sufficient salt is present to cause flocculation. Gelation does occur, however, at salt concentrations below the flocculation point, if the concentration of clay is high enough. The reason is that at high concentrations, the clay platelets are spaced so close together that their diffuse double layers interfere, and they must orient themselves in positions of minimum free energy. Thus, X-rays scattering curves showed no preferred orientation in 2% suspensions of sodium montmorillonite (Hight et al., 1962), but did show parallel orientation in 10% suspensions. As shown in Fig. 4.24, the spacing increased with increase in water content.

THE MECHANISM OF GELATION

Various linkages and plate orientations proposed to account for gel structure may be summarized as follows:

1. Cross-linking between parallel plates, through positive edge to negative surface linkages, to form a house-of-cards structure (Schofield and Samson, 1954; van Olphen, 1977, pp. 103–105).

2. Edge-to-edge association, to form intersecting ribbons (M'Ewen and Pratt, 1957). The basis for this theory is, briefly, that because of the relatively high repulsive potential between the basal surfaces, the preferred platelet orientation will be parallel with edge-to-edge association.
3. Parallel association of plates, held together by the quasicrystalline water between them (Leonard and Low, 1961).

It may well be that all of these mechanisms are operative, and that their relative significance depends on such factors as clay concentration and the strength and sign of the double layer potentials on the platelet edges and surfaces.

In trying to visualize gel structures, one should bear in mind that in a crowded system of platelets up to 10,000 Å wide, spaced somewhere around 300 Å apart, orientation of the particles is restricted by spatial considerations. Furthermore, the platelets are not the rigid little rectangles that we like to draw in schematic diagrams, but flexible films of diverse shapes and sizes, as shown in Fig. 4.7.

In a concentrated suspension, one would expect to see local groups of platelets aligned roughly parallel under the influence of the basal repulsive forces, but not uniform alignment throughout the suspension. Neighboring platelets may flex according to their relative positions and the relative magnitude of their surface and edge potentials. Thus, when the edges are positive, the platelets will flex toward a negative basal surface, as shown in Fig. 4.25 (Norrish and Rausell-Colom, 1961). When the edges are negative, the stronger basal repulsive potential will cause the platelets to align parallel, when not prevented from doing so by mechanical interference. Addition of a thinner reverses positive edge potentials, and increases the repulsive forces between the edges.

It is also to be expected that the layers of crystalline water on the basal surfaces will tend to maintain parallel orientation of the platelets, the only question being to what distance from the surface the water structure is effective.

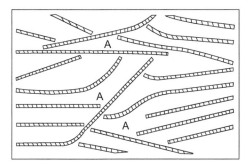

FIGURE 4.25 Schematic representation of edge-to-face bonds. *From Norrish, K. and Rausell-Colom, J.A., 1961. Low-angle X-ray diffraction studies of the swelling of montmorillonite and vermiculite. Clays Clay Miner. 10, 123–149. Courtesy of Pergammon Press Ltd.*

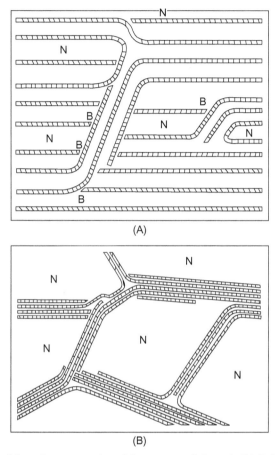

FIGURE 4.26 Schematic representation of the structure of clay gel. (A) Before freezing and (B) after freezing. N, ice crystallization nuclei; B, edge-to-face bonds. *From Norrish, K. and Rausell-Colom, J.A., 1963. Effect of freezing on the swelling of clay minerals. Clay Miner. Bull. (5), 9–16. Courtesy of Clay Mineral Group, Mineralogical Society.*

Attempts have been made from time to time to deduce gel structure by flash-freezing a gel, and then drying it under a vacuum. The process leaves a skeleton framework, presumed to be that of the original gel (Borst and Shell, 1971). The validity of this assumption is dubious, because X-ray diffraction measurements, made while the gel was being flash frozen in the diffractometer, have shown that the interplatelet spacing contracts at the moment of freezing (Norrish and Rausell-Colom, 1963), forming large pores, as shown in Fig. 4.26.

POLYMERS

The polymers discussed in this chapter are organic colloids. They are composed of unit cells, called repeating units (*monomers*), such as the cellulose

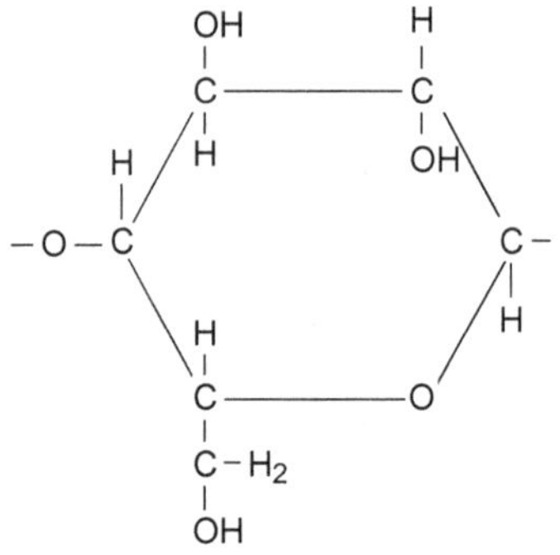

FIGURE 4.27 Cellulose unit.

cell shown in Fig. 4.27, linked together either in straight or branched chains to form macromolecules. A single macromolecule may contain hundreds, even thousands, of unit cells, and thus is well within the colloidal size range.

Polymers are classified as either natural, modified natural, or synthetic. Natural polymers are derived from nature (starch, guar gum, xanthan gum) and lightly processed for industrial uses, but are not modified by chemical reactions. Modified natural polymers (CMC/PAC, CM-starch, HP-guar) are chemically processed to achieve a degree of functionality needed.

Polymers are used in drilling muds when the desired properties cannot be obtained with colloidal clays. For instance, starch, which was the first colloid used in drilling muds, was introduced to provide filtration control in saltwater muds (Gray et al., 1942) because starch is stable in saltwater, whereas clays are not. Modified starches have been developed to enhance their dispensability in different brines base fluids and to impart higher temperature stability.

Xanthan gum was developed to greatly enhance the shear-thinning capabilities of a WBM which was shown in lab studies to dramatically increase the drill rate (Deily et al., 1967). Compared to clay muds, many polymer suspensions have low yield points relative to plastic viscosity, and no real gel strength. Structural properties are obtained, however, with xanthan biopolymer. The structure of xanthan gum is shown in Fig. 11.10. Its high molecular weight ($>2,000,000$) and entangled polymeric structure give the highly non-Newtonian nature of drilling fluids to which they are added. Other polymers that rely on entanglement are scleroglucan, diutan, and some modified starches.

Numerous synthetic polymers have been developed for various purposes, and new ones are constantly being introduced. Only the main types and the principles governing their behavior will be discussed in this chapter. Further details may be found in Chapter 5, Polymers, Chapter 11, Completion, Workover, Packer and Reservoir Drilling Fluids and in reviews in the technical literature (Carico and Bagshaw, 1978; Chatterji and Borchardt, 1981; Lauzon, 1982).

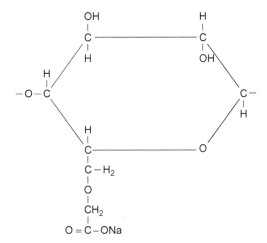

FIGURE 4.28 Sodium carboxymethylcellulose unit.

Carboxymethylcellulose (CMC/PAC) is made by reacting cellulose with chloracetic acid and NaOH, substituting $CH_3COO^-\ Na^+$ for H, as shown in Fig. 4.28. Note that there are three OH groups on each cellulose unit, each capable of substitution. The term *degree of substitution* (DS) refers to the average number of carboxy groups on the chain per unit cell. The difference between CMC and PAC is the DS.

The carboxy group has two important functions: First, it imparts water solubility (strictly speaking, *water dispersibility*) to the otherwise insoluble cellulose polymer. Second, dissociation of Na^+ creates negative sites along the chain. Mutual repulsion between the charges causes the randomly coiled chains to stretch linearly, thereby increasing the viscosity. Soluble salts, especially polyvalent salts, repress dissociation, allowing the chains to coil. Polymers that carry electrostatic charges are termed *polyelectrolytes*. Because its charges are negative, CMC is an anionic polyelectrolyte.

The number of monomers in a macromolecule determines its molecular weight, and is usually called the *degree of polymerization* (DP). By varying the DP and the DS, modified natural polymers are synthesized to suit various purposes. A high DP results in a high viscosity. A high DS also gives a high viscosity (a phenomenon known as the *electroviscous effect*) and increases the resistance to soluble salts. CMC is used as a filtration control agent (Kaveler, 1946). Three grades covering a range of apparent viscosity are available.

A polyacrylamide-acrylate copolymer is another anionic polyelectrolyte. It is made by converting some of the amides on a polyacrylamide chain to carboxylates (see Fig. 4.29), a process called *hydrolysis*. The degree of hydrolysis depends on the end use. A 70% hydrolyzed copolymer is used for filtration control (Scanley, 1959); a 30% one for preserving hole stability

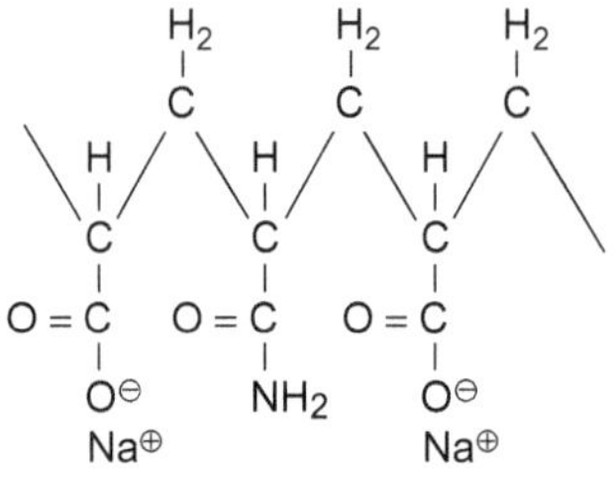

FIGURE 4.29 Acrylamide-sodium acrylate copolymer.

(Clarke et al., 1976) (see chapter: The Surface Chemistry of Drilling Fluids); and a 10% one for clarifying clear water drilling fluids (Gallus, 1962; La Mer and Healey, 1963).

The most likely explanation of the shale stabilizing action of the 30% hydrolyzed copolymer is that it coats the shale surfaces exposed on the sides of the hole, thereby inhibiting disintegration. Similarly, it coats and protects shale drill cuttings, a process known as encapsulation. The coating is believed to result from the attraction between the negative sites on the polymer chain and the positive sites on the edges of the clay platelets. The reason the 30% hydrolyzed copolymer is the most effective for shale preservation is probably that the charged sites on the chain match the spacing of the clay platelets.

The mechanism of water clarification is interesting. The copolymer does not by itself cause flocculation; at least enough salt must be present to initiate flocculation, and then the polymer chains will bind the flocculated particles together (Ruehrwein and Ward, 1952). This point may be demonstrated by changing the order in which salt and the copolymer are added to a suspension of clay in fresh water (see Fig. 4.30A). If the copolymer is added first, the chains are adsorbed around the edges of individual platelets, and are therefore not available for linking between platelets when the salt is subsequently added. Consequently, the platelets separate if the salt concentration is diluted below the flocculation value. On the other hand, if the salt is added first, then the chains can link between adjacent platelets and hold the flocs together when the suspension is diluted.

In order to act as flocculation aids, polyelectrolytes must be added in very small quantities, between about 0.001% and 0.01%. If added in larger amounts, they act as protective colloids and raise the flocculation value of the system. It is believed that at concentrations above 0.01%, the edges of the clay platelets are saturated with polyelectrolyte chains and thus have a strong repulsive charge (see Fig. 4.30B).

Another acrylic copolymer, vinyl acetate-maleic acid, is used as a bentonite *extender* to increase the yield of commercial bentonite (Park et al., 1960). Between 0.1% and 2% of the copolymer is added to the bentonite. When the bentonite is dispersed in fresh water, the copolymer chains form links between the dispersed platelets, increasing the viscosity and yield point.

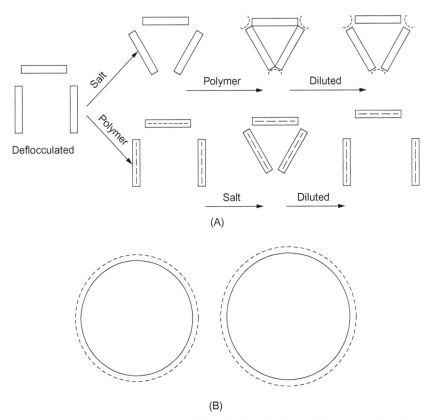

FIGURE 4.30 (A) Effect of order of addition of salt and polymer (schematic). (B) Platelets repel each other when the edges are saturated with polyelectrolyte (schematic, plan view).

A number of nonionic polymers are used for various purposes in drilling fluids. Nonionic polymers have no dissociable inorganic radical, and therefore carry no electrostatic charge. Consequently, they have greater stability in high-salinity fluids. Starch is nonionic, and as previously mentioned, is used for filtration control in salt water muds. It has the advantage of being inexpensive, but the disadvantage of being biodegradable, and a biocide must be used with it.

Other nonionic polymers include hydroxyethylcellulose (HEC) and guar gum. Like CMC, HEC is made from cellulose, but its functional group is an ethylene oxide chain, $(CH_2-O-CH_2)_n$. HEC has two great advantages: it is stable in polyvalent brines, and it is almost completely soluble in acid. Consequently it is used a great deal in completion and workover fluids (see chapter: Introduction to Completion Fluids). Guar gum and HP-guar are used in fracturing fluids and sometimes in workover fluids. They can degraded by enzymes instead of acid. The colloidal activity of natural gums is reduced by

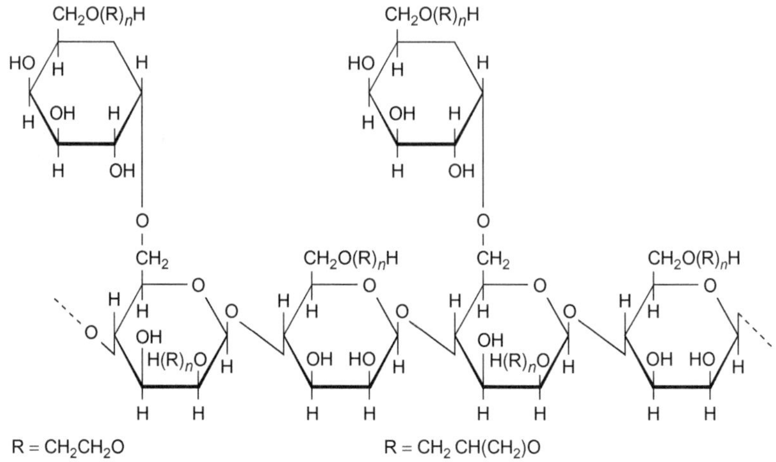

FIGURE 4.31 Hydroxyalkyl guar gum molecule. *Courtesy of Stein Hall Company.*

high concentrations of monovalent salts, and eliminated in polyvalent brines. However, gums that have been reacted with ethylene oxide or propylene oxide (see Fig. 4.31) are stable even in saturated polyvalent brines.

There are also cationic polymers that have positive sites along the chain, created by the dissociation of an inorganic anion. They are used in drilling fluids mostly as emulsifiers and wetting agents, as discussed in Chapter 7, The Filtration Properties of Drilling Fluids. However, a proprietary polyamine has recently been introduced for the purpose of stabilizing formation clays during completion and workover operations (Williams and Underdown, 1981). The cationic group is strongly adsorbed on the clay, reducing its negative charge, and thus inhibiting swelling and dispersion.

Thermal decomposition is a major factor limiting the use of organic polymers in drilling fluids. It may be compensated for by the addition of fresh polymer, but the rate of decomposition increases with temperature, and above a certain temperature becomes excessive. See Chapter 12, Introduction to Fracturing Fluids, for further discussions on polymer temperature stability.

REFERENCES

Barclay, L.M., Thompson, D.W., 1969. Electron microscopy of montmorillonite. Nature 222, 263.
Borst, R.L., Shell, F.J., 1971. The effect of thinners on the fabric of clay muds and gels. J. Petrol. Technol. 23 (10), 1193–1201.
Bradley, W.F., 1940. The structure of attapulgite. Mineralogist 25, 405–410.
Brindley, G.W., Roy, R., 1957. Fourth Progress Report and First Annual Report. API Project 55.
Caenn, R., Darley, H., Gray, G., 2011. Composition and Properties of Drilling and Completion Fluids. Gulf Professional Publishing, Elsevier, Inc, Amsterdam.

Carico, R., 1976. Suspension properties of polymer fluids used in drilling and workover operations. Paper 5870, presented at the California Regional Meeting of the SPE, Long Beach, CA.

Carico, R. and Bagshaw, F., 1978. Description and use of polymers used in drilling, workover and completions. Paper 7747 presented at the AIME Production Technology Symposium, Hobbs, NM.

Carney, L.L., Meyer, R.L., 1976. A new approach to high temperature drilling fluids. SPE Paper 6025. Annual Meeting, New Orleans.

Chatterji, J., Borchardt, J.K., 1981. Application of water-soluble polymers in the oilfield. J. Petrol. Technol. November, 2042–2056.

Clarke, R.N., Scheuerman, R.F., Rath, H., van Laar, H., 1976. Polymerylamide-potassium chloride mud for drilling water-sensitive shales. J. Petrol. Technol. June, 719–727, Trans. AIME 261.

Darley, H.C.H., 1957. A test for degree of dispersion in drilling muds. Trans. AIME 210, 93–96.

Deily, F.H., Lindblom, P., Patton, J.T., Holman, W.E., 1967. New biopolymer low-solids mud speeds drilling operations. Oil Gas J., June 26, 62–70.

Dyal, R.S., Hendricks, S.B., 1950. Total surface areas of clays on polar liquids as a characteristic index. Soil Sci. 69 (June), 421–432.

Engelmann, W.H., Terichow, O., Selim, A.A., 1967. Zeta potential and pendulum sclerometer studies of granite in a solution environment. U.S. Bur. Mines. Report of Investigations 7048.

Gallus, J.P., 1962. Method for drilling with clear water. U.S. Patent No. 3,040,820 (June 26).

Garrison, A.D., ten Brink, K.C., 1940. Some phases of chemical control of clay suspensions. Trans. AIME 136, 175–194.

Gray, G.R., Foster, J.L., Chapman, T.S., 1942. Control of filtration characteristics of salt water muds. Trans. AIME 146, 117–125.

Grim, R.E., 1953. Clay Mineralogy. McGraw Hill Book Co, New York.

Grim, R.E., 1962. Applied Clay Mineralogy. McGraw Hill, New York.

Hauser, E.A., Reed, C.E., 1937. The thixotropic behavior and structure of bentonite. J. Phys. Chem. 41, 910–934.

Hendricks, S.B., Jefferson, M.E., 1938. Structure of kaolin and talc-pyrophyllite hydrates and their bearing on water sorption of the clays. Am. Mineral. 23, 863–875.

Hendricks, S.B., Nelson, R.A., Alexander, L.T., 1940. Hydration mechanism of the clay mineral montmorillonite saturated with various cations. J. Amer. Chem. Soc. 62, 1457–1464.

Hight, R., Higdon, W.T., Darley, H.C.H., Schmidt, P.W., 1962. Small angle scattering from montmorillonite clay suspension. J. Chem. Phys. 37, 502–510.

Hofmann, U., Endell, K., Wilm, O., 1933. Kristalstructur und quellung von motmorillonit. Z. Krist 86, 340–348.

Jackson, M.L., 1957. Frequency distribution of clay minerals in major great soil groups as related to factors of soil formation. Clays Clay Miner 6, 133–143.

Jessen, F.W., Johnson, C.A., 1963. The mechanism of adsorption of lignosulfonates on clay suspensions. Soc. Petrol. Eng. J. 3 (3), 267–273, Trans. AIME 228.

Kahn, A., 1957. Studies in the size and shape of clay particles in aqueous suspension. Clays Clay Miner. 6, 220–235.

Kaveler, H.H., 1946. Improved drilling muds containing carboxymethylcellulose. API Drill. Pred. Prac. 43–50.

La Mer, V.K., Healey, T.W., 1963. The role of filtration in investigating flocculation and redispersion of colloidal suspensions. J. Phys. Chem. 67, 2417–2420.

Lauzon, R.V., 1982. Water-soluble polymers for drilling fluids. Oil Gas J April 19, 93–98.
Leonard, R.A., Low, P.F., 1961. Effect of gelation on the properties of water in clay systems. Clays Clay Miner. 10, 311–325.
Loomis, A.G., Ford, T.E., Fidiam, J.F., 1941. Colloid chemistry of drilling fluids. Trans. AIME 142, 86–97.
Low, P.F., 1961. Physical chemistry of clay-water interaction. Adv. Agron. 13, 323.
Marshall, C.E., 1935. Layer lattices and base exchange clays. Z. Krist. 91, 433–449.
Marshall, C.E., 1949. The Colloid Chemistry of Silicate Minerals. Academic Press, New York.
McAtee, J.L., 1956. Heterogeneity in montmorillonite. Clays Clay Miner. 5, 279–288.
McAtee, J.L., Smith, N.R., 1969. Ferrochrome lignosulfonates (1) X-ray adsorption edge fine structure spectroscopy; (2) interaction with ion exchange resin and clays. J. Colloid Interface Sci. 29 (3), 389–398.
Melrose, J.C., 1956. Light scattering evidence for the particle size of montmorillonite. Symp. Chemistry in Exploration and Production of Petroleum, ACS Meeting Dallas, pp. 19–29.
Mering, J., 1946. On the hydration of montmorillonite. Trans. Faraday Soc. 42 B, 205–219.
M'Ewen, M.B., Pratt, M.I., 1957. The formation of a structural framework in sols of Wyoming bentonite. Trans. Faraday Soc. 53, 535–547.
Norrish, K., 1954. The swelling of montmorillonite. Discuss. Faraday Soc. 18, 120–134.
Norrish, K., Rausell-Colom, J.A., 1961. Low-angle X-ray diffraction studies of the swelling of montmorillonite and vermiculite. Clays Clay Miner. 10, 123–149.
Norrish, K., Rausell-Colom, J.A., 1963. Effect of freezing on the swelling of clay minerals. Clay Miner. Bull. 5, 9–16.
Park, A., Scott, P.P., Lummus, J.L., 1960. Maintaining low solids drilling fluids. Oil Gas J. May 30, 81–84.
Ross, C.S., Hendricks, S.B., 1945. Minerals of the montmorillonite group. Professional Paper 205B, U.S. Dept. Interior, p. 53.
Ruehrwein, R.A., Ward, D.W., 1952. Mechanism of clay aggregation by polyelectrolytes. Soil Sci. 73, 485–492.
Scanley, C.S., 1959. Acrylic polymers as drilling mud additives. World Oil, July, 122–128.
Schofield, R.K., Samson, H.R., 1954. Flocculation of kaolinite due to oppositely charged crystal faces. Discuss. Faraday Soc. 18, 135–145.
van Olphen, H., 1977. An Introduction to Clay Colloid Chemistry, second ed. John Wiley & Sons, New York.
Weaver, C.E., Pollard, L.D., 1973. The Chemistry of Clay Minerals. Elsevier Scientific Publ. Co., New York.
Williams, F.J., Neznayko, M., Weintritt, D.J., 1953. The effect of exchangeable bases on the colloidal properties of bentonite. J. Phys. Chem. 57, 6–10.
Williams, L.H., Underdown, D.R., 1981. New polymer offers effective, permanent clay stabilization treatment. J. Petrol. Technol. July, 1211–1217.

Chapter 5

Water-Dispersible Polymers

INTRODUCTION

There are a confusing number of mud additives for water-based drilling fluids that are called "polymers." They are usually called "water-soluble" but, in actuality, they are water-dispersible and colloidal in nature. One definition used to classify oilfield drilling and completion fluid polymers is: *an organic chemical of above 200 molecular weight with greater than eight repeating units.* The repeating units make up a main, long-chain of material called the backbone. Attached to the backbone are chemical side chains unique to that polymer. If the backbone of a polymer is broken, the functionally of the polymer is destroyed. If the side chain is changed, by temperature or chemical reaction, partial functionality may remain or be changed/lost. The side chain chemistry is crucial to the complexity of the polymer and to determine the polymers chemical reactivity, i.e., precipitation, flocculation, cross-linking, etc.

Polymers vary greatly in functionality and basic properties. Variations in fluid functionality include:

- Rheology modification; bulk viscosity, low-shear rheology, bentonite extension
- Fluid loss control; filter cake modification, seepage control, minimizing lost circulation
- Chemical reactivity; surface coating, clay attachment, flocculation/deflocculation, contaminant removal

Variations in molecular properties include:

- Temperature stability
- Chemical stability
- Biological stability
- Charge density and type
- Molecular weight
- Surface activity

POLYMER CLASSIFICATION

The polymers discussed in this chapter are fluid additives for any system, freshwater or brine water, to control a specific property. A so-called polymer mud is a drilling fluid system that primarily relies on water-dispersible polymers to replace bentonite, chemical thinners, and other common drilling fluid chemicals. It is important to use the proper polymer to control the appropriate function. For example, do not use a fluid loss polymer to control low-shear rate viscosity. Further information on polymer functionality can be found in Chapter 6, The Rheology of Drilling Fluids, Chapter 7, The Filtration Properties of Drilling Fluids, Chapter 9, Wellbore Stability, and Chapter 10, Drilling Problems Related to Drilling Fluids.

Polysaccharides

Polymers can be classified as either natural materials (sometimes called gums), modified-natural materials, or manufactured synthetic materials. Natural polymers are derived from plant material, bacterial or fungus fermentation, or living organisms. They are polysaccharides, meaning composed of many carbohydrate/sugar molecules. Saccharide is a synonym for sugars.

Wikipedia's definition of saccharides and polysaccharides shows the differences among the naturally produced polymers:

> *Natural saccharides are generally of simple carbohydrates called monosaccharides with general formula $(CH_2O)n$ where n is three or more. Examples of monosaccharides are glucose, fructose, and glyceraldehyde. Polysaccharides, meanwhile, have a general formula of $C_x(H_2O)y$ where x is usually a large number between 200 and 2500. Considering that the repeating units in the polymer backbone are often six-carbon monosaccharides, the general formula can also be represented as $(C_6H_{10}O_5)n$ where $40 \leq n \leq 3000$.*
>
> *Polysaccharides contain more than ten monosaccharide units. Definitions of how large a carbohydrate must be to fall into the categories polysaccharides or oligosaccharides vary according to personal opinion. Polysaccharides are an important class of biological polymers. Their function in living organisms is usually either structure- or storage-related. Starch (a polymer of glucose) is used as a storage polysaccharide in plants, being found in the form of both amylose and the branched amylopectin. In animals, the structurally similar glucose polymer is the more densely branched glycogen, sometimes called 'animal starch'. Glycogen's properties allow it to be metabolized more quickly, which suits the active lives of moving animals.*
>
> *Cellulose and chitin are examples of structural polysaccharides. Cellulose is used in the cell walls of plants and other organisms, and is said to be the most abundant organic molecule on Earth. It has many uses such as a significant role in the paper and textile industries, and is used as a feedstock for the production of rayon cellulose acetate, celluloid, and nitrocellulose. Chitin has a similar structure, but has nitrogen-containing side branches, increasing*

its strength. It is found in arthropod exoskeletons and in the cell walls of some fungi. It also has multiple uses, including surgical threads.

Modified-natural polymers use a nature-produced material, which is modified by manipulating the side chains, to gain enhanced functionality or chemical stability. Synthetic polymers are, as the name implies, constructed in a chemical plant, usually with petroleum feedstock.

Natural and modified-natural polymers are, by nature, more chemically and structurally complex than the synthetics. Their backbone and side chains are comprised of various sugar molecules, thus they are called polysaccharides. Chemists have not been able to duplicate nature in constructing the complex chemical structure of the natural gums. On the other hand, synthetic polymers can be designed and manufactured for specific functionality, such as temperature and chemically stability as well as reactivity. The backbone of most natural materials will be broken, depending on the gum, at temperatures from 200°F (100°C) to less than 350°F (150°C). Synthetics can be constructed to withstand temperatures up to 600°F (315°C).

Polymer temperature stability—Most of the temperatures listed for the polymers in this chapter are based on the polymer dispersed in freshwater, with no attempt to extend the temperature degradation of the material. The polymer temperature degradation point is extended by oxygen scangers, brine waters, solids, and other additives. In addition, while circulating the polymer does not experience the maximum bottom-hole temperature. From a practical standpoint, most polymers can be used with bottom-hole temperatures much higher than listed in this chapter. Numerous studies have shown that formate base fluids protect the polymeric backbones aloowing for higher temperature stability while drilling (Annis, 1967, Downs 1991).

TYPES OF POLYMERS

The following is a listing of the current polymer types in use in the oilfield. Oilfield uses of the polymers are primarily in drilling fluids for rheology or fluid loss control and in wellbore stabilization. Some polymers are used in cements and fracturing fluids, in reservoir drill-in fluids and for waste management (solids flocculation).

Starch—Starch is a natural polymer produced from either corn or potatoes, but can be made from other starchy plants. It is supplied as a pregelatinized (water-dispersible) powder. Natural starch is usually treated with a preservative. Modified starches can be either cationic, anionic or nonionic, the most common drilling fluid additive being anionic.

> *Examples*: starch, cm-starch, hp-starch, cmhp-starch
> *Uses*: Fluid loss control for all types of mud systems, particularly useful in salt water systems. Natural starch requires a bactericide while drilling. Modified starches are used in reservoir drill-in fluids.

Guar gum—Derived from the seed of the Guar plant. Regular guar contains residue left from processing the guar bean, which can cause formation damage. HP-guar is further processed with hydroxypropyl side chains and cleaned of excess residue.

Uses: Not used in most drilling fluids (solids reactivity). Has been used in top-hole drilling for quick viscosity. HP-Guar—primary fracturing fluid viscosifier and fluid loss controller

Biopolymers—Polysaccharides manufactured from bacterial or fungal fermentation. They have extremely complex structures with high molecular weights (500 to 2 million+). Their side chains are slightly anionic.

Examples: xanthan gum, wellan gum, diutan, scleroglucan
Uses: Rheology control. Develops high, low-shear-rate viscosities for suspension and carrying capacity.

CMC—A polysaccharide linear polymer based on a plant cellulose backbone modified with carboxymethyl (CM) side chains for water dispersion. Carboxylic acid side chains are anionic. Its functionality depends on the degree of substitution (DS), the number of CM side chains, and molecular weight (MW).

Examples: High MW: Regular CMC or Hi viscosity CMC
Low MW: Low viscosity CMC
Tech grade: Usually high MW but contains up to 40% salt contamination.
PAC (polyanionic cellulose): Higher DS than regular CMC.
Uses: Fluid loss control in drilling; high MW is a bulk viscosifiers with minimal low shear-rate viscosity and minimal suspending ability.

HEC—A polysaccharide linear polymer based on a plant cellulose backbone modified with hydroxyethyl side chains. Its side chains are nonionic. Its functionality depends upon its molecular weight. Usually supplied as a high molecular weight product, >250,000.

Uses: Bulk viscosifier for high density brine fluids, such as the saturated chlorides and bromides. Not normally used in drilling fluids, but used in clear completion fluids, gravel packs, and fracturing fluids. It has no solids suspending ability.

Synthetics—A multitude of synthetic polymers can be designed in chemical manufacturing plants. The two most common synthetics in drilling fluids are the acrylates and polyacrylamides.

Acrylates—Synthetic materials manufactured from acrylic acid. Not as complex structurally as the natural polymers. Usually has a straight-chain carbon backbone but can have a multitude of different side chains, depending on the end product desired. It is usually anionic.

TABLE 5.1 Summary of Polymer Types

Summary of Polymers

Natural	Modified Natural	Synthetic
Starch	CMC/PAC	Polyacrylates
Guar	HP guar	Polyacrylamide
Xanthan gum	CM/HP starch	Vinyl copolymer
Wellan gum	HEC	Styrene copolymer
Diutan		AMPS copolymer
Scleroglucan		

Examples: polyacrylates, vinyl polymers, copolymers, vinyl acetate, maleic anhydride,

Uses: Low molecular weight (<1000)—thinners, deflocculants

    Medium molecular weights—fluid loss, flocculent, shale stabilizer

    High molecular weights—bentonite extender, flocculent

Polyacrylamide—A copolymer of acrylic acid and acrylamide in various ratios. Usually called a partially hydrolyzed polyacrylamide (PHPA). Usually anionic for drilling fluids; can be anionic, nonionic, or cationic for use as a dewatering flocculent.

Uses: flocculants, increased bulk viscosity, shale stabilizer

Cationic polymers—A copolymer, many times with acrylates or acrylamides with ammonium (amides,amines) cationic side chains.

Uses: flocculants, Shale stabilizer

Table 5.1 shows a summary of oilfield polymer types and Table 5.2 is a summary of polymer usage.

GUAR GUM

Fig. 5.1 shows a schematic diagram of the guar gun repeating unit. Guar gum is a polysaccharide, comprised of the sugars galactose and mannose, called a galactomannan. Galactomannans have a mannose backbone with galactose side groups (more specifically, a (1-4)-linked beta-D-mannopyranose backbone with branchpoints from their 6-positions linked to alpha-D-galactose, i.e., 1-6-linked alpha-D-galactopyranose) (Wikipedia 2016).

TABLE 5.2 Primary Functions of the Various Polymer Types

Polymers as Additives By Polymer Primary Function

Rheology	Fluids Loss	Chemical Activity
Modified starch	CMC/PAC	Polyacrylates (high MW)
Xanthan gum	Polyacrylate (medium MW)	Polyacrylamide
Scleroglucan	Starch	Synthetics (high temperature)
Diutan	Modified starch	
HEC	Wellan gum (cements)	
Modified guar		
Polyacrylate (low MW)		

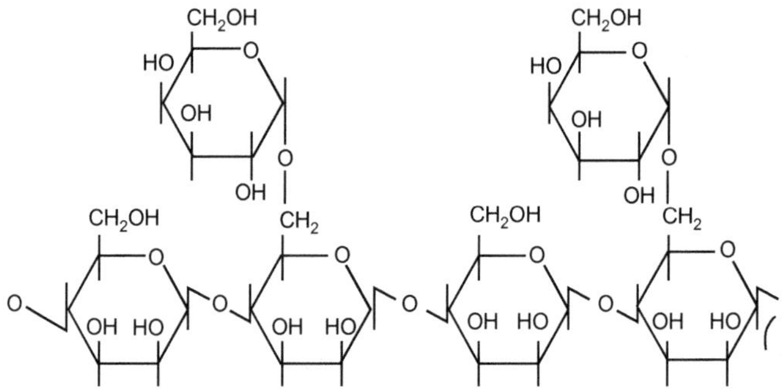

FIGURE 5.1 Schematic illustration of the guar gum monomer.

In order of increasing number of mannose-to-galactose ratio:

fenugreek gum, mannose:galactose $\sim 1:1$
guar gum, mannose:galactose $\sim 2:1$
tara gum, mannose:galactose $\sim 3:1$
locust bean gum or carob gum, mannose:galactose $\sim 4:1$

The largest producer of guar gum is India, with other production from Pakistan, the United States, Australia, and Africa.

Guar gum is not normally used in most drilling fluids (solids reactivity). Has been used in top-hole drilling for quick viscosity. HP-guar—primary fracturing fluid viscosifier and fluid loss controller. Guar and HP-guar are usually cross-linked to gel the fracturing fluid for proppant suspension and placement and to minimize formation damage.

Guar molecules have a tendency to aggregate during the hydraulic fracturing process, mainly due to intermolecular hydrogen bonding. These aggregates are detrimental to oil recovery because they clog the fractures, restricting the flow of oil. Cross-linking guar polymer chains prevents aggregation by forming metal−hydroxyl complexes. The first cross-linked guar gels were developed in the late 1960s. Several metal additives have been used for cross-linking, among them are chromium, aluminum, antimony, zirconium, and the more commonly used, boron. Boron, in the form of $B(OH)_4$, reacts with the hydroxyl groups on the polymer in a two-step process to link two polymer strands together to form bis-diol complexes. Lower concentrations of guar gelling agents are needed when linear guar chains are cross-linked. See Chapter 12, Introduction to Fracturing Fluids, for more information on fracturing fluids.

STARCH

Fig. 5.2 shows the two polysaccharide structures amylose and amylopectin.

Starch was first used for fluid loss control in the late 1930s (Gray et al., 1942). Many vegetable crops contain a starch component, especially root crops. Most oilfield grade starches are from processed corn or potato feedstocks. Regular starch is not dispersible in cold water. It must be pregelatenized to make it dispersible. Also, starches are highly susceptible to bacterial degradation. Oilfield grade starches are pregelatenized and treated with a

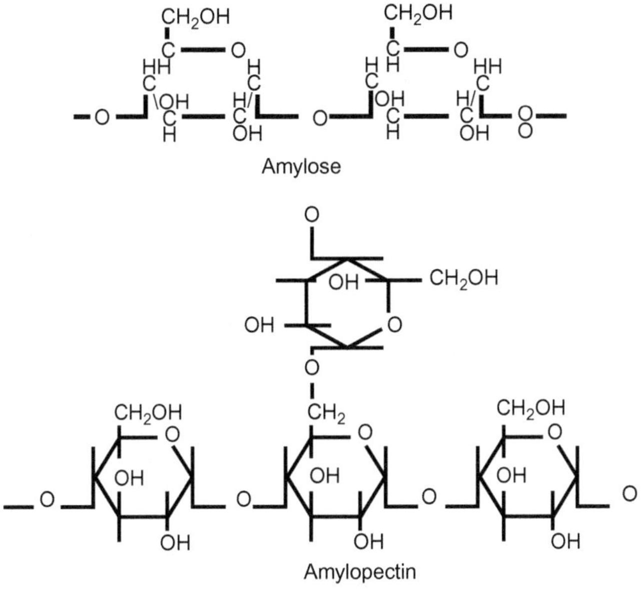

FIGURE 5.2 Schematic illustration of starch monomers amylose and amylopectin.

bactericide before supplied to the rig site. Modified starches, carboxymethyl- and hydroxypropyl-, have been processed during manufacture. Also, the modified starches do not need a bactericide while the drilling fluid is being actively circulated. They do need a preservative if stored.

Temperature stability—natural starch has a relatively low temperature stability, around 200–212°F (~100°C). Modified starches have a higher temperature stability of up to 250–300°F (125–150°C). Higher temperature stability is obtained in saturated brines, especially in formate brines.

The molecular weights of regular and modified starches are 100,000 to 500,000.

Their primary use is for fluid loss control, either to augment bentonite in a freshwater system, or to replace bentonite in a brine system. The modified starches can enhance the low shear rate viscosity (LSRV) of most water-based systems.

FERMENTATION BIOPOLYMERS

Fermentation biopolymers are manufactured as a batch process in fermentation vessels, similar to the manufacture of beer and wine. In the case of oilfield biopolymers, a living organism is grown with the proper temperature, oxygen, and food sources (primarily corn syrup). The organism then builds a polymeric coating around itself, which is then harvested, dried, and ground to a powder for use in a variety of food and industrial applications, one being oilfield drilling and completion fluids.

Xanthan Gum (XC)—Since its introduction in 1964 xanthan gum has been used extensively in the oil industry as a viscosifier for different applications due to its unique rheological properties (Kang and Petit, 1993). These applications include drilling, drill-in, completions, coiled tubing, and fracturing fluids. The unique property is as a rheology modifier, imparting significantly increased LSRV for hole cleaning and suspension properties. The increased LSRV characteristic of these biopolymers is due to the complex nature of the polymer molecules, both from complex side chains and high molecular weights. In addition, when hydrated in either fresh or brine water base fluids, they tend to associate and the long-chain backbones wrap around each other, thereby forming relatively massive dispersed packets. These packets greatly slow down any particle settling that tends to occur.

Fig. 5.3 is a representation of the XC molecule. It is a high molecular weight polysaccharide from bacterial (*Xanthomons campestris*) fermentation. Its molecular weight is greater than 2,000,000. The side chain is obviously much more complex than the plant-based polymers—starch, guar, and cellulose. When dispersed, XC forms a double helix molecular arrangement.

XC exhibits excellent shear and temperature resistance. XC was first used to replace bentonite to enhance the rate of penetration while drilling (Eckel, 1967). At very high salt concentrations complete hydration of

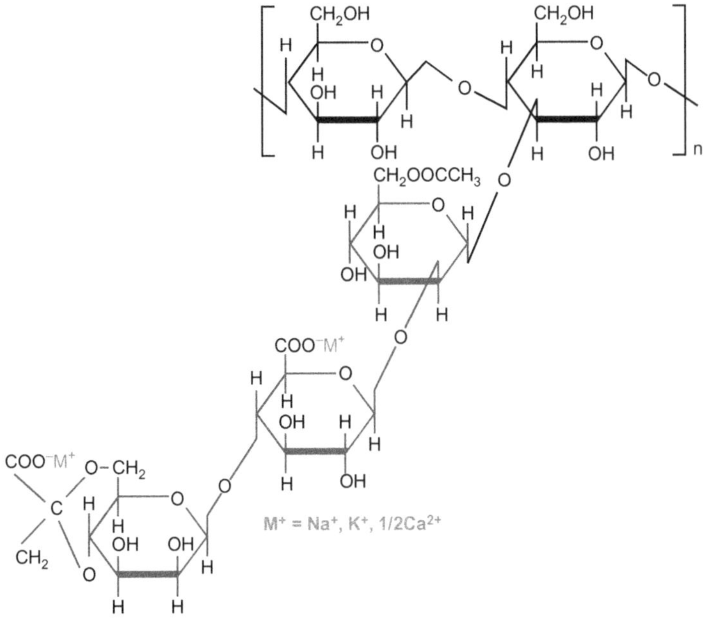

FIGURE 5.3 Schematic representation of the xanthan gum monomer.

xanthan is not achieved (CP Kelco, 2008), but it does hydrate rapidly at temperatures above 250°F (120°C) and high shear (Sinha and Shah, 2014).

XC has moderate temperature stability—250–300°F (120–150°C). A preservative is not normally needed while circulating. XC is susceptible to precipitation in high hardness waters at pHs above 10.5 (cement contamination, saturated calcium brine). There is nonpyruvylated xanthan (NPX) for viscosifying high density $CaCl_2$ brines (Fig. 5.4). NPX is similar in structure to xanthan gum in all other respects aside from the absence of the pyruvic acid group which reduces its anionic character.

Diutan—diutan is a biofermented polymer produced by a newly isolated naturally-occurring bacterial strain of the *Sphingomonas* genus. The chemical structure of the monomer is shown in Fig. 5.5.

The diutan structure is closer to that of welan gum (Fig. 5.8) than that of xanthan gum (Fig. 5.3). However, there are important differences. Diutan has an average molecular weight of 5,000,000, which is much higher than those of welan and xanthan. This is why the length of the diutan molecule is larger than that of welan or xanthan, giving it higher LSRV. Diutan also forms a double helix when dispersed. Figs. 5.6 and 5.7 show viscosity profile graphs of diutan versus a clarified XC.

Diutan gives higher viscosities and is more shear thinning than XC (Navarrete et al., 2001). It also has a higher temperature stability of about 320°F (165°C).

Welan—is an exopolysaccharide used as a rheology modifier in industrial applications such as cement manufacturing and for leakoff control in oilfied

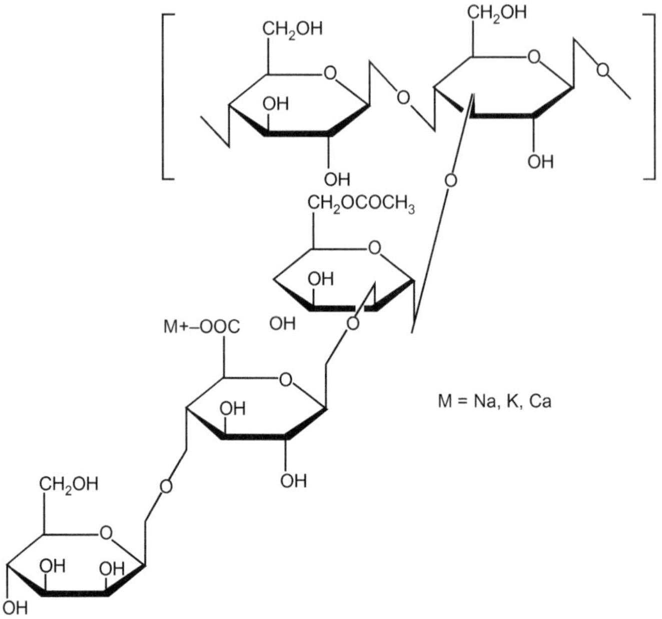

FIGURE 5.4 Schematic representation of the xanthan NPX monomer.

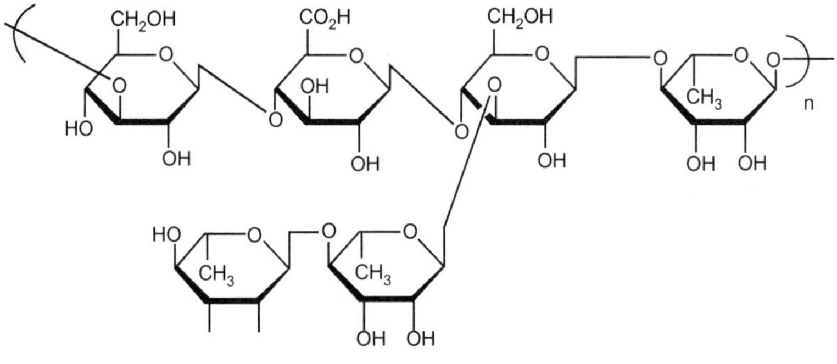

FIGURE 5.5 Schematic representation of the diutan monomer.

cementing operations. It has not been used in drilling or completion fluids. It is produced by fermentation of of a bacteria of the genus *Alcaligenes*. The molecule consists of repeating tetrasaccharide units with single branches of L-mannose or L-rhamnose (Fig. 5.8). In solution, the gum exhibits viscosity retention at elevated temperature, and is stable in a wide pH range, in the presence of calcium ion, and with high concentration of glycols. Fig. 5.9 compares the relative viscosities of XC, diutan, and welan at different concentrations.

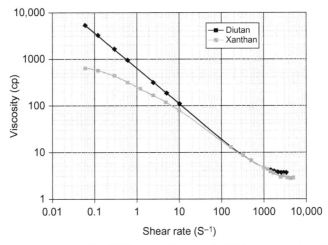

FIGURE 5.6 Viscosity profile of XC versus Diutan at 0.5 lb/bbl concentration. *Courtesy of Kelco Oil Field Group.*

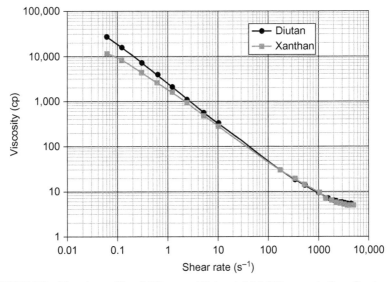

FIGURE 5.7 Viscosity profile of XC versus Diutan at 1.0 lb/bbl concentration. *Courtesy of Kelco Oil Field Group.*

Scleroglucan—scleroglucan is a biopolymer produced by fermentation of a plant pathogen fungus of genus *Sclerotium*. It has found wide applications for enhanced oil recovery (Galino et al., 1996). Fig. 5.10 shows a monomer structure diagram of the scleroglucan polysaccharide and an artist

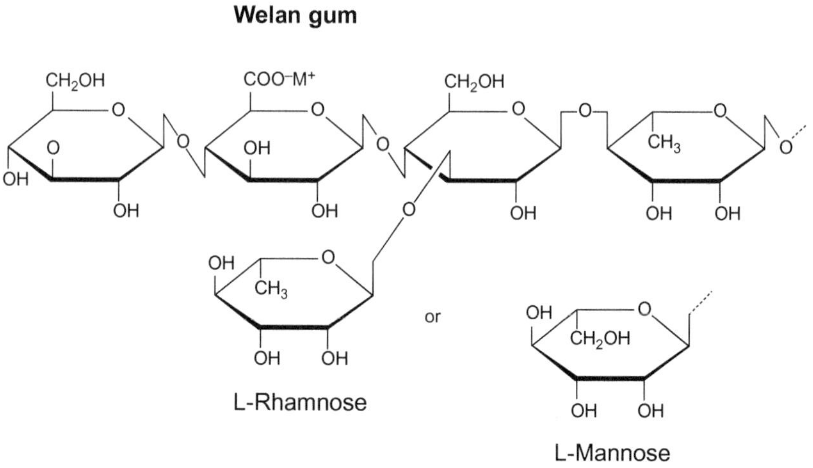

Welan gum

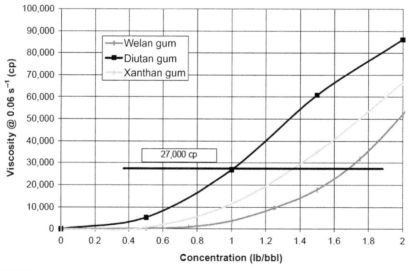

FIGURE 5.8 Schematic representation of the welan monomer.

FIGURE 5.9 Comparison of the relative viscosities of xanthan gum, diutan, and welan at various concentrations. *Courtesy of Kelco Oil Field Group.*

rendering of the molecule's configuration as a water-dispersed colloid. Even though the monomer structure is not as complex as XC, its dispersion configuration results in higher LSRV. It also has a higher temperature stability of up to 230°F (180°C). It is thought that scleroglucan forms a triple helix when dispersed.

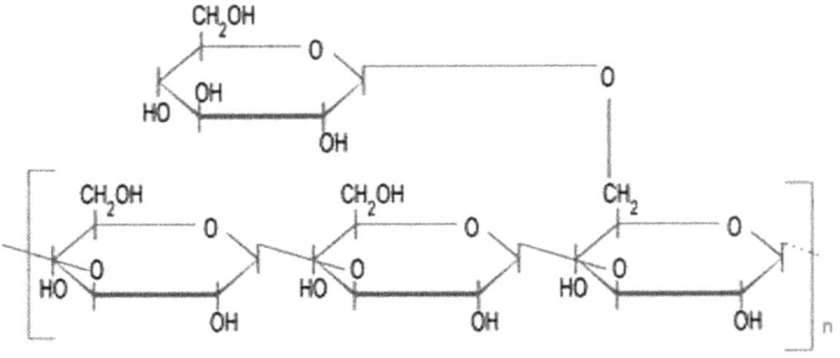

FIGURE 5.10 Schematic representation of the scleroglucan monomer.

CELLULOSICS

Carboxymethylcellulose (CMC) was first use in the 1940s as a fluid loss additive (Kaveler, 1946) and bulk fluid viscosifier. Oilfield cellulosic products are manufactured from plant natural cellulose modified with added short petrochemical side chains—carboxymethyl (CM), hydroxypropyl (HP), and hydroxyethyl (HE). The modification is necessary because natural cellulose is insoluble in water.

Cellulosics can by made in various molecular weight ranges and degrees of substitution (number of side chains). The maximum DS is three side chains per saccharide unit. CMCs that have a higher DS, above approximately 2.0, are called PACs, polyanionic cellulose. Fig. 5.11 is a schematic of the cellulose backbone with CM side chains.

Higher cellulose MWs result in a higher bulk viscosity in the fluid. CMCs and PACs are available as low viscosity (LV), medium viscosity (MV), and high viscosity (HV) products. The PACs, in general, perform better in sauturated brines, including saturated calcium chloride. Most cellulosics have a moderate temperature stability $\sim$250°F (120°C).

CMCs are not as shear thinning as the fermentation biopolymers and show little to no suspending ability. They are excellent fluid loss control materials. HECs are exclusively as a bulk viscosifer of started brines, such as chlorides and bromates, for completion fluids. The have no suspending ability and low particle carrying capacity in annular flow regimes. Table 5.3 lists the various oilfield cellulosic products.

SYNTHETICS

Synthetics are manufactured designer chemicals from petroleum monomers. They can be made in a wide range of molecular weights and charge

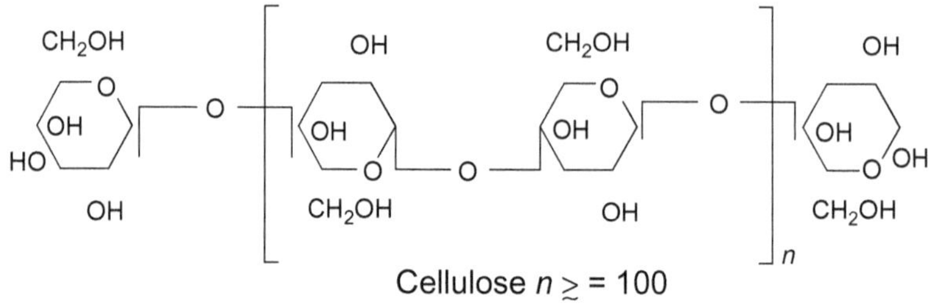

FIGURE 5.11 Schematic representation of the cellulose monomer (Galino et al., 1996).

TABLE 5.3 Modified Cellulosic Products: Differences and Uses

Oilfield Cellulosic		
CMC	PAC	HEC
Low DS	High DS	High DS
Anionic	Anionic	Nonionic
Subject to precipitation with calcium	Can tolerate calcium	Viscosifies high density, saturated brines
Routine fluid loss control	Fluid loss control in fresh and Brine water base fluids.	Bulk viscosifiers for saturated high density brines.

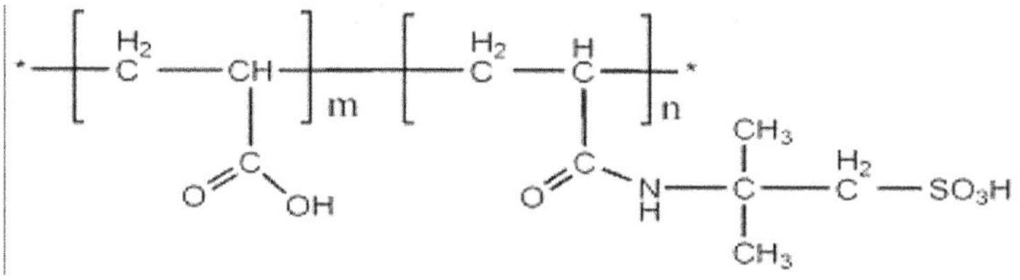

FIGURE 5.12 Schematic representation of a common drilling fluid synthetic monomer.

densities, with both anionic and cationic surface charges. They can be made to withstand very high temperatures, up to 400–600°F (200–300°C). Amine/amide fatty acid copolymers are common synthetics used as wellbore stability additives. Fig. 5.12 shows the monomer of the AA-AMPS oilfield synthetic polymer.

Acrylates and acrylate copolymers—the most common water-based mud synthetic materials are the acrylate copolymers, made from the acrylic acid monomer.

- Low molecular weight, 10,000—thinners/deflocculates, liquid or powder
- Medium molecular weight, 100,000—fluid loss control, usually powder
- High molecular weight, 1,000,000—bentonite extender for low solids, unweighted drilling fluids

Maleic anhydride and styrene copolymers was developed for use in geothermal drilling operations. It is stable to temperatures above 600°F (315°C).

Polycrylamides—first used in the 1960s as a rheology modifier or bentonite extender and then as a wellbore stabilizer in the Shell polymer mud system (Clark et al., 1976). Fig. 5.12 shows a schematic of the polyacrylamide monomer. Polyacrylamides are characterized by their degree of hydrolysis—replacing the amide group with an OH group—and in various MWs and charge type and charge density. Their temperature stability is in the >300°F (150°C) range (Fig. 5.13).

In addition to their use as a film forming material for wellbore stability in KCl drilling fluid, they are used as flocculating agents for cuttings waste management operations—closed-loop or zero-discharge drilling.

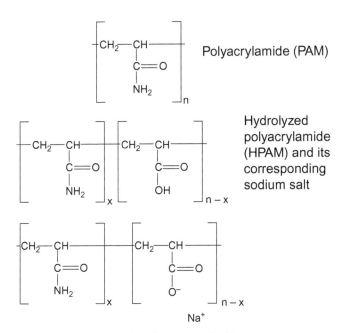

FIGURE 5.13 Schematic representation of the polyacrylamide monomer.

REFERENCES

Annis, M.R., 1967. High temperature properties of water-base drilling fluids. J. Petrol. Technol. 1074–1080, Trans AIME 240.

Clark, R., Scheurmann, R., Rath, H., van Laar, H., 1976. Polyacrylamide/potassium chloride mud for drilling water-sensitive shales. J. Petro. Technol. 719–727, Trans AIME 261.

CP Kelco 2008. Xanthan handbook, eighth ed., 2001–2008.

Downs, J., 1991. High Temperature Stabilisation of Xanthan, in Drilling Fluids by the Use of Formate Salts. Published in Physical Chemistry of Colloids and Interfaces in Oil Production, Editions TechNet, Paris.

Eckel, J.R., 1967. Microbit studies of the effect of fluid properties and hydraulics on drilling rate. J. Petrol. Technol. 541–546, Trans AIME 240.

Gallino, G., Guarneri, A., Poli, G., Xiao, L., 1996. Scleroglucan Biopolymer Enhances WBM Performances. Paper 36426 presented at the SPE Annual Technical Conference Denver, Colorado, October 6–9.

Gray, G.R., Foster, J.L., Chapman, T.S., 1942. Control of filtration characteristics of salt water muds. Trans. AIME 146, 117–125.

Kang, K.S., Petit, D.J., 1993. Xanthan, Gellan, Welan, and Rhamsan. In: Whistler, R.L., BeMiller, J.N. (Eds.), Industrial Gums, Third ed. Academic Press, New York, NY, pp. 341–397.

Kaveler, H.H., 1946. Improved drilling muds containing carboxymethylcellulose. API Drill. Prod. Pract. 43–50.

Navarrețe, R.C., Seheult, J.M., Coffey, M.D., 2001. "New BioPolymers for Drilling, Drill-In, Competions, Spacer Fluids and Coiled Tubing Applications," paper IADC/SPE 62790, presented at the 2000 IADC/SPE Asia Pacific Drilling Technology Conference, Kuala Lumpur, Malaysia, September 11–13.

Sinha, V., Shah, S.N., 2014. Rheological Performance of Polymers in Heavy Brines for Workover and Completion. AADE Fluids Technical Conference and Exhibition held at the Hilton Houston North Hotel, Houston, Texas, April 15–16.

Chapter 6

The Rheology of Drilling Fluids

The science of rheology is concerned with the deformation of all forms of matter, but has had its greatest development in the study of the flow behavior of suspensions in pipes and other conduits. A rheologist is interested primarily in the relationship between flow pressure and flow rate, and in the influence of that relationship of the flow characteristics of the fluid. There are two fundamentally different relationships:

1. The *laminar flow regime* prevails at low flow velocities. Flow is orderly, and the pressure−velocity relationship is a function of the viscous properties of the fluid.
2. The *turbulent flow regime* prevails at high velocities. Flow is disorderly, and is governed primarily by the inertial properties of the fluid in motion. Flow equations are empirical.

The laminar flow equations relating flow behavior to the flow characteristics of the fluid are based on certain mathematical flow models, namely, the *Newtonian*, the *Bingham plastic*, the *Power Law*, the *Power Law with yield stress*, and the *dilatant*. Only the first four are of interest in drilling and completion fluid technology.

Most drilling fluids do not conform exactly to any of these models, but drilling fluid behavior can be predicted with accuracy sufficient for practical purposes by one or more of them. Flow models are usually visualized by means of consistency curves, which are plots either of flow pressure versus flow rate, or of shear stress versus shear rate (see Fig. 6.1).

In the first part of this chapter we discuss the theoretical flow behavior of the four flow models of interest in drilling fluid technology; first in laminar flow, and then in turbulent flow. In this part no units are given for the flow equations because any set of consistent units will serve. Because in practice the terms for pressure, stress, and density are often written in units of mass, whereas those for viscosity are written in units of force, the conversion factor for gravity is included in the equations when appropriate. Practical hydraulic equations for determining flow pressures and velocities in drilling wells are given in the second part of the chapter. Finally, some of the problems related

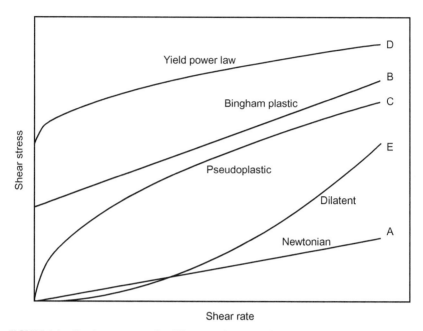

FIGURE 6.1 Consistency curves for different mathematical flow models.

to drilling fluid rheology, such as hole-cleaning, suspension, and swab and surge pressures, are discussed.

Note that this chapter is concerned only with the rheology of liquids. The rheology of gas or foam drilling fluids presents somewhat different problems, and is discussed in Chapter 8, The Surface Chemistry of Drilling Fluids.

LAMINAR FLOW REGIME

Laminar Flow of Newtonian Fluids

Laminar flow is easiest understood by imagining a deck of cards resting on a plane surface. If a force, F, is applied to the end of the top card (see Fig. 6.2), and if, because of friction, the velocity of each successive lower card decreases by a constant amount, dv, from v to zero, then

$$F/A = \tau = -\mu(dv/dr) \qquad (6.1)$$

where A is the area of the face of a card, r the thickness of the deck, dv the difference in velocity between adjoining cards, and dr the distance between them. μ is the frictional resistance to movement between the cards, or, in rheological terms, the *viscosity*. τ is the *shear stress*, dv/dr is the *shear rate*, or *velocity gradient*, defined by the slope of the *velocity profile*.

The Rheology of Drilling Fluids Chapter | 6 153

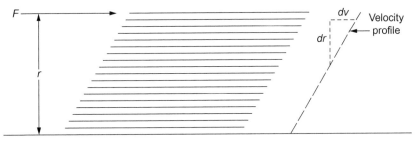

FIGURE 6.2 Schematic illustration of laminar flow of a Newtonian liquid.

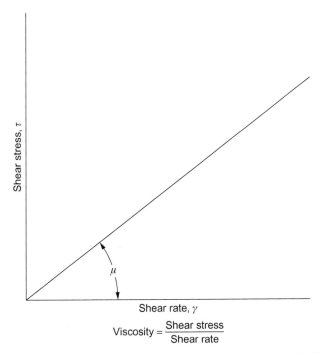

FIGURE 6.3 Consistency curve of a Newtonian fluid. Viscosity = shear stress divided by shear rate.

The consistency curve (often called the flow model) of a Newtonian fluid is a straight line passing through the origin (see Fig. 6.3). The slope of the curve defines the viscosity, so that

$$\mu = \tau/\gamma \tag{6.2}$$

where γ is the rate of shear. Since μ does not change with rate of shear, it is the only parameter needed to characterize the flow properties of a Newtonian liquid.

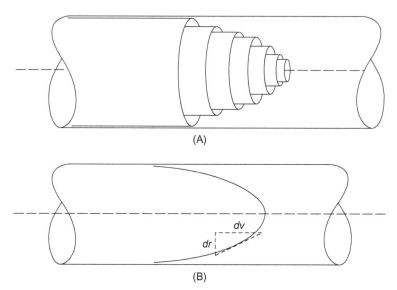

FIGURE 6.4 (A) Schematic representation of laminar flow of a Newtonian fluid in round pipe. Fluid velocity increase from 0 at the wall to a maximum at the axis of the pipe. The velocity profile of the fluid is shown in Fig. 6.4B. The shear rate at any point is the slope of the profile at that point.

The unit of viscosity is Pascal-second (Pa·s) in the SI system. One Pa·s equals 10 poise (P) or 1000 centipoise (cP). The common viscosity unit used in the oilfield is the cP.

Laminar flow of a Newtonian fluid in a round pipe may be visualized as a telescope of concentric cylinders, as shown in Fig. 6.4A. The velocity of the cylinders increases from zero at the pipe wall to a maximum at the axis of the pipe, resulting in a parabolic *velocity profile* (Fig. 6.4B). The shear rate at any point on the radius is given by the slope of the profile at that point. Note that it is maximum at the pipe wall, and zero at the axis.

The relationship between pressure and flow rate is derived as follows: if a fluid flows in a pipe of length L and radius R, the force on the end of a cylinder of radius r will be the pressure difference between the ends of the pipe, P, multiplied by the cross-sectional area of the cylinder, and the shear stress will be

$$\tau = \frac{F}{A} = \frac{\pi r^2 P}{2\pi r L} = \frac{rP}{2L} \tag{6.3}$$

Substituting for τ in Eq. (6.1) gives

$$\frac{rP}{2L} = -\mu \frac{dv}{dr}$$

which leads to Poiseulle's equation for laminar flow of Newtonian liquids in round pipes:

$$Q = \frac{g\pi R^4 P}{8L\mu} \tag{6.4}$$

where Q is the volumetric flow rate, and R is the radius of the pipe.

It is often more convenient to express Poiseulle's equation in terms of V, the average velocity, and the diameter of the pipe, D. Since $Q = V\pi R^2$, Eq. (6.4) becomes

$$P = \frac{32V\mu L}{gD^2} \tag{6.5}$$

For flow in concentric annuli of inner and outer diameters D_1 and D_2, respectively, Eq. (6.3) may be written

$$\tau = \frac{\pi/4(D_2^2 - D_1^2)P}{\pi(D_2 + D_1)L} = \frac{(D_2 - D_1)P}{4L}$$

$(D_2 - D_1)/4$ is called the *mean hydraulic radius*, and may be substituted for $D/4$ in many hydraulic equations. Poiseulle's equation then becomes

$$P = \frac{48V\mu L}{g(D_2 - D_1)^2} \tag{6.6}$$

The viscosity of a Newtonian fluid is determined in a glass capillary viscometer by measuring the time of discharge of a standard volume under a standard gravity head. The viscosity may be calculated from Eq. (6.4), or by calibration with liquids of known viscosity, or from a constant supplied by the manufacturer of the viscometer. Various capillary sizes are available, so that a wide range of viscosities may be conveniently measured.

THE BINGHAM PLASTIC FLOW MODEL

Plastic fluids were first recognized by Bingham (1922), and are therefore referred to as *Bingham plastics*, or *Bingham bodies*. They are distinguished from Newtonian fluids in that they require a finite stress to initiate flow. Fig. 6.5A shows the consistency curve for an ideal Bingham plastic, the equation for which is

$$\tau - \tau_0 = -\mu_p \frac{dv}{dr} \tag{6.7}$$

where τ_0 is the stress required to initiate flow (yield stress), and μ_p is the *plastic viscosity*, which is defined as the shear stress in excess of the yield stress that will induce unit rate of shear. Thus

$$\mu_p = \frac{\tau - \tau_0}{\gamma} \tag{6.8}$$

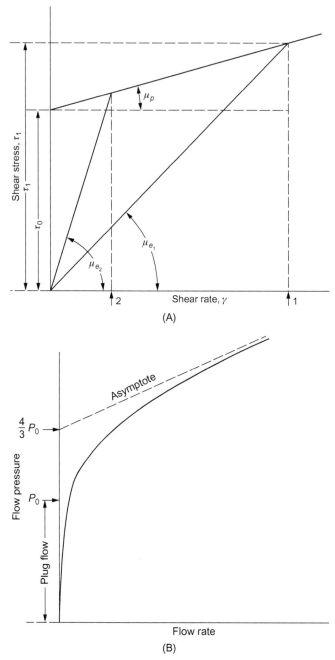

FIGURE 6.5 (A) Consistency curve of an ideal Bingham plastic. Note: The effective viscosity at shear rate 2 is much greater than at shear rate 1. (B) Observed consistency curve of a Bingham plastic. P_0 is the actual yield point, neglecting creep. $4/3P_0$ is the apparent yield point.

The total resistance to shear of a Bingham plastic may be expressed in terms of an effective viscosity, at a specified rate of shear. *Effective viscosity* is defined as the viscosity of a Newtonian fluid that exhibits the same shear stress at the same rate of shear. Fig. 6.5A shows that effective viscosity at shear rate γ_1 is given by

$$\mu_{e1} = \frac{\tau_1 - \tau_0}{\gamma_1} + \frac{\tau_0}{\gamma_1}$$
$$= \mu_p + \frac{\tau_0}{\gamma_1} \qquad (6.9)$$

Thus effective viscosity may be considered as comprising two components: plastic viscosity, which corresponds to the viscosity of a Newtonian fluid, and *structural viscosity*, which represents the resistance to shear caused by the tendency for the particles to build a structure. As shown in Fig. 6.5A, τ_0/γ forms a decreasing proportion of the total resistance to shear as the shear rate increases, so that the effective viscosity decreases with increase in shear rate.

Note particularly that the value of effective viscosity is meaningless unless the rate of shear at which it is measured is specified. Furthermore, as shown in Fig. 1.5, it is not a reliable parameter for comparing the viscous properties of two fluids; for that purpose, at least two parameters are necessary. Nevertheless, effective viscosity is a very useful parameter in many hydraulic equations when the shear rate is known, as will be discussed later.

Plastic flow, as shown in Fig. 6.5A, is never observed in practice; at pressures below the yield point, a slow creep is observed, as shown in Fig. 6.5B. By examining the flow of a suspension in a glass capillary under a microscope, Green (1949) showed that no shearing action was involved in this type of flow. The suspension flowed as a solid plug lubricated by a thin film of liquid at the capillary wall, the particles being held together by attractive forces between them. However low the pressure, there was always some flow, although it might be as low as one cubic centimeter in one hundred years. He therefore concluded that there was no absolute yield point, and redefined the *Bingham yield point* as the shear stress required to initiate laminar flow in the suspension.

Green showed that the flow of a Bingham plastic in a round pipe is as follows: If the pressure is gradually increased from zero, the suspension at first flows as a plug (as described above) and the velocity profile is a straight line normal to the axis of the pipe (Fig. 6.6A). Since shear stress is equal to $rP/2L$ (Eq. 6.3), laminar flow starts at the wall of the pipe when

$$\frac{RP_0}{2L} = \tau_0 \qquad (6.10)$$

where P_0 is the pressure required to initiate plastic flow. At pressures greater than P_0, laminar flow progresses towards the axis of the pipe, so that flow consists of a plug in the center of the pipe surrounded by a zone of laminar flow, and the velocity profile is as shown in Fig. 6.6B. No matter how great the pressure, this plug can never be entirely eliminated because as r becomes very small, P must become very large (since $rP/2L$ must equal τ_0), and infinite when $r = 0$. Thus the consistency curve for flow of a Bingham plastic in a round pipe is, strictly speaking, always nonlinear no matter how great the shear rate. However, an approximate relationship between pressure and flow rate may be derived from the asymptote to the curve, which, as shown in Fig. 6.5B, intercepts the pressure axis at $4/3P_0$. Buckingham (1921) derived this relationship from the intercept and Poiseulle's equation as

$$Q = \frac{\pi R^4}{8\mu_p L}\left(P - \frac{4}{3}P_0\left(1 - \frac{P_0^3}{4P^3}\right)\right) \qquad (6.11)$$

Substituting values for P_0 from Eq. (6.10) and (6.11) may be arranged more conveniently as

$$V = \frac{D^2 P}{32\mu_p L}\left(1 - \frac{4}{3}\left(\frac{4\tau_0 L}{DP}\right) + \frac{1}{3}\left(\frac{4\tau_0 L}{DP}\right)^4\right) \qquad (6.12)$$

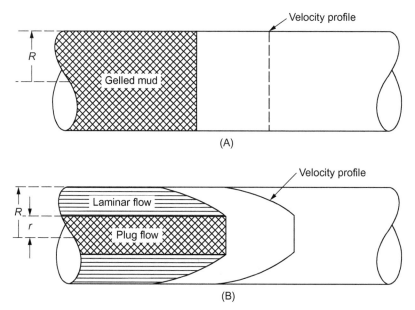

FIGURE 6.6 (A) Plug flow of a Bingham plastic in round pipe. $RP/2L < \tau_0$. (B) Mixed flow of a Bingham plastic in round pipe. $RP/2L < \tau_0$, $rP/2L = \tau_0$.

The last term, $\frac{1}{3}\left(4\frac{\tau_0 L}{DP}\right)^4$, represents the contribution of the area between the full curve and the asymptote. At high flow rates it may be omitted, but for exact solutions the abbreviated form should not be used unless $\frac{DP/4L}{4L}$ exceeds the yield stress, τ_0, by a factor of at least four (Hanks and Trapp, 1967).

At very low flow rates an expression for plug flow should be included in Eq. (6.12). According to Buckingham, plug flow is described by

$$V = \frac{\pi D k P}{2\mu L} \quad (6.13)$$

where k is a constant and μ is the viscosity of the lubricating layer at the wall. The effect of plug flow on overall flow rate is ordinarily insignificant in drilling applications.

The hydraulic radius is defined as $(D_2 - D_1)/4$. It is usually substituted for $D/4$ in Eq. (6.12) when determining flow relations in the annulus. However, this procedure does not give an exact solution (van Olphen, 1950). If such is required, it needs to be calculated from a computer program.

THE CONCENTRIC CYLINDER ROTARY VISCOMETER

The plastic viscosity and yield point of a Bingham plastic are best determined in a concentric cylinder rotary viscometer. The outstanding advantage of this instrument is that, above a certain rotor speed, plug flow is eliminated, and the consistency curve becomes linear.

The essential elements of the viscometer are shown in Fig. 6.7. An outer cup rotates concentrically around an inner cylinder or bob, which is

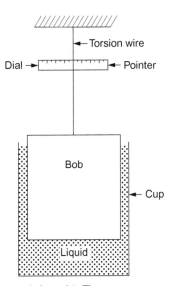

FIGURE 6.7 Torsion viscometer (schematic). The cup rotates.

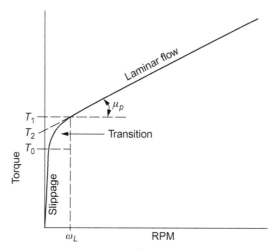

FIGURE 6.8 Consistency curve of Bingham plastic in a direct-indicating viscometer. Flow is laminar throughout annulus at rpm above ω_L.

suspended from a torsion wire. The annulus between the bob and the cup is narrow, about 1 mm. A dial affixed to the wire and a fixed pointer enable the angle through which the wire has turned to be measured.

When the cup is rotated, the bob rotates with it (apart from minor slippage) until the torque in the wire creates a shear stress at the surface of the bob greater than the shear strength of the plastic structure. At that point

$$\frac{T_0}{2\pi R_b^2 h} = \tau_0 \qquad (6.14)$$

where T_0 (see Fig. 6.8) is the torque at the yield point, R_b is the radius of the bob, and h is the effective height of the bob, i.e., the actual height of the bob plus a correction for the end effect of the bottom of the bob. Laminar flow now starts at the surface of the bob and progresses outwards with continued rotation until all the fluid in the annulus is in laminar flow, at which point

$$\frac{T_1}{2\pi R_c^2 h} = \tau_0 \qquad (6.15)$$

where T_1 is the critical torque, and R_c is the inner radius of the cup. With continued rotation at constant speed, the torque increases to an equilibrium value, which depends on the rheological characteristics of the fluid.

The relationship between torque and rotor speed for any speed great enough to maintain all the fluid in the annulus in laminar flow is linear, and is defined by the Reiner-Riwlin (Reiner, 1929) equation:

$$\varpi = \frac{T}{4\pi h \mu_p}\left(\frac{1}{R_b^2} - \frac{1}{R_c^2}\right) - \frac{\tau_0}{\mu_p}\ln\frac{R_c}{R_b} \qquad (6.16)$$

where ϖ is the angular velocity (in radians per second), and T is the corresponding torque. The yield point, τ_0, in Eq. (6.16) is defined by the intercept. T_2, of the extrapolation of the linear portion of the curve on the torque axis, T_0 is given by Eq. (6.16) when $\varpi = 0$, because then

$$\frac{T_2}{4\pi h}\left(\frac{1}{R_b^2} - \frac{1}{R_c^2}\right) = \tau_0 \ln \frac{R_c}{R_b} \tag{6.17}$$

The slope of the line above the critical angular velocity, ω_1, defines the plastic viscosity, μ_p.

The value of the torque in the wire may be obtained from the deflection of the dial and the wire constant, C:

$$T = C\theta \tag{6.18}$$

where T is the torque in dyne centimeters, and θ is the dial deflection in degrees. Usually the wire constant is supplied by the manufacturer of the viscometer; if not, it may be obtained by calibration with Newtonian fluids. Since such fluids have no yield point, C may be determined from Eqs. (6.16) and (6.18):

$$\varpi = \frac{C\theta}{4\pi h \mu}\left(\frac{1}{R_b^2} - \frac{1}{R_c^2}\right) \tag{6.19}$$

A number of commercially-available concentric cylinder rotary viscometers that are suitable for use with drilling muds are described in Chapter 3, Evaluating Drilling Fluid Performance. They are similar in principle to the viscometer shown in Fig. 6.7, but use a spring instead of a torsion wire. All are based on a design by Savins and Roper (1954), which enables the plastic viscosity and yield point to be calculated very simply from two dial readings, at 600 and 300 rpm, respectively. They will be referred to in this chapter as the *direct-indicating viscometer*.

According to Savins and Roper, the underlying theory is as follows: Eq. (6.16), the Reiner-Riwlin equation, is simplified to

$$\mu_p = \frac{A\theta - B\tau_0}{\omega}$$

where A and B are constants that include the instrument dimensions, the spring constant, and all conversion factors; and ω is the rotor speed in rpm.

Then,

$$\mu_p = \overline{PV} = A\left(\frac{\theta_1 - \theta_2}{\omega_1 - \omega_2}\right) \tag{6.20}$$

where θ_1 and θ_2 are dial readings taken at ω_1 and ω_2 rpm. $\overline{PV}$ is the conventional oilfield term for *plastic viscosity* thus measured.

$$\tau_0 = \overline{YP} = \frac{A}{B}\left[\theta_1 - \left(\frac{\omega_1}{\omega_1 - \omega_2}\right)(\theta_1 - \theta_2)\right] \tag{6.21}$$

$\overline{YP}$ is the conventional oilfield term for *yield point* thus measured. The numerical values of A, B, ω_1, and ω_2 were chosen so that

$$A = B = \omega_1 - \omega_2$$

and

$$\omega_1 = 2\omega_2$$

Under these conditions

$$\frac{A}{\omega_1 - \omega_2} = 1, \quad \frac{A}{B} = 1, \quad \frac{\omega_1}{\omega_1 - \omega_2} = 2$$

Eqs. (6.20) and (6.21) then simplify to

$$\overline{PV} = \theta_1 - \theta_2 \tag{6.22}$$

$$\overline{YP} = \theta_2 - \overline{PV} \tag{6.23}$$

To satisfy these requirements, R_b and R_c were chosen so that, with an annulus width of about 1 mm, the value of $A = B$ would be 300, ω_2 was therefore 300 rpm, and ω_1, 600 rpm. For a value of $A = 300$, a spring constant of 387 dyne centimeters per degree was required. These specifications enabled Eq. (6.22) to give the plastic viscosity in centipoises, and Eq. (6.23) to give the yield point almost exactly in pounds per 100 ft².

Effective viscosity may be calculated from the Savins-Roper viscometer reading as follows:

$$1° \text{ dial reading} = 1.067 \text{ lb per 100 ft}^2$$

$$= 5.11 \text{ dynes/cm}^2 \text{ shear stress}$$

$$1 \text{ rpm} = 1.703 \text{ reciprocal seconds, shear rate}$$

$$\mu_e = \frac{\tau}{\gamma} = \frac{5.11}{1.073} \text{ poise per degree per rpm} \tag{6.24}$$

$$= 300 \text{ centipoise per degree per rpm}$$

$$= \frac{300 \times \theta}{\omega}$$

where θ is the dial reading at ω rpm.

In evaluating drilling muds, it was common practice to report the apparent viscosity at 600 rpm, although this measurement is currently not being used. This quantity, defined as the *apparent viscosity (AV)*, and is given by

$$\overline{AV} = \frac{300 \times \theta_{600}}{600} = \frac{\theta_{600}}{2} \tag{6.25}$$

Note that the term *apparent viscosity* is used by some authorities in a more general sense as an alternative for *effective viscosity*. The basis for the determination of plastic viscosity, yield point, and apparent viscosity is shown graphically in Fig. 6.9.

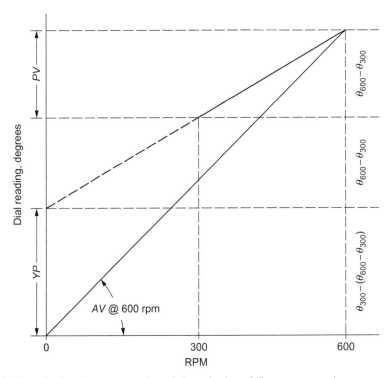

FIGURE 6.9 Graphical interpretation of determination of flow parameters in a two-speed direct-indicating viscometer.

BEHAVIOR OF DRILLING FLUIDS AT LOW SHEAR RATES

As discussed in the preceding section, the consistency curve of a Bingham plastic in a rotary viscometer should be linear at rotor speeds above that required to keep all the fluid in the annulus in laminar flow. But drilling fluids are not ideal Bingham plastics, and they deviate from linearity at low shear rates. This deviation becomes apparent when their behavior is examined in a multispeed viscometer instead of in the two-speed model discussed in the last section. For an example, see the consistency curves of the four bentonite suspensions shown in Fig. 6.10. According to Savins and Roper (1954), flow in the annulus is fully laminar when

$$\frac{T}{2\pi R_c^2 h} > \overline{YP} \tag{6.26}$$

Substituting Eq. (6.26) in Eq. (6.16)

$$\overline{\omega}_L = \frac{\overline{YP}}{2PV}\left(\frac{R_c^2}{R_b^2} - 1 - 2\ln\frac{R_c}{R_b}\right) \tag{6.27}$$

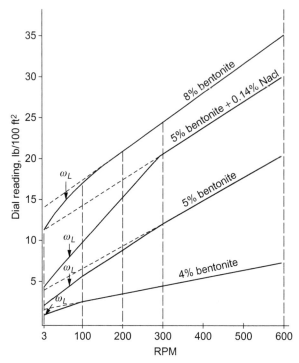

FIGURE 6.10 Behavior of bentonite suspensions in multispeed direct-indicating viscometer. ω_L marks the rpm above which the curves of a Bingham plastic should be linear. *Data courtesy of Brinadd Company.*

Substituting in the instrumental constant

$$\omega_L = 20.62 \frac{\overline{YP}}{\overline{PV}} \qquad (6.28)$$

where ω_L is the critical rpm for full laminar flow.

The critical rpm for each of the four bentonite suspensions was calculated, assuming that they were Bingham plastics, according to Eq. (6.16) and is marked on their curves in Fig. 6.10 by the symbol ω_L. It may be seen that the curves become linear only at considerably higher rpm, and also that the actual yield point is well below that indicated by the extrapolation of the 600 and 300 rpm dial readings.

The behavior of drilling muds may be explained as follows: the consistency curves for ideal Bingham plastics, such as those shown in Figs. 6.5A and 6.8, were based on the behavior of suspensions that had a high concentration of approximately equidimensional particles, such as printing inks and paints. The concentration of solids in such suspensions is high enough to build a structure by grain-to-grain contact. This structure resists shear because of interparticle friction, enhanced, to some degree, by interparticle attractive forces. Once the yield point is exceeded, and laminar flow commences, the particles are presumed to interact no longer, and to influence viscosity merely by the volume they occupy. The effective viscosity is then given by *Einstein's equation*, viz

$$\mu_e = \mu + 2.5\phi \tag{6.29}$$

where μ is the viscosity of the liquid medium and ϕ is the volume fraction of the solids.

As we saw in Chapter 4, Clay Mineralogy and the Colloid Chemistry of Drilling Fluids, the clay particles present in drilling fluids are highly anisodimensional, and can build a structure at very low solid concentrations because of interaction between attractive and repulsive surface forces. At low shear rates the behavior of the clay particles is still influenced by these forces, and, in consequence, the viscosity is relatively high but, as the shear rate increases, the particles gradually align themselves in the direction of flow, and the viscosity then becomes largely dependent on the concentration of all solids present in the mud.

Because of these phenomena, the degree of deviation from linearity in the consistency curves of drilling muds in the rotary viscometer differs from mud to mud, depending on the concentration, size, and shape of the particles. It is most marked with low-solid muds containing a high proportion of clay particles, or long-chain polymers, and least marked with high-solid muds containing silt and barite. It is also influenced by the electrochemical environment, which, as discussed in Chapter 4, Clay Mineralogy and the Colloid Chemistry of Drilling Fluids, determines the interparticle forces. Note, for example, in Fig. 6.10 that the curve for the bentonite flocculated by sodium chloride showed the greatest deviation from linearity.

Unfortunately, there is no way of determining the linearity of the consistency curves, other than by measurement in a multispeed rotary viscometer. Therefore, the usefulness of the rheological parameters $\overline{PV}$ and $\overline{YP}$ is limited. In practice, the most common use of these quantities is for the wellsite evaluation of drilling mud performance, particularly as a guide to maintenance treatments. Thus, the $\overline{PV}$ is sensitive to the concentration of solids, and therefore indicates dilution requirements; the $\overline{YP}$ is sensitive to the electrochemical environment, and hence indicates the need for chemical treatment. The use of $\overline{PV}$ and $\overline{YP}$ for these purposes is justified, since in this case the shape of the consistency curve is of no consequence.

$\overline{PV}$ and $\overline{YP}$ may also be substituted for μ_p and τ_0, respectively, in Eq. (6.12) for predicting laminar flow behavior in pipes, but only for flow at high shear rates. When predicting flow behavior at low rates of shear, it is better to calculate the effective viscosity at the rate of shear prevailing in the pipe, which may then be substituted in Poiseuille's equation (Eq. 6.5). Th required value of effective viscosity is best determined from the power law, which is discussed later in this chapter.

As already mentioned, $\overline{YP}$ does not represent the real yield point. Actually, because of slippage, the consistency curve approaches the stress axis asymptotically, so that the real yield point as defined by Green (1949) (i.e., the stress required to initiate laminar flow) is indeterminable. For practical purposes the initial gel strength in the two speed viscometer, i.e., the maximum dial deflection observed when the cup is rotated by hand immediately after flow has ceased, is probably the best measure of the real yield point. The initial gel in a multi-speed viscometer is taken at 3 rpm.

Because centrifugal effects become significant at high rotor speeds, rotary viscometers cannot be used to determine rheological properties at very high rates of shear. For this purpose a pressurized capillary viscometer, which permits viscosity to be determined over a wide range of shear rates, must be used. This instrument is particularly useful for determining power law parameters, as will be discussed later in this chapter.

THE EFFECT OF THIXOTROPY ON DRILLING MUDS

If the gel strength of a mud is measured immediately after being sheared, and repeatedly after increasingly longer periods of rest, the values obtained will be generally found to increase at a decreasing rate until a maximum value is reached. This behavior is a manifestation of the phenomenon of *thixotropy*, originally defined by Freundlich (1935) as a reversible isothermal transformation of a colloidal sol to a gel. In the case of drilling muds, the phenomenon is caused by the clay platelets slowly arranging themselves in positions of minimum free energy (see Chapter 4, Clay Mineralogy and the Colloid Chemistry of Drilling Fluids) to satisfy electrostatic surface charges. After a period of rest, a thixotropic mud will not flow unless the applied stress is greater than the strength of the gel structure. In other words, the gel strength becomes the yield point, τ_0. If subjected to a constant rate of shear, the clay platelet associations gradually adjust to the prevailing shear conditions, and the effective viscosity decreases with time until a constant value is reached, at which point the structure-building and structure-disrupting forces are in equilibrium. If the rate of shear is increased, there is a further decrease in viscosity with time until an equilibrium value typical of that rate of shear is reached. If the rate of shear is then decreased to the first rate, the viscosity slowly builds up until the equilibrium value for that rate of shear is again reached. Because of these phenomena, Freundlich's original definition of thixotropy has been extended to cover a reversible isothermal change in viscosity with time at constant rate of shear (Goodeve, 1939).

Thixotropy must not be confused with plasticity. As we have already seen, the effective viscosity of a Bingham plastic depends on the rate of shear because its structural component forms a decreasing proportion of the

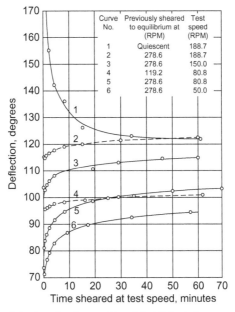

FIGURE 6.11 Flow behavior of a clay mud in a MacMichael viscometer. *From Jones, P.H., Babson, E.C., 1935. Evaluation of rotary drilling muds. Oil Wkly. 25–30.*

total resistance to shear as the shear rate increases. The viscosity of a thixotropic fluid depends on time of shearing, as well as rate of shear, because the structural component changes with time according to the past shear history of the fluid. For this reason, thixotropic fluids are said to be "fluids with a memory." Bingham plastics may or may not be thixotropic, depending on composition and electrochemical conditions. A quick test for thixotropy may be made in a viscometer fitted with an x-y recorder, by increasing and then decreasing the rotor speed. If a hysteresis loop is obtained on the recorder the fluid is thixotropic. The opposite of thixotropy is *rheopexy*. The viscosity of a rheopectic fluid increases with time at constant shear rate. Rheopexy in drilling fluids has not been reported.

The effect of thixotropy on the evaluation of the rheological parameters of drilling muds was first investigated by Jones and Babson (1935). They observed the change in torque with the passage of time, when thixotropic muds were sheared at constant rate in a MacMichael viscometer. Curve 1 in Fig. 6.11 shows the result when a gelled mud was sheared at a constant rate of 189 rpm. The torque decreased sharply during the first 15 min, then decreased gradually until equilibrium was reached after about one hour. Curve 2 shows the behavior of the mud after the shear rate was increased to 279 rpm, maintained there until equilibrium was reached, and then brought back to 189 rpm. Note that the torque gradually built back to the equilibrium value of Curve 1. Curves 4 and 5 show that approximately the same

168 Composition and Properties of Drilling and Completion Fluids

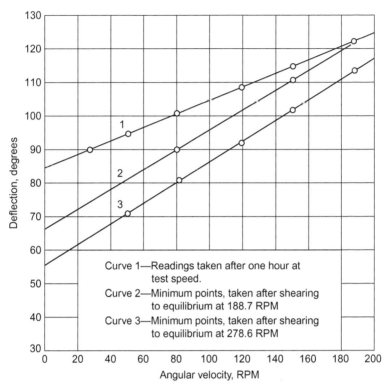

FIGURE 6.12 Equilibrium and instantaneous flow curves of the clay mud shown in Fig. 6.11. Plastic viscosity and yield point are defined by instantaneous Curves 2 and 3. *From Jones, P.H., Babson, E.C., 1935. Evaluation of rotary drilling muds. Oil Wkly. 25–30.*

equilibrium value was obtained at 81 rpm, regardless of whether the mud was presheared at 119 or 279 rpm. These results confirm that thixotropic muds have an equilibrium value that is typical of the shear rate at which it is measured, and which is independent of shear history.

The equilibrium viscosities of the mud at rates ranging from 189 to 21 rpm are shown by Curve 1 in Fig. 6.12. Jones and Babson emphasized that the concepts of plastic viscosity and yield point cannot be applied to this curve because each point represents a different degree of structural breakdown. In other words, the equilibrium curve relates stress to both rate of shear and to the effect of time, whereas flow equations relate stress only to rate of shear. Meaningful values of plastic viscosity and yield point can only be obtained by shearing to equilibrium at a specified speed, and then making torque readings as quickly as possible at lower rpm before any thixotropic change takes place. Curves 2 and 3 in Fig. 6.12 show these instantaneous values after preshearing at 189 and 279 rpm, respectively. For each preshear rate, the area between the equilibrium and instantaneous curves defines the flow conditions at any lower shear rate at any point in time.

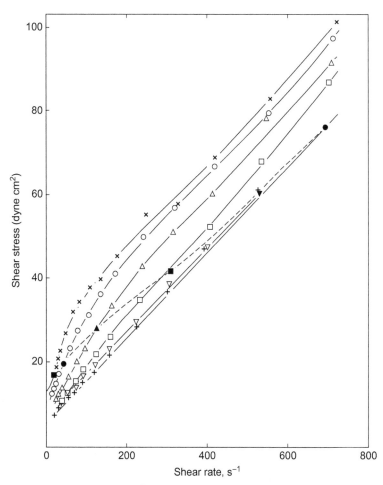

—+— Instantaneous curves after preshearing to an equilibrium viscosity of 11.0 cp.
—▽— Instantaneous curves after preshearing to an equilibrium viscosity of 11.4 cp.
—□— Instantaneous curves after preshearing to an equilibrium viscosity of 13.6 cp.
—△— Instantaneous curves after preshearing to an equilibrium viscosity of 22.4 cp.
—○— Instantaneous curves after preshearing to an equilibrium viscosity of 47.2 cp.
—×— Instantaneous curves after preshearing to an equilibrium viscosity of 90.4 cp.
— — — Equilibrium curve

FIGURE 6.13 Equilibrium and instantaneous curves for a 4.8% bentonite suspension. *From Cheng, D.G.H., Ray, D.J., Valentin, F.H.H., 1965. The flow of thixotropic bentonite suspensions. Through pipes and pipe fittings. Trans. Inst. Chem. Eng. (Lond.) 43, 176–186. Courtesy of the Institute of Chemical Engineers.*

The effect of shear history on viscosity was also shown by Cheng et al. (1965). They presheared bentonite suspensions to equilibrium at rates varying from 700 to 20 reciprocal seconds. In each case the instantaneous curves were obtained at shear rates ranging downward from 700 reciprocal seconds. The results for a 4.8% bentonite suspension, shown in Fig. 6.13, indicate that

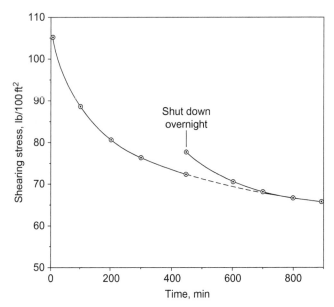

FIGURE 6.14 Slow breakdown of the gel structure caused by the shearing of a 4% slurry of pure sodium montmorillonite contaminated with 50 meq/L of NaCl. Constant shear rate = 480 s^{-1}. *From Hiller, K.H., 1963. Rheological measurements of clay suspensions at high temperatures and pressures. J. Petrol. Technol. 779–789. Trans AIME 228. Copyright 1963 by SPE-AIME.*

the instantaneous effective viscosity at 700 rpm varied from 11 to 63 centipoises depending on the rate of preshearing.

The principles and experimental results discussed above show that the effect of shear history must be taken into account when determining the flow parameters of thixotropic muds. For example, muds must be presheared to equilibrium at a standard rate when comparing the flow properties of different muds. When the flow parameters are to be used for the purpose of calculating pressure drops in a well, the mud must be presheared to a condition corresponding to that prevailing at the point of interest in the well.

Note that the time required for preshearing to equilibrium may be longer or shorter than the one hour reported by Jones and Babson. Fig. 6.14 shows an extreme case where a very long time was required to preshear a flocculated monoionic montmorillonite suspension. Muds brought into the laboratory from a well may also require long preshearing times to reduce them to a state of shear similar to that in the well.

Silbar and Paslay (1962) developed a set of constitutive equations containing five physical parameters that can be used to predict the effect of shear history on the flow of thixotropic materials. They obtained good correlation between their predictions and the experimental results of Jones and Babson (1935).

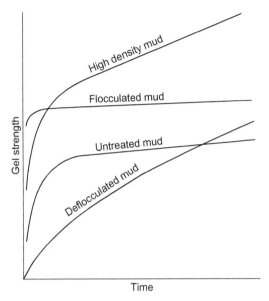

FIGURE 6.15 Increase in gel strength of various mud types with time (schematic).

The high gel strengths developed by thixotropic muds after prolonged periods of rest create another problem for the drilling engineer. The long-term gel strength is a major factor in the pressure required to break circulation after a round trip, and in the magnitude of swab and surge pressures. Unfortunately, the relation between gel strength and time varies widely from mud to mud, depending on composition, degree of flocculation, etc. (as shown in Fig. 6.15), and there is no well-established means of predicting long-term gel strengths. The only important step in this direction was taken by Garrison (1939), who developed the following equation from the observed gelling rates of Californian bentonites:

$$S = \frac{S'kt}{1+kt} \quad (6.30)$$

where S is the gel strength at any time t, S' is the ultimate gel strength, and k is the gel rate constant. Eq. (6.30) may be written as

$$\frac{t}{S} = \frac{t}{S'} + \frac{1}{S'k}$$

which shows that the plot of t/S versus t should be a straight line, the slope of which gives k and the intercept at zero time gives $1/S'k$. Fig. 6.16 shows plots of gel strength versus time for several bentonite suspensions, and Fig. 6.17 shows t/S versus time for the same suspensions. Table 6.1 shows the ultimate gel strength and the gel rate calculated according to Eq. 6.30.

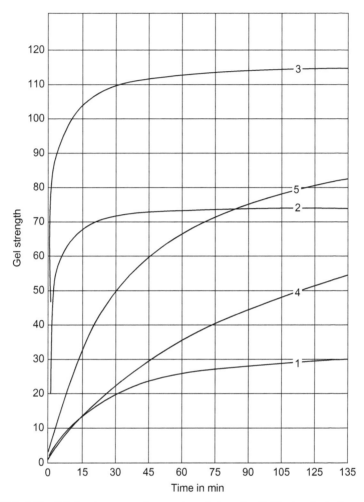

FIGURE 6.16 Relation between gel strength and time for Californian bentonites. *From Garrison, A.D., 1939. Surface chemistry of shales and clays. Trans. AIME 132, 423–436. Copyright 1939 by SPE-AIME.*

There is little evidence in the literature to show whether or not Eq. (6.30) applies to muds other than the bentonites tested by Garrison. Weintritt and Hughes (1965) measured the gel strengths of some field muds containing calcium sulfate and ferrochrome lignosulfonate in a rotary viscometer for rest periods up to one day. Application of their data to Eq. (6.30) shows apparent compliance—although there was considerable scatter of the points—for rest periods up to two hours, but major deviations thereafter.

The Rheology of Drilling Fluids Chapter | 6   **173**

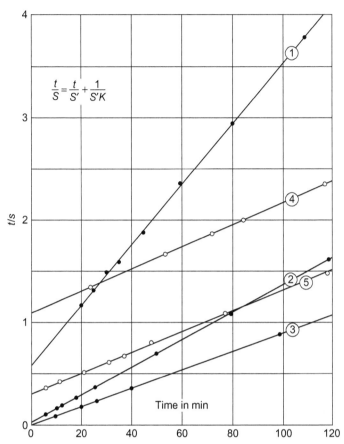

FIGURE 6.17 Gel rate constants of Californian bentonites. *From Garrison, A.D., 1939. Surface chemistry of shales and clays. Trans. AIME 132, 423–436. Copyright 1949 by SPE-AIME.*

TABLE 6.1 Gel Rate Constants Calculated from Fig. 6.16

Curve	Suspension Composition	S^1	k	pH
1	4.5% bentonite	34.4	0.047	9.2
2	5.5% bentonite	74.4	0.75	9.2
3	6.5% bentonite	114	0.79	9.2
4	5.5% bentonite 1% Na tannate	104	0.0089	9.2
5	5.5% monionic Na montmorillonite	99.7	0.033	9.9

Data from Garrison, A.D., 1939. Surface chemistry of shales and clays. Trans. AIME 132, 423–436.

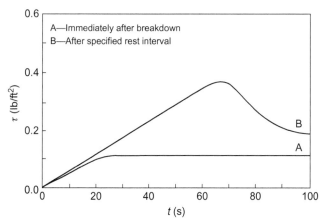

FIGURE 6.18 Typical shear stress loading curves. *From Lord, D.L., Menzie, D.E., 1943. Stress and strain rate dependence of bentonite clay suspension gel strengths, SPE Paper 4231, 6th Conference Drill. & Rock Mech., January 22–23, Austin, pp. 11–18. Copyright 1973 by SPE-AIME.*

The work of these investigators showed the inadequacy of the current method of evaluating gel strength after 10-s and 10-min rest periods. For example, Fig. 6.16 shows that the gel strength may increase very rapidly immediately after the cessation of shearing, so that the initial gel strength (as commonly determined) is very sensitive to time, and meaningful values are therefore hard to obtain. Fig. 6.16 also shows that the 10-min gel strength is not a reliable indication of the ultimate gel strength. For instance, Curves 1 and 4 show approximately the same 10-min gel strengths, but Table 6.1 shows that the ultimate gel strength of the Curve 1 mud was 34 and that of the Curve 4 mud was 104.

One obvious source of scatter in gel strength determinations is variation in the rate of application of load. The importance of this factor was demonstrated by Lord and Menzie (1943), who measured the gel strengths of a 10% bentonite suspension in a modified multispeed rotary viscometer, at rates that varied from 0.5 to 100 rpm, and recorded the change in stress with time. Fig. 6.18 shows the type of curves obtained. The maximum recorded stress was taken to be the gel strength, and the slope of the first part of the curve to be the rate of application of the load. Fig. 6.19 shows that the observed gel strengths (designated Y) increased sharply with increase in stress load rate (designated τ). Similarly, in some tests using a pipe viscometer, Lord and Menzie observed that the breakdown pressure of a gelled mud increased with rate of application of pump pressure.

It appears, therefore, that there is a need for a method of predicting long-term gel strengths, so that circulation breakdown pressures can be estimated more accurately.

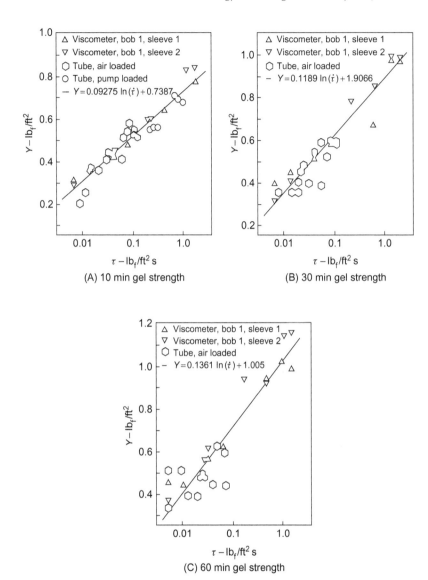

FIGURE 6.19 Increase in gel strength (*Y*) with rate of application of load (τ). *From Lord, D.L., Menzie, D.E., 1943. Stress and strain rate dependence of bentonite clay suspension gel strengths, SPE Paper 4231, 6th Conference Drill. & Rock Mech., January 22–23, Austin, pp. 11–18. Copyright 1973 by SPE-AIME.*

PSEUDOPLASTIC FLUIDS

Pseudoplastic fluids have no yield point; their consistency curves pass through the origin. The curves are nonlinear, but approach linearity at high shear rates. Thus, if stress readings taken at high shear rates are extrapolated

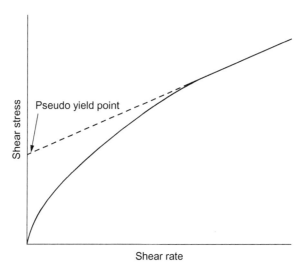

FIGURE 6.20 Consistency curve of a pseudoplastic fluid.

back to the axis, there appears to be a yield point similar to that of a Bingham plastic, hence the name *pseudoplastic* (see Fig. 6.20).

Suspensions of long-chain polymers are typical pseudoplastics. At rest, the chains are randomly entangled, but they do not set up a structure because the electrostatic forces are predominately repulsive. When the fluid is in motion, the chains tend to align themselves parallel to the direction of flow; this tendency increases with increase in shear rate, so that the effective viscosity decreases.

The consistency curve of the pseudoplastic flow model is described by an empirical equation, known as the *power law* (Metzner, 1956, p. 87), viz:

$$\tau = K \left(\frac{dv}{dr}\right)^n \quad (6.31)$$

The n exponent is dimensionless, and the consistency index, K, has the units of Pa-s^n in the SI system and lbf·s^n/100 ft^2 in oilfield units. The yield point for Bingham fluids is characterized in units of lbf/100 ft^2, and plastic viscosity is given in cP. K is the viscosity at a shear rate of 1 s^{-1}, and logically should be expressed in dyne cm/s, but in the oil industry, since it intersects the consistency curve at 1.0 s^{-1} it is expressed as lb/100 ft^2. The exponent n is the flow behavior index, and indicates the degree of shear thinning—the less the value of n, the greater the shear-thinning characteristic.

Actually, the power law describes three flow models, depending on the value of n:

1. Pseudoplastic, $n < 1$, the effective viscosity decreases with shear rate.
2. Newtonian, $n = 1$, the viscosity does not change with shear rate.
3. Dilatant, $n > 1$, the effective viscosity increases with shear rate.

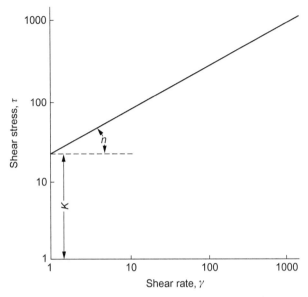

FIGURE 6.21 Logarithmic plot of consistency curve of an ideal power law fluid.

Since Eq. (6.31) may be written

$$\log \tau = \log K + n(\log \gamma) \tag{6.32}$$

a logarithmic plot of shear stress versus shear rate is linear for a pseudoplastic fluid. As shown in Fig. 6.21, the slope of the curve defines n, and the intercept on the stress axis at $\gamma = 1$ defines K (since $\log 1 = 0$).

K and n may either be measured directly from the plot or calculated from two values of stress, as follows:

$$n = \frac{\log \tau_1 - \log \tau_2}{\log \gamma_1 - \log \gamma_2} \tag{6.33}$$

$$\log K = \log \tau_1 = n \log \gamma_1$$

or

$$K = \frac{\tau_1}{\gamma_1^n} \tag{6.34}$$

For example, if dial readings are taken at 600 and 300 rpm in a direct-indicating viscometer, then

$$n = \frac{\log \theta_{600} - \log \theta_{300}}{\log 1022 - \log 511}$$

$$= 3.32 \log \frac{\theta_{600}}{\theta_{300}} \tag{6.35}$$

$$\log K = \log \theta_{600} - 3.0094n$$

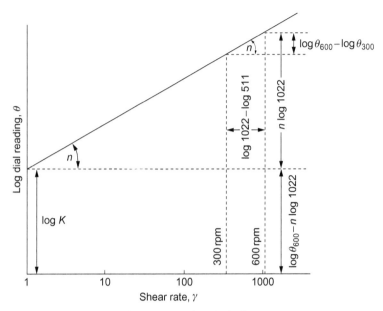

FIGURE 6.22 Determination of n and K in a direct-indicating viscometer.

or

$$K = \frac{\theta_{600}}{(1022)^n} = \text{lb}/100 \text{ ft}^2 \qquad (6.36)$$

A graphical interpretation is given in Fig. 6.22. The effective viscosity of a power law fluid is given by

$$\mu_e = \frac{\tau}{\gamma} = \frac{gK(\gamma)^n}{\gamma} = gK(\gamma)^{n-1} \qquad (6.37)$$

when K is dynes/cm^2, γ in reciprocal seconds, and μ_e is in poises.

For the special case of Newtonian fluids, the slope of the consistency curves on a logarithmic plot is always 45 degrees, since $n = 1$. If the stress is plotted in absolute units, the intercept on the stress axis at $\gamma = 1$ gives the viscosity in poises, as shown in Fig. 6.23.

The velocity profile of pseudoplastic fluids has a central flattish portion, something like the profile of a Bingham plastic (Fig. 6.6B), although pseudoplastics do not have a finite yield point, and hence no central core of unsheared material. The flattening occurs because the local shear rate decreases towards the center of the pipe, and the local viscosity increases accordingly. The degree of flattening increases with decrease in n, according to the following equation (Metzner, 1956, p. 107):

$$\frac{v}{V} = \frac{1+3n}{1+n}\left(1 - \left(\frac{r}{R}\right)^{n+1/n}\right) \qquad (6.38)$$

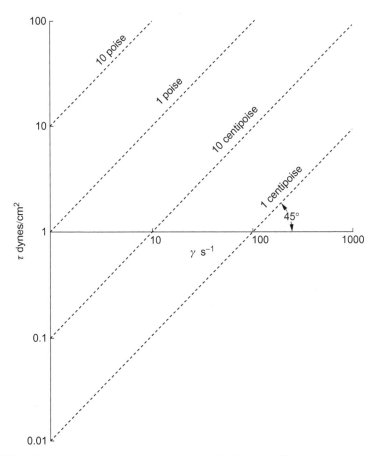

FIGURE 6.23 Logarithmic plot of consistency curves for Newtonian fluids.

where v is the velocity at radius r, and V is the mean velocity. Fig. 6.24 shows velocity profiles for several values of n, calculated from Eq. (6.38).

The equation for the flow of pseudoplastic fluids in round pipes is derived by integrating the power law, Eq. (6.31), with respect to r, which gives the expression (Metzner, 1956, p. 97).

$$P = 4K\left(\frac{6n+2}{2}\right)^n \frac{V^n L}{D^{n+1}} \tag{6.39}$$

This equation is limited to ideal power law fluids that give linear logarithmic consistency curves.

THE GENERALIZED POWER LAW

The consistency curves of most drilling fluids are intermediate between the ideal Bingham plastic and the ideal pseudoplastic flow models. Thus,

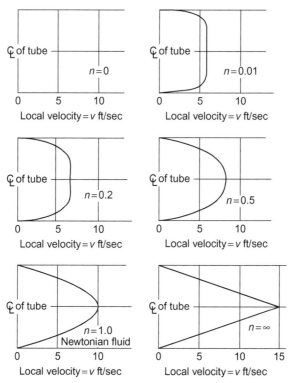

FIGURE 6.24 Dependence of velocity profiles upon flow behavior index. Average velocity of 5 ft/s in all cases. *From Metzner, A.B., 1956. Non-Newtonian technology: Fluid mechanics and transfers. Advances in Chemical Engineering. Academic Press, New York, NY. p. 87. Courtesy of Academic Press.*

Fig. 6.9 shows that the arithmetic plots of multispeed viscometer data are not linear at low rotor speeds, contrary to the Reiner-Riwlin equation, and Fig. 6.25 shows that (except for one polymer-brine fluid) the logarithmic plots are also not linear, contrary to the ideal power law. Similarly, Speers (1984) found that a linear regression technique for calculating n and K from the 600 and 300 rpm readings did not give valid results with a bentonite suspension because of the nonlinearity of the log−log plot. Low-solid and polymer fluids, clay muds heavily treated with thinners, and oil-base muds all tend towards pseudoplastic behavior; high-solid muds, and untreated and flocculated clay muds act more like Bingham plastics.

The generalized power law extends the power law (Eq. 6.31) to cover the flow behavior of these diverse fluids. The nonlinearity of their logarithmic consistency curves shows that n and K are not constant with rate of shear, as required by the power law, and therefore Eq. (6.39) cannot be used to determine the flow behavior of such fluids in pipes. Metzner and Reed (1955)

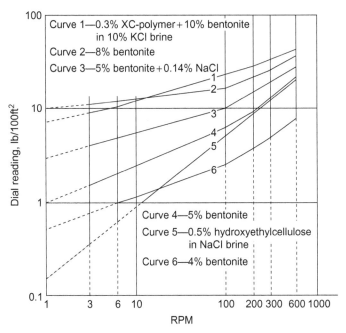

FIGURE 6.25 Consistency curves of typical drilling fluids in a multispeed direct-indicating viscometer. *Data courtesy of Brinadd Co.*

developed the generalized power law to avoid this difficulty. Their work was based on concepts originally developed by Rabinowitsch (1929) and also by Mooney (1931), who showed that for laminar flow of any fluid whose shear stress is a function only of shear rate, the flow characteristics are completely defined by the ratio of the shear stress at the wall of the pipe to the shear rate at the wall of the pipe. Metzner and Reed rearranged the Rabinowitsch-Mooney equation to

$$\left(-\frac{dv}{dr}\right)_w = \frac{3n'+1}{4n'}\frac{8V}{D} \qquad (6.40)$$

where $\left(-\frac{dv}{dr}\right)_w$ is the shear rate at the wall and

$$n' = \frac{d\left(\log\frac{DP}{4L}\right)}{d\left(\log\frac{8V}{D}\right)}$$

By analogy, the power law then becomes

$$\tau_w = \frac{DP}{4L} = K'\left(\frac{8V}{D}\right)^{n'} \qquad (6.41)$$

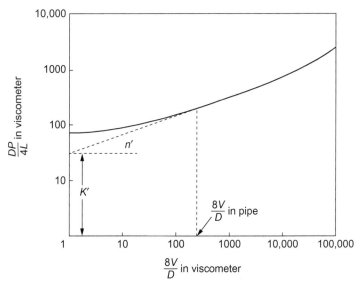

FIGURE 6.26 Determination of n' and K' from capillary viscometer data.

n' is numerically equal to n, and

$$K' = \left(\frac{3n+1}{4n}\right)^n K \tag{6.42}$$

The advantage of Eq. (6.41) over Eq. (6.31) is that the former is already integrated, and therefore it is inconsequential whether or not n' and K' are constants. The two parameters may be obtained from plots of log $DP/4L$ versus log $8\,V/D$. When the curve is nonlinear, n' and K' are obtained from the tangent to the curve at the point of interest, as shown in Fig. 6.26.

The pressure of a fluid flowing in a round pipe at a given velocity may be predicted from data obtained in a pressurized capillary viscometer. The discharge rates are measured over a range of pressures, and plotted as $8\,V/D$ versus $DP/4L$, where D is the diameter of the capillary, and L is its length. The required values of n' and K' are then given by the tangent to the curve at the point where $8\,V/D$ in the pipe equals $8V/D$ in the viscometer, as shown in Fig. 6.26.

Determination of exact values of n' and K' from rotary viscometer data is more involved. Savins (1958) has described a method, based on the relationship of K' in the pipe to K' in the viscometer, but, in practice, it has been found more convenient to determine n and K, instead of n' and K', and then to calculate the pressure loss in the pipe from the generalized power law which, from Eq. (6.41), may be written in the form

$$\tau_w = \frac{DP}{4L} = K\left(\frac{3n+1}{4n}\right)^n \left(\frac{8V}{D}\right)^n \tag{6.43}$$

n and K must, of course, be determined at the shear rates prevailing in the well. A trial and error procedure is therefore necessary, but, in practice, the appropriate speed range on the multispeed viscometer is usually obvious, so all that is necessary is to determine n and K accordingly, as discussed later in this chapter. As a check, the rate of shear at the wall of the drill pipe may be calculated from Eq. (6.40), and if it is found significantly different from the rate at which n and K were evaluated, n and K should be reevaluated, and the pressure loss recalculated.

The effective viscosity in the drill pipe is given by

$$\mu_e = \frac{\tau_w}{\gamma_w} = g \frac{K\left(\frac{3n+1}{4n}\right)^n \left(\frac{8v}{D}\right)^n}{\frac{3n+1}{4n} \frac{8V}{D}} = gK\left(\frac{3n+1}{4n}\right)^{n-1}\left(\frac{8V}{D}\right)^{n-1} \quad (6.44)$$

For flow in the annulus, the shear rate at the wall is given by Savins (1958):

$$\gamma_w = \frac{2n+1}{3n}\frac{12V}{D_2 - D_1} \quad (6.45)$$

so that Eq. (6.43) becomes

$$\frac{(D_2 - D_1)P}{4L} = K\left(\frac{2n+1}{3n}\right)^n \left(\frac{12V}{D_2 - D_1}\right)^n \quad (6.46)$$

and

$$\mu_e = gK\left(\frac{2n+1}{3n}\right)^{n-1}\left(\frac{12V}{D_2 - D_1}\right) \quad (6.47)$$

When n and K are not constant with rate of shear, pressure loss and effective viscosity in the annulus must be determined at the shear rates prevailing in the annulus, and n and K must therefore be determined in a multispeed viscometer. In the API recommended multispeed viscometer, the 100 to 6 rpm $(170 - 10.2 \text{ s}^{-1})$ range is used for this purpose; n is then given by

$$\frac{\log \theta_{100} - \log \theta_6}{\log 170 - \log 10.2} = 0.819 \log \frac{\theta_{100}}{\theta_6}$$

and K is given by

$$\frac{\theta_{100}}{170^n} \text{ lb}/100 \text{ ft}^2$$

or

$$\frac{\theta_{100}}{170^n} \times 4.788 \text{ dyne}/\text{cm}/s$$

Shear rates in the drill pipe are usually covered by the 600–300 rpm range on the multispeed viscometer; equations for determining n and K in this range were given earlier in this chapter. Sample calculations, based on the foregoing equations, are given later in this chapter.

Robertson and Stiff (1976) proposed a three-constant flow model by means of which the effective viscosity at the wall of the drill pipe, or of the annulus, may be calculated from rotary viscometer data. The pressure loss is then calculated by substituting the effective viscosity in Poiseulle's equation.

TURBULENT FLOW REGIME

Turbulent Flow of Newtonian Liquids

A fluid in turbulent flow is subject to random local fluctuations both in velocity and direction, while maintaining a mean velocity parallel to the direction of flow. The average local velocity increases from zero at the pipe wall to a maximum at the axis. Since turbulent flow commences when a certain critical velocity is exceeded, there are three separate flow regimes across the diameter of the pipe: laminar flow next to the wall, where the velocity is below the critical value; a central core of turbulent flow; and a transitional zone between the two.

Fig. 6.27 shows the velocity profile of a Newtonian liquid in turbulent flow. Note particularly that this profile represents the average local velocity at points on the pipe diameter. Because the actual local velocity fluctuates randomly, the slope of the profile does not represent the shear rate. The actual shear rate is indeterminable, and so the flow pressure-rate relationship cannot be obtained from the change of shear stress with shear rate, as is done in the case of the laminar flow regime. Instead, turbulent flow behavior is usually described in terms of two dimensionless groups, namely:

1. The Fanning friction factor:

$$f = \frac{gDP}{2V^2L\rho} \quad (6.48)$$

2. The Reynolds number:

$$N_{Re} = \frac{DV\rho}{\mu} \quad (6.49)$$

The Fanning friction factor expresses the resistance to flow at the pipe wall. It is related to the Reynolds number by an equation originally proposed by von Karran (Dodge and Metzner, 1959):

$$\sqrt{\frac{1}{f}} = A \log\left(N_{Re}\sqrt{f}\right) + C \quad (6.50)$$

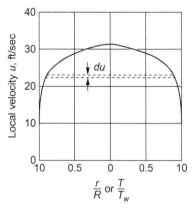

FIGURE 6.27 Newtonian turbulent velocity distribution (N_{Re} = 20,000, d = 1 inch water at 20°C). *From Dodge, D.W., Metzner, A.B., 1959. Turbulent flow of non-Newtonian systems. AIChE. J. 5(2), 194. Courtesy of AIChE.*

The value of the constants A and C depend upon the roughness of the pipe walls, and must be determined experimentally. Fig. 6.28 shows curves based on the von Karman equation for several grades of pipe. Turbulent flow pressures of Newtonian liquids may be predicted by calculating the Reynolds number for the system, finding the corresponding value of f from Fig. 6.28, and then calculating the pressure loss from Eq. (6.48). Note that the viscosity affects the flow pressure only to the extent that it determines the Reynolds number.

It may be deduced from Eqs. (6.48) and (6.49) and Poiseulle's equation (Eq. 6.4) that the Fanning friction factor is related to the Reynolds number in laminar flow by the following equation:

$$f = \frac{16}{N_{Re}} \qquad (6.51)$$

The roughness of the pipe wall does not influence laminar flow behavior, so the $f - N_{Re}$ relationship is the same for all grades of pipe. Laminar flow pressures may be predicted if desired, from Eqs. (6.51) and (6.48).

It has been found experimentally that the change from laminar to turbulent flow always occurs at approximately the same Reynolds number. With Newtonian fluids, transition to turbulent flow begins when the Reynolds number for the system is approximately 2100. Above 3000, flow is fully turbulent.

Turbulent Flow of Non-Newtonian Fluids

The Fanning friction factor and the Reynolds number may also be used to determine turbulent flow behavior of non-Newtonian fluids, provided that

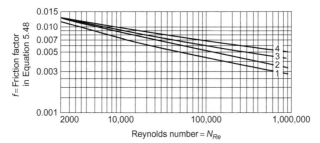

1. Lowest values for drawn brass or glass tubing. (Walker, Lewis and Mc Adams)
2. For clean internal flush tubular goods. (Walker, Lewis and Mc Adams)
3. For full hole drill pipe or annuli in cased hole. (Piggott's data)
4. For annuli in uncased hole. (Piggott's data)

FIGURE 6.28 The relationship between the Fanning friction factor and the Reynolds number. *Data from Bobo, R.A., Hoch, R.A., 1958. Keys to Successful Competitive Drilling. Gulf Publishing Co., Houston (Bobo and Hoch, 1958).*

suitable flow parameters are used. In the past, there has been some question as to what parameter to use for viscosity in the Reynolds number. This question does not arise with Newtonian fluids, because viscosity does not vary with rate of shear, so the viscosity determined in laminar flow may be used in turbulent flow. But as we have seen, viscosity does vary with rate of shear in the case of non-Newtonian fluids, and the rate of shear in turbulent flow cannot be determined. Metzner and Reed (1955) showed that the difficulty may be avoided by deriving a value for effective viscosity from the general power law constants, n' and K', which may be determined from capillary viscometer data, without reference to rate of shear, as shown in Fig. 6.26. Substitution of the effective viscosity so obtained in the Reynolds number gives

$$N'_{Re} = \frac{DV\rho}{gK'\left(\frac{8V}{D}\right)^{n-1}} = \frac{D^{n'} \cdot V^{2-n'}}{gK'8^{n-1}} \qquad (6.52)$$

Note that flow in the capillary viscometer must be laminar. To establish the validity of the generalized Reynolds number, Metzner and Reed evaluated it from the results of a large number of pipe flow experiments, made with non-Newtonian fluids by various investigators, and plotted the values obtained versus the Fanning friction factor (see Fig. 6.29). They found excellent agreement with the classical $f = 16/N_{Re}$ relationship for Newtonian fluids, and fair agreement with the value of 2100 for the critical Reynolds number, but poor agreement with the von Karman equation for turbulent flow.

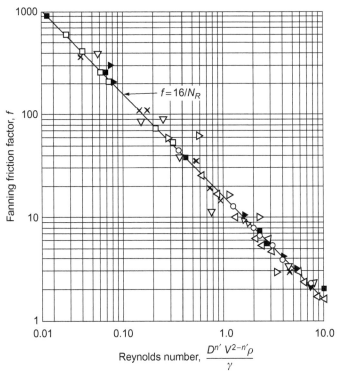

FIGURE 6.29 Friction factor—Reynolds number correlation for non-Newtonian fluids—medium range. *From Metzner, A.B., Reed, J.C., 1955. Flow of non-Newtonian fluids—correlation of laminar, transitional and turbulent flow regimes. AIChE J. 1, 434–440. Courtesy of AIChE.*

In order to reconcile the turbulent flow of non-Newtonian fluids with the von Karman equation, Dodge and Metzner (1959) generalized the von Karman equation as follows:

$$\sqrt{\frac{1}{f}} = A_{1n} \log\left(N'_{Re}(f)^{1-\frac{n'}{2}}\right) + C'_n \qquad (6.53)$$

where A_{1n} and C'_n are dimensionless functions of n. Note that for Newtonian fluids $n = 1$ and Eq. (6.53) reduces to the von Karman equation (Eq. 6.50). The value of the constants was found by determining n' and K' for a number of ideal power law fluids in a capillary viscometer and then finding the corresponding values of f in a pipe viscometer. The value of A_{1n} was found to be $\frac{4.0}{(n')^{0.75}}$ and C'_n to be $\frac{-0.40}{(n')^{1.2}}$.

Eq. (6.53) was derived for ideal power law fluids and does not apply rigorously to Bingham plastic and intermediate fluids because of the variation of n' with shear rate. However, Dodge and Metzner showed that the mean

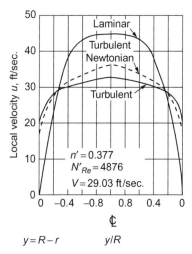

FIGURE 6.30 Typical predicted velocity profiles. All curves are calculated for the same average velocity. *From Dodge, D.W., Metzner, A.B., 1959. Turbulent flow of non-Newtonian systems. AIChE J. 5(2), 194. Courtesy of AIChE.*

flow velocity in the pipe was not greatly influenced by changes in n' in the central section of the pipe where the velocity profile is flat (see Fig. 6.30); therefore, Eq. (6.53) gave a very good approximation of the turbulent flow behavior of such fluids, provided that n' and K' were evaluated at the stresses prevailing at the wall of the pipe.

To verify this conclusion, it was necessary to adopt a system of trial and error in order to find the wall stress at which n' and K' should be evaluated (Metzner, 1956, p. 105). This procedure is not cumbersome, since with most power law fluids there are large regions in which n' and K' are nearly constant, and, further, they vary in opposite directions, so that N'_{Re} is rather insensitive to the shear stress at which n' and K' are evaluated.

The experimental procedure consisted of making a large number of flow tests in three sizes of pipe (1/2, 1, and 2 inches in diameter), with both power law and nonpower law fluids. Fig. 6.31 compares the values of f predicted from capillary viscometer data and Eq. (6.53) with those measured in the pipe flow experiments. The close agreement (maximum deviation 8.5%, mean deviation 1.9%) between the attapulgite clay suspensions and the (ideal power law) polymeric gels shows that Eq. (6.53) is valid for clay suspensions if the flow parameters are evaluated at the wall stresses prevailing in the pipe.

For convenience, Dodge and Metzner plotted the relationship between f, N'_{Re}, and n as shown in Fig. 6.32, and this figure offers the best means of determining f when making hydraulic calculations for flow of mud in drilling wells. n' and K' are best evaluated in a capillary viscometer, but satisfactory results can be obtained with a concentric cylinder viscometer provided that n' and K' do not vary greatly with rate of shear, which is

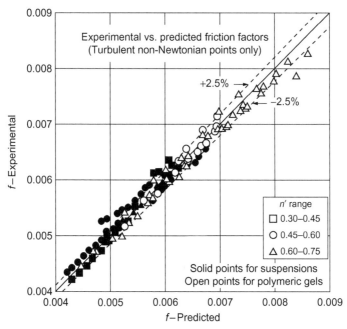

FIGURE 6.31 Comparison of experimental friction with those predicted. *From Dodge, D.W., Metzner, A.B., 1959. Turbulent flow of non-Newtonian systems. AIChE J. 5(2), 194. Courtesy of AIChE.*

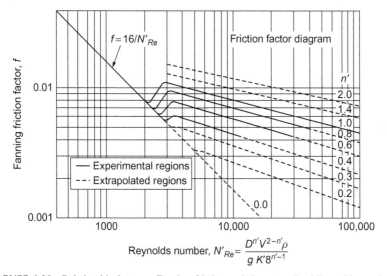

FIGURE 6.32 Relationship between Fanning friction and the generalized Reynolds number. Note that, for a given Reynolds number, f is strongly dependent on the value of n'. *From Dodge, D.W., Metzner, A.B., 1959. Turbulent flow of non-Newtonian systems. AIChE J. 5(2), 194. Courtesy of AIChE.*

usually the case with drilling muds at shear rates above about 300 reciprocal seconds. A convenient procedure is to evaluate n and K from the 600 and 300 rpm readings (Eqs. 6.35 and 6.36), and calculate the effective viscosity from Eqs. (6.44) and (6.47). The generalized Reynolds number for flow in the drill pipe is then given by

$$N'_{Re} = \frac{DV\rho}{\mu_e} = \frac{DV\rho}{gK\left(\frac{3n+1}{4n}\right)^{n-1}\left(\frac{8V}{D}\right)^{n-1}} \quad (6.54)$$

and for flow in the annulus by

$$N'_{Re} = \frac{(D_2 - D_1)V\rho}{\mu_e} = \frac{(D_2 - D_1)V\rho}{gK\left(\frac{2n+1}{3n}\right)^{n-1}\left(\frac{12V}{D_2-D_1}\right)^{n-1}} \quad (6.55)$$

The validity of rotary viscometer data is supported by some field tests made by Fontenot and Clark (1974), in which they determined the effective viscosity from the 600 and 300 rpm readings, calculated N'_{Re} from Eq. (6.54), and f from Eq. (6.53). The pressure loss in the drill pipe thereby predicted agreed well with values measured by downhole pressure gauges. Their results are discussed in greater detail later in this chapter.

Dodge and Metzner's findings were at variance with the prior art and, since old ideas die hard, have not yet been universally accepted by the oil industry. It was previously assumed, implicitly or explicitly, that the f-N_{Re} relationship was the same for Newtonian and non-Newtonian fluids (Melrose et al., 1958; Piggott, 1941; Hedstrom, 1952; Hughes Tool Company, 1954; Dunn et al., 1947; Caldwell and Babbit, 1949; Binder and Busher, 1946; Koch, 1953; Havenaar, 1954). It has been shown that such an assumption can lead to substantial errors in the value of f, especially at Reynolds numbers just above the critical (Metzner and Reed, 1955).

The use of the Newtonian f-N_{Re} relationship when interpreting data from pipe flow experiments led to various misconceptions about the viscosity of non-Newtonian fluids in turbulent flow, such as that it equaled the plastic viscosity, or some multiple thereof, or that it equaled the viscosity of the liquid phase, or that it was a function of the concentration of solids. The values thus obtained were generally too low, and did not appear to change with flow velocity. The latter observation led some authorities to recommend the use of a constant turbulent viscosity of three centipoises in hydraulic calculations. Actually, according to Dodge and Metzner, their experimental data showed that turbulent flow viscosity may vary threefold over a fivefold range of Reynolds numbers. Fortunately, such large variations in viscosity do not lead to equally large errors in the value of f, because f is relatively insensitive to the value of N'_{Re}, as shown in Fig. 6.32.

The Onset of Turbulence

Fig. 6.32 shows that turbulence in a non-Newtonian fluid with an n value of 0.4 would not start until the Reynolds number reached a value of 2900, compared to a value of 2100 for a Newtonian fluid. The difference is significant because it means that the flow velocity would have to be 38% higher, all other factors being equal. These figures emphasize the importance of using the generalized Reynolds number when determining whether or not a non-Newtonian fluid is in turbulent flow. The generalized Reynolds number for the system may be determined from Eq. (6.54) or (6.55), the critical value may be determined from Fig. 6.32, and the value of n for the fluid.

An alternative criterion, the *Z stability parameter*, which determines the point at which turbulence is initiated, has been introduced by Ryan and Johnson (1959). Its chief advantage is that the critical value is always the same, regardless of the value of n. Ryan and Johnson theorized that turbulent flow will start at a point on the radius of the pipe, r/R, at which the Z parameter is maximum, and that the local Reynolds number at that point will be 2100. The maximum value of Z is 808 for all fluids. With Newtonian fluids r/R at Z maximum is $1/\sqrt{3}$; with non-Newtonian fluids r/R increases with decrease in the value of n. Thus the critical Z number of 808 is independent of the value of n, but the mean flow velocity required for turbulence increases with a decrease in the value of n, which is in accordance with the findings of Dodge and Metzner. Ryan and Johnson substantiated their theory with experimental data.

Unfortunately, determination of the Z parameter involves complex calculations, but Walker (1976a,b) has published an approximate method for flow in the annulus, which takes the rotation of the drill pipe into account. His equation is

$$Z = \frac{(D_2 - D_1)^n V^{2-n}}{K} \Psi \tag{6.56}$$

where Ψ is a function of the drill pipe rpm, as shown in Fig. 6.33.

FRICTION REDUCERS

Certain long-chain polymers have the remarkable property of apparently reducing the turbulent viscosity of water. For example, Fig. 6.34 shows that the friction factor curve of a 0.3% suspension of carboxymethylcellulose fell well below that of a pseudoplastic fluid with the same flow-behavior index. Note that the actual amount of the reduction varied with the diameter of the pipe and the Reynolds number, the maximum being about 50%. Similar effects have been observed with other long-chain polymers, many of them commonly used in drilling muds (e.g., the gums, the polyacrylamides, xanthan gum, and hydroxyethylated cellulose) (Savins 1964; Darley and Hartfiel, 1974). In consequence, these polymers dramatically reduce the flow

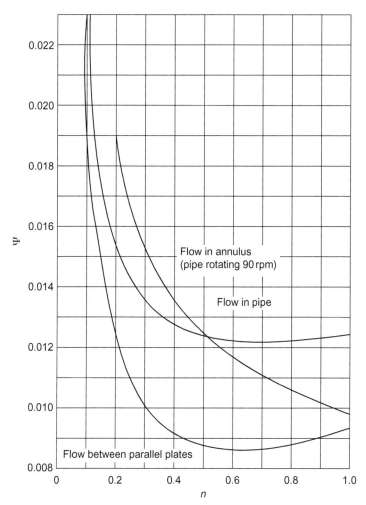

FIGURE 6.33 Determination of Z parameter in annular flow. *From Walker (1976a). Courtesy of Oil and Gas J.*

pressure of water and brines in drilling and workover operations. The amount of the reduction depends on the molecular structure of the polymer, its concentration, the flow velocity, and the pipe diameter. The pressure loss may be determined by making tests with the polymer of interest in a pipe viscometer under turbulent flow conditions, establishing the f-N_{Re} relationship, and then calculating the pressure loss for the specified conditions in the well by means of Eq. (6.48). Such tests have shown that friction reducers can reduce pressure losses by as much as a factor of three (Darley and Hartfiel, 1974). Unfortunately, the beneficial effect of the polymer is lost if the fluid becomes contaminated with substantial amounts of clay.

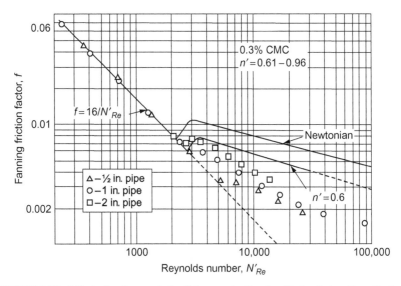

FIGURE 6.34 Effect of carboxymethyl cellulose on the Fanning friction factor. *From Dodge, D.W., Metzner, A.B., 1959. Turbulent flow of non-Newtonian systems. AIChE J. 5(2), 194. Courtesy of AIChE.*

Based on data derived from tests made with full-scale laboratory equipment, Randall and Anderson (1982) developed equations relating f to N_{Re} and to viscometer 600 and 300 readings. These equations were found to predict the pressure losses of polymer fluids—as well as other muds—in the field with acceptable accuracy.

Friction reduction must be distinguished from shear thinning; they are two quite different phenomena. Shear thinning, as we have seen, results from a reduction in structural viscosity. The mechanism of friction reduction is not known for certain, but it appears to result from the elastic properties of the long-chain polymers, which enable them to store the kinetic energy of turbulent flow (Savins, 1964).

INFLUENCE OF TEMPERATURE AND PRESSURE ON THE RHEOLOGY OF DRILLING FLUIDS

The rheological properties of drilling muds under downhole conditions may be very different from those measured at ambient pressures and temperatures at the surface. At depth, the pressure exerted by the mud column may be as much as 20,000 pounds per square inch (1400 kg/cm^2). The temperature depends on the geothermal gradient, and may be more than 500°F (260°C) at the bottom of the hole during a round trip. Fig. 6.35 shows estimated mud temperatures during a normal drilling cycle in a 20,000 foot (6100 m) well (Raymond 1969). Even quite moderate temperatures can have a significant, but largely unpredictable

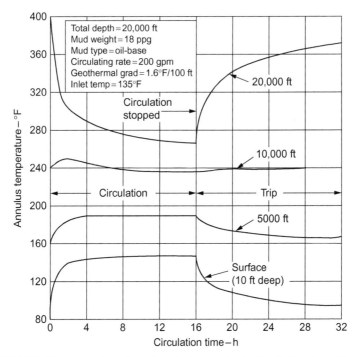

FIGURE 6.35 Temperature trace for various depths in a simulated well. *Courtesy of Raymond, L.R., 1969. Temperature distribution in a circulating fluid. J. Petrol. Technol. 333–341. Trans AIME 246. Copyright 1969 by SPE-AIME.*

influence on the rheological properties. Muds may be thicker or thinner downhole than indicated at the surface, and an additive that reduces viscosity at the surface may actually increase the viscosity downhole (Kelly and Hawk, 1961).

Elevated temperatures and pressures can influence the rheological properties of drilling fluids in any of the following ways:

1. *Physically.* An increase in temperature decreases the viscosity of the liquid phase; an increase in pressure increases the density of the liquid phase, and therefore increases the viscosity.
2. *Chemically.* All hydroxides react with clay minerals at temperatures above about 200°F (94°C). With low-alkalinity muds, such as those treated with caustic tannate or lignosulfonate, the effect on their rheological properties is not significant, except to the extent that the loss of alkalinity lessens the effectiveness of the thinner. But with highly alkaline muds the effect may be severe, depending on the temperature and the species of metal ion of the hydroxide (Darley and Generes, 1956). In the notorious case of high-solid lime-treated muds, hydrated aluminosilicates were formed, and the mud set to the consistency of cement at temperatures above about 300°F (150°C) (Gray et al., 1952).

3. *Electrochemically*. An increase in temperature increases the ionic activity of any electrolyte, and the solubility of any partially soluble salts that may be present in the mud. The consequent changes in the ionic and base exchange equilibria alter the balance between the interparticle attractive and repulsive forces, and hence the degree of dispersion and the degree of flocculation (see Chapter 4, Clay Mineralogy and the Colloid Chemistry of Drilling Fluids). The magnitude and direction of these changes, and their effect on the rheology of mud, varies with the electrochemistry of the particular mud.

Because of the large number of variables involved, the behavior of drilling fluids at high temperature, particularly water-base drilling fluids, is unpredictable, and, indeed, not yet fully understood. Even quite small differences in composition can make considerable differences in behavior (Bartlett, 1967), so that it is necessary to test each mud individually in order to obtain reliable data.

Various types of viscometers may be used to investigate mud rheology at high temperatures and pressures. The consistometer (Weintritt and Hughes, 1965; Sinha, 1970) was the first of these. It measures the time of transit of a magnetically controlled bob through a mud sample. It is a useful instrument for comparing the effect of a large number of variables, but, since there is no means of determining the shear rate, the data obtained are only empirical. To determine the Bingham or power law flow parameters required for hydraulic calculations, it is necessary to use a capillary (Combs and Whitmire, 1960), pipe (Kelly and Hawk, 1961), or rotary viscometer (Annis, 1967; Hiller, 1963; McMordie, 1969), modified for high temperatures and pressure. Rotary viscometers require elaborate instrumentation for temperatures above the boiling point of the liquid phase, and pressures above atmospheric.

Annis (1967) studied the rheology of water-base muds at high temperature. Hiller (1963) studied the effect of high pressure as well, but found that it was minor. The results of their studies showed that the effect of temperature is as follows:

If a suspension is fully deflocculated, the plastic viscosity and yield point decrease with temperature up to 350°F (177°C), whereas if the mud is flocculated, only the plastic viscosity declines and the yield point increases sharply at temperatures above the boiling point of water. For example, compare the behavior of deflocculated monoionic sodium montmorillonite in Fig. 6.36A with the same suspension when flocculated (Fig. 6.36B). Similarly, if such a suspension has been deflocculated by a thinner, the yield point does not increase with temperature, provided that the thinner itself does not degrade, or that the reaction between the clay minerals and the caustic soda does not reduce the pH below the level required to solubilize the thinner.

The plastic viscosity of a clay suspension decreases with temperature at high shear rates because the viscosity of the water decreases. Thus, Fig. 6.37 shows that a plot of the plastic viscosity of a bentonite suspension

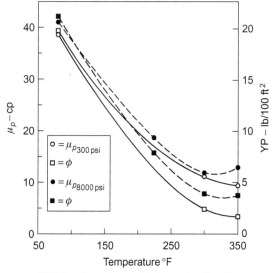

(A) When flocculated with 5 me q/L of NaOH.

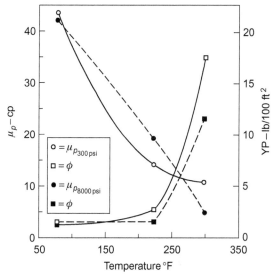

(B) When flocculated with 50 me q/L of NaOH.
Note floc point is 15 me q/L NaOH.

FIGURE 6.36 Effect of temperature on the plastic viscosity, μ_p, and yield point (YP), on 4% monoionic sodium montmorillonite suspensions. *From Hiller, K.H., 1963. Rheological measurements of clay suspensions at high temperatures and pressures. J. Petrol. Technol. 779–789. Trans AIME 228. Copyright 1963 by SPE-AIME.*

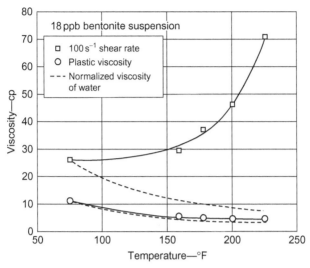

FIGURE 6.37 Effect of temperature and shear rate on the viscosity of an 18 ppb bentonite suspension. *From Annis, M.R., 1967. High temperature properties of water-base drilling fluids. J. Petrol. Technol. 1074–1080. Trans AIME 240. Copyright 1967 by SPE-AIME.*

versus temperature coincides almost exactly with that of the normalized viscosity of water, i.e., the viscosity of water at the specified temperature multiplied by the initial viscosity of the suspension. On the other hand, the effective viscosity of the same suspension at low shear rates increases over the same temperature range. The explanation is, of course, that high temperatures caused an increase in the interparticle attractive forces—as shown by the increase in gel strengths in Fig. 6.38—and the effective viscosity is influenced by interparticle forces at low shear rates, but not at high shear rates (as discussed earlier in this chapter in the section on the shape of the consistency curve).

There is considerable evidence that the degree of dispersion increases when muds are aged dynamically. Thus, Fig. 6.39 shows that the effective viscosity of a bentonite suspension increased at both high and low shear rates after the suspension was rolled at high temperatures. The increase in viscosity at high shear rates must be ascribed to an increase in the degree of dispersion: the greater increase at low shear rates was caused by increases in both the degree of flocculation and the degree of dispersion.

The behavior of suspensions of calcium clays at high temperatures is different from that of sodium clay suspensions, and considerably more complex. The interparticle repulsive forces of calcium clays are much weaker than those of sodium clays; consequently, the effect of high temperature on the degree of flocculation is much stronger, and thus even the plastic viscosity increases, as shown in Fig. 6.40.

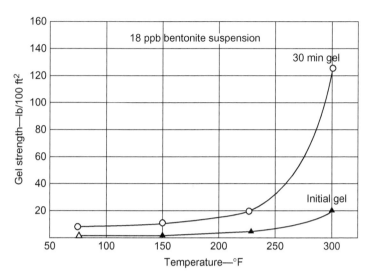

FIGURE 6.38 Effect of temperature on initial and 30-min gel strengths. *From Annis, M.R., 1967. High temperature properties of water-base drilling fluids. J. Petrol. Technol. 1074–1080. Trans AIME 240. Copyright 1967 by SPE-AIME.*

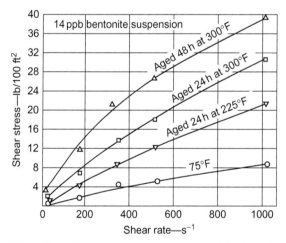

FIGURE 6.39 Effect of hot rolling on shear rate–shear stress relationships. *From Annis, M.R., 1967. High temperature properties of water-base drilling fluids. J. Petrol. Technol. 1074–1080. Trans AIME 240. Copyright 1967 by SPE-AIME.*

The effect of high temperature flocculation on viscosity and gel strength increases with increase in clay concentration. Fig. 6.41 shows the increase in the 10-min gel strength at 300°F (150°C) versus clay concentration for untreated bentonite suspensions and for suspensions treated with the

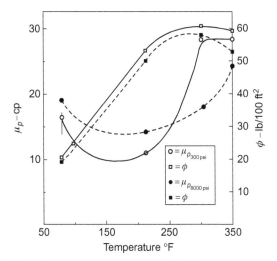

FIGURE 6.40 Effect of temperature and pressure on the yield point ϕ, and plastic viscosity μ_p, of a 13% suspension of pure calcium montmorillonite to which 5 meq/L of $CaCl_2$ have been added. *From Hiller, K.H., 1963. Rheological measurements of clay suspensions at high temperatures and pressures. J. Petrol. Technol. 779–789. Trans AIME 228. Copyright 1963 by SPE-AIME.*

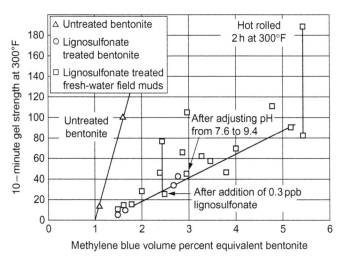

FIGURE 6.41 Effect of clay concentration on the 10-min gel strengths of bentonite suspensions and of lignosulfonate field muds. *From Annis, M.R., 1967. High temperature properties of water-base drilling fluids. J. Petrol. Technol. 1074–1080. Trans AIME 240. Copyright 1967 by SPE-AIME.*

optimum amount of lignosulfonate. Ten-minute gel strengths at 300°F are also shown for a number of lignosulfonate-treated field muds versus their clay content, expressed as equivalents of bentonite, as measured by the methylene blue test (see Chapter 3, Evaluating Drilling Fluid Performance). Note that the correlation is quite good. The slight scatter is probably due to the field muds not being fully deflocculated.

High-temperature behavior varies widely with the type of mud. For example, salt water muds are comparatively stable because the high electrolyte content prevents the clays from dispersing. The behavior of gyp-CLS muds is similar to that of the calcium montmorillonite suspension shown in Fig. 6.40. Lime muds, as already mentioned, develop high gel strengths because of the reaction between the hydroxide and the clay minerals, but calcium surfactant muds remain quite stable at temperatures up to 350°F.

The investigations of Hiller and Annis showed that accurate rheological parameters for water-base drilling muds at elevated temperature can only be obtained by direct measurement at the temperature of interest. However, the results shown in Fig. 6.41 suggest that, for each mud type, correlations in the laboratory might be obtained that would enable approximate subsurface values for that type of mud to be predicted from wellsite tests at ambient temperatures.

Oil-base drilling fluids deteriorate less at high temperatures than do water-base muds, and can withstand higher temperatures. In contrast to water-base muds, however, their viscosities are substantially influenced by pressure, as shown in Fig. 6.42.

The effect of temperature and pressure on the rheology of nonaqueous muds is almost entirely physical, and changes in the subsurface properties

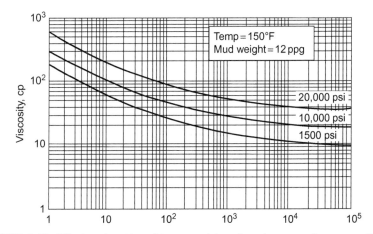

FIGURE 6.42 Effective viscosities of invert emulsion oil muds at several pressures. *From Combs, G.D., Whitmire, L.D., 1960. Capillary viscometer simulates bottom hole conditions. Oil Gas J. 108–113. Courtesy of Oil and Gas J.*

can largely be accounted for by the effect of temperature and pressure on the viscosity of the nonaqueous continuous phase. In the case of diesel oil, Combs and Whitmire (1960) measured the effective viscosities in a capillary viscometer at several temperatures and pressures, and found that, when the viscosities were normalized to the viscosity of diesel oil at the same temperature and pressure, the points all fell on a single curve for each temperature (see Fig. 6.43). They ascribed the small difference between the curves to a decrease in the degree of emulsification at the higher temperature. These results show that the subsurface viscosities of this type of oil mud can be predicted from viscosities measured at ambient temperatures by means of a correction factor, based on the viscosity of diesel oil at the temperature and pressure of interest, provided that the muds remain substantially stable.

Oil muds based on colloidal suspensions of asphalt show a somewhat more complex behavior. Figs. 6.44A and B show the changes in plastic viscosity and yield point, as measured in a high-temperature and pressure rotary viscometer by McMordie (1969), of a suspension of asphalt in diesel oil. There appears to be a synergism between the effect of temperature and pressure, since the increase in viscosity and yield point with pressure is greater at high temperatures than it is at low temperatures.

McMordie et al. (1975) showed that the behavior of oil-base muds (asphaltic/diesel systems) at temperature and pressure may be described by a modification of the power law:

$$\ln \tau = \ln K' + n \ln \gamma + Ap + B/T \tag{6.57}$$

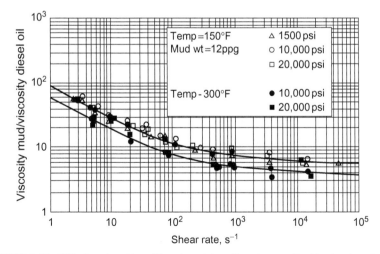

FIGURE 6.43 Effective viscosities of invert emulsion oil muds normalized to the viscosity of diesel oil. *From Combs, G.D., Whitmire, L.D., 1960. Capillary viscometer simulates bottom hole conditions. Oil Gas J. 108−113. Courtesy of Oil and Gas J.*

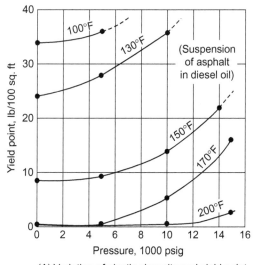

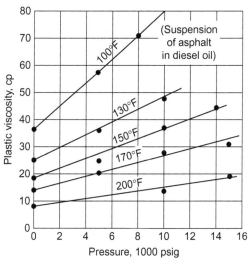

FIGURE 6.44 Variation of plastic viscosity. *From McMordie, W.C., 1969. Viscometer tests mud to 650°F. Oil Gas J. 81–84. Courtesy of Oil and Gas J.*

where A is the pressure constant and B the temperature constant, both of which must be determined separately for each mud. Fig. 6.45 shows the excellent correlation between the experimentally determined relationship of shear stress to shear rate with that calculated from Eq. (6.57) by means of a computer program. Table 6.2 lists the effective viscosities at two shear rates,

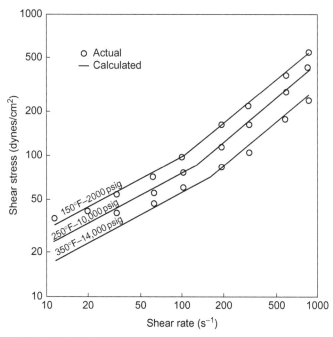

FIGURE 6.45 Comparison of actual with calculated flow properties of a 16 lb/gal asphaltic oil-base mud. *From McMordie Jr., W.C., Bennett, R.B., Bland, R.G., 1975. The effect of temperature and pressure on oil base muds. J. Petrol. Technol. 884–886. Copyright 1975 by SPE-AIME.*

calculated according to Eq. (6.57), for three different oil mud compositions at increasing temperatures and pressures. Note that a mud with the lowest viscosity at the surface may not have the lowest viscosity downhole.

APPLICATION OF FLOW EQUATIONS TO CONDITIONS IN THE DRILLING WELL

The rigorous flow equations that were given in the first part of this chapter were based on two assumptions: first, that the temperature of the fluid remained constant throughout the system, and second, that the rheological properties were not thixotropic. Both these assumptions are violated in the drilling well, but both may be satisfied if the rheological parameters are determined under the flow conditions prevailing at the point of interest in the well. The difficulty lies in ascertaining the flow conditions. The temperature of the mud is constantly changing, as shown in Fig. 6.35, and its precise value at a particular point in the circuit and at a particular time in the drilling cycle depends on a number of variables. Also the rate of shear undergoes drastic changes at several points in the circuit, and there is a considerable time lag before the state of shear reaches even approximate equilibrium; indeed, it may never do so (see "The Effect of Thixotropy on Drilling

TABLE 6.2 Calculated Viscosities of 14 lb/gal Oil-Base Muds

	Asphaltic Formula 1	Asphaltic Formula 2	Oleophilic Inorganic
Viscosity at 500 s^{-1}, cp			
150°F, 0 psig	75	56	37
200°F, 3633 psig	51	44	34
250°F, 7266 psig	38	36	32
300°F, 10,899 psig	31	31	32
350°F, 14,532 psig	28	39	33
Average viscosity	45	39	34
Viscosity at 50 s^{-1}, cp			
150°F, 0 psig	170	118	110
200°F, 3633 psig	124	99	110
250°F, 7266 psig	96	87	112
300°F, 10,899 psig	79	80	115
350°F, 14,532 psig	67	75	120
Average viscosity	107	92	113

SPE/AIME. From McMordie Jr., W.C., Bennett, R.B., Bland, R.G., 1975. The effect of temperature and pressure on oil base muds. J. Petrol. Technol. 884–886. Copyright 1975 by SPE-AIME.

Muds" earlier in this chapter). Along with these uncertainties, there are several unknown factors, such as the width of the annulus in enlarged sections of the hole and the effect of rotation of the drill pipe.

Because of these limitations, the flow pressures or velocities in a drilling well can never be predicted with the precision that is possible in the piping systems in an industrial plant. The question therefore arises: Are these rigorous equations and the computer programs that are, in some cases, necessary to solve them justifiable in terms of time and expense, given the uncertainty of the input data? Might not some simpler, less exact equations, give equally good results? In this section, we shall endeavor to show that the answer to these questions depends on the section of the flow circuit under consideration, the purpose of the investigation, and where the investigation is being carried out—in the laboratory or at the drilling well.

Flow Conditions in the Well

Flow in the drill pipe is usually turbulent, and is therefore only influenced by the viscous properties of the mud to a minor extent. The effective shear rate at the pipe wall, as determined from capillary viscometer data and Eq. (6.40), is generally between 200 to 1000 reciprocal seconds. The conduit dimensions are known accurately, so that the pressure loss can be calculated quite accurately. The only uncertain factor is the roughness of the pipe walls. The pressure loss in the drill pipe is about 20 to 45% of the pressure loss over the whole circuit, i.e., of the standpipe pressure.

Flow velocity through the bit nozzles is extremely high, corresponding to shear rates of the order of 100,000 reciprocal seconds. The pressure loss across the nozzles can be calculated accurately because it depends on the coefficient of discharge, which is essentially independent of the viscous properties of the mud. The pressure loss is about 50 to 75% of the standpipe pressure.

Flow in the annulus is normally laminar, and is therefore a function of the viscous properties of the mud. Shear rates generally lie between 50 and 150 reciprocal seconds. Although the pressure loss from the bit to the surface comprises only 2−5% of the standpipe pressure, a knowledge of the pressure and flow velocity in the various sections of the annulus is very important when dealing with such problems as hole cleaning, induced fracturing, and hole erosion. Unfortunately, accurate prediction of flow relationships is usually difficult and often impossible, owing to various unknown factors and uncertainties. Perhaps the greatest of these unknowns is the diameter of the hole, which may be as much as twice the nominal diameter in enlarged sections of the hole, thereby decreasing the rising velocity by a factor of at least five (see Fig. 6.46).

The influence of the rotation of the drill pipe on velocity profiles is also difficult to account for. There are equations for helical flow (Savins and Wallick, 1966), but these were derived for drill pipe rotating concentrically in a vertical hole, whereas, in practice, the drill pipe whips around in a randomly deviated hole. Also, equations for flow in eccentric annuli show that annular velocity is greatly reduced when the drill pipe lies against the low side of the hole, as in directionally drilled wells, and that equations based on concentric annuli would be seriously in error (Iyoho and Azar, 1981). Finally, there is no practical way to account for the influence of thixotropy on the viscosity of the mud as it rises in the annulus. The high shear rates in the drill pipe and bit nozzles reduce the structural component of the viscosity to a very low value. The shear rates in the annulus are far lower, but change in each section, depending on the drill collar, the drill pipe, and the casing diameters, and on the degree of hole enlargement. The viscosity adjusts to each shear rate, but takes time to do so, and may never reach the equilibrium value except in long sections of gauge or of cased hole.

To summarize, accurate pressure losses in the drill pipe and bit can be reliably predicted, but pressure losses in the annulus are much more questionable. However, quite accurate pressure losses for the whole circuit can

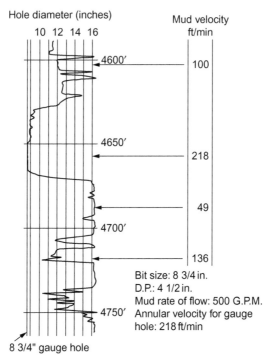

FIGURE 6.46 Typical hole enlargement in shale section. *From Hopkin, E.A., 1967. Factors affecting cuttings removal during rotary drilling. J. Petrol. Technol. 807–814. Trans AIME 240. Copyright 1967 by SPE-AIME.*

be predicted because the annular loss forms such a small percentage of the total. The results of the field tests of Fontenot and Clark (1974) support these conclusions. Fig. 6.47 compares the predicted pressure losses (curves) for a water-base mud in a well in Utah with the values measured by subsurface pressure gauges (points). Note that the agreement between the predicted and measured standpipe pressures is quite good, regardless of whether the calculations were based on constant mud properties (i.e., determined at 115°F (45°C)) or variable mud properties (i.e., determined at estimated downhole temperatures). On the other hand, agreement between predicted and measured annular pressure losses is much poorer, and there is a considerable difference between predictions based on constant and on variable mud properties. Better agreement between annular pressure losses was obtained with an invert emulsion oil mud in a well in Mississippi (see Fig. 6.48), probably because downhole viscosities of this type of mud are easier to predict (as already discussed) and because such muds are less thixotropic.

Politte (1985) developed an equation for predicting the overall circulating pressure of invert emulsion muds by normalizing downhole plastic

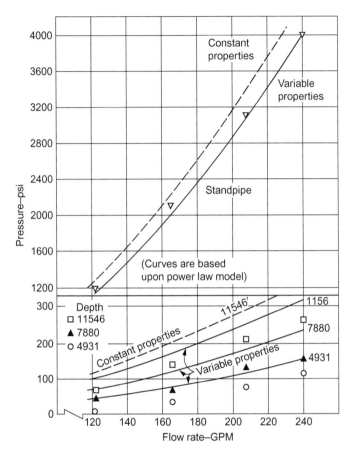

FIGURE 6.47 Comparison of measured and calculated standpipe and annular pressures for test 3—Utah well. *From Fontenot, J.E., Clark, R.K., 1974. An improved method for calculating swab, surge, and circulating pressures. Soc. Petrol. Eng. J. 451–462. Copyright 1974 by SPE-AIME.*

viscosities to changes in the viscosity of diesel oil with temperature and pressure. He also found good agreement between calculated and standpipe pressures, and that the use of downhole viscosities made little difference.

Houwen and Geehan (1986) investigated the changes in rheological properties of invert emulsion muds at temperatures up to 284°F (140°C) and pressures up to 14,500 psi (1019 kg/cm^2) in a viscometer similar to that described by McMordie (1969). They calculated the rheological parameters for the Bingham, Hershel-Bulkey, and Casson flow models, and found that the Casson flow model fitted best. For field use a knowledge of two temperature coefficients, which are specific for a given type of mud, is necessary, and the viscosity of the mud must be measured at two or more temperatures.

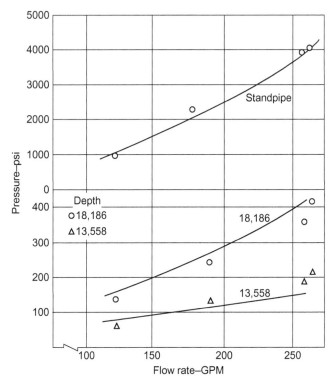

FIGURE 6.48 Comparison of measured and calculated standpipe and annular pressure for test 3—Mississippi well. *From Fontenot, J.E., Clark, R.K., 1974. An improved method for calculating swab, surge, and circulating pressures. Soc. Petrol. Eng. J. 451–462. Copyright 1974 by SPE-AIME.*

Hydraulic Calculations Made at the Well

The rigorous flow equations and testing procedures discussed in the first part of this chapter are suitable for laboratory investigations made for such purposes as relating drilling mud properties to hydraulic performance in the well, but not for hydraulic calculations made at the well. In the laboratory, ample technical personnel, sophisticated equipment, and, above all, time, are available.

In contrast, when hydraulic calculations are made at the wellsite, time is pressing: an immediate answer is required. Equipment is usually limited to the two-speed rotary viscometer. Furthermore, wellsite tests are made to solve a particular problem, and for this purpose it is necessary to know downhole conditions with some degree of certainty. As we have seen, this is by no means always possible. Under these circumstances, it is obviously desirable to use the simplest and quickest test procedures that will give meaningful answers, and to use equations no more complex than the input data justifies.

A number of methods for making wellsite hydraulic calculations have been published (Randall and Anderson, 1982; Willis et al., 1973; Walker and

Korry, 1974; Moore, 1974; Crowley, 1976; Walker, 1976a,b; API Subcommittee 13, 2010), the complexity of which varies with each author's opinion on what degree of accuracy is justifiable. After reviewing these procedures, the following procedures are recommended for making annular pressure calculations.

When drilling in formations that enlarge significantly, calculate the pressure loss in the drill pipe and in the bit nozzles, and subtract the sum from the standpipe pressure to get the pressure loss in the annulus (Moore, 1973; Carlton and Chenevert, 1974). This method is open to the objection that the pressure loss in the drill pipe and in the bit nozzles form such a large proportion of the standpipe pressure that a small error will cause a large percentage error in the annular pressure loss. However, good correlations between annular pressure losses thus calculated and subsurface pressure gauge readings were obtained in a field test by Carlton and Chenevert (1974). They found the method especially useful when circulation is broken after a round trip and when, because of temperature differences and thixotropic effects, bottomhole mud properties differ greatly from those at the surface.

The pressure loss in the open hole section alone may be determined by calculating the pressure loss in the cased hole section using the method described hereunder, and subtracting that loss from the total annular loss. The empirical equations developed by Randall and Anderson (1982) should be used for calculating pressure losses of low-solid polymer fluids in drill pipe.

For direct calculation of effective viscosity and Reynolds number either in the annulus when the annular width is known with sufficient accuracy, or in the drill pipe, the generalized power law model is to be preferred over the Bingham plastic model because it is simpler and it is valid for all flow models, Newtonian and non-Newtonian. In order to obtain meaningful results, the following points should be observed:

1. If the power law parameters, n and K, vary with the rate of shear, they should be determined at the approximate shear rate at the wall of the pipe or hole. For this purpose, it is preferable to use a multispeed viscometer, or better still a variable continuous speed viscometer which covers the annular shear rate range more evenly. If only a two-speed viscometer is available, then n and K for annular calculations should be determined from a line between the 300 rpm reading and the initial gel strength.
2. The Fanning friction factor should be determined from the appropriate value of n and the curves relating n, N'_{Re}, and f in Fig. 6.32. The procedure is just as simple as determining f from the curve for Newtonian fluids, and is obviously much more accurate.
3. The rheological parameters measured at ambient temperatures should be adjusted to estimated subsurface temperatures by regional correlations established in the laboratory for each mud type. Nonaqueous mud parameters should be adjusted for temperature and pressure.

TABLE 6.3 Factors for Converting Common Field Units into Coherent Metric Units (cgs) for Use in Hydraulic Calculations

Field Unit	Multiplied by	Equals Metric Unit
Inch (in.)	2.54	centimeters (cm)
Feet (ft)	30.48	centimeters (cm)
Gallons, US (gal)	0.0037	cubic meters (m^3)
Barrels, US (bbl)	0.159	cubic meters (m^3)
Feet per second (ft/s)	3.048	meter/second (m/s)
Feet per minute (ft/min)	0.35	meter/minute (m/min)
Pounds (mass) per square inch (psi)	68900	dynes/cm^2
Pounds (mass) per square inch per foot (psi/ft)	2262	dynes/cm^2/cm
Pounds (mass) per 100/ft^2 (lb/100 ft^2)	4.78	dynes/cm^2
Pounds (mass) per gallon (lb/gal)	0.120	grams/cm^3 (g/cm^3)
Pounds (mass) per cubic foot (lb/ft^3)	0.0162	grams/cm^3 (g/cm^3)
Centipoise (cp)	0.01	Poise
Gallons per minute (gpm)	63.09	cm^3/second (cm^3/s)
Barrels per minute (bbl/min)	0.265	m^3/minute (m^3/min)

The API Bulletin 13D (API Subcommittee 13 2010) recommends converting all data into consistent metric units before making hydraulic calculations, and converting the results of the calculations back to whatever units are preferred. This procedure is to be commended because it simplifies the calculations and lessens the possibility of mistakes. Conversion factors for units most commonly used are listed in Table 6.3.

A sample calculation, based on the above recommendations, is given below. The properties of a 6% bentonite suspension, weighted to 11 pounds per gallon (specific gravity 1.32), are used in the calculations. These properties, as determined in a multispeed viscometer, are listed in Table 6.4. The well data were assumed to be: drill pipe internal diameter, 3 in. (7.62 cm); external diameter, 3.25 in. (8.225 cm); hole diameter, 8 in. (20.32 cm); circulation rate, 300 gal/min (18,927 cm^3/s).

Determining Pressure Gradient in the Drill Pipe

1. Find the mean velocity:

$$V = Q \times \frac{4}{\pi D^2} = \frac{18977 \times 4}{\pi \times 7.62^2} = 415 \text{ cm/s}$$

TABLE 6.4 Rheological Data of a 6% Bentonite Suspension, 11 lb/gal, as Determined by a Multispeed Viscometer

RPM	Shear Rate (γ) = rpm × 1.703	Dial Reading (lb/100 ft^2)	Stress (τ) dyne/cm^2 = Dial Reading × 5.11
600	1022	30	153
300	511	19.5	100
200	340	16	82
100	170	13	66
6	10.2	7.3	37
3	5.1	7.0	36

2. Plot the shear stress versus the shear rate on log-log paper as shown in Fig. 6.49. Find n from the slope of 600–300 rpm curve:

$$n = 0.65$$

3. Find the rate of shear at the pipe wall from Eq. (6.40):

$$\gamma = \frac{3n+1}{4n} \times \frac{8V}{D}$$

$$= \frac{3 \times 0.65 + 1}{4 \times 0.65} \times \frac{8 \times 415}{7.62} = 494 \text{ s}^{-1}$$

4. Determine the effective viscosity at 494 s^{-1}. From Fig. 6.49, τ at 494 s^{-1} = 98 dyne/cm^2:

$$\mu_e = \frac{\tau}{\gamma} = \frac{98}{494} = 0.198 \text{ poise}$$

5. Determine N'_{Re} to see if flow is turbulent.
  Note: Pounds (mass) are sometimes referred to as pounds (weight); poundals are sometimes referred to as pounds (force).

$$\text{poundal} = \frac{\text{pounds (mass)}}{32.17}$$

From Eq. (6.54): $N'_{Re} = \dfrac{DV\rho}{\mu_g}$

$$= \frac{7.62 \times 415 \times 1.32}{0.198} = 21,100$$

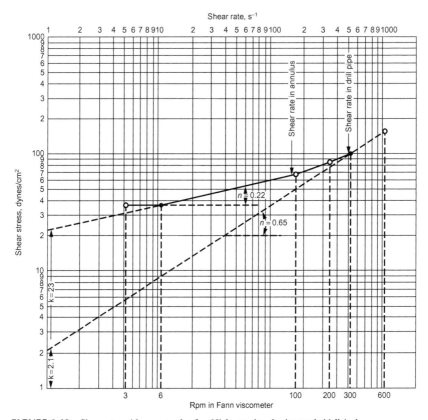

FIGURE 6.49 Shear stress/shear rate plot for 6% bentonite–barite mud, 11 lb/gal.

From Fig. 6.32, when $n = 0.65$, N'_{Re} crit. $= 2600$. Therefore, flow is turbulent.

6. Find the pressure gradient from Eq. (6.48):

$$\frac{gP}{L} = \frac{2fV^2\rho}{D}$$

From Fig. 6.32, f at N'_{Re} 21,100 = 0.005:

$$\frac{gP}{L} = \frac{2 \times 0.005 \times 415^2 \times 1.32}{7.62}$$

$$= 298 \text{ dynes/cm}^2/\text{cm}$$

$$= \frac{298}{2262} = 0.132 \text{ psi/ft}$$

Note: If flow had been laminar find $\frac{gP}{L}$ either from Eq. (6.48), where $f = \frac{16}{N_{Re}}$, or from Poiseulle's equation (Eq. 6.5):

$$\frac{gP}{L} = \frac{32V\mu_e}{D^2}$$

Determining Pressure Gradient in the Annulus

1. Find the mean velocity in the annulus:

$$V = Q \times \frac{4}{\pi(D_2^2 - D_1^2)}$$

$$= 18,927 \times \frac{4}{\pi(20.32^2 - 8.225^2)} = 69.8 \text{ cm/s}$$

2. Find n from the slope of the 100–6 rpm curve in Fig. 6.49:

$$n = 0.22$$

3. Find the shear rate at the wall of the hole from Eq. (6.45):

$$\gamma = \frac{2n+1}{3n} \times \frac{12V}{D_2 - D_1}$$

$$= \frac{(2 \times 0.22) + 1}{3 \times 0.22} \times \frac{12 \times 69.8}{12.1} = 151 \text{ s}^{-1}$$

4. Determine the effective viscosity at 151 s^{-1}.
   From Fig. 6.49, τ at 151 s^{-1} = 64 dynes/cm^2:

$$\mu_e = \frac{\tau}{\gamma} = \frac{64}{151} = 0.424 \text{ poise}$$

5. Determine N'_{Re} to see if flow is turbulent. From Eq. (6.55),

$$N'_{Re} = \frac{(D_2 - D_1)V\rho}{\mu_e}$$

$$= \frac{12.1 \times 69.8 \times 1.32}{0.424} = 2629$$

Fig. 6.32 shows N'_{Re} is critical at ∼4000 when $n = 0.22$. Therefore, flow in the annulus is laminar.

Note: If the Newtonian criterion of 2100 for the critical Reynolds number had been used, the flow would have been judged, wrongly, to be turbulent.

6. Find the pressure gradient either from Eq. (6.48),

$$\frac{gP}{L} = \frac{2fV^2\rho}{D_2 - D_1}, \quad \text{where} \quad f = \frac{24}{N'_{Re}} = 0.0091:$$

$$\frac{gP}{L} = \frac{2 \times 0.0091 \times 69.8^2 \times 1.32}{12.1} = 9.7 \text{ dynes/cm}^2/\text{cm}$$

or from Poiseulle's equation,

$$\frac{gP}{L} = \frac{48V\mu_e}{M^2} = \frac{48 \times 69.8 \times 0.424}{(20.32 - 8.225)^2} = 9.7 \text{ dynes/cm}^2/\text{cm}$$

$$= \frac{9.7}{2.262} = 0.00428 \text{ psi/ft}$$

To show the importance of evaluating n at the shear rate prevailing at the wall of the hole in annular calculations, suppose the above calculations were made with the value of n given by the 600 and 300 rpm readings, i.e., $n = 0.65$.

$$\gamma = \frac{(2 \times 0.65) + 1}{3 \times 0.65} \times \frac{12 \times 69.8}{12.1}$$

$$= 81.6 \text{ s}^{-1}, \text{ instead of } 151 \text{ s}^{-1}$$

and τ at 81.6 s^{-1} = 56 dynes/cm^2:

$$\mu_e = \frac{56}{81.6} = 0.686 \text{ poise, instead of } 0.424 \text{ poise.}$$

Thus, there would have been a 46% error in the rate of shear, and a 62% error in the effective viscosity.

The shear rates in the drill pipe usually approximate to those prevailing in the rotary viscometer between 600 and 300 rpm, and the shear rates in the annulus usually approximate to those between 100 and 6 rpm in the viscometer. If the rate of shear in the well thus calculated falls below the shear rate range in the viscometer at which n was determined, then another calculation should be made using the value of n indicated by a lower shear rate range. Similarly, if the calculated rate falls above the viscometer range, then a higher range should be used to determine n. Of course, many muds will exhibit linear or near linear plots, in which case it is possible to use the value of n given by the 600 to 300 rpm range in the viscometer without undue error.

For maximum accuracy the pressure loss in the various sections of the annulus should be calculated separately according to their annular width. The total pressure loss in the well is given by the sum of the pressure losses in the drill pipe, through the bit nozzles, and in the annulus. The pressure loss through the bit nozzles is given by Schuh (1964):

$$P = \frac{\rho}{2g}\left(\frac{Q}{CA}\right)^2 \quad (6.58)$$

where C is the nozzle constant, which may be taken as 0.95, and A is the total area of the nozzles.

The total annular pressure gradient, static plus hydraulic, is commonly expressed in terms of the *equivalent circulating density*, $\overline{ecd}$:

$$\overline{ecd} = \rho + \frac{gP}{L} \quad (6.59)$$

Thus in the hydraulic calculation given above, the predicted pressure gradient was

$$\frac{gP}{L} = 9.7 \text{ dynes}/\text{cm}^2/\text{cm}$$

and the mud density was 1.32 g/cm³.
Therefore,

$$\overline{ecd} = 1.32 + \frac{9.7}{980}$$

$$= 1.33 \text{ g}/\text{cm}^3$$

$$1.33 \times 8.345 = 11.09 \text{ lb}/\text{gal}.$$

The total pressure exerted by the mud at a given depth may be obtained by multiplying the equivalent circulating density by the depth. Thus, at a depth of 1000 m:

$$P = 1.33 \times \frac{1000 \times 100}{1000} = 133 \text{ kg}/\text{cm}^2$$

or at a depth of 10,000 ft

$$P = 1.33 \times 10,000 \times 0.433 = 5758 \text{ psi}$$

(0.433 is the pressure gradient of water in psi/ft).

RHEOLOGICAL PROPERTIES REQUIRED FOR OPTIMUM PERFORMANCE

The drilling engineer controls mud properties to

1. Minimize pumping costs
2. Maximize bit penetration rates
3. Lift drill cuttings efficiently

4. Lower swab and surge pressures, and lower pressure required to break circulation
5. Separate drill solids and entrained gas at the surface
6. Minimize hole erosion

The rheological requirements for these diverse purposes often conflict, so that it is necessary to optimize the mud properties in order to obtain the best overall performance. The properties required for each purpose are discussed separately below.

Pumping Capacities

Pump capacity must be large enough to maintain a rising velocity in the widest section of the annulus sufficient to lift the drill cuttings efficiently. The pump horsepower required to do this will depend almost entirely on flow conditions in the drill pipe and through the bit nozzles. The pressure loss through the bit nozzles is not affected by the rheological properties, and the pressure loss in the drill pipe is only affected to a minor extent because, there, flow is usually turbulent. As far as rheology is concerned, there are only two possible ways to lower the pressure loss in the drill pipe. One is to increase the carrying capacity of mud (as discussed later in this chapter) so that the circulation rate can be lowered. The other is to use a low-solids polymer mud, whose friction reducing properties will minimize turbulent pressure losses.

Effect of Mud Properties on Bit Penetration Rate

This subject is fully discussed in Chapter 10, Drilling Problems Related to Drilling Fluids. Suffice it to say here that maintaining the viscosity at a low value is a major factor in promoting fast penetration rates (Eckel, 1967). The relevant viscosity is the effective viscosity at the shear rate prevailing at the bit, which is of the order of 100,000 reciprocal seconds.

Hole Cleaning

Before discussing the optimum rheological properties required for lifting drill cuttings, it is first necessary to review the basic mechanisms involved. The rate at which a rising column of fluid will carry solid particles upwards depends on the difference between the velocity of the fluid and the tendency of the particle to fall through the fluid under the influence of gravity. In a still liquid, a falling particle soon acquires a constant downward velocity, known as the *terminal settling velocity*, which depends on the difference in density between the particle and the liquid, the size and shape of the particle, the viscosity of the liquid, and whether or not the rate of fall is sufficient to cause turbulence in the immediate vicinity of the particle.

In the case of spheres falling through a Newtonian liquid, the Reynolds number is given by

$$N_{Re,p} = \frac{d_p v \rho_f}{\mu} \qquad (6.60)$$

where d_r is the diameter of the sphere, v_t the terminal settling velocity, ρ_f the density of the fluid, and μ its viscosity. Under laminar flow conditions the terminal flow velocity is given by Stokes' law:

$$v_t = \frac{2gd_p^2}{36} \times \frac{\rho_p - \rho_f}{\mu} \qquad (6.61)$$

where ρ_p is the density of the particle. Under turbulent flow conditions the terminal settling velocity is given by Rittinger's formula (Piggott, 1941):

$$v_t = 9\sqrt{\left(\frac{d_p(\rho_p - \rho_f)}{\rho_f}\right)} \qquad (6.62)$$

Predicting the terminal velocity of drill cuttings is much more difficult. For one thing, there is the wide range of particles sizes and the particles have irregular shapes; for another, there is the non-Newtonian nature of most drilling fluids.

Terminal velocities in turbulent fall are somewhat easier to predict because the rate of fall is not affected by the rheological properties. Walker and Mayes (1975) proposed the following equation for flat particles falling face down (which is the normal orientation for turbulent fall):

$$v_t = \sqrt{\frac{2gd_p(\rho_p - \rho_f)}{1.12\rho_f}} \qquad (6.63)$$

Terminal velocities predicted by this equation correlated well with experimental data obtained with artificial cuttings of uniform size and shape.

However, the simplest procedure is to determine the terminal settling velocity of the drill cuttings of interest by direct experiment. When determining settling velocities in mud, a layer of a transparent liquid of greater density than the mud should be placed at the bottom of the settling column so that the particles may be seen when they reach the bottom. Fig. 6.50 shows some settling velocities of shale drill cuttings falling through water (Hopkin, 1967).

In a drilling well, cuttings fall under still settling conditions whenever circulation is stopped. In a Newtonian fluid the settling velocity is finite, no matter how viscous the fluid, but, because of the enormous length of the settling column, only a small proportion of the cuttings reach the bottom unless the viscosity approaches that of water. In a non-Newtonian fluid the settling velocity depends on the difference between the stress (τ) created by the difference in gravity ($\rho_p - \rho_f$) and the gel strength of the mud (S). When $\tau < S$,

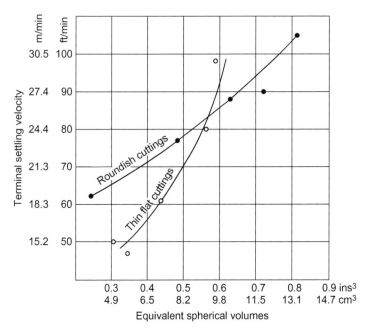

FIGURE 6.50 Terminal settling velocity of shale cuttings in water. *From Hopkin, E.A., 1967. Factors affecting cuttings removal during rotary drilling. J. Petrol. Technol. 807–814. Trans AIME 240. Copyright 1967 by SPE-AIME.*

then v_t is zero, and the cutting is suspended. The initial gel strength of most muds is too low to suspend large cuttings, and suspension depends on the increase of gel strength with time.

In a rising column of fluid, a particle will move upward if the velocity of the fluid is greater than the settling velocity of the particle. However, the particle slips in the rising column, so that the upward velocity of the cutting is less than the annular velocity. Sifferman et al. (1974) defined hole-cleaning efficiency in terms of a *transport ratio*, derived as follows:

$$v_c = v_a - v_s$$

where v_c is the net rising velocity of the cutting, v_a is the annular velocity, and v_s is the slip velocity of the cutting. Dividing both sides of the equation by v_a gives

$$\frac{v_c}{v_a} = \text{transport ratio} = 1 - \frac{v_s}{v_a}$$

Sifferman et al. (1974) measured the decrease in the transport ratio of artificial drill cuttings with increase in annular velocity under simulated well conditions. Fig. 6.51 shows that the ratio tends to level off as the annular velocity increases, and that, at a given velocity, the ratio is strongly dependent on the

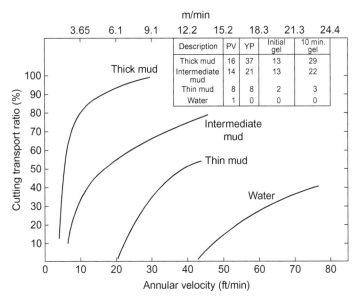

FIGURE 6.51 Cutting removal at low annular velocities (medium cutting, no rotation, 12–3½ inch annulus, 12 lb/gal mud). Artificial cutting 1/8 in. × 1/4 in. × 1/8 in. (0.317 × 0.635 × 0.317 cm) Transport ratio = V_c/V_a. From Sifferman, T.R., Myers, G.M., Haden, E.L., Wall, H.A., 1974. Drill-cutting transport in full scale vertical annuli. J. Petrol Technol. 1295–1302. Copyright 1974 by SPE-AIME.

thickness of the mud. Similar tests, but using actual drill cuttings, by Hussaini and Azar (1983) showed that the rheological properties of the mud had a significant effect on cutting transport only at annular velocities less than about 120 ft/min. Since annular velocities are usually designed to be greater than 120 ft/min, assuming the hole to be to gauge, it follows that rheological properties affect cutting transport only in enlarged sections of the hole.

Hussaini and Azar's tests were conducted with muds having apparent viscosities ranging from 20 to 40 cp. Zeidler (1981) found that annular velocities of 164 ft/min were necessary to clean holes drilled in the Swan Hill and Ferrier fields in Canada with clear water. Evidently, Hussaini and Azar's results do not apply to very low viscosities.

One reason for poor transport efficiency was shown experimentally by Williams and Bruce (1951) to be that flat cuttings tend to recycle locally, as shown in Fig. 6.52. This recycling action is presumed to be caused by the parabolic shape of the laminar velocity profile, which subjects a flat cutting to unequal forces (see Fig. 6.53). In consequence, they turn on edge and migrate to the sides of the annulus, where they descend some distance before migrating back towards the center. The downward descent is caused partly by the low velocity prevailing at the walls, and partly by the edgeways orientation of the cutting.

FIGURE 6.52 Discs recycling in the annulus, drill pipe stationary. *From Williams, C.E., Bruce, G.H., 1951. Carrying capacity of drilling fluids. Trans. AIME 192, 111–120. Copyright 1951 by SPE-AIME.*

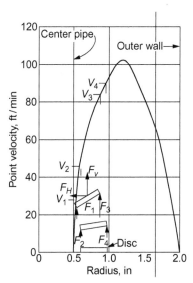

FIGURE 6.53 Unequal forces on flat disc when mud flow is laminar. *From Williams, C.E., Bruce, G.H., 1951. Carrying capacity of drilling fluids. Trans. AIME 192, 111–120. Copyright 1951 by SPE-AIME.*

The Rheology of Drilling Fluids Chapter | 6 221

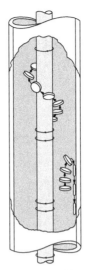

FIGURE 6.54 Helical motion of discs when the drill pipe is rotating. *From Williams, C.E., Bruce, G.H., 1951. Carrying capacity of drilling fluids. Trans. AIME 192, 111–120. Copyright 1951 by SPE-AIME.*

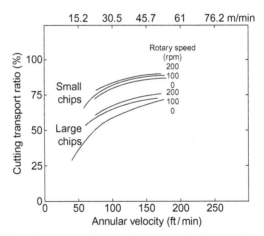

FIGURE 6.55 Effect of rotary speed on cutting transport (medium cutting, 8 × 4 in. annulus, 12 lb/gal mud). Small cutting—1/16 in. × 1/8 in. × 1/3 in. (0.159 × 0.317 × 0.317 cm). Large cuttings—1/4 in. × 1/2 in. × 1/8 in. (0.635 × 1.27 × 0.317 cm). *From Sifferman, T.R., Myers, G.M., Haden, E.L., Wall, H.A., 1974. Drill-cutting transport in full scale vertical annuli. J. Petrol Technol. 1295–1302. Copyright 1979 by SPE-AIME.*

In general, rotation of the drill pipe improves the transport ratio because it imparts a helical motion to the cuttings in the vicinity of the drill pipe (see Fig. 6.54), but the effect was shown by Sifferman et al. (1974) to be rather small (see Fig. 6.55). Theoretically, turbulent flow

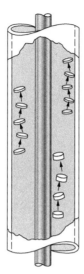

FIGURE 6.56 Discs transported in turbulent flow (center pipe stationary). *From Williams, C.E., Bruce, G.H., 1951. Carrying capacity of drilling fluids. Trans. AIME 192, 111–120. Copyright 1951 by SPE-AIME.*

should improve the transport ratio because the flatter profile eliminates the turning moment (Fig. 6.56), but experimental evidence on this point is not consistent (Sifferman et al., 1974; Williams and Bruce, 1951; Hall et al., 1950; Zeidler, 1972), possibly because of differences in experimental conditions, such as the size and shape of the cuttings. Recent tests in a model wellbore by Thomas et al. (1982) showed that increasing rotary speed increases cutting transport at low-rising velocities, but the effect becomes negligible at high velocities.

Optimum Annular Velocity

Although any velocity greater than the settling velocity of the largest cutting will theoretically lift all the cuttings to surface eventually, too low an annular velocity will lead to an undesirably high concentration of cuttings in the annulus. Because of slip, the concentration of cuttings depends on the transport ratio as well as the volumetric flow rate and the rate of cuttings generation by the bit. Experience has shown that cutting concentrations of more than about 5% by volume cause tight hole, or stuck pipe, when circulation is stopped for any reason (Piggott, 1941; Hopkin, 1967). Fig. 6.57 shows the theoretical upward cutting velocity necessary to maintain a cutting concentration of less than 5% for several hole geometries. Fig. 6.58 shows the minimum annular velocities required to keep the cuttings concentration less then 4%, as calculated by Zamora (1974),

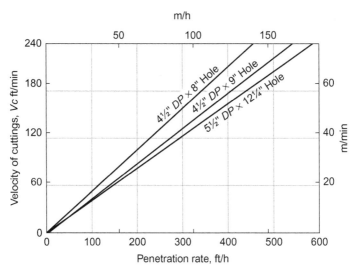

FIGURE 6.57 Rising velocity of cuttings required to keep concentration of cuttings in the annulus less than 5%. Example: With 4½″ DP in 8″ hole and a penetration rate of 100 ft/h, velocity of cuttings must be 55 ft/min. *From Hopkin, E.A., 1967. Factors affecting cuttings removal during rotary drilling. J. Petrol. Technol. 807–814. Trans AIME 240. Copyright 1967 by SPE-AIME.*

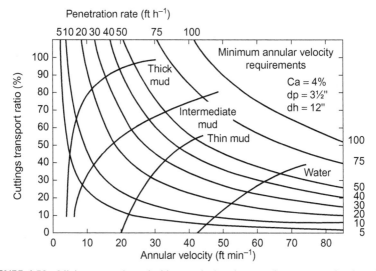

FIGURE 6.58 Minimum annular velocities required to keep cutting concentration less than 4%. Example: with a thin mud and a penetration rate of 50 ft/h, an annular velocity of 42 ft/min is necessary. *From Zamora, M., 1974. Discussion of the paper by Sifferman et al. J. Petrol. Technol. 1302. Copyright 1974 by SPE-AIME.*

for the various muds shown in Fig. 6.58. Zeidler (1974) developed semi-empirical equations for predicting the rising particle velocity necessary to maintain equilibrium between cutting generation and cutting transport, but his work was based on the behavior of drill cuttings in a clear resin fluid. However, subsequent work by others (Hussaini and Azar, 1983; Thomas et al., 1982) has shown that Zeidler's equations are valid for drilling fluids under most conditions.

Since penetration rate decreases with increase in viscosity, it is preferable to maintain adequate hole cleaning by raising annular velocities rather than mud viscosities. However, annular velocities are limited by the following considerations:

1. The rate of increase of the transport ratio falls off increasingly at high annular velocities (as shown in Fig. 6.51).
2. High circulation rates involve disproportionately high pumping costs because the pressure loss in the drill pipe increases with the square of the velocity in turbulent flow.
3. High annular velocities may cause hole erosion. However, high flow rates per se will not cause hole erosion in competent formations, even if the flow is turbulent. The shear stresses exerted by the fluid on the hole walls are in the order of *pounds per hundred square feet*, whereas the shear strengths of competent rocks and shales is in the order of *thousands of pounds per square inch* (see Table 10.3). Thus, in some geologic regions, holes can be drilled with water in highly turbulent flow without causing significant hole enlargement. Unfortunately, in most places where oil wells are drilled the formations are either fractured by tectonic movements, or gravely weakened by physicochemical reactions with the mud. Under such circumstances, hole enlargement will increase with increase in annular velocity (Darley, 1969).

It follows from the above considerations that the optimum annular velocity and mud viscosity depend on whether or not a gauge hole can be maintained. A gauge hole can be kept clean either by water or by minimum viscosity muds, but if the hole enlarges significantly, the fluid velocity in the enlarged sections decreases, so there may be a wide variation in velocity between the gauge and enlarged sections (see Fig. 6.46). Increasing the volumetric flow rate to obtain velocities sufficient to clean the enlarged sections may result in excessive velocities in the gauge sections. Under these conditions the rheological properties of the mud must be adjusted to increase the transport ratio.

Optimum Rheological Properties for Hole Cleaning

On general principles, a mud with predominantly structural viscosity—as indicated by a high ratio of yield point to plastic viscosity, or a low flow

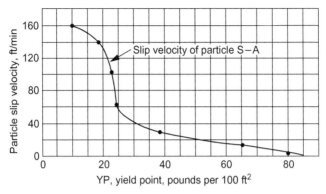

FIGURE 6.59 Particle slip velocity versus yield point for 0.954-in. OD sphere. *From Hopkin, E.A., 1967. Factors affecting cuttings removal during rotary drilling. J. Petrol. Technol. 807–814. Trans AIME 240. Copyright 1967 by SPE-AIME.*

behavior index, n—is desirable for hole-cleaning purposes. Such a mud will be a shear thinning mud, so that the effective viscosity will increase in the enlarged sections, where fluid velocities are low, and decrease in gauge hole sections, where fluid velocities are high.

A mud with a high structural viscosity component would be expected to lift cuttings more efficiently than a Newtonian or near Newtonian fluid, but experimental evidence on this point is contradictory. Hopkin (1967) found that cuttings slip velocity correlated better with yield point than with any other rheological parameter (see Fig. 6.59). Field experience reported by O'Brien and Dobson (1985) seems to confirm Hopkin's results: they found that troublesome cavings could not be cleaned out of large diameter holes in Oklahoma unless the yield point was increased to 30–40 lb/100 ft^2. They also found, when drilling in granite sections, that the size of the cuttings increased as the yield point was increased. For example, the maximum particle dimension was 0.5 in. when the yield point was 19 lb/100 ft^2, 1.1 in. when it was 55, and 1.6 in. when it was 85. On the other hand, Sifferman et al. (1974) found that muds with a yield point of about 20 lb/100 ft^2 (102 dynes/cm^2) had no better transport ratios than Newtonian oils of equivalent viscosity. In one respect, however, their experiments did not fully simulate well conditions: in their tests the mud was pumped into the base of the column by a centrifugal pump, which must have reduced the structural viscosity to a very low value, as occurs at the bit. In the well, the structural viscosity rebuilds as the mud rises in the annulus, but there would have been little time for it to do so in the comparatively short experimental annulus. Hussaini and Azar (1983) found that the YP/PV ratio, the effective viscosity, the yield point, and the initial gel were the controlling rheological factors (but, as previously mentioned, these factors had no significant effect at rising velocities above 120 ft/min).

It has been suggested (Walker 1971) that a mud with a low value of n would also be advantageous because it would have a comparatively flat velocity profile (see Fig. 6.24), and therefore, low shear rates and high local viscosities would prevail over the greater part of the annular radius. There appears to be no good experimental evidence—in which n was varied and all other factors held constant—to support this theory. One would certainly expect flat profiles to be an advantage in so far as they reduce recycling action observed by Williams and Bruce (1951), but, on the other hand, reducing n reduces the mean effective viscosity, since $\mu_e = K(\gamma)^{n-1}$.

Inclined Holes

The behavior of cuttings in high angle deviated wells is very different from that in vertical or low-angle wells (Tomren, 1979; Iyoho, 1980; Becker, 1982; Tomren et al., 1986; Okrajni and Azar, 1986; Gavignet and Sobey, 1986; Martin et al., 1987). In a vertical hole, slip velocity acts parallel to the axis of the hole. In inclined holes, slip velocity has two components, axial and radial. The axial component decreases as the angle of the hole increases, and is zero in a horizontal hole. On the other hand, the radial component increases with the angle of the hole. Consequently, cuttings tend to form on the low side of high-angle holes. High annular mud velocities are necessary in order to limit cutting bed formation.

Extensive laboratory tests under simulated well conditions by Okrajni and Azar (1986) showed that:

1. In holes with inclinations up 45 degrees, laminar flow provided better cutting transport than turbulent, whereas turbulent was better at inclinations above 55 degrees. There was little difference between the two regimes at inclinations between 45 and 55 degrees.
2. When laminar flow was maintained, the higher the yield value of the mud the better the cuttings transport at inclinations below 45 degrees, but yield value had little or no effect at inclinations above 55 degrees.
3. High YP/PV ratios provided better transport at all hole inclinations.
4. The effects of yield value and YP/PV ratios are greater at low annular velocities.
5. As would be expected, the rheological properties of a mud in turbulent flow had little effect on cuttings transport.

Note that although turbulent flow would provide better cuttings transport at high hole inclinations, it could not be used in many applications because of its adverse influence on borehole enlargement. Under such conditions, the highest annular velocity that will not cause turbulent flow should be used. A mud with a high effective viscosity will increase the critical Reynolds number, and thus enable a higher annular velocity to be maintained without causing turbulent flow.

Gavignet and Sobey (1986) developed a model, based on a momentum balance, which may be used to calculate a critical flow rate above which a bed of cuttings will not form. Its value is dependent on pipe, hole, and particle size (see Fig. 6.60A and B). The eccentricity of the drill pipe has a major effect. Even when the flow is turbulent, a bed of cuttings will form at flow rates below a critical value. Calculations based on this model are in fairly good agreement with the results of experiments with water and carbopol by Iyoho (1980) (see Fig. 6.60A and C).

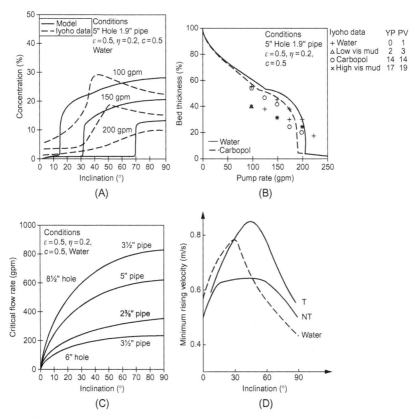

FIGURE 6.60 (A) Effect of pump rate on bed formation and comparison with data of Iyoho (1980). (*From* Gavignet, A.A., Sobey, I.J., 1986. A model for the transport of cuttings in highly deviated wells, SPE Paper 15417, Ann. Tech. Conference, October 5–8, New Orleans, LA. *Copyright 1986 by SPE-AIME.*) (B) Effect of hole size and pipe size and bed formation. (*From* Gavignet, A.A., Sobey, I.J., 1986. A model for the transport of cuttings in highly deviated wells, SPE Paper 15417, Ann. Tech. Conference, October 5–8, New Orleans, LA. *Copyright 1986 by SPE-AIME.*) (C) Comparison of model with data of Iyoho (1980). (*Copyright 1986 by SPE-AIME.*) (D) Minimum rising velocity required to transport cuttings vs inclination and rheology. T = thixotropic fluid; NT = nonthixotropic fluid. After Martin, M., Georges, C., Bisson, P., Konirsch, O., 1987. Transport of cuttings in directional wells, SPE/IADC Paper 16083, Drill. Conference, March 15–18, New Orleans, LA. (*Copyright 1987 by SPE-AIME.*)

Martin et al. (1987) investigated the influence of mud properties, flow rate, and drill string rotation in cylindrical tubes and annuli in the laboratory, and checked their results against field data. Their results showed that

1. The average minimum rising velocity required to transport the cuttings (V_{lim} in Fig. 6.60D) reaches a maximum at inclinations between 30 and 60 degrees.
2. Thixotropic muds are highly undesirable at inclinations above 10° (see Fig. 6.60D) because they almost immobilize the layer of cuttings next to the wall.
3. High apparent viscosities and high YP/PV ratios favor cuttings transport in vertical wells, but are unfavorable at inclinations above 10 degrees. This finding is at variance with Okrajni and Azar's results discussed above.
4. High mud densities enhance cuttings transport.
5. Drill string rotation helps cutting transport because it knocks cuttings from the layer back into the mud stream.

The minimum rising velocities reported by Martin et al. are generally less than those deduced from data in the literature.

Release of Cuttings and Entrained Gas at the Surface

Structural viscosity impedes the separation of drilled solids and entrained gas at the surface. In a quiescent fluid, particles will not fall nor gas bubbles rise, unless the stress created by the difference in density between the particles or bubbles and the fluid is greater than the gel strength of the mud. The high shear rates prevailing in mechanical separators and degassers promotes the release of solids and gas by reducing structural viscosity.

Transient Borehole Pressures

So far, we have discussed pressure arising from steady state flow, but various transient pressures that must be minimized occur during the normal drilling cycle. These transient pressures affect the safety of the well. Fig. 6.61 depicts typical transient pressures in terms of equivalent mud density of a mud whose hydrostatic density is 11.8 lb/gal (specific gravity 1.41) (Clark, 1956). A positive pressure surge occurs each time a stand of drill pipe is run in the hole, because the pipe acts like a loose-fitting piston, forcing mud out of the hole. When the bit reaches bottom, the pressure required to break circulation causes another surge. The greatest surge in the drilling cycle occurs when reaming down rapidly with the pump on, prior to making a connection. Finally, negative surge pressures (or swab pressures as they are usually called) occur when pulling pipe, because of a swabbing effect. It is not necessary for fluid to be physically swabbed out of the annulus, or carried out inside the drill pipe, but the pressure reduction is greater when such actions occur.

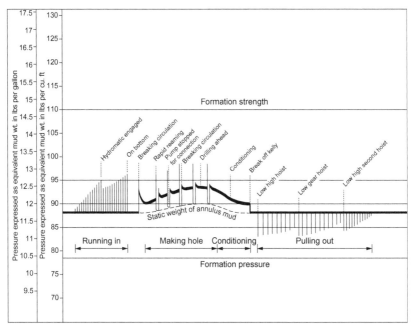

FIGURE 6.61 Transient pressures occurring in the course of a normal drilling cycle. *From Clark, 1956. Courtesy of API.*

In normal or moderately geopressured formations, such as those shown in Fig. 6.61, transient pressures are not hazardous, but, in highly geopressured formations, pore pressures are high, allowing only a small operating margin between the mud density required to control formation fluids and that which would fracture the formation (see Chapter 10, Drilling Problems Related to Drilling Fluids). Under this condition, swab pressures may be sufficient to cause blowouts, and positive surges may cause induced fracturing with consequent loss of circulation. Correlations of swab pressures with actual blowouts, and surge pressures with loss of circulation, have been firmly established by studies of case histories (Cannon, 1934; Goins et al., 1951).

Burkhardt (1961) identified three factors contributing to pressure surges: the pressure required to break the gel, the acceleration or deceleration of the mud, and the viscous drag of the mud. In some field tests, he measured pipe velocities and corresponding subsurface pressures. His results (Fig. 6.62) showed that the maximum positive surge pressure occurred at the maximum pipe velocity and, therefore, that viscous drag caused the peak surge pressures.

Burkhardt derived equations relating surge pressures to pipe velocity, based on the premise that the difference in pressure caused by moving a pipe through a stationary liquid is the same as that caused by flowing the liquid through a stationary pipe at the same velocity. Equations for pressure surges

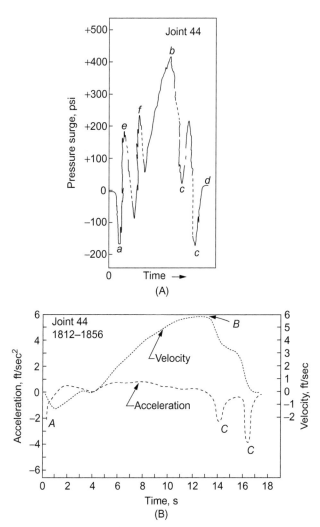

FIGURE 6.62 (A) Typical pressure surge pattern measured as a joint of casing was lowered into a wellbore. (B) Typical pipe velocities and accelerations measured as a joint of casing was lowered into a wellbore. Note: Maximum surge pressure, point b in Fig. 6.62A, occurred at peak pipe velocity, point B in Fig. 6.62B. *From Burkhardt, J.A., 1961. Wellbore pressure surges produced by pipe movement. J. Petrol. Technol. 595–605. Trans. AIME 222. Copyright 1961 by SPE-AIME.*

in a drilling well are complicated by the fact that both the pipe and the liquid move. Thus, the mud has one velocity with respect to the moving pipe, and another with respect to the stationary wall of the hole. For convenience, Burkhardt calculated a single effective velocity.

Another complication is that if the drill pipe is open, part of the mud flows up the annulus and part flows up the drill pipe, in proportion to the relative

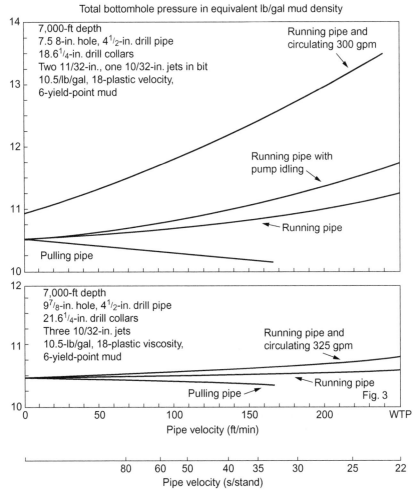

FIGURE 6.63 Predicted pressure surges for alternative well programs. Pressure surges will be much less if 9⅞ hole is drilled. *From Schuh, F.J., 1964. Computer makes surge-pressure calculations useful. Oil Gas J. 96–104. Courtesy of Oil and Gas J.*

resistance offered by each path. For the same effective mud velocity, the pressure—velocity relationship is the same for positive and negative pressure surges.

Exact solutions of these equations require the use of a computer. Programs have been written by Burkhardt (1961), using the Bingham plastic flow model; by Schuh (1964), using the power law model; and by Fontenot and Clark (1974), using the power law and downhole mud properties. Although such programs are unsuitable for wellsite engineering, they are useful for well planning; for example, in determining the hole size necessary for bringing swab and surge pressures within safe limits, as shown in Fig. 6.63.

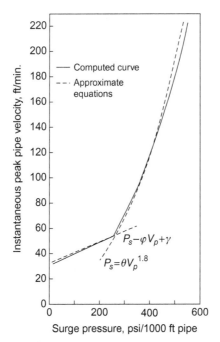

FIGURE 6.64 Rigorous and approximate theoretical relationship between the viscous drag pressure surge and pipe velocity. *From Burkhardt, J.A., 1961. Wellbore pressure surges produced by pipe movement. J. Petrol. Technol. 595–605. Trans. AIME 222. Copyright 1961 by SPE-AIME.*

Burkhardt derived approximate solutions for his equations, which are given here to illustrate the influence of the mud's rheological parameters. For laminar flow the equation is

$$P_s = \frac{B\mu_p V_p + \tau_0}{0.3(D_2 - D_e)} \quad (6.64)$$

and for turbulent flow it is

$$P_s = A\mu_p^{0.21} \rho^{0.806} V_p^{1.8} \quad (6.65)$$

where V_p is the velocity of the pipe in feet per minute; μ_p is the plastic viscosity in centipoises; ρ is the density in pounds per gallon; τ_0 is the yield point in pounds per 100 ft^{-2}; D_2 is the diameter of the hole in inches; A, B, and D_e are form-fitting parameters; and P_s is the surge pressure in pounds per square inch per 1000 ft of pipe. A, B, and D_e are different for open and closed pipe. Exact and approximate solutions are compared in Fig. 6.64.

Although the viscous drag of the mud determines the maximum surge pressure, the downhole gel strength of the mud is an important factor because it determines the value of τ_0 in Eq. (6.64). Burkhardt used the three-min gel strength as measured at the surface for τ_0 in his field tests. In practice, downhole gel strengths will vary quite considerably from this value, depending on the depth of the hole, the temperature gradient, and the thixotropic behavior of the mud. When the drill pipe is started back in the hole after a trip, the mud will have been at rest for several hours. Since the mud at the bottom of the hole will have been undisturbed for much longer than that near the surface, and furthermore will have been subjected to higher temperatures, there will be a gradual increase in gel strength with depth.

Each time a stand is run in, the column of mud above the bit is disturbed for some 15−30 s, but the gel partially rebuilds while the next stand is being made up. When a large number of stands have been run, the gel strength of the mud near the surface will have been reduced considerably, whereas the gel strength of the undisturbed mud at the bit will be that which has developed since the bit was pulled past the point on the way out of the hole. Thus, when the bit reaches bottom, the average gel strength of the mud column will be much greater than the gel strength of the mud coming out of the hole at the surface. Indeed, even after circulation is started, the average effective viscosity remains high until the mud from the bottom of the hole reaches the surface. For example, Fig. 6.65 shows that the annular pressure loss, as measured by Carlton and Chenevert (1974) in field tests, remained above normal until returns from the bottom reached the surface, marked by the rise in gel strength about 95 min after circulation started.

The maximum positive surge pressure to which a formation at a given depth will be subjected occurs when the bit reaches that depth. The value of τ_0 for use in Eq. (6.64) will lie somewhere between the actual gel strength of the mud at that depth and the initial gel strength of the mud coming out of the hole. The undisturbed gel strength is best evaluated by heating a sample of the mud in a closed container for the appropriate time and at the appropriate temperature, and then determining the gel strength with a shearometer.

Swab pressures are lower than positive surge pressures, partly because pipe speeds are lower when pipe is being pulled, and partly because τ_0 is at a minimum when starting out of the hole. Nevertheless, field tests by Cannon (1934) showed that gel strength as measured by a bottomhole pressure gauge was a more important factor than viscosity in determining swab pressures (see Fig. 6.66).

Eq. (6.65) shows that under turbulent flow conditions the mud properties will have little effect on surge pressures, which will depend mainly on pipe speed. Fig. 6.64 indicates that with typical mud and well conditions, turbulent flow will commence at pipe speeds a little above

234 Composition and Properties of Drilling and Completion Fluids

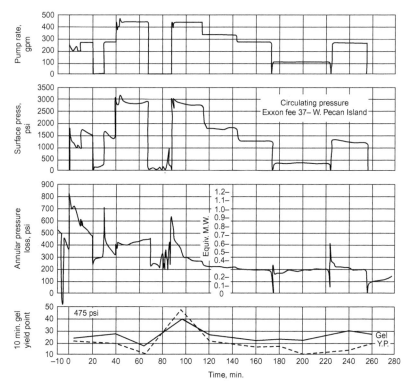

FIGURE 6.65 Decrease in annular pressure loss with time after breaking circulation. Annular pressure loss remained high until gelled mud from the bottom of the hole was circulated out of the hole at 95 min. *From Carlton, L.A., Chenevert, M.E., 1974. A new approach to preventing lost returns, SPE Paper No. 4972, Annual Meeting, October 6–9, Houston. Copyright 1974 by SPE-AIME.*

200 ft/min (60 m/min). This value for the critical pipe speed is supported by the results of field tests by Goins et al. (1951), which showed a marked increase in positive surge pressures at pipe velocities above about 200 ft/min, as shown in Fig. 6.67.

Lal (1984a,b) postulated a flow model for swab and surge pressures based on unsteady flow and a compressible fluid. A computer program enables the change in pressure with time to be predicted. The program is quick to run so that maximum safe pipe speed for a specified mud pressure overbalance can be determined while actually tripping.

Sample computer runs based on various hypothetical conditions indicated that the two variables having a major effect on surge pressures were maximum pipe velocity and the yield point of the mud. Pipe acceleration, mud density, and plastic viscosity had little effect. The effect of thixotropy on downhole mud properties was not taken into account. Lal does not provide a comparison between calculated pressures and actual downhole pressures as

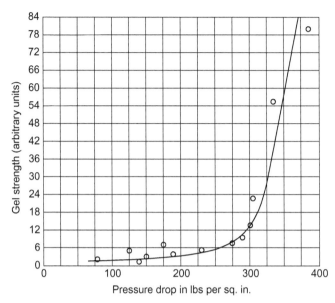

FIGURE 6.66 Relation of gel strength of mud to pressure drop due to swabbing in 7-in. casing. *Data by Cannon, G.E., 1934. Changes in hydrostatic pressure due to withdrawing drill pipe from the hole. API Drill. Prod. Prac. 42–47. Courtesy of API.*

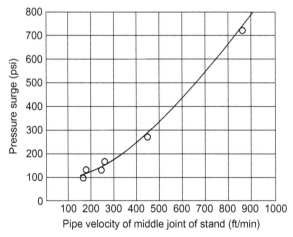

FIGURE 6.67 Variation of positive surge pressure with pipe velocity. High surge pressures above pipe velocities of 200 ft/min are caused by turbulent flow. *From Goins, W.C., Weichert, J.P., Burba, J.R., Dawson, D.D., Teplitz, A.J., 1951. Down hole pressure surges and their effect on loss on circulation. API Drill. Prod. Prac. 125–131. Courtesy of APL Drill. Prod. Prac., 1951.*

measured in field tests. A dynamic swab/surge model that takes into account the elasticity of the pipe, formation, and cement has been developed by Mitchell (1987). Mud properties as a function of temperature and pressure are also considered.

PRACTICAL DRILLING FLUID RHEOLOGY

Techniques for controlling the rheological requirements in the field are beyond the scope of this book. The exact parameter adjustment for optimum drilling and to minimize nonproductive time (NPT) depends on the drilling situation at the time. The drilling fluid technologist observes the operation, the annular flow, the cuttings amount and shape on the shale shaker, bottoms up after a trip, and any anomalies when making connections or during a trip. The technologist then tests the mud and recommends property changes as appropriate. In general, the following rheological guidelines are recommended.

Water-Based Muds

In fresh water, clay based systems high YP/PV ratios are best obtained by lowering the plastic viscosity rather than by increasing the yield point. Maintain the lowest possible plastic viscosity by mechanical removal of drilled solids at the surface. The yield point is controlled by adding or withholding chemical thinners when drilling in colloidal clays, and by adding bentonite when drilling in other formations. The gels need to be kept low and nonprogressive. Also, enhanced control of annular and suspension viscosities can be achieved with polymer additives such as xanthan gum and cm-starch. In saltwater muds, including seawater, annular and suspension viscosities are best controlled by polymer additives and keeping the solids content as low as possible. The Yield Power Law model is best used to calculate annular hydraulics and suspension properties, i.e., LSRV.

Nonaqueous Drilling Fluids

Plastic viscosity, yield point ratios are not an accurate guideline for manipulated the fluid properties for hole cleaning or suspensions. Nonaqueous fluids more closely follow the Power Law or the Yield Power Law model.

Dynamic weight material SAG—A problem unique to nonaqueous fluids is dynamic weight material SAG (Savari et al., 2013; Hanson et al., 1990; Saasen et al., 1991, Jamison and Clements, 1990). Minimizing barite sag is a function of the LSRV characteristics. Increasing the 3 and 6 RPM dial readings are the most important readings for sag control. Another parameter that has been developed to monitor potential sag problems called the low shear rate yield point. The following equation calculates the LSR-YP.

$$\text{LSR-YP} = 2 \times \Theta_3 \text{rpm} - \Theta_6 \text{rpm} \quad (6.66)$$

Annular Effective Viscosity

A relatively easy method of estimating the viscosity in the annulus can be done either by calculation or graphically using the annular velocity and data from a multispeed viscometer,

The API recommended multispeed viscometer is such that the shear rate at each RPM is fixed, as shown in Table 6.5. The number in the Dial Reading Factor column is used to convert the viscometer dial reading from lbf/100 ft^2 to cP viscosity. We use this viscometer data to construct a viscosity profile of the mud. An example for one particular mud is shown in Table 6.5. It is important to realize that the viscometer dial reading is not viscosity—it is shear stress. Newton's definition of viscosity, Eq. (6.2), is used to calculate the viscosity at each rpm. It can be observed that the value of the dial readings decrease with slower rpms, while viscosity increases. A shear thinning mud, then, is one that gets thinner at the higher shear rates.

Annular shear rates are routinely between the 6 and 100 RPM shear rates.

The viscosity data from the multispeed viscometer can plotted on a viscosity profile graph as shown in Fig. 6.68. Fig. 6.69 shows some example viscosity profile plots from a six-speed viscometer for three different muds. This plot clearly shows the shear thinning nature of these muds.

TABLE 6.5 Factors for Converting the API Recommended Viscometer Dial Readings to Viscosities

RPM	Shear Rate, s^{-1}	Dial Reading Factor	Dial Reading. Lbf/100 ft^2	Viscosity. cP
600	1022	0.5	45	22.5
300	510	1.0	29	29
200	340	1.5	21	31.3
100	170	3.0	15	45
60	102	5.0	10	50
30	51	10	8	80
6	10.2	50	6	300
3	5.1	100	4	400

238 Composition and Properties of Drilling and Completion Fluids

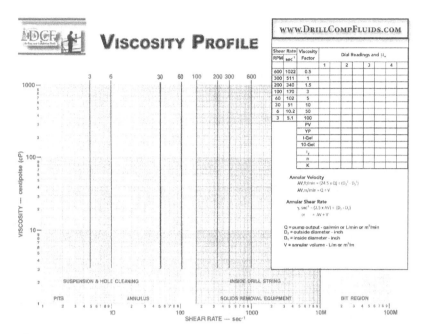

FIGURE 6.68 Viscosity profile graph for. Blank log-log paper to input either 6 or 8 speed viscometer data to plot a viscosity profile for current rig conditions.

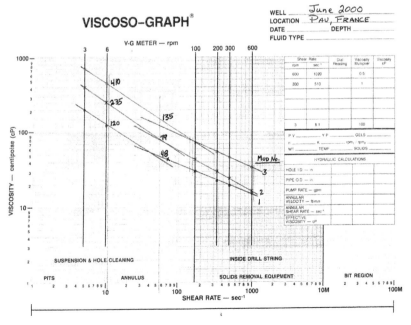

FIGURE 6.69 Example viscosity profile showing 6-speed viscometer data for three drilling fluids and the resulting annular viscosities at two annular shear rates.

The Rheology of Drilling Fluids Chapter | 6   239

Once we have a viscosity profile, we can use the following two equations to determine the annular shear rate.

$$\text{Annular Velocity, ft/min} = 24.4Q/(Do^2 - Di^2) \qquad (6.67)$$

$$\text{Annular Shear Rate, } s^{-1} = 2.4\,V/(Do - Di) \qquad (6.68)$$

Where:

Q = pump rate, gal/min
V = annular velocity, ft/min
Do = outside diameter, in.
Di = inside diameter, in.

On Fig. 6.69 two vertical lines drawn at 11 and 58 s^{-1}, show the different annular viscosities at those two shear rates. Muds 1 and 2 are the same, except that 0.25 lb/bbl xanthan gum was added to the mud. This treatment lowered the n factor, more shear thinning, and increased the effective viscosity dramatically. At 11 s, the viscosity nearly doubled, and undoubtedly the mud had more suspending power and greater hole cleaning capacity with the annular velocity unchanged.

NOTATION

$\overline{AV}$ apparent viscosity at 600 rpm in the rotary viscometer
D diameter
D_1 outer diameter of drill pipe or drill collars
D_2 diameter of hole or inner diameter of casing
F force
f fanning friction factor
g gravitational constant, 32.2 ft/s^2 or 980 cm/s^2
h corrected height of bob in rotary viscometer
K consistency index of a power law fluid
K' generalized consistency index
L length
M mean hydraulic radius
n flow-behavior index of a power law fluid
n' generalized flow-behavior index
N_{Re} Reynolds number
N_{Re}' generalized Reynolds number
P pressure
P_o pressure to initiate flow of a Bingham plastic
$\overline{PV}$ plastic viscosity as determined in a two-speed rotary viscometer
Q volumetric flow rate
R radius
r local radius
R_h radius of bob in rotary viscometer

R_c	inner radius of rotor in rotary viscometer
S	gel strength
T	torque
t	time
V	flow velocity
v	local velocity
v_t	terminal velocity of a particle in a still liquid
v_u	net velocity of a particle in a rising column of liquid
v_s	slip velocity of a particle in a rising column of liquid
$\overline{YP}$	yield point, as determined in a two-speed rotary viscometer
γ	shear rate
γ_w	shear rate at pipe wall
θ	dial reading on rotary viscometer
μ	Newtonian viscosity
μ_p	plastic viscosity of a Bingham plastic
μ_e	effective viscosity of a non-Newtonian fluid
ρ	density
τ	shear stress
τ_θ	shear stress at yield point
τ_w	shear stress at wall of pipe
φ	volume of solids divided by bulk volume of the suspension
ϖ	angular velocity in radians per second
ω	revolutions per minute (rpm)

REFERENCES

Annis, M.R., 1967. High temperature properties of water-base drilling fluids. J. Petrol. Technol.1074–1080, Trans AIME 240.

API Subcommittee 13, 2010. The Rheology of Oil-Well Drilling Fluids, API Bulletin, 13D. American Petroleum Institute, Washington, DC.

Bartlett, L.E., 1967. Effect of temperature on the flow properties of drilling muds, SPE Paper 1861, Annual Meeting, October 6–9, Houston.

Becker, T.E., 1982. The Effect of Mud Weight and Hole Geometry Variations on Cuttings Transport in Directional Drilling, MS Thesis. University of Tulsa, Tulsa, Okla.

Binder, R.C., Busher, J.E., 1946. Study of flow of plastics through pipes. J. Appl. Mech. 13 (2), A-101.

Bingham, E.C., 1922. Fluidity and Plasticity. McGraw-Hill, New York, NY.

Bobo, R.A., Hoch, R.A., 1958. Keys to Successful Competitive Drilling. Gulf Publishing Co, Houston.

Buckingham, E., 1921. On plastic flow through capillary tubes. Proc. ASTM 21, 1154–1156.

Burkhardt, J.A., 1961. Wellbore pressure surges produced by pipe movement. J. Petrol. Technol.595–605, Trans. AIME 222.

Caldwell, D.H., Babbit, H.E., 1949. Laminar flow of sludges with special reference to sewage disposal. Bull. Univ. Illinois (12).

Cannon, G.E., 1934. Changes in hydrostatic pressure due to withdrawing drill pipe from the hole. API Drill. Prod. Prac. 42–47.

Carlton, L.A., Chenevert, M.E., 1974. A new approach to preventing lost returns, SPE Paper No. 4972, Annual Meeting, October 6–9, Houston.

Cheng, D.G.H., Ray, D.J., Valentin, F.H.H., 1965. The flow of thixotropic bentonite suspensions. Through pipes and pipe fittings. Trans. Inst. Chem. Eng. (Lond.) 43, 176–186.

Clark, E.H., 1956. A graphic view of pressure surges and lost circulation. API Drill. Prod. Prac. 424–438.

Combs, G.D., Whitmire, L.D., 1960. Capillary viscometer simulates bottom hole conditions. Oil Gas J. 108–113.

Crowley, M., 1976. Procedures updates drilling hydraulics. Oil Gas J. 59–63.

Darley, H.C.H., 1969. A laboratory investigation of borehole stability. J. Petrol. Technol. 883–892, Trans. AIME 246.

Darley, H.C.H., Generes, R.A., 1956. The use of barium hydroxide in drilling muds. Trans. AIME 207, 252–255.

Darley, H.C.H., Hartfiel, A., 1974. Tests show that polymer fluids cut drill-pipe pressure losses. Oil Gas J. 70–72.

Dodge, D.W., Metzner, A.B., 1959. Turbulent flow of non-Newtonian systems. AIChE. J. 5 (2), 194.

Dunn, T.H., Nuss, W.F., Beck, R.W., 1947. Flow properties of drilling muds. API Drill. Prod. Prac. 9–21.

Eckel, J.R., 1967. Microbit studies of the effect of fluid properties and hydraulics on drilling rate. J. Petrol. Technol. 541–546, Trans AIME 240.

Fontenot, J.E., Clark, R.K., 1974. An improved method for calculating swab, surge, and circulating pressures. Soc. Petrol. Eng. J. 451–462.

Freundlich, H., 1935. Thixotropy. Herman & Cie., Paris, p. 3.

Garrison, A.D., 1939. Surface chemistry of shales and clays. Trans. AIME 132, 423–436.

Gavignet, A.A., Sobey, I.J., 1986. A model for the transport of cuttings in highly deviated wells, SPE Paper 15417, Ann. Tech. Conference, October 5–8, New Orleans, LA.

Goins, W.C., Weichert, J.P., Burba, J.R., Dawson, D.D., Teplitz, A.J., 1951. Down hole pressure surges and their effect on loss on circulation. API Drill. Prod. Prac. 125–131.

Goodeve, C.F., 1939. A general theory of thixotropy and viscosity. Trans. Faraday. Soc. 25, 342.

Gray, G.R., Neznayko, M., Gilkeson, P.W., 1952. Some factors affecting the solidification of lime-treated muds at high temperatures. API Drill. Prod. Prac. 73–81.

Green, H., 1949. Industrial Rheology and Rheological Structures. John Wiley & Sons, New York, NY, pp. 13–43.

Hall, H.N., Thomson, H., Nuss, F., 1950. Ability of drilling mud to lift bit cuttings. Trans. AIME 189, 35–76.

Hanks, R.W., Trapp, D.R., 1967. On the flow of Bingham plastic slurries in pipes and between parallel plates. Soc. Petrol. Eng. 342–346, Trans. AIME 240.

Hanson, P.M., Trigg, Jr., T.K., Rachal, G., and Zamora, M., Investigation of barite 'sag' in weighted drilling fluids in highly deviated wells, SPE paper No. 20423, 65th Annual Technical Conference, New Orleans, Sept. 23–26, 1990.

Havenaar, I., 1954. The pumpability of clay-water drilling muds. J. Petrol. Technol.49–55, Trans. AIME 201, 287-293.

Hedstrom, B.O.A., 1952. Flow of plastic materials in pipes. Ind. Eng. Chem. 44, 651–656.

Hiller, K.H., 1963. Rheological measurements of clay suspensions at high temperatures and pressures. J. Petrol. Technol. 779–789, Trans AIME 228.

Hopkin, E.A., 1967. Factors affecting cuttings removal during rotary drilling. J. Petrol. Technol. 807–814, Trans AIME 240.

Houwen, O.H., Geehan, T., 1986. Rheology of oil-base muds, SPE Paper 15416, Ann. Tech. Conference, October 5–8, New Orleans, LA.
Hughes Tool Company, 1954. Hydraulics in Rotary Drilling, Bull., 1A (Rev.). Hughes Tool Co., Houston.
Hussaini, S.M., Azar, J.J., 1983. Experimental study of drilled cuttings transport using common drilling muds. Soc. Petrol. Eng. J. 11–20.
Iyoho, A.W., 1980. Drilled-Cuttings Transport by Non-Newtonian Fluids Through Inclined, Eccentric Annuli, PhD Dissertion. University of Tulsa, Tulsa, OK.
Iyoho, A.W., Azar, J.J., 1981. An accurate slot-flow model for non-Newtonian fluid flow through eccentric annuli. Soc. Petrol. Eng. J. 565–572.
Jamison, D.E., and Clements, W. R., A new test method to characterize setting/sag tendencies of drilling fluids used in extended reach drilling, ASME 1990 Drilling Tech Symposium, PD Vol. 27, pp. 109–13, 1990.
Jones, P.H., Babson, E.C., 1935. Evaluation of rotary drilling muds. Oil Wkly. 25–30.
Kelly Jr., J., Hawk, D.E., 1961. Mud additives in deep wells. Oil Gas J. 145–155.
Koch, W.M., 1953. Fluid mechanics of the drill string. J. Petrol. Technol. 9–11.
Lal, M., 1984a. Better surge/swab pressures are available with new model. World Oil 81–88.
Lal, M., 1984b. Study finds effects of swab and surge pressures. Oil Gas J. 137–143.
Lord, D.L., Menzie, D.E., 1943. Stress and strain rate dependence of bentonite clay suspension gel strengths, SPE Paper 4231, 6th Conference Drill. & Rock Mech., January 22–23, Austin, pp. 11–18.
Martin, M., Georges, C., Bisson, P., Konirsch, O., 1987. Transport of cuttings in directional wells, SPE/IADC Paper 16083, Drill. Conference, March 15–18, New Orleans, LA.
McMordie, W.C., 1969. Viscometer tests mud to 650°F. Oil Gas J. 81–84.
McMordie Jr., W.C., Bennett, R.B., Bland, R.G., 1975. The effect of temperature and pressure on oil base muds. J. Petrol. Technol. 884–886.
Melrose, J.C., Savins, J.G., Parish, E.R., 1958. Utilization of theory of Bingham plastic flow in stationary pipes and Annuli. Trans. AIME 213, 316–324.
Metzner, A.B., 1956. Non-Newtonian technology: Fluid mechanics and transfers. Advances in Chemical Engineering. Academic Press, New York, NY.
Metzner, A.B., Reed, J.C., 1955. Flow of non-Newtonian fluids–correlation of laminar, transitional and turbulent flow regimes. AIChE J. 1, 434–440.
Mitchell, R.F., 1987. Dynamic surge/swab pressure predictions, SPE/IADC Paper 16156, Drill. Conference, March 15–18, New Orleans, LA.
Mooney, M., 1931. Explicit formulas for slip and fluidity. J. Rheol. 2, 210–222.
Moore, P.E., 1974. Drilling Practices Manual. The Petroleum Publishing Co., Tulsa, pp. 205–228.
Moore, P.L., 1973. Annulus loss estimates can be more precise. Oil Gas J. 111–113.
O'Brien, T.B., Dobson, M., 1985. Hole cleaning: some field results, SPE/IADC paper 13442, Drilling Conf., March 5–8, New Orleans, LA.
Okrajni, S.S., Azar, J.J., 1986. Mud cuttings transport in directional well drilling, SPE Paper 14178, Annual Meeting, Las Vegas, September 22–25; and SPE Drill. Eng. (August), 291–296.
Piggott, R.J.S., 1941. Mud flow in drilling. API Drill. Prod. Prac. 91–103.
Politte, M.D., 1985. Invert mud rheology as a function of temperature and pressure, SPE/IADC Paper 13458, Drilling Conference, March 5–8, New Orleans, LA.
Rabinowitsch, B., 1929. On the viscosity and elasticity of sols. Zeit. Physik. Chem. 145A, 1–26.

Randall, B.V., Anderson, D.B., 1982. Flow of mud during drilling operations. J. Petrol. Technol. 1414–1420.

Raymond, L.R., 1969. Temperature distribution in a circulating fluid. J. Petrol. Technol. 333–341, Trans AIME 246.

Reiner, M., 1929. The theory of plastic flow in the rotational viscometer. J. Rheol. 1, 5–10.

Robertson, R.E., Stiff Jr., H.A., 1976. An improved mathematical model for relating shear stress to shear rate in drilling fluids and cement slurries. Soc. Petrol. Eng. J.31–36, Trans. AIME 261.

Ryan, N.W., Johnson, N.W., 1959. Transition from laminar to tubulent flow in pipes. AIChE J. 5 (4), 433–435.

Saasen, A., Marken, C., Sterri, N., Jakobsen, J., 1991. Monitoring of barite SAG important in deviated drilling. Oil & Gas J. August 26, 1991.

Savins, J.G., 1958. Generalized Newtonian (pseudoplastic) flow in stationary pipes and annuli. Trans. AIME 213, 325–332.

Savins, J.G., 1964. Drag reduction characteristics of solutions of macromolecules in turbulent pipe flow. Soc. Petrol. Eng. J.203–214, Trans. AIME 231.

Savins, J.G., Roper, W.F., 1954. A direct indicating viscometer for drilling fluids. API Drill. Prod. Prac. 7–22.

Savins, J.G., Wallick, G.C., 1966. Viscosity profiles, discharge rates, pressure and torques for a rheologically complex fluid in helical flow. AIChE J. 12 (2), 357–363.

Schuh, F.J., 1964. Computer makes surge-pressure calculations useful. Oil Gas J.96–104.

Sifferman, T.R., Myers, G.M., Haden, E.L., Wall, H.A., 1974. Drill-cutting transport in full scale vertical annuli. J. Petrol Technol. 1295–1302.

Silbar, A., Paslay, P.R., 1962. The analytical description of the flow of thixotropic materials. Symp. on Second-order Effects in Elasticity, Plasticity, and Fluid Dynamics, Internat. Union of Theoretical and Applied Mechanics, Haifa, pp. 314–330.

Sinha, B.K., 1970. A new technique to determine the equivalent viscosity of drilling fluids under high temperatures and pressures. Soc. Petrol. Eng. J. 33–40.

Savari, S., Kulkarni, S., Maxey, J., Teke, K. 2013. A Comprehensive Approach to Barite Sag Analysis on Field Muds. 2013 AADE National Technical Conference and Exhibition held at the Cox Convention Center, Oklahoma City, OK, February 26–27.

Thomas, R.P., Azar, J.J., Becker, T.E., 1982. Drillpipe eccentricity effect on drilled cuttings behavior in vertical wellbores. J. Petrol. Technol. 1929–1937.

Tomren, P.H., 1979. The Transport of Drilled Cuttings in an Inclined Eccentric Annulus, MS Thesis. University of Tulsa, Tulsa, OK.

Tomren, P.H., Iyoho, A.W., Azar, J.J., 1986. An experimental study of cuttings transport in directional wells, SPE Paper 12123, Annual Meeting, San Francisco, CA, October 5–8, and SPE Drill. Eng. (February), 43–46.

van Olphen, H., 1950. Pumpability, rheological properties and viscometry of drilling fluids. J. Inst. Pet. (Lond.) 36, 223–234.

Walker, R.E., 1971. Drilling fluid rheology. Drilling 43–58.

Walker, R.E., 1976a. Annular calculations balance cleaning with pressure loss. Oil Gas J. 5, 82–88.

Walker, R.E., 1976b. Hydraulics limits are set by flow restrictions. Oil Gas J. 86–90.

Walker, R.E., Korry, D.E., 1974. Field method of evaluating annular performance of drilling fluids. J. Petrol. Technol. 167–173.

Walker, R.E., Mayes, T.M., 1975. Design of muds for carrying capacity. J. Petrol. Technol. 893–900.

Weintritt, D.J., Hughes, R.G., 1965. Factors involved in high temperature drilling fluids. J. Petrol. Technol. 707–716, Trans. AIME 234.

Williams, C.E., Bruce, G.H., 1951. Carrying capacity of drilling fluids. Trans. AIME 192, 111–120.

Willis, H.C., Tomm, W.R., Forbes, E.E., 1973. Annular flow dynamics, SPE Paper 4234, 6th Conference, Drill. & Rock Mech., Austin, January 22–23, pp. 43–54.

Zamora, M., 1974. Discussion of the paper by Sifferman et al. J. Petrol. Technol. 1302.

Zeidler, H.U., 1972. An experimental analysis of transport of drilled particles. Soc. Petrol. Eng. J. 39–48.

Zeidler, H.U., 1974. Fluid and Particle Dynamics Related to Drilling Mud Carrying Capacity, PhD Dissertation. University of Tulsa, Tulsa, OK.

Zeidler, H.U., 1981. Better understanding permits deeper clear water drilling. World Oil 167–178.

Chapter 7

The Filtration Properties of Drilling Fluids[1]

INTRODUCTION

To prevent formation fluids from entering the borehole, the hydrostatic pressure of the mud column must be greater than the pressure of the fluids in the pores of the formation. Consequently, mud tends to invade the permeable formations. This does not cause massive losses and lost circulation, but is a steady filtration phenomena into permeable formations. The drilling fluid is dewatered as the mud solids are screened onto permeable formations, forming a cake of relatively low permeability through which only filtrate can pass. Muds must be treated to keep cake permeability as low as possible in order to maintain a stable borehole and to minimize filtrate invasion of, and damage to, potentially productive horizons. Furthermore, high cake permeabilities result in thick filter cakes, which reduce the effective diameter of the hole and cause various problems, such as excessive torque when rotating the pipe, excessive drag when pulling it, and high swab and surge pressures. Thick cakes may cause the drill pipe to stick by a mechanism known as differential sticking, which may result in an expensive fishing job.

Two types of filtration are involved in drilling an oilwell: *static filtration*, which takes place when the mud is not being circulated and the filter cake grows undisturbed; and *dynamic filtration*, which takes place when the mud is being circulated and the growth of the filter cake is limited by the erosive action of the mud stream. Dynamic filtration rates are much higher than static rates, and most of the filtrate invading subsurface formations does so under dynamic conditions. The filtration properties of drilling fluids are usually evaluated and controlled by the API filter loss test (API Recommended Practice 13B-1 2009 and API Recommended Practice B-2 2014), which is a static test and is therefore not a reliable guide to downhole filtration unless the differences between static and dynamic filtration are appreciated, and the test results interpreted accordingly.

---

1. A glossary of notation used in this chapter will be found immediately following this chapter's text.

STATIC FILTRATION

The Theory of Static Filtration

If unit volume of a stable suspension of solids is filtered against a permeable substrate, and x volumes of filtrate are expressed, then $1 - x$ volumes of cake (solids plus liquid) will be deposited on the substrate. Therefore, if Q_c is the volume of the cake, and Q_w the volume of filtrate:

$$\frac{Q_c}{Q_w} = \frac{1-x}{x} \qquad (7.1)$$

and the cake thickness (h) per unit area of cake in unit time will be

$$h = \frac{1-x}{x} Q_w \qquad (7.2)$$

Now, Darcy's law states

$$\frac{dq}{dt} = \frac{kP}{\mu h} \qquad (7.3)$$

where k = permeability in darcys, P = differential pressure in atmospheres, μ = viscosity of the filtrate in centripoises, h = thickness in centimeters, q = volume of filtrate in cubic centimeters, and t = time in seconds.
Therefore,

$$\frac{dq}{dt} = \frac{kP}{\mu Q_w} \times \frac{x}{1-x}$$

Integrating,

$$Q_w^2 = \frac{2kP}{\mu} \times \frac{x}{1-x} t \qquad (7.4)$$

From Eqs. (7.1) and (7.4),

$$Q_w^2 = \frac{2kP}{\mu} \times \frac{Q_w}{Q_c} t \qquad (7.5)$$

If the area of the filter cake is A,

$$Q_w^2 = \frac{2kPA^2}{\mu} \times \frac{Q_w}{Q_c} t \qquad (7.6)$$

This is the fundamental equation governing filtration under static conditions.

Relationship Between Filtrate Volume and Time

Larsen (1938) found that if a mud was filtered through paper at constant temperature and pressure, Q_w was proportional to $\sqrt{t}$, apart from a small zero error. It followed that, for a given mud, Q_w/Q_c and k in Eq. (7.6) were constant with respect to time. Although this finding is not strictly true for all muds (von Engelhardt and Schindewolf, 1952), it is close enough for practical purposes, and forms the basis for the mechanics of static filtration as presently interpreted.

Fig. 7.1 shows a typical plot of cumulative filtrate volume versus time plotted on a square root scale. The intercept on the y-axis marks the zero error. The zero error, commonly called the *mud spurt*, is largely caused by the tendency of the finer mud particles to pass through the filter paper until its pores become plugged. Thereafter only filtrate is expressed, and the curve becomes linear. With most muds the zero error is small, and is often neglected, but it can be substantial when filtration takes place against porous rocks. Some muds plug filter paper almost instantly, in which case the zero error is negative, and represents the volume between the paper and the discharge nipple.

Larsen's experimental results showed that for a given pressure, Eq. (7.6) may be written as

$$Q_w - q_0 = A\sqrt{(C \times t)} \qquad (7.7)$$

where q_a is the zero error, and C is a constant given by

$$C = \frac{2kP}{\mu}\frac{Q_w}{Q_c} \qquad (7.8)$$

Thus the filtration properties of diverse muds can be evaluated by measuring the filtrate volume accumulating in a standard time and under standard conditions. The conditions recommended by the API are:

Time: 30 min
Pressure: 100 psi (6.8 atmospheres, 7 kg/cm^2)
Area of cake: approximately 7 in.^{2} (45 cm^2)

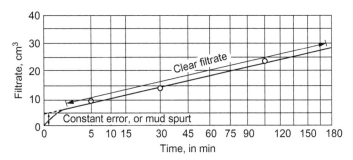

FIGURE 7.1 Relation of filtrate volume to the square root of time.

The filtrate volume that would accumulate in 30 min can be predicted from the volume, Q_w observed at time t_1 from the equation

$$Q_{w_{30}} - q_0 = (Q_{w_1} - q_0)\frac{\sqrt{t_{30}}}{t_1}$$

For example, the 30 min filtrate volume is sometimes predicted by measuring the filtrate volume at 7.5 min, and doubling the value obtained, since $\sqrt{30}/\sqrt{7.5} = 2$

For QA/QC processes when manufacturing additives for fluid loss control, it is somewhat more acurate to discard the first 7.5 min of filtrate volume, then double the volume acquired during the time from 7.5 to 30 min.

Relationship Between Pressure and Filtrate Volume

According to Eq. (7.6), Qw should be proportional to $\sqrt{P}$, and a log–log plot of Q_w versus P should yield a straight line with a slope of 0.5, assuming all factors remained constant. Actually, this condition is never met because mud filter cakes are to a greater or lesser extent compressible, so that the permeability is not constant, but decreases with increase in pressure (Larsen 1938; von Engelhardt and Schindewolf, 1952). Thus

$$Q_w \alpha P^x$$

where the exponent x varies from mud to mud, but is always less than 0.5, as shown in Fig. 7.2.

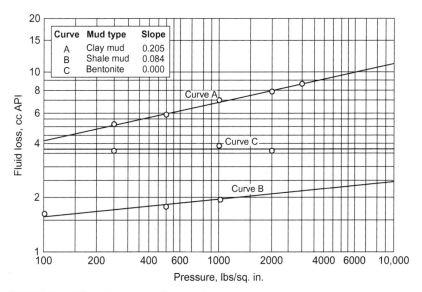

FIGURE 7.2 Effect of pressure on filtrate volume.

The value of the exponent x depends largely on the size and shape of the particles composing the cake. Bentonite cakes, for example, are so compressible that x is zero, and Q_w is constant with respect to P. The reason for this behavior is that bentonite is almost entirely composed of finely divided platelets of montmorillonite, which tend to align more nearly parallel to the substrate with increase in pressure. Thus the permeability of the cake is reduced to a much greater extent than would be the case with a cake composed of, e.g., rigid spheres. With other drilling mud clays it has been found experimentally that the x exponent varies from zero to about 0.2, so it appears that filtration rate is relatively insensitive to changes in pressure.

Outmans (1963) developed a theoretical equation that may be used to predict changes in filtrate volume with filtration pressure, if the compressibility of the cake is known. In practice, it is usually simpler to make the filtration test at the pressure of interest. In the case of nonaqueous muds, another factor comes into play: the increase in the viscosity of the filtrate (the base fluid) with increase in absolute pressure (Simpson, 1974), which tends to reduce the filter loss according to the equation

$$Q_{w_1} = Q_{w_2} \sqrt{\frac{\mu_1}{\mu_2}} \qquad (7.9)$$

where μ_1 and μ_2 are the viscosities at the filtration pressures in the tests for Q_{w_1} and Q_{w_2}, respectively.

Relationship Between Temperature and Filtrate Volume

An increase in temperature may increase the filtrate volume in several ways:

1. Viscosity reduction
2. Flocculation/aggregation status
3. Additive degradation

In the first place, it reduces the viscosity of the filtrate, and therefore the fitrate volume increases according to Eq. (7.9). The viscosity of water and of 6% brine are shown over a range of temperatures in Table 7.1, and over an extended range, for water only, in Fig. 7.3. It is evident that changes in temperature may have a substantial effect on filtrate volume because of changes in filtrate viscosity. For example, the filtrate volume at 212°F (100°C) would be about $\sqrt{\frac{1}{0.284}} = 1.88$ times as large as the volume at 68°F (20°C).

Changes in temperature may also affect filtrate volume through changes in the electrochemical equilibria that govern the degree of flocculation and aggregation, thus altering the permeability of the filter cake. As a result of such effects, filtrate volumes may be higher or lower than predicted from Eq. (7.9), but usually they are higher (Milligan and Weintritt, 1961). For instance, Byck (1939) found that, of the six muds he tested, three had 8% to 58% greater filter

TABLE 7.1 The Viscosity of Water and 6% Sodium Chloride Brine at Various Temperatures

Temperature (°C)	Temperature (°F)	Viscosity Water (Centipoise)	Viscosity Brine (Centipoise)
0	32	1.792	–
10	50	1.308	–
20.2	68.4	1.000	1.110
30	86	0.801	0.888
40	104	0.656	0.733
60	140	0.469	0.531
80	176	0.356	0.408
100	212	0.284	–

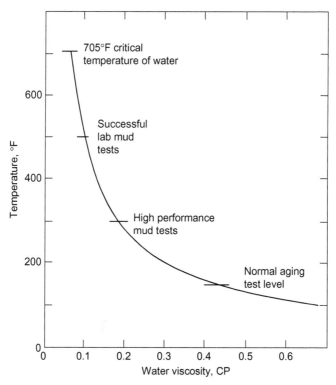

FIGURE 7.3 Viscosity of water at various temperatures. *Data by Milligan, D.J., Weintritt, D.J., 1961. Filtration of drilling fluid at temperatures of 300°F, and above. API Drill. Prod. Prac. 42–48. Courtesy API.*

loss at 175°F (70°C) than had been predicted from the filter loss at 70°F (21°C) by substituting the changes in filtrate viscosity in Eq. (7.9). The permeability of the cakes increased correspondingly, with the maximum change being from 2.2 to 4.5×10^{-3} md—an increase of over 100%. Filtration rates of the other three muds deviated from the predicted values by only ± 5%, and the permeabilities of the cakes remained essentially constant. In more extensive tests, Schremp and Johnson (1952) showed that there is no way that filter losses at high temperature can be predicted from measurements made at a lower temperature. Therefore, it is necessary to test each mud separately at the temperature of interest in a high-temperature cell.

Chemical degradation of one or more components of the mud is a third mechanism by which high temperatures can affect filtration properties. Many organic filtration control agents start to degrade significantly at temperatures above about 212°F (100°C), and the rate of degradation increases with further increase in temperature, until filtration properties cannot be adequately maintained. This subject is discussed further in the section on high-temperature muds in Chapter 9, Wellbore Stability.

THE FILTER CAKE

Cake Thickness

Although cake thickness is the vital factor in problems associated with tight hole, pipe torque and drag, and differential sticking, little attention has been paid to it in the drilling fluid literature. Cake thickness is assumed to be proportional to filter loss, and therefore only filter loss needs to be specified. Actually, although cake thickness *is* related to filter loss, the specific relationship varies from mud to mud, because the value of Q_w/Q_c in Eq. (7.6) depends on the concentration of solids in the mud and on the amount of water retained in the cake. The filter loss decreases with the increase in the concentration of solids, but the cake volume increases, as shown in Fig. 7.4. If an operator adds extra clay to a mud to reduce filter loss, he may believe that he is also reducing cake thickness, but he is actually increasing it.

The amount of water retained in the cakes of muds with different clay bases depends on the swelling properties of the clay minerals involved. Bentonite, for example, has strong swelling properties, and bentonitic cakes therefore have a comparatively high ratio of water to solids, and the $Q_w Q_c$ ratio is correspondingly low. Table 7.2 compares the percent water retained in the cakes of three muds with the amount of water adsorbed by the dry clays in swelling tests. Note that the amount of water in the cakes is only slightly less than that in the swollen clays, and is virtually independent of the percentage of solids in the suspension. Indeed, the percent water in the cake is quite a good measure of the swelling properties of the clay base.

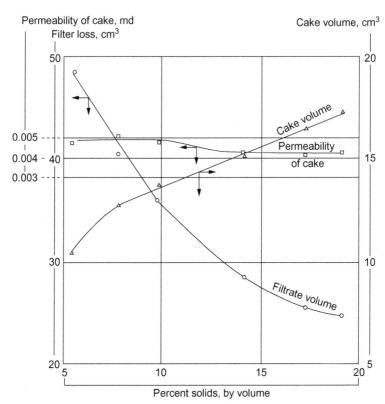

FIGURE 7.4 Variation of filtrate volume, cake volume, and permeability with concentration of solids in a suspension of Altwarmbüchen clay. *Data from von Engelhardt, W., Schindewolf, E., 1952. The filtration of clay suspensions. Kolloid Z 127, 150–164.*

TABLE 7.2 Correlation Between the Amount of Water Adsorbed by Dry Clay and Water Retained in the Filter Cake[a]

Type of Clay	Swelling Test % Solids in Swollen Clay	Filtration Test		
		% Solids in Suspension	100 psi % Solids in Cake	1000 psi % Solids in Cake
Bentonite	14.4	3.17	16.5	
		4.53	15.0	16.7
		6.8	15.2	
Mojave Desert Clay	62.0	23.4	64.5	64.5
		30.2	64.8	
		39.6	65.4	
West Texas clay	40.7	2.42	45.1	53

[a]*Data from Larsen (1938), in* Petroleum Engineer, *Nov. 1938.*

To a lesser extent, cake thickness is determined by particle size and particle-size distribution. These parameters control the porosity of the cake, and therefore the bulk volume relative to the grain volume. The magnitude of these effects was shown by Bo et al. (1965), who measured the porosities of filter cakes formed by mixing nine size grades of glass spheres. Their results may be summed up as follows:

1. Minimum porosities were obtained when there was an even gradation of particle sizes (i.e., a linear particle size distribution curve, as shown in Fig. 7.5A and B), because the smaller particles then packed most densely in the pores between the larger particles.
2. Mixtures with a wide range of particle sizes had lower porosities than mixtures with the same size distribution but narrower size range (see Fig. 7.5A and B).
3. An excess of small particles resulted in lower porosities than did an excess of large particles.

We may expect the inert solids—which are comparatively isodimensional in shape—in drilling muds to exhibit similar phenomena. The behavior of the colloid fraction depends more on particle shape and on electrostatic forces, as discussed in the sections dealing with cake compressibility and cake permeability.

The thickness of the filter cake is difficult to measure accurately, largely because it is not possible to distinguish the boundary between the mud and the upper surface of the cake precisely. The problem arises because the cake is compacted by the hydraulic drag of the filtrate flowing through its pores. The hydraulic drag increases with depth below the surface of the cake, and the local pore pressure decreases correspondingly from the pressure of the mud on the surface of the cake to zero at the bottom of the cake. The compacting pressure (and the resulting intergranular stress) at any point is equal to the mud pressure less the pore pressure, and is therefore equal to zero in the surface layer, and to the mud pressure in the bottom layer of the cake. The distribution of intergranular stress and of density (expressed as porosity) with respect to distance from the bottom of the cake is shown in Fig. 7.6 for theoretical and experimental values determined by Outmans (1963) with a suspension of ground calcium carbonate. Note that the distributions shown do not change with increase in thickness of the cake, so the average porosity of the cake remains constant with respect to time.

Cake Thickness Test

When accurate values of static cake thickness are required, it is advisable to use the method developed by von Engelhardt and Schindewolf (1952), which is as follows: only a limited amount of mud is put in the filtration cell, and filtration is stopped at the moment all of the mud is used up, so that only filter cake remains in the cell. The critical moment to stop filtration is determined

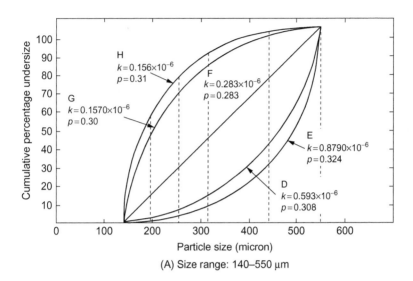

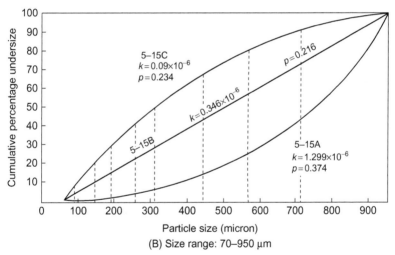

FIGURE 7.5 Permeabilities and porosities of filter cakes of glass spheres. k = permeability in darcys, p = porosity. *From Bo, M.K., Freshwater, D.C., Scarlett, B., 1965. The effect of particle-size distribution on the permeability of filter cakes. Trans. Inst. Chem. Eng. (Lond.) 43, T228–T232. Courtesy of the Institute of Chemical Engineers.*

by observing the filtrate volume at short time intervals, and concurrently plotting the volume versus the square root of the intervals. Filtration is stopped immediately when the curve departs from linearity. The total volume of mud filtered is calculated from the combined weight of the filtrate plus cake, divided by the density of the original mud. The cake volume is then obtained by subtracting the filtrate volume from the volume of mud filtered.

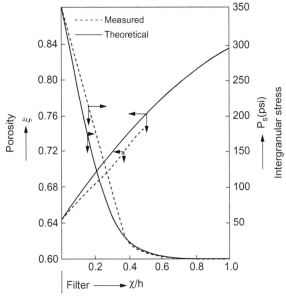

FIGURE 7.6 Distribution of porosity (ξ) and effective stress (P_s) in a filter cake from a suspension of chalk. Filtration pressure = 350 psi. *From Outmans, H.D., 1963. Mechanics of static and dynamic filtration in the borehole. Soc. Petrol. Eng. J. 3, 236—244. Trans. AIME 228. Copyright 1963 by SPE-AIME.*

The Permeability of the Filter Cake

The permeability of the filter cake is the fundamental parameter that controls both static and dynamic filtration. It more truly reflects downhole filtration behavior than does any other parameter. As a parameter for evaluating the filtration properties of muds with different concentration of solids, it has the advantage over filtrate volume in being independent of solids concentration, as shown in Fig. 7.4 (the slight rise in permeability at low solids concentration shown in Fig. 7.4 is probably due to settling of the coarser particles). Furthermore, cake permeability provides useful information on the electrochemical conditions prevailing in the mud.

Cake permeabilities were measured by early investigators of filtration behavior. Williams and Cannon (1938) obtained values between 0.2 and 0.6×10^{-3} md at 8 atmospheres pressure for Gulf Coast field muds, and 72×10^{-3} md for a West Texas mud. Byck (1940) measured permeabilities between 0.46 and 7.42×10^{-3} md at 34 atmospheres pressure with California muds. By making some determinations on cores of permeabilities from 10 to 14,000 md, he showed that the filtration rate was dependent only on the permeability of the cake—at least as long as it was several orders of magnitude lower than the permeability of the formation. Gates and Bowie

(1942) measured the permeabilities of 20 field muds and 40 laboratory muds, and obtained values from 0.31 to over 250×10^{-3} md at 100 psi (7.8 atmospheres).

The above investigators measured cake permeability from the filtration rate at the end of the test and from the cake thickness. Gates and Bowie (1942) mentioned the difficulty of measuring cake thickness accurately. This problem can be avoided by using the method of von Engelhardt and Schindewolf (1952), described above, to determine cake volume, and then calculating permeability from Eq. (7.6), rearranged as follows:

$$k = Q_w Q_c \frac{\mu}{2tPA^2} \qquad (7.10)$$

When Q_w and Q_c are expressed in cm³, t in seconds, P in atmospheres, A in cm², μ in centipoises, Eq. (7.10) then becomes, with the standard API test,

$$k = Q_w Q_c \mu \times 1.99 \times 10^{-5} \text{ md} \qquad (7.11)$$

This method is suitable for laboratory studies of static filtration. At the wellsite, where accuracy is not so important, it is more convenient to measure the filter cake manually (see the last paragraph of this chapter), and to use Eq. (7.6) in the form

$$k = \frac{Q_w h \mu}{2tPA} \qquad (7.12)$$

If h is expressed in millimeters,

$$k = Q_w h \mu \times 8.95 \times 10^{-3} \text{ md} \qquad (7.13)$$

Note: With freshwater muds μ is approximately one centipoise at 68°F (20°C).

The Effect of Particle Size and Shape on Cake Permeability

Krumbein and Monk (1943) investigated the permeabilities of filter cakes of river sand by separating the sand into 10 size fractions and recombining them to obtain two sets of mixtures. In one set, the mixtures had increasingly large mean particle diameters, but all had the same range of particle sizes, which were defined in terms of a parameter phi as shown in Fig. 7.7. In the other set, all the mixtures had the same mean particle diameter, but increasingly wider ranges of particle sizes. The results showed that cake permeability decreased (1) with mean particle diameter, and (2) with increasing width of particles size range (see Fig. 7.8).

One might expect minimum cake permeabilities with an even gradation of particle sizes. However, the experiments of Bo et al. (1965), already referred to, showed that minimum permeabilities were obtained when there was an excess of particles at the fine end of the scale, and not when the size

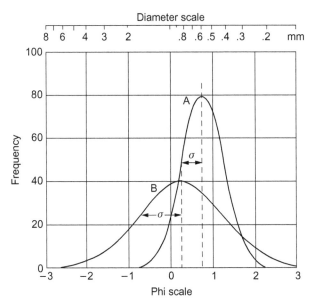

FIGURE 7.7 Examples of a narrow (Curve A) and a wide (Curve B) particle size range. *From Krumbein, W.C., Monk, G.D., 1943. Permeability as a function of the size parameters of unconsolidated sand. Trans. AIME 151, 153–163. Copyright 1943 by SPE-AIME.*

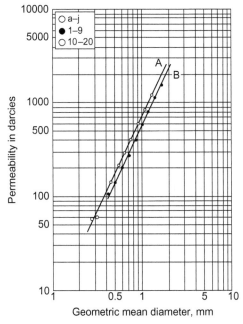

FIGURE 7.8 Variation of permeability of sand filter cake with mean particle diameter and with particle size range. Curve A = individual size fractions; Curve B = mixtures of fractions, all having the same size range but increasing diameter; small vertical circles = mixtures of fractions, all having the same mean diameter. Permeability decreases with increase in size range. *From Krumbein, W.C., Monk, G.D., 1943. Permeability as a function of the size parameters of unconsolidated sand. Trans. AIME 151, 153–163. Copyright 1943 by SPE-AIME.*

distribution curves were linear (see Fig. 7.5A and B). It would appear therefore that a uniform gradation of particle sizes is of secondary importance, but obviously there must be no major gaps, or the finer particles would pass through the pore openings between the larger ones.

Krumbein and Monk (1943) showed that cake permeabilities decreased sharply with particle size. Drilling fluids contain substantial amounts of colloids, whose size may range down to less than 10^{-5} μm. It is not surprising, therefore, that the permeabilities of their filter cakes depend almost entirely on the proportion and properties of the colloidal fraction. Although the data obtained by Gates and Bowie (1942) showed only a broad correlation between particle size and cake permeability—because no allowance was made for different degrees of flocculation—the group of muds with the largest colloidal fraction (see Fig. 7.9A) had cake permeabilities ranging from 1.5 to 0.31×10^{-3} md, whereas the group of muds with no colloids (see Fig. 7.9B) had cake permeabilities so high that they could not be measured.

Cake permeability is, of course, influenced by the kind of colloid as well as by the amount and particle size. For instance, filter cakes of bentonite suspensions in freshwater have exceptionally low permeabilities because of the flat, filmy nature of the clay platelets, which enables them to pack tightly normal to the direction of flow. Solids suspensions in brine fluids are likely to be flocculated, increasing the permeability.

A flat orientation of particles would appear to be contradicted by the work of Hartmann et al. (1986), who investigated the structure of freeze dried filter cakes by means of a scanning electron microscope and observed an open honeycomb structure in freshwater bentonite cakes. However, their finding may be discounted because the honeycomb structure was undoubtedly formed as the gel was being flash frozen, as discussed in the section on the "Mechanism of Gelation" in Chapter 4, Clay Mineralogy and the Colloid Chemistry of Drilling Fluids.

Polymer Fluid Loss Agents

Organic macromolecules, such as the starches, owe their effectiveness to the deformability of the hydrolyzed cells, as well as to their small size. Polyelectrolytes, such as carboxymethylcelluose (Scanley, 1962), are partly adsorbed on the clay particles and partly trapped in the pores, and impede flow both physically and by the electroviscous effect (see Chapter 4, Clay Mineralogy and the Colloid Chemistry of Drilling Fluids). More detail is given about polymers in Chapter 13, Drilling Fluid Components.

Nonaqueous Fluid Loss

In the case of asphaltic dispersions in nonaqueous drilling fluids, filtration control is achieved only if the asphalt is in the colloidal state. Control is lost

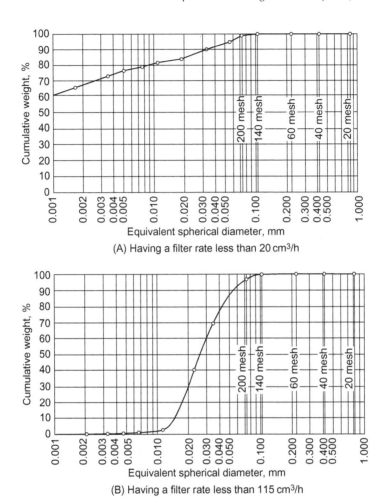

FIGURE 7.9 Particle size distribution of selected muds. (A) Having a filter rate less than 20 cm³/h. (B) Having a filter rate greater than 115 cm³/h. *From Gates, G.L., Bowie, C.P., 1942. Correlation of Certain Properties of Oil-Well Drilling Fluids with Particle Size Distribution. U.S. Bureau of Mines Report of Investigations, No. 3645 (May).*

if the aromatic content of the suspending oil is too low—aniline point above about 150°F (65°C)—because the asphalt coagulates. Control is also lost if the aromatic content of the suspending oil is too high—aniline point below about 90°F (32°C)—because the asphalt passes into a true solution.

In invert emulsions, filtration control is obtained by organophyllic-clays, various synthetic oil-dispersible copolymers (See Chapter 13, Drilling Fluid Components), and finely dispersed water in oil emulsions as well as particle size distribution. These tiny, highly stabilized water droplets act like deformable solids, yielding low-permeability filter cakes.

An alternate method to reduce either WBM or NADF cake permeabilities is adding sized calcium carbonate particles. The optimum particle size distribution is critical and is usually determined by lab testing. This is discussed further in the Reservoir Drill-In Fluid discussion in Chapter 2, Introduction to Completion Fluids.

Effect of Flocculation and Aggregation on Cake Permeability

As explained in Chapter 4, Clay Mineralogy and the Colloid Chemistry of Drilling Fluids, flocculation of muds causes the particles to associate in the form of a loose, open network. This structure persists to a limited extent in filter cakes, causing considerable increases in permeability. The higher the filtration pressure, the more this structure is flattened, so both porosity and permeability decrease with increase in pressure. The greater the degree of flocculation, the greater the interparticle attractive forces, resulting in a stronger structure and more resistance to pressure (Fig. 7.10). The structure is even stronger if flocculation is accompanied by aggregation, because it is then built of thicker packets of clay platelets. For example, Suspension 1 in Fig. 7.10 contained only 0.4 g/L of chloride in the filtrate, sufficient only to cause a weak floc structure. Suspension 2 was obtained by adding 35 g/L of sodium chloride to Suspension 1, sufficient to cause strong flocculation plus aggregation. Consequently, the cake permeabilities and porosities of Suspension 2 were considerably greater than those of Suspension 1, even at high filtration pressures.

Conversely, deflocculation of a mud by the addition of a thinning agent causes a decrease in cake permeability. Moreover, most thinners are sodium salts, and the sodium ion may displace the polyvalent cations in the base exchange positions on the clay, thereby dispersing the clay aggregates and further reducing cake permeability.

Thus, the electrochemical conditions prevailing in a mud are a major factor in determining the permeability of its filter cake. As a generalization it may be said that cake permeabilities of flocculated muds are in the order of 10^{-2} md, those of untreated freshwater muds are in the order of 10^{-3} md, and those of muds treated with thinning agents are in the order of 10^{-4} md.

The Bridging Process

As already discussed, there is a mud spurt at the start of a filter test made on paper before filtration proper begins, and, thereafter, filtrate volume becomes proportional to the square root of the time interval. In the drilling well, mud spurts may be much larger when filtration takes place against the more permeable rocks; in fact, they can be infinite (i.e., circulation is lost) unless the mud contains particles of the size required to bridge the pores of the rock,

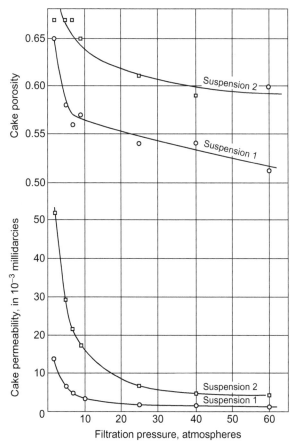

FIGURE 7.10 Reduction of cake porosity and permeability with increase in filtration pressure. Suspension 1 = 0.3004 g of Altwarmbüchen clay per g of suspension, Suspension 2 = 0.2836 g of Altwarmbüchen clay + 35 g/L of NaCl. Filtration time = 60 min. *Data from von Engelhardt, W., Schindewolf, E., 1952. The filtration of clay suspensions. Kolloid Z 127, 150–164.*

and thus establish a base on which the filter cake can form. Only particles of a certain size relative to the pore's size can bridge. Particles larger than the pore opening cannot enter the pore, and are normally swept away by the mud stream; particles considerably smaller than the opening invade the formation unhindered, but particles of a certain critical size stick at bottlenecks in the flow channels, and form a bridge just inside the surface pores. Once a primary bridge is established, successively smaller particles, down to the fine colloids, are trapped, and thereafter only filtrate invades the formation. The mud spurt period is normally very brief, a matter of a second or two at the most (Darley, 1965).

As a result of the process just described, three zones of mud particles are established on or in a permeable formation (see Fig. 7.11):

1. An external filter cake on the walls of the borehole
2. An internal filter cake, extending a couple of grain diameters into the formation
3. A zone invaded by the fine particles during the mud spurt period, which normally extends about an inch into the formation (Abrams, 1977; Darley, 1975; Glenn and Slusser, 1957; Young and Gray, 1967). Experimental results reported by Krueger and Vogel (1954) suggest that these fine particles do not initially cause much permeability impairment, but may do so after filtration has proceeded for some hours, presumably because of migration and consequent pore blocking.

When adequate bridging particles are lacking, the API filter test may give grossly misleading results. A mud might give a neglible loss on filter paper, but give a large one on a permeable formation downhole. The point was well illustrated by experimental data obtained by Beeson and Wright (1952), extracts from which are shown in Table 7.3. Note that the discrepancy between the gross loss on paper and that on the porous media was greater with unconsolidated sand than with consolidated rocks, even when the permeability of the latter was higher. Note also that the discrepancies between the net filter loss on paper and on porous media increased with increase in spurt loss. Evidently the mud spurt plugs the cores to such an extent that the pressure drop within the core becomes significant, thereby reducing the drop across the cake, and reducing cake compaction. Similarly, Peden et al. (1982) got higher dynamic filtration rates on coarse synthetic filter media than on natural sandstones.

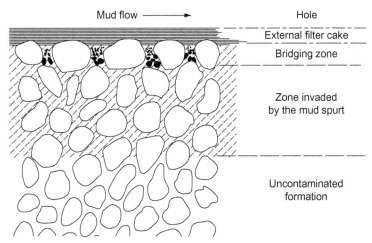

FIGURE 7.11 Invasion of a permeable formation by mud solids (schematic). *From the Sept. 1975 issue of Petroleum Engineer International. Publisher retains copyrights.*

TABLE 7.3 Effect of Filtration Medium on Mud Spurt[a]

Effect per Mud Type	API Test Whatman 50 Filter Paper	Filtration Medium		
		Unconsolidated Sand, 219 md to 299 md	Consolidated Rock, 520 md	Consolidated Rock, 90 md
Bentonite, 1.04 SG				
Gross filter loss	11.5	53	23.4	15.3
Mud spurt	–	29	5.6	3
Net filter loss	11.5	24	17.8	12.3
Native Clay, 1.15 SG				
Gross filter loss	11.5	17		
Mud spurt	–	6		
Net filter loss	11.5	11.0		
Oil Base, 0.93 SG				
Gross filter loss	0	12.3		
Mud spurt	0	9.6		
Net filter loss	0	2.7		

[a]Data from Beeson and Wright (1952).

Haberman et al. (1992) collected data from four wells in the Mississippi delta. The wells were systematically shut in, and the fluid loss from the annulus was measured both with only mud in the annulus and after cement placement. The wells had extensive intervals of permeable sands, making them well-suited for fluid-loss tests. The results indicated a much lower fluid loss per unit area of permeable sands both before and after cementing compared with routine laboratory testing. Results also indicated that cement fluid loss was controlled mainly by filtration properties of the drilling-fluid mud cake.

With regard to the critical size required for bridging, it was shown by Coberly (1937) that because of jamming, particles down to one-third the size of a circular screen opening would bridge that opening. Abrams (1977) showed that particles whose median diameter was about one-third the median pore size of a 5 darcy sand pack would bridge that pack. In order to form an effective base for a filter cake, a mud must therefore contain primary bridging particles ranging in size from slightly less than the largest pore opening in the formation about to be drilled, down to about one-third that size. In addition, there must be smaller particles ranging down to colloidal size, to bridge the smaller formation pores and the interstices between the coarser bridging particles.

The best way to determine primary bridging sizes is to make trial and error tests on cores of the formation of interest. When this is not possible, some guidance may be obtained from published data (Abrams, 1977; Darley, 1975; Glenn and Slusser, 1957) which relates bridging size to permeability. To summarize, it was found that particles less than 2 μm in diameter will bridge rocks of permeability less than 100 md; 10 μm particles will bridge consolidated rocks of permeability between 100 and 1000 md; and 74 μm (200-mesh) particles will bridge sands up 10 darcys. A mud containing a suite of particle sizes up to a maximum of 74 μm should bridge and form a filter cake on all formations except those with macro-openings, such as gravel beds and formations with open fractures, which are discussed in the section on loss of circulation in Chapter 10, Drilling Problems Related to Drilling Fluids.

The greater the concentration of bridging particles, the quicker bridging will occur, and the less will be the mud spurt. On consolidated rocks with permeabilities in the range of 100 to 1000 md, about 1 lb/bbl (2.8 kg/m^3) of the required size range is sufficient to prevent the mud spurt from invading further than an inch into the rock (Glenn and Slusser, 1957). On unconsolidated sands about 5−10 lb/bbl (14−28 kg/m^3) may be required.

Bridging particles in the size ranges and amounts specified above will normally be present in any mud that has been used to drill more than a few feet, except that there will be a shortage of coarse particles if extensive use is made of desanders and desilters when drilling unconsolidated sands. Bridging particles must, however, be added to reservoir drill-in fluids and muds that are formulated for production repair jobs when no drilling is involved (see Chapter 11).

DYNAMIC FILTRATION

Under the condition of dynamic filtration, the growth of the filter cake is limited by the erosive action of the mud stream. When the surface of the rock is first exposed, the rate of filtration is very high, and the cake grows rapidly. However, the growth rate decreases as time passes, until eventually it is equal to the erosion rate; thereafter the thickness of the cake is constant. Under equilibrium dynamic conditions, therefore, the rate of filtration depends on the thickness and permeability of the cake, and is governed by Darcy's law (Eq. 7.3), whereas under static conditions cake thickness increases ad infinitum, and the rate of filtration is governed by Eq. (7.6).

Dynamic filter cakes differ from static cakes in that the soft surface layers of the static cake are not present in the dynamic cake, because its surface is eroded to an extent that depends on the shear stress exerted by the hydrodynamic force of the mud stream relative to the shear strength of the cake's upper layers.

The various stages of dynamic filtration are shown in Fig. 7.12. From T_0 to T_1, the filtration rate decreases and the cake thickness increases. From T_1 to T_2 the thickness of the cake remains constant, but the filtration rate continues to decrease, because, according to Outmans (1963), the filter cake continues to compact (presumably, therefore, the rate of deposition equals the rate of compaction). Another explanation is given by Prokop (1952), who suggested that the permeability of the cake decreases because of a classifying action as the mud stream erodes and redeposits particles in the cake's surface. At time T_2, equilibrium conditions are reached, and both the filtration rate and the cake thickness remain constant. The filtration rate is then given by the following equation (Outmans 1963):

$$Q = \frac{k_1(\tau/f)^{-v+1}}{\mu\delta(-v+1)} \quad (7.14)$$

where k_1 is the cake permeability at 1 psi pressure, δ is the shear stress exerted by the mud stream, f is the coefficient of internal friction of the

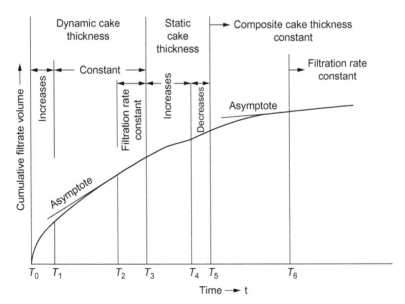

FIGURE 7.12 Relative static and dynamic filtration in the borehole. *From Outmans, H.D., 1963. Mechanics of static and dynamic filtration in the borehole. Soc. Petrol. Eng. J. 3, 236–244. Trans. AIME 228. Courtesy Soc. Petrol. Eng. J. Copyright 1963 by SPE-AIME.*

TABLE 7.4 Equilibrium Cake Thickness Under Dynamic Filtration[a]

Mud Base (all Muds Treated With Lime, Caustic Soda, and Quebracho)	API Filter Loss (cc in 30 min)	Equilibrium Cake Thickness		Mud Velocity Equilibrium	
		(in.)	(mm)	(ft/min)	(m/min)
Bentonite	19	1/32	0.8	125	38
Calcium bentonite and barite	8	3/32	2.4	48	15
Calcium bentonite	10	6/32	4.7	72	22
Attapulgite and bentonite	85	19/32	15.1	220	67
Attapulgite and bentonite	148	21/32	16.6	530	161

Mud circulated through a 2 in. (5.08 cm) diameter hole in consolidated sand. Turbulent flow. Filtration pressure 350 psi (24.6 kg/cm²).
[a] From Prokop (1952). Copyright 1952 by SPE-AIME.

cake's surface layer (see Chapter 9, Wellbore Stability, for definition of this parameter), δ is the thickness of the cake subject to erosion, and $(-v + 1)$ is a function of the cake compressibility.

Prokop (1952) measured dynamic filtration rates in a laboratory tester, in which mud flowed through a concentric hole in a cylindrical artificial core. Table 7.4 shows the equilibrium cake thickness thus obtained with a large number of laboratory muds.

Ferguson and Klotz (1954) obtained some excellent data on dynamic filtration rates in a model well that duplicated field geometry. Holes were drilled in artificial sandstone blocks with 5¼ in. and 5⅜ in. bits. Figs. 7.13–7.16 show the change in dynamic filtration rates with time for four muds at various circulating velocities. Note that the dynamic rates were considerably greater than the static rates as extrapolated from the API filter loss tests. Note also that the time to reach constant dynamic filter rates varied from 2 hours to over 25 h, depending on the type of mud and on the flow velocity. Fig. 7.17 shows the increase in filtration rate with increase in flow velocity. The API filter losses have been marked on the curves of their respective muds to show the lack of correlation with the dynamic rate.

A marked increase in filtration rates under dynamic conditions was also observed by Vaussard et al. (1986) in the cell shown in Fig. 7.18. (Note: For filtration experiments the pipe was centered.) Annular conditions simulated 6¼ in. pipe in an 8½ in. hole.

In contrast to the above two findings, Peden et al. (1982, 1984) observed that dynamic filtration rates decreased with increase in annular velocity

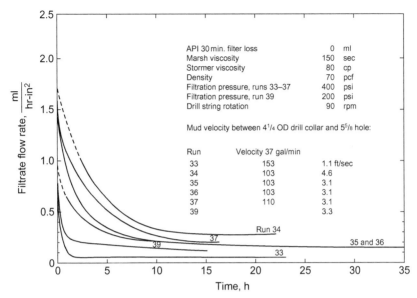

FIGURE 7.13 Dynamic filtration from oil-base mud. *From Ferguson, C.K., Klotz, J.A., 1954. Filtration from mud during drilling. J. Petrol. Technol. 6(2), 29–42. Trans AIME 201. Copyright 1954 by SPE-AIME.*

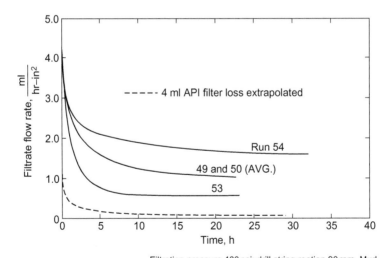

FIGURE 7.14 Dynamic filtration from emulsion mud. *From Ferguson, C.K., Klotz, J.A., 1954. Filtration from mud during drilling. J. Petrol. Technol. 6(2), 29–42. Trans AIME 201. Copyright 1954 by SPE-AIME.*

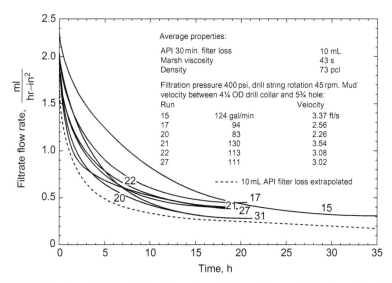

FIGURE 7.15 Dynamic filtration from bentonite mud. *From Ferguson, C.K., Klotz, J.A., 1954. Filtration from mud during drilling. J. Petrol. Technol. 6(2), 29–42. Trans AIME 201. Copyright 1954 by SPE-AIME.*

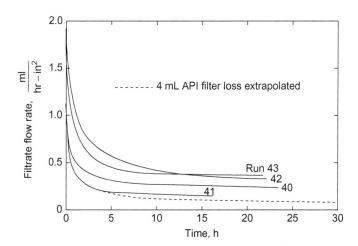

FIGURE 7.16 Dynamic filtration from lime-starch mud. *From Ferguson, C.K., Klotz, J.A., 1954. Filtration from mud during drilling. J. Petrol. Technol. 6(2), 29–42. Trans AIME 201. Copyright 1954 by SPE-AIME.*

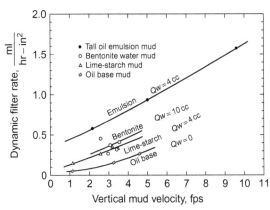

FIGURE 7.17 Dependence of equilibrium dynamic filtration rate on mud flow velocity at 75°F. *From Ferguson, C.K., Klotz, J.A., 1954. Filtration from mud during drilling. J. Petrol. Technol. 6(2), 29–42. Trans AIME 201. Copyright 1954 by SPE-AIME.*

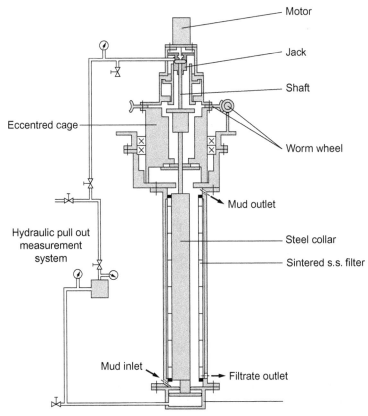

FIGURE 7.18 Filtration and sticking simulator. *From Vaussard, A., Konirsch, O.H., Patroni, J.-M., 1986. An Experimental Study of Drilling Fluids Dynamic Filtration. Annual Tech. Conference, New Orleans, LA, October 5–8, SPE Paper 15412. Copyright 1986 by SPE-AIME.*

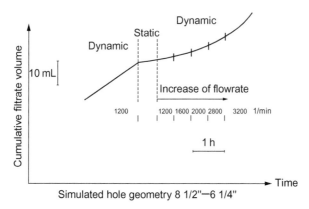

FIGURE 7.19 Filtration versus flow changes. *From Vaussard, A., Konirsch, O.H., Patroni, J.-M., 1986. An Experimental Study of Drilling Fluids Dynamic Filtration. Annual Tech. Conference, New Orleans, LA, October 5–8, SPE Paper 15412. Copyright 1986 by SPE-AIME.*

when circulating a KCl polymer mud in a model wellbore. They suggest that the reason was that high annular velocities tended to erode the coarser particles and increase the deposition of the polymer chains, thus decreasing the permeability of the filter cake. It should be noted that flow was turbulent at their highest annular velocity, and that the shear rates at the two lower velocities were much higher than the shear rates normally prevailing at the wall of a wellbore.

Other findings by Vaussard et al. (1986) were

1. The dynamic rate was reduced by a period of static filtration, but increased if the annular flow rate was increased—markedly so at the onset of turbulence, which occurred at about 1800 L/min (see Fig. 7.19).
2. Invert emulsion mud cakes were easily erodable, resulting in higher dynamic rates than would have been expected from their API filter losses. Spurt losses were high when the solids content was low.
3. The dynamic rates were independent of rock properties, except for coarse sintered media.

FILTRATION IN THE BOREHOLE

The Filtration Cycle in a Drilling Well

In a drilling well, filtration takes place under dynamic conditions while the mud is circulating, and under static conditions when circulation is stopped to make a connection, change bits, etc. A static cake is thus laid down on top of a dynamic one, so the filtration rate decreases and the filter cake thickness increases, as shown between T_3 and T_4 in Fig. 7.12. The amount of filtrate invading the formation under this condition can be calculated approximately

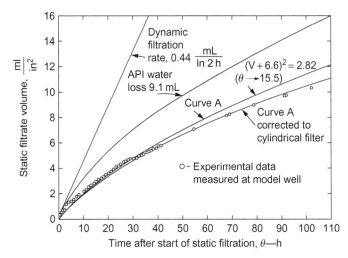

FIGURE 7.20 Static filtration for bentonite mud. *From Ferguson, C.K., Klotz, J.A., 1954. Filtration from mud during drilling. J. Petrol. Technol. 6(2), 29—42. Trans AIME 201. Copyright 1954 by SPE-AIME.*

from Eq. (7.6) by assuming that the dynamic cake was laid down under static conditions, and by obtaining the values of Q_w and t corresponding to the dynamic cake thickness from static test data. Such calculations show that the amount of filtrate invading the formation under static conditions is comparatively small, even during prolonged shutdowns, as shown in Fig. 7.20.

When circulation is resumed, the soft upper layers of the static cake are eroded, and the thickness of the cake decreases (see T_4 to T_5 in Fig. 7.12), but most of the static cake remains. From T_5 to T_6, the thickness of the cake again remains constant while the filtration rate decreases to a new equilibrium value. Thus the thickness of the cake increases every dynamic—static cycle, but the amount of increase is small. The growth of the filter cake is limited by mechanical wear when the drill pipe is rotating, and by abrasion when pulling or running pipe, but these effects cannot be quantified.

Filtration Beneath the Bit

Very little filter cake forms on the bottom of the hole because the action of the mud jets is highly erosive, and because every time a bit tooth strikes, a fresh surface of rock is exposed. At one time it was thought that major filtrate invasion took place beneath the bit, but several investigations (Cunningham and Eenik, 1959; Ferguson and Klotz, 1954; Horner et al., 1957; Young and Gray, 1967) have shown that beneath the bit, filtration is severely restricted by an internal mud cake that forms in the pores of the rock just ahead of the bit. Indeed, even if the drilling fluid is water, filtration is restricted (although to a lesser degree) as drilled solids plug the pores (Young and Gray, 1967). In their tests in the model well, Ferguson and Klotz (1954) measured liquid losses through the movements around the bottom of the hole corresponding to filtrate invasion of from 0.04 to 0.64 well radii, as compared to a potential invasion of from 0.3 to 14.3 well radii if no plugging had occurred.

Ferguson and Klotz (1954) also estimated the amount of filtrate invasion that would take place during the various stages of drilling and completing a hypothetical well, assuming sand at 7000 ft and a total depth of 7500 ft. The results (see Table 7.5) showed that some 95% of the invasion would take place under dynamic conditions while drilling, and only 6% under static conditions while tripping and completing.

TABLE 7.5 Drilling Schedule and Filtrate Invasion[a]

Operation	Time (h)	Filtrate Volume (mL/in^2)	Invasion Radius (in.)	Invaded Zone Thickness (in.)
Drill through zone at 5 fph		7.3	3.5	
Drill below zone at 5 fph	50	120	18.4	14.6
Round trip to replace bit	8	3.5	18.6	14.8
Drill below zone at 5 fph	50	61.5	21.1	17.3
Pull pipe, log well, run pipe	12	2.9	21.3	17.5
Condition Hole to Run Casing				
Circulate drilling mud	2	2.9	21.5	17.7
Pull drill pipe	4			
Run casing	12	2.9	21.7	17.9
Cement casing, end of mud filtration	—	—	—	—
Total mud filtration	138	192	21.7	17.9

[a]From Ferguson and Klotz (1954). Copyright by SPE-AIME 1954.

TABLE 7.6 Comparison Between Calculated and Experimental Bottomhole Filtration Rates[a]

Mud	V_{30} (mL)	C (s/cm^2)	Rate of Drilling (fph)	Bottomhole Filtration Rate (Q = cm^3/s)	
				Calc	Exp
Field	10.1	7.2×10^4	11.3	1.6	3.7
Gel	10.5	6.7×10^4	11.6	1.6	3.6
Gel	10.5	6.7×10^4	6.2	1.6	2.5
Oil base	0.2	1.8×10^8	32	0.04	0.52
Lime starch	4.1	4.4×10^5	19	0.73	0.60
Lime starch	4.1	4.1×10^5	43	0.73	0.6–4

[a]From Havenaar (1956). Copyright by SPE-AIME 1956.

Havenaar (1956) derived the following equation for filtration through the bottom of the hole while drilling:

$$Q = \frac{\pi D^2}{4} \sqrt{\left(\frac{nm}{C}\right)} \qquad (7.15)$$

where Q is the filtration rate in cm^3/s, n the number of cones on a bit rotating at m times per second, and C is determined from Eq. (7.7), using data from the API filtration test. Table 7.6 compares fiiltration rates calculated by this equation with the experimental data of Ferguson and Klotz. The poor correlation obtained with oil-base mud is probably because cakes of oil-base muds are easily eroded, and Eq. (7.15) neglects erosion by the mud jets.

Hassen (1982) has developed a set of equations describing the various stages of filtration downhole. The parameters for these equations may be determined from dynamic tests in the laboratory. The equations can then be used to calculate downhole filtration rates under the relevant conditions. The method is particularly useful for determining the depth of penetration of a water-base mud filtrate into a particular formation. Data obtained by this method were in good agreement with data derived from electric logs.

As might be expected from the discussion on Fig. 7.17, the filtration rate beneath the bit bears no relation to the API filter loss. This lack of correlation was clearly shown by Horner et al. (1957), who measured dynamic filtration rates during microbit drilling tests under conditions where nearly all the filtrate came from beneath the bit (see Fig. 7.21). Their results also

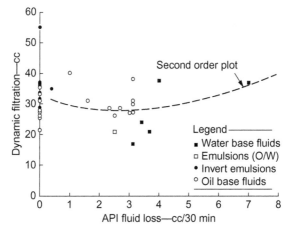

FIGURE 7.21 Filtration rate below the bit versus API fluid loss. *From Horner, V., White, M.M., Cochran, C.D., Deily, F.H., 1957. Microbit dynamic filtration studies. J. Petrol. Technol. (June) 183–189. Trans. AIME 210. Copyright 1957 by SPE-AIME.*

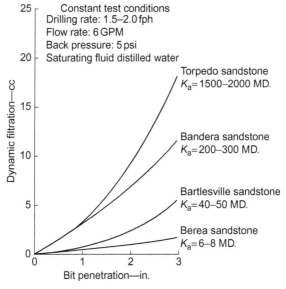

FIGURE 7.22 Effect of rock permeability on dynamic filtration rate below the bit. *From Horner, V., White, M.M., Cochran, C.D., Deily, F.H., 1957. Microbit dynamic filtration studies. J. Petrol. Technol. (June) 183–189. Trans. AIME 210. Copyright 1957 by SPE-AIME.*

showed that filtration rates beneath the bit, unlike filtration rates through the borehole walls, are influenced by the permeability of the formation (see Fig. 7.22). Likewise, microbit drilling tests by Lawhon et al. (1967) showed no correlation with API filter loss.

Evaluation of Downhole Filtration Rates

The lack of correlation between the API filter loss and the dynamic filtration rate shown in Fig. 7.17 casts doubts upon the validity of the API test as a criterion for downhole filtration rates. Experimental work by Krueger (1963) substantiated these doubts. Krueger added increasing amounts of various common filtration-reducing agents to samples of a standard clay mud and measured changes in equilibrium dynamic filtration rates as compared to the API filter loss. In the dynamic tests, the mud was circulated past the faces of sandstone cores mounted on a cylindrical cell containing a concentric mandrel. Filtration pressure was 500 psi (36 kg/cm^2), the temperature was 170°F (77°C), and fluid velocity was 110 ft/min (33 m/min). The results showed that there was a different relationship between dynamic filtration rate and API filter loss for each agent (see Fig. 7.23). Furthermore, the API loss decreased continuously with each addition of starch, CMC, and polyacrylate, but the dynamic rates decreased to a minimum, and then increased. In contrast, the API filter loss decreased very little with additions of lignosulfonate and quebracho, but the dynamic rates were almost as low as those obtained with starch, and much lower than those obtained with CMC and polyacrylate. Much the same rate relationships were obtained when dynamic rates were measured after the deposition of a static cake (Fig. 7.24), and when similar

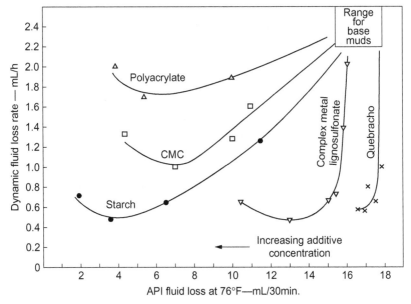

FIGURE 7.23 Comparison of dynamic filtration on 1 in. diameter sandstone cores with API fluid loss. Clay gel-base mud treated with various additives, equilibrium dynamic mud cake. *From Krueger, R.F., 1963. Evaluation of drilling-fluid filter-loss additives under dynamic conditions. J. Petrol Technol. 15(1), 90−98. Trans. AIME 228. Copyright 1967 by SPE-AIME.*

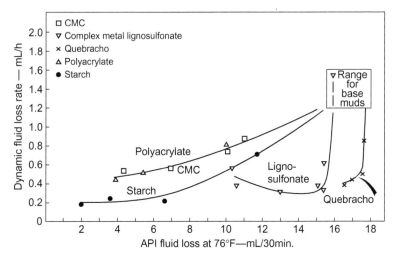

FIGURE 7.24 Comparison of dynamic filtration on 1-inch diameter sandstone cores with API fluid loss. Clay gel-base mud treated with various additives, static mud cake deposited on dynamic cake. *From Krueger, R.F., 1963. Evaluation of drilling-fluid filter-loss additives under dynamic conditions. J. Petrol Technol. 15(1), 90–98. Trans. AIME 228. Copyright 1963 by SPE-AIME.*

dynamic rates were compared with API filter losses measured at 500 psi and 170°F (Fig. 7.25). Krueger also found that emulsification of diesel oil into the mud decreased the API filter loss appreciably, but sharply increased the dynamic filtration rate. Black et al. (1985) found that emulsifying oil decreased the API filter loss but increased the dynamic rate while drilling Berea sandstone in a full-scale model borehole.

In appraising the above results, bear in mind that the treating agents were added to one base clay only. The results should not be regarded as, and were not intended to be, a rating of the effectiveness of the treating agents.

Chesser et al. (1994) developed a stirred fluid loss tester to compare static API filtrates to dynamic rates of field muds from south Louisiana. Figure 7.30 shows a schematic of the stirred fluid loss tester. Figures 7.28 and 7.29 show the accumulated filtrates for a 14 lbm/gal nonaqueous fluid and the dynamic rates at continuously increasing depths.

It seems likely that the poor correlation between API filter loss and dynamic filter rates is due to two factors:

1. Differences in the erodability of the filter cakes. Comparatively high dynamic rates were observed in all tests with oil muds, and the filter cakes of such muds are soft, i.e., f in Eq. (7.14) is low. On the other hand, comparatively low dynamic rates were observed in tests of muds containing lignosulfonates and quebracho. As we saw in Chapter 4, Clay Mineralogy and the Colloid Chemistry of Drilling Fluids, these additives are strongly adsorbed on clay particles.

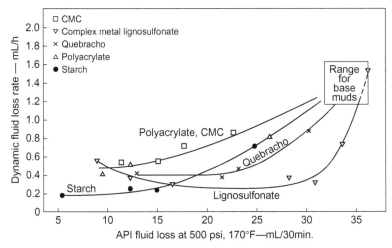

FIGURE 7.25 Comparison of dynamic filtration on 1-inch diameter sandstone cores with API fluid loss at 500 psi, 170°F. Bentonite mud treated with various additives, static mud cake deposited on dynamic cake. *From Krueger, R.F., 1963. Evaluation of drilling-fluid filter-loss additives under dynamic conditions. J. Petrol Technol. 15(1), 90–98. Trans. AIME 228. Copyright 1963 by SPE-AIME.*

2. Differences in the ratio of filtrate volume to cake volume, which affect the API filter loss (see Eq. 7.6), but not the dynamic filtration rate. The only mud-related variables which affect dynamic filtration rates are cake permeability and thickness, and for a given permeability the equilibrium thickness of the cake depends only on the erodability of the cake. For example, if the concentration of clay in a suspension is increased, the API filter loss decreases, but the dynamic filtration remains the same.

Outmans (1963) has suggested that differences in viscosity may also be responsible for the poor correlation, because viscosity influences the shear stress (τ in Eq. 7.14) exerted by the mud stream on the surface of the cake. However, neither Prokop (1952) nor Horner et al. (1957) found any significant relationship between viscosity and dynamic filtration rate.

The danger of relying on the API filter loss as a criterion for downhole dynamic filtration rates is obvious. A treating agent that was recommended on the basis of API test results might give higher rates downhole than another agent that gave a higher API filter loss. Worse still, an agent that reduced the API loss might increase the downhole filtration rate.

Wyant et al. (1985, 1987) compared dynamic and static high temperature, and high pressure filtration rates of a 13 lb/gal invert oil emulsion mud containing increasing amounts of various filtration control agents. The dynamic rates were tested in the cell shown in Fig. 7.26. Two such cells replaced the dynamic cells in the multifunctional circulating system shown in Fig. 7.14. The results showed that the high-temperature, high-pressure

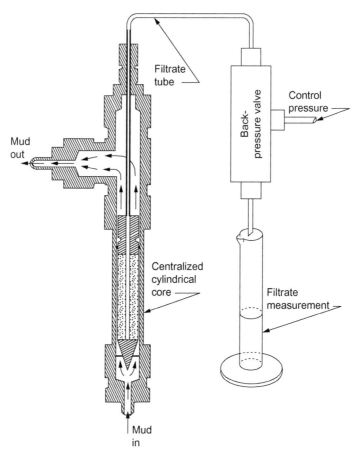

FIGURE 7.26 Dynamic filtration cell and filtrate collection system.

(HTHP) static loss correlated quite well with the short-term (30 min) dynamic loss, but had virtually no relationship with the long-term (equilibrium) dynamic loss (Figs. 7.27–7.30, Chesser et al., 1994).

Despite its shortcomings, the API or some similar static test is the only practical test for control of filtration at the wellsite. However, results should be interpreted in the light of correlations made in the laboratory between API filter loss and dynamic filtration rate (such as those made by Krueger, 1963), but one should use local muds and treating agents. Interpretation of the data would be greatly assisted if cake permeabilities were determined. Cake permeability can be calculated very simply from the API filter loss, cake thickness, and Eq. (7.13). In measuring cake thickness, the soft surface layers should be excluded, since these are not present in the dynamic cake. It is, of course, impossible to know the thickness of the downhole cake, but comparative data would be obtained if some standard procedure were adopted. One method would be to use a cake thickness gauge (Cook, 1954),

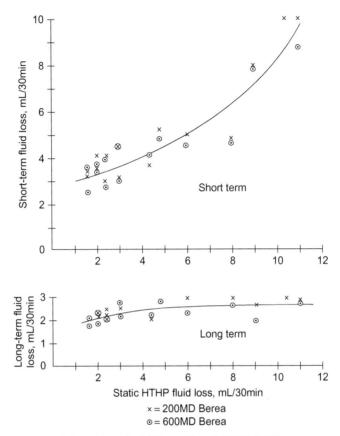

FIGURE 7.27 Correlation of dynamic fluid loss with static HTHP fluid loss.

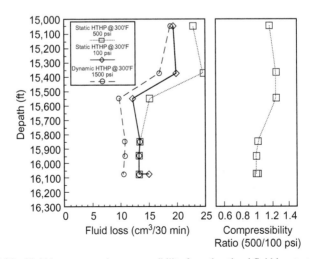

FIGURE 7.28 Fluid loss versus cake compressibility from the stirred fluid loss tester.

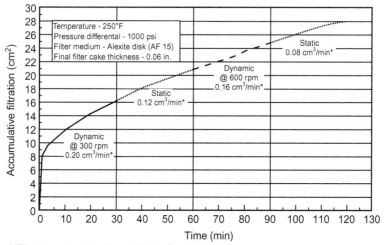

FIGURE 7.29 Fluid loss accumulation for a 14.0 lb/gal (1.67 SG) mud from the stirred fluid loss tester.

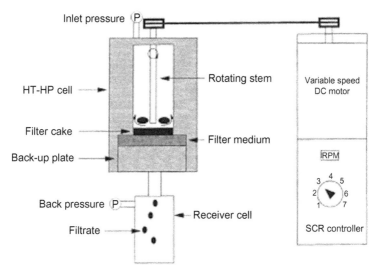

FIGURE 7.30 Stirred dynamic fluid loss tester.

the piston of which would be weighted to penetrate layers of only a certain consistency. The required degree of penetration could be established in the laboratory by correlating static and dynamic cake thicknesses. Cake permeabilities thus determined would not be quantitative, but should be a better guide to downhole dynamic filter rates than is the API filter loss. Remember, though, that static cake thickness is the best guide when concerned with problems such as tight hole, differential sticking, etc. Most of these problems occur when the fluid is not circulating, in which case the gross thickness of the cake, including the soft surface layers, should be measured.

NOTATION

A	area of filter
C	constant defined by Eq. (6.8)
f	coefficient of internal friction
h	thickness of filter cake
k	permeability
md	millidarcies
P	filtration pressure
Q	rate of filtration
Q_ν	volume of filter cake
Q_ω	cumulative filtrate volume
q_δ	volume of mud spurt
t	time
δ	thickness of filter cake subject to erosion
μ	viscosity of filtrate
τ	hydrodynamic shear stress

REFERENCES

Abrams, A., 1977. Mud design to minimize rock impairment due to particle invasion. J. Petrol. Technol. 29 (5), 586–592.

Beeson, C.M., Wright, C.W., 1952. Loss of mud solids to formation pores. Petrol. Eng. August, B40–B52.

Black, A.D., Dearing, H.L., Di Bona, B.G., 1985. Effects of pore pressure and mud filtration on drilling rate on permeable sandstone. J. Petrol. Technol. 37 (9), 1671–1681.

Bo, M.K., Freshwater, D.C., Scarlett, B., 1965. The effect of particle-size distribution on the permeability of filter cakes. Trans. Inst. Chem. Eng. (Lond.) 43, T228–T232.

Byck, H.T., 1939. Effect of temperature on plastering properties and viscosity of rotary drilling muds. Petrol Technol. of AIME November, 1116.

Byck, H.T., 1940. The effect of formation permeability on the plastering behavior of mud fluids. API Drill. Prod. Prac.40–44.

Chesser, B.G., Clark, D.E., Wise, W.V., 1994. Dynamic and Static Filtrate-Loss Techniques for Monitoring Filter-Cake Quality Improves Drilling-Fluid Performance. Society of Petroleum Engineers publication Drilling & Completion, pp. 189–192., September 1994.

Coberly, C.J., 1937. Selection of screen openings for unconsolidated sands. API Drill. Prod. Prac.189–201.

Cook, E.L., 1954. Filter Cake Thickness Gage. U.S. Patent No. 2,691,298. (October 12).

Cunningham, R.A., Eenik, J.E., 1959. Laboratory study of effect of overburden, formation, and mud column pressures on drilling rate of permeable formations. J. Petrol. Technol. January, 9–17, Trans. AIME 216.

Darley, H.C.H., 1965. Design of fast drilling fluids. J. Petrol. Technol. April, 465–470, Trans. AIME 284.

Darley, H.C.H., 1975. Prevention of productivity impairment by mud solids. Petrol. Eng. 47 (10), 102–110.

Ferguson, C.K., Klotz, J.A., 1954. Filtration from mud during drilling. J. Petrol. Technol. 6 (2), 29–42, Trans AIME 201.

Gates, G.L., Bowie, C.P., 1942. Correlation of Certain Properties of Oil-Well Drilling Fluids with Particle Size Distribution. U.S. Bureau of Mines Report of Investigations, No. 3645 (May).

Glenn, E.E., Slusser, M.L., 1957. Factors affecting well productivity. II. Drilling fluid particle invasion into porous media. J. Petrol. Technol. May, 132–139, Trans AIME 210.

Haberman, J., Delestatius, M., Hines, D., Daccord, G., Baret, J-F., 1992. Downhole Fluid-Loss Measurements From Drilling Fluid and Cement Slurries. Journal of Petroleum Technology, August, 44. http://dx.doi.org/10.2118/22552-PA.

Hartmann, A., Özerler, M., Marx, C., Neumann, H.J., 1986. Analysis of Mudcake Structures Formed under Simulated Borehole Conditions. Annual Tech. Conference, New Orleans, LA, October 5–8, SPE Paper 15413.

Hassen, B.R., 1982. Solving filtrate invasion with clay-water base systems. World Oil. 195 (6), 115–120.

Havenaar, I., 1956. Mud filtration at the bottom of the borehole. J. Petrol. Technol. May, 64, Trans. AIME 207, 312.

Horner, V., White, M.M., Cochran, C.D., Deily, F.H., 1957. Microbit dynamic filtration studies. J. Petrol. Technol. June, 183–189, Trans. AIME 210.

Krueger, R.F., 1963. Evaluation of drilling-fluid filter-loss additives under dynamic conditions. J. Petrol Technol. 15 (1), 90–98, Trans. AIME 228.

Krueger, R.V., Vogel, L.C., 1954. Damage to sandstone cores by particles from drilling fluid. API Drill. Prod. Prac. 158–168.

Krumbein, W.C., Monk, G.D., 1943. Permeability as a function of the size parameters of unconsolidated sand. Trans. AIME 151, 153–163.

Larsen, D.H., 1938. Determining the filtration characteristics of drilling muds. Petrol. Eng. September, 42–48. (November), 50–60.

Lawhon, C.P., Evans, W.M., Simpson, J.P., 1967. Laboratory drilling rate and filtration studies of clay and polymer drilling fluids. J. Petrol. Technol. 19 (5), 688–694, Also Laboratory drilling rate studies of emulsion drilling fluids. J. Petrol Technol. (July), 943–948. Trans. AIME 240.

Milligan, D.J., Weintritt, D.J., 1961. Filtration of drilling fluid at temperatures of 300°F, and above. API Drill. Prod. Prac. 42–48.

Outmans, H.D., 1963. Mechanics of static and dynamic filtration in the borehole. Soc. Petrol. Eng. J. 3, 236–244, Trans. AIME 228.

Peden, J.M., Avalos, M.R., Arthur, K.G., 1982. Analysis of Dynamic Filtration and Permeability Impairment Characteristics of Inhibited Water Based Muds. 5th Symposium Formation Damage, Lafayette, LA, March 24–25, SPE Paper 10655.

Peden, J.M., Arthur, K.G., Avalos, M., 1984. The Analysis of Filtration Under Static and Dynamic Conditions. 6th Symposium Formation Damage, Bakersfield, CA, February 13–14, SPE Paper 12503.

Prokop, C.L., 1952. Radial filtration of drilling mud. J. Petrol. Technol. January, 5–10, Trans. AIME 195.

Scanley, C.S., 1962. Mechanism of Polymer Action in Control of Fluid Loss from Oil Well Drilling Fluids. Amer. Chem. Soc. Meeting, Washington, DC.

Schremp, F.W., Johnson, V.L., 1952. Drilling fluid filter loss at high temperatures and pressures. J. Petrol. Technol. June, 157–162, Trans. AIME 195.

Simpson, J.P., 1974. Drilling Fluid Filtration under Simulated Downhole Conditions. Symposium on Formation Damage, New Orleans, February 7-8, SPE Paper 4779.

Vaussard, A., Konirsch, O.H., Patroni, J.-M., 1986. An Experimental Study of Drilling Fluids Dynamic Filtration. Annual Tech. Conference, New Orleans, LA, October 5–8, SPE Paper 15412.

von Engelhardt, W., Schindewolf, E., 1952. The filtration of clay suspensions. Kolloid Z 127, 150–164.

Williams, M., Cannon, G.E., 1938. Filtration properties of drilling muds. API Drill. Prod. Prac. 20–28.

Wyant, R.E., Reed, R.L., Sifferman, T.R., Wooten, S.O., 1985. Dynamic Fluid Loss Measurement of Oil Mud Additives. Annual Meeting, Las Vegas, September 22–25; and SPE Drill. Eng. (March), 63–74. SPE Paper 14246.

Young Jr., F.S., Gray, K.E., 1967. Dynamic filtration during microbit drilling. J. Petrol. Technol. 19 (9), 1209–1224, Trans AIME 240.

Chapter 8

The Surface Chemistry of Drilling Fluids

INTRODUCTION

We have shown how surface forces largely control the behavior of clay suspensions, but surface forces also affect many other aspects of drilling fluid technology, such as the formation of emulsions and foams, bit balling by plastic clays, wellbore stability, and formation damage by drilling mud filtrates. In this chapter, therefore, we will discuss the basics of surface chemistry as it relates to drilling fluid systems.

SURFACE TENSION

The interface between a liquid and a gas behaves as if it were a stretched elastic membrane. The contractile force of this imaginary membrane is known as *surface tension*. Surface tension also occurs at the boundary between a solid and a gas, between a solid and a liquid, and between two immiscible liquids. In the last instance, the tension is referred to as *interfacial tension*.

To illustrate the contractile nature of surface tension, consider two glass plates separated by a thin film of water (Fig. 8.1). Force must be applied to pull the two plates apart because of the contractile meniscus around the periphery of the plates. This phenomenon is known as *capillary attraction*.

Measurement of Surface Tension[1]

Surface tension is defined as the force acting perpendicular to a section of the surface one centimeter long, and is expressed in dyne/cm units. Absolute surface tension must be measured in a vacuum, but it is more convenient to measure it in an atmosphere of its own vapor, or in air (Bikerman, 1958a).

---

[1]. A glossary of notation used in this chapter will be found immediately following this chapter's text.

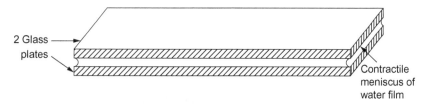

FIGURE 8.1 Illustration of capillary attraction. Contractile force of meniscus around periphery holds plates together.

Surface scientists use a goniometer/tensiometer to measure contact angle, surface tension, and interfacial tension. From the Wikipedia (2016a) listing for tensiometers, the following devices are discussed:

Goniometer/Tensiometer—Surface scientists sometimes use a Goniometer/Tensiometer to measure the surface tension and interfacial tension of a liquid using the pendant or sessile drop methods. A drop is produced and captured using a CCD camera. The drop profile is subsequently extracted, and sophisticated software routines then fit the theoretical Young-Laplace equation to the experimental drop profile. The surface tension can then be calculated from the fitted parameters. Unlike other methods, this technique requires only a small amount of liquid making it suitable for measuring interfacial tensions of expensive liquids.

Du Noüy-Padday method. This method uses a rod which is lowered into a test liquid. The rod is then pulled out of the liquid and the force required to pull the rod is precisely measured. This is a rather novel method which is accurate and repeatable. The Du Noüy-Padday Rod Pull Tensiometer will take measurements quickly and unlike the ring and plate methods, will work with liquids with a wide range of viscosities.

Du Noüy ring method. This type of tensiometer uses a platinum ring which is submersed in a liquid. As the ring is pulled out of the liquid, the force required is precisely measured in order to determine the surface tension of the liquid. This method requires that the platinum ring be nearly perfect; even a small blemish or scratch can greatly alter the accuracy of the results. A correction for buoyancy must be made. This method is considered less accurate than the plate method but is still widely used for interfacial tension measurement between two liquids.

Wilhelmy Plate Tensiometer. The Wilhelmy Plate tensiometer requires a plate to make contact with the liquid surface. It is widely considered the simplest and most accurate method for surface tension measurement.

Another method is to measure the height to which a liquid will rise spontaneously in a capillary tube (Fig. 8.2). At equilibrium the contractile force

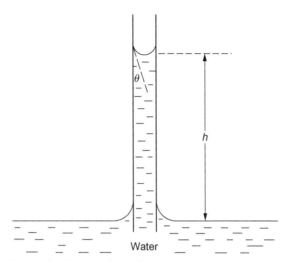

FIGURE 8.2 Rise of a liquid in a capillary tube.

of the meniscus is balanced by the hydrostatic head of the column of liquid, so that

$$\pi r^2 gh\rho = 2\pi r\gamma \cos\theta \tag{8.1}$$

where g is the gravitational constant, h the equilibrium height, ρ the density of the liquid, γ the surface tension, θ the contact angle, and r the radius of the capillary. The surface tension is therefore given by

$$\gamma = \frac{gh\rho r}{2\cos\theta} \tag{8.2}$$

The surface tensions of various substances are given in Table 8.1. Eq. (8.1) may be written

$$gh\rho = \frac{2\gamma \cos\theta}{r} \tag{8.3}$$

The term $gh\rho$ is the force driving the liquid into the capillary and is called the *capillary pressure*. The same equation applies to a drop of mercury on a glass surface.

WETTABILITY

Wetting refers to the ability of a liquid to stick to a solid surface. In drilling fluids, the primary concern is that the base fluid, oil or water, is able to maintain contact with any solids to which they touch, i.e., weight material, drilled solids, drill pipe, etc.

TABLE 8.1 Surface Tensions of Various Substances against Air

Substance	Temperature (°C)	Surface Tension (dynes/cm)
Water	0	75.6
	20	72.7
	50	67.9
	100	58.9
Ethyl alcohol	20	22.3
n-Hexane	20	18.4
Toluene	20	28.4
Mercury	15	48.7
Saturated NaCl brine	20	8.3
Oleic acid	20	32.5
Sorbitan ester	25	40
Aluminum	700	840
Zinc	590	708
Teflon[a]		18.5
Polyethylene[a]		31
Epoxy resin[a]		47

[a]*Critical surface tension for wetting.*

When a drop of liquid is placed on a surface it may or may not spread on (i.e., it may or may not wet) the surface, depending on the balance of forces shown in Fig. 8.3A. Water wets glass because Bikerman (1958c).

$$\gamma_{1,3} > \gamma_{1,2} + \gamma_{2,3} \cos \theta \qquad (8.4)$$

where $\gamma_{1,3}$ is the surface tension between the air and the glass, $\gamma_{1,2}$ is that between the water and the glass, $\gamma_{2,3}$ is that between air and water, and θ is the contact angle between the air–water interface and the glass.

Mercury does not wet glass because its surface tension is too high, and water does not wet Teflon because the surface tension of Teflon is so low. In both cases (see Fig. 8.3B):

$$\gamma_{1,3} + \gamma_{2,3} \cos \theta < \gamma_{1,2} \qquad (8.5)$$

Note that when the liquid does not wet, θ is greater than 90°, and cos θ is therefore negative.

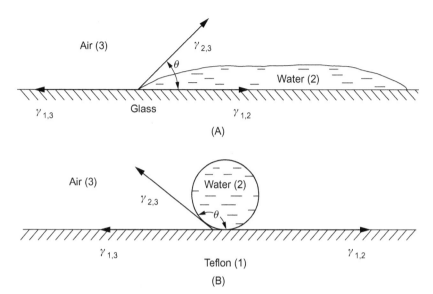

FIGURE 8.3 (A) Forces at an air–water interface. The water spreads on the glass because $\gamma_{1,3} > \gamma_{1,2} + \gamma_{2,3} \cos \theta$. (B) Forces at an air–water–Teflon® interface. The water does not spread because $\gamma_{1,3} + \gamma_{2,3} \cos \theta < \gamma_{1,2}$. The same equation applies to a drop of mercury on a glass surface.

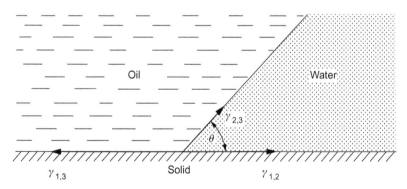

FIGURE 8.4 Preferential wettability. θ is less than 90°; therefore the solid is preferentially water-wet.

If two immiscible liquids are juxtaposed on a surface, one liquid has preference over the other in wetting the surface, depending on the relative tensions between the liquids and the solid, and on the interfacial tension. Eq. (8.4) applies, $\gamma_{1,3}$ and $\gamma_{1,2}$ being the respective tensions between the liquids and the solid, and $\gamma_{2,3}$ being the interfacial tension. The angle θ indicates the preferential wettability. In Fig. 8.4, the solid becomes decreasingly water-wet as θ

increases from 0 to 90°; at 90° the wettabilities are equal, and above 90° the solid is increasingly preferentially oil-wet. Adhesive forces between a liquid and solid cause a liquid drop to spread across the surface. "The degree of wetting (wettability) is determined by a force balance between adhesive and cohesive forces. Cohesive forces within the liquid cause the drop (Fig. 8.3B) to ball up and avoid contact with the surface" (Wikipedia, Wettability, 2016b).

Capillarity and Wettability

Water rises spontaneously in a glass capillary tube because glass is wet by water in the presence of air. Liquids (such as mercury) that do not wet glass will *not* rise spontaneously, and pressure must be applied for them to do so. In other words, the capillary pressure is negative.

Similarly, glass is preferentially water-wet relative to most oils. Water will spontaneously displace oil from a glass capillary, whereas a pressure must be applied in order for oil to displace water. As will be shown in Chapter 2, Introduction to Completion Fluids, these phenomena have an important bearing on the relative permeability of oil and gas reservoirs, and on productivity impairment by mud filtrates.

SURFACE FREE ENERGY

Surface phenomena may also be discussed in terms of surface free energy. Free energy exists at all surfaces because of the lack of balance of the charges around the molecules at the interface. If a liquid wets a surface, it lowers the free energy, and therefore, it performs work, as, for instance, when water rises in a glass capillary. On the other hand, creation of new surfaces involves an increase in free energy. For example, when a solid is split, chemical bonds are broken and an electrostatic surface charge thereby created. Therefore, work must be supplied to split a solid, or to create new surfaces in other ways. *Specific surface energy* is defined as the ratio of work to area. Its dimensions are the same as surface tension: it is expressed in dynes/cm, and the values are numerically equal.

ADHESION

A liquid will adhere to a solid if the attraction of the molecules to the solid surface is greater than their attraction to each other. In other words, the work of adhesion must be greater than the work of cohesion. Thermodynamically this criterion may be expressed as follows (Sharpe et al., 1964):

$$W_{adh} = F_s + F_l - F_{sl} \qquad (8.6)$$

where W_{adh} is the work of adhesion; F_s the surface free energy of the solid; F_l that of the liquid; and F_{sl} that of the newly formed interface. The work of

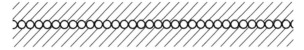

FIGURE 8.5 Microscopic section of two solid surfaces showing small area of contact (schematic).

cohesion is the work of a liquid spreading on itself, and is shown by Eq. (8.6) to be equal to $2F_l$. The criterion for adhesion therefore is

$$W_{adh} - W_{coh} = F_s - F_t - F_d \qquad (8.7)$$

where W_{coh} is the work of cohesion. Therefore the liquid will adhere if

$$F_s > F_l + F_{sl} \qquad (8.8)$$

Attractive forces also exist between two solid surfaces but the surfaces do not adhere when pressed together, because the attractive forces are extremely short range (a matter of a few angstroms), and the area of intimate contact is very small. Even two smooth, highly polished surfaces have microscopic irregularities, and contact is only between highs, as shown in Fig. 8.5. Adhesives bond solid surfaces together because they fill the irregularities while in the liquid state, and then develop sufficient cohesive strength by drying or setting. Solids can also be bonded together if they are ductile enough to be forced into intimate contact. For instance, the oldtime blacksmith welded two steel bars together by heating them white hot, and hammering them together. For the same reason, shales adhere to a bit or to drill collars if they are plastic enough to be forced into intimate contact by the weight of the drill string (see section on bit balling in Chapter 10, Drilling Problems Related to Drilling Fluids).

SURFACTANTS

The term *surfactant* is the standard contraction for *surface active agent*, so called because these agents are adsorbed on surfaces and at interfaces, and lower the surface free energy thereof. They are used in drilling fluids as emulsifiers, wetting agents, foamers, defoamers, and to decrease the hydration of clay surfaces. Many surfactants are discussed in various chapters of this book. The following discussion in this chapter is general in nature, while subsequent sections will cover specific surfactants as appropriate.

Surfactants may be either cationic, anionic, or nonionic. *Cationics* dissociate into a large organic cation and a simple inorganic anion. They are usually the salt of a fatty amine or polyamine, e.g., trimethyl dodecyl ammonium chloride:

$$\left[C_{12}H_{25} - \underset{\underset{CH_3}{|}}{\overset{\overset{CH_3}{|}}{N}} - CH_3 \right]^+ [Cl]^-$$

Anionics dissociate into a large organic anion and a simple inorganic cation. The classic example is a soap, such as sodium oleate:

$$[C_8H_{17}CH:CH(CH_2)_7COO]^- [Na]^+$$

Nonionic surfactants are long chain polymers which do not dissociate, for example, phenol 30-mol ethylene oxide:

$$C_6H_5 - O - (CH_2CH_2O)_{30}H$$

which is known in the drilling industry as DMS (Burdyn and Wiener, 1957).

Since clay minerals and most rock surfaces are negatively charged, the eletrostatic attraction causes the cationic surfactants to be more strongly adsorbed thereon, because of the electrostatic attraction. Anionic surfactants are adsorbed at the positive sites at the ends of clay crystal lattices and at oil−water interfaces. Nonionics, such as DMS, compete with water for adsorption on the basal surfaces of clay crystals (Foster and Waite, 1956), thereby limiting the expansion of swelling clays such as bentonite.

Other nonionics are adsorbed at oil−water interfaces. These compounds consist of an oil-soluble (lipophilic) chain of atoms linked to a water-soluble (hydrophilic) chain. The lipophilic portion dissolves on the oil side of the interface, and the hydrophilic on the water side. An enormous number of such compounds can be synthesized from polyhydric alcohol anhydrides and polyoxyethylene to suit various applications. Two factors help determine their suitability for particular application: the chemical identity of the two chains and the *HLB number* (Griffin, 1954). The latter is defined as the ratio by weight of the hydrophilic part of the molecule to the lipophilic; the greater the HLB number, the more water-soluble is the molecule. Fig. 8.6 shows how the molecule is pulled more into the water phase as the length of the polyoxyethylene chains, and consequently the HLB number, increases.

HLB of a surfactant is also related to solubility. A low HLB will likely be oil-soluble, while a high HLB will be water-soluble. This is a generality and solubility can depend on other factors. From experience (Table 8.2; Croda, 2012), emulsifiers can be classified by HLB range.

Note that the two factors—HLB number and chemical identity—serve only as guides in selecting nonionic surfactants. Final selection must be based on experimental evidence.

Many surfactants perform dual functions, e.g., they may act as an emulsifier and as a wetting agent. Alternatively, blends of compatible surfactants may be used to accomplish several purposes.

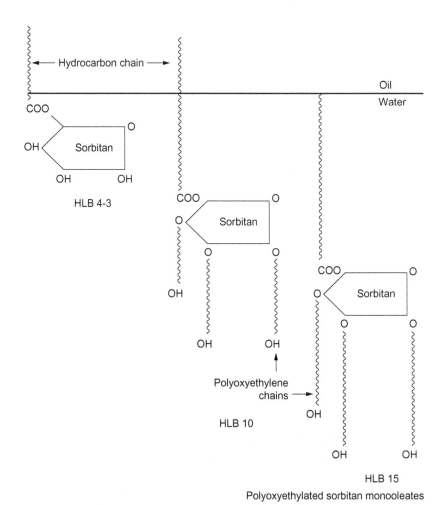

FIGURE 8.6 Effect of hydrophilic–lipophilic balance on the solubility of polyoxyethylated sorbitan monooleates.

EMULSIONS

The interfacial tension between oil and water is very high, so if the liquids are mixed together mechanically, they separate immediately as the agitation ceases, to minimize the interfacial area. Lowering the interfacial tension with a surfactant enables one liquid to form a stable dispersion of fine droplets in the other. The lower the interfacial tension, the smaller the droplets and the more stable the emulsion. The interfacial tension between mineral oil and water is about 50 dynes/cm, and a good emulsifier will lower it to about 10 dynes/cm. In most emulsions, oil is the *dispersed phase* and water is the *continuous phase*

TABLE 8.2 General Range of HLB Characteristics Versus Usage (Croda 2012)

HLB Range	Use
4−6	W/O emulsifiers
7−9	Wetting agents
8−18	O/W emulsifiers
13−15	Detergents
10−18	Solubilizers

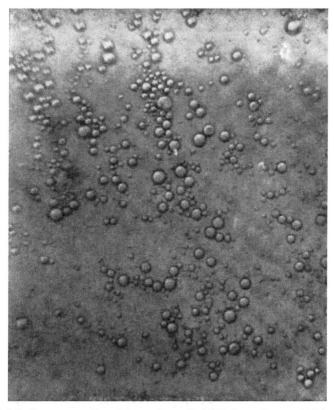

FIGURE 8.7 Two percent oil emulsion mud magnified 900 times. *Courtesy M-I Swaco (formerly Dresser Magcobar).*

(see Fig. 8.7), but drilling fluid invert emulsions, in which water is the dispersed phase, can be made with a suitable emulsifier. Besides lowering interfacial tension, the emulsifier stabilizes the emulsion, because its molecules are adsorbed at the oil−water interfaces, forming a skin around the droplets (see

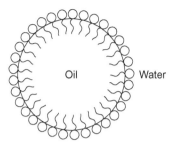

FIGURE 8.8 Protective skin of surfactant molecules around oil droplet (schematic).

Fig. 8.8). This skin acts as a physical barrier, preventing the droplets from coalescing when they collide.

Emulsion droplets may carry a small electrostatic charge. The consequent mutual repulsion contributes to the stability of the emulsion, but the charge can only be maintained in low-salinity (i.e., low-conductivity) water. The stability of an emulsion increases with increase in viscosity of the continuous phase because the number of collisions between the droplets is decreased. Similarly, the stability decreases with increase in temperature because the number of collisions increases.

The viscosity of an emulsion increases with the increase in the proportion of the dispersed phase, but major increases only take place when the proportion of the dispersed phase exceeds 40%. For example, an oil-in-water emulsion had a viscosity of 12 MPa•s when the oil content was 40%, 35 mPa•s when it was 60%, and 59 mPa•s when it was 70%. From a drilling fluid standpoint this results in an increasing API Funnel Viscosity and API Plastic Viscosity.

Although the maximum packing of spheres of uniform diameter is 74% of the gross volume, the proportion of the dispersed phase can exceed this limit, partly because the droplets are not uniform in size, but also because they are deformable. However, it becomes increasingly difficult to form a stable emulsion as the proportion of the dispersed phase is increased above 75%.

Because the creation of new surfaces involves work, some degree of mechanical agitation is required to form an emulsion, but the lower the interfacial tension, the less the amount of work required. In certain cases the emulsifier lowers the interfacial tension so much that merely pouring the constituents together provides sufficient agitation to form the emulsion. In other cases, a high input of energy is required. The most important requirement of a mixer is a high mixing shear rate, which may be obtained either by a high-speed rotor with a narrow clearance between the blades and the housing, or by forcing the components through a small orifice under high pressure (See mixing equipment in Chapter 3, Evaluating Drilling Fluid Performance).

Whether an oil-in-water (O/W) or a water-in-oil (W/O) emulsion is formed depends on the relative solubility of the emulsifier in the two phases. Thus, a preferentially water-soluble surfactant, such as sodium oleate, will form an O/W emulsion because it lowers the surface tension on the water side of the oil−water interface, and the interface curves towards the side with the greater surface tension, thereby forming an oil droplet enclosed by water. On the other hand, calcium and magnesium oleates are soluble in oil, but not in water, and thus form W/O emulsions. Similarly, a nonionic surfactant with a large hydrophilic group (HLB number 10 to 12) will be mostly soluble in the water phase, and thus form an O/W emulsion, whereas a nonionic surfactant with a large lipophilic group (HLB number about 4) will form a W/O emulsion (Griffin, 1954).

Typical O/W emulsifiers historically used in freshwater muds are alkyl aryl sulfonates and sulfates, polyoxyethylene fatty acids, esters, and ethers. A polyoxyethylene sorbitan tall-oil ester (Mallory et al., 1960), sold under various trade names, is used in saline O/W emulsions, and an ethylene oxide derivative of nonylphenol, $C_9H_{19}C_6H_4O(CH_2-CH_2-O)_{30}H$, known as DME, is used in calcium-treated muds (Burdyn and Wiener, 1957). Fatty acid soaps, polyamines, amides, or mixtures of these are used for making W/O emulsions.

An O/W emulsion can be broken by adding a small amount of a W/O emulsifier, and vice versa. In either case, the emulsion will be reversed if too much of the contrasting emulsifier is added.

Whether a given emulsion is oil-in-water or water-in-oil can easily be determined by adding some of the emulsion to a beaker of water. Because its continuous phase is water, an O/W emulsion will disperse readily in the water, whereas a W/O emulsion will remain as a separate phase.

Stable emulsions can be formed without the presence of a surfactant by the adsorption of finely divided solids, such as clays, drilled solids, weight material, CMC, starch, and other colloidal materials, at the oil−water interfaces. A skin of solid particles is thus formed around the dispersed droplets, which prevents their coalescence. Since the particles do not significantly lower the interfacial tension, they are known as *mechanical* emulsifiers or a Pickering-stabilized emulsion (Pickering, 1907).

Strongly water-wet particles will remain wholly in the water phase, and strongly oil-wet particles will remain wholly in the oil phase, so that in neither case will they act as mechanical emulsifiers. In order to form stable emulsions, the particles must be somewhere between slightly oil-wet and slightly water-wet, so that they remain partly in each phase, as shown in Fig. 8.9. Ideally the most stable emulsion is formed when the contact angle is 90°.

Most aqueous drilling fluids contain the finely divided particles required to form mechanical emulsions, and the electrochemical conditions are such that the particles are adsorbed at oil−water interfaces. Dispersed clays and various colloidal additives, especially lignosulfonates in alkaline solution

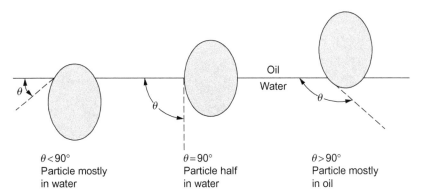

FIGURE 8.9 Idealized diagram showing effect of contact angle on immersion of particle. The most stable emulsion is formed when $\theta = 90°$ (schematic).

(Browning, 1955), act as mechanical emulsification agents, so that quite stable O/W emulsions are formed merely by adding oil and providing sufficient mechanical agitation. As a rule, however, mechanical emulsions are not as stable as chemical ones. When sufficient stability is not achieved, the emulsion can be stabilized by adding small quantities of a suitable chemical emulsifier.

OIL-WETTING AGENTS

Nitrogen compounds with long hydrocarbon chains are the most frequently used oil-wetting agents. N-alkyl trimethylene diamine chloride is a typical example. The salt ionizes in water to yield a large organic cation and two chloride anions:

$$\left[\begin{array}{ccccccccc} & H & & H & & H & & H & & H \\ & | & & | & & | & & | & & | \\ R & - & N & - & C & - & C & - & C & - & N & - & H \\ & | & & | & & | & & | & & | \\ & H & & H & & H & & H & & H \end{array} \right]^{++} [Cl]^- [Cl]^-$$

where R is an alkyl chain with 18 carbon atoms. The diamine acts as an O/W emulsifier as well as a wetting agent. The cation's nonpolar tail dissolves in the oil phase, and the nitrogen polar end extends into the water, thus imparting a positive charge to the oil droplet. Because most metal and mineral surfaces carry a negative surface charge, the oil droplets are attracted to the surface, where they break, depositing a film of oil as shown in Fig. 8.10. Many other hydrocarbon-nitrogen compounds may be used as combined wetting agents and emulsifiers, provided the length of the hydrocarbon chain exceeds 10 carbon atoms. A balance must be struck between the wetting and the emulsifying actions—too strong an emulsifying action and the oil will

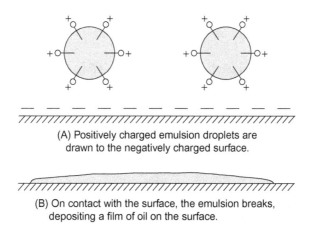

(A) Positively charged emulsion droplets are drawn to the negatively charged surface.

(B) On contact with the surface, the emulsion breaks, depositing a film of oil on the surface.

FIGURE 8.10 Mechanism of oil-wetting by an O/W emulsion (schematic).

not wet; too strong a wetting action and excessive oil is lost. The relative strength of the two actions depends upon various conditions, such as the nature of the surface to be coated, electrochemical conditions in the water phase, pH, temperature, etc., and the surfactant must be selected accordingly. For example, if oil-wetting is excessive, use one of the ethylene oxide adducts of diamines which reduces wetting in proportion to the number of moles of ethylene oxide added. Alternatively, a secondary emulsifier may be used, either a cationic or a nonionic one. Insufficient wetting may be corrected by adding an oil-soluble surface tension reducer, such as the salt of a fatty acid and a polyamine.

Wetting agents are sometimes added to O/W emulsion muds in order to oil-wet the drill string, and thus reduce torque, prevent bit balling, or inhibit corrosion. The problem with this approach is that the oil also coats the mud solids, and if considerable amounts of finely divided clays are present, the amount of oil lost may be excessive.

Oil-wetting agents are added to all oil-base muds to prevent the water-wetting of barite and drilled solids. Lecithin is effective for this purpose (Simpson et al., 1961). The structure of lecithin is

$$\left[\begin{array}{l} C_{17}H_{35}\text{-COO-}CH_2 \\ \quad | \\ C_{17}H_{35}\text{-COO-}CH \quad O \\ \quad | \quad\quad\quad\quad | \\ \quad\quad CH_2\text{-O-P-O}-(CH_2)_2 \\ \quad\quad\quad\quad\quad \| \\ \quad\quad\quad\quad\quad O \end{array} \right]^{-} [N(CH_3)_3]^{+}$$

The phosphate group carries a negative charge, and the quaternary amine a positive one. The nonpolar forked hydrocarbon tail dissolves in the oil phase, and the polar head in the water phase. If excess of lecithin is added, the interfacial tension is reduced.

Reaction products of clay and oil-wetting agents are used to prepare oil-dispersible additives for oil-base muds (Simpson et al., 1961). Clays that are used in water-base muds, such as bentonite and attapulgite, may be made oil-dispersible by base exchange reactions with an onium salt. The onium salt ionizes to yield

$$[C_{14}H_{29}NH_3]^+[Cl]^-$$

The inorganic cations on the clay surface are replaced by the organic cations of the onium salt, thus rendering the clay oil-dispersible. When added to an oil mud, such organophilic clays greatly improve the suspending properties (see Fig. 8.11). Similarly, lignite may be made oil-dispersible and used to reduce the filtration of oil muds.

Oil-wetting agents may also be used to make a mechanical invert emulsion with finely divided chalk. Chalk particles are normally water-wet, but by adding a small quantity of a fraction of tall oil, which consists mostly of oleic and linoleic acid, the particles may be made partially oil-wet (Darley, 1972). Upon agitation the particles are adsorbed at the oil—water interfaces around the water droplets, resulting in a mechanical emulsion whose continuous phase is oil.

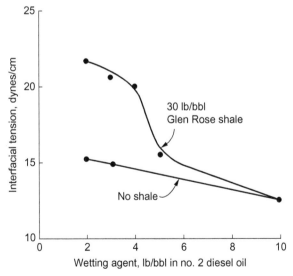

FIGURE 8.11 Adsorption of wetting agent by shale particles in oil as indicated by interfacial tension of diesel oil against water. *From Simpson, J.P., Cowan, J.C., Beasley, A.E., 1961. The new look in oil-mud technology. J. Petrol. Technol. 1177–1183. Copyright 1961 by SPE-AIME.*

FOAMS

When water is encountered in air or gas drilling, foaming agents are added to facilitate its removal. Foam formation is quite a simple matter: it merely requires the injection into the airstream of a surfactant that sufficiently lowers the surface tension of the water. Foams, however, tend to collapse in a short time because the surface free energy of the system is thereby reduced. In drilling, foam longevity must be considered when selecting a foamer.

Foams and Mists

Foams and mists are colloidal systems in which the two phases are a gas and a liquid. Distribution of the two phases depends on the relative amounts of each present. This ratio is usually expressed as either the volume fraction of the gas (foam quality) (David and Marsden, 1969) or the volume fraction of the liquid (LVF) (Beyer et al., 1972). In the foam quality range from 0 to about 0.54 the foam consists of independent bubbles dispersed in the gas (see Fig. 8.12); in the range from 0.54 to 0.96 the system is analogous to an emulsion with gas as the internal phase and the liquid as the external phase. Above a quality of 0.96 the system consists of ultramicroscopic droplets of water dispersed in the gas, and is termed a *must* or an *aerosol*.

The factors governing the formation and stability of foams is not well understood. Obviously, since foaming involves a huge increase in surface area, the reduction of surface tension by the addition of a surfactant is

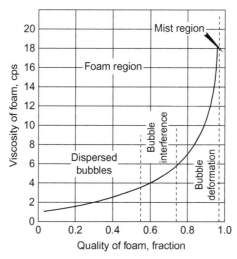

FIGURE 8.12 Effect of quality of foam on foam viscosity. *From Mitchell, B.J., 1971. Test data fill theory gap on using foam as a drilling fluid. Oil Gas J. 96–100. Courtesy of Oil and Gas J.*

essential. However, reduction of surface tension is not the only pertinent factor. The molecular structure of the surfactant also appears to be significant. One theory is that the anions are oriented normal to the surface of the film, and the cations are distributed in the solution between the walls (Bikerman, 1958b). Thus, the walls carry an electrostatic charge, and repulsion between these charges hinder coalescence. Because of the lack of basic theory, foaming agents must be evaluated by empirical tests (see Chapter 3, Evaluating Drilling Fluid Performance).

Foams are used for three purposes in the drilling industry: (1) to remove formation water that enters the borehole when drilling with air; (2) as a low-density fluid to remove drill cuttings and other solids when completing or working over wells in depleted reservoirs (Anderson et al., 1966); and (3) as an insulating medium in Arctic wells (Anderson, 1971).

If a water-bearing sand is encountered when drilling with air, the inflowing water accumulates in the bottom of the hole and creates a back pressure that increases air volume requirements and reduces the rate of penetration. In addition, it causes cuttings to ball up and adhere to the bit and drill string. Injection of a suitable foamer into the airstream enables the air to carry the water and the cuttings out of the hole as a foam. Maximum efficiency is achieved when all the inflowing water is converted into foam as it enters the hole, and the foam remains stable just long enough to reach the surface. The choice of surfactants depends on the salinity of the water and on whether or not oil is present. Suitable surfactants include anionic soaps, alkyl polyoxyethylene nonionic compounds, and cationic amine derivatives, all of which are commercially available.

The cutting-carrying capacity of foam depends on the square of the annular velocity and on the rheological properties of the foam. The rheological properties depend mainly on the viscosity of the air and the liquid, and on the quality of the foam (Mitchell, 1971) (see Figs. 8.12 and 8.13). When the foam quality is between 0.60 and 0.96, foam behaves as a Bingham plastic (Beyer et al., 1972; Krug and Mitchell, 1972). Buckingham's equation (see Eq. (6.12) in Chapter 6, The Rheology of Drilling Fluids) may be used to determine flow pressure/flow rate relations if modified to allow for slippage at the pipe wall and for changes in air/water ratios (and hence in foam viscosity) with pressure. Beyer et al. (1972) have determined the relationship between slippage, shear stress at the wall, and LVF, and between viscosity and LVF, by means of pilot-scale experiments. From these relationships and Buckingham's equation, they developed a mathematical model that describes the flow of foam in vertical tubes and annuli. Computer programs based on this flow model may be used to determine optimum gas and liquid flow rates, pressures, circulation times, and solids lifting capacity in prospective workover jobs (Millhone et al., 1972). Detailed analysis should be made whenever jobs in new fields or under new conditions are being undertaken (Figures 8.14–8.18).

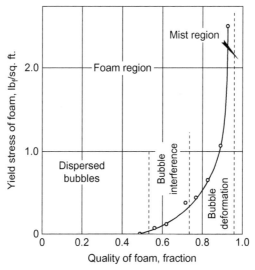

FIGURE 8.13 Effect of quality of foam on yield stress of foam. *From Mitchell, B.J., 1971. Test data fill theory gap on using foam as a drilling fluid. Oil Gas J. 96–100. Courtesy of Oil and Gas J.*

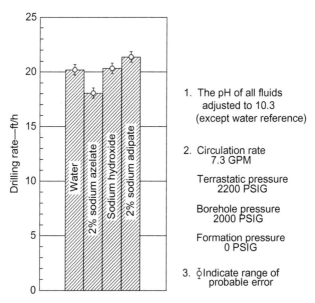

FIGURE 8.14 Comparison of effects of chemical additives with water on drilling rate of Indiana limestone with a 2-cone rock microbit at 50 rpm and 1000 lb bit weight. *From Robinson, L.H., 1967. Effect of hardness reducers on failure characteristics of rock. Soc. Petrol. Eng. J. 295–300. Trans. AIME 240. Copyright 1967 by SPE-AIME.*

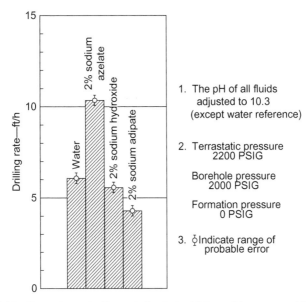

FIGURE 8.15 Comparison of effects of chemical additives with water on drilling rate of Indiana limestone with a 3-blade micro drag bit at 50 rpm and 500 lb bit weight. *From Robinson, L.H., 1967. Effect of hardness reducers on failure characteristics of rock. Soc. Petrol. Eng. J. 295–300. Trans. AIME 240. Copyright 1967 by SPE-AIME.*

The rheological properties of the foam in normal air-drilling operations are of no great concern because annular velocities sufficient to clean the hole are economically feasible. However, high annular velocities are undesirable when completing easily erodable formations in workover wells where minimum bottomhole pressures are advantageous, and in very large diameter holes. Under these circumstances, a preformed stiff foam, made from a surfactant plus bentonite or a polymer, is used (Anderson et al., 1966; Hutchison and Anderson, 1972).

DEFOAMERS

The action of defoamers is also not well understood, and they, too, must be selected empirically. High molecular weight alcohols and aluminum stearate are commonly used. Gas-cut muds may be thought of as a type of foam, but in this case the action is largely mechanical. The gas bubbles are trapped by the gel structure, and may be removed by breaking the gel by agitation or by the addition of thinning agents. The effect is enhanced if the operation is carried out under reduced pressure.

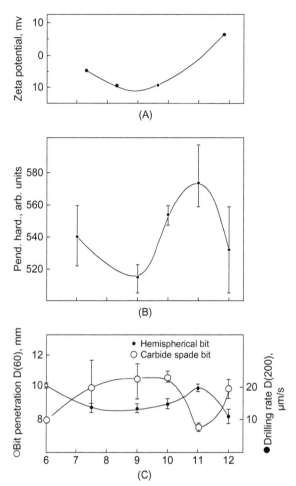

FIGURE 8.16 Variation of (A) zeta potential, (B) pendulum hardness, and (C) rate of penetration for calcite in water buffered environments. Note: The rate of penetration of the spade bit is given as total depth penetrated in the first 60 s. *From Jackson, R.E., Macmillan, N.H., Westwood, A.R.C., 1974. Chemical enhancement of rock drilling. In: Proceedings of the 3rd Cong. Internat. Soc. Rock Mechanics, September 1–7, Denver.*

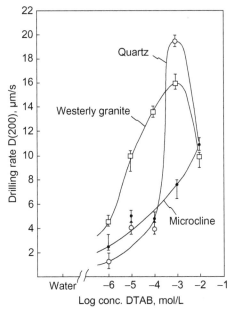

FIGURE 8.17 Variation in rate of drilling quartz, microcline, and Westerly granite with a diamond core bit in aqueous DTAB environments (2200 rpm). *From Jackson, R.E., Macmillan, N. H., Westwood, A.R.C., 1974. Chemical enhancement of rock drilling. In: Proceedings of the 3rd Cong. Internat. Soc. Rock Mechanics, September 1−7, Denver.*

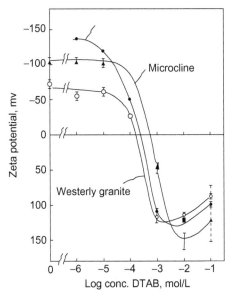

FIGURE 8.18 Zeta potentials of quartz, microcline, and Westerly granite in aqueous DTAB environments. *From Jackson, R.E., Macmillan, N.H., Westwood, A.R.C., 1974. Chemical enhancement of rock drilling. In: Proceedings of the 3rd Cong. Internat. Soc. Rock Mechanics, September 1−7, Denver.*

NOTATION

F_l surface free energy of a liquid
F_{sl} surface free energy at a solid/liquid boundary
g gravitational constant
h height of liquid rise in a capillary
r capillary radius
W_{coh} work of cohesion
W_{adh} work of adhesion
γ surface tension
ρ density
θ contact angle

REFERENCES

Anderson, G.W., 1971. Near-gauge hole through permafrost. Oil Gas J.129–142.

Anderson, G.W., Harrison, T.F., Hutchison, S.O., 1966. The use of stable foam as a low-pressure completion and sand-cleanout fluid. API Drill Prod. Pract. 4–13.

Beyer, A.H., Millhone, R.S., Foote, R.W., 1972. Flow behavior of foam as a well circulating fluid. SPE Paper No. 3986, Annual Meeting, October 8–11, San Antonio.

Bikerman, J.J., 1958a. Surface Chemistry; Theory and Applications, second ed. Academic Press, New York, pp. 8–13.

Bikerman, J.J., 1958b. Surface Chemistry; Theory and Applications, second ed. Academic Press, New York, p. 99.

Bikerman, J.J., 1958c. Surface Chemistry; Theory and Applications, second ed. Academic Press, New York, p. 340.

Browning, W.C., 1955. Lignosulfonate stabilized emulsions in oil well drilling fluids. J. Petrol. Technol. 9–15.

Burdyn, R.F., Wiener, L.D., 1957. Calcium surfactant drilling fluids. World Oil 101–108.

CRODA, Inc 2012. The HLB System.

Darley, H.C.H., 1972. Chalk emulsion–a new completion fluid. Petrol. Eng. 45–51.

David, A., Marsden, S.S., 1969. The rheology of foam. SPE Paper No. 2544, Annual Meeting, September 1, Denver.

Foster, W.R., Waite, J.M., 1956. Adsorption of polyoxyethylated phenols on some clay minerals. Amer. Chem. Soc., Symp. on Chemistry in the Exploration and Production of Petroleum, April 1956, Dallas, pp. 8–13.

Griffin, W.C., 1954. Calculation of HLB values of non-ionic surfactants. J. Soc. Cosmetic Chem. 5 (4), 1–8.

Hutchison, S.O., Anderson, G.W., 1972. Preformed stable foam aids workover drilling. Oil Gas J. 74–79.

Jackson, R.E., Macmillan, N.H., Westwood, A.R.C., 1974. Chemical enhancement of rock drilling. In: Proceedings of the 3rd Cong. Internat. Soc. Rock Mechanics, September 1–7, Denver.

Krug, J.A., Mitchell, B.J., 1972. Charts help find volume, pressure needed for foam drilling. Oil Gas J. 61–64.

Mallory, H.E., Holman, W.E., Duran, R.J., 1960. Low-solids mud resists contamination. Petrol. Eng. B25–B30.

Millhone, R.S., Haskin, C.A., Beyer, A.H., 1972. Factors affecting foam circulation in oil wells. SPE Paper 4001, Annual Meeting, October 8, San Antonio.
Mitchell, B.J., 1971. Test data fill theory gap on using foam as a drilling fluid. Oil Gas J. 96–100.
Pickering, S.U., 1907. Emulsions. J. Chem. Soc. 91.
Robinson, L.H., 1967. Effect of hardness reducers on failure characteristics of rock. Soc. Petrol. Eng. J. 295–300, Trans. AIME 240.
Sharpe, L.H., Schohorn, H., Lynch, C.J., 1964. Adhesives. Int. Sci. Technol. 26–37.
Simpson, J.P., Cowan, J.C., Beasley, A.E., 1961. The new look in oil-mud technology. J. Petrol. Technol. 1177–1183.
Wikipedia 2016a. Tensiometers. <www.Wikipedia.com>.
Wikipedia 2016b. Wettability. <www.Wikipedia.com>.

Chapter 9

Wellbore Stability[1]

INTRODUCTION

Maintaining a stable borehole is one of the major problems encountered in drilling oil wells. When the hole is not kept open, an additional string of casing may be required. The number of strings of casing that can be set in a well is limited. Also, each unplanned casing set adds to the cost of the well, as well as increasing the nonproductive time (NPT) effecting drilling efficiency.

Hole instability takes several forms: soft, ductile formations that are squeezed into the hole; hard, brittle formations that spall under stress; and, most common, shales that slump and cave, with consequent hole enlargement, bridges of cavings, and fill during connections and trips. These problems add greatly to drilling time and costs, and may result in disasters such as blowouts, stuck pipe, and sidetracked holes.

THE MECHANICS OF BOREHOLE STABILITY

Wellbore stability has two main considerations. First, the mechanical aspects of stability covering the stresses and pressures acting on the walls of the hole and the ability of the hole to resist such stresses (Maury and Sauzay, 1987); and second, instability resulting from physicochemical interactions between the drilling fluid and exposed shale (Mondshine and Kercheville, 1966). Instability may be caused by excessive stress alone, by physicochemical effects alone, or by a combination of both factors. Failure of the borehole in tension, i.e., induced fracturing, is covered in the section on loss of circulation in Chapter 10, Drilling Problems Related to Drilling Fluids.

Brief Review of the Geology and Geophysics of Sedimentary Basins

Sedimentary basins, in which the great majority of oil reservoirs are found, are formed by the deposition of sediments carried down by rivers and settled

---

1. A glossary of notation used in this chapter will be found immediately following this chapter's text.

under bodies of water. Sands are deposited near shore, silts further out, and clays in the deep lake or ocean water. When first deposited, the sediments are soft and have a high water content. As they are buried by succeeding sediments, water is squeezed out by compaction. At shallow depths the effects of compaction are fully reversible, i.e., cores or drill cuttings recovered at the surface can readily be dispersed to individual grains. Such sediments are termed *unconsolidated*. At greater depths the sediments gradually become *consolidated* by compaction and diagenesis. Diagenesis refers to mineralogical and chemical changes that take place under the influence of subsurface temperatures, pressures, and electrochemical environment. Eventually the grains are cemented together by deposition of minerals from siliceous or calcareous formation waters, forming sandstones, indurated shales, and claystones. When thus fully consolidated, rocks cannot be dispersed except by direct mechanical action such as grinding. Formations composed of sands and silts are termed *arenaceous*. Those composed of clays and mixtures of clays and silts are termed *argillaceous*.

As the sediments are buried, more water is moved out and the rock grains support more of the earth's overburden pressure. This called normal compactions. If there is a trapping mechanism, the water cannot be expelled and the liquid in the pore spaces will carry more of the overburden pressure and the rocks grains will support less.

Rapid sedimentation takes place under geosynclinal conditions, i.e., in regions where the earth's surface is slowly sinking (as is occurring at present in the Gulf of Mexico). Because of tectonic forces in the earth's crust, the surface may go through cycles of elevation and depression. Thus, old sedimentary basins may be raised above sea level, and may be overlain by or intermingled with other types of sediments, such as carbonates, sulfates, and rock salt, which are laid down by evaporation of saturated solutions in lagoons and inland lakes.

The Geostatic Gradient

A sediment at depth must support the weight of the sediments above it. The total stress, S, imposed by the overburden (solids plus water) is termed the *geostatic* (or lithostatic) pressure, and is given by the equation

$$S = \rho_B Z \qquad (9.1)$$

where ρ_B is the bulk density of the sediments, and Z is the depth. ρ_B is usually taken as 144 lb/ft^3 (19.2 lb/gal, 2.3 SG) so that the geostatic gradient $S/Z = 1$ psi/ft (0.23 kg/cm^2/m). But, as mentioned earlier, in recent sedimentary basins the bulk density is low at the surface and increases with depth. Fig. 9.1 shows the average geostatic gradient plotted against depth for the Gulf Coast region (Eaton, 1969).

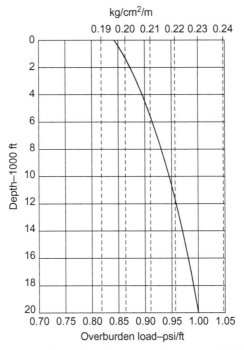

FIGURE 9.1 Composite overburden load for all normally compacted Gulf Coast formations. *From Eaton, B.A., 1969. Fracture gradient prediction and its application in oilfield operations. J. Petrol. Technol. 246, 1353–1360. Trans. AIME 246. Copyright 1969 by SPE-AIME.*

Hydrostatic Pore Pressure Gradients

Once a sediment has been compacted sufficiently for grain to grain contact to be established, the overburden load is carried independently by the solid matrix and the fluid in the pores (Hubbert and Willis, 1957), so that

$$S = \sigma + p_f \tag{9.2}$$

where σ is the intergranular stress between the grains, and p_f is the pressure exerted by the column of fluid within the pores, commonly referred to as the *formation pressure* or the *pore pressure*. p_f increases with depth and the density of the fluid so that

$$p_f = \rho_f Z \tag{9.3}$$

where ρ_f is the density of the pore fluid. The pore fluid in the Mid-Continent region is freshwater, so $p_f/Z = 0.433$ psi/ft (0.1 kg/cm^2/m) (Freeze and Frederick, 1967). In the Gulf Coast region, the salinity of the pore water is around 80,000 ppm and $p_f/Z = 0.465$ psi/ft (0.107 kg/cm^2/m) (Dickinson, 1953). In the Williston Basin, North Dakota, the salinity is about 366,000 ppm, and $p_f = 0.512$ psi/ft (0.118 kg/cm^2/m) (Finch, 1969).

Abnormal or Geopressured Gradients

Abnormal pressures occur when fluids expelled by a compacting sediment cannot migrate freely to the surface (Dickinson, 1953). This condition typically occurs in a thick argillaceous series, because clays develop very low permeabilities when compacted. For example, bentonite has a permeability of about 2×10^{-6} md when compacted at a pressure 8500 psi (600 kg/cm^2), corresponding to a depth of burial of 8500 ft (2600 m) (Darley, 1969). The permeabilities of shales are of the same order of magnitude. Sandstones are mostly in the range of $1-10^3$ md.

When a clay formation is in contact with a sand layer that provides a permeable path to the surface, water is at first freely expelled from the clay as the clay compacts. However, a layer of low-permeability compacted clay develops adjacent to the sand, restricting flow from the remainder of the clay body. Thus, in a thick clay formation, the rate of expulsion is not able to keep pace with the rate of compaction, and the pore pressure therefore increases above that normal for the depth of burial. Such shales are said to be *geopressured* (Stuart, 1970), or *abnormally pressured*. Any sand body, either interbedded in, or contiguous with the shale will also be geopressured if it is isolated from the surface either by pinch-out or faulting (see Fig. 9.2). Geopressures will, in the course of geologic time, decline to normal pressures, but the thicker the shale, the longer the time required.

Geopressures may have any value up to the total weight of the overburden, and the density of the mud must be increased accordingly. Thus, densities greater than 19.2 lb/gal (2.3 SG) may be required to control formation fluids—oil, gas, or water—in geopressured formations. Fig. 9.3 shows typical subsurface pressures and stresses in the Gulf Coast region. Because of the abnormally high water content of geopressured shales, a plot of bulk density versus depth will identify geopressured zones and indicate their magnitude, as shown in Fig. 9.4.

Geopressures may be encountered at quite shallow depths, e.g., at about 4000 ft (1200 m) in the Forties Field, North Sea (Koch, 1976). Highly geopressured formations, however, are only found at considerable depths, and the geopressuring is usually associated with the diagenesis of montmorillonite to illite. Illite contains much less water of hydration than montmorillonite, so the diagenesis is accompanied by the expulsion of water from the clay crystal, thereby increasing the geopressure. It has been established beyond doubt that the geopressures found in the Gulf Coast at depths of about 10,000 ft (3,000 m) are associated with diagenesis (Burst, 1969; Powers, 1967; Perry and Hower, 1972; Schmidt, 1973). It is known that the conversion of montmorillonite to illite takes place under pressure at temperatures of about 200°F (94°C), if the electrochemical environment is suitable. The requisite conditions are present in the Gulf Coast from about 10,000 ft down. Furthermore, some experiments with artificial diagenesis, in which sediments from Gulf Coast

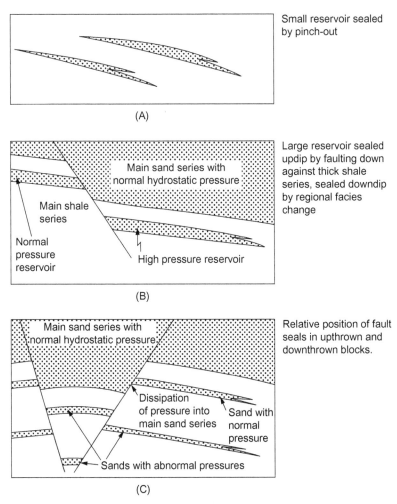

FIGURE 9.2 Types of reservoir seals necessary to preserve abnormal pressures (A) Pinchout, (B) Down fault, (C) Upthrown/Downthrown faults. *From Dickinson, G., 1953. Geological aspects of abnormal pressures in the Gulf Coast, Louisiana. AAPG Bull. 37 (2), 410−432. Proc. 3rd World Petrol. Cong. 1, 1−17. Courtesy AAPG.*

wells were heated with simulated seawater in pressure bombs, checked well with field data (Hiltabrand et al., 1973). Finally, montmorillonite is the dominant clay mineral found in cores down to 10,000 ft, but then decreases and is almost completely replaced by illite at 14,000 ft (4300 m).

Abnormally high pressures may also be found in formerly normally pressured formations that have been elevated above sea level by tectonic forces, and some surface layers then eroded. Isolated sand bodies within such formations will then have high pore pressures relative to their depth below the surface.

314 Composition and Properties of Drilling and Completion Fluids

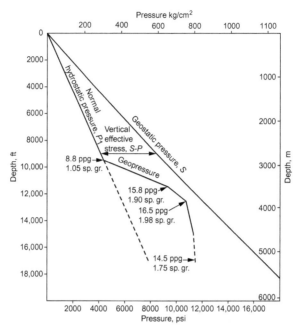

FIGURE 9.3 Geostatic and pore pressures in a typical Gulf Coast well. Effective stress is the intergranular or matrix stress, $S - P_f$. Figures on the pore pressure curve are equivalent mud densities.

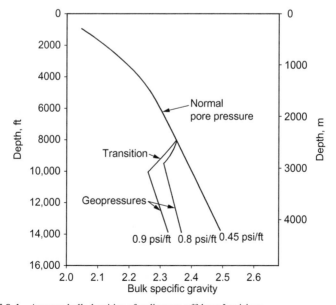

FIGURE 9.4 Average bulk densities of sediments, offshore Louisiana.

The Behavior of Rocks Under Stress

The behavior of rocks under stress may be studied in a triaxial tester, such as that shown in Fig. 9.5. A cylindrical specimen is enclosed in a flexible jacket, an axial load is applied by a piston, and an external confining pressure is exerted by a liquid surrounding the jacket. An internal pore pressure may also be applied. The usual procedure is to increase the axial load while maintaining constant confining and pore pressures, and to measure the resulting axial deformation (called the *strain*).

Results of such tests (Robinson, 1956; Hubbert and Willis, 1957) have shown that the deformation of the rock depends on the stress between the grains, and is independent of the pore pressure. Thus the *effective, intergranular*, or *matrix* stress is equal to the applied load less the pore pressure. The difference between the axial and confining intergranular stresses induces a *shear stress* in the specimen. Fig. 9.6 shows the relationship between the shear stress and the strain for three typical rocks. The relationship is linear, indicating elastic deformation, until the elastic limit is reached. The shear stress at this point

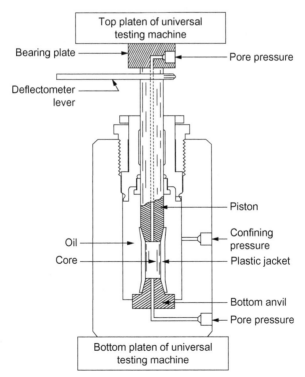

FIGURE 9.5 Triaxial test cell. *From Robinson, L.H., 1956. Effect of pore and confining pressure on failure characteristics of sedimentary rocks. Trans. AIME 216, 26−32. Copyright 1969 by SPE-AIME.*

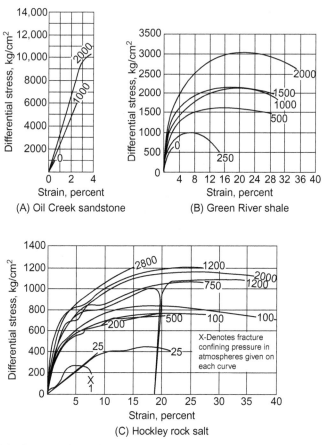

FIGURE 9.6 Stress–strain relationships for a sandstone, a shale, and a rock salt from triaxial tests of jacketed specimens. (A) Oil Creek sandstone. (B) Green River shale. (C) Hockley rock salt. *From Handin and Hager. Courtesy Shell Development Co.*

is called the *yield stress*. Above the yield stress two types of deformation may occur:

1. In *brittle failure*, the rock shatters suddenly. This type of failure is exhibited by hard, indurated rocks, such as sandstones, as shown in Fig. 9.6A.
2. *Plastic deformation* causes rapidly increasing strain with little increase in stress, or a decrease in stress, until the specimen eventually shatters. This type of failure is exhibited by ductile rocks, such as rock salt and shales, as shown in Fig. 9.6B and C.

Note particularly the difference between the *ultimate strength*, the *ultimate strain*, and failure. Ultimate strength is defined as the maximum stress on the stress–strain curve. Failure occurs when the ultimate strain is reached, and the rock shatters. In the case of brittle failure, the ultimate strength and the ultimate strain are reached at virtually the same stress. With both brittle and ductile rocks, the ultimate strength and the ductility increase with increase in confining pressure. Therefore, the strength and ductility of rocks in the earth's subsurface increase with depth of burial.

The Subsurface Stress Field

As already mentioned, a subsurface rock must bear the weight of the overburden, i.e., the solid matrix plus interstitial fluids. Eq. (9.4) shows that the effective stress gradient resulting from this load is

$$\frac{\sigma}{z} = \frac{S - p_f}{z} \qquad (9.4)$$

Since rocks are viscoelastic, the vertical stress generates horizontal components. According to Eaton (1969), the horizontal components are equilateral, and may be determined from *Poisson's ratio*, which is equal to unit change in transverse dimension divided by unit change in length. However, this thesis involves the assumption that the sediments are rigidly confined, so that no lateral movement takes place. That lateral movements do, in fact, take place is shown by the extensive faulting observed in the earth's crust.

Hubbert and Willis (1957) have pointed out that the horizontal stresses are modified by the tectonic forces that have acted throughout geologic history. They resolve the effective stresses acting on a subsurface rock into three unequal principal components acting at right angles to each other. Thus, σ_1 is the greatest principal stress, regardless of its direction; σ_2 is the intermediate principal stress, and σ_3 is the least principal stress. The three possible arrangements of these stresses are shown in Fig. 9.7. When the

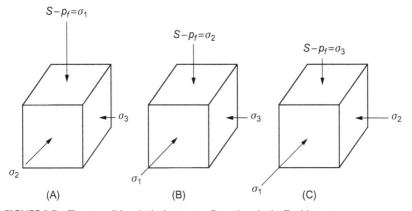

FIGURE 9.7 Three possible principal stress configurations in the Earth's crust.

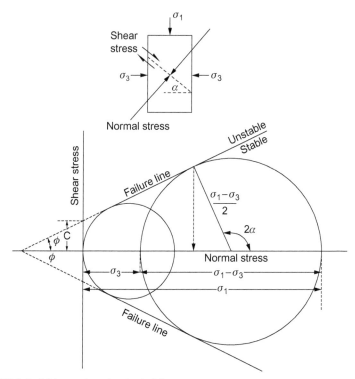

FIGURE 9.8 Mohr envelope for plastic failure.

difference between σ_1 and σ_3 exceeds the strength of the rock, faulting takes place, and the stress is relieved, but then gradually builds up again.

The conditions for faulting may be determined by constructing a Mohr diagram, using data from a triaxial test on the rocks of interest. Referring to Fig. 9.8, the axial and confining stresses (representing σ_1 and σ_3, respectively) at the maximum stress on the stress—strain curve are plotted as abscissae, and a circle drawn around them. The procedure is repeated at several confining pressures. The area enclosed by the tangents to these circles defines the conditions for stability. The intercept of the failure line on the y axis gives the cohesive strength of the rock, C, and the slope of the line, ϕ, is the angle of internal friction, which is a measure of the ductility.

From the geometry of the Mohr diagram it may be shown that the least principal stress at faulting is given by

$$\sigma_3 = \sigma_1 \left(\frac{1 - \sin \varphi}{1 + \sin \varphi} \right) - \frac{2c \cos \varphi}{1 + \sin \varphi} \qquad (9.5)$$

Hubbert and Willis (1957) showed that c for unconsolidated sand is 0, and φ is 30° (Fig. 9.9), in which case Eq. (9.5) reduces to $\sigma_3 = \frac{1}{3}\sigma_1$, and that

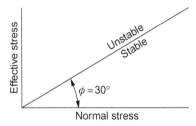

FIGURE 9.9 Mohr envelope for an unconsolidated sand. Cohesive strength is 0. Faulting occurs when $\sigma_3 = 1/3\sigma$ (schematic).

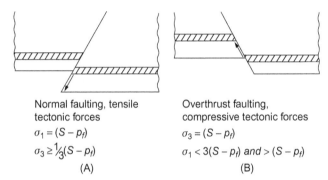

Normal faulting, tensile tectonic forces
$\sigma_1 = (S - p_f)$
$\sigma_3 \geq \tfrac{1}{3}(S - p_f)$
(A)

Overthrust faulting, compressive tectonic forces
$\sigma_3 = (S - p_f)$
$\sigma_1 < 3(S - p_f)$ and $> (S - p_f)$
(B)

FIGURE 9.10 (A) Normal faulting, tensile tectonic forces. (B) Overthrust faulting, compressive tectonic forces.

approximately the same relationship holds for sandstones and anhydrite. They concluded that in regions such as the Gulf Coast where tectonic forces are relaxed and tension faulting (see Fig. 9.10A) is prevalent, σ_3 would be horizontal and that its value would vary between ⅓ and ½ σ_1, depending upon the stress history. In regions where there were active compressive tectonic forces, as indicated by overthrust faulting (see Fig. 9.10B), σ_1 would be horizontal, and σ_3 would be vertical. In that case, the horizontal stress might be three times as great as the vertical, and the land would be rising.

Bear in mind that the relationships between σ_1 and σ_3 given by Hubbert and Willis (1957) are valid only for the stated values of c and φ. For rocks with substantially different values of c and φ the relation between the two stresses may be deduced from Eq. (9.5). For example, with unconsolidated clays, φ is 0 (see Fig. 9.11), and Eq. (9.5) reduces to

$$\sigma_3 = \sigma_1 - 2c \tag{9.6}$$

Stresses Around a Borehole

When a hole is drilled into a subsurface rock the horizontal stresses are relieved, and the borehole contracts until the radial stress at its walls is equal

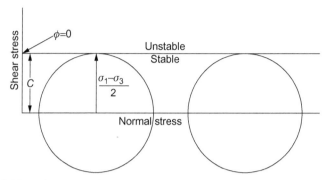

FIGURE 9.11 Mohr envelope for unconsolidated clay. The angle of internal friction, ϕ, is 0 in a perfectly ductile material. Failure occurs when $\sigma_1 - \sigma_3$ is twice the cohesive strength, C (schematic).

to the pressure of the mud column, p_w, minus the pore pressure, p_f. The load is transferred to a zone of hoop stresses that create tangential shear stresses around the borehole. The hoop stress is maximum at the wall, and decreases with radial distance into the formation. If the strain caused by the relief of radial stress does not exceed the elastic limit of the rock, deformation will be elastic and usually will be negligible. However, under certain conditions—e.g., deep holes, holes drilled closely to gauge—the deformation may be sufficient to cause pinched, or even stuck, bits. Once the deformation is reamed off, no further difficulties are experienced.

If the strain exceeds the elastic limit, the resulting deformation will be plastic because of the high confining stresses prevailing at depths of interest in the subsurface. A ring of plastically deformed rock therefore forms around the borehole (see Fig. 9.12). The radius of the borehole decreases and the outer radius of the plastic zone increases until the radial stress at the borehole walls equals $p_w - p_f$.

If the ultimate strain is not exceeded, the hole is then stable. The inner and outer radii of the plastic zone at stability depend upon the ductility of the rocks, φ; the cohesive strength of the rock, c; and the stress distribution in both the plastic and elastic zones. Because both the stresses in both zones increase with the depth of the hole, the width of the plastic zone required for stability also increases with depth. The outer radius of the plastic zone required for stability at the stated virgin horizontal stresses is shown for three rock types in Fig. 9.13. If the ultimate strain of the rock is exceeded before the required width of the plastic zone is reached, the borehole collapses (Broms, pers. comm.). Note that because of the vast mass involved, subsurface rocks deform slowly at a decreasing rate, a phenomenon known as creep.

With respect to the stability of the hole, the tangential shear stress, σ_θ, is the greatest principal stress; and the effective confining pressure, p_c, is the least principal stress. p_c is a function of the principal earth effective stresses,

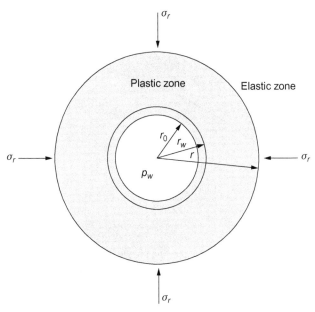

FIGURE 9.12 Plastic deformation of a bore hole. r_w = original radius of the hole, r_0 = hole radius after deformation, r = outer radius of plastic zone, p_w = pressure of the borehole fluid.

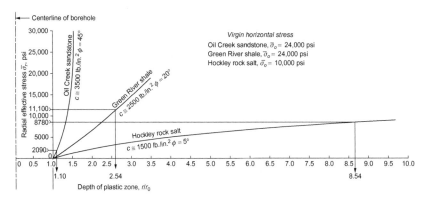

FIGURE 9.13 Determination of the depth of plastic zone for stability of a borehole through a sandstone, a shale, and a rock salt. *Note:* ϕ and c were determined from Mohr diagrams, using data from triaxial tests shown in Fig. 9.6. *From Broms, B., Personal communication to H.C.H. Darley. Courtesy Shell Development Co.*

σ_1, σ_2, σ_3, and $(p_w - p_f)$. Thus, the stability of the hole depends on the difference between σ_θ and p_c, and on their distribution in the surrounding formation. If σ_θ is plotted against p_c, the hole will be stable if the points fall below the Mohr failure curve for the formation in question, and unstable if they fall above (see Fig. 9.14). The Mohr failure curve is obtained from triaxial tests

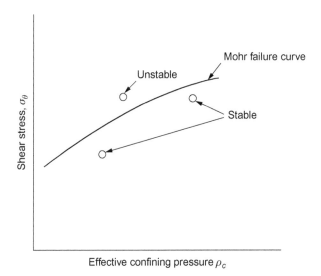

FIGURE 9.14 Criterion for compressive failure. σ_θ and p_c are derived from the three principal earth stresses and $p_w - p_f$. The Mohr failure curve is established from triaxial tests on cores from the formation in question.

made on specimens of the formation. Note that since an increase in $(p_w - p_f)$ increases the confining pressure, raising the density of the mud increases the stability of the hole. However, excessive increase in density will cause the hole to fail in tension, with consequent loss of circulation (see the section on induced fractures in Chapter 10, Drilling Problems Related to Drilling Fluids).

Fig. 9.15 shows the stress distribution around a borehole for the conditions that the virgin horizontal stresses are equal, $p_w = p_f$, and there is no fluid flow to or from the formation. Assuming these conditions, Broms calculated the elastic limit and the ultimate strain of the three rocks shown in Fig. 9.6. In assessing the results shown in Tables 9.1 and 9.2, bear in mind that the values in both tables were derived under rather unrealistic assumptions, and in practice the critical depths might deviate substantially from those shown.

The mechanics of borehole stability under more complex conditions have been reviewed by various authors (Westergaard, 1940; Topping, 1949; Galle and Wilhoite, 1962; Paslay and Cheatham, 1963; Remson, 1970; Desai and Reese, 1970; Daemen and Fairhurst, 1970; Gnirk, 1972; Nordgren, 1977; Fairhurst, 1968; Bradley, 1979a; Cheatham, 1984). Desai and Reese (1970) obtained essentially the same value for the Green River shale as Broms, but they used the finite element method.

Mitchell et al. (1987) used the finite element method to analyze borehole movements in sandstones and limestones and obtained quantitative agreement with field data. Maury and Sauzay (1987) modified the Mohr-Coulomb theory

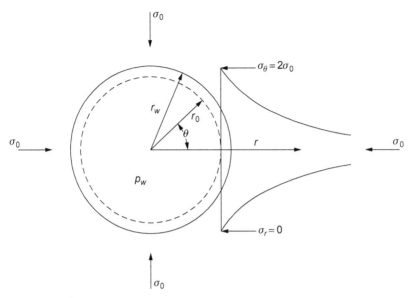

FIGURE 9.15 Stresses around a borehole when $p_w - p_f = 0$. r_w = radius of gauge hole; r_0 = radius of hole after deformation; σ_0 = virgin horizontal effective stress, assuming $\sigma_2 = \sigma_3$; σ_θ = hoop stress; σ_r = horizontal stress at radius r.

TABLE 9.1 Horizontal Virgin Effective Stresses, and Equivalent Depths, at Elastic Limit of Borehole Walls

Rock	σ_0 at Elastic Limit	Equivalent Depths Assuming	
		Horizontal Stress = 1/3 Vertical	Horizontal Stress = Vertical × 3
Oil creek sandstone	17,000 psi	94,000 ft	10,500 ft
	1190 kg/cm²	28,600 m	3200 m
Green river shale	7160 psi	40,000 ft	4400 ft
	500 kg/cm²	12,200 m	1340 m
Hockley salt	820 psi	4550 ft	500 ft
	60 kg/cm²	1390 m	150 m
Assumed:	Pore pressure gradient 0.46 psi/ft		
	Equilateral horizontal effective stresses		
	No mud cake, fluid flow, or temperature effects		
	Air in borehole		

Source: Data from Broms, courtesy Shell Development Co.

TABLE 9.2 Horizontal Virgin Effective Stresses, and Equivalent Depths, for Collapse of Borehole Walls

Rock	Ultimate Strain at Zero Confining Pressure	Corresponding σ_0	Equivalent Depths Assuming	
			Horizontal Stress = 1/3 Vertical	Horizontal Stress = Vertical × 3
Oil Creek sandstone	0.7	24,000 psi	133,000 ft	14,800 ft
		1687 kg/cm²	40,525 m	4510 m
Green River shale	8.3	24,000 psi	133,000 ft	14,800 ft
		1687 kg/cm²	40,525 m	4510 m
Hockley Salt	16.1	6000 psi	33,000 ft	3700 ft
		422 kg/cm²	10,055 m	1127 m

Assumed: Pore pressure gradient 0.46 psi/ft
Equilateral horizontal effective stresses
No mud cake, fluid flow, or temperature effects
Air in borehole

Source: Data from Broms, courtesy Shell Development Co.

to allow for the various stress conditions that occur around boreholes. They postulated 8—10 failure parameters, but found that any given drilling problem depended only on two or three of them. The various modes of failure are shown in Fig. 9.16. The top left-hand model corresponds to the spalling observed in the model borehole (see Fig. 9.17) and the top right-hand model corresponds to the anisotropic failure shown in Fig. 9.18.

Nordgren (1977) obtained expressions for σ_θ and p_c under the condition that the principal horizontal stresses are not equal, which occurs in regions of active faulting, as discussed earlier in this chapter. If the horizontal stresses are not equal, then σ_θ is not evenly distributed around the circumference of the hole—it is greatest parallel to the least horizontal stress, and least parallel to the greatest horizontal stress. Thus, when faulting is normal (tensile), σ_θ is greatest in the direction of the dip (see Fig. 9.19), and in the case of overthrust (compressive) faulting, it is greatest in the direction of the strike. Fig. 9.20 shows the variation in σ_θ on the circumference of the hole for three assumed ratios of principal horizontal stresses σ_A and σ_B. Note that significant hoop stresses extend only a few hole diameters into the formation.

Unfortunately, the value of the greatest horizontal stress is usually not known, and that of the least is only known in developed regions when it can be deduced from induced fracturing data. However, Hottman et al. (1978) were able to deduce both stresses in two wildcat wells in the Gulf of Alaska by noting the effect of various changes in mud density and pore pressure on hole stability, and plotting tentative values of σ_θ and p_c together with a Mohr failure curve, which was established from triaxial tests on the formation in question. The region is one of active overthrust faulting, and they found both horizontal stresses to be greater than the overburden:

$\sigma_1 = 1.3$ to 1.4 psi/ft (0.299 to 0.322 kg/cm^2/m)
$\sigma_2 = 1.03$ to 1.3 psi/ft (0.237 to 0.299 kg/cm^2/m)

When the hole is deviated significantly, the stability depends on the angle of the hole with respect to the principal horizontal stresses and on the value of those stresses. Bradley (1979a) has developed the concept of a stress cloud to enable the effect of the large number of variables to be demonstrated graphically, as shown in Fig. 9.21. Any portion of the stress cloud that crosses the Mohr failure curve indicates unstable conditions. Fig. 9.22 shows the effect of mud density and hole angle on stability. Bradley (1979b) also showed that the stress cloud method could be used to demonstrate the stress field in the vicinity of a salt dome.

The Influence of Hydraulic Pressure Gradient on Hole Stability

So far we have neglected the effect of fluid flowing into or out of the formation on the stress field around the borehole. Nevertheless, it has been theoretically (Paslay and Cheatham, 1963) and experimentally (Darley, 1969)

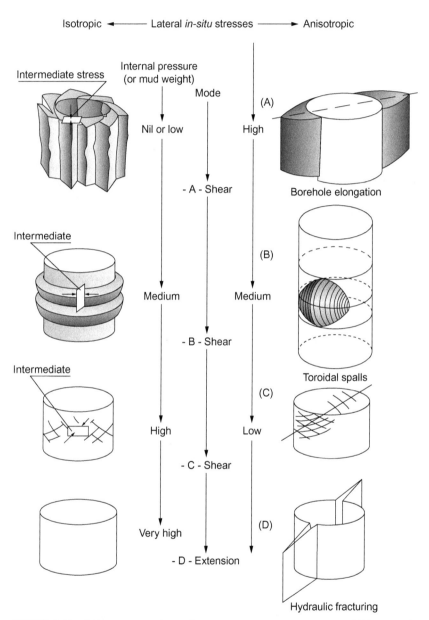

FIGURE 9.16 Failure modes at the wall. *From Maury, V.M., Sauzay, J.M., 1987. Borehole instability: case histories, rock mechanics approach, and results. SPE/IADC Paper 16051, Drill. Conf., New Orleans, LA, March 15–18. Copyright 1987 by SPE-AIME.*

Wellbore Stability Chapter | 9 327

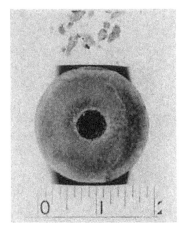

FIGURE 9.17 Spalling of reconstituted shale specimen in model borehole. Atoka shale, bulk density of specimen 2.52 g/cm^3, maximum applied stress 4400 psi (309 kg/cm^2), yield stress 2800 psi (190 kg/cm^2), air in the hole. Note typical tangential fracture pattern caused by excessive hoop stresses. *From Darley, H.C.H., 1969. A laboratory investigation of borehole stability. J. Petrol. Technol. 883–892. Trans. AIME 246. Copyright 1969 by SPE-AIME.*

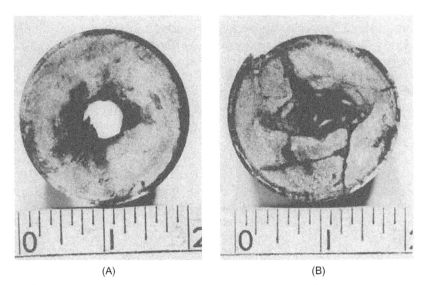

(A)　　　　　　　　　　　　(B)

FIGURE 9.18 Failure caused by stress concentrations. (A) Initiation of spalling at points of maximum stress. (B) Collapse with time at same stress. Natural specimen cut from core of Mitchell shale. *Courtesy of Shell Development Co.*

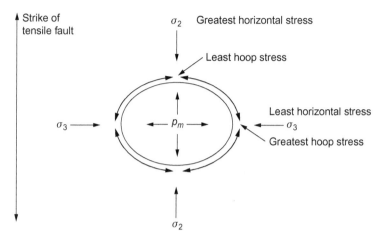

FIGURE 9.19 Stress concentrations in a region of tensile faulting. In a region of compressive faulting, the location of the greatest and least stresses with respect to the strike would be reversed.

shown that the hydraulic pressure gradient thus created can materially influence hole stability. The magnitude and direction of flow in the pores of the formation is a function of the pressure differential, Δp, between the pressure exerted by the mud column, p_w, and the formation pore pressure, i.e., $\Delta p = p_w - p_f$.

At the instant of penetration Δp is applied at the face of the borehole, but in the course of time a pressure gradient is established in the pores of the formation. When equilibrium conditions have been reached, the pore pressure, p_r, at any radius, r, from the center of the hole is given by the well-known radial flow equation

$$p_r = \frac{\mu q}{2\pi k} \ln\left(\frac{r}{r_w}\right) \tag{9.7}$$

where μ is the viscosity of the fluid, q is the rate of flow per unit vertical thickness, k is the permeability of the formation, and r_w the diameter of the borehole.

At the wall of the hole $\Delta p = -p_f$ in the case of a hole being drilled with air. In high-permeability formations the resulting inflow of fluid will be excessive, and air drilling must be discontinued, but in low-permeability formations the rate of inflow may be tolerable. The resulting distribution of p_r at two time intervals is shown in Fig. 9.23A. The effect of this hydraulic gradient is to increase the hoop stress gradient, and hence to destabilize the hole. In air-drilled holes such hydraulic gradients are of no great consequence, because inflow rates are low, but similar gradients are established in producing wells,

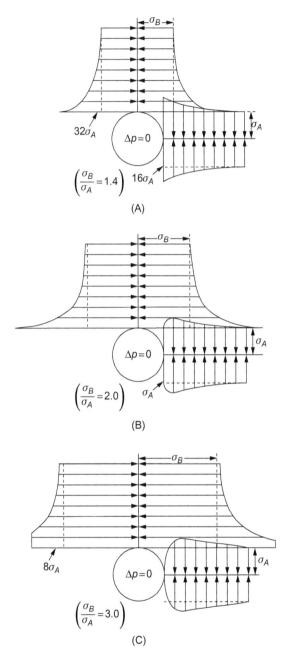

FIGURE 9.20 Stress states about a bore hole for regional-stress ratios σ_B/σ_A of (A) 1.4, (B) 2.0, and (C) 3.0. *From Hubbert, M.K., Willis, D.G., 1957. Mechanics of hydraulic fracturing. J. Petrol. Technol. 210, 153–166. Trans. AIME, 210. Copyright 1959 by SPE-AIME.*

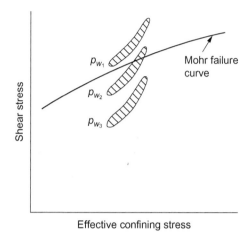

FIGURE 9.21 Stress clouds for three borehole stresses, $p_{w3} > p_{w2} > p_{w1}$, other conditions remaining the same. *After Bradley, W.B., 1979a. Mathematical concept—stress cloud—can predict borehole failure. Oil Gas J. 77, 92–97. Courtesy of the Oil and Gas Journal.*

and, where inflow rates are high, may be a major factor affecting hole stability (Broms, pers. comm.).

In wells drilled with liquids, the density of the mud column is normally maintained great enough to ensure that p_w exceeds p_f by a safe margin, so that the flow of fluid is from the hole to the formation, thereby reducing the hoop stress. If no mud cake is established, as when drilling with brine, the hydraulic pressure gradient decreases with r, as shown in Fig. 9.23B, but the pressure drop across an element on the wall of the hole is small. If, however, there is mud in the hole, a filter cake is established and virtually the whole pressure drop is then on the face of the formation thus preventing the caving of sand grains and fractured shale fragments. While drilling low permeability shales, a filter cake, similar to higher permeability producing zones, is not formed. Water invasion into these shale formations is by either capillary action or osmosis, discussed later in this chapter.

Occurrence of Plastic Yielding in the Field

Pure plastic yielding as discussed earlier in this chapter occurs only with two types of formation: rock salt and soft unconsolidated shales. Except for anomalies in the earth's stress field in the vicinity of salt domes, there is little problem in drilling salt beds encountered at shallow depths. Even if the yield stress is exceeded, the rate of deformation is slow, and the hole may be kept to gauge by reaming. Reaming does not significantly affect the strength of the borehole, because the amount reamed off is small relative to the width of the plastic zone, and readily compensated for by a slight increase in the outer radius. Because of the high ductility of salt, slow deformation (creep) will continue as long as the salt formation is exposed.

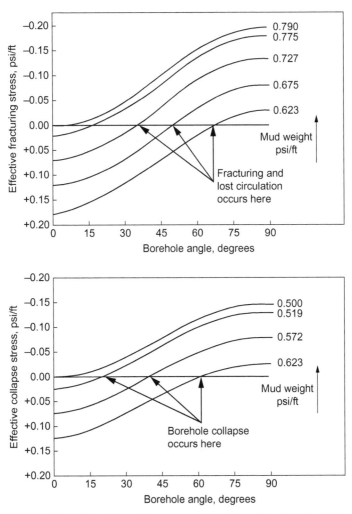

FIGURE 9.22 Effect of borehole angle on compressive failure. *From Bradley, W.B., 1979a. Mathematical concept—stress cloud—can predict borehole failure. Oil Gas J. 77, 92–97. Courtesy of the Oil and Gas Journal.*

Much greater problems are experienced when *deep* salt beds are drilled. As shown in Fig. 9.6, the ultimate strength of salt increases comparatively little with increase in confining pressure. On the other hand, the strength decreases markedly with temperature, and the ductility increases. At depths below about 10,000 ft (3000 m) the effect of confining pressure is more than offset by the effect of temperature, and the strength actually decreases with increase in depth. Because of its high ductility, the salt may transmit almost the full weight of the overburden, and mud densities in excess of 19 lb/gal

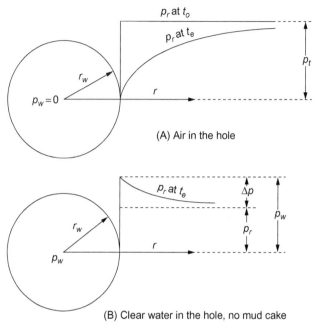

FIGURE 9.23 Pressure gradients of fluids flowing to or from borehole when drilled with air (A) and clear water (B). p_r at t_o = pressure gradient at the instant of penetration. p_r at t_e = pressure gradient when equilibrium conditions have been established.

(2.3 SG) may be necessary to control plastic flow (see Fig. 9.24) (Leyendecker and Murray, 1975).

Salt beds may be drilled either with saturated salt muds or with oil-base muds. The use of saturated salt fluids is tricky, because the brine may be saturated at the surface, but undersaturated at the high temperatures prevailing subsurface. Slight undersaturation is beneficial because the brine dissolves salt off the walls of the hole, thereby counteracting creep. However, if the salinity is too low, too much salt is dissolved and the hole enlarges excessively (Sheffield et al., 1983). An additional complication is that some salt formations contain substantial amounts of potassium, calcium, and magnesium chlorides that dissolve in saturated sodium chloride brines. These salts are present in the Zechstein salt beds that are encountered in North Sea wells, but the problem has been overcome by the use of saturated brines from wells producing from the Zechstein formation (Moore, 1977).

Oil-base fluids whose internal phase is saturated brine do not dissolve salt formations; therefore, they eliminate hole enlargement. On the other hand, they increase the risk of stuck pipe, but the pipe can usually be freed by circulating about 20 barrels of freshwater. Problems associated with water-wetting of barite and other mud constituents are sometimes experienced, necessitating

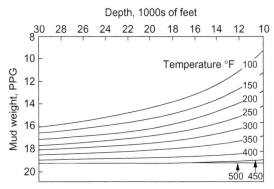

FIGURE 9.24 Density of mud required to control plastic flow of salt formations. Mud weights given are the theoretical densities that will permit a creep rate no greater than 0.1% per hour at the specified temperatures at the depth of interest. *From Leyendecker, E.A., Murray, S.C., 1975. Properly prepared salt muds aid massive salt drilling. World Oil 180, 93–95.*

treatment with oil-wetting agents and gelling agents (Leyendecker and Murray, 1975).

Plastic yielding may also occur when drilling at shallow depths in recent argillaceous sediments, such as are encountered in the Gulf Coast and North Sea. The predominant clay mineral in these sediments is sodium montmorillonite, and the sediments have a high water content, as will be discussed in the next section. Because of their soft, putty-like nature, these sediments are usually referred to as *gumbo shales*.

Unconsolidated shales have an angle of internal friction of zero, and a Mohr diagram shows that they deform plastically whenever the shear stress exceeds their cohesive strength, as shown in Fig. 9.11. Plastic yielding of this kind is illustrated in Fig. 9.25A (Darley, 1969). The specimen shown was made by compacting a slurry of Miocene drill cuttings to a bulk specific gravity of 2.0. The specimen was deformed in the model borehole shown in Fig. 9.26 and yielded under an isotropic load of 1700 psi (120 kg/cm^2). Plastic yielding may also occur with gumbo shales even if their yield stress is not exceeded. In this case, uptake of water from a freshwater mud causes the sides of the hole to swell and deform as shown in Fig. 9.25B.

In a drilling well, plastic deformation of gumbo shales gives rise to large volumes of cuttings and spallings, sometimes in sufficient quantity to blank out the shaker screen. Sometimes they agglomerate when rising in the annulus, forming mud rings large enough to block the flow line. Frequent reamings are necessary to avoid stuck pipe. The tendency of the cuttings and spallings to swell and disperse when rising in the annulus aggravates mud problems.

Gumbo shales can be drilled extremely fast, but hole cleaning considerations limit drilling rates. The maximum permissible rate is dependent on the circulation rate. Even with drilling rates thus limited, the time of exposure before protective casing is set is sufficiently short to enable the interval to be

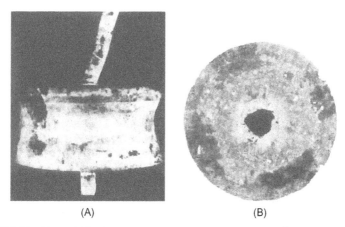

(A) (B)

FIGURE 9.25 Plastic yielding in model borehole. (A) bulk density 2.0 g/cm^3, air in the hole, stress at failure 1700 psi (120 kg/cm^2). (B) Bulk density 2.22 g/cm^3, 5% bentonite mud in hole, applied stress 1000 psi (70 kg/cm^2). *From Darley, H.C.H., 1969. A laboratory investigation of borehole stability. J. Petrol. Technol. 883–892. Trans. AIME 246. Copyright 1969 by SPE-AIME.*

drilled with low-density inhibited muds (Moore, 1977). The antiaccretion additives discussed in the bit balling section of Chapter 10, Drilling Problems Related to Drilling Fluids, have in many cases been successful in minimizing the effect of gumbo shales. The types of inhibited muds for gumbo drilling are discussed later in this chapter.

Lastly, plastic yielding may be experienced in geopressured shales. By definition, geopressured shales have a high water content relative to their depth of burial. For example, Fig. 9.4 shows that the bulk density at 10,000 ft of a shale with a pore pressure gradient of 0.9 psi/ft is the same as that of a normally pressured shale at about 5000 ft. The geopressured shale has to withstand an overburden load of 10,000 ft, and is therefore liable to deform plastically into the hole. Deformation is normally prevented because the mud density is raised to prevent the inflow of formation fluids from interbedded high permeability sands: the increase in density also prevents the inflow of shale. But when the shale contains no interbedded sands, the geopressure may not be recognized and the density of the mud not raised, in which case shale will be squeezed into the hole. Gill and Gregg (1975) state that in North Sea wells an underbalance of a few hundred psi—1 lb/gal or less in mud density—may permit plastic flow.

Brittle-Plastic Yielding

We stated earlier that deformation of all rocks at depth was plastic because of the high confining pressure. However, penetration by the bit allows the radial stress at the wall of the hole to fall until it equals $(p_w - p_f)$. Therefore,

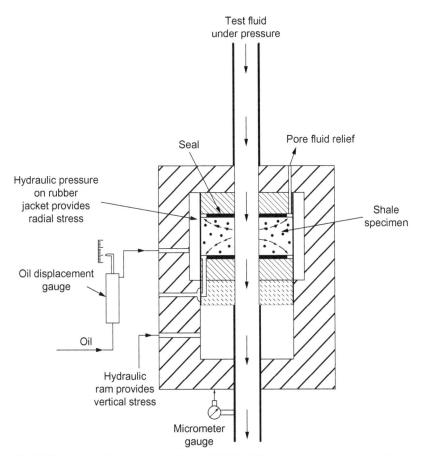

FIGURE 9.26 Model borehole. *From Darley, H.C.H., 1969. A laboratory investigation of borehole stability. J. Petrol. Technol. 883–892. Trans. AIME 246. Copyright 1969 by SPE-AIME.*

deformation may be brittle in the immediate vicinity of the hole and plastic further into the formation as the confining pressure increases. This type of yielding is most likely to occur in an air-drilled hole because $(p_w - p_f)$ is negative. This phenomenon has been demonstrated in the model borehole (Fig. 9.26). Specimens of consolidated shales, both natural and artificial, with air in a predrilled hole, were subjected to increasing triaxial stress. When the stress exceeded the yield stress, spalling developed around the hole, as shown in Fig. 9.17. When the specimen was sectioned at the conclusion of the experiment, concentric slip rings indicating plastic deformation could be seen around the spalled zone. In another test, a specimen was subjected to stress well in excess of the yield stress, and a hole then drilled. The rate of radial deformation decreased with time, as shown in Fig. 9.27. Other tests showed that creep continued indefinitely.

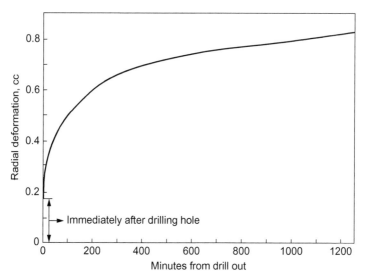

FIGURE 9.27 Decay of deformation with time, in model borehole with 2-inch specimen, 1/2-inch hole. *From Darley, H.C.H., 1969. A laboratory investigation of borehole stability. J. Petrol. Technol. 883–892. Trans. AIME 246. Copyright 1969 by SPE-AIME.*

As previously discussed, in a tectonically stressed region the shear stress on the wall of the hole will be maximum parallel to the least principal horizontal stress; one would therefore expect spalling to be at those points. This hypothesis was confirmed by Hottman et al. (1978), who reported that a high resolution, four arm caliper survey in the previously mentioned well in the Gulf of Alaska showed that, at the depth where spalling occurred, the diameter of the hole was 18½ in (47 cm) parallel to the least horizontal stress and to gauge (12¼ in, 31 cm) 90 degrees to it.

Stress concentrations may also be caused by out-of-round holes, e.g., key seats, doglegs, and anisotropic rock properties, such as bedding planes and oriented fracture systems. Spalling at specific points creates out-of-round holes, thereby increasing the stress concentrations. This self-perpetuating mechanism was demonstrated in the model borehole with specimens cut from a core of Mitchell shale, which has anisotropic rock properties because of old parallel fracture lines. The specimens were left in the tester under the same stress conditions, but for increasing lengths of time. Spalling increased in severity with time of exposure, and specimens left for several days eventually collapsed (see Fig. 9.18). The points of spalling correlated with the old fracture lines.

The obvious way to cure spalling is to raise the density of the mud. The problem is that there is seldom enough data available to determine whether the spalling is due to excessive stress, to weakening of the formation by the mud filtrate, water imbibition, or to a combination of each. Under these circumstances it is usually preferable to first try improving the shale stabilizing properties of the mud along the lines suggested later in this chapter. Spalling due to stress concentrations at doglegs and key seats can often be avoided if close attention is paid to drill-string design, weight-on-bit, and rotary speed.

Hole Enlargement

The most common form of borehole instability is the hole enlargement that occurs when the yield stress is not exceeded until the formation is weakened by water being imbibed into the shale. The water invades a layer around the wellbore from the filtrate in high permeability zones and by capillary action or osmosis in low permeability zones. This then causes spalling and caving, exposing fresh surfaces to invasion, and the hole gradually enlarges. The only remedy is to use a shale stabilizing mud, either nonaqueous or brine water-based, discussed later in this chapter.

Formations with No Cohesive Strength

Fig. 9.9 shows that when a formation (such as unconsolidated sand) has no cohesive strength, the Mohr failure envelope passes through the origin. These formations have very high permeability. Therefore, when drilled with air or a clear fluid that exerts no confining stress on the walls of the hole, the sand sloughs into the hole. However, when drilled with a mud that has good filtration properties, the pressure drop across the filter cake imparts cohesive strength and the cake provides pore throat plugging action, and a gauge or near-gauge hole is obtained. The mud must contain enough bridging solids (see the section on the bridging process, in Chapter 7, The Filtration Properties of Drilling Fluids) to enable a cake to form quickly. Otherwise, the hole will be washed out to considerable distance by the turbulent flow conditions prevailing around the bit.

A similar but more difficult situation is often encountered when drilling through active fault zones. In this case, the grinding action of the fault has shattered the formation into loose rubble. Sometimes these zones consist of highly polished fragments of shale, in which case the shale is called *slickensided*.

Because of the lack of cohesive strength, sloughing of rubble zones is difficult to prevent. Use of a mud with good filtration properties is essential because rubble zones have *fracture permeability*. Sealing the fracture openings with a low-permeability cake enables the mud pressure overbalance to be applied at the face of the formation. Special sealing agents are often added to the mud (See Chapter 10, Drilling Problems Related to Drilling Fluids on Lost Circulation).

Good drilling practice is important when drilling through rubble zones. Keep annular velocities low to avoid fluid erosion and do not hang up and circulate with bit in a rubble zone. Reaming down fast with pumps on will push mud into the formation, and pulling up fast will pull it out again, bringing the shale with it. Also, increase the carrying capacity of the mud as described in Chapter 6, The Rheology of Drilling Fluids.

Coal Seams

Coal is a very brittle material with a low compressive strength, often containing many natural fractures. In regions of high tectonic stress, such as the Rocky Mountain foothills in Canada, coal seams almost explode into the hole when the horizontal stress is relieved by the bit, often causing stuck pipe (Willis, 1978). Large chunks and slivers of coal are recovered at the surface. Caliper logs often show undergauge hole in a coal section, indicating creep. The best technique for drilling coal seams appears to be to drill very slowly through them, with a good hole-cleaning mud. Some success has been seen with mixed-metal hydroxide, also called mixed-metal silicate, muds (see chapter: Drilling Fluid Components). These fluids have extremely high low-shear rate viscosity and K values. High-density muds cannot be used because of low fracture gradients.

HOLE INSTABILITY CAUSED BY INTERACTION BETWEEN THE DRILLING FLUID AND SHALE FORMATIONS

Adsorption and Desorption of Clays and Shales

The various forms of hole instability resulting from interaction between the drilling fluid and argillaceous formations are all related to hydration phenomena. As discussed in Chapter 4, Clay Mineralogy and the Colloid Chemistry of Drilling Fluids, water is adsorbed on clays by two mechanisms: adsorption of monomolecular layers of water on the planar surfaces of clay crystal lattices (commonly referred to as crystalline swelling or surface hydration), and osmotic swelling resulting from the high concentration of ions held by electrostatic forces in the vicinity of the clay surfaces. Crystalline swelling is exhibited by all clays. Swelling pressures are high, but the increase in bulk volume is comparatively small. Interlayer osmotic swelling occurs only with certain clays of the smectite group (notably sodium montmorillonite), and causes large increases in bulk volume, but swelling pressures are low.

If a dry clay is confined, but given access to free water, it develops a swelling pressure. Similarly, if a clay that has been allowed to equilibrate with free water is compacted and water is expelled, swelling pressures develop. The swelling pressure at any given water content is related to the vapor pressure of the clay at the same water content by the equation (Chenevert, 1969):

$$p_s = -\frac{RT}{V}\ln\frac{p}{p_0} \qquad (9.8)$$

Where p_s is the swelling pressure in atmospheres; T is the absolute temperature (°K); V is the partial molar volume of water, L/mole; R is the gas constant (L atm/mole °K); and p/p_0 is the relative water vapor pressure at equilibrium with the shale and is approximately equal to the activity of the water in the shale. Thus, the potential swelling pressure of a compacted shale whose water content is known can be predicted from adsorption or desorption isotherms of that shale (see Fig. 9.28) (Chenevert, 1970b). Isotherms are determined by equilibrating specimens of the shale with water vapor in atmospheres of known humidity and constant temperature. Fig. 9.29 shows that the swelling pressure of the layer of crystalline water adjacent to a clay surface is extremely high, but that of succeeding layers decreases rapidly (Powers, 1967).

The relationship between swelling and compacting pressure may be studied experimentally in a compaction cell such as that shown in Fig. 9.30 (Darley, 1969). Fig. 9.31 plots equilibrium water content versus effective stress for samples of sodium and calcium bentonites cut from outcrops. The test specimens were cut normal to the bedding planes. Since sodium montmorillonite exhibits osmotic swelling, but calcium montmorillonite does not, the curves suggest that the high water content of the sodium clay at stresses less than about 2000 psi (140 kg/cm^2) was due to osmotic swelling. At higher stresses both clays were being desorbed of crystalline water.

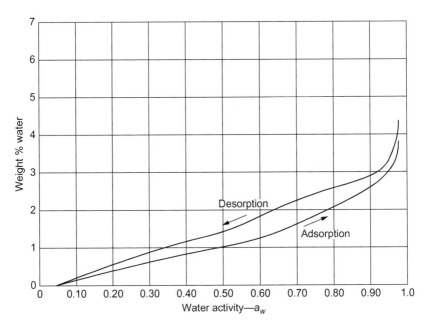

FIGURE 9.28 Adsorption/desorption isotherm. Wolfcamp shale. $T = 75°F$ (24°C). *From Chenevert, M.E., 1970b. Shale control with balanced activity oil-continuous muds. J. Petrol. Technol. 22, 1309–1316. Trans. AIME 249. Copyright 1970 by SPE-AIME.*

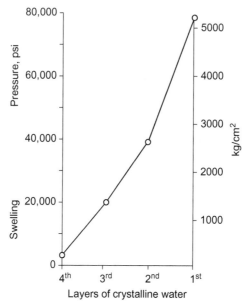

FIGURE 9.29 Approximate crystalline swelling pressure of montmorillonite. *After Powers, M. C., 1967. Fluid release mechanisms in compacting marine mudrocks and their importance in oil exploration. AAPG Bull. 51 (7), 1240–1254. Courtesy AAPF.*

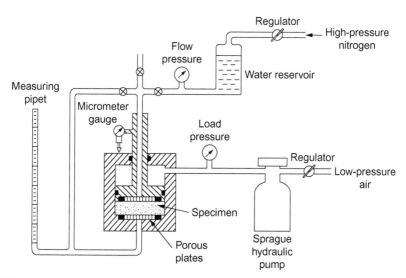

FIGURE 9.30 Compaction cell. *From Darley, H.C.H., 1969. A laboratory investigation of borehole stability. J. Petrol. Technol. 883–892. Trans. AIME 246. Copyright 1969 by SPE-AIME.*

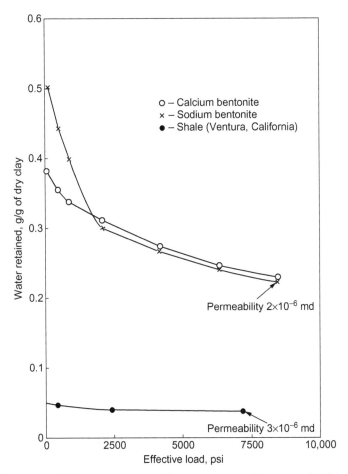

FIGURE 9.31 Equilibrium water content of clays versus compacting pressure. Load applied by hydraulic ram as shown in Fig. 9.29. Water expelled at 0 pore pressure. *From Darley, H.C.H., 1969. A laboratory investigation of borehole stability. J. Petrol. Technol. 883–892. Trans. AIME 246. Copyright 1969 by SPE-AIME.*

If the sample in the compaction cell discussed above had consisted of pure montmorillonite, and all the clay crystals had sedimented with their basal surfaces parallel to the bedding planes, the swelling pressures would have equaled the compacting pressures when equilibrium conditions were obtained. Actually, swelling pressures were less than compacting pressures, as shown by Fig. 9.32, which compares bulk densities calculated from adsorption isotherms with bulk densities calculated from the compaction data. A similar plot of compaction data obtained by Chilingar and Knight (1960) with a sample of commercial bentonite, which had first been equilibrated with an excess of distilled water, is also shown. Evidently, in both

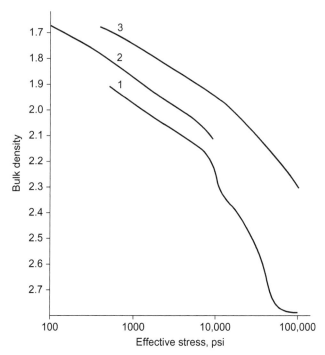

FIGURE 9.32 Comparison of desorption and compaction data. Curve 1 was calculated from the desorption isotherm. Curve 2 was calculated from the compaction curve in Fig. 9.30. Curve 3 was calculated from the compaction data of Chilingar and Knight. Grain density of bentonite taken as 2.8 g/cm^3.

compaction experiments the clay crystals were to some extent randomly oriented, and the compacted specimens contained pore water as well as water of hydration.

Hydration of the Borehole with Water-Based Drilling Fluids

Where argillaceous sediments are compacted by the weight of overlying sediments, water adsorbed by clay minerals is expressed along with pore water. The amount of water remaining in the subsurface sediments depends on the depth of burial, the species and amounts of clay minerals present in the sediment, the exchange cations thereon, and the geologic age of the formation. Average bulk densities for the various ages are shown in Fig. 9.33 (Dallmus, 1958). When the shale is penetrated by the bit, the horizontal earth stresses on the walls of the hole are relieved and the shale is contacted by the drilling fluid. Water is then drawn by capillary action and osmosis in or out of the formation depending on the pore throat openings and the activity of the water in the shale relative to that in the mud. The activity of water in a compacted

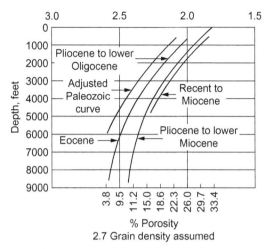

FIGURE 9.33 Effect of age and depth on shale density. *Dallmus, K.F., 1958. Mechanics of basic evaluation and its relation to habitat of oil in the basin, habitat of oil. AAPG Bull. 42, 883–931. Courtesy AAPG.*

shale is reduced by hydrogen bonding to the clay crystal surfaces, and by hydration of the counterions held in the double layer by electrostatic forces (see Chapter 4, Clay Mineralogy and the Colloid Chemistry of Drilling Fluids). The activity of the water and pore throat opening sizes decrease with depth because interlayer spacing decreases with increased compaction.

Both adsorption and desorption may destabilize the hole: adsorption will cause the hole to yield if the resultant swelling pressure exceeds its yielding stress (see Fig. 9.25B). Desorption causes shrinkage cracks to develop around the hole. Caving occurs when the fluid from the hole invades the cracked zone, equalizing the pressure in the cracks with that in the hole. A clear brine invades the cracks freely, and the hole rapidly enlarges (see Fig. 9.34A). A mud with bridging material tends to plug the cracks, greatly reducing the rate of pressure equalization, and allowing a large proportion of the mud pressure overbalance to be applied at the wall of the hole. Consequently, caving is greatly reduced (see Fig. 9.34B).

Brittle Shales

Caving and hole enlargement are frequently experienced in the older, consolidated shales that contain no montmorillonite. It was formerly believed that swelling was not the cause of caving in these so-called brittle shales, because the cavings were hard and showed no obvious signs of swelling. Chenevert (1970a), however, showed that these shales can develop extremely high swelling pressures when confined and contacted with water. In a drilling well, the swelling pressure increases the hoop stresses around the hole. When the hoop

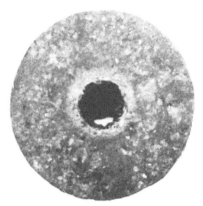

(A) Borehole fluid saturated sodium chloride brine.   (B) Same as (a), but with 2% hydrolyzed starch.

FIGURE 9.34 Caving of montmorillonitic shales in model borehole. Reconstituted Miocene shale specimens, bulk density 2.22 g/cm^3, applied stress 1000 psi (70 kg/cm^2), flow rate 150 ft/min (46 m/min). (A) Saturated NaCl, (B) Saturated NaCl plus 2% hydrolyzed starch. *From Darley, H.C.H., 1969. A laboratory investigation of borehole stability. J. Petrol. Technol. 883–892. Trans. AIME 246. Copyright 1969 by SPE-AIME.*

stress at the wall of the hole exceeds the yield stress of the shale, hydrational spoiling occurs. In the laboratory, Chenevert observed that the swelling pressure increased with time and eventually caused an explosive enlargement of the hole. Similarly, in the field, it has often been observed that severe caving does not occur until several days after the shale is penetrated by the bit.

Many shales contain old fracture lines or invisible microfractures. Time and high confining pressures have partially healed these fractures, so that a specimen recovered at the surface appears quite competent. When contacted with water, however, the water penetrates along the fracture lines, the resulting swelling pressures break the adhesive bonds, and the shale falls apart (see Figs. 9.35–9.37). A similar process undoubtedly takes place downhole and facilitates destabilization of the borehole.

Control of Borehole Hydration

If the mud density is correct, the borehole hydration is, in many cases, the prime cause of hole instability, and in many other cases a contributing factor, thus every effort must be made to control it. The introduction of silicate muds, which consisted of sodium silicate and saturated sodium chloride brine marked the first attempt to do so (Baker and Garrison, 1939). These muds were so successful at controlling hydration and dispersion that drill cuttings of gumbo shale were recovered at the surface with bit tooth marks still visible. Unfortunately, the rheological properties of silicate muds were so difficult to control that their use was discontinued at that time. New generations

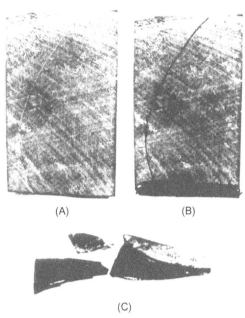

FIGURE 9.35 Destabilization of shale specimen by hydration of a healed fracture. (A) Shale specimen after machining from core. (B) After placing water at base. (C) Specimen falls apart. *From Darley, H.C.H., 1969. A laboratory investigation of borehole stability. J. Petrol. Technol. 883–892. Trans. AIME 246. Copyright 1969 by SPE-AIME.*

FIGURE 9.36 Destabilization of shale specimen by penetration of water along microfractures. Apparently competent specimen brought up in junk basket in air-drilled hole.

of silicate systems and reduced solids content are being used successfully for wellbore stability. Silicate systems are discussed in Chapter 13, Drilling Fluid Components.

Nonaqueous Fluids and Wellbore Stability

Since then, the muds that have been most successful at preventing hydration of shale formations have been nonaqueous fluids with concentrated brines as

FIGURE 9.37 Specimen pictured in Fig. 9.35A shortly after contacting with water. *Note*: Only exterior surfaces of fragments were wet, interior was dry. *Courtesy of Shell Development Co.*

the internal phase. As originally hypothesized by Mondshine and Kercheville (1966), if the salinity of the aqueous phase was equal to the salinity of the water in the pores of the shale, hydration would be prevented. Subsequently Mondshine (1969) modified this method to allow for the swelling pressure, which he determined approximately from the effective stress on the shale at the depth of interest. Chenevert (1970a), however, pointed out that the essential factor is the activity of the water in the shale (as determined in the laboratory by the vapor pressure of preserved cores), since it determines the potential swelling pressure, as shown by Eq. (9.8). Therefore, swelling will be prevented if the activity of the water in the internal phase of the mud is equal to the in situ activity of the water in the shale formation. This requirement may be accomplished by determining the in situ water content of the shale from density logs (see the section on induced fractures, in Chapter 10, Drilling Problems Related to Drilling Fluids) or, less accurately, from Fig. 9.33. The in situ activity of the water can then be read from the adsorption isotherm of that shale at the appropriate water content (see Fig. 9.28). The activity of the aqueous phase of the mud is then adjusted to the same value by the addition of sodium or calcium chloride. Field results indicated that corrections for the difference between the activity at laboratory and at downhole temperatures are usually not necessary.

Chenevert's method has proved successful in the field in preventing instability due to hydration, but his original requirement that the activity of the mud should match but not be less than the activity of the water in the shale has been found to be unnecessary. Field experience has shown that the hole is not destabilized by excessive mud salinity when the continuous phase is oil, and laboratory experiments in the model borehole have shown that desorption actually strengthens the shale. The reason probably is that the continuous phase of the mud is oil, and oil cannot enter incipient cracks because of the high capillary pressure (see Chapter 8, The Surface Chemistry of Drilling Fluids). Thus the fluid pressure in the cracked zone remains at the formation pore pressure, and the whole of the mud pressure overbalance, $(p_w - p_f)$, is applied on the walls of the hole, whereas with water-base muds pressure equalization takes place, albeit slowly.

Although controlled activity nonaqueous drilling fluids are best for preventing hydration of formation clays, their cost is high, it is sometimes difficult to obtain satisfactory formation evaluation with them, and they have various other disadvantages (discussed in Chapter 1, Introduction to Drilling Fluids). Over the years special water-base muds for maintaining hole stability have been developed (Denny and Shannon, 1968; Shell, 1969; Darley, 1976; Gray and Gill, 1974; Clark et al., 1976; O'Brien and Chenevert, 1973; Steiger, 1982; Mondshine, 1974; Jones, 1981; Walker et al., 1983; Walker et al., 1984; Lu, 1985; Chaney and Sargent, 1985; Fraser, 1985; Fleming, 1986; Holt et al., 1986; Chesser, 1986; Griffin et al., 1986; Simpson et al., 1998; McDonald et al., 2002).

In some areas these muds provide adequate hole stability and mud costs are lower than with oil muds, but sometimes the savings in mud costs are more than offset by higher drilling costs.

Soluble salts are used in water-base muds to control swelling, and various polymers provide rheological properties and control dispersion. Salts control swelling by two mechanisms: lowering the activity of the water and cation exchange. In water-based muds, lowering the activity of the water can be used to stabilize shales only to a limited extent. Mud salinities below the balancing salinity will reduce osmotic swelling, but when the continuous phase is water, salinities above balancing will cause shrinkage cracks and consequent destabilization, as already discussed. Maintaining a balanced activity is not a practical proposition.

Cation Exchange Reactions

An adequate degree of shale stabilization can usually be achieved by cation exchange reactions, usually the replacement of Na^+ by K^+. Table 9.3 and Fig. 9.38 show that potassium chloride (KCl) is more effective at reducing linear swelling than equivalent concentrations of other salts, and Bol (1986) showed the same phenomenon in volumetric swelling tests on confined specimens of Pierre shale. The potassium ion is more effective because of its low hydration energy and its small size, which enables it to fit into the holes in the silica layers in the clay crystal, thus reducing interlayer swelling (see Table 4.4).

The concentration of KCl required to suppress swelling depends on the cation exchange capacity of the shale, and the exchange constants of the ions involved. Steiger (1982) found that shales high in montmorillonite required up to 90 lb/bbl (256 kg/m^3), whereas illitic shales required only 20 lb/bbl (57 kg/m^3). Because of its stability in high-salinity brines, polyanionic

TABLE 9.3 Effect of Concentration of Salts on Linear Swelling of Shales

Shale	Clay Minerals in the Shale (%)	Brine (%)	Reduction of Swelling Relative to Swelling in Water (%)
Anahuac	40 montmorillonite	3 KCl	19
	5.5 illite	3 NaCl	8
Midway	35 illite	Freshwater	21
	15 interlayered		
	15 chlorite	3 KCl	64
		5 KCl	69
		Sat. KCl	79
		3 NaCl	36
Wolfcamp	15 illite	3 KCl	57
	3 chlorite	3 NaCl	21

Note: Linear swelling was measured by means of a strain gauge. All brines contained 1 ppb xanthan gum.
Source: From O'Brien, D.E., Chenevert, M.E., 1973. Stabilizing sensitive shales with inhibited potassium-based drilling fluids. J. Petrol. Technol. 189–1100. Trans. AIME 255. Copyright 1973 by SPE-AIME.

cellulose is commonly used to provide filtration control in KCl muds. Starch is also used, and Steiger (1982) reports good results with a synergistic mixture of polyanionic cellulose and starch. Xanthan gum or prehydrated bentonite is used to provide cutting carrying capacity.

The performance of KCl muds is greatly increased by the addition of certain long chain anionic polymers that coat shale surfaces, and thus protect the walls of the hole from disintegration (see Fig. 9.39). The most likely explanation of the coating action is that the negative sites on the polymer chain are attracted to the positive sites on the clay crystal edges. Tests in the model borehole by Clark et al. (1976) showed that partially hydrolyzed polyacrylamide-polyacrylate copolymer (PHPA) is the best polymer for maintaining hole stability in an illitic shale (Atoka shale). 10.5 lb/bbl (3%) KCl was added to prevent swelling (see Table 9.4). Tests by Bol (1986) in a similar machine showed that PHPA was the best polymer to prevent erosion of a montmorillonitic shale (Pierre shale), but 10% KCl was necessary to prevent swelling. Strain gauge tests have shown that PHPA does not inhibit swelling (Steiger, 1982)—that is the function of the KCl. As already mentioned, the optimum concentration of KCl depends on the CEC of the shale. Hole stability can often be maintained with KCl alone, but considerably higher concentrations are required if the PHPA is omitted.

The degree of hydrolysis of the PHPA is important (see Table 9.4), presumably because it spaces the negative sites along the chain so that they match the positive sites on the clay platelet edges.

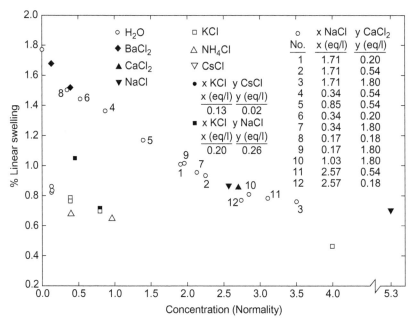

FIGURE 9.38 Effect of cation concentration and species on linear swelling. Clay mineral analysis of shale: 9.2% montmorillonite, 11.2% mixed layer, 35% illite, 5.5% chlorite, 4.4% kaolinite. *From Steiger, R.P., 1982. Fundamentals and use of potassium/polymer drilling fluids to minimize drilling and completion problems associated with hydratable clays. J. Petrol. Technol. 34, 1661–1670. Copyright 1982 by SPE-AIME.*

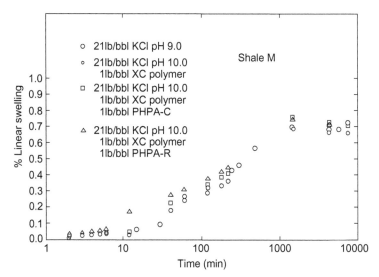

FIGURE 9.39 Effect of KCl and polymer solutions on linear swelling (same shale as in Fig. 9.36). *From Steiger, R.P., 1982. Fundamentals and use of potassium/polymer drilling fluids to minimize drilling and completion problems associated with hydratable clays. J. Petrol. Technol. 34, 1661–1670. Copyright 1982 by SPE-AIME.*

TABLE 9.4 Polymers Evaluated for Shale Protection[a]

Polymer	Test Concentration (ppb)	Salt	Test Concentration (ppb)	Test[b] Time (min)	Sample Erosion (%)
Polyacrylamide: 30% hydrolyzed, M. W. $> 3 \times 10^6$	0	KCl	10.5	5	20.7
	0.063	KCl	10.5	1403+	<1.0
	0.50	KCl	10.5	1420+	<1.0
	0.50[c]	NaCl	10.5	8640+	1.8
Polyacrylamide: 5% hydrolyzed, M. W. $\sim 1 \times 10^6$	0.50	KCl	10.5	104	12.2
Polyacrylamide: <1% hydrolyzed, M. W. $\sim 12 \times 10^6$	0.50	KCl	10.5	50	—
Sodium polyacrylate	4.0[c]	NaCl	10.5	172+	—
Modified starch	5.0	KCl	10.5	110	8.6
	17.5	KCl	10.5	1370+	<1.0
Polyanionic cellulose	0.50	KCl	10.5	38	9.1
Xanthan gum	0.50	KCl	10.5	21	10.7
Polyethylene oxide	0.50	KCl	10.5	55	11.6
Vinyl ether-vinyl pyrolidone copolymer	1.11	KCl	10.5	6	16.7

[a] Test conditions: Atoka shale, 3500 psi stress, 800 ft/min flow rate.
[b] Designates no sample failure at test termination.
[c] Run on different Atoka shale, 2500 psi stress.
Source: From Clark, R.K., Scheuerman, R.K., Rath, H., van Laar, H., 1976. Polyacrylamide/Potassium-chloride mud for drilling water-sensitive shales. J. Petrol. Technol. 261, 719–727. Trans. AIME 261. Copyright 1976 by SPE-AIME.

Another important benefit of formulating KCl muds with PHPA is that it coats the drill cuttings (usually referred to as *encapsulation*) thereby inhibiting their dispersion and incorporation into the mud. Table 9.5 shows the effect of various additives on the rate of dispersion of Pierre shale cuttings, expressed as a percentage of cuttings retaining their original size after hot rolling. Note that PHPA is by far the most effective polymer, and that KCl alone had little effect.

Densities up to 10 lb/gal (SG 1.20) may be obtained with KCl; barite must be added if higher densities are necessary. Small additions of lignosulfonate or other thinner may be necessary to control gelation at densities above 16 lb/gal (SG 1.92). Potassium hydroxide (KOH) should always be used instead of NaOH for pH control to avoid the dispersive action of the sodium ion.

KCl muds have been successful at maintaining hole stability in hard brittle shales, usually at considerable savings in mud and drilling costs. In soft montmorillonitic shales results are more questionable. Very high concentrations of KCl are required, and maintenance costs are high because of rapid depletion of KCl and polymers. In some cases overall well costs were as high as with nonaqueous muds, and the KCl mud was less effective at maintaining hole stability. However, Clark and Daniel (1987) report reduced overall drilling costs in over 300 Louisiana and Texas Gulf Coast wells. Benefits included higher rates of penetration, less stuck pipe, less cuttings dispersion, and better rheological properties than with other water-base systems.

TABLE 9.5 Dispersion Tests on Pierre Shale Cuttings

Base Mud	PHPA (%)	KCl (%)	% Recovery of Cuttings of Original Size
2.86% bentonite	0	0	12
	0.21	0	76
	0	2.86	21
	0	10	36
	0.21	2.86	84
	0.21	10	86
Freshwater	0.21% PHPA		90
	0.75% xanthan gum		63
	1.5% CMC-HV		23
	0.6% PAC		22

Source: After Bol, G.M., 1986. The effect of various polymers and salts on borehole and cutting stability in water- base drilling fluids. IADC Paper No. 14802, Drilling Conference, Dallas, TX, February.

Lime-lignosulfonate or gyp-lignosulfonate muds are often used for drilling dispersible montmorillonitic clays and shales. Their beneficial action depends solely on cation exchange, primarily the substitution of Ca^{++} for Na^+. Fig. 9.31 shows that osmotic swelling is much less in calcium bentonite than in sodium bentonite, and that crystalline swelling is not significantly affected by the calcium ion. Thus, lime muds help stabilize shales that exhibit osmotic swelling, but have no effect on shales, such as illites, that exhibit only crystalline swelling. Osmotic swelling is the cause of dispersion (see section on clay swelling mechanisms in chapter: Clay Mineralogy and the Colloid Chemistry of Drilling Fluids); therefore, lime muds inhibit hole enlargement in dispersible shales. Also, they inhibit dispersion of drill cuttings, and thus help maintain low viscosities and faster drilling rates.

A developed potassium lime mud (KLM) represents a considerable improvement over conventional lime muds (Walker et al., 1983, 1984). The substitution of KOH for NaOH improves the shale stabilizing action of the mud for reasons already mentioned. Also, the use of a polysaccharide deflocculant instead of lignosulfonate reduces the dispersion of the cuttings. In laboratory rolling tests there was much less dispersion of tablets of southern bentonite in muds containing KOH instead of NaOH. Wells drilled with the KLM mud experienced less hole enlargement and lower maintenance costs than did offsetting wells drilled with conventional water-base muds. No tendency to high-temperature gelation was noted in wells with bottomhole temperatures above 300°F (149°C). The KLM mud achieved a notable success under adverse conditions when used to drill a well in the Navarin Basin in the Bering Sea (Holt et al., 1986).

Other formulations for drilling dispersible shales consist of various polymers and potassium lignite (Mondshine, 1974) or KOH−lignosulfonate (Jones, 1981; Fraser, 1985) or KOH−calcium lignite (Fleming, 1986), or a potassium-base derivative of humic acid (Pruett, 1987). When low salinities are required for logging purposes, a mud containing diammonium phosphate, polyanionic cellulose, and bentonite (DAP−PAC) may be used (Denny and Shannon, 1968). Another low-salinity inhibitive mud consists of 10×10^6 MW PHPA (compare with 3×10^6 MW PHPA used by Clark et al., 1976) and as little as 1% KCl (Chesser, 1986). A freshwater PHPA ($>15 \times 10^6$ MW) with small amounts of KOH has been used to replace chrome lignosulfonate muds in south Texas. Improved hole stability was obtained due to less mechanical erosion and reduced hydroxyl concentration. Mud costs were higher but total drilling costs were lower because of higher rates of penetration and longer bit life.

Diffusion Osmosis and Methyl Glucoside

Simpson and Dearing (2000) noted that prior studies have documented two driving forces involved in the transfer of water into shales. One is the hydraulic pressure differential between the drilling fluid and shale pore fluid. A second is a chemical osmotic force dependent upon the difference between the water activity (vapor pressure) of the drilling fluid and that of the shale pore fluid under downhole conditions. They hypothesized another driving force called *diffusion osmosis*. This is determined by the difference in concentrations of the solutes in the drilling fluid and shale pore fluid. Diffusion osmosis then results in transfer of solutes and associated water from higher to lower concentration for each species, opposite to the flow of water in chemical osmosis. If the diffusion osmotic force exceeds the chemical osmotic force, invasion of ions and water can increase the pore pressure and water content of the shale near the borehole surface. Additionally, the invading ions can cause cationic reactions that alter the clay structure in the shale. All of these effects tend to destabilize the shale.

Simpson and Dearing propose that these destabilizing ionic reactions within a shale can be minimized by using a suitable nonionic polyol (such as methyl glucoside) to reduce the activity of a freshwater drilling fluid. In certain situations the addition of salt to such a freshwater drilling fluid to obtain further reduction of water activity can cause an increase in the diffusion osmotic force that offsets some, or all, of the desired increase in chemical osmotic force.

Simpson and Dearing state that chemical osmotic effectiveness can be improved by emulsification of a nonaqueous phase in the drilling fluid. A freshwater drilling fluid containing methyl glucoside for activity control and emulsified pentaerythritol oleate prevented hydration and maintained stability of Pleistocene shale from the Gulf of Mexico. Drill cuttings from such a drilling fluid should be environmentally acceptable for discharge at offshore or land locations.

Selection of Mud Type for Maintaining Borehole Stability

This chapter should have made it clear that hole instability is a complex problem, the nature of which depends on the borehole environment. The type of drilling fluid that will provide maximum hole stability therefore varies from area to area; no one fluid is best for all areas. Some investigators (Mondshine, 1969; Steiger, 1982; Kelly, 1968; Allred and McCaleb, 1973) have attempted to base the choice of drilling mud on a classification of shales according to clay mineral composition and texture. The difficulty with this approach is that too many variables exist for shales to be placed in a few simple categories. Also, hole stability is influenced by other factors, such as tectonic stress, pore pressure, dip of the formations, and degree of compaction. For example, the Atoka shale in southeast Oklahoma is notoriously unstable in the vicinity of the overthrust Choctaw fault, yet comparatively few problems are experienced with the same shale in undisturbed areas a few miles further north.

Before formulating a mud to minimize borehole problems, the first step should be to collect as much information as possible on the geology, the stress history, and the fault patterns of the region. Temperature gradients, pore pressure gradients, and in situ water contents of shales should be obtained from logs in the nearest wells. Samples of problem shales should be obtained for laboratory testing. The best source of samples is a well preserved core, but if none is available drill cuttings must be used. Cores are to be preferred because much valuable information may be obtained from the lithology, structure, presence of fractures, degree of hydration, etc. Coring costs are high, but if cores are taken early in the life of a field, and suitable tests are made, the coring costs will be repaid many times over by savings in subsequent drilling costs. The objection to drill cuttings is that they are likely to have been altered by hydration and base exchange reactions with the mud on the way up the hole. Dust from air-drilled wells avoids the contamination problem but not hydrational changes.

Laboratory Tests

The following laboratory tests should be made:

1. *Clay mineral analysis by X-ray diffraction, cation exchange capacity, and exchange cations* (see section on ion exchange in Chapter 4). Where facilities for these tests are not available, the methylene blue test may be used instead (see section on clay mineral identification in Chapter 4). This test permits a rough estimate to be made of the amount of montmorillonite present. As already mentioned, the concentration of KCl required for the control of swelling depends largely on the amount of montmorillonite present.
2. *Balancing salinity*. A test to determine the salinity required to balance the in situ activity of subsurface shales is necessary. Obviously, it is a waste of money to use an oil-base mud unless the salinity of its aqueous phase is kept greater than the required balancing salinity. The test as described by Chenevert (1970b) is made by placing chips of dried shale in desiccators containing saturated solutions of various salts covering a wide range of vapor pressures (see Table 9.6). After one day, 90% of equilibrium is reached; the chips are then removed and weighed, and the water content is calculated and plotted against relative humidity. The in situ activity of the water in the shale is then given by the intercept of its in situ water content on the isotherm. This value is an indication of the potential swelling pressure should the shale pick up water from the drilling mud—the lower the in situ activity, the greater the maximum possible swelling pressure. The salt content of an oil-base mud required to balance the shale activity may be calculated from graphs such as those shown in Figure 9.40. Note that a reliable estimate of the in situ water content cannot be obtained from the density of the drill cuttings, but can be obtained from a specimen cut from the *center* of a *well-preserved*

TABLE 9.6 Water Activities of Various Saturated Salt Solutions

Desiccator Number	Salt Employed	p/p_0
1	$ZnCl_2$	0.100
2	$CaCl_2$	0.295
3	$Ca(NO_3)_2$	0.505
4	NaCl	0.755
5	$(NH_4)_2SO_4$	0.800
6	$Na_2C_4H_4O_6 \cdot 2H_2O$	0.920
7	KH_2PO_4	0.960
8	$K_2Cr_2O_7$	0.908

Source: From O'Brien, D.E., Chenevert, M.E., 1973. Stabilizing sensitive shales with inhibited potassium-based drilling fluids. J. Petrol. Technol. 189–1100. Trans. AIME 255. Copyright 1973 by SPE–AIME.

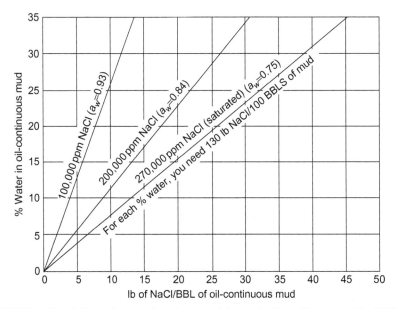

FIGURE 9.40 NaCl requirements for balanced activity muds. *From Chenevert, M.E., 1970b. Shale control with balanced activity oil-continuous muds. J. Petrol. Technol. 22, 1309–1316. Trans. AIME 249. Copyright 1970 by SPE-AIME.*

core, or can be estimated from density logs. Instead of equilibrating the chips in desiccators containing saturated solutions of various salts, it is sometimes convenient to equilibrate them against a series of solutions with increasing concentrations of sodium chloride, and against another

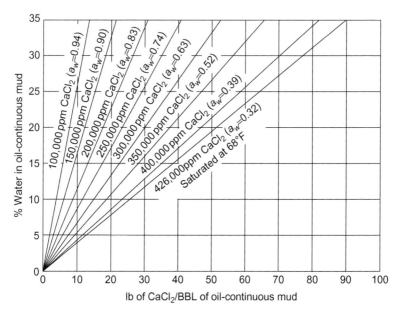

FIGURE 9.41 CaCl$_2$ requirements for balanced activity muds. *From Chenevert, M.E., 1970b. Shale control with balanced activity oil-continuous muds. J. Petrol. Technol. 22, 1309–1316. Trans. AIME 249. Copyright 1970 by SPE-AIME.*

series with increasing concentrations of calcium chloride. If the equilibrium water contents are then plotted against the salinities, the salinity of the mud required to balance the swelling pressure of the shale may be read directly from the curve at the intercept with the in situ water content (see Fig. 9.41). As already discussed, it is not possible to formulate a balanced-salinity water-base mud. Nevertheless the balancing salinity test should be made because the value obtained is useful for diagnostic purposes.

3. *Swelling measurements.* In order to measure physical swelling, the shale sample must be immersed in the test fluid so that cation exchange can take place. Linear swelling may be measured with a strain gauge (Chenevert, 1970b). Volumetric swelling may be measured by confining a specimen in a cylinder with a piston, and observing the displacement of the piston as fluid is sucked through a permeable disc in the bottom of the cylinder (Ritter and Gerault, 1985). Alternatively, swelling may be measured by the linear displacement of a piston in the apparatus shown in Fig. 9.42 (Osisanya and Chenevert, 1987).

4. *Dispersion tests.* This test compares the degree to which cuttings will be dispersed into the mud by the drilling process. A weighed sample of dried cuttings, or core fragments, in a coarse size range is rolled for a standard time—the length of which depends on the dispersibility of the cuttings—

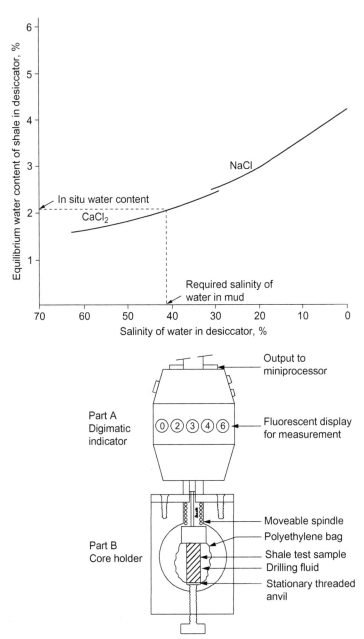

FIGURE 9.42 Direct determination of mud salinity requirements for balanced activity muds (schematic). Direct reading digimatic swelling indicator. *From Osisanya, S.O., Chenevert, M.E., 1987. Rigsite shale evaluation techniques for control of shale-related wellbore instability problems. SPE/IADC Paper 16054, Conf., New Orleans, LA, March 15–18. Copyright 1987 by SPE-AIME.*

at the temperature of interest. The mud is then screened through a fine screen, and the cuttings remaining on the screen are dried and weighed. The percentage loss of weight is taken as a measure of dispersibility. The test is empirical, and any set of conditions may be chosen to suit local shales or problems. For instance, if the problem is an increase in viscosity caused by dispersion of the cuttings, four to 10 mesh cuttings might be chosen, rolled in the mud for 8 hours and sieved through a 325-mesh screen. Another procedure, found necessary by Nesbitt et al. (1985) when they observed that the active material in the shale was located in a network of fractures, was to roll large pieces of core and determine the amount remaining on a 5-mesh sieve. The dispersion test is a valuable aid to mud selection; it is simple and quick, and has the great advantage that a number of candidate fluids may be run simultaneously.

5. *Rigsite tests*. Osisanya and Chenevert (1987) describe six tests that may be made at the rig site to aid in drilling troublesome shales, namely swelling, dispersion, cation exchange capacity, hydration capacity, hydrometer, and capillary suction time. The first two are by far the most important. They state that all the tests are dependent on the particle size of the shale used.

Laboratory performance tests are helpful when choosing between several alternative muds, or in determining the optimum formulation of a particular mud. Such tests should be made under simulated subsurface conditions. Performance tests that consist merely of immersing shale specimens in candidate muds may give misleading results, because unconfined shale can disintegrate without developing any significant swelling pressure, whereas downhole the shale will not disintegrate unless swelling pressure develops sufficiently to increase the hoop stresses above the yield stress.

Performance tests may be carried out either in a microbit drilling machine or a model borehole such as that shown in Fig. 9.26. In applying stresses, remember that the effective stress is the load minus the pore pressure, so that it is convenient to set the pore pressure at zero, and the vertical stress, the confining stress, and the mud pressure at their respective values at the depth of interest in the well, minus the pore pressure at that depth.

Specimens may be either cut from cores or reconstituted in a compaction cell (such as that shown in Fig. 9.30) from slurries of powdered shale. Natural specimens represent underground conditions more truly, but—since no two natural specimens are exactly alike—each experiment must be repeated on a number of specimens, and the results averaged. Reconstituted specimens have far better reproducibility, but give only qualitative results, because conditions that have come about over millions of years underground cannot be reproduced in a matter of days in the laboratory.

The time required to compact a specimen increases sharply with increase in height of the specimen, because the basal surfaces compact first, greatly restricting the outflow of water from the center. A 2-inch-high homogenous specimen in approximate equilibrium with the compacting pressure can be made in one day. Performance tests may be used to compare candidate muds with respect to their effect on (1) the mode of failure, whether it be plastic yielding or caving in hard fragments; (2) hydration of the borehole walls, which may be determined by taking a sample of the hydrated zone around the hole and comparing its water content with that of the original specimen; and (3) borehole diameter, which may be determined by measuring the volume of oil required to fill the hole. If the hole has enlarged so much that the specimen has collapsed, the time to collapse may be used as a parameter.

Membrane Efficiency

It is estimated that wellbore instability problems results in several billions in excess cost in worldwide drilling operations each year. The overwhelming majority of wellbore instability problems arise from shale instability induced by exposure to drilling fluids. Researchers are working to develop a real-time approach to manage stability problems in the field. A number of field and operational parameters significantly influences borehole instability (Table 9.7). To better manage borehole stability problems and reduce the associated learning curve, operators must understand the interrelationships among the various parameters and integrate these variables into well planning. Although these factors are considered in the drilling plan, field engineers still must constantly monitor the overall fluid system and make adjustments as needed (Ewy and Morton, 2008; Schlemmer et al., 2002).

Another key issue is the perpetual search for a truly effective water-based drilling fluid replacement for nonaqueous fluids. Nonaqueous drilling fluids are widely recognized as superior to water-based fluids for stabilizing shales

TABLE 9.7 Field and Operational Parameters That Influence Borehole Stability

Drilling Fluid	Rock Properties	Drilling Operations	*In-situ* Stresses	Drill String
Composition	Strength	Hole orientation	Overburden and horizontal stresses	Bottomhole assembly
Pressure	Permeability and porosity	Open hole time	Pore pressure	Vibrations
Flow rate and rheology	Mud-rock interactions	Tripping		
Temperature				

Source: Tare and Mody, 2002.

due to establishing a semipermeable membrane on the wellbore (Santos et al., 2001). One strand of this work involves the search for materials to enhance membrane efficiencies in water-based fluids as a way to slow or stop the deterioration of shales exposed to drilling fluids. The pore pressure transmission (PPT) test (Patel et al., 2002) continues to be a viable method for comparing the relative stabilizing effects of drilling fluids by measuring the membrane efficiency produced on a shale. Fig. 9.43 shows a schematic of a typical PPT device. Preserved shale cores about 25.4 mm in diameter and 6−8 mm in length are used. Multiple cores can be tested at one time. Comparisons of the PPT tests versus traditional shale stability testing are shown in Table 9.8.

Traditional shale stability tests are always suspect in that they cannot simulate downhole wellbore conditions. The PPT test, on the other hand, more nearly represents wellbore/mud interactions and the effect produced on in situ shales. It is recognized that there are three different membranes developed by shale/mud interactions (Tare and Mody, 2002):

- Type 1, shale stabilizing additives, such as saccharides and derivatives, carbohydrates, acrylic acid copolymers, siloxanes, high concentrations of lignosulfonate, glycols/polyglycols and derivatives (Santos et al., 2001);
- Type 2, impermeable precipitates, such as silicate and some aluminum additives (Boyd et al., 2002; Bland et al., 2002).
- Type 3, separation of the aqueous phase of the mud from water-filled shale pores as provided by nonaqueous muds (Bland et al., 2002).

In related work, the Drilling Engineering Association's joint industry project, DEA-113, has been completed. This study looked at drilling fluid/shale interactions with an eye toward shale membrane enhancement using

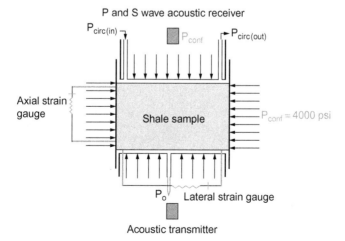

FIGURE 9.43 Pore pressure transmission tester. *From Tare, U.A., Mody, F.K., 2002. Managing borehole stability problems: on the learning, unlearning and relearning curve. AADE-02-DFWM-HO-31, AADE 2002 Technology Conference, Houston, TX, April 2−3.*

TABLE 9.8 Shale Test Characteristics

Traditional Shale Stability Testing	PPT Shale Membrane Tester
Hardness	Chemical osmotic flow volume
Moisture content	Chemical osmotic pressure development
Dimensions	Hydraulic flow volume
Tensile or compression strength	Hydraulic pressure development
Extrudability	Net direction and volume of flows
Abrasion disintegration	Shale permeability
Salinity change	Shale/fluid conductivity
Hot roll with drilling fluid	Water/oil content

the wellbore simulator operated by OGS Laboratories in Houston. A water mud containing soluble silicates was the best performer in those tests. Silicate muds, which form a Type 2 membrane, have continually showed well as borehole stabilizing fluids. They have not met with a good deal of success on the Gulf Coast, however, because of perceived difficulties in handling the mud-making shales in the Gulf. Silicate muds are continually being used in many areas of the United States and Canada as a nonaqueous-based mud replacement (Bland et al., 2002; Ewy and Morton, 2008).

Properties for Wellbore Stability by Water-Based Fluids

Most companies have been actively researching how best to make a water-based mud for wellbore stability (Bland et al., 2002). In general, the findings have shown that shale stability times can be enhanced by maintaining the following properties:

- *Correct density*. Optimizing mud weight for shale pore pressure and fracture gradient.
- *Osmotic control*. Water phase salinity/activity by use of chloride salts or formate brines.
- *Clay fixation*. Salt type, such as K^+, Ca^{2+}, Al^{3+}, silicates.
- *Accretion control*. Adding amphoteric ROP enhancers or surfactant/wetting agents, such as detergents, amines/amides chemicals, silicates.
- *Membrane former*. Reducing the shale permeability and/or pore space size with formates, glycols, polyglycerines, synthetic oils, silicates, gilsonite, asphalts.
- *Encapsulation*. Coating shale surfaces with polymers (PHPA) and surface active agents.

Planning mud programs is most difficult—but also most important—in newly discovered areas where much of the necessary information, such as lithology and pore and fracture gradients, is unavailable, and shale samples may be hard to obtain. Intensive sampling (preferably by coring) and laboratory testing should be done in early wells, and the results correlated with field experience. The information accumulated will save much time and money in later wells, and also reduce the amount of laboratory testing required.

No mud will maintain a stable hole unless its properties are kept up to specifications. Frequent mud checks at the rig and remedial treatments based thereon are therefore essential. When drilling with polymer muds it is particularly important that the polymer concentration be maintained at the required level. Loss of polymer by adsorption on drill cuttings proceeds rapidly, especially when drilling rates are fast. As the polymer concentration drops, the rate of dispersion of cuttings increases, further increasing the rate of polymer adsorption. Polymer adsorption continues until the polymer concentration approaches zero, with consequent major destabilization of the borehole.

Good drilling practices are also essential to maintenance of hole stability. Experience has shown that much less hole enlargement occurs in troublesome shales if a straight hole is drilled and doglegs are avoided. Pipe speeds when tripping should be kept low in order to minimize transient pressures. High fluid velocities in the annulus will cause hole enlargement when drilling through rubble zones, and through highly stressed, spalling formations, and will exacerbate enlargement caused primarily by the interactions between shales and the drilling fluid, as discussed above. Erosion will be much more severe if the mud is in turbulent flow. Velocities in the annulus may be lowered either by decreasing pump speed or by installing smaller bit nozzles while maintaining the same pump pressure. It may be necessary to adjust the rheological properties of the mud to increase the cutting-carrying capacity, or to change the flow from turbulent to laminar (see Chapter: The Rheology of Drilling Fluids).

NOTATION

c cohesive strength
K tensile strength
p_m hydrostatic pressure of the mud column
p_w hydrostatic pressure of mud plus any hydraulic or transient pressures in the annulus
p_f pressure of fluid in formation pores
Δ_p $p_w - p_f$
r radius at point of interest
r_w radius of gauge hole
r_0 radius of deformed hole
S overburden load, solids plus pore fluids
Z depth
μ viscosity

ρ_b bulk density
ρ_f density of formation water
σ effective intergranular stress
σ_0 effective horizontal stress in virgin rock
σ_1 greatest principal stress
σ_2 intermediate principal stress
σ_3 least principal stress
σ_θ hoop stress (or tangential or circumferential stress)
σ radial confining stress around borehole
Ø angle of internal friction

REFERENCES

Allred, R.B., McCaleb, S.B., 1973. Rx for Gumbo shale. SPE Paper No. 4233, 6th Conf. Drill. and Rock Mech., Austin, TX, January 22–25, pp. 35–42.

Baker, C.L., Garrison, A.D., 1939. The chemical control of heaving shale. Part 1. Petrol. Eng.50–58, Part 2. Petrol. Eng. (Eng.), 102–110.

Bland, R.G., Waughman, R.R., Tomkins, P.G., Halliday, W.S., 2002. Water-based alternatives to oil-based muds: do they actually exist? IADC/SPE 74542, IADC/SPE Drilling Conference Dallas, Texas, February 26–28.

Bol, G.M., 1986. The effect of various polymers and salts on borehole and cutting stability in water- base drilling fluids. IADC Paper No. 14802, Drilling Conference, Dallas, TX, February.

Boyd, J.P., McGinness, T., Bruton, J., Galal, M., 2002. Sodium silicate fluids improve drilling efficiency and reduce costs by resolving borehole stability problems in Atoka shale. AADE-02-DFWM-HO-35, AADE 2002 Technology Conference Houston, Texas, April 2–3.

Bradley, W.B., 1979a. Mathematical concept—stress cloud—can predict borehole failure. Oil Gas J. 77, 92–97.

Bradley, W.B., 1979b. Predicting borehole failure near salt domes. Oil Gas J. 77, 125–130.

Broms, B., Personal communication to H.C.H. Darley.

Burst, J.F., 1969. Diagenesis of Gulf Coast clayey sediments and its possible relation to petroleum migration. AAPG Bull. 53 (1), 73–93.

Chaney, B.P., Sargent, T.L., 1985. Low colloid polymer mud provides cost effective prevention of wellbore enlargement in the Gulf of Mexico. SPE Paper No. 14243, Annual Meeting, Las Vegas, NV, September 1985; and SPE Drill. Eng. 1986, 466–470.

Cheatham, J.B., 1984. Wellbore stability. J. Petrol. Technol. 36, 889–896.

Chenevert, M.E., 1969. Adsorptive pressures of argillaceous rocks. 11th Symp. Rock Mech., Berkeley, CA, June, pp. 16–19.

Chenevert, M.E., 1970a. Shale alteration by water adsorption. J. Petrol. Technol. 22, 1141–1148.

Chenevert, M.E., 1970b. Shale control with balanced activity oil-continuous muds. J. Petrol. Technol. 22, 1309–1316, Trans. AIME 249.

Chesser, B.G., 1986. Design considerations for an inhibitive and stable water-base mud system. IADC/SPE Paper No. 14757, Drilling Conference, Dallas, TX, February.

Chilingar, G.V., Knight, L., 1960. Relationship between pressure and moisture content of kaolinite, illite, and montmorillonite. AAPG. Bull. 44, 101–106.

Clark, D.E., Daniel, S., 1987. Protective colloid system improves solids removal efficiency, drilling fluid stability, and drilling performance. SPE/IADC Paper 16081, Drill. Conf. 15–18, New Orleans, LA, March.

Clark, R.K., Scheuerman, R.K., Rath, H., van Laar, H., 1976. Polyacrylamide/Potassium-chloride mud for drilling water-sensitive shales. J. Petrol. Technol. 261, 719–727, Trans. AIME 261.

Daemen, J.J., Fairhurst, C., 1970. Influence of inelastic rock properties on the stability of a wellbore. SPE Paper No. 3032, Annual Meeting, Houston, TX, 4–7.

Dallmus, K.F., 1958. Mechanics of basic evaluation and its relation to habitat of oil in the basin, habitat of oil. AAPG Bull. 42, 883–931.

Darley, H.C.H., 1969. A laboratory investigation of borehole stability. J. Petrol. Technol. 883–892, Trans. AIME 246.

Darley, H.C.H., 1976. Advantages of polymer muds. Petrol. Eng. 16, 46–52.

Denny, J.P., Shannon, J.L., 1968. New way to inhibit troublesome shale. World Oil 189, 111–117.

Desai, C.S., Reese, L.C., 1970. Stress-deformation analysis of deep boreholes. In: Proceedings of the 2nd International Soc. Rock Mech. Congr., Belgrade, September 2–26, pp. 475–484.

Dickinson, G., 1953. Geological aspects of abnormal pressures in the Gulf Coast, Louisiana. AAPG Bull. 37 (2), 410–432, Proc. 3rd World Petrol. Cong. 1, 1–17.

Eaton, B.A., 1969. Fracture gradient prediction and its application in oilfield operations. J. Petrol. Technol. 246, 1353–1360, Trans. AIME 246.

Ewy, R.T., Morton, E.K., 2008. Wellbore stability performance of water base mud additives. SPE 116139, SPE SPE ATC Denver, September 21–24.

Fairhurst, C., 1968. Methods of determining rock stresses at great depth. TRI-68 Missouri River Div., Corps of Engineers, February.

Finch, W.C., 1969. Abnormal pressure in antelope field North Dakota. J. Petrol. Technol. 246, 821–826.

Fleming, C.N., 1986. Moderate pH, potassium, polymer treated mud reduces washout. IADC/SPE Paper No. 14758, Drilling Conference, Dallas, TX, February.

Fraser, L., 1985. KOH/Lignosulfonate mud controls Laffan, Nahr Umr/Wasia shales off Abu Dhabi. Oil Gas J. 84, 124–127.

Freeze, G.I., Frederick, W.S., 1967. Abnormal drilling problems in the Andarko Basin. API Paper 851–41–L, Div. of Prodn., Mid-Continent Meeting, March.

Galle, E.M., Wilhoite, J.C., 1962. Stresses around a wellbore due to internal pressue and unequal principal geostatic stresses. Soc. Petrol. Eng. J. 2, 145–155, Trans. AIME 225.

Gill, J.A., Gregg, D.N., 1975. Complex sediments hard to interpret. Oil Gas J. 73, 63–66.

Gnirk, P.F., 1972. The mechanical behavior of uncased wellbores in elastic/plastic media under hydrostatic stress. Soc. Petrol. Eng. J. 12, 49–59, Trans. AIME 253.

Gray, G.R., Gill, J.A., 1974. Stabilizing boreholes in Australian offshore drilling. Petrol. Eng. 46, 49–58.

Griffin, J.M., Hyatdavoudi, A., Ghalambor, A., 1986. Design of chemically balanced drilling fluid leads to a reduction in clay destabilization. SPE Drill. Eng. 1, 31–42.

Hiltabrand, R.R., Ferrell, R.E., Billings, G.K., 1973. Experimental diagenesis of Gulf Coast argillaceous sediment. AAPG Bull 57 (2), 338–348.

Holt, C.A., Brett, J.E., Johnson, J.B., Walker, T.O., 1986. The use of potassium-lime drilling fluid system in Navarin Basin drilling. IADC/SPE Paper No. 14755, Drilling Conference, Dallas, TX, February.

Hottman, C.E., Smith, J.H., Purcell, W.R., 1978. Relationship among earth stresses, pore pressure and drilling problems, offshore Gulf of Alaska. SPE Paper No. 7501, Annual Meeting, Houston, TX, October 1–3.

Hubbert, M.K., Willis, D.G., 1957. Mechanics of hydraulic fracturing. J. Petrol. Technol. 210, 153–166, Trans. AIME, 210.

Jones, R.D., 1981. Potassium fluids to control shale. Oil Gas J. 197−204.
Kelly Jr., J., 1968. Drilling problem shales, Part 1. Classification simplifies mud selection. Oil Gas J. 66, 67−70.
Koch, R.D., 1976. Forties Field involves high angle, Gumbo. Oil Gas J. 74, 70−78.
Leyendecker, E.A., Murray, S.C., 1975. Properly prepared salt muds aid massive salt drilling. World Oil 180, 93−95.
Lu, C.F., 1985. A new technique for the evaluation of shale stability in the presence of polymeric drilling fluid. SPE Paper No. 14249, Annual Meeting, Las Vegas, NV, September.
Maury, V.M., Sauzay, J.M., 1987. Borehole instability: case histories, rock mechanics approach, and results. SPE/IADC Paper 16051, Drill. Conf., New Orleans, LA, March 15−18.
McDonald, M., Reifsnyder, R., Sidorkiewicz, V., LaPlant, D., 2002. Silicate based drilling fluids: a highly inhibitive mud system offering HS&E benefits over traditional oil based muds. AADE- 02-DFWM-HO-29, AADE 2002 Technology Conference, Houston, TX, April 2−3.
Mitchell, R.F., Goodman, R.A., Wood, E.T., 1987. Borehole stresses: plasticity and the drilled hole effect. SPE/IADC Paper 16053, Drill. Conf., New Orleans, LA, March 15−18.
Mondshine, T.C., 1969. New technique determines oil-mud salinity needs in shale drilling. Oil Gas J. 67, 70−75.
Mondshine, T.C., 1974. Tests show potassium-mud versatility. Oil Gas J. 72, 120−130.
Mondshine, T.C., Kercheville, J.D., 1966. Successful gumbo shale drilling. Oil Gas J. 64, 194−203.
Moore, W.D., 1977. North Sea chemistry is unique. Oil Gas J. 75, 132−150.
Nesbitt, L.E., King, G.P., Thurber, N.E., 1985. Shale stabilization principles. SPE Paper No. 14248, Annual Meeting, Las Vegas, NV, September.
Nordgren, R.P., 1977. Strength of well completions. In: Proceedings of the 18th Symposium of Rock Mechanics, Keystone, CO, June, pp. 4A3−1−9.
O'Brien, D.E., Chenevert, M.E., 1973. Stabilizing sensitive shales with inhibited potassium--based drilling fluids. J. Petrol. Technol. 225, 189−1100, Trans. AIME 255.
Osisanya, S.O., Chenevert, M.E., 1987. Rigsite shale evaluation techniques for control of shale-related wellbore instability problems. SPE/IADC Paper 16054, Conf., New Orleans, LA, March 15−18.
Paslay, P.R., Cheatham, J.B., 1963. Rock stresses induced by flow of fluids into boreholes. Soc. Petrol. Eng. J. 3, 85−94, Trans. AIME 228.
Patel, A., Stamatakis, E., Young, S., Cliffe, S., 2002. Designing for the future — a review of the design, development and testing of a novel, inhibitive water-based drilling fluid. AADE-02-DFWM-HO-33, AADE 2002 Technology Conference, Houston, TX, April 2−3.
Perry Jr., E.A., Hower, J., 1972. Late-stage dehydration in deeply buried peletic sediments. AAPG Bull. 56 (10), 2013−2021.
Powers, M.C., 1967. Fluid release mechanisms in compacting marine mudrocks and their importance in oil exploration. AAPG Bull. 51 (7), 1240−1254.
Pruett, J.O., 1987. A potassium base derivative of humic acid proves effective in minimizing wellbore enlargement in the Ventura Basin. SPE/IADC Paper 16080, Drill. Conf., New Orleans, LA, March 15−18.
Remson, D., 1970. Applying soil mechanics to well repair, completions. Oil Gas J. 68, 54−57.
Ritter, A.J., Gerault, R., 1985. New optimization programs for reactive shale formations. SPE Paper No. 14247, Annual Meeting, Las Vegas, NV, September.
Robinson, L.H., 1956. Effect of pore and confining pressure on failure characteristics of sedimentary rocks. Trans. AIME 216, 26−32.

Santos, H., Petrobras, S.P.E., Perez, R., 2001. What have we been doing wrong in wellbore stability? SPE 69493 SPE Latin American and Caribbean Petroleum Engineering Conference, Buenos Aires, Argentina, March 25–28.

Schlemmer, R., Friedheim, J.E., Growcock, F.B., M-I, L.L.C., Bloys, J.B., Headley, J.A., Polnaszek, S.C., 2002. Membrane efficiency in shale – an empirical evaluation of drilling fluid chemistries and implications for fluid design. IADC/SPE 74557 IADC/SPE Drilling Conference, Dallas, TX, February 26–28.

Schmidt, G.W., 1973. Interstitial water composition and geochemistry of deep Gulf Coast shales and sandstones. APPG Bull. 57 (2), 321–337.

Sheffield, J.S., Collins, K.B., Hackney, R.M., 1983. Salt drilling in the rocky mountains. Oil Gas J. 81, 70–76.

Shell, F.J., 1969. New mud improves productivity. Drilling 9, 47–50.

Simpson, J.P, Dearing, H.L., 2000. Diffusion osmosis - an unrecognized cause of shale instability. IADC/SPE 59190, IADC/SPE Drilling Conference, New Orleans, LA, February 23–25, 3–6.

Simpson, J.P., Walker, T.O., Aslakson, J.K., 1998. Studies dispel myths, give guidance on formulation of drilling fluids for shale stability. IADC/SPE 39376, IADC/SPE Drilling Conference, Dallas, TX, March 3–6.

Steiger, R.P., 1982. Fundamentals and use of potassium/polymer drilling fluids to minimize drilling and completion problems associated with hydratable clays. J. Petrol. Technol. 34, 1661–1670.

Stuart, C.A., 1970. Geopressures. Supplement to proceedings of the 2nd symp. abnormal subsurface pressure, Louisiana State Univ., Baton Rouge, LA, January 30.

Tare, U.A., Mody, F.K., 2002. Managing borehole stability problems: on the learning, unlearning and relearning curve. AADE-02-DFWM-HO-31, AADE 2002 Technology Conference, Houston, TX, April 2–3.

Topping, A.D., 1949. Wall collapse in oilwells as a result of rock stress. World Oil 112–120.

Walker, T.O., Dearing, H.L., Simpson, J.P., 1983. Potassium modified muds improve shale stability. World Oil 81, 93–100.

Walker, T.O., Dearing, H.L., Simpson, J.P., 1984. The role of potassium in lime muds. SPE Paper No. 13161, Annual Meeting, Houston, TX, September 16.

Westergaard, H.M., 1940. Plastic state of stress around a deep well. J. Boston Soc. Civil Eng. 27 (1), 2.

Willis, C., 1978. Rocky mountain coal seams call for special drilling techniques. Oil Gas J. 76, 143–152.

Chapter 10

Drilling Problems Related to Drilling Fluids

INTRODUCTION

In oil well drilling, various problems related to drilling fluid properties arise. Some of these problems, such as slow drilling rate or excessive drill pipe torque, merely render the drilling less efficient. Others, such as stuck drill pipe or loss of circulation, may interrupt the drilling progress for weeks and sometimes lead to abandonment of the well. Collectively, the problems that lead to the suspension of "drilling ahead" are called nonproductive time (NPT) (York et al., 2009).

NPT can result from the following issues

- Stuck Pipe
- Twist Off
- Flow check
- Kick
- Directional Correction
- Cement Squeeze
- Mud/Chemical Problems/Logistics
- Lost Circulation
- Wellbore Instability
- Waiting on Weather
- Casing or Wellhead Failure
- Rig Failure
- Equipment Failure
- Shallow Gas or Water Flows

The costs associated with drilling NPT can be extraordinary (Fig. 10.1).

This chapter covers the mud technical issues that can increase costs by delaying optimum drilling; stuck pipe, lost circulation, slow drill rate, and torque and drag. Wellbore instability is covered in Chapter 9, Wellbore Stability. Issues involving drilling fluids include that cause NPT are logistical delays, conditioning the mud to correct mud properties, unexpected weight ups, and/or inefficient displacement issues. Other issues covered

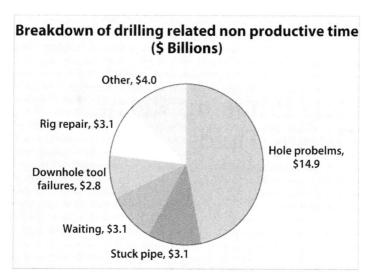

FIGURE 10.1 Drilling costs related to nonproductive time (York et al., 2009).

in this chapter are the effect of ultrahigh temperatures on drilling fluid properties and additives and corrosion caused by drilling fluids.

DRILL STRING TORQUE AND DRAG

Because no hole is truly vertical, and because the drill string is flexible, the rotating drill pipe bears against the side of the casing and wellbore at numerous points. The frictional resistance thus generated may require considerable extra torque than otherwise required to turn the bit. Similarly, considerable frictional resistance to raising and lowering the pipe may occur—a problem referred to as drag. Under certain conditions—highly deviated holes, holes with frequent changes in direction, undergauge hole, or poor drill string dynamics—torque and drag can be large enough to cause an unacceptable loss of power. The addition of certain lubricating agents to the mud can alleviate this power loss.

In general engineering practice, friction is reduced by interposing a film of oil or grease between moving metal parts. These lubricants are evaluated by their effect on the coefficient of friction, which is defined as the ratio of the frictional force parallel to the contact surface to the force acting normal to the contact surface. Expressed mathematically (see Fig. 10.2):

$$u = \frac{F}{W_1} \qquad (10.1)$$

where u is the coefficient of friction, F the force parallel to the surface, and W_1 the force normal to the surface. Note that u is constant for homogenous

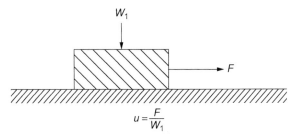

FIGURE 10.2 Measurement of coefficient of friction. $u = F/W$.

FIGURE 10.3 Lubricity tester for drilling muds. *Courtesy of NL Baroid Petroleum Services.*

surfaces; thus, for a given value of W_1, F is independent of the area of contact.

Lubricants

To evaluate lubricants for torque reduction, Mondshine (1970) used the apparatus shown in Fig. 10.3. The steel test block simulating the wall of the hole is pressed against the test ring by a torque arm. F is measured by the amperes required to turn the ring at a given rpm when immersed in the test mud. To get repeatable results, a steel test block was used. Mondshine found that although the coefficients of friction measured with steel blocks differed from those measured with sandstone or limestone blocks, the relative results obtained with different muds were substantially the same.

Results obtained with this machine were not in accord with some previously held notions. For example, it was thought that bentonite reduced torque because of its slippery nature. Test results showed that it did so only at low loads (less than 100 psi), and sharply increased the coefficient of friction at high loads. Similarly, it was believed that oil emulsified into a mud with an oil-wetting surfactant reduced torque. Test results showed that oil lightly stirred into the mud reduced friction considerably, but had no effect when tightly emulsified.

Table 10.1 shows the effect various water additives have on the coefficient of friction of water and two freshwater muds. These results were obtained under the standard conditions of 60 rpm and 150 inch-pounds (720 psi) load, which were judged to be representative of field conditions. The table shows that many of the agents reduced the coefficient of friction with water; some did so to a lesser extent with a simple bentonite mud, but only a fatty acid, a sulfurized fatty acid, and a blend of triglycerides and alcohols reduced friction in all the muds. These additives also reduced friction in a seawater mud.

The triglyceride mixture is one of the commercial water-soluble lubricity agents now commonly used in water-base muds to reduce torque. Oil muds are excellent torque reducers, presumably because of their oil-wetting properties; however, cost and potential pollution prevent the use of oil muds where the only advantage is torque reduction.

The fatty acid compounds referred to above are extreme pressure (EP) lubricants that were originally introduced by Rosenberg and Tailleur (1959) to decrease the wear of bit bearings. The action of EP lubricants differs from that of ordinary lubricants. Under EP, ordinary lubricants are squeezed out from between the bearing surfaces, and the resulting metal to metal contact causes galling and tearing. According to Browning (1959), the lubricating properties of EP lubricants are due to the lubricants reacting chemically with the metal surfaces at the high temperatures generated by metal-to-metal contact. The reaction product forms a film that is strongly bonded to the metal surface and acts as a lubricant.

Glass or plastic beads have been shown to reduce torque and drag (Lammons, 1984). For instance, in a field test, 4 lb/bbl of 44–88 micron beads reduced drag from 37,000 to 25,000 lb. The action is not clearly understood. The beads may act as ball bearings, or they may become embedded in the filter cake and provide a low friction bearing surface.

The effect of mud composition on wear and friction between tool joints and casing has been investigated by Bol (1985). He found that small-scale lubricity testers, such as the API tester, do not represent casing/tool joint contact conditions. He therefore developed a full-scale tester, using tool joints and oilfield casing. Tests with this instrument showed that wear was

TABLE 10.1 Comparison of Various Mud Lubricants (Mondshine, 1970)

Lubricant	Concentration, lb/bbl[a]	Water	Lubricity Coefficient Mud A[b]	Mud B[b]
None		0.36	0.44	0.23
Diesel oil	0.1	0.23	0.38	0.23
Asphalt	8	0.36	0.38	0.23
Asphalt and diesel oil	8			
Diesel oil	0.1	0.23	0.38	0.23
Graphite	8	0.36	0.40	0.23
Graphite and diesel oil	8			
Diesel oil	0.1	0.23	0.40	0.23
Sulfurized fatty acid	4	0.17	0.12	0.17
Fatty acid	4	0.07	0.14	0.17
Long-chained alcohol	2	0.16	0.40	0.23
Heavy metal soap	2	0.28	0.40	0.23
Heavy alkylate	4	0.17	0.36	0.23
Petroleum sulfonate	4	0.17	0.32	0.23
Mud detergent, brand X	4	0.11	0.32	0.23
Mud detergent, brand Y	4	0.23	0.32	0.23
Mud detergent, brand Z	4	0.15	0.38	0.23
Silicate	4	0.23	0.30	0.26
Commercial detergent	4	0.25	0.38	0.25
Chlorinated paraffin	4	0.16	0.40	0.25
Blend of modified triglycerides and alcohols	4	0.07	0.06	0.17
Sulfonated asphalt	8	0.25	0.30	0.25
Sulfonated asphalt and diesel oil	0.1	0.07	0.06	0.25
Walnut hulls (fine)	10	0.36	0.44	0.26

[a] Concentration in lb/bbl except for diesel oil which is given in bbl/bbl.
[b] Mud A: 15 g bentonite in 350 mL water; Mud B: 15 g bentonite, 60 g Glen Rose shale, 3 g chrome lignosulfonate, 0.5 g caustic soda in 350 mL water.

very high with bentonite suspensions, but decreased with the addition of barite. Additions of 0.5–2% of commercial lubricants all reduced wear by about the same amount.

Coefficients of friction were calculated from the test data. The results may be summarized as follows: oil emulsion muds, 0.15; unweighted water-base muds, 0.35–0.5; weighted water-base muds, 0.25–0.35. Polymeric additives, diesel oil, and glass beads had no effect. Results with lubricants were erratic, but, in general, they reduced the coefficient of friction of unweighted and low-density muds to about 0.25.

In assessing Bol's results, bear in mind that they do not necessarily apply to torque and drag in the open hole. In that case, the determining factor is the friction between the pipe and casing, (metal to metal contact), pipe and formation (metal to shale/sand), and metal to mud cake in which the lubricating additives are incorporated.

A device that more likely simulates wellbore and casing contact is the fully-automated Lubricity Evaluation Monitor (LEM-NT) (Slater and Amer, 2013) incorporating new technology shown in Fig. 10.3. This device is a modification of the lubricity tester shown in Fig. 3.24. The LEM-NT evaluates relative lubricity by measuring the friction coefficient between a rotating bob and a side-loaded, simulated wellbore immersed in the fluid of interest. Torque is measured by a noncontact rotational torque meter and fluid is circulated around the bob and sample. Computer-controlled, pneumatically-applied side loads are periodically relieved to allow fluid refreshment between the formation and the fluid on the bob. Simulated wellbores include ceramic cylinders, sandstone, and casing material. The apparatus accepts multiple diameters. Other wellbore materials can be used providing they are competent enough to be shaped to the proper dimensions and not disintegrate during testing.

DIFFERENTIAL STICKING OF THE DRILL STRING

Stuck pipe is one of the commonest hazards encountered in drilling operations. Sometimes the problem is caused by running or pulling the pipe into an undergauge section of the hole, or into a key seat or a bridge of cavings. In such cases, the driller is usually able to work the pipe free. A more intractable form of stuck pipe, known as *differential sticking*, characteristically occurs after circulation and rotation have been temporarily suspended, as when making a connection. The phenomenon was first recognized by Hayward in 1937, and the mechanism was demonstrated by Helmick and Longley (1957) in the laboratory.

Mechanism of Differential Sticking

The mechanism of *differential sticking*, stated briefly, is as follows: A portion of the drill string lies against the low side of a deviated hole.

While the pipe is being rotated, it is lubricated by a film of mud, and the pressure on all sides of the pipe is equal. When rotation of the pipe is stopped, however, the portion of the pipe in contact with the filter cake is isolated from the mud column, and the differential pressure between the two sides of the pipe causes drag when an attempt is made to pull the pipe. If the drag exceeds the pulling power of the rig, the pipe is stuck. Thus, increasing drag when pipe is being pulled warns that the pipe is liable to differential sticking.

Outmans (1958) has made a rigorous analysis of the mechanism of differential sticking which may be summarized as follows: The weight distribution of the drill string is such that the drill collars will always lie against the low side of the hole, and, therefore, differential sticking will always occur in the drill-collar section of the hole. When the pipe is rotating, the collars bear against the side of the hole with a pressure equal to the component of the weight of the collars normal to the sides of the hole. Thus, the depth the collars penetrate into the cake depends on the deviation of the hole, and on the rate of mechanical erosion beneath the collars relative to the rate of hydrodynamic erosion by the mud stream over the rest of the hole. Unless the hole is highly deviated or horizontal, or the rate of rotation very high, the collars will penetrate only a short distance into the filter cake, as shown in Fig. 10.4A.

When rotation is stopped, the weight of the pipe compresses the isolated mud cake zone, forcing its pore water out into the formation. As explained in the section on cake thickness in Chapter 7, The Filtration Properties of Drilling Fluids, the effective stress in a mud cake increases with decrease in local pore pressure. The effective stress therefore increases as pore water is forced out of the cake, and the friction thereby created between the pipe and the cake is the fundamental cause of differential sticking. After very long set times the pore pressure in the cake becomes equal to the pore pressure in the formation and the effective stress is then equal to the difference between the pressure of the mud in the hole and the formation pore pressure, i.e., $p_m - p_f$. The force required to pull the pipe is then given by

$$F = A(p_m - p_f)u \qquad (10.2)$$

where F is the force, A is the contact area, and u is the coefficient of friction between the collars and the cake. Because the ultimate value of F is not reached under normal field conditions, Outmans computed the value of F^1, which he defined as half the ultimate value of F. He found that as well as increasing with u, $(p_m - p_f)$, and A, F^1 increased with the compressibility and thickness of the filter cake, the hole deviation, and the diameter of the drill collars. It decreased with increase in diameter of the hole. In the drilling

---

1. A lubricity tester in which the shaft can, if desired, bear against a filter cake is shown in Figure 3.24.

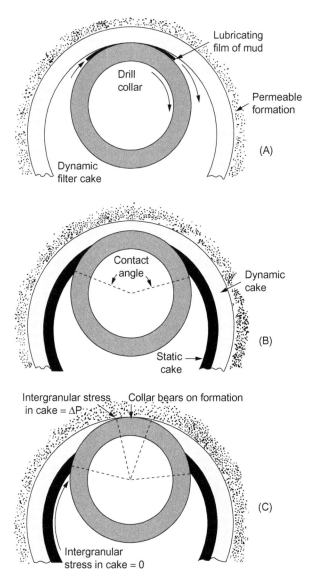

FIGURE 10.4 Differential pressure sticking mechanisms. (A) Pipe rotating and collar penetrates only a short distance into cake. (B) Pipe stationary and collar pushed into cake by differential pressure. (C) Highly deviated hole, pipe stationary and the pressure between the cake and collar varies from 0 to the differential pressure.

well, the pull-out force also increases with set time because filtration continues under static conditions. Thus, a static cake builds up around the collars, thereby increasing the angle of contact between the cake and the collars (see Fig. 10.4B).

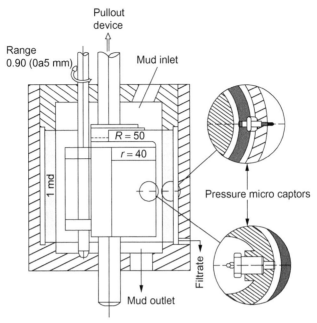

FIGURE 10.5 Pipe sticking tester. The pipe simulator is gradually pushed into the cake by turning the eccentric. *From Courteille, J.M., Zurdo, C., 1985. A new approach to differential sticking. In: SPE Paper No. 14244, Annual Meeting, Las Vegas, NV, 22–25 September. Copyright 1985 by SPE-AIME.*

Courteille and Zurdo (1985) investigated pipe-sticking phenomena in the apparatus shown in Fig. 10.5. Among other things, they measured fluid pressures at the cake/mud interface and at various points in the cake and filter media. They found that

1. The major pressure drop occurred across the internal mud cake (see Fig. 10.6).
2. With a thin filter cake (2 mm API) there was no change in pore pressure at the cake/pipe interface during or after embedment (see Fig. 10.7).
3. With thicker filter cakes (4–6 mm) the pressure at the cake/pipe interface fell with time after maximum embedment, but never reached the value in the pores of the uncontaminated porous medium (see Fig. 10.7).

These results indicate that the pressure differential across a stuck pipe will never reach the full value of $p_m - p_f$

Outmans' assumption that differential sticking always occurs in the drill collar section of the drill string is not substantiated by field experience. In a study of 56 recovery logs, Adams (1977) found that in 31 cases the drill pipe was stuck, and in the remaining cases either the drill collars only or the drill collars and the drill pipe were stuck. These results in no way invalidate the

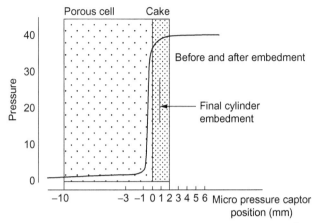

FIGURE 10.6 Pore pressure distribution before and after embedment. $40 \times 10^{5\text{Pa}}$ (580 psi) mud pressure. Atmospheric pore pressure in the uncontaminated filter medium. Cake thickness was 2 mm. *From Courteille, J.M., Zurdo, C., 1985. A new approach to differential sticking. In: SPE Paper No. 14244, Annual Meeting, Las Vegas, NV, 22–25 September. Copyright 1985 by SPE-AIME.*

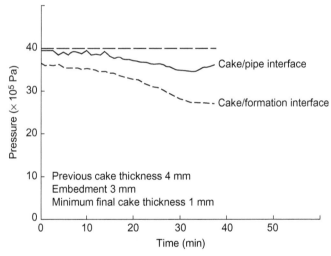

FIGURE 10.7 Change in pore pressure with time after embedment. Initial cake thickness was 4 mm. Minimum thickness after embedment was 1 mm. *From Courteille, J.M., Zurdo, C., 1985. A new approach to differential sticking. In: SPE Paper No. 14244, Annual Meeting, Las Vegas, NV, 22–25 September. Copyright 1985 by SPE-AIME.*

basic mechanisms postulated by Outmans; they merely show that sticking may occur at any point in the drill string where it bears against a permeable formation with a filter cake thereon. The chances that sticking will occur in the drill collar section are increased by the weight distribution of the drill string, which ensures that the collars will always lie against the low side of the hole, but are decreased by the circumstance that the filter cake will be much thinner because of erosion caused by the high shear rates prevailing in the narrow annulus around the drill collars.

The compaction of the filter cake and its effect on the coefficient of friction postulated by Outmans was confirmed experimentally by Annis and Monaghan (1962), who measured the friction between a flat steel plate and a filter cake in the apparatus shown in Fig. 10.8. They found that it increased with time up to a maximum (as shown in Fig. 10.9), and thereafter became constant. Moreover, no more filtrate was produced after the coefficient of friction reached its maximum.

As discussed in the section on adhesion in Chapter 8, The Surface Chemistry of Drilling Fluids, two solid surfaces will stick together if brought into sufficiently intimate contact. Hunter and Adams (1978) showed experimentally that adhesive forces contribute to the total frictional force involved in pulling stuck pipe, and that lubricants reduce adhesion and thus reduce pull-out force. Apparently some surfactants also reduce adhesion. For instance, when measuring friction in the apparatus shown in Fig. 7.24, Wyant et al. (1985) observed that when a styrene butadiene block copolymer was added to an invert oil emulsion mud, the plate would not adhere to the cake. An example of the magnitude of adhesive force may be deduced from the work of Courteille and Zurdo (1985). They measured the pull-out force when the pipe simulator shown in Fig. 10.5 was fully embedded. Table 10.2 shows that when the mud pressure was zero, the pull-out force per unit area of pipe-to-cake contact (which may then be taken as a measure of adhesion) was almost the same—about $0.1\,\mathrm{daN/cm^2}$ (1.40 psi)—for all three cake

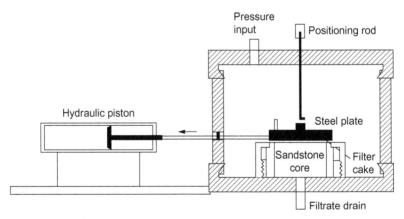

FIGURE 10.8 Apparatus for measurement of sticking coefficient. *From Annis, M.R., Monaghan, R.H., 1962. Differential pipe sticking—laboratory studies of friction between steel and mud filter cake. J. Pet. Technol. 537–542. Trans AIME. 225. Copyright 1962 by SPE-AIME.*

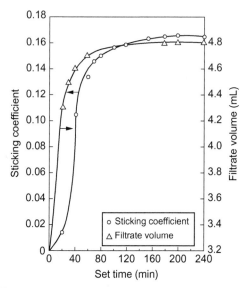

FIGURE 10.9 Effect of set time on sticking coefficient. *From Annis, M.R., Monaghan, R.H., 1962. Differential pipe sticking—laboratory studies of friction between steel and mud filter cake. J. Pet. Technol. 537–542. Trans AIME. 225. Copyright 1962 by SPE-AIME.*

TABLE 10.2 Pull-Out Force after Maximum Embedment (Courteille and Zurdo, 1985)

	Pull-Out Pressure daN/cm² (psi)	
Mud	Mud Pressure: Zero	40×10^5 Pa (580 psi)
Low filter loss, 10cc, 2 mm API	0.09 (1.305)	0.13 (2.45)
Medium filter loss, 70cc, 4 mm API	0.1 (1.45)	0.23 (3.33)
High filter loss, 120cc, 6 mm API	0.1 (1.45)	0.30 (4.35)

thicknesses. When mud pressure was applied, the pull-out force increased with increasing cake thickness because of decreasing pore pressure at the cake/pipe interface.

Differential sticking is particularly liable to occur when drilling high-angle or horizontal holes. In such circumstances, the weight component of the collars normal to the wall of the hole, and erosion under the collars, may be so high that no external filter cake can form (see Fig. 10.4C). The weight of the collars is then borne by the formation, and the cake in the fillet between the collars and the formation will not be compacted when rotation ceases. The frictional forces acting on the collar will then derive in part from the friction between the collars and the formation, and in part from the effective stress between the cake in the fillet and the collars. In the section on inclined holes in Chapter 6, The Rheology of Drilling Fluids, it was shown that a bed of cuttings tends to form on the low side of the hole when annular velocities are low. Evidence obtained by Wyant et al. (1985) leads them to suggest that such cuttings will become incorporated in the filter cake, greatly increasing sticking tendencies.

Prevention of Differential Sticking

One way to prevent differential sticking is to minimize the contact area by suitable drill string design. Noncircular collars, fluted or spiraled drill collars, and drill pipe stabilizers have been used for this purpose. Long drill collar sections and oversized (*packed hole*) collars increase the contact area, and for that reason increase the chances of stuck pipe. This effect may be offset by drilling the hole straighter.

Another approach is to maintain suitable mud properties. Outmans (1958) showed that the pull-out force increased with differential pressure, contact area, thickness of the filter cake, and coefficient of friction. The differential pressure may be minimized by keeping the mud density as low as possible, consistent with well safety. To minimize contact area and cake thickness, cake permeability should be kept low, and drilled solids content should be reduced by rigorous desilting. Remember that cake thickness does not necessarily correlate with filter loss (see section on cake thickness in chapter: The Filtration Properties of Drilling Fluids), and therefore should always be directly measured when concerned with differential sticking. The coefficient of friction of the filter cake depends on the composition of the mud. A number of papers on this subject have been published (Annis and Monaghan, 1962; Haden and Welch, 1961; Simpson, 1962; Krol, 1984; Kercheville et al., 1986), but correlation between them is difficult because of differences in the testing methods and procedures. Some investigators measured the torque or pull-out force, while others measured the coefficient of friction in the lubricity tester shown in Fig. 10.3. The latter tests are of doubtful value because the composition of the mud is unlikely to have the same effect on the friction between two steel surfaces as on the friction between steel and the mud filter cake. Results may be summarized as follows:

1. Oil-base muds have much lower coefficients of friction than water-base muds. Since they also lay down very thin filter cakes, they are much better muds for avoidance of differential sticking. This conclusion was strikingly confirmed in the field study by Adams (1977), who found that out of 310 cases of stuck pipe in southern Louisiana, only one occurred when oil-base mud was in the hole.

2. Sufficient evidence is not available for establishing which class of water-base muds yields filter cakes with the lowest coefficient of friction. The best (though rather limited) data were obtained by Simpson (1962), who measured the torque required to free a disc embedded in a cake of standard thickness at elevated temperatures, under conditions of both static and dynamic filtration (see Fig. 10.10 and Table 10.3).
3. Barite content increases the friction coefficient of all muds.
4. Emulsification of oil or the addition of friction reducers to a water-base mud reduces the force required to free stuck pipe. Krol (1984) compared the effect of diesel oil and various commercial additives on pull-out force in an apparatus closely simulating downhole conditions. He found that 2% diesel oil reduced the force required by 33%, and that larger amounts had no further effect. Only a few commercial additives reduced the pull-out force significantly more than 33%, and some reduced it considerably less. The amount of additive required to achieve maximum reduction depended on the amount of solids in the mud and the degree of dispersion. For example, increasing the density from 11.75 lb/gal (1.41 SG) to 16 lb/gal (1.92 SG) with barite increased the amount of additive required from 2 to 4%. Krol concluded that for maximum effect additives must improve filtration properties, coat solids, and wet metal surfaces. Kercheville et al. (1986) measured the effect of diesel oil and environmentally acceptable mineral oil and friction reducers on the torque required to free a steel disc stuck in a filter cake for increasing lengths of time. The results showed that the mineral oil was just as effective as the diesel oil, and the nonpolluting additives were more effective than the oils.

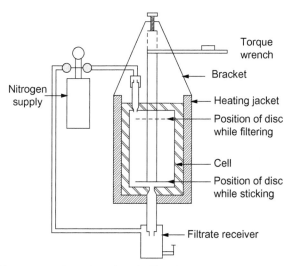

FIGURE 10.10 Apparatus for static testing of differential pressure sticking. *From Simpson, J.P., 1962. The role of oil mud in controlling differential-pressure sticking of drill pipe. In: SPE Paper 361, Upper Gulf Coast Drill and Prod. Conf., Beaumont, TX, 5 April. Copyright 1962 by SPE-AIME.*

TABLE 10.3 Differential Pressure Sticking Tests (Simpson, 1962)

Mud (All Muds Weighted to 14 lb/gal [1.68 SG] with Barite)	Torque after 30-Minute Set Time (in lb)
Laboratory-Prepared Muds	
Freshwater chrome lignosulfonate	75
Gyp-chrome lignosulfonate	0
Invert emulsion oil-base mud	0
Field Muds	
Freshwater chrome lignosulfonate	115
Gyp-chrome lignosulfonate	63
Gyp-chrome lignosulfonate + 8% oil	4
Chrome lignite/chrome lignosulfonate	10
Invert emulsion oil-base mud	0
Asphaltic oil-base mud	0

Note: Filter cake deposited dynamically at 200°F [93°C] 1/32 inch (0.8 mm) thick. Differential pressure 500 psi (35 kg/cm^2).

Freeing Stuck Pipe

One way to free stuck pipe is to reduce the pressure differential either by reducing the density of the mud, or by setting a drill-stem tester (Sartain, 1960). The more usual way is to spot oil around the stuck section. The capillary pressure of oil on aqueous filter cake runs into thousands of pounds, so the pressure of the mud column (or at least a good part of it) on the face of the filter cake compresses it, and reduces the contact angle (see Fig. 10.11). It was formerly thought that the oil freed the pipe by penetrating between the pipe and the cake, thereby reducing friction. The work of Annis and Monaghan (1962) casts doubt on this theory; they found spotting oil was ineffective. Obviously the oil did not penetrate between the cake and plate, and if it had done so, it would not have reduced the area of contact because of the geometry of their apparatus. However, it is possible that under downhole conditions, the oil penetrates along the fillet as the cake is compressed, particularly if the pipe is being worked, and oil thus possibly assists in the freeing process. Note that excess of oil-wetting surfactants should not be used in spotting fluids, since they reduce capillary pressure.

Before spotting oil, the depth at which the pipe is stuck must be determined. The usual method is to measure the amount of pipe stretch produced

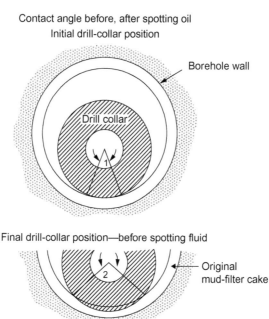

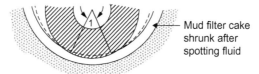

FIGURE 10.11 Contact angle before and after spotting oil. *From Outmans, H.D., 1974. Spot fluid quickly to free differentially stuck pipe. Oil Gas J. 65–68 (Outmans, 1974). Courtesy Oil and Gas J.*

by a given amount of pull. The length of free pipe can then be calculated from charts which list stretch versus pull for the various pipe sizes and weights. Several logging devices are also available for locating the sticking point. The best of these is the drill pipe recovery log, which can define free and stuck sections by means of sound attenuation, even when there are multiple stuck sections (Adams, 1977).

The oil slug must be weighted to the same density as the mud, if it is to stay in place. Based on field experience, Adams (1977) recommends using an excess volume of spotting fluid, and waiting at least 12 h for the cake to be compressed.

SLOW DRILLING RATE

Laboratory Drilling Tests

It has long been obvious that the properties of the drilling mud have a profound effect on the rate of penetration of the bit. Changing from drilling with air to drilling with water always results in a marked drop in penetration rate; changing from water to mud produces another sharp drop. Furthermore, there is a correlation between the penetration rate decrease and the depth, even when allowance is made for the older, harder rocks that are encountered as the depth of the hole increases. To understand why drilling fluids have such a profound effect on drilling rate, a review of basic drilling mechanisms is necessary. The principal factors involved were established by a number of investigators (Murray and Cunningham, 1955; Eckel, 1958; Cunningham and Eenik, 1959; Garnier and van Lingen, 1959; van Lingen, 1962; Maurer, 1962; Young and Gray, 1967) using micro- or full-scale laboratory drilling machines. In these tests jacketed rock specimens were subjected to simulated subsurface vertical and horizontal stresses. Pore pressures were sometimes applied but usually set at zero, which was permissible since, as we have seen, the effective stress is the load less the pore pressure (see the section on the behavior of rocks under stress in chapter: Wellbore Stability). These tests established beyond doubt that the critical factor governing penetration rate was a function of mud column pressure, not the stresses to which the rocks were subjected. The pressure of the mud column affected the penetration rate by holding the chips (created by the bit) on the bottom of the hole, as discussed below.

Static Chip Hold-Down Pressure

The differential pressure between the mud column and the formation pore pressure ($p_m - p_f$) causes the mud to filter into the formation beneath the bit, but the filtration rate bears no relation to the API filter loss nor to the filtration rate into the sides of the hold (see the section on filtration below the bit in chapter: The Filtration Properties of Drilling Fluids). Under the bit, the filter cake is continuously being removed by the bit teeth. If a tri-cone bit is rotating at 100 rpm, a tooth strikes the same spot about every 0.2 s. In time intervals of this magnitude, filtration is still in the mud spurt stage, and the amount of fluid that invades the formation depends on the concentration of bridging particles and their size relative to the size of the rock pores (see the section on the bridging process in chapter: The Filtration Properties of Drilling Fluids) rather than the colloidal characteristics of the mud (Darley, 1965).

The bridging particles establish an internal filter cake in the pores of the rock immediately below the bottom of the hole, and the finer particles invade somewhat further. This process is repeated over and over as successive layers of rock are removed. The magnitude of the resulting pressure gradients ahead of the bit was measured by Young and Gray (1967) in a micro drilling machine. In their apparatus, the invading fluid was constrained to flow downwards, and the change in pore pressure as the bottom of the hole approached a pressure tap was observed. Also, the distribution of permeability in the rock was calculated from Darcy's law. Fig. 10.12 shows that the pressure gradient was very high in the first millimeter or so below the bottom of the hole, corresponding to the internal mud cake, and then decreased to almost zero in the invaded zone. Since a bit tooth will penetrate well into the invaded zone, practically the full pressure differential $(p_m - p_f)$ will act across a chip (Fig. 10.13).

That the pressure drop across the internal mud cake is one of the major factors affecting the penetration rate was confirmed experimentally by Black et al. (1985) in a full-scale model borehole. They measured the filtration rate while drilling Berea sandstone and while circulating only. The difference between the two rates gave the rate of filtration through the bottom of the hole while drilling. The pressure drop in the uncontaminated sandstone was calculated from this rate and the permeability of the sandstone by means of Darcy's law. This pressure drop subtracted from the total pressure drop across the sandstone gave the pressure drop across the internal mud cake. Tests were conducted on four water-base muds at constant borehole mud pressure and various back pressures in the sandstone. Fig. 10.14 shows the relationship between the penetration rate and the pressure drop across the internal mud cake for two bit weights. Similar results were obtained with the other muds. No relationship with the API filter loss, either on filter paper or Berea sandstone discs, was observed.

Vaussard et al. (1986) investigated the effect of differential pressure on penetration rate when drilling through limestone rocks in a full-scale model borehole. They found that the penetration rate increased with differential pressure with a 10 md limestone but remained constant with a 0.5 md one.

Garnier and van Lingen (1959) have shown that this static CHDP limits penetration rates by two mechanisms (see Fig. 10.15). In the first place, it acts as a confining pressure and therefore strengthens the rock. The importance of this action varies with the type of rock, and is most pronounced with unconsolidated sand. For example, though loose sand drills without any measurable resistance at atmospheric pressure, it drills like hard rock at 5000 psi (351 kg/cm^2) (Cunningham and Eenik, 1959), as shown in Fig. 10.16.

In the second mechanism, the differential pressure acting across the chip opposes its dislodgement. The magnitude of this effect varies with the type of bit. It is greatest with drag bits—which drill with a scraping action—and the effect increases with the increase in the cutting angle (*rake*). It is least with roller bits, which drill mostly by crushing.

Laboratory tests have shown that static CHDP occurs even when drilling with water, although the magnitude of the effect is considerably less (see Fig. 10.17). The phenomenon is caused by fine particles generated by the action of the bit

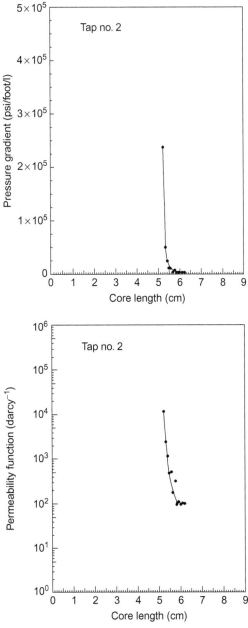

FIGURE 10.12 Permeability function and pressure gradient versus core length for high fluid loss mud (vertical Berea sandstone, $P_b = 1000$ psig). *From Young Jr., F.S., Gray, K.E., 1967. Dynamic filtration during microbit drilling. J. Pet. Technol. 1209–1224. Trans AIME. 240. Copyright 1967 by SPE-AIME.*

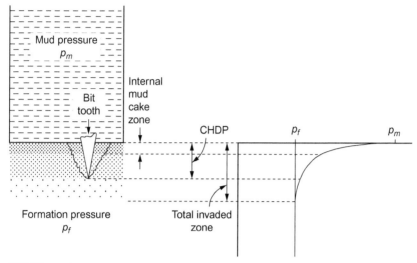

FIGURE 10.13 Pressure distribution ahead of the bit (schematic). Chip hold-down pressure (CHDP) is virtually equal to $p_m - p_f$.

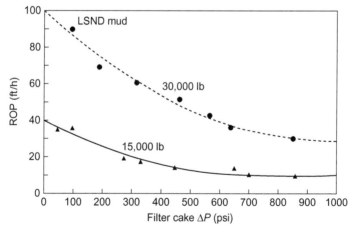

FIGURE 10.14 Penetration rate versus pressure drop across the internal mud cake for weights on bit of 15,000 and 30,000 lb. Low-solid nondispersed mud. *From Black, A.D., Dearing, H.L., Di Bona, B.G., 1985. Effects of pore pressure and mud filtration on drilling rate in permeable sandstone. J. Pet. Technol. 1671–1681. Copyright by SPE-AIME.*

being carried into the surface pores of the rock, thus forming a pseudo internal mud cake (Cunningham and Eenik, 1959; Young and Gray, 1967).

That the penetration rate in permeable rocks depends on $p_m - p_f$, and not on p_m, was shown by Cunningham and Eenik (1959), who determined

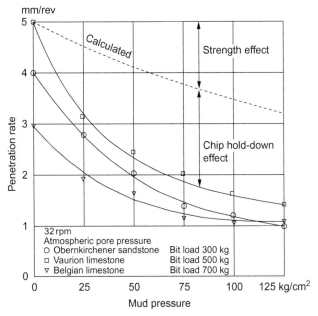

FIGURE 10.15 Penetration rate as a function of mud pressure at atmospheric pore pressure. *From Garnier, A.J., van Lingen, N.H., 1959. Phenomena affecting drilling rates at depth. J. Pet. Technol. 232–239. Copyright 1959 by SPE-AIME.*

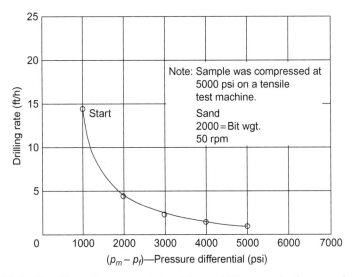

FIGURE 10.16 Effect of pressure differential on drilling rate in loose sand. *From Cunningham, R.A., Eenik, J.G., 1959. Laboratory study of effect of overburden, formation, and mud column pressure on drilling rate of permeable formations. Trans AIME. 216, 9–17. Copyright 1959 by SPE-AIME.*

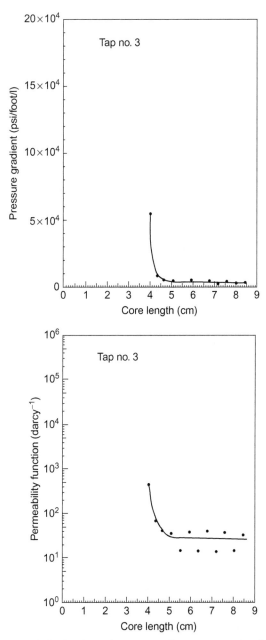

FIGURE 10.17 Permeability function and pressure gradient versus core length for water (vertical Berea sandstone, p_b = 1000 psig). *From Young Jr., F.S., Gray, K.E., 1967. Dynamic filtration during microbit drilling. J. Pet. Technol. 1209–1224. Trans AIME. 240. Copyright 1967 by SPE-AIME.*

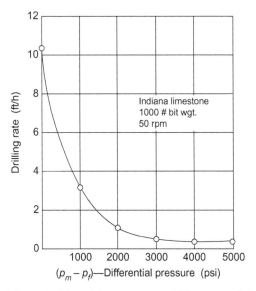

FIGURE 10.18 Effect of differential pressure on drilling rate of Indiana limestone. Overburden ($\sigma_0 = 6000$ psi). *From Cunningham, R.A., Eenik, J.G., 1959. Laboratory study of effect of overburden, formation, and mud column pressure on drilling rate of permeable formations. Trans AIME. 216, 9–17 Copyright 1959 by SPE-AIME.*

penetration rates at various values of p_m and p_f, and found that the results correlated with $p_m - p_f$ (Fig. 10.18).

The results discussed above show that filtration through the bottom of the hole causes a pressure drop between the top and bottom of a chip. In rocks of low permeability, initial filtration rate is low, so static CHDP does not have time to fully develop. In rocks of very low permeability, such as shales, no filtrate enters the formation and no filter cake is formed, so there is no pressure drop across a chip. The only mechanism affecting penetration rate is the strength effect caused by $p_m - p_f$ acting on the bottomhole surface. Warren and Smith (1985) have shown that $p_m - p_f$ will be greater in shales than in permeable rocks under similar conditions. In both cases the rocks expand as the overburden pressure is relieved by the bit, but, in the case of shales, pore pressure equalization will not occur in the available time, because of the low permeability of the shale. Hence, p_f is reduced, and $p_m - p_f$ is greater than in the case of permeable rocks.

Dynamic Chip Hold-Down Pressure

Garnier and van Lingen (1959) have shown that as the static CHDP declines with decreasing rock permeability, another type of hold-down pressure, which they call the *dynamic CHDP*, develops. This pressure

390 Composition and Properties of Drilling and Completion Fluids

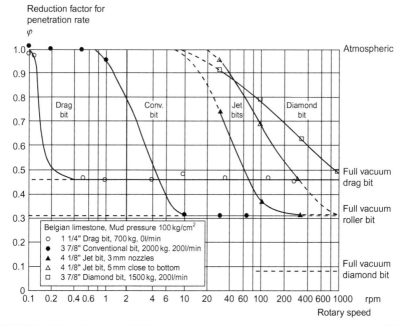

FIGURE 10.19 Dynamic hold-down versus rotary speed. *From van Lingen, N.H., 1962. Bottom scavenging—a major factor governing penetration rates at depth. J. Pet. Technol. 187–196. Trans AIME. 225. Copyright 1962 by SPE-AIME.*

develops because, although a bit tooth penetrates the rock and creates fractures when it strikes, in order for the chips to fly out of the crater, fluid must flow in to take their place. Because of the low permeability of the rock, the fluid can only flow in through the fractures, the width of which is initially very small. If the fluid does not flow in fast enough, a vacuum is created under the chips, while the full weight of the mud column acts on top of the chip.

The magnitude of the dynamic chip hold-down pressure depends on the rpm of the bit, the permeability of the rock, and the type of bit (van Lingen, 1962). For a given rock and bit, the dynamic CHDP increases—and the rate of penetration decreases—as the rpm increases, until a full vacuum is pulled under the chips, and then remains constant with further increase in rpm. The maximum dynamic CHDP is thus one atmosphere plus the pressure of the mud column, p_m. Fig. 10.19 shows the rate of penetration of various bits versus rpm when drilling in Belgian limestone, permeability 0.5 md. The differences in the performance of the various bit types are related to the velocity of the mud across the hole bottom (note, for instance, the superior performance of jet bits) which suggests that the filtration characteristics of the mud influence dynamic CHDP, presumably because they control rate of penetration of fluid into the cracks.

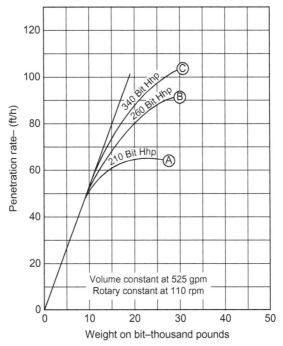

FIGURE 10.20 Influence of lack of hydraulic horsepower on "balling up" and penetration rate. *From Speer, J.W., 1958. A method for determining optimum drilling techniques. Oil Gas J. 90–96. Courtesy of Oil and Gas J.*

Bottomhole Balling

If the drill cuttings are not removed from beneath the bit as fast as they are generated, they will be reground, and a layer of broken rock will build up between the bit and the true hole bottom. From data obtained in the field, Speer (1958) showed that a plot of penetration rate versus weight was linear up to a certain bit weight, and then fell off rather rapidly. Since the decrease in rate was less at higher bit hydraulic horse power (see Fig. 10.20), Speer concluded that the phenomenon was caused by inadequate scavenging of the cuttings from beneath the bit. In laboratory tests at high bit weights, a layer of crushed rock mixed with mud solids was noticed on the bottom of the hole at the conclusion of the tests (Cunningham and Eenik, 1959).

The importance of bottomhole cleaning has been shown by Maurer (1962). He deduced that under conditions of perfect cleaning, the rate of penetration would be given by

$$R = \frac{CNW^2}{D^2 S^2} \qquad (10.3)$$

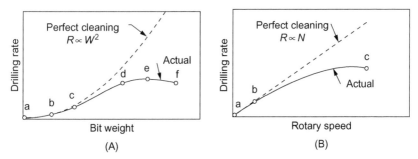

FIGURE 10.21 Rate/weight/speed relationships for imperfect cleaning. *From Maurer, W.C., 1962. The "perfect cleaning" theory of rotary drilling. J. Pet. Technol. 1270–1274. Trans AIME. 225. Copyright 1962 by SPE-AIME.*

where R is the rate of penetration; C, a drillability constant; N, the bit rpm; W, the weight on the bit; D, the diameter of the bit; and S, the drillability strength of the rock. In laboratory experiments with full-scale bits and near perfect cleaning conditions—atmospheric pressure, water as the drilling fluid, and impermeable Beeksmantown dolomite—he found this formula to be valid, except that the R/N ratio was not linear at rotary speeds greater than 300 rpm, obviously because of dynamic CHDP. On the other hand, when drilling with a mud pressure equivalent to a mud column of 3000 ft (914 m), all other conditions being the same, he obtained R/N and R/W curves very different from the theoretical, as shown in Fig. 10.21a,b. Since most drilling in the field is done with bit weights as shown from c to d on the curves, it is evident that inadequate bottomhole scavenging is a major factor restricting penetration rates in the field.

Bit Balling

Like bottom balling, bit balling occurs at high bit weights. In hard formations the teeth become partially clogged with cuttings. How far such clogging restricts penetration rate is obscure because the effect is indistinguishable from bottom balling. Garnier and van Lingen (1959) postulated that when the cuttings are pressed against the bit surfaces, they adhere because of the difference between pore pressure in the cuttings and the mud pressure. Thus, they are held by a mechanism similar to differential sticking of drill pipe. Eventually the cuttings are released because filtration neutralizes the pressure differential.

A much worse type of bit balling occurs in soft shales, particularly gumbo shales and swelling shales that adsorb water from the mud. In this case, a ball of compacted shale may build up and cover the whole bit, preventing further drilling progress. The driller must then either try to spud the ball off, or pull a green bit. To avoid bit balling, gumbo shales are often drilled with less weight than would otherwise be optimum.

The severity of bit balling in soft shales is caused by two factors: (1) the differential pressures postulated by Garnier and van Lingen are magnified by the hydrational forces of compressed subsurface shales (see section on hydration of the borehole in chapter: Wellbore Stability); and (2) adhesive forces become significant because the ductile shales deform, and are forced into intimate contact with the bit surfaces. As discussed in the section on surface free energy in Chapter 8, The Surface Chemistry of Drilling Fluids, short range attractive forces become effective when contact between solids is intimate. In addition, soft shales—or shales that become soft on contact with aqueous drilling fluids—have low internal cohesion and, as already mentioned, adhesion depends on the difference between the adhesive and cohesive forces.

The mechanism of adhesion (accretion) in the case of bit balling is probably hydrogen bonding, extending from the molecular layers of water adsorbed on the shale surfaces to the layer of water adhering to the steel surface. From soil mechanics technology (Sashikumar et al., 2011) shale cuttings can be classified as to their accretion potential. When first cut by the bit the cutting is relatively dry clay. Shales adsorb water to relive their internal stresses. As water is adsorbed, a "plastic limit" is reached and the sticking tendency is very high. NADF differ from WBMs in the very high capillary pressures in shales that prevent uptake of the NABF. This helps inhibit accretion. In addition, steel surfaces have a nonaqueous film that will hinders clay adherence.

A laboratory test procedure was developed to screen materials for accretion control (Metath et al., 2011) called the Rolling Bar Accretion Test Method. A mild steel bar is placed in the center of a stainless steel test cell which is half full of the drilling fluid. Sized shale pieces are evenly distributed around the bar. The cell is then placed in a roller oven and rolled for a selected time at room temperature. Accreted solids are scrapped off the bar and weighed. Various possible antiaccretion agents are mixed into the drilling to see the relative effects of preventing or minimizing balling. Most promising antiaccretion agents are based on ammonia chemistry, i.e., amines and amides as copolymers with fatty acid derivitives.

Many additives used to formulate HPWBM for wellbore stability are also effective as antiaccretion agents. Brines are also effective, especially potassium chlorides and formates (see chapter: Wellbore Stability). Chesser and Perricone (1983) recommend the use of an aluminum lignosulfonate chelate[2] to prevent bit balling. This complex is adsorbed on the surface of the shale through linkages between the aluminum and the oxygen atoms in the silicate

---

2. A chelate is a heterocyclic ring of organic molecules having coordinate bonds with a metal ion.

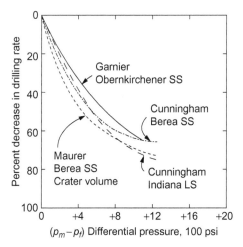

FIGURE 10.22 Laboratory data. Percent decrease in drilling rate as differential pressure increases. *From Vidrine, D.J., Benit, B.J., 1968. Field verification of the effect of differential pressure on drilling rate. J. Pet. Technol. 676–682. Copyright 1968 by SPE-AIME.*

layer on the shale surface. These linkages thereby disrupt the hydrogen bonding. Because the aluminum is chelated, the concentration of aluminum ions in the aqueous phase is very low, so the mud is not flocculated.

The Effect of Mud Properties on Drilling Rate

Density is the most important mud property affecting penetration rate. For any given formation pressure, the higher the density, the greater will be the differential pressure, and, consequently, the greater the static chip hold down, and likelihood of bottomhole and bit balling. Fig. 10.22 summarizes the effect of pressure differential on drilling rate as observed in the laboratory by the various investigators discussed above; Fig. 10.23 shows similar results obtained by Vidrine and Benit (1968) in a controlled field study. Note that in both figures, drilling rate decreased by over 70% when the differential pressure increased from 0 to 1000 psi (70 kg/cm^2). In addition, decreasing mud density decreases dynamic chip hold down, permitting faster rpm, and, by decreasing pressure losses in the drill pipe, increases hydraulic horsepower available at the bit.

It follows that the lowest possible mud density should always be carried. Wherever possible, use air, gas, foam, or an underbalanced mud column. In normally pressured formations, keep differential pressures no higher than necessary to establish a filter cake on unconsolidated sands (100–200 psi). Note that, for a constant mud density and formation pressure gradient, the differential pressure increases with depth. For example, a 10 lb/gal mud would exert a differential pressure of 70 psi at 1000 ft and 700 psi at

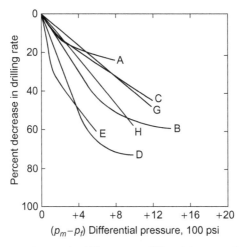

FIGURE 10.23 Percent decrease in drilling rate as differential pressure increases. The letters refer to different wells. *From Vidrine, D.J., Benit, B.J., 1968. Field verification of the effect of differential pressure on drilling rate. J. Pet. Technol. 676–682. Copyright 1968 by SPE-AIME.*

10,000 ft if the formation pressure gradient remained unchanged at 0.45 kpsi per foot. In geopressured formations, safety of the well must be the first consideration, but do not increase the mud density unnecessarily; for instance, do not increase the density because of gas swabbed into the hole on a trip—reduce the gel strength or lower the hoisting speeds instead. Lower differential pressures can be carried if the mud returns are continuously monitored for gas, or if drill string telemetry is installed.

Viscosity is another mud property that materially influences penetration rate. Low viscosity promotes fast rates mainly because of good scavenging of cuttings from under the bit. The relevant viscosity is the effective viscosity at the shear rate prevailing under the bit, not the plastic or funnel viscosity. The determination of viscosity at high shear rates was discussed in the section on the generalized power law in Chapter 6, The Rheology of Drilling Fluids. Eckel (1967) obtained a rather good correlation between kinematic viscosity (viscosity/density) and drilling rate as shown in Fig. 10.24. Note, however, that worthwhile increases in drilling rate were only obtained at viscosities less than 10 cs. The tests were made in a microbit drilling machine with a wide variety of liquids: water, salt solutions, glycerine, oils, and water-base and oil-base muds. Viscosities were measured at the shear rates prevailing in the bit nozzles.

Low viscosities are particularly important at high rotational speeds because of lower dynamic chip hold down. When the bit tooth strikes, the fractures are at first exceedingly small and the viscosity of the filtrate is probably the relevant factor, but as the chips move out, the viscosity of the mud becomes relevant (Mettath et al., 2011).

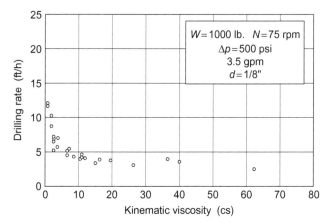

FIGURE 10.24 Drilling rate versus kinematic viscosity at 3½ gal/min. *From Eckel, J.R., 1967. Microbit studies of the effect of fluid properties and hydraulics on drilling rate. J. Pet. Tech. 541–546. Trans AIME. 240. Copyright 1958 by SPE-AIME.*

Because their viscosities are higher than aqueous muds, conventional oil muds tend to give comparatively slow penetration rates. Simpson (1978) describes a special low-viscosity invert emulsion mud that gave drilling rates as fast as or faster than did aqueous muds in offset wells. Low viscosities are obtained by using not more than about 15% water in the dispersed phase, a low-viscosity oil, and a minimum amount of additives such as bentones, oil-dispersible lignites, tall oil soaps, and asphaltenes (Simpson, 1979; Carney, 1980).

The concentration of particulate solids is another mud property that affects drilling rate. High concentrations of solids reduce drilling rates because they increase mud density and viscosity. Use of weighting agents with specific gravities higher than that of barite, such as itabirite and ilmenite (see chapter: Drilling Fluid Components), has enabled faster drilling rates to be obtained (Rupert et al., 1981; Blattel and Rupert, 1982; Scharf and Watts, 1984; Simpson, 1985; Montgomery, 1985) because the volume of solids required for a given mud density is less, and hence the viscosity is lower. Also, these materials are harder than barite, so there is less attrition in the course of drilling, and thus less increase in viscosity.

Much higher penetration rates are experienced as the percentage of solids approaches zero than can be explained by the negligible decrease in viscosity. The reason is reduced CHDP, as discussed later. The actual solids concentration that it is possible to maintain depends on well conditions and the type of mud being used. When drilling in low-permeability sandstones and carbonates that remain true to gauge, it is possible to drill with clear water. Emphasis is placed on the word *clear*, because field experience has shown that remarkably small amounts of solids can greatly reduce penetration rates

(Gallus et al., 1958; Lummus, 1965). For this reason, small amounts of a flocculation aid, such as 10% hydrolyzed polyacrylamide copolymer, are added at the flow line. If proper settling facilities are provided, clear water is obtained at the suction. The settling facilities usually consist of large earthen pits, because the flocs are voluminous and their specific gravity is only slightly greater than water. Consequently, their settling rates are low, and they are sensitive to stray currents. Settling efficiency is greatly increased if the pits are baffled to distribute the flow evenly over the whole surface of the pit, and if a weir is placed at the discharge end of the pit to skim clear water from the surface. Because of hindered settling, settling efficiency falls off very rapidly if solids content is allowed to rise much above 1% by volume.

Another type of ultralow-solid fluid (commonly known as *milk emulsion*) used to drill hard formations, consists of water or brine in which 5% diesel oil is emulsified with an oil-wetting surfactant. The emulsified oil is believed to provide a small measure of filtration control, and to protect the drill cuttings from disintegration by the oil-wetting action. Fig. 10.25 shows the faster drilling times and decreased bit wear that have been achieved with this type of fluid in West Texas (Mallory, 1957).

In most holes, filtration control is necessary, and therefore the drilling fluid must have a colloidal base, which makes maintenance of a low solids content more difficult. Some decrease in rate of penetration is inevitable, but quite high rates can be obtained if the bridging solids content can be kept below 4% by weight (Darley, 1965). If the concentration of bridging solids can be kept low enough, the amount of mud spurt that can invade the formation between successive tooth strikes will be unable to establish an effective internal filter cake, and therefore the static CHDP is minimized. However, unlimited time is available for bridging on the sides of the hole, which are therefore protected by a normal filter cake. Fig. 10.26 shows the increase in static CHDP (expressed as a percentage of the applied pressure) as increasing amounts of bridging solids were added to a 2% starch suspension, all other factors being held constant. Note that the CHDP increases rapidly as the solids increase to 4%, and much more slowly thereafter. Fig. 10.26 also shows that the penetration rate, as measured in a microbit drilling machine with the same starch suspensions, correlates fairly well with the CHDP. Confirmation of these results was obtained by Doty (1986) who made a comprehensive series of full-scale drilling tests under simulated downhole conditions. He found that the penetration rate through Berea sandstone with a clear 13 lb/gal $CaCl_2/CaBr_2$ polymer fluid was 5−10 times as fast as with the same fluid plus 6% fine solids. Tests were also conducted with two other fluids, a conventional lignosulfonate mud, and a low viscosity oil mud. Filtration rates through the sandstone were measured while drilling and when only circulating; the difference between the two gave a measure of the filtration rate through the bottom of the hole while drilling. Typical results are

398 Composition and Properties of Drilling and Completion Fluids

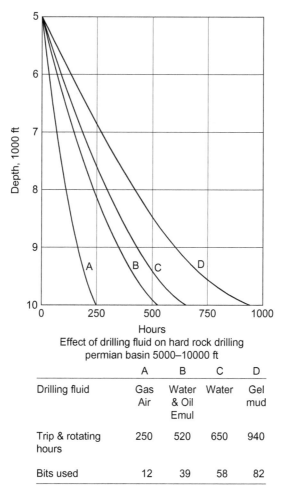

FIGURE 10.25 Effect of oil-in-water emulsion on total drilling time. *From the April 1957 issue of Petroleum Engineer International. Publisher retains copyright.*

shown in Table 10.4. The correlation between the penetration rate and the filtration rate through the bottom of the hole shows that CHDP is the controlling mechanism. Note that the filtration rate while circulating with the clear brine fluid was much the same as with the other fluids, indicating that an effective filter cake was deposited on the sides of the hole. No correlation was found between the API fluid loss and the penetration rate.

In order to obtain the ultrafast drilling rates possible with clear brine polymer drilling fluids, virtually all the drilled solids must be removed when they reach the surface. When drilling in hard rock formations, this requirement can be achieved by desanders and desilters, if efficiently operated, or

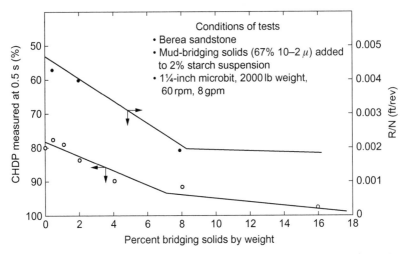

FIGURE 10.26 Influence of bridging solids on drilling rate and CHDP. *From Darley, H.C.H., 1965. Designing fast drilling fluids. J. Pet. Technol. 465–470. Trans AIME. 234. Copyright 1965 by SPE-AIME.*

TABLE 10.4 Filtration Rates and Penetration Rates When Drilling in Berea Sandstone (Doty, 1986)

	Filtration Rate cc/s		
Drilling Fluid	Circulating	Drilling	Penetration Rate (ft/h)
Clear brine/polymer	1.97	28.3	70
Same + 6% solids	0.95	4	10
Lignosulfonate	2.5	6.9	15
Invert oil emulsion	2.17	3.9	6

by sedimentation in large earth pits, as described above in the section on clear brines. As long as gauge hole can be maintained, the sole function of the polymer is to limit filtration into the walls of the hole. Obviously, a nonviscous, zero yield point polymer should be selected since a viscous polymer will prevent efficient separation of drilled solids. Also, low viscosity will maximize bottomhole cleaning and minimize dynamic CHDP. Another consideration in selecting polymers is that certain ones act as friction reducers in turbulent flow (see Figs. 6.33 and 10.27) and thus reduce pressure losses in the drill pipe and increase the hydraulic horsepower at the bit.

Densities up to 10 lb/gal (1.2 SG) are usually obtained with KCl or NaCl, up to 11.5 lb/gal (1.38 SG) with $CaCl_2$, and up to 15.2 lb/gal (1.82 SG) with

CaCl$_2$ and/or CaBr$_2$, but the higher densities are expensive and may not be economically justifiable. Note that only polymers such as hydroxyethylcellulose or hydroxyalkyl gums can be used with the polyvalent brines.

Clear polymer fluids also permit fast drilling rates in shales. For example, Clark et al. (1976) obtained penetration rates 50–100% faster than those in offset wells when drilling in brittle shales in the Canadian foothills with the KCl-polyacrylamide clear fluid mentioned in Chapter 9, Wellbore Stability. Again, in the full-scale drilling tests mentioned above, Doty (1986) found that the clear brine drilling fluid drilled the fastest through Pierre shale (a soft montmorillonitic shale) under all conditions except high bit weight combined with high mud pressure. But the use of clear brine polymer drilling fluids in shales is limited by two problems:

1. Dispersion of drill cuttings: This makes it difficult to maintain a low enough solids content. Use of a polymer that coats (encapsulates) the drill cuttings will keep the solids content within bounds when drilling in consolidated shales (Darley, 1976), but in soft, unconsolidated shales excessive build-up of solids is very difficult to prevent.
2. Hole enlargement: If the hole enlarges significantly, a viscous polymer, such as xanthan gum or prehydrated bentonite, must be added to clean the hole. High viscosities and yield points prevent adequate separation of

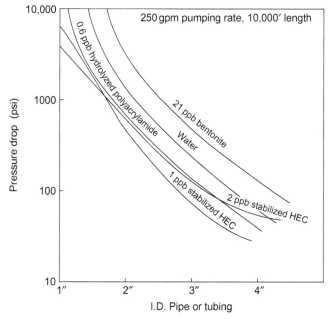

FIGURE 10.27 Effect of polymers on pressure loss of liquids in turbulent flow. *Courtesy of Brinadd Co.*

Drilling Problems Related to Drilling Fluids Chapter | 10   401

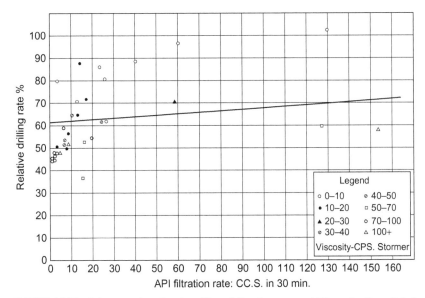

FIGURE 10.28 Laboratory data showing effect of filtration rate on drilling rate. *From Eckel, J.R., 1954. Effect of mud properties on drilling rate. API Drill. Prod Prac. 119–124. Courtesy of API.*

solids at the surface, and again, the solids content becomes excessive. Therefore, every effort should be made to prevent hole enlargement, both by good drilling practice and by the use of one of the shale stabilizing KCl-polymer fluids mentioned in Chapter 9, Wellbore Stability.

When drilling with conventional high-solid muds, mud properties have much less influence on penetration rates. Variations in viscosity and solids content have a comparatively small effect on penetration rate, as shown in Figs. 10.24 and 10.26. There is no direct correlation with filtration properties (see Fig. 10.28) (Eckel, 1954), although there is a tendency for both to move in the same direction, because both decrease with increase in the colloidal fraction. The only mud property that has a major effect on drilling rate in high-solid muds is density (discussed earlier in this section).

When drilling in shales, drilling rates may be improved by the use of muds or mud additives that inhibit bit balling. Cunningham and Goins (1957) found that emulsification of oil increased penetration rates in microbit tests with Vicksburg and Miocene shales (see Fig. 10.29), presumably because of decreased bit balling. See discussion on bit balling in this chapter.

LOSS OF CIRCULATION

Circulation in a drilling well can be lost into fractures induced by excessive mud pressures, into preexisting open fractures, or into large openings with

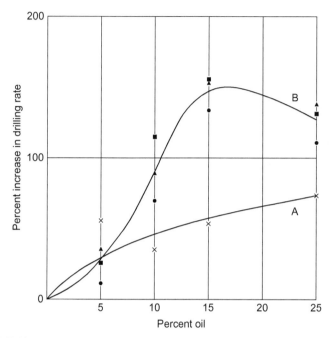

FIGURE 10.29 Increase in drilling rate versus percent oil. A = Vicksburg shale; B = Miocene shale, La. *From Cunningham, R.A., Goins Jr., W.C., 1957. Laboratory drilling of gulf coast shales. API Drill. Prod. Prac. 75–85. Courtesy of API.*

structural strength (such as large pores or solution channels). The conditions under which each of these types of openings can occur are discussed in the following three sections.

Induced Fractures

The mechanism of induced fracturing in a drilling well is similar to hydraulic fracturing during well completion, the only difference being that the latter is deliberate and desirable, while the former is unintentional and most unwelcome. A fracture is induced whenever the difference between the mud pressure and the formation pore pressure, $p_m - p_f$, exceeds the tensile strength of the formation plus the compressive stresses surrounding the borehole. Since the tensile strength of rocks is usually small compared to the compressive stresses, it is generally (though not always justifiably) left out of the calculation. The direction of the fracture must be normal to the least principal stress. Except in regions of active mountain building, the least principal stress is horizontal, and therefore an induced fracture is vertical (Hubbert and Willis, 1957). As discussed in the section on stresses around the borehole in Chapter 9, Wellbore Stability, the least principal stress, σ_3, is related to the overburden effective stress, $S - p_f$, by a factor k_1, the value of which depends on the tectonic history of the geologic region. Therefore, a fracture will be induced when

$$p_w - p_f > (S - p_f) \qquad (10.4)$$

p_w includes the hydrostatic pressure of the mud column, p_m, plus the pressure loss in the rising column if the well is being circulated, plus any transient pressures such as surge pressures when pipe is being run into the hole.

A distinction must be made between the pressure required to initiate a fracture and that required to extend it. As discussed in Chapter 9, Wellbore Stability, there is a concentration of hoop stresses around the borehole. The value of these stresses depends on the ratio between the two principal horizontal stresses, σ_2 and σ_3, and they may not be distributed evenly round the circumference of the borehole (Hubbert and Willis, 1957). In order to initiate a fracture, $p_w - p_f$ must be greater than the minimum hoop stress, and in order to extend it beyond the hoop stress zone, it must be greater than σ_3. The minimum hoop stress may vary from zero to twice σ_3, so the initiation pressure may be greater or smaller than the extension pressure, as shown in Fig. 10.30.

In a drilling well, mud density must be kept great enough to control formation fluids, but not so great as to induce a fracture. In a normally pressured formation, an ample margin of safety ensures that no problem arises. In geopressured zones, however, the difference between the fracture pressure and the formation fluid pressure becomes very small as the geopressuring increases. Ability to predict formation and fracture pressures then becomes important so that mud and casing programs may be planned to minimize the risks.

Formation pore pressures may be determined from shale resistivity logs, or from acoustic logs in nearby wells. Shale bulk density is directly related to shale resistivity and to a function of shale transit time (Hottman and Johnson, 1965). Thus a plot of either shale resistivity or a function of transit time reveals anomalies in bulk density (see Fig. 10.31) which are related to pore fluid pressures. The precise relationship depends on the geologic region, and must be determined empirically by measurement of formation fluid pressures in interbedded sand lenses. Fig. 10.32 shows an example of fluid pressure gradient versus a shale acoustic parameter. Once such a relationship is established, it may be used to predict formation fluid pressures in future wells.

A less accurate but more convenient method depends on plotting drilling rate versus depth (Jorden and Shirley, 1966). Drilling rate is related to $p_m - p_f$, as shown in the previous section. The rate normalized for drilling variables is expressed in the following equation:

$$\frac{R}{N} = a \left(\frac{W}{D} \right)^d \qquad (10.5)$$

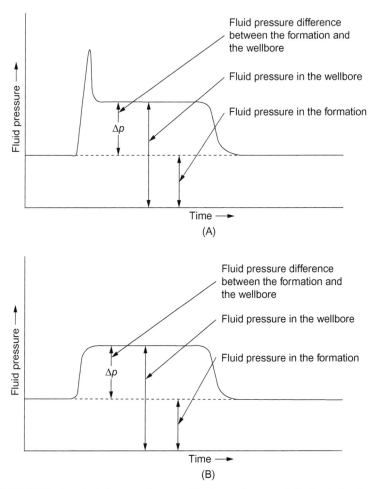

FIGURE 10.30 Idealized diagram of two possible types of pressure behavior during fracture treatment depending upon various underground conditions. *From Hubbert, M.K., Willis, D.G., 1957. Mechanics of hydraulic fracturing. J. Pet. Technol. 153–166. Trans AIME. 210. Copyright 1957 by SPE-AIME.*

where a and d are constants. The relationship between the exponent d and $p_m - p_f$ is established for a particular region by direct measurement (see Fig. 10.33). Field experience and modifications thereof have been discussed by Fontenot and Berry (1975).

An estimate of pore pressures in sand bodies in shales may be made from a mathematical model relating the pressure directly to resistivity and density log data, thus eliminating the need for first establishing a correlation (Stein, 1985).

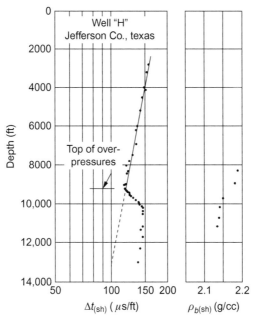

FIGURE 10.31 Shale travel time and bulk density versus burial depth. *From Hottman, C.E., Johnson, R.K., 1965. Estimation of pore pressure from log-derived shale properties. J. Pet. Technol. 717–722. Trans AIME. 234. Copyright 1965 by SPE-AIME.*

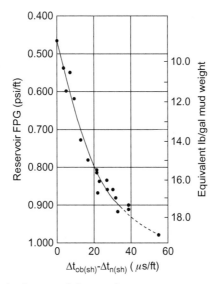

FIGURE 10.32 Relation between shale acoustic parameter $t_{ob(sh)} - t_{n(sh)}$ and reservoir fluid pressure gradient (FPG). *From Hottman, C.E., Johnson, R.K., 1965. Estimation of pore pressure from log-derived shale properties. J. Pet. Technol. 717–722. Trans AIME. 234. Copyright 1965 by SPE-AIME.*

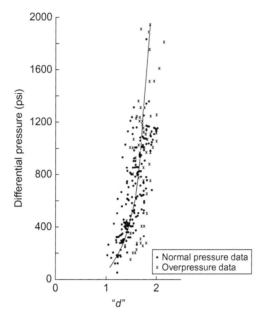

FIGURE 10.33 d exponent–differential pressure relationship, bit run data. *From Jorden, J.R., Shirley, O.J., 1966. Application of drilling performance data to overpressure detection. J. Pet. Technol. 1387–1394. Copyright 1966 by SPE-AIME.*

Pore pressures may also be determined from gamma ray logs (Zoeller, 1984). These logs reflect the degree of shaliness of a formation; consequently, there is a regional correlation between gamma ray curves and shale compaction. As discussed above, shale density can be correlated by experience with pore pressure. Therefore, gamma ray curves can be similarly correlated, and any departure from the normal regional gamma ray-depth curve can be related to a change in pore pressure. The gamma ray values can be measured while drilling (MWD) and thus provide real time information.

Fracture pressure at any depth of interest may be predicted by substituting pore pressures, determined as outlined above in Eq. (10.4), provided that the regional value of k_1 is known. Methods for determining k_1 empirically have been published by Eaton (1969) and Matthews and Kelly (1967). Although their rationales differ, both methods are based essentially on the degree of compaction of shales, as determined from resistivity or sonic logs, and the relation of the degree of compaction to observed fracture gradients. Fig. 10.34 shows the fracture *initiation pressure* (p_{frac}) gradients predicted by the two methods for the Gulf Coast region. Remember that these curves do not apply to other regions because k_1 varies according to local subsurface stresses. Curves for fracture *injection pressure* gradients based on the theoretical limits of σ_3 as postulated by Hubbert and Willis (1957) are also shown in Fig. 10.33.

As mentioned previously, fracture initiation pressures are determined by the minimum hoop stress, which may be more or less than the least horizontal stress. To extend a fracture beyond the hoop stress zone the injection pressure must be greater than the least horizontal stress. The instantaneous shut-in pressure (ISIP) is the best measure of the least horizontal stress. Breckels and van Eekelen (1982) therefore developed equations relating the least horizontal stress to depth, based on observed ISIPs. The correlation

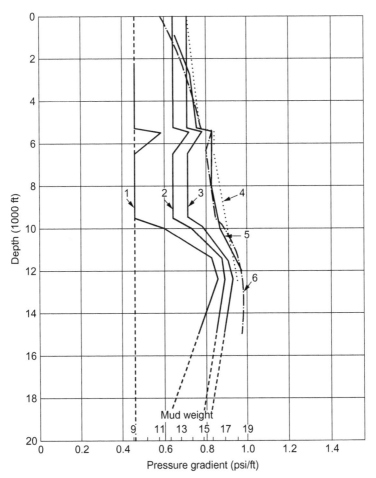

FIGURE 10.34 Pore and fracture pressure gradients in the Gulf Coast. Curve 1: Pore pressure gradient. Curve 2: Hubbert and Willis fracture injection pressure gradient when $\sigma_3 = \frac{1}{3}(S - p)$. Curve 3: Hubbert and Willis fracture injection pressure gradient when $\sigma_3 = \frac{1}{2}(S - p)$. Curve 4: Matthews and Kelley fracture injection pressure gradient. Curve 5: Goldsmith and Wilson fracture gradient (derivation not published). Curve 6: Eaton fracture initiation pressure gradient. *From Eaton, B.A., 1969. Fracture gradient prediction, and its application in oilfield operations. J. Pet. Technol. 1353–1360. Trans AIME. 246. Copyright 1969 by SPE-AIME.*

depends on pore pressure, and is not the same for normal and abnormal pressures, but Breckels and van Eekelen developed combined correlations, which are as follows:

For the Gulf Coast, depths less than 11,500 ft:

$$S_{Hmin} = 0.197Z^{1.145} + 0.46(p_f - p_{fn})$$

Depths greater than 11,500 ft:

$$S_{Hmin} = 1.167Z - 4,596 + 0.46(p_f - p_{fn})$$

For Venezuela, depths between 5900 and 9200 ft:

$$S_{Hmin} = 0.210Z^{1.145} + 0.56(p_f - p_{fn})$$

For Brunei, depths less than 10,000 ft:

$$S_{Hmin} = 0.227Z^{1.145} + 0.49(p_f - p_{fn})$$

where S_{Hmin} = the total minimum horizontal stress (effective stress plus pore pressure) and p_{fn} = the normal pore pressure, assuming a gradient of 0.465 psi/ft.

Measurements of the vertical distribution of S_{Hmin} in the lower Mesa Verde formation in Colorado have been made by Warpinski et al. (1985). Accurate, reproducible ISIPs were obtained by conducting repeated, small volume "minifracs" through perforations. Results showed that the value of S_{Hmin} depended on the lithology. In shales the horizontal stress approached that of the vertical, and fracture gradients were greater than 1 psi/ft, whereas in sandstones the gradients were 0.85–0.9 psi/ft.

Theoretical values of S_{Hmin} were calculated from the equation

$$S_{Hmin} = \frac{v}{1-v}(S - p_f) + p_f$$

where v is Poisson's ratio, which was obtained from the long space sonic logs and experimentally determined Young moduli. The calculated values deviated substantially from the values measured in shales, indicating that the shales did not behave elastically—probably because of creep.

Daines (1982) developed a method for predicting fracture pressures from data obtained in the first fracture test in a wildcat well. His model for horizontal stress contains two components: an expression for the compressive stress created by the pressure of the overburden, and a superimposed stress, σ_1, created by tectonic forces, if any. A fracture then occurs when

$$p_{frac} = \sigma_t + \sigma_1\left(\frac{v}{1-v}\right) + p_f$$

After the first fracture test, σ_t is obtained by subtracting the elastic component from the fracture pressure. Poisson's ratio is obtained from the experimentally determined values shown in Table 10.5. Lithology is determined

TABLE 10.5 Suggested Poisson's Ratios for Different Lithologies (Daines, 1982)

Clay, very wet	0.50
Clay	0.17
Conglomerate	0.20
Dolomite	0.21
Greywacke	
Coarse	0.07
Fine	0.23
Medium	0.24
Limestone	
Fine, micritic	0.28
Medium, calcarenitic	0.31
Porous	0.20
Stylolitic	0.27
Fossiliferous	0.09
Bedded fossils	0.17
Shaly	0.17
Sandstone	
Coarse	0.05
Coarse cemented	0.10
Fine	0.03
Very fine	0.04
Medium	0.06
Poorly sorted, clayey	0.24
Fossiliferous	0.01
Shale	
Calcareous (<50% $CaCO_3$)	0.14
Dolomitic	0.28
Siliceous	0.12
Silty (<70% silt)	0.17
Sandy (<70% sand)	0.12
Kerogenaceous	0.25
Siltstone	0.08
Slate	0.13
Tuff	
Glass	0.34

from drill cuttings, and pore pressure by the usual means, previously discussed. Assuming that the ratio σ_t/σ_1 remains constant with depth, the fracture pressure at any subsequent depth can be calculated from the lithology and pore pressure at that depth. Use of the method is limited because σ_t/σ_1 will only be constant with depth when the formations are close to horizontal and when the structure does not change significantly with depth.

Daneshy et al. (1986) have measured ISIPs and fracture orientation during drilling operations by creating microfractures on the bottom of the hole. A packer is set a few feet above the bottom and small volumes of mud pumped at very low rates through the tubing. Fracture orientation is determined by subsequent coring. The ISIPs did not show any observable relation to Young's modulus, Poisson's ratio, or tensile strength.

In a highly geopressured formation (as mentioned earlier), the operating margin between a blowout and an induced fracture is small. For example, if the formation pore pressure gradient is 0.95 psi/ft and k_1 is 0.5, the fracture pressure gradient will be

$$\frac{pw}{z} = 0.5(1 - 0.95) + 0.95 = 0.975 \text{ psi/ft}$$

where Z is the depth. Thus, the mud column must exert a gradient of at least 0.95 psi/ft in order to prevent a blowout, but must be less 0.975 psi/ft in order to avoid loss of circulation. At 10,000 ft (3047 m) the operating margin would be 250 psi (17.5 kg/cm^2). Some extra margin of safety might be provided by the tensile strength of the rock (see Table 10.6) (Topping, 1949)

TABLE 10.6 Average Rock Strengths (Topping, 1949)

Rock	Ultimate Compressive Strength (psi/kg-cm^2)	Ultimate Tensile Strength	Ultimate Shearing Strength	Poisson's Ratio
Sandstone	13,000 psi	450 psi	1300 psi	0.27
	910 kg/cm^2	31 kg/cm^2	91 kg/cm^2	
Marble	9000	300	1350	0.27
	630	21	94	
Limestone	11,000	200	1200	0.23
	770	14	84	
Granite	20,000	650	2000	0.21
	1400	45	140	

Note: Strengths given are averages. Strength varies widely for the same rock, e.g., ultimate compressive strength for limestone varies from 4000 to 20,000 psi (920 to 4600 kg/cm^2).

but if the rock contained natural fractures, the tensile strength would be zero. Be that as it may, it obviously becomes extremely important to minimize circulating pressures and transient pressures in the annulus when drilling in highly geopressured formations.

Up to this point, we have discussed fracturing under the condition that significant amounts of borehole fluid do not invade the formation, as would be the case when a mud cake is present. When there is no mud cake, the outward flow of fluid lessens the stress concentration around the hole, and thus should reduce the breakdown pressure. However, laboratory tests have shown that the effect of fluid invasion depends on the rock type and the stress level (Haimson, 1973). Fluid invasion was found to lower the breakdown pressure of low-porosity rocks, such as shales, at stress levels above 1000–2000 psi (70–140 kg/cm^2), and of high-porosity rocks at stress levels above 5000–10,000 psi (350–700 kg/cm^2).

A fracture, once initiated, will continue to extend, and drilling fluid will continue to be lost, until the pressure at the tip of the fracture falls below the regional fracture injection pressure, either because of friction or filtrate losses along the length of the fracture; or if pumping has been stopped, until the pressure of the mud remaining in the hole falls below the fracture injection pressure.

Natural Open Fractures

Natural open fractures can exist at depth only when one of the principal stresses is tensile. Formerly it was believed that conditions of absolute tension could not exist in the compressive field prevailing subsurface. Secor (1965), however, has shown from the geometry of the combined Griffith and Mohr failure envelopes (see Fig. 10.35) that tension fractures can develop down to a depth where $\sigma_1 = 3K$ (where K is the tensile strength), and be held open to a depth where $\sigma = 8K$.

Effective stresses decrease with pore pressures. Consequently, the maximum permissible depth for open fractures is greater in geopressured than in normal pressured zones. Fig. 10.36 shows the effect of an incremental increase in p_f, and Fig. 10.37 shows the maximum depths at which open fractures may be encountered plotted against p_f/S for several values of K.

In a drilling well the mud pressure, p_m, normally exceeds p_f and therefore if an open fracture is encountered, mud is certain to be lost until p_m is reduced below p_f by the same mechanisms that limit the extension of induced fractures.

Openings With Structural Strength

When no tensile stresses exist in subsurface formations, voids can only exist if they have structural strength great enough to withstand the earth's compressive forces. Examples of such voids are

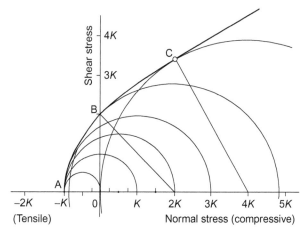

FIGURE 10.35 The composite failure envelope showing a family of Mohr stress circles tangent at point A and one Mohr stress circle tangent at each of points B and C. *From Secor, D.T., 1965. Role of fluid pressure jointing. Am. J. Sci. 263, 633–646. Courtesy Am. J. Science.*

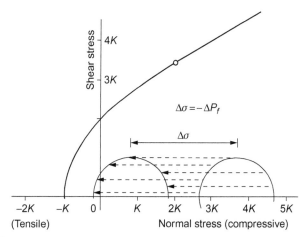

FIGURE 10.36 A failure envelope diagram showing the effect of an incremental increase of fluid pressure on the position of the Mohr stress circle, assuming the total principal stresses remain constant. *From Secor, D.T., 1965. Role of fluid pressure jointing. Am. J. Sci. 263, 633–646. Courtesy Am. J. Science.*

1. Solution channels caused by water percolating through carbonate formations for millions of years. These channels may range in size from pinholes to caverns. Limestones often contain vugs (small cavities) interconnected by solution channels. The structural strength of cavities decreases with their size, so large caverns are only found at shallow depths.
2. Coarse granular beds such as gravels.

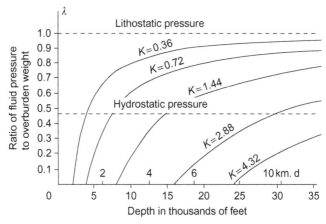

FIGURE 10.37 A graph for the case σ_1 vertical showing the maximum depth in the earth where open fractures can occur for a variety of tensile strengths. The bulk specific gravity of the rock is 2.3. K is the tensile strength in 10^5 lb/ft^2. 1×10^5 lb per square foot = 694 psi = 49 kg/cm^2. *From Secor, D.T., 1965. Role of fluid pressure jointing. Am. J. Sci. 263, 633–646. Courtesy Am. J. Science.*

3. Natural fractures that have been closed by subsequent compressive forces but which retain some permeability because they are propped by irregularities or crystal growths on their sides, or by loose rock fragments. Such fractures may be anywhere from a few microns to several millimeters in width. Some otherwise impermeable formations have appreciable fracture permeability because of the presence of multiple microfractures.

Circulation will be lost into any opening as long as p_m exceeds p_f unless the drilling fluid contains particles large enough to bridge the opening. Openings in a compressive field are distinguished from tensile open fractures in that they do not enlarge unless p_w exceeds p_{frac}.

Lost Circulation Materials

An enormous variety of materials have been used at one time or another in attempts to cure lost circulation. They may be divided into four categories:

1. Fibrous materials, such as micronized cellulose, shredded sugar cane stalks (bagasse), cotton fibers, hoghair, shredded automobile tires, wood fibers, sawdust, and paper pulp. These materials have relatively little rigidity, and tend to be forced into large openings. If large amounts of mud containing a high concentration of the fibrous material are pumped into the formation, sufficient frictional resistance may develop to effect a seal. If the openings are too small for the fibers to enter, a bulky external

filter cake forms on the walls of the hole, and is knocked off when the well is cleaned out.
2. Flaky materials, such as shredded cellophane, mica flakes, plastic laminate and wood chips. These materials are believed to lie flat across the face of the formation and thereby cover the openings. If they are strong enough to withstand the mud pressure, they form a compact external filter cake. If not strong enough, they are forced into the openings, and their sealing action is then similar to that of fibrous materials.
3. Granular materials, such as sized calcium carbonate, ground nutshells, or vitrified, expanded shale particles. The latter are made by firing ground shales at temperatures up to 1800°F (982°C) (Gockel et al., 1987). These materials have strength and rigidity and when the correct size range is used, seal by jamming just inside the openings in a manner similar to that described in Chapter 7, The Filtration Properties of Drilling Fluids for the bridging of normal porous formations, i.e., there must be some particles approximately the size of the opening and a gradation of smaller particles (Goins and Nash, 1957; Gatlin and Nemir, 1961). Experiments by Howard and Scott (1951) showed that the greater the concentration of particles in the mud, the larger the opening bridged (see Fig. 10.38), and that strong granular materials, such as nutshells, bridged larger openings than did fibrous or flaky materials; however, weak granular materials, such as expanded perlite, did not (see Fig. 10.39).
4. Slurries whose strength increases with time after placement, such as hydraulic cement, diesel oil-bentonite-mud mixes, and high filter loss muds. Neat cement is usually used only for squeezes at the casing shoe. If used to plug a channel in open hole, the strength of the cement may be reduced by contamination with mud. Also, a danger of sidetracking exists when drilling out cement left in the hole.

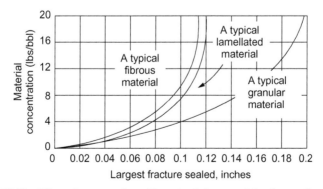

FIGURE 10.38 Effect of concentration of lost circulation materials when sealing fractures. *From Howard, G.C., Scott, P.P., 1951. An analysis of the control of lost circulation. Trans AIME. 192, 171–182. Copyright 1951 by SPE-AIME.*

Material	Type	Description	Concentration lbs/bbl	Largest Fracture Sealed inches (0 – 0.20)
Nut shell	Granular	50% −3/16+10 Mesh 50% −10+100 Mesh	20	
Plastic	"	"	20	
Limestone	"	"	40	
Sulphur	"	"	120	
Nut shell	"	50% −10+16 Mesh 50% −30+100 Mesh	20	
Expanded perlite	"	50% −3/16+10 Mesh 50% −10+100 Mesh	60	
Cellophane	Lamellated	3/4 inch flakes	8	
Sawdust	Fibrous	1/4 inch particles	10	
Prairie hay	"	1/2 inch fibers	10	
Bark	"	3/8 inch fibers	10	
Cotton seed hulls	Granular	Fine	10	
Prairie hay	Fibrous	3/8 inch particles	12	
Cellophane	Lamellated	1/2 inch flakes	8	
Shredded wood	Fibrous	1/4 inch fibers	8	
Sawdust	"	1/16 inch particles	20	

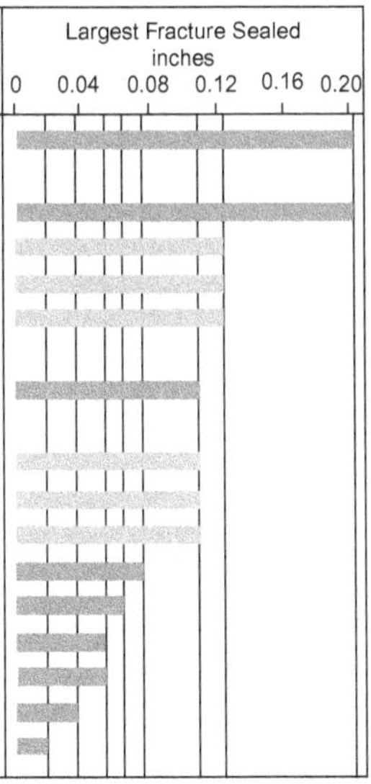

FIGURE 10.39 Summary of material evaluation tests. Analysis and control of lost circulation by Howard and Scott (1951). *Copyright 1951 by SPE-AIME.*

The principle underlying diesel oil-bentonite (DOB) slurries is that large amounts of bentonite—300 lb/bbl (850 kg/m^3)—can be readily mixed with diesel oil. When the DOB slurry is mixed with water or mud, the bentonite hydrates and forms a dense plastic plug whose shear strength depends on the ratio of mud to DOB (Dawson and Goins, 1953). A slurry that sets to a hard plug can be obtained by adding cement along with bentonite to the diesel oil (DOBC).

Messenger (1973) studied the effect of composition of DOB and DOBC slurries and the effect of the ratio of mud added thereto, using an apparatus in which the slurries were pumped between spring loaded discs, thus simulating an induced fracture. The shear strength of the slurries was calculated from the force on the springs. He recommended mud-to-DOB mixes of from 1:1 to 2:1, because slurries that never set hard form a *breathing seal*, that is, if subjected to a pressure surge during subsequent drilling operations, the plug deforms to maintain the seal. Slurries that do not develop shear strengths in excess of 5 psi (0.35 kg/cm^2) can be mixed at the surface and pumped down the drill pipe, and are suitable for sealing small fractures. Slurries that develop shear strengths high enough to seal large fractures must be mixed downhole proximate to the point of loss by pumping the DOB down the drill pipe and the mud down the annulus.

Another type of slurry that thickens downhole is made, typically, by dispersing a polyacrylamide in water and then emulsifying the dispersion in a paraffinic mineral oil, using a polyamine as the emulsifying agent. Bentonite is then added and remains in the external (oil) phase. At normal shear rates there is little contact between the water and the bentonite, so the slurry remains relatively thin while being pumped down the drill pipe. But at the high shear rates prevailing at the bit, the emulsion breaks and the bentonite mixes with the water. Cross-linking by the polyacrylamide results in a semi-solid mass that thickens further as it is pumped into the cracks and fissures in the formation (Hamburger et al., 1985). Besides restoring lost circulation, this technique may also be used to stop unwanted inflows, such as a gas influx. The technique has been used for various purposes 10 times to date, and has been successful in most of them.

High filter loss slurries are suitable for sealing fractures or channels in permeable formations. The rapid loss of filtrate deposits a filter cake, which eventually fills the fracture or a small void. One such slurry is composed of attapulgite, nutshells, and cotton fiber, plus barite if necessary (Ruffin, 1978). The API filter loss of this mixture is around 36 cm^3 but may be increased by adding lime. Another mixture consists of attapulgite, diatomaceous earth, and granular and fibrous lost circulation materials (LCM). (Moore et al., 1963) When flocculated with 10 lb/bbl of sodium chloride, the mixture has a filter loss of several hundred cubic centimeters.

Regaining Circulation

When circulation is lost, the first step should always be diagnosis—why and where is the loss occurring, and what are the characteristics of the formation at that depth? Much information can often be gained from circumstantial evidence. For instance, if the loss occurs while drilling ahead in a normally pressured zone with no change in mud weight, the fluid will almost certainly be lost into a preexisting void which the bit has just encountered. If the loss occurs when pipe is being run into the hole, one may safely assume that a transient surge pressure has induced a fracture.

If the depth at which the loss occurs is not obvious, a survey should be made to determine it. This information must be known if LCM is to be placed correctly. Furthermore, the local geologist may be able to provide useful information on the lithology of the formation at that depth. One survey procedure is to pump cooled mud down the hole and then run a temperature survey (Goins and Dawson, 1953). The inflection in the temperature curve indicates the point of loss, as shown in Fig. 10.40. Another method is to lower a transducer, an instrument that detects movement of the mud downhole (Bardeen and Teplitz, 1956). Downward flow of the mud creates a pressure difference across a diaphragm that causes a transducer to transmit a signal to the surface through the cable that suspends the instrument. A third method is to pump down mud containing a radioactive tracer, and then to run a gamma ray log.

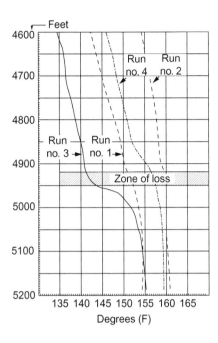

FIGURE 10.40 Lost circulation zone located by temperature survey. *From Goins and Dawson (1975). Courtesy API.*

Observation of the fluid level after the pumps are shut down will also provide useful information, especially if the local formation pore pressure and fracture gradients are known. Fluid levels may be determined by means of an echometer or, less accurately, by counting the pump strokes required to fill the hole.

Diagnosis costs time and money, but usually pays off in the long run: it enables the best remedial treatment to be applied at once, and avoids the costly—and possibly damaging—trial and error approach. The symptoms and suggested treatment of the various types of lost circulation are given hereunder.

Loss Into Structural Voids

When mud is lost into cavities, highly porous formations, propped fractures or other spaces, the mud level in the annulus falls until the hydrostatic head is equal to the formation pore pressure. An idea of the size of openings may be gained from measuring the rate at which the level falls. This rate may be very fast in cavernous limestones, or so slow into small openings that partial returns are obtained when circulating. Other indications are the depth at which the loss occurs; the greater the depth, the smaller the openings are likely to be.

Losses into large caverns occur only at very shallow depths, and usually the bit drops several feet when they are encountered. Losses into such caverns are very difficult to cure; enormous volumes of LCM slurries may be pumped in without appreciable increase in pumping pressure. Unconventional means, such as dropping sacks of cement (without removing the sacks) have occasionally been successful, but often it is necessary to reduce the pressure of the mud column, p_m, below the formation pressure, p_f, by drilling with aerated mud or foam until protective casing can be set. This method is tricky—if the mud density is reduced too much, formations fluids will backflow into the hole, and must be controlled by applying back pressure. Another method sometimes used is to drill blind with water and allow the returns, mixed with drill cuttings, to flow into the zone of loss.

Losses into smaller cavities, too large to be bridged by granular materials, may be plugged by pumping in viscous LCM slurries or time-set slurries that develop high shear strength. High squeeze pressures are unnecessary and undesirable because they introduce the risk of inducing a fracture.

Losses into vugular limestones, gravel beds, and propped fractures are best stopped by circulating granular materials while drilling, until the formation is fully penetrated. The required size and concentration can be established only by experience. It is generally best to start with a low concentration, say 5 lb/bbl (15 kg/m^3), of a fine grade and add larger amounts and coarser grades if necessary. Ground nutshells are preferable to ground carbonates when drilling in nonproductive formations, because they are less abrasive and about two-and-a-half times less dense. Therefore, about two-and-a-half times less by weight is required to provide the same volume of bridging solids, and lower gel strengths are required to suspend the particles. Carbonates must, however, be used in potentially productive formations because their acid solubility permits the removal of productivity impairment (see chapter: Completion, Workover, Packer, and Reservoir Drilling Fluids). Materials and procedures for shutting off losses into structural voids have been discussed in detail by Canson (1985). Problem analysis, treatment materials, and emplacement techniques are discussed.

Losses Induced by Marginal Pressures

When circulation is lost while drilling, but the hold stands full when the pumps are shut down (or will do so after a short waiting period), the loss is caused by the marginal increase in bottomhole pressure due to the hydraulic pressure drop in the annulus. In other words, p_w exceeds p_{frac}, but p_m does not. When the pumps are shut down, the fracture closes and mud solids bridge over the opening. Similarly, when p_m is close to p_{frac}, circulation may be temporarily lost because of pressure surges when running pipe, or because of peak bottomhole pressures when breaking circulation after a trip.

Temporary circulation losses of this sort are best remedied by adjusting mud properties and operating conditions, rather than by the application of lost circulation materials. The following procedures are recommended:

1. Carry the lowest mud density consistent with well safety.
2. Use the lowest circulation rate that will clean the hole adequately.
3. Adjust the rheological properties to give maximum hole cleaning with minimum pressure drop in the annulus (see chapter: The Rheology of Drilling Fluids).
4. Do not drill with a balled bit and drill collars (see the section on bit balling, earlier in this chapter).
5. Run pipe in slowly, and above all, do not ream down rapidly with the pumps on (see chapter: The Rheology of Drilling Fluids).
6. Break circulation several times on the way into the hole. When on the bottom, break circulation slowly, and raise pipe while doing so.
7. Minimize gel strengths.

If mud losses occur even under optimum operating conditions, then it will be necessary to use some type of lost circulation material. Continuous circulation of ground nutshells while drilling appears to be the simplest and best method of control. Howard and Scott (1951) showed that these materials sealed fractures as they occurred, and prevented their extension. An alternative would be to squeeze soft DOB plugs, which, as already mentioned, adjust to seal transient pressures. Repeated applications might, however, be necessary to seal new fractures as they occur.

Losses Caused by the Hydrostatic Pressure of the Mud Column Exceeding the Fracture Pressure

In this case p_m exceeds p_{frac} and the fluid level falls when the pumps are shutdown, until the hydrostatic head of fluid remaining in the hole equals p_{frac}. This kind of loss typically occurs when the density of the mud is raised to control a kick in the lower part of the hole, and the mud pressure gradient thus increased exceeds the fracture pressure gradient somewhere higher up in the hole. Usually, a fracture is induced just below the casing shoe where the difference between p_m and p_{frac} is greatest (see Fig. 10.41). Short of setting another string of casing, the only cure is to pump LCM into the fracture until the squeeze pressure exceeds the maximum transient pressure expected when drilling operations are resumed. Final squeeze pressures up to 1000 psi (70 kg/cm^2) have been reported (Moore et al., 1963). High squeeze pressures strengthen the borehole by widening the fracture, thereby increasing the hoop stresses around the walls (see Fig. 10.42). Note that the squeeze pressure induces and seals new fractures below the initial fracture. The occurrence of such fractures is shown by fluctuations in the injection pressure (Moore et al., 1963).

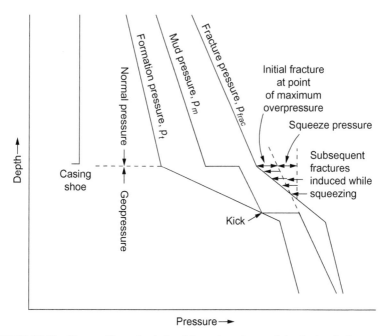

FIGURE 10.41 Diagram illustrating induced fracturing when mud density is raised to control kick.

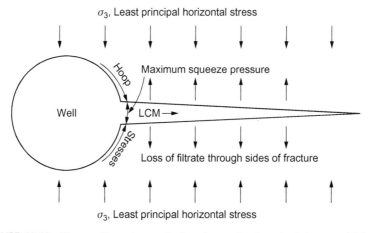

FIGURE 10.42 Diagram illustrating mechanics of squeezing lost circulation material into a vertical fracture.

Fibrous materials, mixtures of fibrous and granular materials, DOBC, and high filter loss slurries may be used for high pressure squeezes, the principal requirement being that they develop sufficient shear strength and do not flow back into the hole when the squeeze pressure is released (Lummus, 1968).

Natural Open Fractures

As already mentioned, open fractures occur when one of the principal stresses is tensile. As there is at present no means of measuring subsurface rock stresses in drilling wells, there is no means of differentiating between natural open fractures and induced fractures. Since the remedy would be the same—high-pressure squeezes to increase compressive hoop stresses round the bore hole—open fractures may well have been encountered but not recognized.

WELLBORE STRENGTHENING

The term "wellbore ballooning" was coined in the 1980s to describe a phenomenon in which the wellbore seemingly loses mud to the formation and then regains the mud later. Many times, the regain looked like the well was flowing. Anytime the well seems to be flowing, rig crews are taught to shut down and check for a formation kick. This can result in significant NPT. It is unfortunate that the term ballooning was used, because it implies that the wellbore is elastic and moves in and out like a balloon, which it does not.

In the 1990s a series of catastrophic lost circulation incidents with synthetic base fluids in deep water drilling operations were encountered. These incidents cost several millions of dollars, both from lost mud and NPT rig costs. Joint industry projects were initiated by the Drilling Engineering Association to look into lost circulation in deep water drilling. It was determined that nonaqueous fluids many times had a lower fracture propagation pressure than water-based systems. In addition, due to the narrow drilling pressure margins (Lee et al., 2012), the equivalent circulating densities (ECD) calculation inaccuracies lead to induced lost circulation. Part of this problem was the varying temperatures experienced by the NADF while circulating—bottom-hole temperatures of 250°F + and mudline riser temperatures of about 40°F.

More accurate computer algorithms were developed based on the power law/yield stress model to more accurately determine ECD. The drilling fluid service companies also developed accurate PVT data for their base fluids to use with the new software, again gaining more accurate ECD information. This then lead to the development of flat rheology NADF to minimize the temperature variation in ECD (Young et al., 2012)

At the same time, using geophysical information on the formation being drilled, Poisson's Ratio, and Young's Modulus, greater accuracy on the fracture initiation and propagation pressures were calculated. Using this information and the hoop stress model discussed earlier (Fig. 10.42), certain lost circulation materials were developed to strengthen the wellbore to have higher ignition and propagation fracturing pressures.

Through these R&D efforts, a variety of materials has been found to bridge and help strengthen the wellbore while drilling, including (Caenn, 2013):

- Deformable graphite;
- Small-sized calcium carbonate;
- Small nut shells;
- Micronized cellulose; and
- Fibers.

Particle size distribution is important to effectively strengthen the wellbore. A laboratory test was developed using the American Petroleum Institute standard particle plugging apparatus (PPA) and slotted metal disks (the PPA is used to determine the ability of particles in the drilling fluid to effectively bridge the pores in the filter medium). Many service companies have developed software that inputs real-time data from measurement-while-drilling services to design the correct particle sizes needed to strengthen the wellbore.

David Beardmore developed the following guidelines for wellbore strengthening (IADC, 2013):

1. Putting Granular loss-circulation materials in the mud before can help stop up fractures in the foundation, preventing loss. This includes materials like ground-up rock, nuts, and paper.
2. Using high fluid loss slurry when you have lost circulation into a fracture, the slurry loses all of its liquid to the formation and leaves behind solids to plug the fracture and prevent losses.
3. Chemical sealant is like an epoxy that is pumped into the wellbore that leaks into the formation and then sets up in the pore space and prevent losses.
4. Heating the wellbore will make it stronger. If you heat the rock around the wellbore, it's going to try and expand, so it increases rock stress, which makes it harder to break down.
5. Smear effect—This happens when drilling with casing. The pipe smears the cuttings and filter cake into the wall, creating a plaster-like sealant that keeps the wellbore from breaking down.
6. Rigid-plug forming treatment are liquids that, when mixed, transform into cross-linked polymers, latex rubber, cement, etc., that set up in the fractures of a well to prevent lost circulation.
7. Ultralow fluid loss muds are used to mitigate losses in permeable formations. By putting things in the mud that reduce the initial spurt loss of mud into the formation, that makes the filter cake nearly impermeable and the wellbore is less likely to break down.

HIGH TEMPERATURES
Geothermal Gradients

The increase in temperature with depth in the earth's crust is called the *geothermal gradient*, and is expressed in °F/100 ft (°C/km). The heat flow in the upper crust is derived from two sources: (1) heat conducted from the lower crust and mantle; and (2) radiogenic heat in the upper crust (Diment et al., 1975). Conducted heat is low in regions of ancient tectonism, e.g., the eastern half of the United States, and high in regions of recent tectonism, e.g., the mountainous regions of the West (see Fig. 10.43). Temperature gradients vary widely within each region depending on:

1. The amount of radiogenic heat in the upper crust.
2. Structural features: gradients are high at structural highs.
3. Thermal conductivity of the formation: gradients are low in conductive formations such as sandstones, and high in low conductivity ones such as shales.
4. Convective flow: in thick permeable beds, water circulates by convection, causing high temperatures at relatively shallow depths.

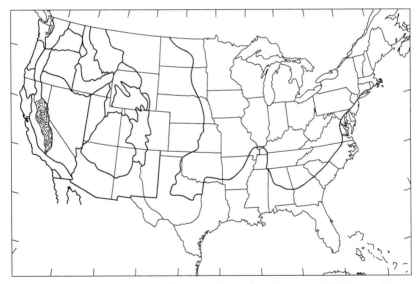

FIGURE 10.43 Map showing probable extent of hot (stippled), normal (white), and cold (dotted) crustal regions of the United States. Physiographic provinces do not necessarily represent heat flow provinces. *From Diment, W.H., Urban, T.C., Sass, J.H., Marshall, B.V., Munroe, R.J., Lachenbruch, A.H., 1975. Temperatures and heat contents based on conductive transport of heat. In: Assessment of Geothermal Resources of the United States, Geological Survey Circular 726. Dept. Interior, Washington, DC, pp. 84–103. Courtesy U.S. Dept. Int.*

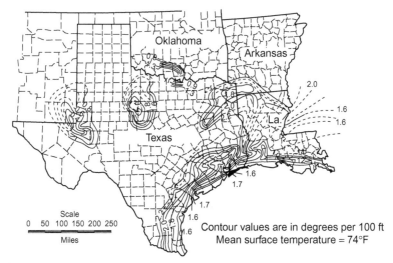

FIGURE 10.44 Contour map of isothermal gradients in the southwest US *From Moses, P.L., 1961. Geothermal gradients. API Drill. Prod. Prac. 57–63. Courtesy of API.*

5. Pore pressure: temperature gradients are higher in geopressured formations (Schmidt, 1973).

Because of these factors, geothermal gradients vary from 0.44°F/100 ft (8°C/km) to 2.7°F/100 ft (50°C/km) according to location in the United States (Kruger and Otte, 1974). In the Gulf Coast, they vary from 1.2°F/100 ft (22°C/km) to 2.2°F/100 ft (40°C/km) (see Fig. 10.44) (Moses, 1961). Very high gradients are found in the steam wells in the Salton Sea area of California. Here, a thick aquifer overlies a local upthrust of igneous rock. Temperatures of water circulating by convections rise as high as 680°F (360°C) at depths of about 5000 ft (1500 m), which equals a gradient of 12.5°F/100 ft (230°C/km) from the surface to the top of the aquifer (Renner et al., 1975).

Detailed surveys have shown that geothermal gradients are not linear with depth, but vary according to the factors listed above, i.e., formation, pore pressure, etc. In Louisiana's Manchester Field, for instance, the gradient is 1.3°F/100 ft (23.6°C/km) in the normally pressured zone above 10,500 ft (3200 m) and 2.1°F/100 ft (38°C/km) in the geopressure zone below (Schmidt, 1973). Elsewhere in the Gulf Coast, gradients as high as 6°F/100 ft (109°C/km) have been observed in the geopressured zone (Jones, 1969).

The bottomhole temperatures of drilling wells are always less than the virgin formation temperature. For example, the maximum temperature logged in a geothermal well drilled to 4600 ft in the Imperial Valley of California was 430°F (221°C) after an 8-h shutdown, but the well subsequently produced steam at a temperature at 680°F (360°C).

The reason for the difference between bottomhole and formation temperature is that the mud, while circulating, cools the formation around the lower part of the hole, transfers the heat to formations around the upper part of the hole, and loses it to the atmosphere at the surface. Fig. 10.45 shows the change in mud and formation temperatures with time of circulation as calculated by Raymond (1969) for a hypothetical well with a virgin formation temperature of 400°F. During a round trip, the temperature of the mud at the bottom of the hole rises, but under normal circumstances does not have time to reach virgin formation temperature (see Fig. 10.46). Note that only the mud in the bottom half of the hole increases in temperature: in the rest of the hole and at the surface, it decreases in temperature. Thus, the average temperature of the mud is always substantially lower than the bottomhole logged temperature (a fact which should be remembered when making and evaluating high-temperature stability tests).

The hydrostatic pressure of the mud column in a well depends on the density of the mud in the hole, which differs from the density at the surface because of increases in temperature and pressure with depth. Therefore, calculating hydrostatic pressures from surface densities will result in error.

McMordie et al. (1982) measured changes in density with temperature and pressure in a variable pressure autoclave. Fig. 10.47 shows the results for a freshwater bentonite mud and a low-viscosity oil mud (85/15 oil/water ratio). Both muds were loaded to three different densities (approximately 11, 14,

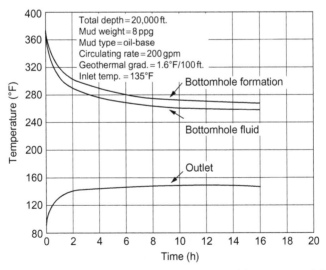

FIGURE 10.45 Effect of time on temperature in a simulated well. *From Raymond, L.R., 1969. Temperature distribution in a circulating drilling fluid. J. Pet. Technol. 333−341. Trans AIME. 246. Copyright 1969 by SPE-AIME.*

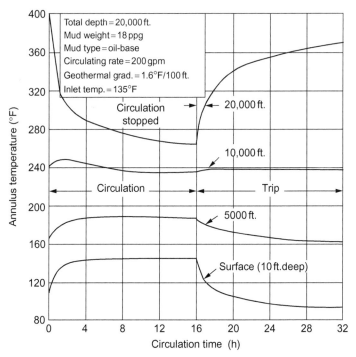

FIGURE 10.46 Temperature trace for various depths in a simulated well. *From Raymond, L.R., 1969. Temperature distribution in a circulating drilling fluid. J. Pet. Technol. 333–341. Trans AIME. 246. Copyright 1969 by SPE-AIME.*

and 18 lb/gal—1.32, 1.68, and 2.16 SG), but changes in density appeared to be unaffected by the initial density.

Hoberock et al. (1982) calculated the pressure at the bottom of the hole from the sum of incremental changes in density with depth. They used a material balance to account for the differences in compressibility between oil and water. Solids were assumed incompressible. Calculations were made to show the effect of total depth, temperature gradient, circulation rate, and mud formulation. Table 10.7 shows some typical results. The pressure difference shown in the last column is the approximate difference between the bottom-hole pressure as calculated from the mud density at the surface and the actual bottomhole pressure. The values shown are for a water-base mud (77% water, no oil). Results for an oil-base mud (70% oil, 7% water) were not substantially different from the water-base mud.

The pressure differences shown in Table 10.7 show that calculating bottom-hole pressures in deep hot holes from the density of the mud at the surface can result in bottomhole underpressures quite sufficient to cause a blowout.

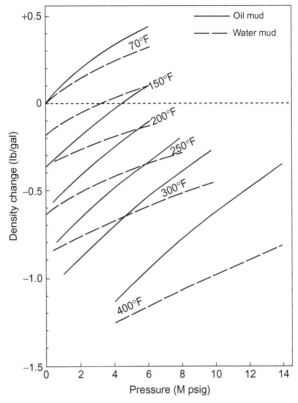

FIGURE 10.47 Effect of temperature and pressure on the density of oil- and water-base muds. *From McMordie, W.C., Bland, R.G., Hauser, J.M., 1982. Effect of temperature and pressure on the density of drilling fluids. SPE Paper No. 11114, Annual Meeting, New Orleans, LA, September. Copyright 1982 SPE-AIME.*

A computer program is available for predicting downhole mud temperatures from mud properties; surface operating data; and pipe, casing, and hole dimensions (Thomson and Burgess, 1985). It is necessary to know the local geothermal gradient and the thermal properties of the downhole formations, at least approximately. Downhole mud temperatures may be determined directly by measurement-while-drilling (MWD) techniques (Gravely, 1983).

High-Temperature Drilling Fluids

In Chapter 7, The Filtration Properties of Drilling Fluids, it was shown that the degree of flocculation of bentonite suspensions starts to increase sharply with increase in temperature above about 250°F (121°C). The consequent increase in yield point can be controlled by the addition of thinning agents, but, unfortunately, thinning agents themselves degrade in the same

TABLE 10.7 Approximate Difference in Bottomhole Pressure Due to Changes in Mud Density (Hoberock et al., 1982)

Total Depth (ft)	Meters	Temperature (°F/100 ft)	Gradient (°C/km)	Circulation Rate (gpm)	(m³/min)	Pressure Difference (psi)	(kg/cm²)
10,000	3030	2.0	37	300	114	0	0
15,000	4545	—	"	"	"	−100	−7
20,000	6060	—	"	"	"	−300	−21
25,000	7575	—	"	"	"	−650	−45
—	"	1.2	22	"	"	−125	−9
—	"	1.6	29	"	"	−350	−24
—	"	2.0	37	0	0	−825	−58
—	"	"	37	150	57	−725	−51

temperature range. Degraded additives can be replaced, but as the rate of degradation increases, costs increase and eventually become excessive. For example, lignosulfonate is commonly used to maintain rheological properties and filtration control in clay muds at high temperatures. Kelly (1965) circulated such muds continuously at temperatures up to 350°F (177°C) in a laboratory unit, adding more lignosulfonate from time to time as required to keep the rheological properties constant. His results showed that the lignosulfonate started to degrade at 250°F (121°C), but its thinning characteristics could be maintained up to 350°F (177°C) by the addition of small amounts of sodium chromate. The filtration properties did not start to degrade, with or without the sodium chromate, at temperatures below 350°F. As mentioned in Chapter 6, The Rheology of Drilling Fluids, the temperature at which degradation becomes excessive may be calculated from reaction rate mechanics. Usually it is established by experience, but it may be determined by laboratory testing.

Lignite is more temperature stable than is lignosulfonate. Muds containing lignite and DMS (see the section on surfactants in Chapter 8: The Surface Chemistry of Drilling Fluids) retained their rheological and filtration properties after heating under static conditions for 352 hours at 400°F (204°C) (Burdyn and Wiener, 1957; Cowan, 1959). Bentonite-lignite-DMS muds are used extensively to drill geothermal wells with bottomhole shut-in temperatures around 450°F (234°C), but gel strengths and costs are high (Anderson, 1961; Cromling, 1973).

Because of their low permeability, basement rocks may usually be drilled with air or water, both of which (especially air) have the advantage of temperature stability and fast drilling rates in hard rocks. Mitchell (1981) has developed an improved method for calculating air temperatures, pressures, and velocities. Water was used to drill two 15,000 ft wells at the Geothermal Test Site in New Mexico (Carden et al., 1985). The wells were inclined at 35° to the vertical below 10,400 ft. 50 bbl sweeps of high-viscosity bentonite pills helped clean the hole, and similar sweeps of pills containing triglycerides and alcohol reduced friction. Formation temperatures exceeded 600°F (316°C).

Oil muds are considerably more temperature stable than water muds. They have been used to drill wells with bottomhole logged temperatures up to 550°F (287°C). Oil muds are best for drilling deep hot wells in the Gulf Coast and Mississippi. Because of high geopressures, mud densities of about 18 lb/gal (2.15 SG) are necessary to drill these wells. If a water mud is used, the combination of high temperatures and high solids content gives rise to high viscosities and gel strengths. If the mud becomes contaminated by salt water or other flocculants, it becomes impossible to control the rheological properties.

One problem with oil muds is that at temperatures above 350°F (177°C) the organophilic clays used to provide structural viscosity degrade, and the

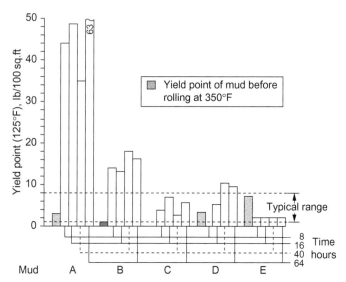

FIGURE 10.48 Yield point of 9 lb/gal water muds after rolling at 350°F, 300 psi. *From Remont, L.J., Rehm, W.A., McDonald, W.J., Maurer, W.C. Evaluation of commercially available geothermal drilling fluids. Sandia Report No. 77–7001, Sandia Laboratories, Albuquerque, NM, pp. 52–75. Courtesy of Sandia Laboratories.*

cutting-carrying capacity of the mud deteriorates. To offset this problem Portnoy et al. (1986) have introduced a lightly sulfonated polystyrene (i.e., a low ratio of sulfonated styrene units to styrene units) polymer (SPS). Laboratory tests have shown that SPS provides good rheological properties at temperatures up to 400°F (204°C), and it has performed satisfactorily in the field in wells with bottomhole temperatures up to 432°F (222°C). Since SPS does not become activated until the temperature reaches 300°F (149°C), it is best used in combination with organophilic clay to provide rheological control at lower temperatures.

The high-temperature properties of some commercial muds have been evaluated by Remont et al. To stimulate their behavior while circulating, the muds were rolled at 350°F (117°C) for periods up to 64 h, and their rheological and filtration properties were then determined. Figs. 10.48 and 10.49 show the yield points and high-temperature high-pressure (HTHP) filtration properties of 9 lb/gal (1.1 SG) water muds. These muds consisted primarily of bentonite and lignite, except Sample E, which contained sepiolite and an unspecified polymer. Note that Sample E had the lowest yield points but the highest filtration rates. Figs. 10.50 and 10.51 show that the HTHP filtration rates of 18 lb/gal oil-base muds are much lower than those of water-base muds after rolling for extended periods at 350°F (note the difference in scale).

Drilling Problems Related to Drilling Fluids Chapter | 10 **431**

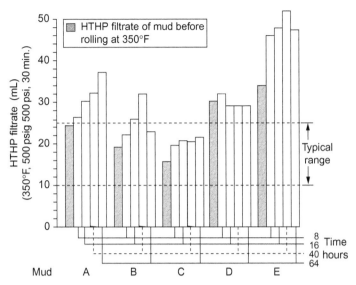

FIGURE 10.49 HTHP filtrate of 9 lb/gal water muds after rolling at 350°F, 300 psi. *From Remont, L.J., Rehm, W.A., McDonald, W.J., Maurer, W.C. Evaluation of commercially available geothermal drilling fluids. Sandia Report No. 77–7001, Sandia Laboratories, Albuquerque, NM, pp. 52–75. Courtesy of Sandia Laboratories.*

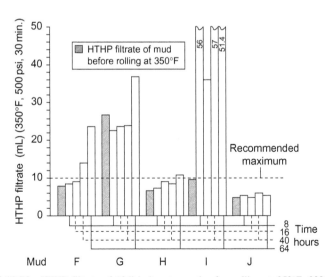

FIGURE 10.50 HTHP filtrate of 18 lb/gal water muds after rolling at 350°F, 300 psi. *From Remont, L.J., Rehm, W.A., McDonald, W.J., Maurer, W.C. Evaluation of commercially available geothermal drilling fluids. Sandia Report No. 77–7001, Sandia Laboratories, Albuquerque, NM, pp. 52–75. Courtesy of Sandia Laboratories.*

432 Composition and Properties of Drilling and Completion Fluids

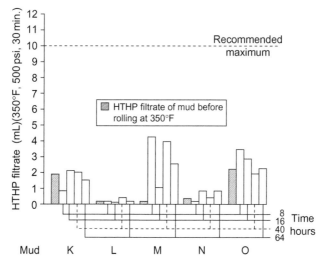

FIGURE 10.51 HTHP filtrate of 18 lb/gal oil muds after rolling at 350°F, 300 psi. *From Remont, L.J., Rehm, W.A., McDonald, W.J., Maurer, W.C. Evaluation of commercially available geothermal drilling fluids. Sandia Report No. 77–7001, Sandia Laboratories, Albuquerque, NM, pp. 52–75. Courtesy of Sandia Laboratories.*

To simulate the behavior of muds left standing at the bottom of the hole for prolonged periods, the muds were aged statically at temperatures up to 500°F (260°C) and pressures up to 15,000 psi (1055 kg/cm^2). Figs. 10.52 and 10.53 show that only three oil muds remained stable under these conditions.

Because of the interest in drilling for geothermal sources of energy, considerable research has been undertaken lately to develop muds that are more stable at high temperatures. When evaluating the results of such work, bear in mind that both high temperature and time of exposure to that temperature are pertinent factors. Claims based on exposure times of a few hours should be treated with caution.

Because of pollution problems, oil muds are not considered suitable for drilling geothermal wells, and research has therefore concentrated on water-base muds. As indicated by the work of Remont et al., sepiolite appears to be a suitable viscosifier. Sepiolite is a fibrous clay mineral similar to attapulgite (see section on attapulgite in chapter: Clay Mineralogy and the Colloid Chemistry of Drilling Fluids). When slurries of sepiolite are subjected to high shear rates, the bundles of fibers separate to innumerable individual fibers. Mechanical interference between these fibers is primarily responsible for the rheological properties, and sepiolite muds are therefore little affected by the electrochemical environment. The suspension properties of the mineral itself are stable up to at least 700°F (371°C) (Carney and Meyer, 1976).

Because of its rod-like shape, sepiolite provides no filtration control. However, Carney and Meyer (1976) state that satisfactory HTHP filtration

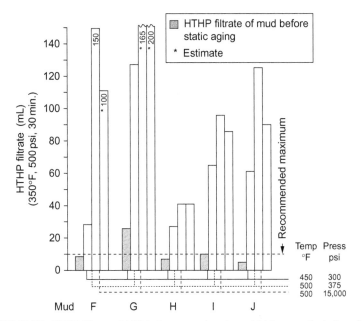

FIGURE 10.52 HTHP filtrate of 18 lb/gal water mud static aged 24 hours at the indicated temperatures and pressures. *From Remont, L.J., Rehm, W.A., McDonald, W.J., Maurer, W.C. Evaluation of commercially available geothermal drilling fluids. Sandia Report No. 77–7001, Sandia Laboratories, Albuquerque, NM, pp. 52–75. Courtesy of Sandia Laboratories.*

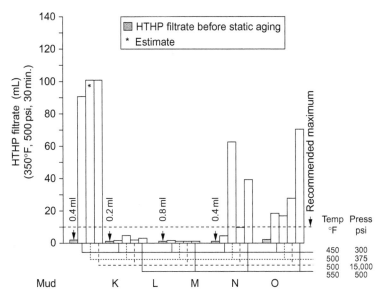

FIGURE 10.53 HTHP filtrate of 18 lb/gal oil muds static aged 24 h at the indicated temperatures and pressures. *From Remont, L.J., Rehm, W.A., McDonald, W.J., Maurer, W.C. Evaluation of commercially available geothermal drilling fluids. Sandia Report No. 77–7001, Sandia Laboratories, Albuquerque, NM, pp. 52–75. Courtesy of Sandia Laboratories.*

TABLE 10.8 Typical Ultra High Temperature Mud Formulation Using Sepiolite in Conjunction with Various Filtration Control Additives (Carney and Meyer, 1976)

15 lb/bbl Sepiolite

5 lb/bbl Wyoming bentonite

2 lb/bbl Polymer A

0.5 lb/bbl Polymer B

0.5 lb/bbl NaOH

TABLE 10.9 Effect of Static Aging at Various Temperatures on the Mud Formulation in Table 10.8 (Carney and Meyer, 1976)

Aging (°F)	—	350	400	450	500	560
Aging (h)	—	24	24	24	24	24
Shear	—	25	28	30	60	0
Plastic viscosity	12	13	13	13	10	9
Yield point	3	7	5	4	5	3
Gel strengths	2/2	2/2	2/2	0/0	0/0	0/16
pH	11.8	9.5	9.3	9.2	8.6	9.2
API filtrate	8.2	8.5	8.8	8.7	8.4	12.7
High temp. filtrate						
300°F at 500 psi	32.0	36.0	37.4	36.2	33.0	34.0
400°F at 500 psi	—	—	—	—	29.8	33.2
450°F at 500 psi	—	—	—	—	31.6	37.4

properties can be obtained with sepiolite muds containing small amounts of bentonite and two unspecified polymers. The proportion of bentonite used is too small to cause high gel strengths. Tables 10.8 through 10.11 show the composition and properties of one of their formulations.

Bannerman and Davis (1978) have described the use of sepiolite muds in the field. Typical compositions are shown in Table 10.12. Use of these muds in geothermal wells in the Imperial Valley of California have resulted in much better rheological control than previously experienced with bentonite-lignite systems. Prolonged shearing was at first found necessary in order to obtain the desired initial viscosity, but this difficulty was overcome by the use of a specially designed high-shear mixer.

TABLE 10.10 Performance of the Mud Formulation in Table 10.8 on the Ultra High Temperature, High-Pressure Thickening Time Tester (Carney and Meyer, 1976)

Time (min.)	Viscosity (uc)[a]	Temp. (°F)	Pressure (psi)
0	5	80	500
15	4	281	2500
30	5	413	5000
45	4	528	10,000
60	6	618	15,000
75	5	632	17,500
90	5	648	20,000
120	5	683	20,000
150	5	700	20,000
180	5	700	20,000

[a]Units of consistency.

TABLE 10.11 Properties of the Mud Formulation in Table 10.8 after Being Taken to 700°F on the Ultra High Temperature, High-Pressure Thickening Time Tester (Carney and Meyer, 1976)

Plastic viscosity	40
Yield point	38
Gel strengths	0/18
pH	8.8
API filtrate	11.9
High temp. filtrate	
350°F at 500 psi	28
400°F at 500 psi	35.2

Note: 4 ppb gilsonite added to the above mud lowered the high-temperature filtrate at 450°F to 22 cm³/30 min.

Two low molecular weight polymers have been shown to be stable at temperatures above 400°F (204°C), and may be used as deflocculants to maintain low rheological properties in high-temperature wells. One is sodium polyacrylate, with a molecular weight less than 2500 (Chesser and Enright, 1980),

TABLE 10.12 Two Sepiolite Drilling Fluids[a] (Bannerman and Davis, 1978)

	System 1	System 2
Ingredients (lb/bbl):		
Sepiolite	15	15
NaOH	1	1
Modified lignite	5	0
Sulfonated lignite	0	5
Sodium polyacrylate	2	2
Drilled solids	25	25
Fresh water		
Apparent viscosity (cp)	66	25
Plastic viscosity (cp)	46	41
Yield point (lb/100 ft^2)	40	34
Gels (lb/100 ft^2)	8/25	6/22
pH	9.5	9.0
API filtrate (cm^3)	10.0	20.0
High temp./high pressure Filtrate (cm^3)	19.0	20.0

[a] Fluids are heat aged at 460°F in a rotating bomb.

which is suitable for freshwater muds but is sensitive to calcium contamination. The other is the sodium salt of sulfonated styrene maleic anhydride copolymer, with a molecular weight between 1000 and 5000 (SSMA), which may be used in fresh or seawater muds. The SSMA chain has a high charge density because it contains three ionizing carboxyls per monomer, i.e., a degree of substitution of three. It is believed that this high charge density enables it to remain adsorbed on the clay particles at high temperatures, whereas lignosulfonates become severely desorbed at temperatures above 350°F (177°C). However, laboratory tests showed that the lowest viscosities in consistometer runs at both 400°F (204°C) and 500°F (260°C) were obtained with a combination of SSMA and lignosulfonates. In some field tests it was found that the addition of SSMA to lignosulfonate muds eliminated difficulties previously experienced in getting logging tools to bottom, even in geothermal wells with bottomhole temperatures up to 700°F (371°C).

Low molecular weight polymers provide good rheological properties at high temperatures, but do not provide adequate filtration control, which must be obtained with long chain polymer. Son et al. (1984) described a vinylamide/vinylsulfonate with a molecular weight between one and two million, which maintained good rheological and filtration properties at 400°F (204°C). Tests also showed that the copolymer prevented flocculation in muds containing 10% each of NaCl and $CaCl_2$ and increased the tolerance of KCl muds to drilled solids. In field tests, good rheological and filtration properties were maintained in two 20,000 ft plus wells with bottomhole temperatures in excess of 400°F (204°C).

Perricone et al. (1986) investigated two other vinyl sulfonated copolymers with molecular weights in the range of 750,000–1,500,000. In laboratory tests, these copolymers were added to freshwater, seawater, and lignosulfonate field muds, and hot rolled for 16 hours. Low HTHP—500 psi (35 kg/cm^2), 300°F (149°C)—filter losses were maintained in all cases, but in the tests with lignosulfonate field muds the rheological properties were higher than with the untreated base mud.

Field results with one of these copolymers have been remarkable: Low HTHP filter losses have been maintained in over 30 geothermal wells, many with bottomhole temperatures in excess of 500°F (260°C). In one such well the HTHP filter loss increased only 2 cc after logging for 72 h. The copolymer was added to a bentonite-lignite deflocculated system, with SSMA as the deflocculant. The other copolymer has been used in a North Sea well, with a bottomhole temperature of 350°F (177°C), to provide filtration control in a nondamaging coring fluid.

CORROSION OF DRILL PIPE

Although the components of water-base drilling fluids are not unduly corrosive, degradation of organic additives by high temperature or bacteria may result in corrosive products. Also, contamination by acid gases (such as carbon dioxide and hydrogen sulfide) and by formation brines can cause severe corrosion. Under adverse conditions, the replacement of corroded drill pipe becomes an economic problem. A more severe problem arises if the corrosion is not detected and the pipe fails while drilling. In this section, the several ways in which corrosion can occur and the necessary corrective measures are briefly discussed.

Electrochemical Reactions

If a metal is placed in a solution of one of its salts, the metal ions tend to pass into solution, leaving the metal negatively charged with respect to the solution. Only a minute amount of cations leaves the metal, and they are held close to the surface by its negative charge. An electrostatic double layer (similar to the electrostatic double layer discussed in chapter: Clay Mineralogy and the Colloid Chemistry of Drilling Fluids) is thus formed, and the negative charge is the Nernst surface potential, the value of which is a function of the identity of the metal and the concentration—or rather the activity—of the salt solution.

Table 10.13 shows the Nernst potential of various metals when immersed in solutions of their own salts at unit activity, and referred to the potential of a hydrogen electrode, also at unit activity. This series is known as the *electromotive series*. The greater the tendency for a metal to ionize, the more negative its potential, and the more reactive it is in an aqueous environment. For instance, potassium reacts violently in pure water and zinc reacts in acid solutions whereas silver is inert even in concentrated acids. The underlying principle is that cations from the metal displace cations lower in the electromotive series from solution. Thus, zinc displaces hydrogen, but silver cannot.

This principle may be used to generate an electric current. For example, if strips of zinc and copper are placed in a solution of copper sulfate and connected by an external conductor, as shown in Fig. 10.54, the zinc passes into solution forming zinc sulfate, while copper is deposited on the copper strip. In passing into solution, the zinc gives up two electrons that flow along the wire to the copper strip where they are accepted by the copper ions, thus forming molecular copper. By this means, chemical energy, as represented by the equation

TABLE 10.13 The Electromotive Series

The Surface Potential of Metals When Immersed in Solutions of Their Own Salts at Unit Activity, Referred to the Standard Hydrogen Electrode

Metal	Volts
Potassium (K^+)	−2.92
Sodium (Na^+)	−2.72
Magnesium (Mg^{++})	−2.34
Aluminum (Al^{+++})	−1.67
Zinc (Zn^{++})	−0.76
Iron (Fe^{++})	−0.44
Lead (Pb^{++})	−0.13
Copper (Cu^{++})	+0.34
Mercury (Hg^{++})	+0.80
Silver (Ag^+)	+0.80
Gold (Au^+)	+1.68

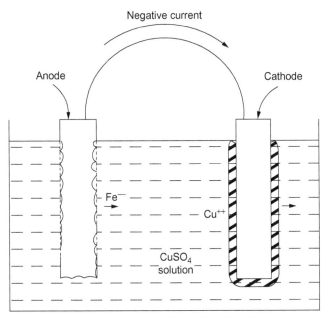

FIGURE 10.54 Schematic representation of an electrochemical cell.

$$Zn^{++} + CuSO_4 = ZnSO_4 + Cu^{++} \rightarrow Cu^0 \qquad (10.6)$$

is converted into electrical energy, which can perform work. The device is known as an *electrochemical cell*; the metal strips are *electrodes*. The zinc is the *anode* and the copper is the *cathode*. Because zinc is higher on the electromotive series than copper, it is negative with respect to the copper.

Similar cells are set up between electrodes of any two metals when in contact with electrolyte solutions and connected by a conductor. Familiar examples are flashlight and automobile batteries. The action of the zinc-copper sulfate-copper cell is reversible, i.e., if a current is sent in the opposite direction by a battery, copper passes into solution and zinc is deposited on the zinc electrode. But if the cell contained, say, sulfuric acid instead of copper sulfate, hydrogen would be discharged at the cathode, and would bubble off as molecular hydrogen. Such a cell is irreversible. Since electrode potential is a function of ionic activity, electrochemical cells are also set up if electrodes of the same metal are immersed in solutions of different ionic activity and connected by a conductor. Such cells are known as *concentration cells*.

Electrochemical activity is the fundamental cause of corrosion. Local differences in surface potential caused by lack of homogeneity of the metal provide sites for anodes and cathodes. The body of the metal is the conductor. Corrosion always takes place at the anode, as in the zinc-copper sulfate-copper cell discussed above, and the corrosion reaction products, such as hydrogen, are discharged at the cathode.

Corrosion cells are set up in drill pipe because steel is an alloy, and contains iron and iron carbide crystals. The iron crystals almost always act as anodes, the iron carbide crystals almost always act as cathodes, and the circuits are completed by aqueous drilling muds, causing general corrosion of the pipe surface (Patton, 1974a,b). Patches of scale or deposits of any sort also provide cathodic sites, causing local corrosion or pitting. Local corrosion may also be caused by concentration cells set up by differences in ionic activity at the bare surface of the pipe and under barriers such as drill pipe protectors and patches of scale (Bush et al., 1966).

Another type of electrochemical cell is set up by differences in oxidizing conditions, fundamentally because oxidation involves a gain of electrons, e.g., when ferrous iron is converted to ferric iron:

$$Fe^{++} \rightarrow 1e + Fe^{+++}$$

Potential differences caused by this mechanism are called *redox potentials*. In a drilling well, oxygen is inevitably entrained in the mud at the surface by mixing and conditioning operations. Downhole, the surface of the pipe is exposed to oxidizing conditions, but reducing conditions prevail under patches of rust or other barriers. An anode is therefore set up under the barrier as follows (Hudgins, et al., 1966):

$$Fe^0 - 2e \rightarrow Fe^{++} \tag{10.7}$$

and a cathode at the surface of the pipe:

$$O_2 + 2H_2O + 4e \rightarrow 4\,OH^- \tag{10.8}$$

depositing ferric hydroxide according to the equation:

$$4Fe^{++} + 6H_2O + 4e \rightarrow 4Fe(OH)_3 \tag{10.9}$$

Corrosion pits are thus formed under the scale, as shown in Fig. 10.55. Note that corrosion will still occur even though there is some oxygen under the barrier. The essential condition is that less oxygen be present under the barrier than at the bare surface of the metal, thus establishing an oxygen concentration cell.

If the products of corrosion accumulate at the cathode, the flow of electrons is impeded, and the corrosion process slows up. The cathode is then said to be *polarized*. For example, cations of hydrogen may polarize a cathode by coating it with a layer of hydrogen atoms. If the hydrogen atoms unite and bubble off as molecular hydrogen, the cathode is said to be *depolarized*. Dissolved oxygen can act as a depolarizer by reacting with the hydrogen to form water, thus accelerating the corrosion process.

Stress Cracking

When a metal is subjected to cyclic stresses, it eventually fails, even though the applied stress may be well below the normal yield stress. Failure is

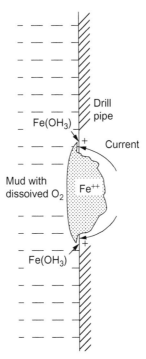

FIGURE 10.55 Oxygen corrosion cell.

caused by cracks, which start at points of high stress concentration in notches or other surface defects and deepen with repeated cycling. Failures of this sort are commonly encountered in metallurgical engineering and are known as *fatigue failures*.

Steel can endure a number of stress cycles before failure. The number decreases with the applied stress, the hardness of the steel, and the corrosiveness of the environment (see Fig. 10.56). In a drilling well, fatigue cracking is greatly accelerated by dissolved salts, oxygen, carbon dioxide, and hydrogen sulfide, because an anode develops at the bottom of the crack, and a cathode at the surface (Bush et al., 1966) (see Fig. 10.57). Thus, the propagation of the crack is accelerated by metal ions going into solution at the bottom of the crack. Corrosion-fatigue cracks are a major cause of washouts and pipe failures.

Another form of stress cracking is known as *hydrogen embrittlement* (Bush et al., 1966; Hudgins et al., 1966; Mauzy, 1973; Hudgins, 1970; Patton, 1974a,b). It was stated previously in this section that hydrogen ions generated in a corrosion cell give up their charge at the cathode and bubble off as molecular hydrogen. However, a certain amount of hydrogen remains in the atomic form and is able to penetrate the steel. Normally, the amount

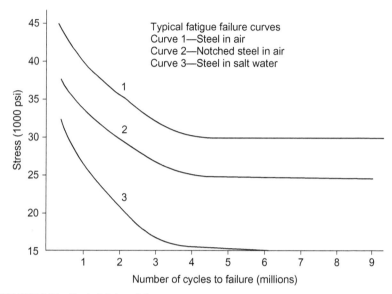

FIGURE 10.56 Typical fatigue failure curves.

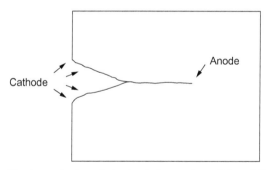

FIGURE 10.57 Development of an electrochemical corrosion cell in a fatigue stress crack. *From Bush, H.E., Barbee, R., Simpson, J.P., 1966. Current techniques for combatting drill pipe corrosion. API Drill. Prod. Prac. 59–69. Courtesy of API.*

that penetrates is so small that no harm is done. In the presence of hydrogen sulfide, however, the formation of atomic hydrogen is enhanced, and a larger amount penetrates the steel, where it concentrates at points of maximum stress. When a critical concentration is reached, the crack grows rapidly and failure occurs. This type of failure is known as *sulfide stress cracking*.

The time to failure depends on the following three variables:

1. The stresses in the steel, either residual or imposed. The higher the stress, the shorter the failure time. Below a certain stress, whose value depends on the strength of the steel, failure will not occur.

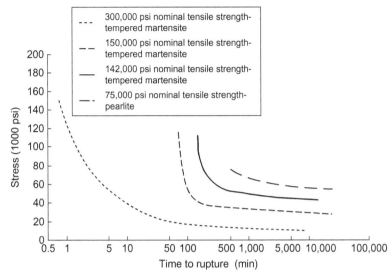

FIGURE 10.58 Effect of applied stress and tensile strength on hydrogen embrittlement failure. *From Mauzy, H.L., 1973. Minimize drill string failures caused by hydrogen sulphide. World Oil 65–70.*

2. The tensile strength or hardness of the steel. Sulfide stress cracking is not usually a problem with steels of tensile strength less than 90,000 psi (6000 kg/cm^2; Rockwell hardness C 22), as shown by Fig. 10.58.
3. The amount of hydrogen present. The more hydrogen, the shorter the failure time.

Severe problems with hydrogen stress cracking are experienced when drilling mud is contaminated with hydrogen sulfide. Fig. 10.59 shows the rapid failures that can occur when the stress in the drill pipe is high. Failure occurred in less than 1 h when the concentration of sulfide was high, and even when the concentration was very low, failure occurred in a week or so.

Control of Corrosion

The simplest and most common method of corrosion control is to use a highly alkaline mud. There are, however, limits to this practice: Notably, degradation of clay minerals by the hydroxyl ion starts at temperatures above 200°F (93°C) when the pH of the mud is above 10. As mentioned in Chapter 7, The Filtration Properties of Drilling Fluids, calcium hydroxide can cause solidification of the mud at temperatures above 300°F (149°C), and all hydroxides cause significant degradation and thickening at such temperatures. The wisest policy is to maintain the pH between 9 and 10, which, in many wells, will keep corrosion within acceptable limits, and at the same

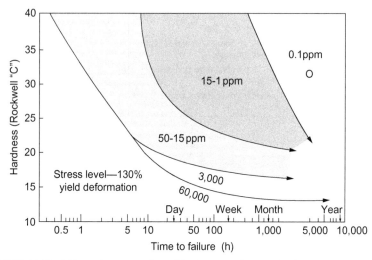

FIGURE 10.59 Effect of concentration of hydrogen sulfide on the failure time of highly stressed steel. *From the 1970 issue of Petroleum Engineer International. Publisher retains copyright.*

time allow tannate and lignosulfonate thinners to operate most efficiently. If previous experience in the locality has shown this procedure to be inadequate for corrosion control, or if excessive corrosion is detected by examination of the steel-ring coupons, which should be included in every drill string (see the section on corrosion tests in chapter: Evaluating Drilling Fluid Performance), then treatments appropriate to the particular type of corrosion involved must be applied. Table 10.14 (Bush, 1974) summarizes the common types of corrosion, their identification in the field, and suggested treatments. Details of the action and control of the various contaminants are as follows:

Carbon dioxide. Carbon dioxide dissolves in water and lowers the pH by forming carbonic acid. Corrosion is best controlled by maintaining the pH between 9 and 10 with sodium hydroxide, but, when the inflow of the gas is large, the formation of excessive amounts of soluble carbonates causes high viscosities. In such cases, calcium hydroxide may be used to neutralize the acid, but the resulting calcium carbonate precipitate tends to form scale, thus setting up corrosion cells. This tendency may be offset by the use of scale inhibitors, and by cleaning the pipe during round trips.

Hydrogen sulfide. Hydrogen sulfide may massively contaminate mud by a sudden inflow of sour gas, or gradually by the gradual degradation of lignosulfonates either by sulfate-reducing bacteria or by high temperatures. The thermal degradation of lignosulfonates starts at about 330°F (165°C) and increases gradually until a major decomposition occurs at

TABLE 10.14 Troubleshooting Chart to Combat Drilling Fluid Corrosion (Bush, 1974)

Cause	Primary Source	Visual Form of Attack	Identification — Corrosion By-product	Identification — Test	Treatment
Oxygen	Water additions	Concentration cell. Pitting under barrier or deposits	Primarily magnetite Fe_3O_4	Not acid soluble 15% HCl	Oxygen scavenger: Initial treating range equivalent of 2.5–10 lb/h of sodium sulfite. Maintain 20–300 mg/L sulfite residual, engineer to reduce air entrapment in pits. Defoam drilling fluid.
Air entrapment	Mixing and solids control equipment	Pits filled with black magnetic corrosion by-products		By-product attracted to magnet	Coat pipe with film forming inhibitors during trips to reduce atmospheric attack and cover concentration cell deposits.
Mineral scale deposits	Formation and mud materials	Corrosion cell pits below deposit	Iron products below mineral deposit	White mineral scale; calcium, barium and/or magnesium compounds	If mineral scale deposits on drill pipe, treat with ªscale inhibitor at 5–15 mg/L or approximately 25–75 lb/day added slowly and continuously. Treatments of scale inhibitor may be reduced after phosphate residual exceeds 15 mg/L. Treatments of 1 gal/1000 bbl mud/day can be used for

(Continued)

TABLE 10.14 (Continued)

Cause	Primary Source	Visual Form of Attack	Identification Corrosion By-product	Test	Treatment
					maintenance treatment under normal drilling conditions. [a]Organic phosphorous compounds
Aerated drilling fluid	Injected air	Severe pitting	Oxides of iron	Black to rust red	Maintain chromate concentration at 500–1000 mg/L with chromate compounds or zinc chromate compounds. Maintain high pH. Keep drill pipe free of mineral scale deposits with scale inhibitor. Coat pipe with filming inhibitors during trips.
Carbon dioxide gas	Formation	Localized to pitting, dark brown to black film	Iron carbonate	Slow effervescence in 5% HCl	Maintain basic pH with caustic soda to neutralize the acid-forming gas.
	Thermally degraded mud products				Mineral scale precipitates can form as a result of the metastable reaction of carbonic acid with calcium. Scale inhibitors treatments of 5–15 mg/L

Hydrogen sulfide gas	Formation	Localized to sharp pitting, dark blue to black film on equipment [a]SSCC failures	Iron sulfides	Acid arsenic solution produces a bright yellow precipitate soluble in 15% HCl, rotten egg odor, dark blue to black film on exposed equipment. Lead acetate test.	Maintain high pH with caustic soda. Remove sulfide ions by precipitation with metal compounds, such as oxides of iron, Fe$_3$O$_4$, and/or zinc compounds ZnCO$_3$ or ZnO. Sulfide 0–100 ppm. 3–5 lb/bbl iron oxide (Fe$_3$O$_4$), 0.1–0.5 lb/bbl zinc compound. The combined treatments of iron oxide and zinc compounds should provide lower sulfide ion contamination in most drilling fluids. (25–75 lb/day) may be required. Filming inhibitors sprayed or float-coated on drill pipe during trips.
Atmospheric	Generalized to localized	Iron oxide rust	Visual	Wash equipment free of salts and mud products and spray with atmospheric filming inhibitors.

[a]Note: SSCC sulfide stress corrosion cracking produces rapid brittle failure of high-strength metal (usually tool joints or bits).

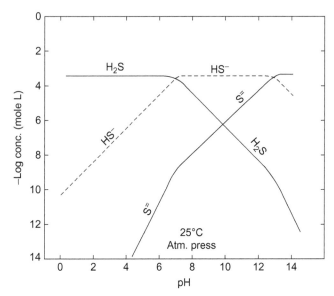

FIGURE 10.60 Equilibrium of the aqueous system H_2S HS^-, S^-, relative concentrations versus pH. *From Garrett, R.L., Clark, R.K., Carney, L.L., Grantham, C.K., 1978. Chemical scavengers for sulphides in water-base drilling fluids. In: SPE Paper 7499, Annual Meeting, Houston, TX, 1–3 October. Copyright 1978 by SPE-AIME.*

450°F (232°C) (Skelly and Kjellstrand, 1966). Reaction products are hydrogen sulfide, carbon dioxide, and carbon monoxide.

Molecular hydrogen sulfide is a poisonous gas, and every possible precaution must be taken to protect rig personnel when it is encountered, even in small quantities. It is weakly acidic when dissolved in water, and attacks iron to form iron sulfides, as approximately described by the equation

$$H_2S + Fe^{++} = FeS_x \downarrow + 2H^+ H_2^0 \uparrow \qquad (10.10)$$

The sulfides are deposited on the pipe as a black powder.
Hydrogen sulfide ionizes in two stages:

$$H_2S \rightleftharpoons H^+ + HS^- \qquad (10.11)$$

$$HS + OH \rightleftharpoons S^{--} + H_2O \qquad (10.12)$$

These reversible reactions are a function of pH, as shown by Fig. 10.60 (Garrett et al., 1978). It may be seen that the sulfide is in the form of H_2S up to about pH 6, as HS^- from pH 8 to 11, and as S^- above pH 12. Since sulfide stress cracking is caused by the atomic hydrogen formed together with HS^- in the first ionization stage (Eq. 10.vi), it follows that maintaining the pH between 8 and 11 is not a viable means of control.

The formation of atomic hydrogen is suppressed when the pH is above 12, but maintaining such high alkalinities is not desirable because it involves the accumulation of S^- in the mud (Eq. 10.vii). Should the pH fall because of a sudden inflow of more hydrogen sulfide, or for any other reason, the ionization reactions would reverse, and large amounts of atomic hydrogen or, possibly, hydrogen sulfide gas, would be generated. A high pH is, of course, also undesirable in high-temperature wells because of the aforementioned degradation of clay minerals. It is preferable, therefore, to combat hydrogen sulfide by the addition of a scavenger rather than by maintaining a high pH.

Formerly, copper salts were used to scavenge hydrogen sulfide until it was realized that these salts caused bimetallic corrosion (Chesser and Perricone, 1970) (i.e., the process shown in Fig. 10.54). To prevent bimetallic corrosion, the cation of the scavenger must be higher on the electromotive series than iron. Zinc meets this qualification, and basic zinc carbonate is now commonly used. Care must be taken to maintain the pH between 9 and 11. At higher or lower pH, the solubility of zinc carbonate increases sharply and the zinc ion flocculates clay suspensions (Garrett et al., 1978). Fig. 10.61 shows the consequent increase in yield point, gel strength, and filtration rate. It is evident that optimum properties are at pH 9, or thereabouts.

Flocculation by the zinc ion may be avoided by the use of a zinc chelate (Carney and Jones, 1974). The zinc is held by coordinate bonds in the chelating agent so that a very low level of zinc ions is maintained in solution, but the zinc is readily available to react with the sulfide as required. Powdered iron minerals also act as scavengers for hydrogen sulfide. For example, hydrogen sulfide reacts with iron oxides to form insoluble iron sulfides. The reaction takes place at the surface so that the efficiency of the material depends on the surface area exposed. A synthetic form of magnetite, Fe_3O_4, which has a high specific surface because of its porous nature, is commercially available. The reaction product is pyrite, but the chemistry involved is complex, and depends on a number of variables such as pH, mud shear rate, and temperature (Samuels, 1974). Reaction time is fastest at low pH, and the material is therefore most effective in neutralizing sudden large influxes of hydrogen sulfide (Ray and Randall, 1978). The ability to operate at low pH is also an advantage in high-temperature wells. Another great advantage of the material is that, being insoluble, it does not affect the rheological and filtration properties of the mud.

Oxygen. Oxygen dissolved in the mud during mixing and treating operations is almost always present in drilling muds, and a few parts per million is sufficient to cause significant corrosion. Pitting, caused by the formation of oxygen corrosion cells (see Fig. 10.55) under patches of rust or scale, is characteristic of oxygen corrosion.

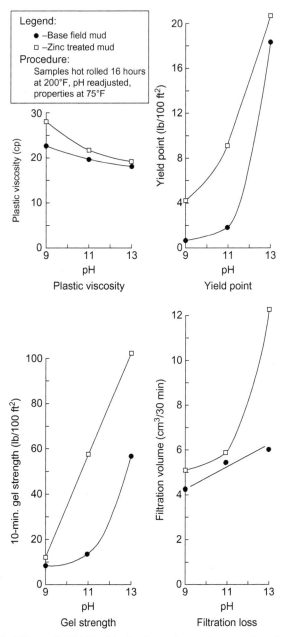

FIGURE 10.61 Effects of 6 lb/bbl of basic zinc carbonate on properties of an 11.8 lb/gal deflocculated field mud at 9–13 pH. *From Garrett, R.L., Clark, R.K., Carney, L.L., Grantham, C.K., 1978. Chemical scavengers for sulphides in water-base drilling fluids. In: SPE Paper 7499, Annual Meeting, Houston, TX, 1–3 October. Copyright 1978 by SPE-AIME.*

Oxygen corrosion increases sharply with increase in temperature, and is influenced by salinity. Brines and saline muds are more corrosive than freshwater muds because of the increase in conductivity, but the corrosion rate is less at very high salinities because the solubility of oxygen is less. The effect of temperature and salinity are illustrated in Fig. 10.62 (Cox, 1974). Low-solid polymer muds are more corrosive than conventional high-solid muds because the tannates and lignosulfonates added to clay muds for rheological control also act as oxygen scavengers. In general, oxygen corrosion decreases with increase in pH up to about pH 12, but then increases above that value (see Fig. 10.63) (Sloat and Weibel, 1970).

The best way to prevent oxygen corrosion is to minimize the entrainment of air at the surface by using only submerged guns in the pits, and arranging for returns from desanders, desilters, etc., to discharge below the mud level. The hopper is a prime source of air entrainment, and should be used only when adding solid conditioning materials. Fig. 10.64 (Bradley, 1970) shows the reduction in oxygen content when all conditioning operations were stopped.

The mud should be continuously monitored for corrosion so that remedial action can be quickly undertaken should unduly high rates develop. Corrosion meters now available can be mounted so that the rig crew can observe the corrosion rate at all times (Bush, 1974). Corrosion by oxygen can also be evaluated by placing a corrosion ring coupon just below the kelly, and another just above the drill collars. Since oxygen is consumed by corrosion reactions as it travels down the pipe, the difference between the weight lost by the upper and lower rings is a measure of the corrosion rate.

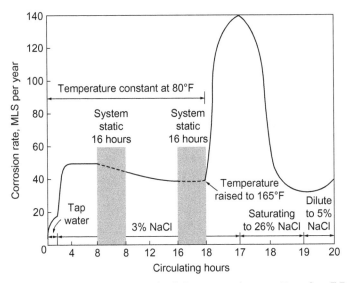

FIGURE 10.62 Effect of temperature and salinity on corrosion rate. *From Cox, T.E., 1974. Even traces of oxygen can cause corrosion. World Oil 110–112.*

452   Composition and Properties of Drilling and Completion Fluids

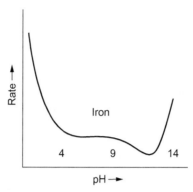

FIGURE 10.63 Variation in corrosion rate with pH value. *From Sloat, B., Weibel, J., 1970. How oxygen corrosion affects drill pipe. Oil Gas J. 77–79. API Div. Prod. Rocky Mountain Dist. Mtg., Denver, CO, April. Courtesy of API.*

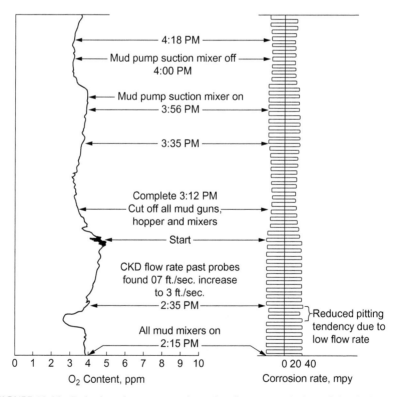

FIGURE 10.64 Reduction of oxygen corrosion at low flow rates and when mixing devices are stopped. *From the 1970 issue of Petroleum Engineer International. Publisher retains copyright.*

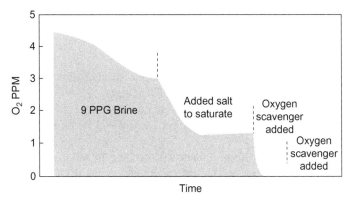

FIGURE 10.65 Reduction in oxygen content by addition of scavenger. *From Cox, T.E., 1974. Even traces of oxygen can cause corrosion. World Oil 110–112.*

If too much oxygen builds up in the mud, it must be removed by a suitable scavenger. Sulfite compounds are commonly used for this purpose (Bush, 1974; Cox and Davis, 1976). They react with oxygen to form sulfates, as, for instance:

$$O_2 + 2Na_2SO_3 \rightarrow 2Na_2SO_4 \qquad (10.13)$$

Fig. 10.65 (Cox, 1974) shows the marked reduction in oxygen when a scavenger is added.

Corrosion can also be reduced by making the pipe oil-wet with amines or amine salts (Bush, 1973). The action of these compounds was discussed in Chapter 9, Wellbore Stability. If the amine is added continuously, most of it will be lost by adsorption on clay particles. A better practice is to add small amounts at regular intervals, say 4 gallons (15 L) of a 10% solution in diesel oil every 30 min. Alternatively, the pipe may be sprayed with a solution during round trips.

Another practice is to coat the pipe internally with a protective film of plastic. This technique works well in high-solid muds because most of the oxygen is consumed by organic treating agents on the way down the hole. However, the oxygen is not so consumed when the drilling fluid is brine or a low-solid polymer mud. Consequently, oxygen is available to corrode the outside of the pipe on the way up the hole.

Microbiological Corrosion

Many species of bacteria exist in water-base drilling fluids and contribute to corrosion by forming patches of slime under which corrosion cells become established (Baumgartner, 1962). More severe damage is caused by a species of bacteria known as *Desulfovibrio*, which develops under impermeable deposits where conditions are completely anaerobic. They reduce sulfates present in the mud to form hydrogen sulfide by reacting with cathodic hydrogen, as follows:

$$SO_4 + 10H^- \rightarrow H_2S + 4H_2O \quad (10.14)$$

Corrosion is caused not only by the hydrogen sulfide thus created, but also by depolarization of the cathode (Baumgartner, 1962; Johnson and Cowan, 1964). Furthermore, by degrading mud additives such as lignosulfonates, the bacteria adversely affect the rheological and filtration properties of the mud.

The extent to which microorganisms contribute to the corrosion of drill pipe is not known for certain. The general opinion is that if the physical properties of the mud are not harmed, bacterial corrosion is not a problem (Bush, 1974). Microorganisms may be controlled by the addition of a biocide. Many biocides are available for general use, but in the case of drilling muds the choice is limited by the requirements that the biocide must not adversely affect the properties of the mud, and must not itself be corrosive. For example, copper salts cannot be used, because of bimetallic corrosion. Chlorinated phenols and paraformaldehyde appear to be the most suitable biocides (Johnson and Cowan, 1964). The amount required depends on the concentration of solids in the mud, and may be as much as 2 lb/bbl (6 kg/m^3).

Use of Oil Muds to Control Corrosion

Oil muds effectively prevent corrosion because they are nonconductive and contain strong oil-wetting agents. Their high cost is fully justified under severe conditions; for instance, when drilling sour gas reservoirs in deep, hot wells. They should also be considered when the drilling fluid is to be left in the hole as a packer fluid (see chapter: Completion, Workover, Packer, and Reservoir Drilling Fluids).

NOTATION

A Area
D Diameter of bit
d Exponent relating R/N to W/D
F Force parallel to a sliding surface
K Tensile strength
k_1 Constant relating vertical to horizontal stress
N Bit rpm
P_f Pressure of fluid in formation pores
P_{frac} p_w pressure required to induce a fracture
p_m Pressure of static mud column
p_w p_m plus pressure drop in annulus, plus any transient pressure
S Overburden pressure, solids plus liquids
S_{Hmin} Minimum horizontal total stress
W Weight on bit

W_1 Force normal to a sliding surface
Z Depth
σ_1 Greatest principal stress
σ_2 Intermediate principal stress
σ_3 Least principal stress

REFERENCES

Adams, N., 1977. How to control differential pipe sticking. Part 3. Pet. Eng. 44−50.
Anderson, E.T., 1961. How world's hottest well was drilled. Pet. Eng. 47−51.
Annis, M.R., Monaghan, R.H., 1962. Differential pipe sticking—laboratory studies of friction between steel and mud filter cake. J. Pet. Technol. 537−542, Trans AIME. 225.
Bannerman, J.K., Davis, N., 1978. Sepiolite muds for hot wells, deep drilling. Oil Gas J. 86, 144−150.
Bardeen, T., Teplitz, A.J., 1956. Lost circulation info. with new tool for detecting zones of loss. J. Petro. Tech. 36−41, Trans AIME. 207.
Baumgartner, A.W., 1962. Microbiological corrosion—what causes it, and how it can be controlled. J. Pet. Technol. 1074−1078, Trans AIME. 225.
Black, A.D., Dearing, H.L., Di Bona, B.G., 1985. Effects of pore pressure and mud filtration on drilling rate in permeable sandstone. J. Pet. Technol. 1671−1681, Trans AIME. 248.
Blattel, S.R., Rupert, J.P., 1982. Effect of weight material type on rate of penetration using dispersed and non-dispersed water-base muds. In: SPE Paper No. 10961, Annual Meeting, New Orleans, LA, September.
Bol, G.M., 1985. Effect of mud composition on wear and friction of casing and tool joints. In: SPE/ IADC Paper No. 13457, Drilling Conf., New Orleans, LA, 5−8 March.
Bradley, B.W., 1970. Oxygen cause of drill pipe corrosion. Pet. Eng. 54−57.
Breckels, I.M., van Eekelen, H.A.M., 1982. Relationship between horizontal stress and depth in sedimentary basins. J. Pet. Technol. 2191−2199, Trans AIME. 245.
Browning, W.C., 1959. Extreme-pressure lubes in drilling muds. Oil Gas J. 67, 213−218.
Burdyn, R.F., Wiener, L.D., 1957. Calcium surfactant drilling fluids. World Oil. 34, 101−108.
Bush, H.E., 1973. Controlling corrosion in petroleum drilling and packer fluids. In: Nathan, C.C. (Ed.), Corrosion Inhibition. NACE, Houston, TX, p. 109.
Bush, H.E., 1974. Treatment of drilling fluid to combat corrosion. In: SPE Paper No. 5123, Annual Meeting, Houston, TX, 6−9 October.
Bush, H.E., Barbee, R., Simpson, J.P., 1966. Current techniques for combatting drill pipe corrosion. API Drill. Prod. Prac. 59−69.
Caenn, R., 2013. Enhancing wellbore stability, strengthening and clean-out focus of fluid system advances. American Oil and Gas Reporter, October 2013.
Canson, B.E., 1985. Loss circulation treatment for naturally fractured, vugular, or cavernous formations. In: SPE/IADC Paper No. 13440, Drilling Conf., New Orleans, LA, March.
Carden, R.S., Nicholson, R.W., Pettitt, R.A., Rowley, R.C., 1985. Unique aspects of drilling hot dry rock geothermal wells. J. Pet. Technol. 821−834, Trans AIME. 248.
Carney, L., 1980. New inverts give good performance but for wrong reasons. Oil Gas J. 88.
Carney, L.J., Meyer, R.L., 1976. A new approach to high temeprature drilling fluids. In: SPE Paper 6025, Annual Meeting, New Orleans LA, 6 March.
Carney, L.L., Jones, B., 1974. Practical solutions to combat detrimental effects of H2S during drilling operations. In: SPE Paper 5198, Symp. on Sour Gas and Crude, Tyler, TX, 11−12 November.

Chesser, B.G., Enright, D.P., 1980. High temperature stabilization of drilling fluids with a low-molecular-weight copolymer. J. Pet. Technol. 243, 950–956.

Chesser, B.G., Perricone, A.C., 1970. Corrosive aspects of copper carbonate in drilling fluids. Oil Gas J. 78, 82–85.

Chesser, B.G., Perricone, A.C., 1983. A physiochemical approach to the prevention of balling of gumbo shales. In: SPE Paper 4515, Annual Meeting, Las Vegas, NV, September.

Clark, R.K., Scheuerman, R.F., Rath, H., van Laar, H., 1976. Polyacrylamide-potassium chloride mud for drilling water-sensitive shales. J. Pet. Technol. 719–727, Trans AIME. **261**.

Courteille, J.M., Zurdo, C., 1985. A new approach to differential sticking. In: SPE Paper No. 14244, Annual Meeting, Las Vegas, NV, 22–25 September.

Cowan, J.C., 1959. Low filtrate loss and good rheology retention at high temperatures are practical features of this new drilling mud. Oil Gas J. 67, 83–87.

Cox, T.E., 1974. Even traces of oxygen can cause corrosion. World Oil. 51, 110–112.

Cox, T., Davis, N., 1976. Oxygen scavengers. Drilling 68.

Cromling, J., 1973. How geothermal wells are drilled and completed. World Oil. 50, 42–45.

Cunningham, R.A., Eenik, J.G., 1959. Laboratory study of effect of overburden, formation, and mud column pressure on drilling rate of permeable formations. Trans AIME 216, 9–17.

Cunningham, R.A., Goins Jr., W.C., 1957. Laboratory drilling of gulf coast shales. API Drill. Prod. Prac. 75–85.

Daines, S.R., 1982. Prediction of fracture pressures in wildcat wells. J. Pet. Technol. 245, 863–872.

Daneshy, A.A., Slusher, G.L., Chisholm, P.T., Magee, D.A., 1986. In-situ stress measurements during drilling. J. Pet. Technol. 249, 891–898.

Darley, H.C.H., 1965. Designing fast drilling fluids. J. Pet. Technol. 465–470, Trans AIME. **234**.

Darley, H.C.H., 1976. Advantages of polymer fluids. Pet. Eng. 48, 46–48.

Dawson, D.D., Goins Jr., W.C., 1953. Bentonite-diesel oil squeeze. World Oil. 30, 222–233.

Diment, W.H., Urban, T.C., Sass, J.H., Marshall, B.V., Munroe, R.J., Lachenbruch, A.H., 1975. Temperatures and heat contents based on conductive transport of heat. Assessment of Geothermal Resources of the United States, Geological Survey Circular 726. Dept. Interior, Washington, DC, pp. 84–103.

Doty, P.A., 1986. Clear brine drilling fluids: a study of penetration rates, formation damage, and wellbore stability in full scale drilling tests. In: SPE/IADC Paper No. 13441, Drilling Conf., New Orleans, LA, 5–8 March; and SPE Drill. Eng. 17–30.

Eaton, B.A., 1969. Fracture gradient prediction, and its application in oilfield operations. J. Pet. Technol. 1353–1360, Trans AIME. **246**.

Eckel, J.R., 1954. Effect of mud properties on drilling rate. API Drill. Prod Prac. 119–124.

Eckel, J.R., 1958. Effect of pressure on rock drillability. Trans AIME 213, 1–6.

Eckel, J.R., 1967. Microbit studies of the effect of fluid properties and hydraulics on drilling rate. J. Pet. Tech 541–546, Trans AIME. **240**.

Fontenot, J.E., Berry, L.M., 1975. Study compares drilling-rate-based pressure-prediction methods. Oil Gas J. 83, 123–138.

Gallus, J.P., Lummus, J.L., Fox, J.E., 1958. Use of chemicals to maintain clear water for drilling. J. Pet. Technol. 70–75, Trans AIME. 213.

Garnier, A.J., van Lingen, N.H., 1959. Phenomena affecting drilling rates at depth. J. Pet. Technol. 209, 232–239.

Garrett, R.L., Clark, R.K., Carney, L.L., Grantham, C.K., 1978. Chemical scavengers for sulphides in water-base drilling fluids. In: SPE Paper 7499, Annual Meeting, Houston, TX, 1–3 October.

Gatlin, C., Nemir, C.E., 1961. Some effects of size distribution on particle bridging in lost circulation and filtration tests. J. Pet. Technol. 575–578, Trans AIME. 222.

Gockel, J.F., Gockel, C.E., Brinemann, M., 1987. Lost circulation: a solution based on the problem. In: SPE/IADC Paper 16082, Drill. Conf., New Orleans, LA, 15–18 March.

Goins Jr., W.C., Dawson, D.D., 1953. Temperature surveys to locate zone of lost circulation. Oil Gas J. 61, 170, 171, 269–276.

Goins, W.C. Jr., Nash, F. Jr., 1957. Methods and composition for recovering circulation of drilling fluids in wells. U.S. Patent No. 2,815,079 (3 December).

Gravely, W., 1983. Review of downhole measurement-while-drilling systems. J. Pet. Technol. 246, 1439–1445.

Haden, E.L., Welch, G.R., 1961. Techniques for preventing differential sticking of drill pipe. API Drill. Prod. Prac. 36–41.

Haimson, B.C., 1973. Hydraulic fracturing of deep wells. 2nd Annual Report of API Project 147. American Petroleum Institute, Dallas, TX, 35–49.

Hamburger, C.L., Tsao, Y.H., Morrison, M.E., Drake, E.N., 1985. A shear thickening fluid for stopping unwanted flows while drilling. J. Pet. Technol. 248, 499–504.

Hayward, J.T., 1937. Cause and cure of frozen drill pipe and casing. API Drill. Prod. Prac. 8–20.

Helmick, W.E., Longley, A.J., 1957. Pressure-differential sticking of drill pipe, and how it can be avoided. API Drill. Prod. Prac. 55–60.

Hoberock, L.L., Thomas, D.C., Nickens, H.V., 1982. Here's how compressibility and temperature affect bottom-hole mud pressure. Oil Gas J. 90, 159–164.

Hottman, C.E., Johnson, R.K., 1965. Estimation of pore pressure from log-derived shale properties. J. Pet. Technol. 717–722, Trans AIME. 234.

Howard, G.C., Scott, P.P., 1951. An analysis of the control of lost circulation. Trans AIME 192, 171–182.

Hubbert, M.K., Willis, D.G., 1957. Mechanics of hydraulic fracturing. J. Pet. Technol. 153–166, Trans AIME. 210.

Hudgins Jr., C.M., 1970. Hydrogen sulphide corrosion can be controlled. Pet. Eng. 33–36.

Hudgins Jr., C.M., McGlasson, R.L., Medizadeh, P., Rosborough, W.M., 1966. Hydrogen sulphide cracking of carbon and alloy steels. Corrosion 238–251.

Hunter, D., Adams, N., 1978. Laboratory test data indicate water base drilling fluids that resist differential-pressure pipe sticking. In: Paper OTC 3239, Offshore Technology Conf., Houston, TX, 8–11 May.

IADC 2013. Advances in high-performance drilling fluids enhance wellbore strength, help curb loss. Drilling Contractor magazine, 9 January, 69.

Johnson, D.P., Cowan, J.C., 1964. Recent Developments in Microbiology of Drilling and Completion Fluids. Am. Inst. Biological Sci., Washington, DC.

Jones, P.H., 1969. Hydrodynamics of geopressure in the northern Gulf of Mexico Basin. J. Pet. Technol. 246, 803–810.

Jorden, J.R., Shirley, O.J., 1966. Application of drilling performance data to overpressure detection. J. Pet. Technol. 243, 1387–1394.

Kelly, J., 1965. How lignosulfonate muds behave at high temperatures. Oil Gas J. 73, 111–119.

Kercheville, J.D., Hinds, A.A., Clements, W.R., 1986. Comparison of environmentally acceptable materials with diesel oil for drilling mud lubricity and spotting formulations. IN: IADC/SPE Paper No. 14797, Drilling Conf., Dallas, TX, 1986.

Krol, D.A., 1984. Additives to cut differential pressure sticking in drillpipe. Oil Gas J. 92, 55–59.

Kruger, P., Otte, C., 1974. Geothermal Energy. Stanford Univ. Press, Stanford, CA, p. 73.

Lammons, R.D., 1984. Field use documents glass bead performance. Oil Gas J. 92, 109−111.
Lee, J., Cullum, D., Friedheim, J., Young, S., 2012. A new SBM for narrow margin extended reach drilling. In: IADC/SPE Drilling Conference and Exhibition, San Diego, CA, 6−8 March.
Lummus, J.L., 1965. Chemical removal of drilled solids. Drill. Contract. 21, 50−54, 67.
Lummus, J.L., 1968. Squeeze slurries for lost circulation control. Pet. Eng. 59−64.
Mallory, H.E., 1957. How low solids fluids can cut costs. Pet. Eng 1321−1324.
Matthews, W.R., Kelly, J., 1967. How to predict formation pressure and fracture gradients. Oil Gas J. 75, 92−106.
Maurer, W.C., 1962. The "perfect cleaning" theory of rotary drilling. J. Pet. Technol. 1270−1274, Trans AIME. 225.
Mauzy, H.L., 1973. Minimize drill string failures caused by hydrogen sulphide. World Oil. 50, 65−70.
McMordie, W.C., Bland, R.G., Hauser, J.M., 1982. Effect of temperature and pressure on the density of drilling fluids. SPE Paper No. 11114, Annual Meeting, New Orleans, LA, September.
Messenger, J.U., 1973. Common rig materials combat severe lost circulation. Oil Gas J. 81, 57−64.
Mettath, S., Stamatakis, E., Young, S. and De Stefano, G. 2011. The prevention and cure of bit balling in water-based drilling fluids. In: 2011 AADE Fluids Conference Paper 11-NTCE-28, Houson, TX, 7−9 April.
Mitchell, R.F., 1981. The simulation of air and mist drilling for geothermal wells. In: SPE Paper No. 10234, Annual Meeting, San Antonio, TX, October.
Mondshine, T.C., 1970. Drilling mud lubricity. Oil Gas J. 78, 70−77.
Montgomery, M., 1985. Discussion of "the drilling mud dilemma−recent examples". J. Pet. Technol. 248, 1230.
Moore, T.F., Kinney, C.A., McGuire, W.J., 1963. How Atlantic squeezes with high water loss slurry. Oil Gas J. 71, 105−110.
Moses, P.L., 1961. Geothermal gradients. API Drill. Prod. Prac. 57−63.
Murray, A.S., Cunningham, R.A., 1955. Effect of mud column pressure on drilling rates. Trans AIME 204, 196−204.
Outmans, H.D., 1958. Mechanics of differential sticking of drill collars. Trans AIME 213. 265−274.
Outmans, H.D., 1974. Spot fluid quickly to free differentially stuck pipe. Oil Gas J. 82, 65−68.
Patton, C.C., 1974a. Corrosion fatigue causes bulk of drill string failures. Oil Gas J. 163−168.
Patton, C.C., 1974b. Dissolved gases are key corrosion culprits. Oil Gas J. 82, 67−69.
Perricone, A.C., Enright, D.P., Lucas, J.M., 1986. Vinyl-sulfonate copolymers for high temperature filtration control of water-base muds. In: Drilling Conf. SPE/IADC Paper 13455, New Orleans, LA; and SPE Drill. Eng. 358−364.
Portnoy, R.C., Lundberg, R.D., Werlein, E.R., 1986. Novel polymeric oil mud viscosifier for high temperature drilling. In: IADC/SPE Paper No. 14795, Drilling Conf., Dallas.
Ray, J.D., Randall, B.V., 1978. Use of reactive iron oxide to remove h2s from drilling fluid. In: SPE Paper 7498, Annual Meeting, Houston, TX, 1−3 October.
Raymond, L.R., 1969. Temperature distribution in a circulating drilling fluid. J. Pet. Technol. 333−341, Trans AIME. 246.
Remont, L.J., Rehm, W.A., McDonald, W.J., 1977. Maurer, W.C. Evaluation of commercially available geothermal drilling fluids. Sandia Report No. 77−7001, Sandia Laboratories, Albuquerque, NM, pp. 52−75.

Renner, J.L., White, D.E., Williams, D.L., 1975. Hydrothermal convection systems. Assessment of Geothermal Resources of the United States, Geological Survey Circular, No. 726. Dept. Interior, Washington, DC, pp. 5–57.

Rosenberg, M., Tailleur, R.H., 1959. Increased drill bit use through use of extreme pressure lubricant drilling fluids. J. Pet. Technol. 195–202, Trans AIME. 216.

Ruffin, D.R., 1978. New squeeze for lost circulation. Oil Gas J. 86, 96–97.

Rupert, J.P., Pardo, C.W., Blattel, S.R., 1981. The effects of weight material type and mud formulation on penetration rate using invert oil systems. In: SPE Paper No. 10102, Annual Meeting, San Antonio, TX, 5 October.

Samuels, A., 1974. H_2S need not be deadly, dangerous, destructive. In: SPE Paper 5202, Symp. on Sour Gas and Crude, Tyler, TX, 11–12 November.

Sartain, B.J., 1960. Drill stem tester frees stuck pipe. Pet. Eng. B86–B90.

Scharf, A.D., Watts, R.D., 1984. Itabirite: an alternative weighting material for heavy oil base muds. In: SPE Paper No. 13159, Annual Meeting, Houston, TX, 16–19 September.

Schmidt, G.W., 1973. Interstitial water composition and geochemistry of deep gulf coast shales and sandstones. AAPG Bull. 57 (2), 321–337.

Secor, D.T., 1965. Role of fluid pressure jointing. Am. J. Sci. 263, 633–646.

Simpson, J.P., 1962. The role of oil mud in controlling differential-pressure sticking of drill pipe. In: SPE Paper 361, Upper Gulf Coast Drill and Prod. Conf., Beaumont, TX, 5 April.

Simpson, J.P., 1978. A new approach to oil muds for lower cost drilling. In: SPE Paper 7500, Annual Meeting, Houston, TX, 1–3 October.

Simpson, J.P., 1979. Low colloid oil muds cut drilling costs. World Oil. 56, 167–171.

Simpson, J.P., 1985. The drilling mud dilemma–recent examples. J. Pet. Technol. 248, 201–206.

Skelly, W.G., Kjellstrand, J.A., 1966. The thermal degradation of modified lignosulfonates in drilling muds. In: API Paper 926–1106, Spring Meeting, Div. of Prodn., Houston, TX, March.

Slater, K., Amer, A. 2013 New automated lubricity tester evaluates fluid additives, systems and their application. In: 11th Offshore Mediterranean Conference and Exhibition paper in Ravenna, Italy, 20–22 March.

Sloat, B., Weibel, J., 1970. How oxygen corrosion affects drill pipe. Oil Gas J. 77–79, API Div. Prod. Rocky Mountain Dist. Mtg., Denver, CO, April.

Son, A.J., Ballard, T.M., Loftin, R.E., 1984. Temperature-stable polymeric fluid-loss reducer tolerant to high electrolyte contamination. In: SPE Paper No. 13160, Annual Meeting, Houston, TX, September.

Speer, J.W., 1958. A method for determining optimum drilling techniques. Oil Gas J. 66, 90–96.

Stein, N., 1985. Resistivity and density logs key to fluid pressure estimates. Oil Gas J. 81–86.

Thomson, M., Burgess, T.M., 1985. The prediction and interpretation of downhole mud temperature while drilling. In: SPE Paper No. 14180, Annual Meeting, Las Vegas, NV, September.

Topping, A.D., 1949. Wall collapse in oil wells as a result of rock stress. World Oil. 26, 112–120.

van Lingen, N.H., 1962. Bottom scavenging—a major factor governing penetration rates at depth. J. Pet. Technol. 187–196, Trans AIME. 225.

Vaussard, A., Martin, M., Konirsch, O., Patroni, J-M, 1986. An experimental study of drilling fluids dynamic filtration. In: SPE Paper 15142, Ann. Tech. Conf., New Orleans, LA, 5–8 October.

Vidrine, D.J., Benit, B.J., 1968. Field verification of the effect of differential pressure on drilling rate. J. Pet. Technol. 231, 676–682.

Warpinski, N.R., Branagan, P., Wilmer, R., 1985. In-situ measurements at US DOE's multiwell experimental site Mesaverde Group, Rifle, Colorado. J. Pet. Technol. 248, 527–536.

Warren, T.M., Smith, M.B., 1985. Bottomhole stress factors affecting drilling rate at depth. J. Pet. Technol. 248, 1523–1533.

Wyant, R.E., Reed, R.L., Sifferman, T.R., Wooten, S.O., 1985. Dynamic fluid loss measurement of oil mud additives. In: SPE Paper No. 13246, Annual Meeting, Las Vegas, NV, 22–25 September; and SPE Drill. Eng. (1987), 63–74.

York, P.A., Prichard, D.M., Dodson, J.K., Dodson, T., Rosenberg, S.M., Gala, D., et al., 2009. Eliminating non-productive time associated with drilling through trouble zones. In: Offshore Technology Conference Paper 20220, Houston, TX, 4–7 May.

Young Jr., F.S., Gray, K.E., 1967. Dynamic filtration during microbit drilling. J. Pet. Technol.1209–1224, Trans AIME. 240.

Young, S., Friedheim, J., Lee, J., Prebensen, O.I., 2012. A new generation of flat rheology invert drilling fluids. IN: SPE Paper 154682, Prepared for Presentation at the SPE Oil and Gas India Conference and Exhibition Held in Mumbai, India, 28–30 March 2012.

Zoeller, W.A., 1984. Determine pore pressure from MWD logs. World Oil. 61, 97–102.

Chapter 11

Completion, Workover, Packer, and Reservoir Drilling Fluids

INTRODUCTION

Oil companies traditionally divide engineering duties between exploration and production disciplines. There are drilling engineers who specialize in designing and implementing drilling programs and there are production or completion engineers who take over from the drilling engineers after a well has been cased and cemented. There are also reservoir evaluation engineers who specialize in evaluating the various rock formations that a well has penetrated to determine if hydrocarbons are present.

The objective of the drilling engineers and drilling operations personnel are often at odds with what is needed by the evaluation and completion engineers. Drilling engineers want to punch a hole down to the objective as quickly and cheaply as possible. Evaluation and completion issues associated with poor borehole conditions are not their problem. Evaluation and completion engineers want a borehole that is in as good condition as possible, i.e., that is not washed out and that has an undamaged producing zone. These engineers may be at odds with each other over how a well is to be drilled, and the resulting drilling program is usually a compromise. Field well drilling programs, since the reservoir parameters are in large part known, are normally not contentious and are developed with cost reduction in mind.

If engineering management and leadership are good, then the drilling program is designed to meet the needs of the evaluation and completion engineers for an exploration well. If it is not, and they are too focused on "saving money," then the resulting well will often be in poor condition making evaluation difficult at best, and in some cases impossible. Completing a well with a bad borehole can be very challenging as well, particularly if there are multiple zones to test. If the wellbore is in bad condition the cement job is usually ineffective and it is not possible to isolate productive zones because the hydrocarbons can flow between the casing and the borehole.

EXPENSE VERSUS VALUE

Over the last few decades, operators have realized that compromises can be made between the "cost" portion of a wellbore (drilling) and the "value" portion (production). Drilling deeper, quicker, and cheaper may be false economy. Drilling and completion engineers now agree that ensuring a nondamaged reservoir is the most important goal for both. The value portion of the hole justifies an increased drilling cost in the reservoir.

The first step to make sure we get the most value from the drilling fluid used to drill the reservoir is to identify and understand the mechanisms of damage to the particular formation which result in reduced permeability. This process can be time-consuming and expensive in itself. It involves gathering information from many sources—seismic, laboratory core evaluations, and offset drilling and production information. With this information we can calculate or estimate the potential damage caused by a particular reservoir drilling fluid (RDF) type. The overall damage is called the skin effect and can be caused by a variety of different mechanisms.

THE SKIN EFFECT

Reservoir engineers have calculated from pressure drawdown data that many wells produce at less than their potential (Hurst, 1953; van Everdingen, 1953). A barrier, or *skin*, appears to exist around the wellbore impairing the flow of oil and gas, as shown in Fig. 11.1. This skin is caused by a zone of reduced permeability around the wellbore, and results from contamination by mud particles or filtrate. In some wells, poor completion practices—such as too widely spaced perforations, or insufficient penetration into the reservoir—also contribute to the observed skin (Matthews and Russell, 1967). In addition, horizontal, shale wells may have a nonuniform skin along their length complicating pressure-transient analysis (Al-Otaibi and Ozkan, 2005).

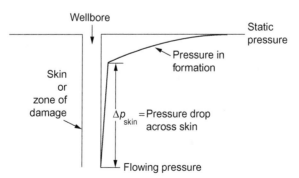

FIGURE 11.1 Pressure distribution in a reservoir with a skin. *From the October 1953 issue of Petroleum Engineer International.*

Although the contaminated zone extends at most only a few feet into the reservoir, it can cause quite a large reduction in well productivity, because the flow is radial, and therefore pressure drop is proportional to ln r/r_w, where r_w is the radius of the well, and r is the radius of interest. Muskat (1949) gives the ratio of productivity of a damaged well to that of an undamaged well as:

$$\frac{Q_d}{Q} = \frac{\ln \dfrac{r_e}{r_w}}{\dfrac{k}{k_d}\ln \dfrac{r_d}{r_w} + \ln \dfrac{r_e}{r_d}} \qquad (11.1)$$

where Q is the undamaged productivity, Q_d is the damaged productivity, r_e is the radius of the drainage area, k and k_d are the virgin and the damaged permeabilities, and r_d is the radius of the damaged zone.

Fig. 11.2 shows that the effect of reduced rock permeability (as measured in a lineal core) on the productivity of a well decreases with distance from the borehole. There are several mechanisms by which mud solids or filtrate may reduce well productivity. These may be summarized as follows:

- Capillary phenomena: relative permeability effects resulting from changes in the relative amounts of water, oil, and/or gas in the pores; wettability effects; and pore blocking by aqueous filtrates.
- Swelling and dispersion of indigenous clays by the mud filtrate.
- Penetration of the formation, and pore plugging by particles from the mud and filter cake.

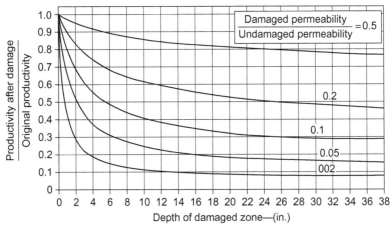

FIGURE 11.2 Reduction in productivity or infectivity due to invasion of damaging fluid around the wellbore. *From Tuttle, R.N., Barkman, J.H., 1974. New non-damaging and acid-degradable drilling and completion fluids. J. Petrol. Technol. 1221–1226. Courtesy J. Petrol. Technol. Copyright 1974 by SPE-AIME.*

- Plugging of gravel packs, liners, and screens by mud filter cake.
- Salt precipitation from mixing the filtrate and formation water.
- Slumping of unconsolidated sands and compression of wellbore surfaces.

In this chapter we first discuss these mechanisms in detail and then describe the various workover and completion fluids, and the best means of minimizing or avoiding formation damage.

Capillary Phenomena

When the filtrate from a water-base mud invades an oil-bearing formation it displaces the oil. Under certain circumstances not all of the water is produced back, and productivity is thereby impaired. This mechanism was the first type of formation damage to be recognized, and is commonly called *waterblock*.

As we saw in Chapter 8, The Surface Chemistry of Drilling Fluids glass is wet by water in preference to air or oil; therefore, water will spontaneously displace air or oil from a glass capillary, but a pressure (known as the threshold pressure) must be applied in order for air or oil to displace water. Permeable rocks are analogous to a bundle of capillary tubes of innumerable different diameters (Purcell, 1949). In reality, of course, the flow paths through a rock are much more complicated, being tortuous and three dimensional, but the concepts of capillary behavior still apply. The actual pore structure of most rocks consists of an irregular, three-dimensional network of pores connected by narrow channels. The capillary properties of these structures may be demonstrated by forcing mercury at increasingly higher pressures into a specimen of the rock, and plotting the percentage of the pore volume occupied by the mercury against the injection pressure, as shown in Fig. 11.3 (Swanson, 1979).

To help visualize the tortuous paths followed by a fluid flowing through a rock, Swanson (1979) forced molten Woods metal—which, like mercury, does not wet rock surfaces—into rock specimens. He then dissolved the matrix of the rock with acid, leaving a pore cast of Woods metal. Figs. 11.4 and 11.5 show scanning electron micrographs of the casts of high- and low-permeability sandstones. Note that many of the large pores are connected by small capillaries.

In most cases, water will spontaneously displace oil or gas from a rock specimen, so it is evident that rocks are normally preferentially water wet. The degree of water wetness can be determined from imbibition tests, in which the amount of oil spontaneously displaced by water from a specimen, or the amount of water forced out by oil under pressure is measured (Amott, 1958).

When two immiscible fluids are flowing simultaneously through a permeable medium, the flow paths of the fluids are controlled by the preferential wettability. In the case of water and oil (or gas) flowing through rocks that

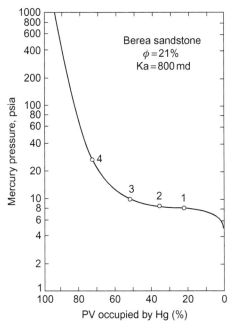

FIGURE 11.3 Mercury capillary-pressure curve applying to pore casts in Fig. 10.4. *From Swanson, B.F., 1979. Visualizing pores and nonwetting phase in porous rock. J. Petrol. Technol. 10–18. Copyright 1979 by SPE-AIME.*

are preferentially water wet, the water flows along the surface of the grains and through the minor capillaries, and the oil flows through the center of the pores and the larger flow channels. The relative permeability to each fluid (i.e., the permeability to the fluid expressed as a ratio of the permeability when only a single fluid—usually air—is present) depends on the wettability of the rock and on the percent saturation of each fluid. Fig. 11.6 shows typical oil and water relative permeabilities plotted against percent water saturation. As would be expected, the relative permeability to oil at a given saturation is greater than the relative permeability to water at the equivalent water saturation. The *residual water saturation* (shown in Fig. 11.6) is the minimum water saturation when only oil is flowing at a given pressure drop, and the *residual oil saturation* is the corresponding value for oil.

A virgin oil reservoir is at residual water saturation for the prevailing reservoir pressure. Invasion by mud filtrate drives the oil towards residual oil saturation, and when the well is brought onto production, the oil drives the filtrate back towards residual water saturation. However, as shown in Fig. 11.6, the relative permeability to water becomes very low as residual water is approached. It may therefore take a considerable length of time before all the filtrate is expelled, and full production is obtained, particularly

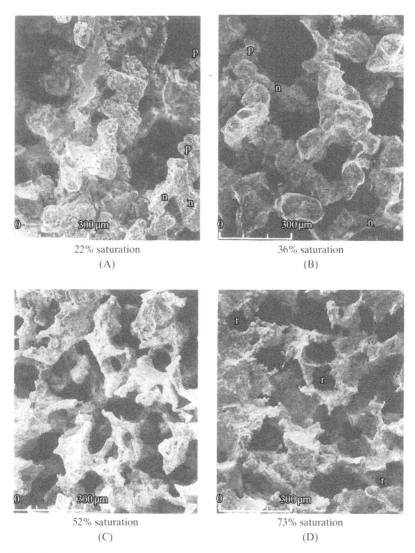

22% saturation
(A)

36% saturation
(B)

52% saturation
(C)

73% saturation
(D)

FIGURE 11.4 Woods metal pore cast of Berea sandstone. *From Swanson, B.F., 1979. Visualizing pores and nonwetting phase in porous rock. J. Petrol. Technol. 10–18. Copyright 1979 SPE-AIME.*

if the oil−water viscosity ratio is low (Ribe, 1960). In most virgin reservoirs, pressures are high enough to expel all the filtrate eventually, so impairment caused by relative permeability effects is temporary. However, in low-pressure and low-permeability reservoirs and in workover wells, capillary pressures become significant. Capillary pressures are inversely proportional to radius, and some capillaries of rocks are so small that capillary pressures

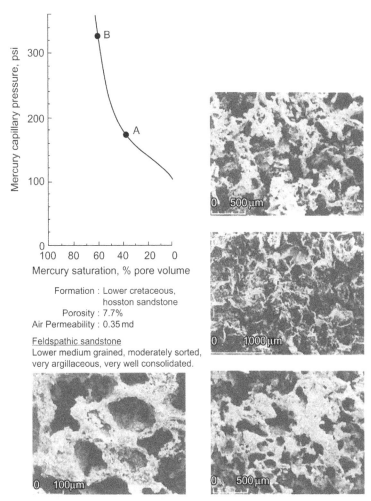

FIGURE 11.5 Mercury capillary pressure curve, rock texture, and pore cast of Hosston sandstone. *From Swanson, B.F., 1979. Visualizing pores and nonwetting phase in porous rock. J. Petrol. Technol. 10–18. Copyright 1979 by SPE-AIME.*

may run into hundreds of pounds per square inch. Capillary pressure promotes the displacement of oil by an aqueous filtrate but opposes the displacement of the filtrate by the returning oil. The pressure drop may not be high enough to drive filtrate out of the finer capillaries, especially in the immediate vicinity of the borehole wall, where the pressure drop at the oil–water interface approaches zero. This mechanism, which we will refer to as *waterblock*, causes permanent impairment—and even complete shutoff—in highly depleted reservoirs. Waterblock in gas reservoirs was formerly called the *Jamin effect* (Yuster and Sonney, 1944).

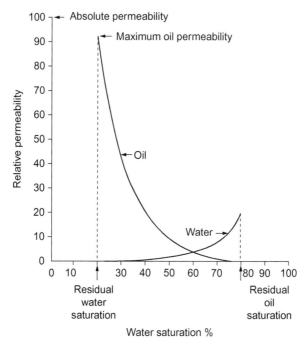

FIGURE 11.6 Relative permeabilities to oil and water in a preferentially water-wet reservoir.

Waterblock may be avoided by the use of oil muds, provided no water is in their filtrates under bottomhole conditions. Oil muds have two limitations: first, they should not be used in drilling dry gas sands, because not all the oil will be produced back, and will thus leave a second residual phase. Second, cationic surfactants used as emulsifiers decrease the degree of water wetness (Amott, 1958) of the grain surfaces, and, if the mud is poorly formulated, may even convert the surfaces to the oil-wet condition. In an oil-wet rock, the shape of the relative permeability curves shown in Fig. 11.6 are interchanged, so that the relative permeability to oil at low water saturations is greatly decreased.

In situ emulsification of interstitial oil is another possible cause of capillary impairment if the filtrate of an oil-in-water emulsion mud contains appreciable quantities of emulsifier. Emulsification is possible because, although the bulk flow rate of the filtrate is low, the rate of shear at constrictions in the flow channels is high. If a stabilized emulsion is formed, the droplets become trapped in the pores and reduce the effective permeability. However, emulsifiers will only be present in the filtrate if excess is present in the emulsion mud. Therefore, in situ emulsification can be avoided if care is taken in formulating and maintaining the emulsion muds.

Permeability Impairment by Indigenous Clays

Nearly all sands and sandstones contain clays, which profoundly influence the permeability of the rock. These clays derive from two possible sources: detrital clays are clays which have sedimented with the sand grains at the time the bed was deposited. Diagenetic clays are clays which have subsequently been precipitated from formation waters or which were formed by the interaction of formation waters and preexisting clay minerals (Almon, 1977). The clays may be present as part of the matrix, as coating on the pore walls, or lying loose in the pores. Diagenetic clays usually occur as a deposit of clay platelets on the pore walls oriented normally to the grain surfaces (see Fig. 11.7A). Clays may also be present as thin layers, or partings, in the

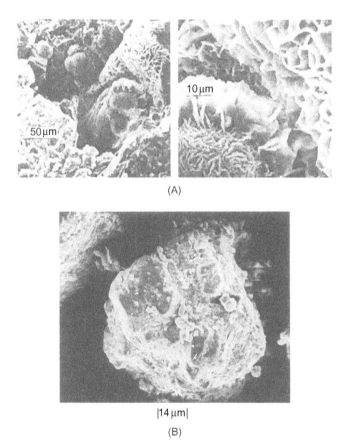

FIGURE 11.7 Examples of clays and other fines coating grain surfaces. (A) Diagenetic clay minerals coating pore surfaces. (*From Almon, W.R., 1977. Sandstone diagenesis is stimulation design factor. Oil Gas J. 56−59. Courtesy of Oil and Gas J.*) (B) Fine particles located on surfaces of larger formation and grains. (*From Muecke, T.W., 1979. Formation fines and factors controlling their movement through porous media. J. Petrol. Technol. 144−150. Copyright 1978 by SPE-AIME.*)

sand beds. Carbonate formations are seldom clay-bearing, and, when clays are present, they are incorporated in the matrix.

The action of aqueous filtrates on indigenous clays can severely reduce the permeability of the rock, but only if the clays are located in the pores. Nowak and Krueger (1951) found that the permeability of a dry core, which contained montmorillonite, was 60 mD to air, but only 20 mD to the interstitial water from the same formation. With other brines, the permeability decreased with decrease in the salinity of the brine; with distilled water, it was only 0.002 mD. Experimental evidence suggested that the reduction in permeability was caused by the swelling and dispersion of the montmorillonite, and the subsequent blocking of the pores was caused by the migrating particles. Formations whose permeability is reduced by aqueous fluids are called *water-sensitive formations*.

Other investigators (Hower, 1974; Dodd et al., 1954; Bardon and Jacquin, 1966) have shown that permeability reduction is greatest when montmorillonite and mixed-layer clays are present. Reduction is less with mite, and least with kaolinite and chlorite. On the basis of petrographic examination, Basan (1985) has classified reservoirs in order of potential impairment according to the nature and location of the clays in the rock pores. Permeability impairment may also be caused by the loose fines of minerals such as micas and quartz (Mungan, 1965). Muecke (1979) reported that fines of unconsolidated Gulf Coast sands consisted of 39% quartz, 32% amorphous materials, and 12% clay. The fines were located on the surface of the larger grains, as shown in Fig. 11.7B.

Mechanism of Impairment by Indigenous Clays

The mechanism whereby aqueous fluids impair water-sensitive rocks has been studied by a number of investigators whose work is discussed below. To simplify interpretation of the results, experiments were made on single-phase systems. Usually, concentrated sodium chloride brine was first flowed through the core or sandpack, followed either by floods of successively lower salinity or by distilled water. The study by Bardon and Jacquin (1966) is particularly illuminating because it separates the reduction in permeability caused by crystalline swelling from that caused by dispersion, and from that caused by deflocculation, and defines the salinity range in which each of these phenomena occur. In some experiments with sandpacks containing montmorillonite, they found that the decrease in permeability with salinity down to 20 g/L of sodium chloride (see Fig. 11.8A) could be correlated quantitatively with the increase in bulk volume of the clay due to crystalline swelling that was reported by Norrish (see the section on clay swelling mechanisms in

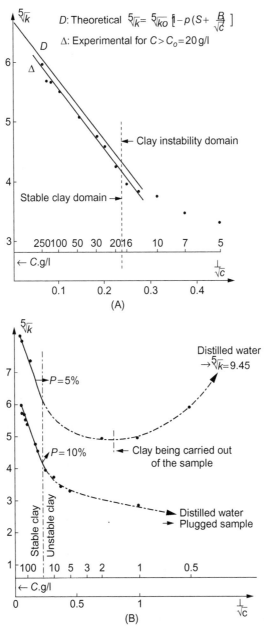

FIGURE 11.8 (A) Decrease in permeability of a sand pack containing 10% montmorillonite with decrease in salinity of sodium chloride brine. (A) Salinities above 20 g/L, impairment caused by crystalline swelling. (B) Salinities less than 20 g/L, impairment caused by dispersion, and by deflocculation at salinities less than 1 g/L. *From Bardon, C., Jacquin, C., 1966. Interpretation and practical application of flow phenomena in clayey media. In: SPE Paper No. 1573, Annual Meeting, Dallas, October 25. Copyright 1966 by SPE-AIME.*

see chapter: Clay Mineralogy and the Colloid Chemistry of Drilling Fluids). This relationship was defined by the equation:

$$5\sqrt{k} = 5\sqrt{k_o}\left[1 - p\left(A + \frac{B}{\sqrt{c}}\right)\right] \quad (11.2)$$

where k is the observed permeability, k_o the permeability when no clay is present in the pack, p is the percent clay, c the concentration of the sodium chloride in grams per liter, and A and B are constants.

At salinities less than 20 g/L the clays became "unstable" (i.e., dispersed). Bardon and Jacquin point out that 20 g/L is almost the identical concentration of sodium chloride at which Norrish observed the sudden expansion in clay lattice spacing, and close to the salinity that marks the change from aggregation to dispersion, as measured by the clay volume and optical density tests (23 g/L, see the section on aggregation and dispersion in see chapter: Clay Mineralogy and the Colloid Chemistry of Drilling Fluids).

Finally, they observed that when distilled water was flowed through the pack, it plugged completely if the pack contained 10% montmorillonite, but if the pack contained only 5% montmorillonite, the permeability increased, and clay was discharged from the end (see Fig. 11.8B). These sharp changes in behavior are understandable when it is remembered that at salinities below about 1 g/L clay particles are deflocculated as well as dispersed, and therefore much more mobile than when flocculated and dispersed (see chapter: Clay Mineralogy and the Colloid Chemistry of Drilling Fluids). Similar tests with natural sandstones containing illite or kaolinite agreed qualitatively with the results obtained with the sand packs and montmorillonite (see Fig. 11.9).

Most natural sands and sandstones contain considerably less clay minerals than did the packs in Bardot and Jacquin's experiments; e.g., sands in Southern California contain from 1% to 2% clay minerals (Somerton and Radke, 1983). Consequently, the decrease in permeability caused by crystalline swelling is much less than that shown in Fig. 11.8A. Crystalline swelling causes the volume of montmorillonite to no more than double (see "Clay Swelling Mechanics," see chapter: Clay Mineralogy and the Colloid Chemistry of Drilling Fluids); if the clay is present only as a thin coating on the pore walls, doubling the volume has little effect on the permeability. On the other hand, the flow paths through natural sand and sandstone formations are more tortuous than those in artificial sandpacks; consequently, dispersed and deflocculated clay particles are usually trapped, with a consequent sharp drop in permeability. Thus the shape of the curves in Fig. 11.9 is typical of those obtained with natural sands and sandstones, even when montmorillonite is present.

Work by other investigators (Gray and Rex, 1966; Muecke, 1979; Jones and Neil, 1960; Holub et al., 1974) leaves no doubt that the decrease in permeability at low salinities is caused by the displacement and dispersion of

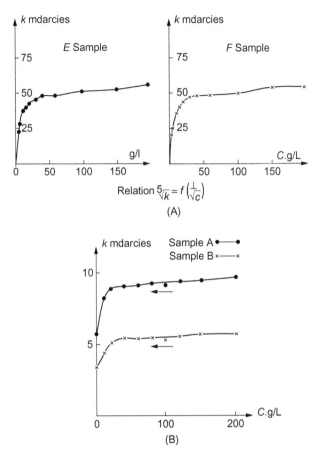

FIGURE 11.9 Decrease in permeability of sandstones with decrease in salinity of sodium chloride brine. (A) Vosges sandstone, containing illite. (B) Hassi-Messaoud sandstone containing kaolinite. *From Bardon, C., Jacquin, C., 1966. Interpretation and practical application of flow phenomena in clayey media. In: SPE Paper No. 1573, Annual Meeting, Dallas, October 25. Copyright 1966 by SPE-AIME.*

the clay or other fines from the pore walls by the invading fluid, and by subsequent trapping at the pore exits (as shown in Fig. 11.10). This mechanism is now commonly referred to as *clay blocking*. The mechanism is analogous to the formation of external mud filter cakes, but in this case multitudinous internal microfilter cakes are formed on the pore exits. The very low permeabilities observed under the deflocculating conditions of distilled water floods are analogous to the low permeabilities of the filter cakes of deflocculated muds compared to those of flocculated muds. The effect of deflocculation can be shown by flooding sand packs containing montmorillonite with sodium chloride brines containing a deflocculant, such as

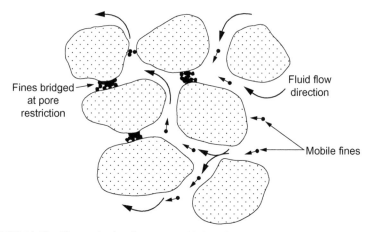

FIGURE 11.10 Clays and other fines move with invading aqueous filtrate and are trapped at pore restrictions. *From Muecke, T.W., 1979. Formation fines and factors controlling their movement through porous media. J. Petrol. Technol. 144–150. Copyright 1978 by SPE-AIME.*

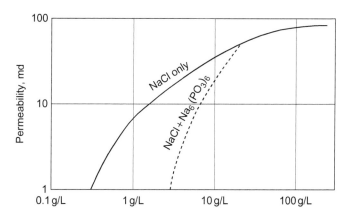

FIGURE 11.11 Salinity of sodium chloride brine with and without sodium hexametaphosphate.

sodium hexametaphosphate. As mentioned in Chapter 4, Clay Mineralogy and the Colloid Chemistry of Drilling Fluids sodium hexametaphosphate raises the flocculation point of montmorillonite in sodium chloride brines to 20 g/L, so that the particles are both deflocculated and dispersed at lower salinities. Consequently, permeabilities are much lower than those obtained when no hexametaphosphate is present (as shown in Fig. 11.11). The point has considerable practical importance because it means that deflocculants such as the complex phosphates and tannates should never be added to dispersed muds when drilling through water-sensitive formations.

Another factor that affects clay blocking is the rate of reduction of salinity. Experiments by Jones (1964) showed that permeability reduction was much

TABLE 11.1 Permeability Reduction of Berea Cores With Continuous Salinity Decrease[a]

	Test 1, $C = C_0 e^{-0.05}$		Test 2, $C = C_0 e^{-0.05}$	
Time[b] (h)	Salinity (ppm)	Permeability (mD)	Salinity (ppm)	Permeability (mD)
0	C_0 = 30,000	190	C_0 = 30,000	190
1	18,200	180	28,500	187
2	11,200	175	27,100	187
4	4050	170	24,500	188
8	550	100	20,100	188
10	200	50	18,200	187
20	1	25	11,000	186
40			4050	183
60			1500	180
80			550	180
100			200	179
150			17	178
210			1	177

[a]From Mungan (1965).
[b]Injection rate = 1 pore volume/h = 120 mL/h.

less if the salinity of the brine floods was reduced gradually in steps, rather than an abrupt change from a concentrated brine to freshwater. Mungan (1965) showed that if the sodium chloride brine was flooded very slowly through a core of Berea sandstone, and the salinity was very gradually and continuously reduced, permeability impairment could be almost completely avoided (see Table 11.1). An abrupt change from 30,000 ppm sodium chloride brine to freshwater reduced the permeability from 190 mD to less than 1 mD after only 1.2 pore volumes had been injected. Evidently the shock of sudden change in the electrochemical environment promotes dispersion of the particles.

Effect of Base Exchange Reactions on Clay Blocking

When more than one species of cation is present in a clay−brine system, the behavior of the clay is determined by the base exchange reactions. It will be remembered from Chapter 4, Clay Mineralogy and the Colloid Chemistry of

Drilling Fluids that the base exchange equilibrium constants favor the adsorption of polyvalent cations over monovalent cations. Further, when polyvalent cations are in the base exchange positions, the clay aggregates do not disperse, even in distilled water. Consequently, when the base exchange ions on the clays in water-sensitive rocks are predominately polyvalent, permeability impairment by freshwater filtrates is avoided. For example, the base exchange ions on the clays in Berea sandstone are predominately calcium, so that when a dry core is flooded with distilled water, no impairment is observed. If, however, the dry core is first flooded with a sodium chloride brine, the clays are converted to the sodium form, and a subsequent flood with distilled water will plug the core. But when the sodium chloride brine contains polyvalent cations in proportion sufficient to keep the clay in the polyvalent form, dispersion is inhibited. Thus, Jones (1964) showed that a ratio of one part of calcium chloride or magnesium chloride to 10 parts sodium chloride prevented clay blocking in Berea sandstone, provided that the salinity was reduced gradually (see Fig. 11.12). The exchange of cations on the clay with those in the filtrate is governed by the law of mass action, and the exchange constant favors the adsorption of polyvalent ions over monovalent ions (see section on ion exchange in see chapter: Clay Mineralogy and the Colloid Chemistry of Drilling Fluids).

Although potassium is monovalent, it is closely held in the clay lattice and thus does not give rise to the high repulsive forces that cause dispersion. Steiger (1982) reports that the polymer-KC1 mud developed for stabilizing shales (see chapter: Wellbore Stability) also greatly reduced formation damage.

The cations in subsurface formations are in equilibrium with those in the interstitial water. The major constituent of formation brines is normally sodium chloride in concentrations above 20 g/L (20,000 ppm), plus significant amounts of calcium and magnesium salts so that the clays are stabilized on the pore walls. The invasion of mud filtrate upsets the existing equilibrium and may or may not cause dispersion and clay blocking, depending on the formulation of the drilling fluid. The large number of variables involved make advisable a laboratory test on cores from the formation of interest, to determine the optimum formulation, as will be discussed later. In general, the filtrate of brine muds with salinities at least as great as those shown in Table 11.2 will cause no impairment other than that caused by crystalline swelling. To control swelling, Griffin et al. (1984) have suggested determining the amounts of the various cations on the formation clay, calculating the activity of the connate water accordingly, then balancing that activity by adding the appropriate amounts of salts of the same cations to a polymer drilling completion fluid. The filtrates of freshwater muds will cause clay blocking, especially if chemical thinning agents such as tannates and complex phosphates are present, but the filtrates of lime and calcium lignosulfonate muds cause no impairment provided the Ca/Na ratio is high enough to repress dispersion. It is probably safe to assume that if this ratio is high

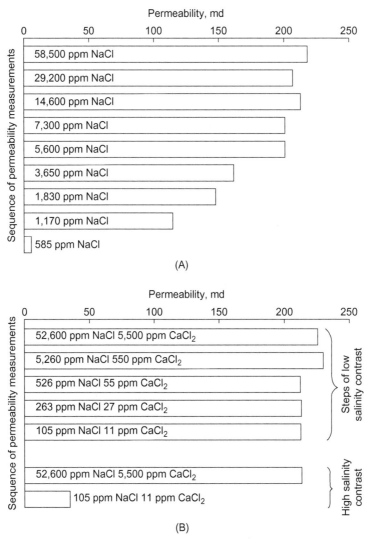

FIGURE 11.12 Decrease in permeability with decrease in salinity. (A) Brine contains sodium chloride only. (B) Brine contains one part calcium chloride to 10 parts sodium chloride. *From Jones, F.O., 1964. Influence of chemical composition of water on clay blocking of permeability. J. Petrol. Technol. 441–446. Trans. AIME 231. Copyright 1964 by SPE-AIME.*

enough to keep the clays in the mud repressed, it will be high enough to keep those in the formation repressed.

It is important to prevent dispersion of formation clays, because once dispersion has occurred, it cannot be reversed by floods with high-salinity brines. Some improvement can be obtained because of the flocculating action of the brines, but laboratory tests have shown that the original permeability

TABLE 11.2 Minimum Salinities of Brines Required to Prevent Clay Blocking in Water-Sensitive Formations

Species of Clay Mineral in the Brine	Sodium Chloride, ppm NaCl	Calcium Chloride, ppm $CaCl_2$	Potassium Chloride, ppm KCl
Montmorillonite	30,000	10,000	10,000
Illite, kaolinite, chlorite	10,000	1000	1000

can only be restored by drying the core, thus condensing the clays back on the pore walls (Slobod, 1969). Mud acid will remove clay blocks, but the effect on permeability is only temporary (Bruist, 1974).

It has been shown that the migration of fines and the consequent permeability impairment can occur even when no cation exchange reactions are involved. This phenomenon was demonstrated by Bergosh and Ennis (1981) by flowing simulated pore water from a reservoir through cores from that reservoir. The migrating fines were believed to have loosened from the matrix by the shearing force of the fluid, or by solution of carbonate cement.

Gruesbeck and Collins (1982) have developed and experimentally confirmed a theory for the entrainment and deposition of mineral fines (such as those shown in Fig. 11.7B) in a porous medium. They showed that the minimum interstitial velocity for entrainment depends on the pore and particle size distributions and on the properties of the reservoir fluids. Sharma and Yortos (1986) developed equations governing the behavior of fines after they have been released. They showed that their subsequent entrapment at pore throats depends on the amount of particles released, the flow rate, and the pore size distribution of the porous medium.

The Effect of pH

Clay dispersion is influenced by pH because it affects the base exchange equilibrium, but its effect on a particular system depends on the electrochemical conditions in that system. The pH of the filtrate may be the cause of impairment by another mechanism if the matrix cement is amorphous silica. Filtrates with a very high pH dissolve the silica, releasing fine particles, which may then block pores (Mungan, 1965).

Clay Blocking in Two-Phase Systems

So far, we have discussed clay blocking in single-phase systems. In practice, we are concerned with two-phase systems in which the factor of interest is

Test no.	Type of mud filtrate	% Interstitial water		% of original oil permeability recovered 0 20 40 60 80 100 120
		Before	After	
6	Clay-water base	34.3	37.6	
7	Fresh water-starch	34.6	45.3	
8	Fresh water-driscose	32.2	36.2	
9	Calcium chloride-starch	32.3	25.7	
10	Lime-starch	28.5	27.4	
11	Lime-tannate	36.2	43.3	
12	Fresh water emulsion	32.0	37.7	
13	Salt water emulsion	28.8	26.6	
14	Oil base	25.2	24.9	

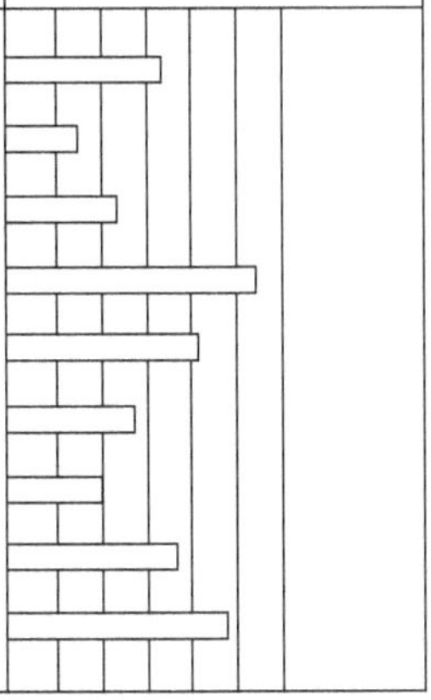

FIGURE 11.13 Effect of field mud filtrates on oil permeability of Paloma field (Stevens sand) cores. *From Nowak, T.J., Krueger, R.F., 1951. The effect of mud filtrates and mud particles upon the permeabilities of cores. API Drill. Prod. Prac. 164−181. Courtesy API.*

the effect of clay blocking on the permeability to oil or gas. In the laboratory this effect is investigated by bringing a core to residual water saturation, determining the permeability to oil (k_1), exposing the core to mud filtrate, and then flowing oil in the opposite direction until constant permeability k: is obtained. The criterion for impairment then is $k_{o2}/k_{o1} \times 100$.

Permeability impairment by this criterion is much less than that observed in single-phase systems. For instance, in Nowak and Krueger's (1951) experiments, 20% of the initial permeability to oil was recovered with a core that had been plugged by distilled water. Considerably higher recoveries were obtained with cores damaged by mud filtrates, especially filtrates from inhibited muds (see Fig. 11.13). Oil must be backflowed for a considerable time before maximum return permeability is obtained. For example, Bertness (1953) found that up to 62 hours were required before the rate stabilizes (see Table 11.3). Such delays may lead to misinterpretation of drill-stem test results.

The wettability of the moveable fines is an important factor in the flow behavior of two-phase systems. The fines are usually water wet, and therefore move only when the water phase is mobile. In a virgin reservoir at residual water saturation, the fines are immobilized in the connate water films around the sand grains, as shown in Fig. 11.14. When an aqueous filtrate invades the reservoir, the water phase becomes mobile, and the fines migrate to form bridges at flow constrictions, as already discussed for single-phase flow. When the well is completed, and oil flows in the reverse direction, the bridges are broken up and the fines carried along with the

TABLE 11.3 Flow Behavior of Stevens Zone Sands, Field A[a]

		Permeabilities, Millidarcys				
Sample Number	To Air (K_a)	To Oil, With Interstitial Water	To Oil, Directly After Freshwater Damage	Total Hours of Oil Flow	Maximum Pressure Gradient, psi/in	Final Oil Permeability (K_o), mD
1	25	6.4	0	44	12	3.8
2	26	5.6	0.4	35	8	5.6
3	36	9.5	0.2	26	8	9.8
4	38	25	1.8	42	12	21
5	82	61	1.3	54	12	58.5
6	88	67	2.8	44	8	69
7	128	99	4.5	51	8	99
8	142	107	6.1	42	8	105
9	234	192	9	56	8	197
10	259	198	11.3	62	6	199
11	299	243	7.4	61	6	241
12	306	203	6.8	61	10	207

[a]From Bertness (1953).

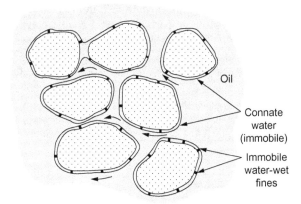

FIGURE 11.14 After oil has replaced aqueous filtrate, remaining fines are immobilized in connate water. *From Muecke, T.W., 1979. Formation fines and factors controlling their movement through porous media. J. Petrol. Technol. 144–150. Copyright 1978 by SPE-AIME.*

oil—water interface. Some of these fines may be carried into the well with the front, some form bridges in the opposite direction, and some remain in the water films left on the grain surfaces. The actual behavior of the fines depends on the properties of the reservoir, and the rate of production. Field studies have shown that productivity decreases at high rates of production (Krueger et al., 1967), apparently because the high rate increases the concentration of fines, which favors bridge formation. In a micromodel study Muecke (1979) observed that once the oil—water interface had passed out of the model, no more fines were discharged. However, if oil and water were flowed simultaneously through the model, the fines continued to migrate indefinitely, because multiphase flow caused local pressure disturbances.

Permeability Impairment by Particles From the Drilling Mud

It is now well established that particles from the mud can invade the formation and cause impairment by blocking constrictions in the flow channels. However, as we showed in Chapter 7, The Filtration Properties of Drilling Fluids mud particles can only penetrate the formation during the mud spurt period, before the filter cake is established. Once the filter cake is fully formed, it filters out the finest colloids because of its structure and very low permeability (around 10^{-3} mD). The permeability may continue to decrease, but the decrease will be caused not by particles passing through the cake, but by transport and rearrangement of particles already carried in by the mud spurt.

It follows that the way to control mud particle damage is to minimize the mud spurt by ensuring that enough bridging particles of the right size are present in the mud. Bridging particles, it will be recalled from Chapter 7, The Filtration Properties of Drilling Fluids fit into and block the surface pores of the rock, thus forming a base on which the filter cake can form. To be effective, the primary bridging particles must be not greater than the size of the pore openings and not less than 1/3 that size, and there must be a range of successively smaller particles down to the size of the largest particles in the colloidal fraction. The greater the amount of bridging particles, and the lower the permeability of the rock, the quicker the particles will bridge, and the smaller will be the mud spurt (Hurst, 1953; Purcell, 1949).

Since the mud spurt occurs when the formation is first exposed by the bit, tests for mud particle penetration should be made under the condition that the external cake is being continuously removed. Even under that condition, mud particles penetrate a remarkably short distance into a rock when adequate bridging particles are present. For example, tests by Glenn and Slusser (1957) showed mud particles penetrated 2−3 cm into alundum cores. Microbit tests by Young and Gray (1967) indicated that mud particles penetrated about 1 cm into Berea sandstone cores having a maximum permeability of 105 mD. Krueger (1973) cites studies indicating particle penetration of 2−5 cm.

Most of the impairment caused by particle invasion is concentrated in the first few millimeters of the rock. For example, Young and Gray (1967) found the permeability of the first centimeter of the Berea sandstone cores to be reduced to about 10^{-2} mD, and the remainder of the core to be essentially undamaged. In more permeable rocks, the permeability of the invaded zone beyond the first centimeter may be reduced to 70−80% of the original permeability.

Impairment that extends only a few centimeters into the formation can be eliminated by gun perforating. Klotz et al. (1974a) have shown that formation damage can thus be eliminated provided that the length of the tunnel—normally about 8″ (20 cm)—exceeds the depth of the damaged zone by at least 50% (see Fig. 11.15). Therefore, we may conclude that damage from mud particles is not a cause for concern, provided that the mud contains adequate bridging particles and that the well is gun perforated (or underreamed) with a nondamaging completion fluid, which will be described later.

On the other hand, deep and irreversible damage will be done if adequate bridging particles are not present. An experiment by Abrams (1977) with a brine that contained no particles large enough to bridge illustrates an extreme case. The brine contained 1% of particles less than 12 μm, and was injected continuously into a 5-darcy sand pack in a radial system. Fig. 11.16 shows the resulting severe impairment plotted as reduction in permeability versus depth of penetration. Backflushing with oil did little to restore permeability, and Abrams calculated that a similar impairment in a well with a drainage radius of 500′ would reduce productivity to 14% of potential, compared to 99% which has been obtained with a fluid containing bridging particles.

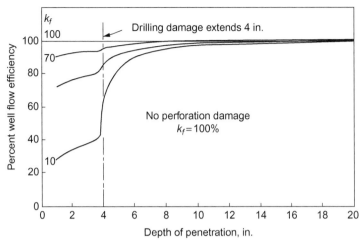

FIGURE 11.15 Effect of depth of gun perforations on well productivity. Note mud damage is virtually eliminated when the depth of the tunnel exceeds the depth of the invaded zone by 50%. k_f is the permeability of the invaded zone as a percent of initial permeability. Percent well flow efficiency is ratio of the productivity of a damaged well to that of an undamaged well expressed as a percent, and calculated for radial flow. *From Klotz, J.A., Krueger, R.F., Pye, D.C., 1974a. Effects of perforation damage on well productivity. J. Petrol. Technol. 1303−1314. Trans. AIME 257. Copyright 1974 by SPEAIME.*

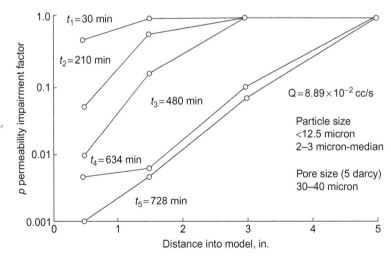

FIGURE 11.16 Productivity impairment by particle invasion when no bridging particles were present. *From Abrams, A., 1977. Mud design to minimize rock impairment due to particle invasion. J. Pet. Technol. 586−592. Copyright 1977 by SPE-AIME.*

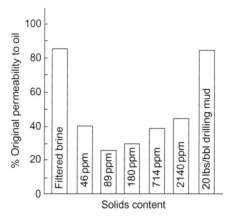

FIGURE 11.17 Permeability impairment in a 3″ Berea core caused by brines containing low concentrations of micron-sized particles, filtered brines, and a KCl-based drilling fluid containing bridging solids.μm *From Krueger, R.F., 1986. An overview of formation damage and well productivity in oilfield operations. J. Petrol. Technol. 131–152. Copyright 1986 by SPE-AIME.*

Krueger (1986) describes an experiment which showed very clearly that to avoid permeability impairment, either a properly filtered brine or a mud containing adequate bridging solids must be used; as shown in Fig. 11.17, dirty brines cause severe impairment.

Occurrence of Mud Particle Damage in the Field

Any mud that has drilled more than a few feet will contain more than 1 lb/bbl (3 kg/m^3) of particles in the size range of 50 to 2 μm, which is all that is required to bridge consolidated rocks of permeability less than about 1 darcy. In such formations, therefore, no special precautions are needed, and the productive interval can usually be drilled with the same mud that was used to drill the upper part of the hole. There are, however, a number of formations and types of operation in which adequate bridging requirements are difficult to estimate, and sizes larger than 50 μm may be required. Under the circumstances discussed below, steps must be taken to assure adequate bridging; when that cannot be assured, it is advisable to use a special completion or workover fluid whose solids degrade or may be dissolved when the operation is completed.

Unconsolidated Sands

Unconsolidated sands often require particles larger than 50 μm to bridge. Because of the wide range of particle and pore sizes and shapes, it is difficult to specify sizes and amounts for bridging, but 5–10 lb/bbl (15–30 kg/m^3) of bridging particles with a maximum size of 150 μm (100-mesh screen) should suffice for all formations except gravel beds and formations with open fractures or channels. Drilling fluids are sometimes deficient in bridging particles larger than 50 μm, and certainly will be if subjected to effective desanding and desilting. Particle size distribution should therefore be watched closely when drilling in unconsolidated sands, and mechanical separators should be adjusted to maintain enough coarse particles for bridging. Coarse grinds of suitable material, such as calcium carbonate, should be added if necessary.

Apart from the productivity damage caused by deep mud particle invasion. lack of sufficient bridging particles causes slumping of unconsolidated sands and hole enlargement. As explained in Chapter 9, Wellbore Stability an unconsolidated sand has a coefficient of cohesion of zero, and will therefore slump into the hole unless a mud cake is formed. The pressure drop across a mud cake increases the cohesion and reduces the compressive stresses around the borehole. It is essential that a mud cake be formed quickly, because the turbulent flow around the bit creates highly erosive conditions, and hole enlargement is rapid. Failure to establish a mud cake quickly will result not only in productivity impairment, but also will lead to sand production, casing buckling, and other production problems associated with enlarged hole (Bruist, 1974). Correlation between particle sizes maintained in the mud and caliper logs will soon establish optimum bridging requirements for a particular reservoir.

Reservoirs With Fracture Permeability

Some reservoirs, notably carbonates, have a very low matrix permeability, and production depends on flow through a network of microfractures. The fractures are mostly less than 10 μm in width, but may be much wider. Because of the uncertainty of fracture size, and because of the geometry involved, bridging fractures is more difficult than bridging porous media. If the fractures are not bridged, fine mud particles invade the fracture and filter internally against the sides of the fracture until it is filled with mud cake. Such internal mud cakes are not removed by backflow, and productivity is greatly impaired. Such reservoirs should therefore be drilled with a fluid whose solids are degradable.

When Gun Perforating

Most wells are drilled, cased, cemented, and then perforated. The damage caused by gun perforating with conventional muds in the hole has been recognized for some time (Allen and Atterbury, 1958). Even if a well is perforated with a completely nondamaging fluid, productivity is decreased by the formation of a zone of crushed rock around the perforation tunnel (see Fig. 11.18). If the productivity of this zone is further reduced by mud solids or filtrate, the productivity of the well will be reduced severely. Based on experimental data, Klotz et a1. (1974a) have calculated loss of well

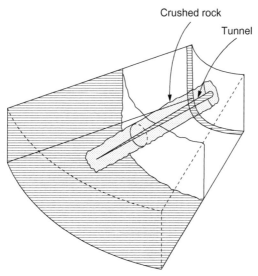

FIGURE 11.18 Schematic representation of a gun perforation, showing zone of crushed rock around tunnel. *From Klotz, J.A., Krueger, R.F., Pye, D.C., 1974a. Effects of perforation damage on well productivity. J. Petrol. Technol. 1303–1314. Trans. AIME 257. Copyright 1974 by SPE-AIME.*

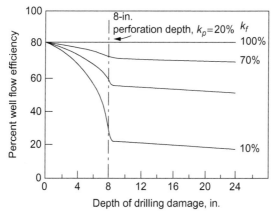

FIGURE 11.19 Effect of damage by mud while drilling on well productivity when perforated with a nondamaging fluid. k_p is permeability of crushed zone around perforation tunnel as a percent of initial permeability. *From Klotz, J.A., Krueger, R.F., Pye, D.C., 1974a. Effects of perforation damage on well productivity. J. Petrol. Technol. 1303–1314. Trans. AIME 257. Copyright 1974 by SPE-AIME.*

productivity resulting both from formation damage by mud solids and filtrate while drilling and from damage to the zone around the tunnel. Fig. 11.19 shows that if a well is perforated with a completely nondamaging fluid and if the permeability of the formation has not been impaired while drilling

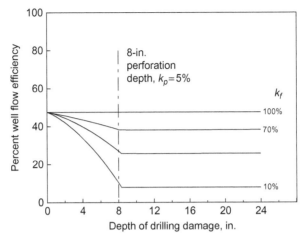

FIGURE 11.20 Effect of mud damage while drilling on well productivity when perforated with a damaging fluid. *From Klotz, J.A., Krueger, R.F., Pye, D.C., 1974a. Effects of perforation damage on well productivity. J. Petrol. Technol. 1303–1314. Trans. AIME 257. Copyright 1974 by SPE-AIME.*

($k_r = 100\%$) permeability of the crushed zone (k_p) is 20% of the original rock permeability, and productivity is 80% of potential. If the permeability of the formation has been impaired during drilling, productivity may be reduced to as low as 20% of potential, depending on the value of k_f and the depth of invasion. Fig. 11.20 shows that if the well is perforated with a damaging fluid in the hole, the permeability of the crushed zone may be reduced to as little as 5% of the original permeability, and the maximum productivity that can then be expected (even if there is no damage during drilling) is 45% of potential. These results show that perforating with a nondamaging fluid in the hole is of the utmost importance. Several authors have shown that underbalanced perforating can minimize gun perforation damage (Bolchover and Walton, 2006; Behrmann, 1996; Walton, 2000).

Workover Wells

Muds used in workover wells differ from drilling muds in that they usually have no opportunity to pick up bridging solids in the hole, so bridging solids must be included in the formulation. The importance of bridging solids was not realized in the past, and workover fluids containing only colloidal materials, such as starch, carboxymethylcellulose (CMC), xanthan gum, and bentonite, were often used in freshwater. These fluids had the required rheological properties, and appeared to have acceptable filtration properties because the tests were made on filter paper. In the hole, however, they penetrated deeply into moderate- and high-permeability formations, causing considerable drops in productivity. If the workover fluid was highly overbalanced to the downhole pressure, seepage losses occurred. Also, circulation was sometimes lost, in which case major losses of productivity occurred. Nowadays, common practice is to add bridging particles to workover fluids, but nevertheless, workover wells are especially liable to formation damage for several reasons. In the first place, the correct size of bridging particles is not known because previous production may have opened up flow channels of unknown size. Furthermore, changes in intergranular stress around the wellbore may have altered the pore structure, particularly if the well has produced sand.

Another problem is that reservoir pressures are usually low in workover wells, sometimes less than hydrostatic, and consequently mud overbalance pressures are apt to be high. A high pressure differential between the mud and the formation increases the mud spurt because of dynamic effects, and also increases the chances of loss of circulation through induced fracturing. Finally, workover wells are liable to impairment arising from damage to previous gun perforations, which are exposed to the workover fluid throughout the whole operation (Klotz et al., 1974a).

Because of these various problems, it is generally advisable to use a degradable fluid in workover operations, such as any of the following:

Calcium carbonate salt fluidsm. Calcium carbonate particles can be dissolved when contacted with acid. Brine solutions are usually used to get the required density without using barite.

Sized salt fluids. These fluids use specially ground sodium chloride crystals, sized to minimize or stop filtrate entrance into the formation. The idea is that the salt crystals in the filter cake will be dissolved by the water produced when flow is restored. Sized salt fluids must be formulated in a saturated sodium chloride fluid. If the fluid becomes undersaturated then the sized salt particles may dissolve, negating their bridging ability.

Mixed metal hydroxide. MMH fluids have been found to be beneficial in milling operations. MMH is a low-solid, flocculated, cationic drilling fluid system which provides excellent hole-cleaning and solids-suspension characteristics. The key product in this system is the cationic inorganic polymagnesium aluminum hydroxyl, the viscosifier that provides the system's unique rheology.

Gravel Pack Operations

Plugging of the gravel by external filter cake left on the face of the formation poses a problem in gravel pack operations. In open-hole gravel packs, the cake will plug the gravel when the well is produced, unless it is readily dispersible in the produced fluids and the maximum particle size of its solids is less than one-third the size of the openings between gravel. Rather than rely on this mechanism, a better practice is to use a mud with degradable solids when underreaming prior to gravel packing. It is obviously desirable that mud not be contaminated with drilled solids when underreaming; therefore, efficient mechanical separation must be provided at the surface, and muds that enhance the removal of the drilled solids while retaining the bridging solids must be used.

Completion of Water Injection Wells

Water injection wells are especially liable to impairment by mud solids because the flow is from the well into the formation when the well is completed. Thus, any solids left in the hole after washing the well will be carried into the formation by the injected water, or will filter out on the face of the borehole. Accordingly, degradable materials are advisable for use in completing or working over a water injection well.

PREVENTION OF FORMATION DAMAGE

The surest way to prevent formation damage by mud solids or filtrate is to operate with an underbalanced mud column, so that no solids or filtrate can invade the formation. Unfortunately, this operation is risky in high-pressure wells. This method requires the use of special equipment and trained crews and may not be economically feasible. In wells with hydrostatic formation pressures, an underbalanced column may be achieved by the use of oil-base fluids, but in drilling wells it is difficult to maintain the necessary low density because of the incorporation of drilled solids into the mud. However, in some types of workover operations, oil-base fluids or crude oil may enable an underbalanced column to be maintained. In very low pressure wells, gas or foam may be used (see the section on foaming in see chapter: The Surface Chemistry of Drilling Fluids).

In most wells, an overbalanced column must be maintained, and prevention of impairment requires the use of a nondamaging fluid. As already mentioned, damage to water-sensitive formations may be prevented by using inhibited muds, or brine fluids. Table 11.2 lists the minimum recommended concentrations for sodium, potassium, and calcium chloride inorganic brines. Note that calcium and potassium chlorides have about the same inhibiting power, but calcium chloride suffers from the disadvantage that it may cause impairment by precipitating carbonates or sulfates, which are often present in formation waters. Recent work has shown a significant advantage for the use of formate brines over the inorganic brines (Fleming et al., 2016).

Therefore, when the brine density required is less than 9.7 lb/gal and calcium ions are judged to be potentially damaging, potassium chloride is generally acceptable as a completion/workover fluid. If the density requirement is greater than 9.7 lb/gal and a calcium-free brine is desirable, then sodium bromide or potassium bromide may be used to obtain densities up to 12.5 lb/gal and potassium formate will give densities up to 13 lb/gal. In some cases, zinc bromide or cesium formate brines have been used to provide high-density calcium-free completion fluids. An alternative to the use of chloride or bromide brines is the use of formate or acetate brines in completion and workover operations. The organic brines have shown a number of advantages as discussed in Chapter 2, Introduction to Completion Fluids.

Completion Workover Procedures

The question of what constitutes a nondamaging fluid depends not only on the properties of the fluid but also on the completion or workover procedure in question. When the well is perforated and/or gravel packed, experience indicates that the wellbore should contain a solids-free or extremely clean fluid (Klotz et al., 1974b; McLeod, 1982; Olivier, 1981). Also, if possible, the perforations should be shot underbalanced. The problem with defining a nondamaging fluid is a trade-off between high seepage loss to the formation, in the case of a solids-free brine, which may disrupt the equilibrium of the rock matrix or carry fine particles into the formation, and plugging of perforations and gravel packs in the case of a solids-laden fluid.

In general, perforations are shot in clear brines. To maintain the clarity of a solids-free brine, filters must be used during the circulation of the brine downhole. The most common approach to brine filtration is the use of a filter press utilizing diatomaceous earth as a filter aid and absolute micron rated filtration cartridges downstream from the press. This arrangement allows any type of brine to be filtered to turbidities of less than 5 NTU. If care is taken to properly clean the casing and tubing, the turbidity of the wellbore effluent can be as low as 5–10 NTU. Note that NTUs do not correlate directly to solids content in ppm, but calibration curves may be prepared according to standard American Petroleum Institute (API) practice (API RP 13J, 2014). Maly (1976) has emphasized the extreme care that must be taken to remove the contaminating solids at the surface, but even if this is done, enough solids may be picked up on the way down the tubing to cause considerable impairment. Tuttle and Barkman (1974) showed that it was necessary to reduce the solids content of Louisiana bay water to less than 2 ppm in order to prevent significant impairment. Such reduction of solids content is possible with currently available filtration equipment as described above.

Given that wholesale loss of brine to the formation should be avoided when possible during completion and workover procedures, what can be done to stop fluid loss? The use of properly sized bridging solids may be the best approach. Remember, any solids in the wellbore during perforating or gravel packing operations may cause considerable damage. Various types of soluble or degradable bridging materials are available commercially, and the choice between them depends on reservoir conditions and type of operation. Sized particles of oil-soluble resins or waxes may be used as bridging agents for oil reservoirs. Any particles left in or on the formation are dissolved when the well is brought into production. Obviously such particles are of no use in dry gas reservoirs or water injection wells. Organic particles have the advantage over mineral bridging agents in that their density is about one-half that of drilled solids. Thus, when drilling or underreaming the productive interval, the drilled solids may be removed at the surface by gravity separation methods without removing the bridging particles.

Ground carbonates (limestone, oyster shells, dolomite) were the first degradable bridging particles to be used in workover fluids, and are still frequently used. On completion of the job, they are removed with acid if necessary. Carbonates are inexpensive, and may be used in any type of reservoir, but suffer from the following disadvantages:

1. Acidization is an extra operation and an additional expense.
2. Acid may dissolve iron on the way down the hole, and iron compounds present in the formation. Then, when the acid is spent, the pH rises, and iron hydroxide is precipitated, causing considerable impairment (Jones, 1964).
3. All of the carbonate particles may not be contacted by the acid, which tends to follow the path of least resistance. To avoid this problem, alternate slugs of acid and diverting agent are necessary.
4. In reservoirs where the matrix cement is calcite, the acid tends to dissolve the calcite, releasing fines.

Regardless of these objections, carbonates are the most suitable degradable particles for use in dry gas reservoirs. Furthermore, the above objections do not apply to carbonate reservoirs that must be acidized in any event.

Long-chain polymers are used in degradable muds to obtain theological properties, and, in some cases, filtration control. Unfortunately, many polymers are, at best, only partially degradable. One sometimes reads in the literature, or in product specifications, of "water-soluble" polymers, which suggests that the polymers are thereby nondamaging. In point of fact, none of these polymers enter into true solution; their particles are in the colloidal size range, and have chain lengths that may exceed $0.1\ \mu m$, which is comparable to the width of a medium sized clay platelet. If the particles penetrate deeply into the formation, they cause considerable impairment. They are difficult to reverse out because they are adsorbed on silica surfaces and on the edges of clay lattices (see chapter: Clay Mineralogy and the Colloid Chemistry of Drilling Fluids). The damage that can be caused by polymers was shown by Tuttle and Barkman (1974), who injected polymer suspensions (containing no bridging agents) into 450 mD sandstone cores. With guar gum, they obtained only 25% of the original permeability after backflow, and 43% when the contaminating polymer was hydroxyethylcellulose (HEC) (see Fig. 11.21A). Of course, in practice, bridging agents would be added to

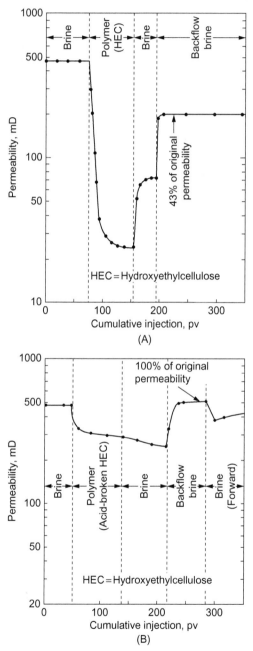

FIGURE 11.21 (A) Effect of unbroken hydroxyethylcellulose solution on permeability of Cypress sandstone. (B) Effect of acid-broken hydroxyethylcellulose solution on permeability of Cypress sandstone. *From Tuttle and Serkman (1974). Copyright 1974 by SPE-AIME.*

the suspension to prevent deep invasion of the polymer, but in the kinds of operations requiring degradable muds, effective bridging is not assured. To guard against the possibility of deep invasion, a degradable polymer should be used. HEC is almost completely soluble in acid, and Tuttle and Barkman obtained 90−100% return permeability after injecting acid into a core contaminated by it (see Fig. 11.21B). Guar gum will degrade over a period of time if an enzyme is incorporated in the formulation, but about 9% residue remains after degradation. This residue is sufficient to cause severe impairment. For example, Tuttle and Barkman obtained only 50% return permeability alter the gum had broken. However, derivatives of guar gum, such as hydroxyethyl and hydroxypropyl guar gum are degradable, leaving only 1−2% residue (Githens and Burnham, 1977). Similarly, although starch itself is not acid soluble, starch derivatives, such as hydroxyalkylated and esterified starches, are acid soluble and can be degraded by enzymes.

SELECTION OF COMPLETION AND WORKOVER FLUIDS

The planning steps for the selection of an appropriate completion and/or a reservoir drill-in fluid involve investigating the following:

- Determine the volume of brine needed for the job, including displacements and possible losses.
- Determine the fluid density required based on the bottomhole pressure and true vertical depth.
- Determine the true crystallization temperature of the selected brine and adjust as necessary.
- Determine any compatibility issues that may occur—corrosion, clay sensitivity, brine/formation incompatibilities, etc.

Deepwater Completion Fluid Selection

Deepwater drilling can impose other criteria that need to be addressed (Jeu et al., 2014). In addition to the above criteria, they recognized the following approach to designing deepwater completion fluids.

- Density required to control formation pressure and the ability to modify density without adverse effect on crystallization and hydrate inhibition.
- Crystallization point at seafloor temperature at the maximum anticipated pressure.
- Hydrate inhibition at seafloor temperature and the maximum anticipated pressure.
- Compatibility with formation, both reservoir rock and shale laminations.
- Compatibility with reservoir fluids—formation water and hydrocarbons.
- Fluid compatibility between completion brine and the following:

- gravel pack or frac pack fluids, stimulation chemicals and acids;
- corrosion inhibitors and packer fluid additives;
- fluid loss control materials and LCM breakers.
- Compatibility with subsea control fluids and elastomeric seals.

The primary differences between land and shelf completions are designing for methane hydrate inhibition and compatibility with control fluids. If using a riser, the abrupt change in temperature while circulating from the bottomhole past the mudline has to be recognized.

Of particular importance is that many brines can cause solid precipitates that can block subsea controls. Jeu found that calcium and zinc chlorides and bromides can plug SCSSV control lines, chemical injection valves or annulus bled-off valves.

Solids-Free Brines

Various brines are used as completion or workover fluids. A summary of these brines is presented in Table 11.4. The density of a brine is adjusted by altering the concentration of the salt or salts in solution. Because these salts are soluble in the water, calculation of brine composition is not straightforward. Hence, brines are prepared with the use of empirical blend charts. Salt blend charts are available from brine manufacturers and suppliers.

Notice that all the brines are made from a few basic materials. These materials include premixed stock brines, such as NaCl, $CaCl_2$, $CaBr_2$, $ZnBr_2$, or Na/K/Cs formate, and the appropriate dry salts as needed to augment the brine solutions. The manner in which these various materials are blended depends largely on the density and crystallization temperature requirements.

TABLE 11.4 Brines Used as Completion or Reservoir Drill-In Fluids

Inorganic Salts	Organic Salts
Monovalent	**Monovalent**
Sodium chloride, NaCl	Sodium formate, $NaCHO_2$
Sodium bromide, NaBr	Potassium format, $KCHO_2$
Potassium chloride, KCl	Cesium formate, $CeCHO_2$
Potassium bromide, KBr	Sodium acetate, NaC_2HO_2
Divalent	
Calcium chloride, $CaCl_2$	Potassium acetate, KC_2HO_2
Calcium bromide, $CaBr_2$	Cesium acetate, CeC_2HO_2
Zinc bromide, $ZnBr_2$	

Crystallization temperature is matched to environmental conditions. Density and crystallization temperature are determined according to standard API procedures (API RP 13J, 2014). Because of the potential for formation damage caused by loss of fluid to the formation (Milhone, 1983), brines used in completing or working over highly permeable zones may require added bridging solids to control loss of fluid. Brines may also be difficult to use when the producing horizon contains unconsolidated sands, because they do not prevent slumping and washouts (Ellis et al., 1981). Slumping can only be prevented by the deposition of a filter cake on the walls of the hole so that the pressure over balance is applied on the face of the formation. Clear brines may require increased overall viscosity or the use of viscous pills when cuttings, millings, etc., must be cleaned out of the hole. Because they have low effective annular viscosities, their carrying capacity is low.

Viscous Brines

Newtonian viscous brines have been used in order to avoid the various disadvantages of acidizing filter cakes as discussed above. To limit invasion into the formation, the brines are thickened to high viscosities (up to several hundred cp) with HEC. These viscous brines contain no bridging particles, therefore no external filter cake is formed and no positive shutoff obtained. However, the rate of invasion is reduced because of the high viscosity. According to Scheuerrman (1983), in some cases a minimum of 4.2 lb/gal (12 kg/m^3) HEC may be required to obtain the necessary viscosity (see Fig. 11.22).

In practice, a pill of the viscous brine is spotted across the loss zone or above the perforations. The volume of the pill should be sufficient to penetrate into the zone at least 3′. A viscosity breaker (cellulose enzyme or oxidizing agent) can be added to the pill to enable backflow out of the formation at the end of the job. However, most commonly used HEC breakers act in a matter of a few hours up to a day or so, depending on the downhole temperature. Thus, they are useful only for short-term jobs. For longer jobs, reliance has traditionally been placed on long-term thermal degradation of the HEC. However, based on the data from Fig. 11.23, breakdown of the HEC within practical time limits in most formations is unlikely. The exceptions would be applications at very high temperatures ($>275°F$) or in low-density brines (<11.6 lb/gal). Remember that even low-viscosity HEC fluids can cause permeability impairment if the polymer remains stable (Fig. 11.21A), but not after acid degradation (Fig. 11.21B).

An approach to the use of viscous pills, which have been used successfully for seepage loss control, is the use of low-density brine as the medium for the pill. The low-density viscous brine is actually squeezed into the formation instead of spotted in the wellbore. This allows it to be used in the

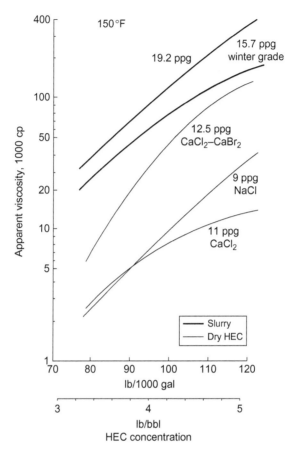

FIGURE 11.22 Effect of poymer concentration on apparent viscosity. *From Scheuerrnan, R.F., 1983. Guidelines for using HEC polymers for viscosifying solids-free completion and workover brines. J. Petrol. Technol. 306–314. Copyright 1983 by SPE-AIME.*

presence of higher density brines without pressure control problems. Also, a cross-linkable HEC polymer has been used to control seepage loss in extremely permeable zones.

HEC is very slow to develop viscosity in brines of density greater than 12 lb/gal (1.4 SG). Even when heated, up to 5 hours mixing time may be required in order to reach maximum viscosity, especially in Na and Ca brines.

HEC is used in these heavy brines in the field by first prehydrating the polymer in an inert solvent, such as isopropyl alcohol, then mixing it into the brine. This process does essentially the same thing as heat. In certain brines containing zinc, HEC will not yield even when heated or prehydrated. For these situations the zinc concentration must be increased to a sufficiently

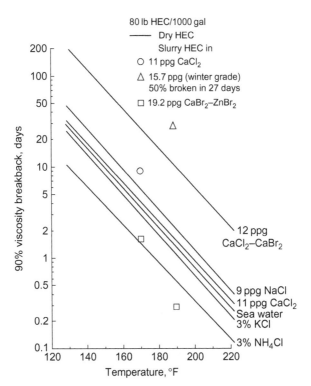

FIGURE 11.23 Viscosity breakback times. *From Scheuerman, R.F., 1983. Guidelines for using HEC polymers for viscosifying solids-free completion and workover brines. J. Petrol. Technol. 306–314. Copyright 1983 by SPE-AIME.*

high level, relative to the other salts, so that the solution will solvate the prehydrated HEC. (Note: The minimum zinc concentration is approximately 7.5–9.0% by weight.)

Water-Base Fluids Containing Oil-Soluble Organic Particles

Several types of fluids use oil-soluble organic particles, such as waxes and resins, to act as bridging agents. In some of these fluids, the particles are deformable at low enough temperatures to act as filtration control agents as well as bridging agents. These systems operate best at temperatures between 150°F and 200°F (65–95°C). The particles become too rigid at temperatures below 150°F, and too soft at temperatures above 200°F. In the system described by Fischer et al. (1974, 1975; Matthews and Russell, 1967; Dodd et al., 1954; Ribe, 1960), the organic particles consist of a blend of wax, surfactants, and an ethylene-vinyl copolymer. Filter losses down to 24 cm API can be obtained with these particles and losses down to 7 cm^3 can be obtained by the addition of a lignite. HEC and xanthan gum are used for rheological control, if required. Densities up to 10 lb/gal (1.20 SG) can be achieved with potassium chloride and to 13 lb/gal with potassium formate.

The thermoplastic resin particles in the system described by Crowe and Cryar (1976) are sufficiently deformable to provide filter losses down to 7 cm without the addition of a supplementary agent, but control is lost on rocks of permeability greater than about 900 mD. This system has the advantage of being stable in all brines up to saturation.

Suman (1976) describes a system in which the thermoplastic resins—which have a much higher softening point (360°F (182°C))—provide bridging requirements only, and do not contribute to filtration control. Starch derivatives, or other polymers, provide filtration control, and HEC is used to obtain carrying capacity when necessary. This system is also stable in all brines up to saturation at temperatures up to at least 300°F (149°C).

Acid-Soluble and Biodegradable Systems

Ground calcium carbonate is commonly used as a bridging agent in acid-soluble and biodegradable systems. It is completely soluble in acid, and is available in a wide range of particle sizes, from several millimeters down to hundredths of a micron, and may be used at any temperature encountered in an oil well. Tuttle and Barkman (1974) found that, if suitable size ranges were selected, suspensions of calcium carbonate alone could be used for short-term remedial work in gun perforated wells. However, for most purposes, it is necessary to add polymers for filtration control and carrying capacity. Polymers that are commonly used include CMC and polyacrylonitrile, which are not acid soluble; xanthan gum (50% acid soluble) and guar gum, which may be degraded with enzymes as previously noted; and starch derivatives, enzyme degradable, and HEC, which are almost completely acid soluble. Note that magnesium oxide must be added to HEC to provide high-temperature stability (Jackson, 1976). Both guar gum and HEC have low annular and low-shear rate viscosities and are nonthixotropic, which is advantageous because of increased efficiency in separating gas and extraneous solids. For applications that require high carrying capacity and suspending properties, xanthan gum, diutan, or scleroglucan are better choices.

Saturated Na/K chloride or formate brines may be weighted up to 15.0 lb/gal (1.80 SG) using finely ground $CaCO_3$. The particle size range of the $CaCO_3$ should be from 1 to 40 μm with an average of 6–10 μm to optimize suspension of the solids and filtration control. Also, $FeCO_3$ can be used to obtain densities beyond 15.0 lb/gal (1.80 SG). However, during acid cleanup, remember the possibility of ferrous hydroxide precipitation if the pH exceeds 7. $FeCO_3$ should be used in a calcium-free brine, because in the presence of calcium ions, the $FeCO_3$ converts to a hydroxide−oxide of iron similar to the mineral Goethite or limonite. Also, the particle size must be very fine to prevent errosion of metal goods such as pump liners and valves.

Fluids With Water-Soluble Solids

A completion and workover fluid has recently been introduced that uses sized grains of sodium chloride as bridging and weighting agents (Mondshine, 1977). The grains are suspended in saturated brine by a polymer and a dispersant (both unspecified). Densities up to 14 lb/gal (1.68 SG) are attainable. When the well is brought into production, the salt grains are removed by formation water, or the well may be cleaned by flushing with undersaturated brines. This fluid would obviously be especially suitable for water injection wells.

An Oil-in-Water Emulsion for Gun Perforating

Earlier, the importance of using a nondamaging fluid when gun perforating overbalanced was emphasized. Priest and Morgan (1957) and Priest and Allen (1958) developed a solids-free emulsion specifically for this purpose. Typically, it consists of 40% oil emulsified in sodium chloride or calcium chloride brine.

The oil phase is either kerosene or carbon tetrachloride, or mixtures of the same, depending on the density required. The maximum density is 12.5 lb/gal (1.50 SG). The emulsion is stable enough to provide filtration control for 24 hours. To minimize costs, only a slug of emulsion is pumped into the well, and spotted opposite the interval to be perforated, the density having previously been adjusted so that it maintains this location. Results from the field showed that the emulsion caused no impairment either when perforating or during workover jobs with exposed perforations

Nonaqueous Fluids

Under most conditions, conventional oil-base muds make excellent fluids for drilling through the productive interval. Indeed, they were first developed for this purpose. Their low mud spurt minimizes particle invasion, and their filtrate, being oil, does not cause waterblock or impair water-sensitive formations. Laboratory and field tests have shown that oil muds cause less impairment in water-sensitive formations than do conventional water-base muds (Nowak and Krueger, 1951; Priest and Allen, 1958; Kersten, 1946; Miller, 1951; Stuart, 1946; Trimble and Nelson, 1960). Limitations on oil muds are that they may cause changes in wettability, and that they are unsuitable for use in dry gas reservoirs.

However, conventional oil-base muds are not readily degradable, and therefore should not be used under the conditions already discussed, which make the use of a degradable mud advisable. Oil muds are designed for maximum stability while drilling; any water that they contain or pick up while drilling is tightly emulsified by powerful surfactants. Thus, there is a danger of emulsion blocking should bridging fail and the whole mud penetrate deeply into the formation. One would expect the low-viscosity oil muds to be less likely to cause emulsion blocking, since they contain less surfactants and have higher oil—water ratios.

In reservoirs with aromatic crude, damage from asphalt in asphaltic oil muds will automatically be removed when the well comes on production, because asphalt is soluble in aromatic oil (Methven and Kemick, 1969). Otherwise, it may be removed by washing with aromatic solvents. Asphalt muds should not, however, be used in condensate reservoirs, because asphalt is precipitated by light hydrocarbons, such as hexane.

Some success in reversing damage caused by all types of oil muds has been reported by Goode et al. (1984). They inject, and subsequently backflow, an aqueous-based fluid containing 2% KCl, diesel oil, a mutual solvent, and a blend of unspecified surfactants.

Lease crude is used in many workover operations. It has the advantages of being cheap and readily available. However, lease crude contains many particulate impurities that may cause impairment (Maly, 1976). Crude oil should not, therefore, be used under conditions such that it is liable to invade the formation in significant quantities.

A degradable water-in-oil emulsion is available for use under conditions that require an oil-base fluid with filtration and rheological properties (Darley, 1972). The emulsion droplets are stabilized by a skin of finely divided chalk particles instead of by organic sufactants. When contacted by acid, the chalk particles dissolve and the emulsion breaks to oil and water, leaving no residue. The composition is particularly suitable for use in workover wells because it is available in sacked form, which may be mixed with lease crude and water or brine to form the finished emulsion.

TESTS FOR POTENTIAL FORMATION DAMAGE BY COMPLETION FLUIDS

The complexities of formation damage make it difficult to formulate a nondamaging completion fluid unless extensive laboratory testing is carried out. Such testing necessarily involves considerable expense, both in cutting a core and in the laboratory time involved. However, the costs are miniscule compared with the money saved if the productivity of a newly discovered field is improved by only a small percentage.

One problem—which cannot be completely avoided—is that the core will be altered by contamination by mud particles and filtrate. In some microbit coring tests, Jenks et al. (1968) showed that the filtrate may expel over 50% of the in-place oil. They found that contamination may be considerably reduced by maintaining mud overbalance pressures no greater than 200 psi (14 kg/cm) and by using muds with low spurt and filtrate losses. Such filtration properties are obtained with oil muds, which have the additional

Completion, Workover, Packer, and Reservoir Drilling Fluids Chapter | 11

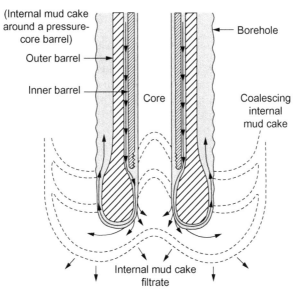

FIGURE 11.24 Penetration of mud particles under pressure core barrel. *From Webb, M.G., Haskin, C.A., 1978. Pressure coring used in Midway-Sunset's unconsolidated cores. Oil Gas J. 52−55. Courtesy Oil and Gas J.*

advantage that their filtrate does not affect water-sensitive clays. Low spurt and fluid losses can also be obtained in water-based muds by judiciously selecting the proper solids distribution and fluid loss polymers. Webb and Haskin (1978) observed bands of particles from mud and the formation in cores cut by rubber sleeve and pressure core barrels (see Fig. 11.24), but not in cores cut by conventional core barrels.

All cores should be wrapped in plastic sheeting and sealed immediately after recovery. If cores are allowed to dry, indigenous clays become coated with residual oil and the properties of the core are irreversibly altered (Keelan and Koepf, 1977). Laboratory test procedures depend on local conditions and individual preference. A few suggestions that may be of use follow.

Cut the test plugs along the diameter of the core. Then, cut off both ends of the plugs at the point where mud contamination is obvious. This procedure will minimize the effect of contamination by the coring fluid, especially if a wide diameter core has been cut.

The following preliminary tests will be useful in planning the testing program:

1. Extract the core with an aromatic solvent, dry, and determine both the permeability to air and the porosity.
2. Make an X-ray analysis to indentify clay minerals, or, at least, make a methylene blue test (see chapter: Clay Mineralogy and the Colloid Chemistry of Drilling Fluids) to evaluate clay mineral activity.

3. On cores from very low permeability reservoirs, make a mercury injection test to determine capillary pressures.
4. Analyze a sample of formation water for soluble salts.

Tests for evaluating prospective completion fluids with respect to formation damage should be made on fresh plugs with the interstitial fluids in place. Drying and extracting alters the wettability of the pore surfaces (Amott, 1958; Keelan and Koepf, 1977). The usual test procedure is as follows:

1. Flow natural or synthetic formation brine through the plugs until constant permeability is obtained.
2. Backflow oil (nitrogen in the case of a gas reservoir) until constant permeability is obtained (k_{o1}).
3. Expose to the test fluid under 500 psi (35 kg/cm^2) differential pressure (assuming the fluid has filter-loss control) until at least one pore volume of mud filtrate has passed through the core. With a 3″ (7.6 cm) plug, an exposure of a day or so may thus be necessary, unless a dynamic filtration cell is used.
4. Backflow oil to constant permeability (k_{o2}).

The criterion for formation damage is then $- k_{o2}/k_{o1} \times 100$.

The following tests will be found useful in diagnosing the cause of any impairment that may have been observed, and also in devising remedial measures:

1. To check for the effect of filtrate on indigenous clays without interference by particles from the mud, extract a large volume of filtrate in a multiple filter press, and repeat the above sequence of tests, but in step 3 of the flow sequence, flood with the filtrate. Adjust the pressure drop to give an appropriate flow rate, and continue flow until constant permeability is obtained.
2. Check the water saturation after the initial and final oil flood. If there is a large difference, make imbibition tests (Amott, 1958) to determine if there has been a change in wettability.
3. To check for impairment by particles from the mud, extract and dry a plug, then fire at 600°C for at least 6 hours to inactive indigenous clays. Repeat the usual sequence of floods, using the test mud in step 3 of the flow sequence. Any permeability loss then reflects the damage caused by mud particles only. To check the depth of invasion, cut off successive slices of the plug, starting with 0.25 cm, and proceed in 1 cm steps, checking the permeability of the remainder of the plug until it becomes constant. This test should be used in determining the optimum size and amount of bridging particles.
4. Check for mutual precipitation by mixing mud filtrate and formation water.

PACKER FLUIDS AND CASING PACKS

Functions and Requirements

When a well is being completed, it is good practice to set a packer between the tubing and the casing above the productive interval, and to fill the annulus with a *packer fluid*. This procedure is simply a safety measure; when it is not followed, the casing head is subjected to the full reservoir pressure in the case of a dry gas well, and to the reservoir pressure less the column of liquid in the annulus in the case of an oil well. The packer fluid also reduces the pressure differential between the inside of the tubing and the annulus, and between the outside of the casing and the annulus. The density of the packer fluid may or may not be great enough for the column of liquid to balance the tubing pressure at the bottom of the tubing, but even when it does so, there is a pressure differential at shallower depths which increases with decrease in depth.

Since a packer fluid remains in place until it is necessary to do remedial work on the well, which may not be for years, it has certain special requirements:

1. It must be mechanically stable, so that solids do not settle on the packer.
2. It must be chemically stable under bottomhole temperatures and pressures, so that high gel strengths, which would prevent the mud from being circulated, do not develop.
3. It should contain materials that would seal any leaks that might develop.
4. It must not itself cause appreciable corrosion, and it must protect the metal surfaces from corrosion by formation fluids that might leak into the annulus.

Since the perforations will be exposed to the packer fluid in the course of completion and remedial operations, the packer fluid must not damage the formation.

Casing packs are fluids left above the cement in the annulus between the borehole walls and the casing, primarily to protect the casing against corrosion by formation fluids. They may also help control formation pressures, and increase the chances of recovering casing, should the need to do so develop. They must have all the properties of packer fluids, and, in addition, they must have good bridging and filtration properties so as to prevent loss of the fluid pack or its filtrate to permeable formations. A summary of the properties of the various types of packer fluids and casing packs is given below. For detailed recommendations, the reader is referred to an informative paper by Chauvin (1976).

Aqueous Packer Fluids

Aqueous Drilling Muds as Packer Fluids

Water-based drilling muds that have been used to drill the well are often left in the hole as packer fluids. Advantages are convenience and economy.

However, they suffer from the great disadvantage that they are inherently corrosive. Tubing and casing in a producing well are subject to the same reactions that corrode the drill pipe (see chapter: Drilling Problems Related to Drilling Fluids for details), but, whereas in a drilling well, remedial treatments can be applied as often as necessary, treating agents added to a packer fluid must last indefinitely, and there is no assurance that they will do so. Therefore, leaving a water-base drilling mud in the hole as a packer fluid may result in development of casing or tubing leaks in the course of time, and the practice is inadvisable except in wells where corrosive conditions are mild. Also, the mud will solidify over time with temperature, and workover costs can be extremely high.

Low-Solid Packer Fluids

Low-solid packer fluids usually consist of a polymer viscosifier, a corrosion inhibitor, and soluble salts for weight control. Bridging particles, filtration control agents, and sealing materials (such as micronized cellulose) are included if needed. These simple systems are easier to control than high-solid drilling muds. There are no problems with high-temperature degradation of lignosulfonates or clay minerals, and corrosion can be inhibited by oil-wetting agents since the loss of inhibitor is greatly reduced by the low-solid content. One unfavorable characteristic is that the common polymers used in casing packs are pseudoplastic and are not thixotropic. Therefore, particulate solids will slowly settle, but there are so few solids to settle—and no barite—that sedimentation seldom creates difficulties. Another problem is that polymers are to various degrees unstable at elevated temperatures. Long-term stability tests under anticipated bottomhole temperatures should therefore be made on polymer fluids before they are put in the well. Mayell and Stein (1973) describe a low-solid packer fluid consisting of attapulgite clay in saturated sodium chloride, with sodium chromate for inhibition of corrosion, and sodium carbonate to raise pH to 10.5. Field tests showed this fluid to have excellent high-temperature stability.

Clear Brines

Clear brines may be used as packer fluids, but not as casing packs because they lack filtration control. If they are rigorously filtered and do not become contaminated with drilling mud left on the sides of the tubing or casing, they are virtually solids free, with all the attendant advantages. Various brines are used as packer fluids, ranging from seawater up to zinc bromide and cesium formate. The decision as to what type and density brine to use is based on several factors, including pressure control, corrosion properties, availability, and cost.

Corrosion is a major concern in designing a packer fluid. In general, brines exhibit low corrosion rates. Except for zinc bromide brines with densities above 18.0 lb/gal, the static equilibrium corrosion rates of various metals in most chloride brines will be less than 10 mils per year (mpy) and formate brines are even lower (Cabot, 2016). Addition of a corrosion inhibitor can yield lower corrosion rates. Inhibitors are generally selected based on the application temperature. Organic chemicals such as amines are used at lower temperature (i.e., below 300°F) and inorganics such as thiocyanates are used for higher temperatures.

Brine Properties

The downhole density of brines, as with any fluid, increases with pressure and decreases with temperature (Schmidt et al., 1983; Adams, 1981; Poole, 1984; Hubbard, 1984). The change can be predicted by means of a model developed by Thomas et al. (1984). Since the effect of pressure is comparatively small, an approximate value of downhole density may be obtained from changes in temperature (Schmidt et al., 1983). However, particularly for critical work and in extremely high density, expensive brines, it is advisable to search the literature methods and include the pressure term in the density correction calculation.

Crystallization in surface facilities is sometimes a problem in handling brines. At densities approaching saturation, salts will crystallize out if the temperature falls below a certain critical value, which depends on the composition of the brine (Milhone, 1983; Adams, 1981; Poole, 1984; Hubbard, 1984). For example, a 14.8-lb/gal (1.77 SG) $CaCl_2/CaBr_2$ brine will crystallize if the temperature of the brine falls below 63°F (17.2°C). Mild crystallization results in deposition of crystals in the surface tanks and lines, with lighter weight brine going downhole. Severe crystallization may cause the entire brine volume to turn to slush or solidify completely.

Oil-Base Packer Fluids and Casing Packs

As discussed in Chapter 10, Drilling Problems Related to Drilling Fluids oil muds are noncorrosive and more thermally stable than water muds. These characteristics make them especially suitable for use as packer fluids, and will usually outweigh their disadvantages of high cost and pollution potential. In deep, hot holes, sedimentation of barite and other solids may become a problem (Hamby and Tuttle, 1975), but can he avoided by the addition of an oil-dispersible bentonite or other organophyllic gelling material (Simpson, 1968).

Oil-base fluids, as well as formates, can be used whenever temperatures are too high for salt brine/water muds, and whenever corrosion is expected to be severe, e.g., when the formation contains hydrogen sulfide. Oil muds make ideal casing packs because of their resistance to corrosion, excellent filtration properties, and also because they facilitate the recovery of easing, should the need to do so arise (Jackson, 1964). They are used in the Arctic to replace water-base muds from the casing annuli in the permafrost zone (Ng, 1975; Remont and Nevins, 1976; Goodman, 1978).

RESERVOIR DRILLING FLUIDS

A RDF, also called a drill-in fluid, is a combination drilling fluid and completion fluid, specially formulated to maximize the productivity index of a producing well. It must maintain the desired properties of a good drilling fluids, i.e., density, viscosity, and fluid loss control, as well as having good completion fluid properties, i.e., nondamaging to the wellbore and the internal pore structure of the producing zone. These special formulations are not usually obtained as off-the-shelf systems, but must be specially formulated to match the specific characteristics of the producing zone. Most of the wellbore is drilled with a regular drilling fluid, either aqueous or nonaqueous, which is then displaced with the special RDF.

As with drilling and regular completion fluids, these special RDFs can take many forms and a variety of products have been developed by the fluids service companies to fit special needs. The basic ingredients for drill-in fluids are

- Nondamaging base fluid compatible with the in situ formation waters; either a brine or a synthetic oil.
- Optimum viscosity and fluid loss control; degradable/breakable polymers.
- Proper particle size distribution of all solids in the fluid; usually carbonates or sized salts for bridging and filter cake control.
- Removable filter cake.
- Fluid weight control by nondamaging products; heavy brines, sized salt, or carbonates or micronized weight material.

In addition to the above criteria, any other chemicals added to the fluid must not cause unwanted wettability or emulsion problems.

Designing the optimum drill-in fluid can involve extensive laboratory testing to select the proper fluid system and additives. These fluids are also usually more expensive than regular muds.

Why a Special Reservoir Drilling Fluid?

Since there are so many fluid choices available and so many cost options, many wonder why we would add more confusion and especially why we would increase the expense of drilling a well. After all, drilling engineers are judged on their ability to produce a wellbore on time and under budget. In the past, operators usually had separate drilling and completion departments staffed with engineers who specialized in those disciplines. They usually had minimal contact with one another and almost no input into each other's design and implementation processes.

In recent years, however, more companies have developed interdisciplinary teams to develop projects based on optimizing the production capability of a given field. This has been brought about, in part, by the downsizing that has swept the industry, as well as the realization that optimum return on investment does require up-front planning and complete cooperation among all parties involved in drilling, completing, and producing a well. The term "well construction" has been introduced to describe this process. The well construction process must also include all the relevant service companies as well as the various in-house engineering, geological, chemical, business, and management staffs.

Cost Versus Value

The driving force behind using special drill-in fluids is the realization that there are two phases involved in drilling an oil and gas well—cost and value phases.

- Drilling above the producing zone = cost.
- Drilling in the producing zone = value.

Usually the majority of the length of a wellbore is an expense item that affects the cash flow and bottom line of a company. It makes sense from an accounting and good engineering economics standpoint to minimize the cost of drilling the well. From a production standpoint, however, cost is not as important as the cost-effectiveness of maximizing or optimizing production. The payout from the well over a period of time determines to a large extent the company's return on investment.

Therefore, in drilling situations in which large sections of the drilled formation is the producing zone, i.e., horizontal wells, it makes more economic sense to focus on cost-effectiveness than on overall cost. Special drilling fluids, usually more expensive, are used to optimize production. The primary means by which this is accomplished is by reducing formation damage in the producing interval.

Reservoir Drilling Fluid Properties

An RDF must have both drilling fluid and completion fluid properties. The properties of fluids used in petroleum operations can be classified into four broad categories: density, viscosity, fluid loss, and reactivity (see chapter: Introduction to Drilling Fluids). For drill-in fluids, the following are the common methods currently used to achieve good fluid properties.

Density

- No solids fluids—clear brines and pure oils.
- Balanced drilling—soluble solids (such as carbonates or sized salts) or insoluble solids (such as barite).
- Underbalanced drilling—pure pneumatic (air, natural gas, nitrogen) and pneumatic with additives (mist, foam, stiff foam).

Viscosity

The thickness or thinness of the fluid is used to optimize hole cleaning and suspension and to minimize fluid invasion. Drill-in fluids are always formulated to enhance low shear rate viscosities with polymer additives, such as xanthan gum, and/or a modified starch, such as carboxymethylstarch (CM-starch). Special processing of the polymers may be necessary to ensure that no fine solids from the manufacturing process are left in the finished product.

Fluid Loss

Fluid loss control is maintained by viscosity and an ultra low permeability filter cake. In addition, the spurt loss must be near zero to minimize base fluid invasion into the reservoir. The cake must be thin, tough, and easily removable. Filtrate viscosity plays an important role in minimizing the depth of invasion of the filtrate. Special situations, such as voids, vugs, and fractures, can require the use of nondamaging lost circulation control materials, such as micronized cellulose, low-viscosity CMC, and CM-starch.

Reactivity

Reactive properties to control include chemical incompatibilities, clay swelling, particle migration, emulsions, and lubricity. Filtrate chemistry must also be adjusted to ensure compatibility with formation clays and connate waters, and to not cause wettability or emulsion problems.

Fluid Types

The most common fluid types currently being used as drill-in fluids are:

- Pneumatic: stiff foam and nitrogen, either with or without additives.
- Water-base systems: sized salt, sized carbonates, clear brines, MMH.
- Nonaqueous: synthetics, either inverts or 100% oil-base fluid.

Fluid Design

Certain information is mandatory to optimize a well's value. Needed essential information includes:

- Downhole properties: formation analysis, pore fluid analysis, logging information, mud type used, and filter cake properties. Also, any drilling

operations synopsis is helpful, especially descriptions of nonproductive times, such as lost circulation, wellbore instabilities, and pressure control problems.
- Laboratory tests: complete core analysis, and return permeability tests.

The more accurate and complete the information available, the more value is added to the well! Three distinct steps are needed to properly select the best drill-in fluid for the well to be drilled. These steps are:

Step 1: Rock and fluid characterization and identification of damage potential.
Step 2: Laboratory testing of potential drill-in fluid systems, compatibility and return permeabilities, stimulation options.
Step 3: Fluid design, modification, and exploiting stimulation options.

The following describes the essential information needed and the laboratory testing to be done.

Downhole Information: Gather as Much Information as Possible from These Sources

- Wireline and mud logs.
- Offset data.
- Cores: recovered, preserved, analyzed.
- In situ fluids: accurate at downhole conditions.
- Reservoir production potential.

Geological Characterization and Mercury Injection

- Perform a geological analysis on core fragments or sister plugs: SEM, XRD, and thin sections.
- CT scan the plugs, solvent clean, and determine gas properties, then saturate with brine and determine porosity and permeability.
- Determine the pore size by mercury injection using the most representative core material after permeability and porosity tests on all plugs.
- Determine capillary pressures (gas/liquid or oil/brine) to determine phase trapping potential.
- Restore the core to reservoir conditions: wettability, fluid saturations, and stress conditions.

Tests at Reservoir Conditions—Core Initialization and Mud Damage Assessment Using Dead Oil

1. Mount the core into a multiport, HTHP core fluid cell.
2. Determine the permeability at reservoir conditions.
3. Flood or centrifuge with pure hydrocarbon to reduce to Swi and obtain keo.
4. Perform stock tank oil flood in both directions.

5. Age 7–41 days.
6. Determine keo in the production direction, degree of polymerization (DP) versus applied pressure.
7. Circulate the mud with a net pressure equivalent to actual conditions.
8. Repeat keo determination, DP versus applied pressure (or flow rate) curves.
9. Compare the two DP versus applied pressure curves.
10. Remove the core and examine the plug using CT scans and thin sections.

Laboratory Tests

There are no recognized standard return permeability or formation damage tests. To get accurate data you must simulate downhole conditions as closely as possible.

- Return the core to downhole conditions for return permeability testing.
- Test fluid incompatibility at downhole conditions; predictions and experience.
- Perform surface plugging at downhole conditions.
- Test return fluid flow at downhole conditions.

Stimulation Options

The results from all the tested mud systems are compared. If a mud system has caused considerable damage, a stimulation option is implemented. The stimulation option to be used is determined by discussion among all interested stake holders, considering operator and service companies, based on mud type, filter cake characteristics, and the likely damage mechanisms.

Result

After going through the above process, the best-performing system is selected for use in the field. After it is used, the results are analyzed and evaluated, and the system modified as necessary (Fig. 11.25).

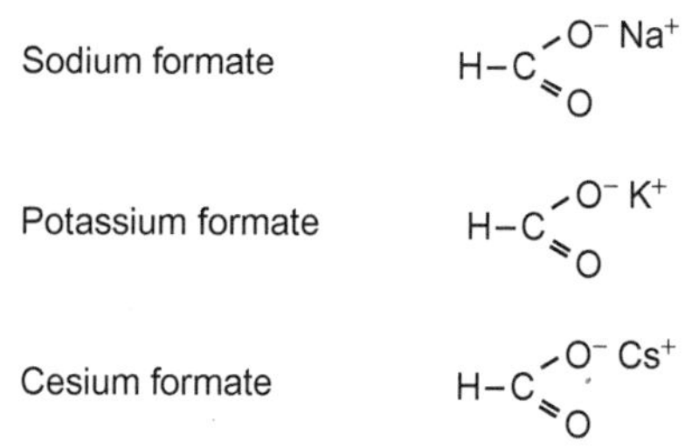

Sodium formate $H-C\begin{smallmatrix}O^- Na^+\\ O\end{smallmatrix}$

Potassium formate $H-C\begin{smallmatrix}O^- K^+\\ O\end{smallmatrix}$

Cesium formate $H-C\begin{smallmatrix}O^- Cs^+\\ O\end{smallmatrix}$

FIGURE 11.25 Chemical structures of formate brines. *Courtesy Cabot Specialty Fluids.*

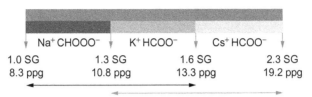

FIGURE 11.26 Specific gravity range of formate brines. *Courtesy Cabot Specialty Fluids.*

FORMATE BRINES

Formate brines are the aqueous solutions of the alkali metal salts of formic acid. These salts are readily soluble in water, yielding high-density brines with low crystallization temperatures. The chemical structures of the three formate salts used in the oilfield are shown in Fig. 11.26.

The formate anion is the most hydrophilic of the family of carboxylic acid anions, yet it retains significant organic characteristics when compared with the halides. This organic character is seen in the solubility of formate salts in organic solvents, such as methanol or ethylene glycol.

The formate anion is also an antioxidant, readily scavenging hydroxyl free radicals. This means formate brines in general can provide thermally sensitive solutes, such as water-soluble polymers, with considerable protection against oxidative degradation at high temperatures (Clarke-Sturman et al., 1986).

The alkali metal cations (Na^+, K^+, and Cs^+) are all monovalent, giving them their unique compatibility with biopolymers while at the same time contributing to their nondamaging behavior in reservoirs. Their molar and weight percentage solubilities in water at 20°C/68°F are shown in Table 11.5.

The formate brines cover the entire fluid density range normally required in drilling and completion. Specific gravity ranges of formate brines are shown in Fig. 11.26. The alkali metal formates in solution also exert a structuring effect on surrounding water molecules, making water more ice-like in nature. This water-structuring behavior has a beneficial effect on the conformation of dissolved macromolecules, making them more ordered, rigid, and stable at high temperatures. The combination of antioxidant and water-structuring properties imparts formate brines with the potential to extend the thermal stability ceiling of many common drilling fluid polymers. An example of this is the commonly used viscosifier xanthan gum, which in concentrated formate brine can be stabilized up to around 356°F (180°C) for 16 hours. This is significantly higher than in any other brines. By adding some other antioxidant chemicals and oxygen scavengers the stability can be raised further to around 400°F (204°C) (Messler et al., 2004).

Compared with other alkali metal cations, the cesium cation is heavier and more electropositive. Cesium is the heaviest of the stable group I

TABLE 11.5 Basic Properties of Sodium, Potassium, and Cesium Formate Brines

Brine	Formula	Molecular Weight (g/mol)	Solubility at 20°C/68°F (moles/L)	Solubility at 20°C/68°F (%wt)	Solution Density SG	Solution Density ppg
Sodium formate	NaCHOO	68.01	9.1	46.8	1.33	11.1
Potassium formate	KCHOO	84.12	14.5	76.8	1.59	13.2
Cesium formate	CsCHOO	177.92	—	—	2.3	19.2
Cesium formate monohydrate	CsCHOO·H$_2$O	195.94	10.7	83	2.3	19.2
Formate ion	CHOO$^-$	45.02	—	—	—	—

elements, with an atomic weight of 132.9. Cesium is also the most electropositive of all the stable elements making cesium formate the most ionic of the formate salts.

Complete mixing formulations and test procedures can be found on the Formate Brines website (Cabot, 2016).

Formate Test Procedures

Table 11.6 shows the recommended changes from standard API protocols that are necessary for formate fluids. The most serious deviation from the API recommended procedures is the solids analysis.

Do not use the API retort with formate fluids. The standard API retort test should never be used with formate fluids because the condensation chamber of the standard retort could get plugged with salt crystals, causing the retort to burst. Even if the retort test could be performed safely, the results are invalid since most solids are formed from formate salts crystallizing out of the highly concentrated brines.

Solids in a formate mud generally comprise of drilled solids and calcium carbonate bridging solids (no weighting material is required in formate muds). Based on this, an alternative solids analysis procedure is detailed next.

Calcium Carbonate

A method has been developed to test for calcium carbonate weighting material in formate fluids. The method, which is based on the standard API total hardness (Ca^{2+}, Mg^{2+}) method, involves removing the carbonate component as carbon dioxide by lowering the pH. The method determines the combined calcium carbonate and magnesium carbonate concentrations, which means that any dolomite-type weighting material is also determined. The method is as follows:

- Add 1 mL formate drilling fluid to a 100 mL volumetric flask.
- Add 9 mL 2N/5N hydrochloric acid.
- Agitate gently to ensure all the calcium carbonate has dissolved.
- Fill the volumetric flask to the 100 mL line with deionized water and shake.
- Take a 10 mL sample from the volumetric flask and place in a smaller conical flask or beaker.
- Add 0.5 mL 8N potassium hydroxide (KOH).
- Check that the pH is at 14 with pH paper and add more potassium hydroxide, if required.
- Add Calver 2 Indicator and titrate with EDTA (0.01 M), recording the volume of EDTA required to change from red to blue.

TABLE 11.6 API Tests in Formate Brines and Fluids

Property	Test Method	Comments
pH	CSF method	CSF method—on sample diluted with deionized water
Density (drilling fluid)	API-13B-1	Use CSF temperature correction method
Density (brine)	API-13J	Use densitometer and CSF temperature correction method
Solids analysis	CSF method	DO NOT run retort with formate fluids. Use alternative CSF method
Chlorides	API-13B-I	No change
Total hardness	API-13B-I	Test is not required in buffered formate fluids. If the test is used on formate fluids without buffer present, the sodium hypochlorite treatment should be avoided
Calcium	API-13B-I	Test is not required in buffered formate fluids. If the test is used on formate fluids without buffer present, the sodium hypochlorite treatment should be avoided
Cation exchange capacity	API-13B-I	No change
Turbidity	API-13J	No change
API fluid loss	API-13B-I	No change. Advisable to run two cells to obtain enough clear filtrate for chemical analyses
HPHT fluid loss	API-13B-I	No change
Rheology	API-13B-1	No change
Alkalinity and lime content (P_f, M_f)	API-13B-1	Cannot be measured in formate brines. See alternative CSF method for buffer concentration
Buffer concentration (CO_3^{2-} and HCO_3^-)	CSF method	Replaces API alkalinity test (P_f, M_f)
K—content	IDF technical manual	Does not work in formates as sodium hypochlorite is an oxidizing

The calcium carbonate concentration can be calculated as:

$$CaCO_3(g/L \text{ or } kg/m^3) = 10 \cdot V \text{ EDTA (mL)} \quad (11.3)$$

where $CaCO_3$ = $CaCO_3$ concentration (kg/m³ or g/L); and V EDTA = volume 0.01 M EDTA (mL).

Completion, Workover, Packer, and Reservoir Drilling Fluids Chapter | 11   515

Example:
If EDTA titration = 5 mL, then the calcium carbonate concentration would be 50 g/L (since the chemical analysis was performed using 10 mL of the 100-mL prepared sample).

Drill Solids

The amount of drill solids in the mud is calculated by determining the total solids in the mud (low-gravity solids comprising drill solids and calcium carbonate), and then subtracting the calcium carbonate portion. The low-gravity solids in the mud can be calculated by measuring the mud and filtrate densities, using the following equation:

$$\text{LGS } (\%v) = \rho\text{mud} - \rho\text{filtrate } \rho\text{LGS} - \rho\text{filtrate} \cdot 100 \qquad (11.4)$$

where ρmud = density or SG of mud; $\rho\text{filtrate}$ = density or SG of filtrate; and ρLGS = density or SG of low-gravity solids.

The density of the mud is measured using a pressurized mud balance and the density of the mud HPHT filtrate is measured using either the densitometer or a 5 mL gravity bottle. If a density bottle is used, the bottle is first weighed empty, then filled with filtrate and reweighed. The density is calculated from the difference between the two weights divided by the filtrate volume, which is inscribed on the gravity bottle. The temperatures are also measured and the densities are corrected to standard temperature (15.6°C/60°F) using DensiCalc II. By assuming that the density of the low-gravity solids is 2.5 s.g./20.84 ppg, the low-gravity solids concentration in the fluid can be calculated as

Metric Units

$$\text{LGS } (g/L) = 25 \cdot \text{LGS } (\%V) \qquad (11.5)$$

Field Units

$$\text{LGS } (\text{lb}/\text{bbl}) = 8.76 \cdot \text{LGS } (\%V) \qquad (11.6)$$

and the drill solids concentration is calculated as follows:

$$\text{DS} = \text{CLGS} - \text{CCaCO}_3 \qquad (11.7)$$

where DS, LGS, and $CaCO_3$ are concentrations of drill solids, low-gravity solids, and calcium carbonate, respectively. This equation is valid for all density units.

REFERENCES

Abrams, A., 1977. Mud design to minimize rock impairment due to particle invasion. J. Pet. Technol.586–592.

Adams, N., 1981. Workover well control conclusion–how to use fluids to best advantage. Oil Gas J.254–275.

Al-Otaibi, A.M., Ozkan, E., 2005. Interpretation of skin effect from pressure transient tests in horizontal wells. In: SPE Paper 93296 Middle East Oil & Gas Show and Conference, Bahrain, March 12–15.

Allen, T.O., Atterbury, T.H., 1958. Effectiveness of gun perforating. Trans. AIME 201, 8–14.

Almon, W.R., 1977. Sandstone diagenesis is stimulation design factor. Oil Gas J.56–59.

Amott, E., 1958. Observations relating to the wettability of porous rock. In: SPE Paper No. 1167, Regional Meeting, Los Angeles, October 16.

API RP 13J, 2014. Testing of Heavy Brines.

Bardon, C., Jacquin, C., 1966. Interpretation and practical application of flow phenomena in clayey media. In: SPE Paper No. 1573, Annual Meeting, Dallas, October 25.

Basan, P.B., 1985. Formation damage index number: a model for the evaluation of fluid sensitivity in shaly sandstones. In: SPE Paper No. 14317, Annual Meeting, Las Vegas, September.

Behrmann, L.A., 1996. Underbalance criteria for minimum perforation damage. In: SPE Conference Paper 30081-PA.

Bergosh, G.L., Ennis, D.O., 1981. Mechanism of formation damage in matrix permeability of geothermal wells. In: SPE Paper No. 10135, Annual Meeting, San Antonio, October 5.

Bertness, T.A., 1953. Observations of water damage to oil productivity. API Drill. Prod. Prac.287–295.

Bolchover, P., Walton, I.C., 2006. Perforation damage removal by underbalance surge flow. In: Society of Petroleum Engineers. SPE Paper 98220, International Symposium on Formation Damage, Lafayette LA, February 15–17.

Bruist, E.H., 1974. Better performance of Gulf Coast wells. In: SPE Paper No. 4777, Symp. on Formation Damage Control, New Orleans, February 28, pp. 83–96.

Cabot Special Fluids, 2016. Formate Brine website. www.FormateBrines.com.

Chauvin Jr., D.J., 1976. Selecting packer fluids: here is what to consider. World Oil87–92.

Clarke-Sturman, A.J., Pedley, J.B., Sturla, P.L., 1986. Influence of anions on the properties of microbial polysaccharides in solution. Int. J. Biol. Macromol. 8, 355.

Crowe, C.W., Cryar Jr., H.B., 1976. Development of oil soluble resin mixtures for control of fluid loss in water base workover and completion fluids. In: SPE Paper No. 5662, Second Symp. on Formation Damage Control, Houston, January 29, pp. 7–17.

Darley, H.C.H., 1972. Chalk emulsion–a new completion fluid. Petrol. Eng.45–51.

Dodd, C.G., Conley, F.R., Barnes, P.M., 1954. Clay minerals in petroleum reservoir sands. Clays Clay Miner. 3, 221–238.

Ellis, R.C., Snyder, R.E., Suman, G.O., 1981. Gravel packing requires clean perforations, proper fluids. World Oil.

Fischer, P.W., Gallus, J.P., Krueger, R.F., Pye, D.S., Simons, F.J., Talley, B.F., 1974. An organic "clay substitute" for nondamaging drilling and completion fluids. In: SPE Paper No. 4651, Symp. on Formation Damage Control, New Orleans, February 7–8, pp. 7–18.

Fischer, P.W., Pye, D.S., Gallus, J.P., 1975. Well completion and workover fluid. U.S. Patent No. 3,882,029 (May 6).

Fleming, N., Moland, L.G., Svanes, G., Watson, R., Green, J., Patey, I., Howard, S., 2016. Formate drilling and completion fluids: evaluation of potential well-productivity impact, valemon. In: SPE Conference Paper 174217-PA, SPE Production & Operations, February, 31. http://dx.doi.org/10.2118/174217-PA.

Githens, C.J., Burnham, J.W., 1977. Chemically modified natural gum for use in well stimulation. Soc. Petrol. Eng. J.5–10.

Glenn, E.E., Slusser, M.L., 1957. Factors affecting well productivity. 11. Drilling fluid particle invasion into porous media. J. Petrol. Technol.132–139, Trans. AIME 210.

Goode, D.L., Berry, S.D., Stacy, A.L., 1984. Aqueous-based remedial treatment for reservoirs damaged by oil-phase drilling muds. In: Paper No. 12501, 6th Symp. Formation Damage Control, Bakersfield, CA, February 13.

Goodman, M.A., 1978. Reducing permafrost thaw around wellbores. World Oil. 71−76.

Gray, D.H., Rex, R.W., 1966. Formation damage in sandstones caused by clay dispersion and migration. Clays Clay Miner. 14, 355−366.

Griffin, J.M., Hayatvoudi, A., Ghalambor, A., 1984. Design of chemically balanced polymer drilling fluid leads to a reduction in clay destabilization. In: SPE Paper No. 12491, 6th Symp. Formation Damage Control, Bakersfield, CA, February 13, p. 185.

Gruesbeck, C., Collins, R.E., 1982. Entrainment and deposition of fine particles in porous media. Soc. Petrol. Eng. J.847−856.

Hamby, T.W., Tuttle, R.N., 1975. Deep, high-pressure sour gas is challenge. Oil Gas J.114−120.

Holub, R.W., Maly, G.P., Noel, R.P., Weinbrandt, R.M., 1974. Scanning electron microscope pictures of reservoir rocks reveal ways to increase oil production. In: SPE Paper No. 4787, Symp. on Formation Damage Control, New Orleans, February 7−8, pp. 187−196.

Hower, W.F., 1974. Influence of clays on the production of hydrocarbons. In: SPE Paper 4785, Symp. on Formation Damage Control, New Orleans, February 7, pp. 165−176.

Hubbard, J.T., 1984. How temperature and pressure affect clear brines. Petrol. Eng.58−64.

Hurst, W., 1953. Establishment of skin effect and its impediment to flow into a wellbore. Petrol. Eng.B6−B16.

Jackson, G.L., 1964. Oil-system packer fluid insures maximum recovery. Oil Gas J.116−118.

Jackson, J.M., 1976. Magnesia stabilized additive for nonclay wellbore fluids. U.S. Patent No. 3,953,335 (April 27).

Jenks, L.H., Huppler, J.D., Morrow, N.R., Salathiel, R.A., 1968. Fluid flow within a porous medium near a diamond core bit. Can. J. Petrol. Technol.172−180.

Jeu, S.J., Foreman, D., Fisher, B., 2014. Systematic approach to selecting completion fluids for deepwater subsea wells reduces completion problems. In: Conference Paper AADE-02-DFWM-HO-02.

Jones, F.O., 1964. Influence of chemical composition of water on clay blocking of permeability. J. Petrol. Technol.441−446, Trans. AIME 231.

Jones, F.O., Neil, J.D., 1960. Clay blocking of permeability. In: SPE Paper No. 1515-G, Annual Meeting, Denver, October 25.

Keelan, D.K., Koepf, E.H., 1977. The role of core analysis in evaluation of formation damage. J. Petrol. Technol.482−490.

Kersten, G.V., 1946. Results and use of oil-base fluids in drilling and completing wells. API Drill. Prod. Proc.61−68.

Klotz, J.A., Krueger, R.F., Pye, D.C., 1974a. Effects of perforation damage on well productivity. J. Petrol. Technol.1303−1314, Trans. AIME 257.

Klotz, J.A., Krueger, R.F., Pye, D.S., 1974b. Maximum well productivity in damaged formations requires deep, clean perforations. In: SPE Paper No. 4792, Formation Damage Symposium, New Orleans, February 7−8.

Krueger, R.F., 1973. Advances in well completion and stimulation during J.P.T.'s first quarter century. J. Petrol. Technol.1447−1461.

Krueger, R.F., 1986. An overview of formation damage and well productivity in oilfield operations. J. Petrol. Technol.131−152.

Krueger, R.F., Vogel, L.C., Fischer, P.W., 1967. Effects of pressure drawdown on clean-up of clay or silt blocked sandstone. J. Petrol. Technol.397−403, Trans. AIME 240.

Maly, G.P., 1976. Close attention to smallest detail vital for minimizing formation damage. In: SPE Paper No. 5702, Second Symp. on Formation Damage Control, Houston, January 24–30, pp. 127–145.

Matthews, C.S., Russell, D.A., 1967. Pressure buildup and flow tests in wells. In: Soc. Petrol Eng. of AIME Monograph, Dallas.

Mayell, M.J., Stein, F.C., 1973. Salwater gel pack fluid for high pressure completions. In: SPE Paper No. 4611, Annual Meeting, Las Vegas, August 30. Reprinted in Drilling, September 1974, pp. 86–89.

McLeod Jr., H.O., 1982. The effect of perforating conditions on well performance. In: SPE Paper No. 10649, Formation Damage Symposium, Lafayette, LA, March 24–25.

Messler, D., Kippie, D., Broach, M., 2004. A potassium formate milling fluid breaks the 400° fahrenheit barrier in deep Tuscaloosa coiled tubing clean-out. In: SPE Paper 86503, Formation Damage Control, Lafayette, LA, March 24–25.

Methven, N.E., Kemick, J.G., 1969. Drilling and gravel packing with an oil base fluid system. J. Petrol. Technol.671–679.

Milhone, R.S., 1983. Completion fluids for maximizing productivity-state of the art. J. Petrol. Technol.47–55.

Miller, G., 1951. Oil-base drilling fluids. In: Brill, E.J. (Ed.), Proceedings of the 3rd World Petrol, Conf., Sec. II, 2, Leiden, pp. 321–350.

Mondshine, T.C., 1977. Completion fluid uses salt for bridging, weighting. Oil Gas J.124–128.

Muecke, T.W., 1979. Formation fines and factors controlling their movement through porous media. J. Petrol. Technol.144–150.

Mungan, N., 1965. Permeability reduction through changes in pH and salinity. J. Petrol. Technol.1449–1453, Trans. AIME 254.

Muskat, M., 1949. Physical Principles of Oil Production. McGraw-Hill Book Company Inc, New York, p. 243.

Ng, F.W., 1975. Process removes water-base mud from permafrost-zone well bores. Oil Gas J.87–92.

Nowak, T.J., Krueger, R.F., 1951. The effect of mud filtrates and mud particles upon the permeabilities of cores. API Drill. Prod. Prac.164–181.

Olivier, D.A., 1981. Improved completion practices yield high productivity wells. Pet. Eng..

Poole, G.L., 1984. Planning is key to completion fluid use. Oil Gas J.90–94.

Priest, G.G., Allen, T.O., 1958. Non-plugging emulsions useful as completion and well servicing fluids. J. Petrol. Technol.11–14.

Priest, G.G., Morgan, B.E., 1957. Emulsions for use as non-plugging perforating fluids. J. Petrol. Technol.177–182, Trans. AIME 210.

Purcell, W.R., 1949. Capillary pressures – their measurements using mercury and the calculation of permeability there from. Trans. AIME 186, 39–48.

Remont, L.J., Nevins, M.J., 1976. Arctic casing pack. Drilling43–45.

Ribe, K.H., 1960. Production behavior of water-blocked well. J. Petrol. Technol.1–6, Trans. AIME 219.

Scheuerrnan, R.F., 1983. Guidelines for using HEC polymers for viscosifying solids-free completion and workover brines. J. Petrol. Technol.306–314.

Schmidt, D.D., Hudson, T.E., Harris, T.M., 1983. Introduction to brine completion and workover fluids-part 1: chemical and physical properties of clear brine completion fluids. Petrol. Eng.80–96.

Sharma, M.M., Yortos, Y.C., 1986. Permeability impairment due to fines migration sandstones. In: SPE Paper No. 14189, Formation Damage Symp., Lafayette, LA, February.

Simpson, J.P., 1968. Stability and corrosivity of packer fluids. API Drill. Prod. Prac.46−52.

Slobod, R.L., 1969. Restoring permeability to clay-containing water damaged formations. In: SPE Paper No. 2683, Annual Meeting, Denver, September 29−October 1.

Somerton, W.B., Radke, C.J., 1983. Role of clays in the enhanced recovery of petroleum from some California sands. J. Petrol. Technol.643−654.

Steiger, R.P., 1982. Fundamentals and use of potassium/polymer fluids to minimize drilling and completion problems associated with hydratable clays. Petrol. Technol.1661−1670.

Stuart, R.W., 1946. Use of oil-base mud at Elk Hills naval petroleum reserve number one. API Drill. Prod. Proc.69−75.

Suman, G.O., 1976. New completion fluids protect sensitive sands. World Oil55−58.

Swanson, B.F., 1979. Visualizing pores and nonwetting phase in porous rock. J. Petrol. Technol.10−18.

Thomas, D.C., Atkinson, G., Atkinson, B.L., 1984. Pressure and temperature effects on brine completion fluid density. In: SPE Paper No. 12489, 6th Symp, Formation Damage Control, Bakersfield, CA, February 13, pp. 165−174.

Trimble, G.A., Nelson, M.D., 1960. Use of invert emulsion mud proves successful in zones susceptible to water damage. J. Petrol. Technol.23−30.

Tuttle, R.N., Barkman, J.H., 1974. New non-damaging and acid-degradable drilling and completion fluids. J. Petrol. Technol.1221−1226.

van Everdingen, A.R., 1953. The skin effect and its influence on well productivity. Trans. AIME 198, 171−176.

Walton, I.C., 2000. Optimum underbalance for the removal of perforation damage. In: SPE Conference Paper 63108-MS.

Webb, M.G., Haskin, C.A., 1978. Pressure coring used in Midway-Sunset's unconsolidated cores. Oil Gas J.52−55.

Young Jr., F.S., Gray, K.E., 1967. Dynamic filtration during microbit drilling. J. Petrol. Technol.1209−1224, Trans. AIME 240.

Yuster, S.T., Sonney, K.J., 1944. The drowning and revival of gas wells. Petrol. Eng.61−66.

Chapter 12

Introduction to Fracturing Fluids

INTRODUCTION

In the early years of oil well drilling with high-solids clay slurries, the production zone would be damaged by solids. Also, many wells were drilled into formations with low permeability that restricted production. These wells were stimulated to improve the flow of oil. Some of the methods used to stimulate these formations was by "shooting" the well with dynamite or nitroglycerin, acidizing, or by bullet gun perforating. After World War II, the use of shaped charges, developed for use in the military's bazooka, became the preferred method of perforation, thereby bypassing any mud damage.

The first experimental hydraulic fracturing treatment in the United States took place in 1947 by Stanolind Oil and Gas Corporation (Later Pan American Petroleum and then Amoco) in the Hugoton gas field in Grant County, Kansas. Floyd Farris of Stanolind developed the first fracturing fluid, "1000 gallons of naphthenic-acid and palm-oil (napalm)-thickened gasoline" ... "and sand from the Arkansas River" (Montgomery and Smith, 2010). A patent was issued to Stanolind in 1949 and an exclusive license to use and develop the hydrofrac process was given to Halliburton Oil Well Cementing Company. Halliburton then performed the first two commercial fracturing treatments in Oklahoma and Texas in 1949.

Fracturing fluids have developed since that time with a variety of base fluids being used. Since the advent of horizontal drilling in the 1990s, fracturing of low permeability formations, including ultralow permeability shales, has increased dramatically. Until shale and/or gas were economically drilled horizontally, fracturing was reserved for vertical and directionally drilled wells through low-permeability sands, coal-bed methane production, and frac-pac completion procedures (Gallegos and Varela, 2012).

Hydraulic fracturing occurs when the pressure in the wellbore gets high enough to split the surrounding formation apart. Unintentional fracturing can lead to wellbore ballooning, lost circulation, a drop in hydrostatic pressure in the wellbore, and possible blowouts. These scenarios lead to increased nonproductive time and increased well costs (see chapter: Drilling Problems Related to Drilling Fluids).

Intentional fracturing (well stimulation) consists of pumping a fluid containing solids, which keep the fracture open (proppants), to facilitate an increase in permeability resulting in higher production rates. The permeability of the surrounding formation is not changed. The fracturing operation creates a high-permeability channel allowing the oil and or gas to flow more easily into the wellbore.

FRACTURING FLUID TYPES

There are usually two stages in a fracture job, the pad stage followed by the slurry stage. The pad stage initiates the fracture, opening the formation. Since the overburden pressure tends to close the fracture, the slurry phase injects the proppant into the pad that was created and props open the fracture.

Purpose of a Fracturing Fluid

The primary purpose of the fracturing fluid is to transmit the pump truck's pressure to the zone to be fractured. There are two distinct pressures involved, the fracture initiation pressure and the fracture propagation pressure. Fracturing fluids with higher viscosities will have higher friction losses while pumping, therefore less pressure is exerted on the formation. Fracturing fluids with higher solids content may cause fracture initiation but not be efficient at fracture propagation and may plug the producing zone, called sand off or screenout.

The secondary purpose, but equally important, is to efficiently transport the propping agent into the fracture so the fracture does not close, maintaining a permeability higher than the native producing zone permeability. These two purposes sometime work against each other.

The following are, in general, the properties that a good fracturing fluid must possess (King, 2010):

- Help create a wide fracture
- Low friction pressure to minimize friction pressure losses during injection
- Good fluid loss control
- Transport the propping agent into the fracture, and release the proppant to stabilize the fracture.
- Compatible with the formation rock and fluid, minimum damage
- Easey or formulate and mix
- Shear stability
- Cost-effective
- All chemical additives are approved by appropriate environmental agencies
- Revert to a low-viscosity fluid for easy cleanup for flowback

FRACTURING FLUIDS COMPOSITION

A comprehensive review of the state of fracturing fluids was presented by Ghaithan in 2014. The following is adapted from the abstract of that paper, *A critical review of hydraulic fracturing fluids over the last decade.* The SPE also presented a webinar of the paper that can be found on their webinar website (Ghaithan Al-Muntasheri, 2015).

Guar-based polymers are still being used in fracturing operations for wells at temperatures less than 300°F (148.9°C). To minimize the damage associated with these polymers, the industry used lower polymer concentrations in formulating the fluid. Another approach was to alter the cross-linker chemistry to generate higher viscosities at lower polymer loadings. Commercial guar contains a minimum of 5 wt% residues causing damage to proppant packs. The industry then shifted towards the use of cleaner guar-based polymers, primarily HP-guar. Fracturing in deeper wells in hotter reservoirs, the need arose for a new class of thermally stable polyacrylamide-based polymers. These synthetic polymers offer sufficient viscosity at temperatures up to 232°C (450°F). To address the challenge of high pressure pumping requirements on the surface, high density brines have been used to increase the hydrostatic pressure by 30%.

New breakers were introduced that would decross-link the gel by reacting with the cross-linker. To minimize the environmental impact of using massive amounts of freshwater and to minimize costs associated with treating produced water, the use of produced water in hydraulic fracturing treatments was implemented.

Advancements in the use of slickwater include the use of drag-reducing agents (polyacrylamide-based polymers) to minimize friction along with the development of breakers to improve the cleanup of these drag reducers.

For foamed fluids, new viscoelastic surfactants (VES) that are compatible with CO_2 have been developed. Nanotechnology was used to develop nanolatex silica, reducing the concentration of boron used in conventional cross-linkers. Another advancement in nanotechnology was the use of 20 nm silica particles suspended in guar gels.

Basic Components and Additives

A fracturing fluid consists of:

- Base fluid—water, oil, gas
- Chemical additives—see Table 12.1
- Proppants—sand, ceramics

Table 12.2 shows the variety of fracturing fluids systems available. Some fluids, such as a slickwater fluid, use very few different additives. Others,

TABLE 12.1 Generic Additives Used in Fracturing Fluids

Fracturing Fluid Additives

Biocides	Fluid-loss additives
Breakers	Gelling agents
Buffers	Nonemulsifiers
Clay stabilizers	Paraffin inhibitors
Diverting agent	Scale inhibitors
Emulsifiers	Proppants
Foamers	Surfactants
Friction Reducers	Temperature Stabilizers

TABLE 12.2 Fracture Fluid Types

Base Fluid	Fluid Type	Components
Water-Based	Slickwater	Water
		Friction reducer (polyacrylamide)
	Linear—gelled water	Guar gum
		HPG (hdroxypropylguar)
		HEC (hydroxyethylcellulose)
		CMHPG—(carboxymethylhydroxypropalguar)
	Cross-linked	Cross-linker—borate, titanium, zirconium
		Guar Gum
		Hydroxypropylguar
		Carboxymethylhydroxypropylguar
		Carboxymethylhydroxyethylcellulose
	Polymer free	Water
		VES (viscoelastic surfactant)

(*Continued*)

TABLE 12.2 (Continued)

Base Fluid	Fluid Type	Components
Gas Nitrogen, CO_2	Water-based foam	Water
		Foam generator
	Acid-based foam	Acid
		Foam generator
		Nitrogen
	Alcohol-based foam	Methanol
		Foam generator
		Nitrogen
	Energized Fluid	Nitrogen, Carbon dioxide
Nonaqueous	Linear	Gelling agent
	Cross-linked	Polyester gels
	Water external emulsions	Water + oil
		Emulsifier

such as a cross-linked system, will contain a number of different additives. In addition, various combinations of systems have been designed for specific purposes. For example the VES systems can be formulated with cross-linked gels, nonaqueous fluids, and foams. No one fracturing fluid will contain all of these additives. Figs. 12.1 and 12.2 show typical differences in additive amounts between a fracture job in Canada's Montney shale and in Pennsylvania's Marcellus shale (All Consulting, 2012).

SELECTING A FRACTURING FLUID SYSTEM

Water-Based

Slickwater—Slickwater fracturing fluids can be prepared from either fresh or saline water. The saline water can be either NaCl or KCl for inhibition of clays present in the formation. They are prepared with a minimal number of additives. A friction reducer is usually added to these fluids. Other additives used are bactericides and possibly scale inhibitors when using a brine-base fluid.

The basic characteristics of slickwater fluids are:

- Low friction loss
- Low viscosity (<50 cP)

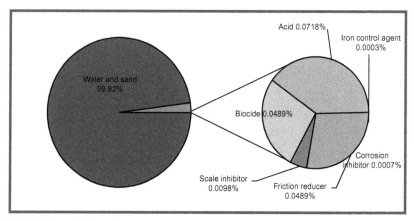

FIGURE 12.1 Volumetric breakdown of a fracture treatment typical of the Montney shale in Canada. Arthur, J., 2009. Modern shale gas development. Presented at Oklahoma Independaent Petroleum Association Mid-Continent OBM & Shale Gas Symposium, Tulsa, OK, 8 December (Arthur, 2009).

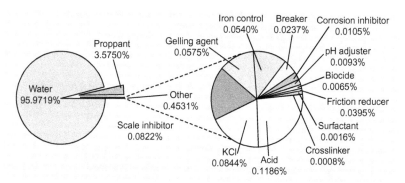

FIGURE 12.2 Volumetric breakdown of a fracture treatment typical of the Marcellus shale in Pennsylvania. Arthur, J., Bohm, B., Layne, M., 2009. Considerations for development of Marcellus Shale gas. World Oil July (Arthur et al., 2009).

- Low residue
- Low proppant transport efficiency

The greatest advantages or these fluids are their low fluid costs and low residue.

The use of low-viscosity fluids in slickwater fracturing pumped at high rates generates narrow, complex fractures using low concentrations of propping agent (PetroWiki, 2016). Pumping rates must be high enough to transport proppants through long distances in horizontal wells, enter the fracture, and then place the proppant into the open fracture to prevent screenouts. With low proppant transport efficiency, there is always the possibility of a screenout.

Water, a Newtonian fluid, pumped at very high pressures is placed in turbulence. The addition of polymers, primarily polyacrylamide friction reducer, tend to suppress turbulence. Friction reducers do not function unless turbulence is achieved.

Disadvantages of slickwater fluids (PetroWiki, 2016):

- Larger volumes of water often required.
- Larger horsepower requirement (to maintain high pump rates, 60−110 bpm).
- Limited fracture-width (due to low proppant concentration).
- Reduced flowback-water recovery, due to the imbibement of the base fluid into the fracture network.
- Limited to fine-mesh propping agents. Larger proppants will settle faster increasing the likelihood of screenouts.

Linear gels—Linear gels can be formulated in either fresh or brine water. The polymers typically used are HP-guar or CMC. The basic characteristics of linear gel fluids are:

- Mild friction pressure
- Variable viscosity (10−60 cP)
- High residue

Linear gel viscosity is several orders of magnitude higher than slickwater and has improved proppant-suspension. The use of uncross-linked gels in late-slurry stages of a fracturing treatment, after the pad and early-slurry stages used slickwater, are often referred to as "hybrid" fracturing treatments. A breaker must be used to facilitate flowback.

Cross-linked gels—These fluids use polymers that can be cross-linked forming an elastic gel structure that can hold the proppant in suspension. The cross-linked structure must be broken prior to flowback. This high viscosity structure, however, results in higher pressure drops during pumping.

The basic characteristics of cross-linked gels are:

- High viscosity (>100 cP)
- Excellent proppant transport
- High residue
- Expensive
- Temperature- and pH-dependent

The most common cross-linking agent is a borate compound. The temperature/time design of these fluids is critical to their success. The common polymers used in cross-linked gel fluids are: Guar Gum, HydroxyPropylGuar, CarboxyMethylHydroxyPropylGuar, and CarboxyMethylHydroxyEthylCellulose.

Gelled nonaqueous fluids—Depending on environmental rules and regulations, the base fluid can be either diesel oil, mineral oil, or synthetic fluids. The basic characteristics of these fluids are:

- Compatibility with reservoir (using produced oil)
- In some cases, economics (using produced oil)
- Lower fracture fluid cleanup costs
- Inherently dangerous
- Decreased hydrostatic head
- Complicated and poorly understood rheology
- Requires intense quality control

Energized fluids—Energized fluids consist of one or more gas components either by itself or with a liquid component. They are generally referred to as foams. Foamed fracturing fluids reduce the amount of water used in the fracturing treatment. These treatments can be costly, however, they can show better performance than nonenergized fluids (Burke et al., 2011; Jacobs, 2016).

Energized fracturing fluids make use of either CO_2-H_2O mixtures, N_2-H_2O, $CO_2-MeOH-H_2O$ mixtures, or regular surfactant/air/water foamed fluids. They are primarily used in water-sensitive formations, depleted underpressured wells and low-permeability gas formations. The main advantages of energized systems are minimizing water flowback, minimal clay/water interactions, and high proppant transport efficiency.

Pure Liquid CO_2 or N_2 gases and liquefied petroleum gas characteristics include (Watts, 2014):

- No other additives required
- Clean system, minimal residue
- Requires expensive special equipment
- Nondamaging to the formation
- Chemistry similar to oil gels

Foam fracturing fluid characteristics include:

- Minimal fluid on formation
- Self-energizing
- Moderate fluid loss control
- Reduced solids—fluid loss additives (FLA) and polymers
  Disadvantages of foams:
- Loss of hydrostatic head
- Expense of energizing medium
- Produced gas can be contaminated for extended period

Foam fracturing base fluid variations include:

- Water (fresh and brine)
- Water−methanol

- Methanol
- Hydrocarbon
- Nitrogen
- CO_2
- CO_2/N_2 mixtures
- Cross-linked gel

At temperatures greater than 88°F (31°C) and pressures typical of those in fracturing treatments (above 1000 psi), CO_2 behaves as a supercritical fluid. In such a case, the energized fluid is considered an emulsion (Gupta and Hlidek, 2010; Arias et al., 2008). Note that CO_2 has more hydrostatic pressure than N_2 and therefore, it is preferred in cases where the formation breakdown pressure is high.

Synthetic HPHT polymers—The natural polymers are limited to less than 300°F (149°C) temperature before degrading. Examples of synthetics developed for higher temperature stability up to 400°F (204°C) are 2-acrylamido-2-methylpropanesulfonic acid (AMPS) and copolymers of partially hydrolyzed polyacrylamide (PHPA)-AMPS-vinyl phosphonate (PAV). These polymers can be used in fresh or brine waters and are natural drag reducers. They can be used in slickwater or in linear and cross-linked gels and with CO_2 as an energized system, although the temperature may be limited to below 260°F (127°C) for CO_2 fluids.

Holtsclaw and Funkhouser (2010) reported the use of a PAM-based polymer as a fracturing fluid to carry proppants. The system is based on a terpolymer of 2-acrylamdio-2-methylpropanesulfonic acid (AMPS) and acrylamide (AM). Funkhouser and Norman (2003) reported that the best performance of this system is obtained upon the use of 60% AMPS, 39.5% amide, and 0.5% acrylate. They also found that sodium bromate is the best breaker for the system. Holtsclaw and Funkhouser (2010) reported that this copolymer cross-linked with Zr^{4+} gave a viscosity of 700 cp for more than 1.5 h at a temperature of 402°F (204°C) at a shear rate of 40 s^{-1}. Under the same conditions, a gel based on CMHPG containing the same concentration of polymer and cross-linker lost significant viscosity after 22 min of shear.

Gupta and Carman (2011) developed a high temperature fracturing gel based on a copolymer of partially hydrolyzed polyacrylamide (PHPA)-AMPS-vinyl phosphonate (PAV). This copolymer was cross-linked by zirconium. They reported that the system maintained a 1000 cP viscosity after shearing at 100 s^{-1} for 2 h at 425°F (218°C).

Gaillard et al. (2013) used three PAM-based polymers for use in HPHT fracturing. These polymers are high molecular weight (5–15 million g/gmol) PAM-based friction reducers. They all contain at least 1.5 mol% of a hydrophobe and have an acrylate content of 10–25 mol%. The third polymer had an AMPS content of 10–25 mol%. The polymers exhibited good proppant settling properties.

VES (viscoelastic surfactant)—These materials are believed to be less damaging to the proppant packs as they leave less residues. In addition, they do not require cross-linkers, simplifying their formulation. A variety of surfactants can be used including quaternary ammonium salts (Samuel et al., 1999, 2000), amidoamine + < 100 nm ZnO (Crews and Huang, 2008; Huang and Crews, 2008; Huang et al., 2010) and zwitterionic (Sullivan et al., 2006) surfactants.

Huang et al. (2010) reported that a VES is a low molecular weight molecule having a hydrophilic head and a long hydrophobic tail and in the presence of salts such as potassium chloride, ammonium chloride, or ammonium nitrate, it forms elongated micellar structures. With sufficient concentration of these micelles, the micellar structures entangle to build viscosity. Because this entanglement is not based on a chemical cross-linking reaction, these fluids have high leakoff rates. The high leakoff rate prevents their use in reservoirs with permeabilities larger than 100 mD.

Although VES fluids can break efficiently in the lab, data indicated that 20% of field treatments utilizing VES-based fluids in the 1990s required the use of remedial actions to revive wells fractured with these fluids (Crews and Huang, 2008). VES-based fluids can break upon contact with hydrocarbons or when the salt content is reduced in the mixing brine. These two conditions may not be met in all cases. For example, when reservoirs do not produce hydrocarbons, then the VES does not break. Crews (2005) reported the use of breakers with VES fluids.

Acidized fracturing fluids (Adapted from PetroWiki, 2016)—The most commonly used acid fracturing fluid is 15% hydrochloric acid (HCl). To obtain more acid penetration and more etching, 28% HCl is sometimes used as the primary acid fluid. On occasion, formic acid (HCOOH) or acetic acid (CH_3COOH) is used because these acids are easier to inhibit under high-temperature conditions. However, acetic and formic acid cost more than HCl. Hydrofluoric acid (HF) should never be used during an acid fracturing treatment in a carbonate reservoir, only in shales or sands.

Typically, a gelled water or cross-linked gel fluid is used as the pad fluid to fill the wellbore and break down the formation. The water-based pad is then pumped to create the desired fracture height, width, and length for the hydraulic fracture. Once the desired values of created fracture dimensions are achieved, the acid is pumped and fingers into the fracture to etch the walls of the fracture creating fracture conductivity. The acid can be gelled, cross-linked, or emulsified to maintain fracture width and minimize fluid leakoff. Because the acid is reactive with the formation, fluid loss is a primary consideration in the fluid design. Large amounts of FLA are generally added to the acid fluid to minimize fluid leakoff. Fluid-loss control is most important in high-permeability and/or naturally fractured carbonate formations.

There are several unique considerations to be understood when designing acid fracture treatments. Of primary concern is acid-penetration distance into the fracture. The pad fluid is used to create the desired fracture dimensions. Then the acid is pumped down the fracture to etch the fracture walls, which creates fracture conductivity. When the acid contacts the walls of the fracture, the reaction between the acid and the carbonate is almost instantaneous, especially if the temperature of the acid is 200°F or greater. As such, the treatment must be designed to create a wide fracture, with minimal leakoff, with viscous fluids. If a wide fracture is created with a viscous acid and minimal fluid loss, then a boundary layer of spent acid products will reduce the rate at which the live acid contacts the formation at the walls of the fracture. However, as the flow in the fracture becomes more turbulent and less laminar, the live acid will contact the walls of the fracture more easily, and the acid will not penetrate very far into the fracture before becoming spent.

Methanol fracturing fluids—Alcohol fracturing fluids can also be formulated in many variations—linear, cross-linked, and foamed fluids are available. Adding methanol to water-based treatments reduces water content, lowers interfacial tension, and enhances the evaporation of the water-based filtrate during reservoir cleanup. The characteristics of alcohol fracturing fluids are:

Advantages

- Low surface tension
- Inherent volatility for removal during flowback
- Enhanced removal of water

Disadvantages

- Inherently dangerous
- Difficult to degrade
- Expensive
- Minimal rheological data available

Concentration of alcohol is critical. The following alcohol content is recommended.

Guar	≤25%
HP-Guar	≤60%
HP-Cellulose	100%

High Density Brines—A formation can only be fractured if the bottomhole pressure exceeds the formation breakdown pressure. Some unconventional reservoirs are located at depths up to 20,000 ft. and the reservoir temperature can be as high as 355°F (179.4°C) (Bartko et al., 2009). Fracturing such wells requires bottomhole treating pressures up to 20,000 psi. There are limitations on the pressure rating of some bottomhole completion equipment. There could also be limitations on the pressure rating of surface equipment and pumping equipment where they do not exceed 15,000 psi (Qiu et al., 2009). The density of standard fracturing gels prepared in field brines is usually around 8.7 lb/gal (Simms and Clarkson, 2008). Thus, the use of high density brines can increase the hydrostatic pressure by 30%.

ADDITIVES

Fluid loss—FLA control leakoff into the fracture matrix or a natural fracture. Without controlling the leakoff, the fracture will be short and wide and depending on the formation characteristics, fluid invasion will occur. With an FLA, the fracture will be long and slender with minimal fluid invasion. FLAs include:

- Bridging agents—Solids—100 mesh silica flour, nano-silica.
- Plastering agents—Soft material—resins, starch, colloidal polymers.
- Multiphase—hydrocarbon emulsions or entrained gases.

Friction reducers—Newtonian fluids (water and glycerine are examples) pump at very high pressures when placed in turbulence. Polymer additions will suppress the entrance into turbulence by controlling molecular migration. High molecular weight, low viscosity materials at low concentrations work best in smooth wall pipes. Viscosity is detrimental in smooth pipes but moderate concentration of gel works best in tubing and casing. Friction reducers do not function unless turbulence is achieved. The most common friction reducers are based on polyacrylamide chemistry.

Bacteria control—Bacteria are unicellular, microscopic organisms. There are estimated to number over 10^{30} individual bacteria worldwide (Whitman et al., 1998). Different bacteria can thrive in a wide range of environmental conditions, such as:

- Oxygen—both anaerobic and aerobic bacteria exist.
- Temperature—Bacteria can function at temperatures ranging from 12°F to 220°F (-11 to 104°C).
- Pressure—Bacteria can function at pressures from 0 to 25,000 psi (172 MP).
- Salinity—Bacteria can function in salinities ranging from 0 to 30%.
- pH—The pH range for bacteria survival is 1.0–10.2.

Specific bacterial problems include:

- Degrade polymers (enzyme secretion).
- Produce hydrogen sulfide, sulfate reducing bacteria (SRBs).
- Corrosion in anaerobic conditions.
- Plugging both downhole and on the surface.

Glutaraldehyde is the most common oilfield bactericide. Other bactericides include:

- Quaternary ammonium
- Amine salts
- Aldehydes
- Chlorinated phenols
- Organosulfur
- Heavy metals
- Thiocyanates
- Carbamates
- Combinations of the above

Breakers—The use of polymers, cross-linking agents, and any wall building material requires a breaker prior to flowback. If any residual materials remain, production from the well might be impeded. Breakers are usually combinations of materials including oxidizing materials, acids, and enzymes.

Oxidizer breakers (Montgomery, 2013) include ammonium persulfate, sodium persulfate, and calcium and magnesium peroxides. The main disadvantage of oxidizing breakers is that both how well they work and how fast they work is a function of the amount of chemical added. A concentration of 0.5 lb/1000 gal of persulfate breaker will break a polymer viscosity back to the viscosity of water but will damage the proppant pack so that only 20% of the original conductivity remains. If we want to get the maximum retained permeability we need to go to concentrations of 10 to 12 lb/1000 gallons which will break the fluid viscosity instantly. To counteract this and retard the release of the persulfate encapsulated breakers were developed (Lo and Miller, 2002). There are two types of encapsulated breakers available. The release rate of the first type is controlled by hydrostatic pressure, elevated temperatures, and the pH of the fracturing fluid. The second method of release is by crushing the capsule coating as the fracture closes.

Enzyme breakers (Montgomery, 2013) are protein molecules that act as organic catalysts that attach to and digest polymers at specific sites along the polymer backbone. Because they are catalysts they are not "used up" during the breaking process and persist until there is no polymer present to digest. Typical enzymes that are used include hemicellulase, cellulose, amylase, and pectinase. These enzymes are susceptible to thermal degradation and denaturing when exposed to very high or very low pH so are limited to mild temperatures below 150°F (66°C) and fluid pHs between 4 and 9. Recent work by Brannon and Tjon-Joe-Pin have developed proprietary GLSE (Guar Linkage-Specific Enzymes) that are reported to work at temperatures higher than 300°F (Brannon and Tjon-Joe-Pin, 2003).

Clay stabilizers—Clay stabilizers inhibit clay swelling and possible dislodgment into the fracture. The following chemicals are clay stabilizers used in fracturing fluids:

- Potassium chloride
- Sodium chloride
- Ammonium chloride
- Calcium chloride
- Acrylamides
- Cationic polymers
- Quaternary compounds

Diverting agents—The purpose of diverting agents is to divert flow to another zone when stimulation of the first zone is complete. Physical diverters, such as ball sealers, packers, and ball and baffle techniques work best. Chemical techniques can be successfully used in matrix treatments, but are less successful in fracturing treatments. Typical diverting chemicals include:

- Karaya powder
- Graded naphthalene
- Oil external emulsion
- High concentration linear gel
- Oyster shells
- Polymer coated sand
- Buoyant particles
- High quality foams
- Flake boric acid

REFERENCES

All Consulting, LLC, 2012. The Modern Practices of Hydraulic Fracturing: A Focus on Canadian Resources. Petroleum Technology Alliance Canada and Science and Community Environmental Knowledge Fund. <http://www.all-llc.com/publicdownloads/ModernPracticesHFCanadianResources.pdf>.

Al-Muntasheri, G., 2014. A critical review of hydraulic fracturing fluids over the last decade. SPE 169552. This Paper Was Prepared for Presentation at the SPE Western North American and Rocky Mountain Joint Regional Meeting held in Denver, CO, 16–18 April.

Al-Muntasheri, Dr. Ghaithan A., 2015. Fluids for Fracturing Petroleum Reservoirs. Web Events. SPE. <https://webevents.spe.org/products/fluids-for-fracturing-petroleum-reservoirs>.

Arias, R.E., Nadezhdin, S.V, Hughes, K., Santos, N., 2008. New viscoelastic surfactant fracturing fluids now compatible with CO_2 drastically improve gas production in rockies. Paper SPE 111431 Presented at the SPE International Symposium and Exhibition on Formation Damage Control, Lafayette, LA, 13–15 February.

Arthur, J., 2009. Modern shale gas development. Presented at Oklahoma Independaent Petroleum Association Mid-Contenent OBM & Shale Gas Symposium, Tulsa, OK, 8 December. <http://www.oipa.com/page_images/1262876643.pdf>.

Introduction to Fracturing Fluids Chapter | 12 535

Arthur, J., Bohm, B., Layne, M., 2009. Considerations for development of Marcellus Shale gas. World Oil July.

Bartko, K., Arocha, C., Mukherjee, T.S., 2009. First application of high density fracturing fluid to stimulate a high pressure and high temperature tight gas producer sandstone formation of Saudi Arabia. Paper SPE 118904 Presented at the SPE Hydraulic Fracturing Technology Conference, The Woodlands, TX, 19−21 January.

Brannon, H., Tjon-Joe-Pin, 2003. Enzyme breaker technologies: a decade of improved well stimulation. SPE 84213 Presented at the SPE Annual Technical Conference, Denver, CO, 5−8 October.

Burke, L.H., Nevison, G.W., Peters, W.E., 2011. Improved unconventional gas recovery with energized fracturing fluids: montney example. Paper SPE 149344 Presented at the SPE Eastern Regional Meeting Held in Columbus, OH, 17−19 August.

Crews, J., 2005. Internal Phase Breaker Technology for viscoelastic surfactant gelled fluids. Paper SPE 93449 Presented at the SPE International Symposium on Oilfield Chemistry, Houston, TX, 2−4 February.

Crews, J., Huang, T., 2008. Performance enhancements of viscoelastic surfactant stimulation fluids with nanoparticles. Paper SPE 113533 Presented at the SPE Europec/EAGE Annual Conference and Exhibition, Rome, 9−12 June.

Funkhouser, G.P., Norman, L., 2003. Synthetic polymer fracturing fluid for high-temperature applications. Paper SPE 80236 Presented at the SPE International Symposium on Oilfield Chemistry, Houston, TX, 5−7 February.

Gaillard, N., Thomas, A., Favero, C., 2013. Novel associative acrylamide-based polymers for proppant transport in hydraulic fracturing fluids. Paper SPE 164072 Presented at the 2013 SPE International Symposium on Oilfield Chemistry, The Woodlands, TX, 08−10 April.

Gallegos, T., Varela, B., 2012. Trends in Hydraulic Fracturing Distributions and Treatment Fluids, Additives, Proppants, and Water Volumes Applied to Wells Drilled in the United States from 1947 through 2010—Data Analysis and Comparison to the Literature By Scientific USGS Investigations Report 2014−5131.

Gupta, D.V.S., Carman, P., 2011. Fracturing fluid for extreme temperature conditions is just as easy as the rest. Paper SPE 140176 Presented at the SPE Hydraulic Fracturing Technology Conference and Exhibition, The Woodlands, TX, 24−26 January.

Gupta, D.V.S., Hlidek, B.T., 2010. Frac-fluid recycling and water conservation: a case history. SPE Prod. Facil. J. 25 (2), 65−69.

Holtsclaw, J., Funkhouser, G.P., 2010. A cross-linkable synthetic-polymer system for high-temperature hydraulic- fracturing applications. SPE Drill. Complet. 25 (4), 555−563.

Huang, T., Crews, J., 2008. Nanotechnology applications in viscoelastic surfactant stimulation fluids. SPE Prod. Oper. 23 (4), 512−517.

Huang, T., Crews, J., Agrawal, G., 2010. Nanoparticle pseudocrosslinked micellar fluids optimal solution for fluid-loss control with internal breaking. Paper SPE 128067 Presented at the International Symposium and Exhibition on Formation Damage Control, Lafayette, LA, 10−12 February.

Jacobs, T, 2016. Shale revolution revisits the energized fracture. JPTFebruary 2016. <http://www.spe.org/jpt/article/10666-downturn-represents-stress-test-for-unconventional-hydraulic-fracture-modeling/>.

King, G., 2010. Thirty years of gas shale fracturing: what have we learned? Presented at the SPE Annual Technical Conference and Exhibition, Florence, 19−22 September. SPE-133456. <http://dx.doi.org/10.2118/133456-MS>.

Lo S.-W., Miller, M., 2002. Encapsulated breaker release rate at hydrostatic pressure and elevated temperatures. SPE 77744 Presented at the Annual Meeting and Exhibition, San Antonio, TX, 29 September–2 October.

Montgomery, C., 2013. Fracturing fluid components, Chapter 12. In: Bunger, A., McLennan, J., Jeffrey, R. (Eds.), Effective and Sustainable Hydraulic Fracturing book. ISBN 978-953-51-1137-5, Published: May 17, 2013 under CC BY 3.0 license, Intech Open Science. <http://dx.doi.org/10.5772/56422> <http://cdn.intechopen.com/pdfs-wm/44660.pdf>.

Montgomery, C., Smith, M., 2010. Hydraulic fracturing. JPT December.

PetroWiki, 2016. Fracturing fluids and additives. <http://petrowiki.org/Fracturing_fluids_and_additives>.

Qiu, X., Martch, E., Morgenthaler, L., Adams, J., Vu, H., 2009. Design criteria and application of high-density brine- based fracturing fluid for deepwater frac packs. Paper SPE 124704 Presented at the SPE Annual Technical Conference and Exhibition, New Orleans, LA, 4–7 October.

Samuel, M., Card, R.J., Nelson, E.B., Brown, J.E., Vinod, P.S., Temple, H.L., et al., 1999. Polymer-free fluid for fracturing applications. SPE Drill. Complet. 14 (4), 240–246.

Samuel, M., Polson, D., Kordziel, W., Waite, T., Waters, G., Vinod, P.S., et al. 2000. Viscoelastic surfactant fracturing fluids: application in low permeability reservoirs. Paper SPE 60322 Presented at the SPE Rocky Mountain Regional/Low Permeability Reservoirs Symposium and Exhibition, Denver, CO, 12–15 March.

Simms, L., Clarkson, B., 2008. Weighted frac fluids for lower-surface treating pressures. Paper SPE 112531 Presented at the SPE International Symposium and Exhibition on Formation Damage Control, Lafayette, LA, 13–15 February.

Sullivan, P.F, Gadiyar, B., Morales, R.H., Hollicek, R., Sorrells, D., Lee, J., et al. 2006. Optimization of a viscoelastic surfactant (VES) fracturing fluid for application in high permeability reservoirs. Paper SPE 98338 Presented at the SPE International Symposium and Exhibition on Formation Damage Control, Lafayette, LA, 15–17 February.

Watts, R., 2014. Hydraulic fracturing fluid systems energized with nitrogen, carbon dioxide optimize frac effectiveness. Am. Oil Gas Rep. 57 (February), 78–88.

Whitman, W.B., Coleman, D.C., Wiebe, W.J., 1998. Prokaryotes: the unseen majority. Proc. Natl. Acad. Sci. USA 95, 6578–6583.

Chapter 13

Drilling Fluid Components

INTRODUCTION

Oil field drilling fluids are mixtures of natural and synthetic chemical compounds and are essential to the successful completion of a drilling operation. An efficient drilling fluid must exhibit numerous characteristics, such as desired rheological properties (plastic viscosity, yield pint, shear-thinning rheology, and gel strengths), fluid loss control, stability under various temperature and pressure conditions, and stability against contaminates such as saltwater, calcium sulfate, cement, alkaline waters, and formation CO_2. In addition, additives are available for controlling the lubrication properties, minimizing torque and drag, wellbore strengthening, eliminating chemical wellbore instabilities, corrosion control, and minimizing formation damage.

This chapter covers many of the chemical entities used as additives in drilling fluids. These additives control fluid properties that, in general, fall under one or more of these headings:

- Mud weight
- Viscosity
- Fluid loss
- Chemical reactivity.

The chemistry of the additive is entirely determined by the base fluid used to make the drilling fluid. These base fluids were covered in Chapter 1, Introduction to Drilling Fluids; water, nonaqueous, or pneumatic. A multitude of configurations can be formulated within each base fluid. The additive used is determined by the property to be adjusted. The property to be adjusted is determined by the current drilling operation and/or problem to be addressed. The chemistry of the additive is determined by the base fluid used. Figs. 13.1–13.3 show most of the additives used to control each of the above functions for water, brine, and nonaqueous base fluid.

SUMMARY OF ADDITIVES

In Table 13.1 a classification of generic functions for drilling fluid additive uses are listed. As mentioned, some additives are used for multiple functions.

538 Composition and Properties of Drilling and Completion Fluids

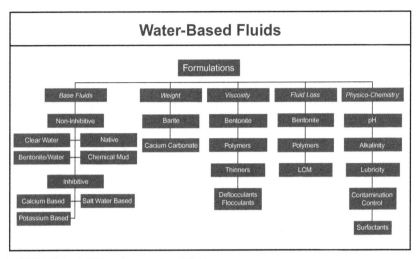

FIGURE 13.1 Additives for water-based fluids.

FIGURE 13.2 Additives for brine-based fluids.

In this chapter, additive's primary use will be covered in detail and secondary functions mentioned.

Many additives have been covered in the preceding chapters of this book. Those will not be covered in detail in this chapter. The additives in this section have been gleaned for various publications, many from patents. Some are in commercial use, others are not.

WEIGHTING AGENTS

Many weighting materials have been used in the drilling industry through the years. The primary material used to increase mud weight since Stroud's

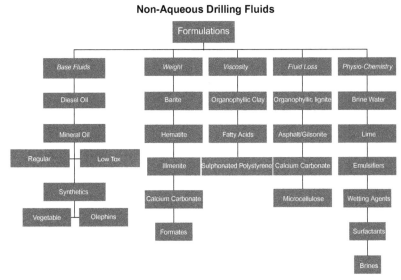

FIGURE 13.3 Additives for nonaqueous fluids.

TABLE 13.1 Summary of Additives

Additive Type	
Alkalinity, pH control	Hydrate suppressants
Bactericides	Lost circulation materials
Corrosion inhibitors	Lubricants/pipe freeing agents
Calcium reducers	Shale control inhibitors
Defoamers	Surface active agents
Emulsifiers	Temperature stability agents
Filtrate reducers	Thinners. dispersants
Flocculants	Viscosifiers
Foaming agents	Weight control material

patent in the 1920s has been barite. Conversely, the mud weight can be reduced by foaming or by the addition of hollow glass particles. The densities of weighting agents used through the years are summarized in Table 13.2.

In some cases, an alternative to barite (or other solid or weight material) is the use of soluble salts. Saturation with sodium chloride (common salt) increases the density of water to 10.0 lb/gal, and with calcium chloride to 11.8 lb/gal. Sodium and potassium formate brines can also be used to

TABLE 13.2 Densities of Weighting Agents

Material	Specific Gravity (g/cm^{-3})
Barite	3.9−4.4
Hematite	4.9−5.3
Iron Oxide (manufactured)	4.7
Siderite	3.7−3.9
Ilmenite	4.5−5.1
Manganese tetroxide	4.7
Lead carbonate	6.6
Galena	7.4−7.7
Calcite	2.6−2.8
Dolomite	2.8−2.9
Celestite	3.7−3.9

increase the mud weight without adding solid weight material. Densities of 19.0 + lb/gal can be reached with formate, chloride, and bromide brines (see chapter: Introduction to Drilling Fluids). The advantage of soluble salts as weighting agents is that density increases are achieved without increasing the solids content of the mud; however, the effect of the salt additions on other mud properties must be considered. Sometimes the desired mud weight is achieved through a combination of brines and other weighting agents. In offshore drilling, seawater, specific gravity about 8.5, is normally used as the makeup fluid.

Barite (BaSO$_4$)

Barite has been used as a weighting agent in drilling fluids since the 1920s. It is preferred over other materials because of its high density, low production costs, low abrasiveness, and ease of handling. Barite producers blend ores from different sources to obtain the desired average density to meet the current API specific gravity specification of 4.1 (API Specification 13A, 2015). Mud densities can be increased to 20 lb/gal or more with barite, if the drilled solids content is kept low.

The barite ore is crushed and ground for use in drilling muds. Particle size is important. Large particles will require thick mud for suspension, and will be removed by the shaker. On the other hand, very fine particles are undesirable since they result in a large total surface area of solids being exposed to the liquid phase. The API specification states that no more than

5% pass through a 325-mesh screen, and no more than 3% should be retained on a 200-mesh screen in the wet screen analysis.

Some barite ores contain alkaline-soluble carbonate minerals that can be detrimental to a drilling fluid, such as iron carbonate (siderite), lead carbonate (cerussite), and zinc carbonate (smithsonite) (Kulpa et al., 1992). API Spec 13A specifies that soluble alkaline earth metals present in the ground barite should not exceed 250 mg/L on the total hardness test. Details of the chemical analysis of barite are shown in API RP 13K (2016).

Most of the barite for drilling fluid use is mined in China. The ore is then shipped to grinding plants located in various parts or the world, close to oil field drilling activities. In 2010, China supplied about 51% of worldwide barite production and India about 14%.

Iron Oxides—Fe_2O_3

Hematite/Itabirite—Research and actual field use has shown hematite to be an excellent alternative to barite. Hematite is an ore composed chiefly of a soft micaceous hematite and small amounts of quartz. The micaceous hematite is very thin tablets or leaves of irregular outline, and the quartz is an aggregate of grains. Hematite closely resembles mica schist. Imagine the mica of such a schist replaced by a substance of mica-like thinness and the metallic luster of polished iron for a good idea of how hematite looks. Hematite is virtually free of contaminants, and is easily mined from a nearly inexhaustible source.

Additionally, hematite has three main advantages over barite: (1) higher specific gravity; (2) better particle distribution; and (3) a low particle attrition rate. For use in the oil field, hematite is refined by removing the quartz to an amount which is less than 1%.

The hematite API specific gravity specification is 4.9. Specific gravity is the key to mud weighting material performance. Less hematite is required to build mud weight because of its higher specific gravity. That reduces both the cost of weighting the mud and the amount of solids added to the system. Fewer solids in the mud reduce the need for expensive treating chemicals. Less solids means a lower plastic viscosity, lower gel strengths, a thinner filter cake, and better displacement by cement. Lower plastic viscosity reduces shear losses in the drill pipe and through the bit nozzles, resulting in more hydraulic horsepower in the jet stream to blast and clean the bottom of the hole. Users report 30% faster drilling rates.

Ilmenite ($Fe\text{-}TiO_2$)

Ilmenite has a specific gravity of up to 5.1. Saasen et al. (2001) reported a higher drilling penetration rate from the use of ilmenite, because a lesser colloidal solids fraction was produced during drilling. Ilmenite was used in drilling operations in the North Sea in 1979 and 1980. The ilmenite was a fairly coarse ground material compared to the presently used ilmenite. The drilling fluid properties were easier to control compared to drilling with barite. Again, this was because ilmenite has a lower tendency of being ground down to finer particles. Environmental aspects suggest replacing barite with ilmenite. However, the use of ilmenite as a weighting material can cause severe abrasion problems. Using ilmenite with a narrow particle size distribution around 10 can reduce the erosion to a level experienced with barite (Saasen et al., 2001).

Galena (PbO)

Galena is a high specific gravity lead oxide weight material (6.5) which drills faster because less solids are required to achieve a given mud weight compared to barite (4.1). A Mohs hardness of 5—6 can result in abrasion problems while using this material. It is only used in special cases when the mud weight needs to be increased to above 19.0 lb/gal while minimizing the total solids content.

Calcium Carbonate—$CaCO_3$

It is reasonable to replace barite- and iron-based weighting material with carbonate, specific gravity about 2.7, if a high degree of weighting the muds is not required by the drilling conditions. Besides being cheaper than barite, a carbonate weighting material is less abrasive, which is especially important when drilling is performed in producing formations, and it is readily soluble in hydrochloric acid. The main shortcomings of carbonate powders are due to the presence of a coarsely divided fraction and noncarbonate impurities (Lipkes et al., 1995).

MANGANESE TETROXIDE

Manganese tetraoxide, Mn_3O_4, has been recently used as a weighting material for water-based drilling fluids. Mn_3O_4 has a specific gravity of 4.8. It is used in muds for drilling deep gas wells. The filter cake formed by this mud also contains Mn_3O_4 (Moajil et al., 2008).

Several articles engaged with the use of manganese tetroxide with other additives in drilling fluid formulations reported negative effects on the reservoir performance. The permeability of reservoirs is reduced when they are contacted with such drilling fluids. Special and expensive stimulation techniques have been proven to be necessary.

Unlike $CaCO_3$, Mn_3O_4 is a strong oxidant (Moajil et al., 2008). Therefore, the use of HCl is not recommended for the removal of the filter cake. Various organic acids, chelating agents, and enzymes have been tested up to 150°C.

Research has been presented that indicates that a drilling fluid formulation containing manganese tetroxide results in a minimal reduction of the permeability of the reservoir (Al-Yami, 2009). These formulations are particularly useful in wells that are otherwise difficult to stimulate. A return permeability of 90% or greater is achieved without the need for acidizing treatments.

Hollow Glass Microspheres

Initially, glass microspheres were used in the 1970s to overcome severe lost circulation problems in the Ural Mountains. The technology has been used in other sites (McDonald et al., 1999). Hollow glass beads reduce the density of a drilling fluid and can be used for underbalanced drilling (Medley et al., 1995, 1997a,b). Field applications have been reported (Arco et al., 2000).

VISCOSITY MODIFIERS

This section covers both thickeners and thinners for drilling fluids. A variety of compounds are useful as thickeners. Chapter 5, Water-Dispersible Polymers, covers the common polymers used for viscosity modification, or mare appropriately rheology modification, see Chapter 6, The Rheology of Drilling Fluids, on rheology. The following discussions cover other miscellaneous materials used as viscosity modifiers.

pH Responsive Thickeners

The viscosity of ionic polymers is sometimes dependent on the pH. In particular, pH responsive thickeners can be prepared by copolymerization of acrylic or methacrylic acid ethyl acrylate or other vinyl monomers and tristyrylpoly(ethyleneoxy)$_x$ methyl acrylate. Such a copolymer provides a stable aqueous colloidal dispersion at an acid pH lower than 5.0 but becomes an effective thickener for aqueous systems upon adjustment to a pH of 5.5−10.5 or higher (Robinson, 1996, 1999).

Mixed Metal Hydroxides/Silicates (MMH/MMS)

By addition of mixed metal hydroxides, typical bentonite muds are transformed to an extremely shear-thinning fluid (Lange and Plank, 1999). At rest these fluids exhibit a very high viscosity but are thinned to an almost waterlike consistency when shear stress is applied. In theory, the shear-thinning rheology of mixed metal hydroxides and bentonite fluids is explained by the formation of a three-dimensional, fragile network of mixed metal hydroxides and bentonite. The positively charged mixed metal hydroxide particles attach themselves to the surface of negatively charged bentonite platelets. Typically, magnesium aluminum hydroxide salts are used as mixed metal hydroxides. Mixed metal hydroxides demonstrate the following advantages in drilling (Felixberger, 1996):

- High cuttings removal
- Suspension of solids during shutdown
- Lower pump resistance
- Stabilization of the borehole
- High drilling rates
- Protection of the producing formation

Mixed metal hydroxide drilling muds have been successfully used in horizontal wells, in tunneling under rivers, roads, and bays, for drilling in fluids, for drilling large-diameter holes, with coiled tubing, and to ream out cemented pipe.

Mixed metal hydroxides can be prepared from the corresponding chlorides treated with ammonium (Burba and Strother, 1991). Experiments done with various drilling fluids showed that the mixed metal hydroxides system, coupled with propylene glycol (Deem et al., 1991), caused the least skin damage of the drilling fluids tested. Thermally activated mixed metal hydroxides, made from naturally occurring minerals, especially hydrotalcites, may contain small or trace amounts of metal impurities besides the magnesium and aluminum components, which are particularly useful for activation (Keilhofer and Plank, 2000).

Mixed hydroxides of bivalent and trivalent metals with a three-dimensional spaced-lattice structure of the garnet type ($Ca_3Al_2[OH]_{12}$) have been described (Burba et al., 1992; Mueller et al., 1997).

Nonaqueous Gelling Agent

A fragile gel is a gel that can be easily disrupted or thinned under shear stress, etc. But it can quickly return to a gel when the stress is alleviated or removed, e.g., when the circulation of the fluid is stopped. Fragile gels may be disrupted by a mere pressure wave or a compression wave during drilling. They break instantaneously when disturbed, reversing from a gel back into a liquid form with minimum pressure, force, and time. Metal cross-linked phosphate esters impart a fragile progressive gel structure to a variety of nonaqueous-based drilling fluids, both at neutral or acidic pH.

The amount of phosphate ester and metal cross-linker used in a drilling fluid depends on the oil type and the desired viscosity of the drilling fluid. Generally, however, more phosphate ester and metal cross-linker is used for gelling or enhancing the viscosity of the fluid for transport than is used for imparting fragile progressive gel structure to the drilling fluid. Thus, metal cross-linked phosphate ester compositions enhance the fluid viscosity for suspending weighting materials in drilling fluids during transport of the fluids (Bell and Shumway, 2009).

Thinners

Phosphate Thinned Water-Based Muds

Phosphates are effective only in small concentrations. The mud temperature must be less than 55°C. The salt contamination must be less than 500 mg/L sodium chloride. The concentration of calcium ions should be kept as low as possible. The pH should be between 8 and 9.5. Some phosphates may decrease the pH, so adding more NaOH is required.

Lignite Muds

Lignite muds are high-temperature resistant up to 450°F (230°C). Lignite can control viscosity, gel strength, and fluid loss. The total hardness must be lower than 20 ppm.

Quebracho Muds

Quebracho is a natural product extracted from the heartwood of the *Schinopsis* trees that grow in Argentina and Paraguay. Quebracho is a well characterized polyphenolic and is readily extracted from the wood by hot water. Quebracho is widely used as a tanning agent. It is also used as a mineral dressing, as a dispersant in drilling muds, and in wood glues. Quebracho is commercially available as a crude hot water extract, either in lump, ground, or spray-dried form, or as a bisulfite treated spray-dried product that is completely soluble in cold water. Quebracho is also available in a bleached form, which can be used in applications where the dark color of unbleached quebracho is undesirable (Shuey and Custer, 1995).

Quebracho-treated freshwater muds are used in shallow depths. It is also referred to as *red mud* because of the deep red color. Quebracho acts as a thinner. Poly(phosphate)s are also added when quebracho is used. Quebracho is active at low concentrations and consists of tannates.

Lignosulfonate Muds

Lignosulfonate freshwater muds contain ferrochrome lignosulfonate for viscosity and gel strength control. These muds are resistant to most types of drilling contamination because of the thinning efficiency of the lignosulfonate in the presence of large amounts of salt and extreme hardness.

LOST CIRCULATION ADDITIVES

Filtration control is an important property of a drilling fluid, particularly when drilling through permeable formations in which the hydrostatic pressure exceeds the formation pressure. It is important for a drilling fluid to quickly form a filter cake which effectively minimizes fluid loss, but which also is thin and erodable enough to allow product to flow into the wellbore during production (Jarrett and Clapper, 2010). A few fluid loss additives for drilling fluids are summarized next for quick reference.

There are a number of methods that have been proposed to help to prevent the loss of a circulation fluid (Messenger, 1981). Some of these methods use fibrous, flaky, or granular materials to plug the pores as the particulate material settles out of the slurry. Lost circulation additives are summarized in Table 13.3. Other methods propose to use materials that interact in the fissures of the formation to form a plug of increased strength.

Water Swellable Polymers

Certain organic polymers absorb comparatively large quantities of water, e.g., alkali metal poly(acrylate) or cross-linked poly(acrylate)s (Green, 2001).

TABLE 13.3 Lost Circulation Additives

Material	References
Encapsulated lime	Walker (1986)
Encapsulated oil-absorbent polymers	Delhommer and Walker (1987a), Walker (1987, 1989)
Hydrolyzed poly(acrylonitrile)	Yakovlev and Konovalov (1987)
Divinylsulfone, cross-linked	
Poly(galactomannan) gum	Kohn (1988)
PU foam	Glowka et al. (1989)
Partially hydrolyzed poly(acrylamide) 30% hydrolyzed, cross-linked with Cr^{3+}	Sydansk (1990)
Oat hulls	House et al. (1991)
Rice products	Burts (1992, 1997)
Waste olive pulp	Duhon (1998)
Nut cork	Fuh et al. (1993), Rose (1996)
Pulp residue waste	Gullett and Head (1993)
Petroleum coke	Whitfill et al. (1990)
Shredded cellophane	Burts (2001)

Such water-absorbent polymers, insoluble in water and in hydrocarbons, can be injected into the well with the objective of encountering naturally occurring or added water at the entrance to and within an opening in the formation. The resulting swelling of the polymer forms a barrier to the continued passage of the circulation fluid through that opening into the formation.

The hydrocarbon carrier fluid initially prevents water from contacting the water-absorbent polymer until such water contact is desired. Once the hydrocarbon slug containing the polymer is properly placed at the lost circulation zone, water is mixed with the hydrocarbon slug so that the polymer will expand with the absorbed water and substantially increase in size to close off the lost circulation zone (Bloys and Wilton, 1991; Delhommer and Walker, 1987; Walker, 1987, 1989). The situation is similar to an oil-based cement. The opposite mechanism is used by a hydrocarbon-swellable elastomer (Wood, 2001).

Anionic Association Polymer

Another type of lost circulation agent is the combination of an organic phosphate ester and an aluminum compound, e.g., aluminum isopropoxide. The action of this system as a fluid loss agent seems to be that the alkyl phosphate ester becomes cross-linked by the aluminum compound to form an anionic association polymer, which serves as the gelling agent (Reid and Grichuk, 1991).

Other lost circulation additives can be encapsulated. The encapsulation is dissolved and the material swells to close fissures. Microbubbles in a drilling fluid can be generated by certain surfactants, and polymers known as *aphrons* are a different approach to reduce the fluid loss (Ivan et al., 2001).

Permanent Grouting

Lost circulation also can be suppressed by grouting permanently, either with cement (Allan and Kukacka, 1995; Cowan and Hale, 1994) or with organic polymers that cure in situ.

LUBRICANTS

During drilling, the drill string may develop an unacceptable rotational torque or, in the worst case, become stuck (see chapter: Drilling Problems Related to Drilling Fluids). When this happens, the drill string cannot be raised, lowered, or rotated. Common factors leading to this situation include:

- Cuttings or slough buildup in the borehole
- An undergauge borehole
- Irregular borehole development embedding a section of the drill pipe into the drilling mud wall cake
- Unexpected differential formation pressure

Differential pressure sticking occurs when the drill pipe becomes imbedded in the mud wall cake opposite a permeable zone. The difference between the hydrostatic pressure and the formation pressure holds the pipe in place, causing the drill pipe to become stuck. Differential sticking may be prevented, and a stuck drill bit may be freed, using an OBM or an oil-based or water-based surfactant composition. Such a composition reduces friction, permeates drilling mud wall cake, destroys binding wall cake, and reduces differential pressure.

Polarized Graphite

Graphite is a classical lubricant (Zaleski et al., 1998). Because of environmental concerns, compositions for solid lubricants have been developed in order to replace molybdenum disulfide. These compositions consisted of graphite, sodium molybdate, and sodium phosphate (Holinski, 1995). Later, such formulations have been addressed as polarized graphite. Polarized graphite can be used as a lubricant additive for rock bits. Unlike graphite, polarized graphite is a unique material that exhibits extremely good load-carrying ability and antiwear performance.

Graphite consists of carbon in a layered structure, and the lack of polarity inhibits graphite powder from forming a lubricant film and adhering to metal surfaces. The polarization of graphite results in the material having good adhesion to metal and forming a lubricant film that can carry extremely high loads without failure. Ordinary graphite has a laminar hexagonal crystal structure and the closed rings of carbon atoms do not normally have any electrical polarization. Hence, graphite has good lubricity in that the layers may slip or shear readily. However, the lack of polarity leads to a poor adhesion to metal surfaces.

Graphite can be treated with alkali molybdates or tungstenates, to impart a polarized layer at the surface of the graphite. Alternating positive and negative charges are formed on the surfaces. The treated graphite shows an extremely good load-carrying capacity and antiwear performance, somewhat similar to molybdenum disulfide. The polarized graphite exhibits a good adhesion of particles on metal surfaces and good film-forming properties (Denton and Lockstedt, 2006).

The adhesion properties of polarized graphite allow it to adhere to metal surfaces and to form a film that serves as a physical barrier between adjacent metal surfaces. Therefore, the polarized graphite, unlike graphite alone, acts as an adhesion promoter to the surface and changes the lubricity of the graphite. Thus, the grease composition may support much heavier loads, with a lower coefficient of friction (Denton and Lockstedt, 2006).

Ellipsodial Glass Granules

The use of ellipsoidal glass granules instead of spherical glass beads is substantiated by the effort to increase the contact surface of antifrictional particles, reduce their ability to penetrate deeply into the mud cake, and increase their breaking strength (Kurochkin and Tselovalnikov, 1994; Kurochkin et al., 1990, 1992a,b).

Paraffins

Purified paraffins are nontoxic and biodegradable (Halliday and Clapper, 1998). Biodegradable purified paraffins may be used as lubricants, rate of penetration enhancers, or spotting fluids for WBM.

Olefins

Olefin isomers containing 8−30 carbon atoms are suitable. However, isomers having fewer than 14 carbon atoms are more toxic, and isomers having more than 18 carbon atoms are more viscous. Therefore olefin isomers having 14−18 carbon atoms are preferred (Halliday and Schwertner, 1997).

Phospholipids

In aqueous drilling fluids, phospholipids are effective lubricating agents (Patel et al., 2006). Phospholipids are naturally occurring compounds, e.g., lecithin belongs to the class of phospholipids. An introduction to phospholipid chemistry has been given by Hanahan (1997). Phospholipids also find use as polymers (Nakaya and Li, 1999).

Because of their ionic nature, some phospholipids are soluble in water. A preferred compound as lubrication additive for aqueous drilling fluids is cocoamidopropyl propylene glycol diammonium chloride phosphate (Patel et al., 2006). Phosphatides or phospholipids are environmentally safe lubricating additives (Garyan et al., 1998).

Alcohols

Silicate-based aqueous drilling fluids have long been known to inhibit formation damage caused by water but have also long been known to have poor lubricity properties. Lubricants commonly known and used in WBMs do not provide good lubricity in silicate muds (Fisk et al., 2006). Chang reported in 2011 a new material that lowers the coefficient of friction in silicate drilling fluids (Chang et al., 2011).

A lubricant composition has been developed for silicic acid-based drilling fluids. This composition comprises 2-octyldodecanol and 2-ethylhexylglucoside

(Fisk et al., 2006). Alternative alcohols include oleyl alcohol, stearyl alcohol, and poly(etherglycol)s. Lubricants for both water-based and OBMs for use at low temperatures have been found as fatty acid partial glycerides.

Instead of alcohols, amino alcohols can be used. For example, a lubricating composition has been synthesized by the reaction of polymerized linseed oil with diethanol amine at 160°C. A product with a viscosity of around 2700 mPas at 40°C is obtained (Argillier et al., 2004). The viscosity can be reduced by adding some methyl oleate to the reaction product.

Synthetic PAOs are nontoxic and effective in marine environments when used as lubricants, return-of-permeability enhancers, or spotting fluid additives for WBM. A continuing need exists for other nontoxic additives for WBM, which serve as lubricants, return-of-permeability enhancers, and spotting fluids. Both poly(alkylene glycol) (PAG) (Alonso-Debolt et al., 1999) and side-chain polymeric alcohols such as poly(vinyl alcohol) (PVA) have been suggested. These substances are comparatively environmentally safe (Penkov et al., 1999; Sano, 1997).

PVAs may be applied as such or in cross-linked form (Audebert et al., 1996). Cross-linkers can be aldehydes, e.g., formaldehyde, acetaldehyde, glyoxal, and glutaraldehyde, to form acetals, maleic acid, or oxalic acid to form cross-linked ester bridges, or dimethylurea, poly(acrolein), diisocyanate, and divinylsulfonate (Audebert et al., 1994, 1998).

ETHERS AND ESTERS

Ethers and Esters

2-Ethylhexanol can be epoxidized with 1-hexadecene epoxide. This additive also helps reduce or prevent foaming. By eliminating the need for traditional oil-based components, the composition is nontoxic to marine life, biodegradable, environmentally acceptable, and capable of being disposed of at the drill site without costly disposal procedures (Alonso-Debolt et al., 1995). There is a growing interest in alternatives with better biodegradability, in particular esters.

The use of esters in water-based systems, particularly under highly alkaline conditions, can lead to considerable difficulties. Ester cleavage can result in the formation of components with a marked tendency to foam, which then introduce unwanted problems into the fluid systems.

Ester-Based Oils

Several ester-based oils are suitable as lubricants (Durr et al., 1994; Genuyt et al., 2001), as are branched chain carboxylic esters (Senaratne and Lilje, 1994). Tall oils can be transesterified with glycols (Runov et al., 1991) or condensed with monoethanolamine (Andreson et al., 1992). The ester class also comprises natural oils, such as vegetable oil (Argillier et al., 1999), spent sunflower oil (Kashkarov et al., 1997, 1998; Konovalov et al., 1993a,b), and natural fats, e.g., sulfonated fish fat (Bel et al., 1998). In WBM systems no harmful foams are formed from partially hydrolyzed glycerides of predominantly unsaturated C_{16} to C_{24} fatty acids.

The partial glycerides can be used at low temperatures and are biodegradable and nontoxic (Mueller et al., 2000). A composition for high-temperature applications is available (Wall et al., 1995). It is a mixture of long-chain poly(ester)s and poly(amide)s (PA)s. In the case of esters from, e.g., neopentylglycol, pentaerythrite, and trimethylolpropane with fatty acids, tertiary amines, such as triethanol amine, together with a mixture of fatty acids, improve the efficiency (Argillier et al., 1997).

Phosphate Esters

It has been found that the inclusion of poly(ether) phosphate esters in combination with PEG can give aqueous drilling fluids that provide good lubricating properties in a wide range of drilling fluids (Dixon, 2009).

Biodegradable Lubricants

A biodegradable lubricating composition has been proposed. This composition is based on an aliphatic hydrocarbon oil and a fatty acid ester (Genuyt et al., 2006). It is important that the hydrocarbon is not aromatic. The composition is used as a continuous oil phase in an invert emulsion in a petroleum drilling fluid or mud. The composition is particularly useful in offshore drilling in deep water or in inclined or long-range drilling. In the case of deep water drilling, the temperature of the water is around 4°C. Therefore, the viscosity of drilling fluids needs to be controlled at these low temperatures.

Starch Olefin Copolymer

Laboratory tests indicated that a starch olefin copolymer lubricant compositions lower both API and HTHP fluid loss values. Results are represented in Table 13.4. The coefficients of friction are up to 45% lower than those of the untreated base muds and are similar to those of OBMs. Only 0.5% of starch lubricant must be added to get satisfactory results (Sifferman et al., 2003).

CLAY AND SHALE STABILIZERS

It is important to maintain the wellbore stability during drilling, especially in water-sensitive shale and clay formations. The rocks within these types of formations absorb the fluid used in drilling. This absorption causes the rock to swell and may lead to a wellbore collapse. The swelling of clays and the

TABLE 13.4 Coefficients of Friction and Fluid Loss Values of Drilling Fluids Containing Starch Lubricant Composites

Composition	Friction		Fluid Loss	
	Coeff.	Reduction	API	HTHP
	k	%	mL	mL
Base mud	0.3126	–	8.0	26
Field mud + 3% lubricant	0.2981	4.6	4.4	14
Base mud + starch composite with 0.5% high MW olefin	0.2732	12.6	3.1	12
Base mud + starch composite with 0.5% base olefin	0.2653	15.1	3.4	11
Base mud + starch composite with 0.5% high MW olefin + ester	0.2551	18.4	3.0	13
Base mud + starch composite with 0.5% base olefin + ester olefin copolymer	0.2473	20.9	3.2	12
Base mud + starch composite with 0.5% poly(butene)	0.1672	46.5	2.7	10

problems that may arise from these phenomena have been reviewed in the literature (Durand et al., 1995a,b; Van Oort, 1997; Zhou et al., 1995). Clay stabilizers are shown in Table 13.5.

Salts

Swelling can be inhibited by the addition of KCl. Relatively high levels are required. Other swelling inhibitors are both uncharged polymers and poly(electrolyte)s (Anderson et al., 2010).

Quaternary Ammonium Salts

Choline salts are effective antiswelling drilling fluid additives for underbalanced drilling operations (Kippie and Gatlin, 2009). Choline is addressed as a quaternary ammonium salt containing the N,N,N-trimethylethanolammonium cation. An example of choline halide counterion salts is choline chloride.

The quaternization of a polymer from dimethyl amino ethyl methacrylate has been described. To an aqueous solution of a homopolymer from dimethyl amino ethyl methacrylate, sodium hydrochloride is added to adjust the pH to 8.9. Then, some water is added again and hexadecyl bromide as alkylation

TABLE 13.5 Clay Stabilizers

Additive	References
Polymer lattices	Stowe et al. (2002)
Partially hydrolyzed poly(vinylacetate)[a]	Kubena et al. (1993)
Poly(acrylamide)[b]	Zaitoun and Berton (1990), Zaltoun and Berton (1992)
Anionic and cationic copolymers	Aviles-Alcantara et al. (2000), Hale and Van Oort (1997), Patel and McLaurine (1993), Smith and Thomas (1995a,b, 1997)
Partially hydrolyzed acrylamide–acrylate copolymer, potassium chloride, and polyanionic cellulose (PAC)	Audibert et al. (1992), Halliday and Thielen (1987)
Cationic starches and PAGs	Branch (1988)
Hydroxyaldehydes or hydroxyketones	Westerkamp et al. (1991)
Pyruvic aldehyde and a triamine	Crawshaw et al. (2002)
In situ cross-linking of epoxide resins	Coveney et al. (1999a,b)
Quaternary ammonium carboxylates[BD,LT]	Himes (1992)
Copolymer of styrene and substituted maleic anhydride (MA)	Smith and Balson (2000)
Potassium salt of carboxymethyl cellulose	Palumbo et al. (1989)
Water-soluble polymers with sulfosuccinate derivative-based surfactants, zwitterionic surfactants[BD,LT]	Alonso-Debolt and Jarrett (1994, 1995a)
Capryloamphoglycinate	Alonso-Debolt and Jarrett (1995b)
Cocoamphodiacetate	Alonso-Debolt and Jarrett (1995b)
Disodium cocoamphodiacetate	Alonso-Debolt and Jarrett (1995b)
Lauroamphoacetate	Alonso-Debolt and Jarrett (1995b)
Sodium capryloamphohydroxypropyl sulfonate	Alonso-Debolt and Jarrett (1995b)
Sodium mixed C_8 amphocarboxylate	Alonso-Debolt and Jarrett (1995b)
Alkylamphohydroxypropyl sulfonate	Alonso-Debolt and Jarrett (1995b)
Partially hydrolyzed poly(acrylamide) and PPG, or a betaine	Patel et al. (1995)
Quaternized trihydroxyalkyl amine	Patel et al. (1995)
Polyfunctional poly(amine)	McGlothlin and Woodworth (1996)
Poly(acrylamide)	Ballard et al. (1994)
Amphoteric acetates and glycinates	Jarrett (1997a)

[MA]Maleic anhydride.
[BD]Biodegradable.
[LT]Low toxicity.
[SF]Well stimulation fluid.
[a]75% Hydrolyzed, 50 kDa.
[b]Shear-degraded, for montmorillonite clay dispersed in sand packs.

agent, and further benzylcetyldimethyl ammonium bromide is added as an emulsifier. This mixture is then heated, with stirring, to 60°C for 24 h (Eoff et al., 2006).

Amine Salts of Maleic Imide

Compositions containing amine salts of imides of MA polymers are useful for clay stabilization. These types of salts are formed, e.g., by the reaction of MA with a diamine such as dimethyl aminopropylamine, in (EG) solution (Poelker et al., 2009). The primary nitrogen dimethyl aminopropylamine forms the imide bond.

In addition, it may add to the double bond of MA. Further, the EG may add to the double bond, but also may condense with the anhydride itself. On repetition of these reactions, oligomeric compounds may be formed.

Finally, the product is neutralized with acetic acid or methanesulfonic acid to a pH of 4. The performance was tested in Bandera sandstone. The material neutralized with methanesulfonic acid is somewhat less then that neutralized with acetic acid. The compositions are particularly suitable for water-based hydraulic fracturing fluids.

Thermally treated carbohydrates are suitable as shale stabilizers (Sheu and Bland, 1992a,b,c). They may be formed by heating an alkaline solution of the carbohydrate, and the reaction product may be reacted with a cationic base. The inversion of nonreducing sugars may be first effected on selected carbohydrates, with the inversion catalyzing the browning reaction.

Potassium Formate

Clay is stabilized in drilling and treatment operations by the addition of potassium formate to the drilling fluid. Further, a cationic formation control additive is added. Potassium formate can be generated in situ from potassium hydroxide and formic acid. The cationic additive is basically a polymer containing quaternized amine units, e.g., polymers of dimethyl diallyl ammonium chloride or AAm (Smith, 2009).

In the clay pack flow test, where the higher volumes at a given time indicate better clay stability, the addition of a small amount of potassium formate increases the volume throughput for a given polymer concentration. For example, 0.1% poly(dimethyl diallyl ammonium chloride) added to the formulation had a volume at 10 min of 112 mL. The same polymer, when combined with potassium formate and treated at 0.05% of the polymer, i.e., half the original polymer concentration, had a volume of 146 mL, indicating better clay stability and a possible synergistic effect from the addition of the potassium formate (Smith, 2009).

Saccharide Derivatives

A drilling fluid additive, which acts as a clay stabilizer, is the reaction product of methyl glucoside and alkylene oxides, such as EO, PO, or 1,2-butylene oxide. Such an additive is soluble in water at ambient conditions, but becomes insoluble at elevated temperatures (Clapper and Watson, 1996). Because of their insolubility at elevated temperatures, these compounds concentrate at important surfaces such as the drill bit cutting surface, the borehole surface, and the surfaces of the drilled cuttings.

Sulfonated Asphalt

Asphalt is a solid, black-brown to black bitumen fraction, which softens when heated and rehardens upon cooling. Asphalt is not water soluble and difficult to disperse or emulsify in water. Sulfonated asphalt can be obtained by reacting asphalt with sulfuric acid and sulfur trioxide. By neutralization with alkali hydroxides, such as NaOH or NH_3, sulfonate salts are formed. Only a limited portion of the sulfonated product can be extracted with hot water. However, the fraction thus obtained, which is water soluble, is crucial for the quality.

Sulfonated asphalt is predominantly utilized for water-based drilling fluids but also for those based on oil (Huber et al., 2009). Apart from reduced filtrate loss and improved filter cake properties, good lubrication of the drill bit and decreased formation damage are important features assigned to sulfonated asphalt as a drilling fluid additive (Huber et al., 2009). In particular, clay inhibition is enhanced by sulfonated asphalt in the case of water-based drilling fluids. If swellable clays are not inhibited, undesirable water absorption and swelling of the clay occurs, which can cause serious technical problems, including the instability of the borehole or even a stuck pipe.

The mechanism of action of sulfonated asphalt as a clay inhibitor in a drilling fluid is explained by the fact that the electronegative sulfonated macromolecules attach to the electropositive ends of the clay platelets. Thereby, a neutralization barrier is created, which suppresses the absorption of water into the clay. In addition, because the sulfonated asphalt is partially lipophilic, and therefore water repellent, the water influx into the clay is restricted by purely physical principles. As mentioned already, the solubility in water of the sulfonated asphalt is crucial for proper application. By the introduction of a water-soluble and an anionic polymer component, the proportion of water-insoluble asphalt can be markedly reduced.

In other words, the proportion of the water-soluble fraction is increased by introducing the polymer component. Especially suitable are lignosulfonates as well as sulfonated phenol, ketone, naphthalene, acetone, and aminoplasticizing resins (Huber et al., 2009).

Grafted Copolymers

The clay stabilization of copolymers of styrene and MA grafted with PEG have been investigated (Smith and Balson, 2004). The amounts of shale recovery from bottle rolling tests have been used to measure the shale inhibition properties. The tests were done using Oxford clay cuttings and a water-sensitive shale, sieved to 2−4 mm. The swelling is performed in 7.6% aqueous KCl. The grafted copolymer used is an alternating copolymer of styrene and MA. The polymer is grafted with PEG with different molecular weight. The amount of shale recovery with various PEG types is shown in Table 13.6.

It seems that there is an optimum, with respect to the molecular weight of the grafted PEG. Further, the results in the lower part of Table 13.6 indicate that increasing the amount of styrene in the backbone increases also the amount of shale recovered.

TABLE 13.6 Amount of Shale Recovery

Sample	KCl [%]	Shale Recovery [%]
KCl only	7.6	25
PEG	7.6	38
SMAC MPEG 200	7.6	54
SMAC MPEG 300	7.6	87
SMAC MPEG 400	7.6	85
SMAC MPEG 500	7.6	72
SMAC MPEG 600	7.6	69
SMAC MPEG 750	7.6	70
SMAC MPEG 1100	7.6	66
SMAC MPEG 1500	7.6	49
KCl only	12.9	27
PEG	12.9	53
SMAC MPEG 500	12.9	85
SMAC 2:1 MPEG 500	12.9	95

(Smith and Balson, 2004).
SMAC Styrene and MA copolymer.
SMAC 2:1 Styrene and MA copolymer, 2 styrene units for every MA.
MPEG Poly(ethylene glycol) monomethyl ethers, the number refers to the molecular weight.

Poly(oxyalkylene Amine)s

One method to reduce clay swelling is to use salts in drilling fluids. Salts generally reduce the swelling of clays. However, salts flocculate the clays resulting in both high fluid losses and an almost complete loss of thixotropy. Further, increasing the salinity often decreases the functional characteristics of drilling fluid additives (Patel et al., 2007).

Another method for controlling clay swelling is to use organic shale inhibitor molecules in drilling fluids. It is believed that the organic shale inhibitor molecules are adsorbed on the surfaces of clays, with the added organic shale inhibitor competing with water molecules for clay reactive sites and thus serving to reduce clay swelling. Poly(oxyalkylene amine)s are a general class of compounds that contain primary amino groups attached to a poly(ether) backbone. They are also addressed as poly(ether amine)s. They are available in a variety of molecular weights, ranging up to 5 kDa.

Poly(oxyalkylenediamine)s have been proposed as shale inhibitors. These are synthesized from the ring-opening polymerization of oxirane compounds in the presence of amino compounds. Such compounds have been synthesized by reacting Jeffamine with two equivalents of EO. Alternatively, PO is reacted with an oxyalkyldiamine (Patel et al., 2007). The poly(ether) backbone is based either on EO, PO, or a mixture of these oxirane compounds (Patel et al., 2007). A typical poly(ether amine) is shown in Fig. 13.4. Such products belong to the Jeffamine product family. A related shale hydration inhibition agent is based on an N-alkylated 2,2′-diaminoethylether.

Anionic Polymers

Anionic polymers may be active as the long chain with negative ions attaches to the positive sites on the clay particles or to the hydrated clay surface through hydrogen bonding (Halliday and Thielen, 1987). Surface hydration is reduced as the polymer coats the surface of the clay. The protective coating also seals, or restricts the surface fractures or pores, thereby reducing or preventing the capillary movement of filtrate into the shale. This stabilizing process is supplemented by PAC. Potassium chloride enhances the rate of polymer absorption onto the clay.

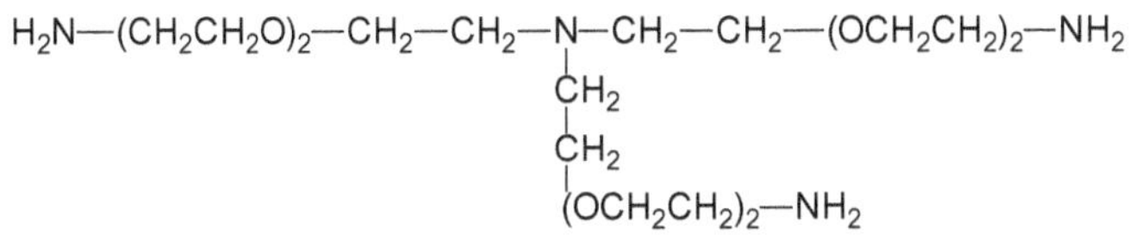

$$H_2N-(CH_2CH_2O)_2-CH_2-CH_2-\underset{\underset{\underset{(OCH_2CH_2)_2-NH_2}{|}}{\underset{CH_2}{|}}}{N}-CH_2-CH_2-(OCH_2CH_2)_2-NH_2$$

FIGURE 13.4 Poly(ether amine) (Klein and Godinich, 2006).

Shale Encapsulator

A shale encapsulator is added to a WBM in order to reduce the swelling of the subterranean formation in the presence of water. A shale encapsulator should be at least partially soluble in the aqueous continuous phase in order to be effective.

A conventional encapsulator is a quaternary PAM, which is preferably a quaternized PVA. Suitable examples of anions that are useful include halogen, sulfate, nitrate, formate, etc. (Patel et al., 2009).

By varying the molecular weight and the degree of amination, a wide variety of products can be tailored. It is possible to create shale encapsulators for the use in low salinity, including freshwater (Patel et al., 2009). The repeating units of quaternized etherified poly(vinyl alcohol) and quaternized PAM are shown in Fig. 13.5.

Membrane Formation

In order to increase the wellbore stability, it is possible to provide formulations for water-based drilling fluids, which can form a semipermeable osmotic membrane over a specific shale formation (Schlemmer, 2007). This membrane allows a comparatively free movement of water through the shale, but it significantly restricts the movement of ions across the membrane and thus into the shale.

The method of membrane formation involves the application of two reactants to form in situ a relatively insoluble Schiff base, which deposits at the shale as a polymer film. This Schiff base coats the clay surfaces to build a polymer membrane. The first reactant is a soluble monomer, oligomer, or polymer with ketone or aldehyde or aldol functionalities or precursors to those. Examples are carbon hydrates, such as dextrin, a linear of branched starch. The second reactant is a primary amine. These compounds are reacting by a condensation reaction to form an insoluble cross-linked polymerized product. The formation of a Schiff base is shown in Fig. 13.6.

Fig. 13.6 illustrates the reaction of a dextrine with a diamine, but other primary amines and poly(amine)s will of course react in the same way. Long-chain amines, diamines, or poly(amine)s with a relatively low amine ratio may require supplemental pH adjustment using materials such as

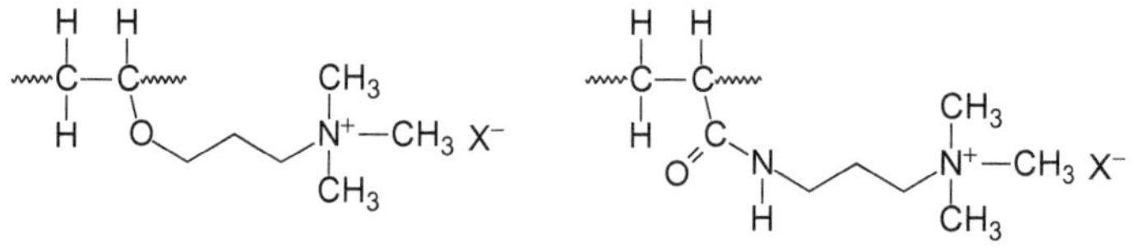

FIGURE 13.5 Quaternized etherified poly(vinyl alcohol) and quaternized poly(acrylamide) (Patel et al., 2009).

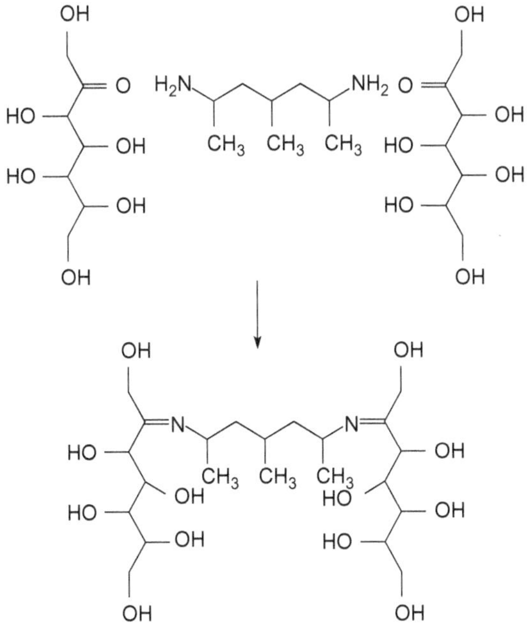

FIGURE 13.6 Formation of a Schiff base (Schlemmer, 2007).

sodium hydroxide, potassium hydroxide, sodium carbonate, potassium carbonate, or calcium hydroxide (Schlemmer, 2007). The Schiff base formed in this way must be essentially insoluble in the carrier brine in order to deposit a sealing membrane on the shale during drilling of a well.

By carefully selecting the primary polymer and the cross-linking amine, the relative concentrations of these components, together with the adjustment of pH, cross-linking and polymerization and precipitation of components occurs, which effectively forms an osmotically effective membrane on or within the face of the exposed rock. The polymerization and precipitation of the osmotic membrane on the face of the exposed rock significantly retards water or ions from moving into or out of the rock formation, typically shale or clay. The ability to form an osmotic barrier results in an increased stability in the clays or minerals, which combine to make the rock through which the borehole is being drilled (Schlemmer, 2007).

FORMATION DAMAGE PREVENTION

Formation damage due to invasion by drilling fluids is a well-known problem in drilling. Invasion of drilling fluids into the formation is caused by the differential pressure of the hydrostatic column, which is generally greater than the formation pressure, especially in low-pressure or depleted zones

(Audibert et al., 1999; Whitfill et al., 2005). Invasion is also caused by openings in the rock and the ability of fluids to move through the rock. When drilling depletes sands under overbalanced conditions, the mud will penetrate progressively into the formation unless there is an effective flow barrier present at the wellbore wall.

The presence of a mobile water phase can cause the migration of fines and subsequent formation damage. Therefore, it is desirable to minimize the migration of the fines, since fines block flow paths, choking the potential production of the well, as well as causing damage to downhole and surface equipment (Nguyen et al., 2010).

Horizontal drilling may also drill across highly fractured or permeable, low-pressure, or depleted zones, which increases the probability of the drill pipe getting stuck due to lying on the low side of the borehole. The exposure of numerous fractures or openings having low formation pressures has increased the problems of lost circulation and formation invasion (Whitfill et al., 2005). Poly(acrylate)s are often added to drilling fluids to increase the viscosity and limit formation damage.

SURFACTANTS

Surfactants are used to change the interfacial properties. Suitable surfactants are given in Table 13.7. Methyl-diethyl-alkoxymethyl ammonium methyl sulfate has high foam extinguishing properties (Fabrichnaya et al., 1997). Alkylpoly(glucoside)s (APGs) are highly biodegradable surfactants (Nicora and McGregor, 1998). The addition of APGs, even at very low concentrations, to a polymer mud can drastically reduce the fluid loss even at high temperatures. Moreover, both fluid rheology and temperature resistance are improved.

TABLE 13.7 Surface Active Agents for Drilling Muds

Compound	References
Alkylpolyglycosides	Lecocumichel and Amalric (1995)
Amphoteric surfactants	Dahanayake et al. (1996)
Acetal or ketal adduct hydroxy polyoxyalkylene ether[a]	Felix (1996)
Amphoteric anion ethoxy and propoxy units	Hatchman (1999a)
Alkanolamine	Hatchman (1999b)

[a]Controlling foam formation, drilling muds.

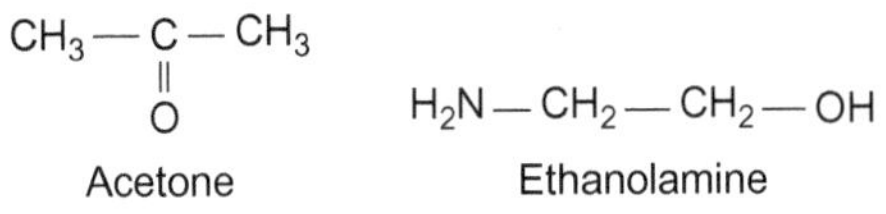

$$CH_3-C-CH_3$$
$$\overset{\|}{O}$$
Acetone

$$H_2N-CH_2-CH_2-OH$$
Ethanolamine

FIGURE 13.7 Acetone, ethanolamine.

There are special shale stabilizing surfactants consisting of nonionic alkanolamides (Jarrett, 1996, 1997b), e.g., acetamide monoethanolamines and diethanol amines. Acetone and ethanolamine are shown in Fig. 13.7.

EMULSIFIERS

Emulsions play an important role in fluids used for oil field applications. These include, most importantly, drilling and treatment fluids. Actually in these fields of applications, the emulsions are not addressed as such, rather, they are addressed, e.g., as OBMs or WBMs, which are essentially emulsions from the view of physics. Oil field emulsions are sometimes classified based on their degree of kinetic stability (Kokal and Wingrove, 2000; Kokal, 2006):

- Loose emulsions: Those that will separate within a few minutes. The separated water is sometimes referred to as the free water.
- Medium emulsions: They will separate in some 10 min.
- Tight emulsions: They will separate within hours, days, or even weeks. Thereafter, the separation may not be complete.

Emulsions are also classified by the size of the droplets in the continuous phase. When the dispersed droplets are larger than $0.1\,\mu$, the emulsion is a macroemulsion (Kokal, 2006).

From the purely thermodynamic view, an emulsion is an unstable system. This arises because there is a natural tendency for a liquid−liquid system to separate and reduce its interfacial area and thus its interfacial energy (Kokal and Wingrove, 2000).

A second class of emulsions is known as microemulsions. Such emulsions are formed spontaneously when two immiscible phases are brought together with extremely low interfacial energy. Microemulsions have very small droplet sizes, less than 10 nm, and are stable from the view of thermodynamics. Microemulsions are fundamentally different from macroemulsions in their formation and stability.

INVERT EMULSIONS

Invert emulsions have a continuous phase that is an oleaginous fluid and a discontinuous phase that is a fluid, which is at least partially immiscible in the oleaginous fluid. Actually, invert emulsion are water-in-oil emulsions.

Invert emulsions may have desirable suspension properties for particulates like drill cuttings. As such, they can easily be weighted if desired. It is well known that invert emulsions can be reversed to regular emulsions by changing the pH or by protonating the surfactant. In this way, the affinity of the surfactant for the continuous and discontinuous phases is changed (Taylor et al., 2009). For example, if a residual amount of an invert emulsion remains in a wellbore, this portion may be reversed to a regular emulsion to clean out the emulsion from the wellbore. Invert emulsion compositions can be used, where the organic phase is gelled. For example, diesel can be gelled with decanephosphonic acid monoethyl ester and a Fe^{3+} activator (Taylor et al., 2009). Polymers are often used to increase the viscosity an aqueous fluid. The polymer should interact with this fluid as it should show a tendency to hydrate. Microemulsions may be helpful to achieve this target (Jones and Wentzler, 2008).

Breakers

As breakers for invert emulsions, polymerized linseed oil reacted with diethanol amine has been proposed (Audibert-Hayet et al., 2007). Breakers are described in more detail in Chapter 11, Completion, Workover, Packer, and Reservoir Drilling Fluids, with drill-in fluids.

Drilling Fluid Systems

Invert emulsion fluid systems tend to exhibit high performance with regard to shale inhibition, borehole stability, and lubricity. However, invert emulsion fluid systems have a high risk of loss of circulation (Xiang, 2010). Latex additives can counterbalance this drawback, but since water must be added, an oil-base drilling fluid system becomes an unbalanced invert emulsion system with different rheological properties. It is possible to rebalance an unbalanced invert emulsion fluid system. Conventionally, an unbalanced invert emulsion fluid system would either have to be rebalanced in the field or transported offsite to be rebalanced.

There are special formulations that avoid this drawback. The latex particles are not dispersed in the oil-base continuous phase. Rather, the latex particles are dispersed in the emulsified aqueous phase. In fact, one of the advantages of invert emulsion fluids is that they can achieve at least some of the benefits of having an aqueous phase without requiring the aqueous phase to be in direct contact with the borehole wall (Xiang, 2010).

APHRONS

Aphrons are of importance in fluids used for oil field applications (Belkin et al., 2005; Growcock et al., 2007). The first use of aphrons in a drilling fluid application was described in 1998 (Brookey, 1998). An aphron is a phase that is surrounded by a tenside-like film, e.g., a soap bubble. The term originates from the Greek αφροσ for foam. The topic has been described by Sebba, who obviously coined the term (Sebba, 1984, 1987). Aphrons have been also termed as biliquid foams. In short, an aphron is a foam, where the liquid skin is built up from two phases. So, in contrast to a conventional air bubble, which is stabilized by a surfactant monolayer, the outer shell of an aphron consists of a much more robust surfactant trilayer.

Colloidal gas aphrons as proposed by Sebba are made up from a gaseous inner core surrounded by a thin aqueous surfactant film composed of two surfactant layers. In addition, there is a third surfactant layer that stabilizes the structure (Watcharasing et al., 2008). The basic structure of a colloidal gas aphron is shown in Fig. 13.8.

An aphron drilling fluid is similar to a conventional drilling fluid, but the drilling fluid system is converted to an energized air-bubble mud system before drilling (Kinchen et al., 2001).

When a drilling fluid migrates into a loss zone, aphrons move faster than the surrounding liquid phase and quickly form a layer of bubbles at the front of the fluid. This bubble front and the radial flow pattern of the fluid rapidly reduce the shear rate and raise the fluid viscosity, resulting in diminishing the invasion of the fluid. Further, aphrons exhibit only little affinity for each other or for the mineral surfaces of the pores or fractures. Consequently, the sealings they form are soft. Their lack of adhesion enables them to be

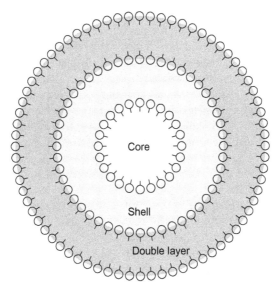

FIGURE 13.8 Basic structure of an aphron (Watcharasing et al., 2008).

flushed out easily whenever desired (Growcock et al., 2007). The aphrons reduce the density of a drilling fluid and provide bridging and sealing of the formations contacted by the fluid as the bubbles expand to fill the openings exposed while drilling. Low shear rate polymers strengthen the microbubbles and act as fluid loss agents (Brookey, 2004). In this way, lost circulation is prevented.

Moreover, the IFT between the base fluid and produced oils or gases is low, so the fluids have no tendency to formation damage. Depleted wells, which are very expensive to drill underbalanced or with other remediation techniques, have been drilled overbalanced with the aid of aphron drilling fluids (Growcock et al., 2007).

In a foam of the water-lamella type small globules of oil are encapsulated in a surfactant-stabilized film and separated from one another by a further thin lamella of water. Biliquid foams are essentially of two types (Sebba, 1984):

- Oil-lamella
- Water-lamella

An oil-lamella biliquid foam consists of aqueous cells coated with an oil film and separated from one another by an oil lamella. In this type, the oil phase corresponds to the aqueous phase in a conventional gas foam, and the water globules correspond to the gas cells. The second type of biliquid foam is one in which the discontinuous phase is the oil or a nonpolar liquid and the encapsulating phase is water or a hydrogen-bonded liquid that contains a soluble surfactant. The encapsulating film as well as the foam lamella are stabilized by the surfactant. In both types, the cells are held together by capillary pressures, just as are the soap bubbles in a gas foam. A water-lamella biliquid foam must be distinguished from oil-in-water emulsions, in which the discontinuous oil phase is separated from the continuous aqueous phase by a single interface. At moderate gas concentrations, the stability of bubbles in an aqueous medium depends primarily on the viscosity of the bulk fluid and the IFT. More specifically, the stability is determined by the rate of mass transfer between the viscous water shell and the bulk phase. This transfer is known as Marangoni convection (Bjorndalen and Kuru, 2008). Carlo Marangoni published his results in 1865 in the course of his doctoral thesis at the University of Pavia.

If, by some disturbing effects, e.g., a temperature gradient, a gradient in the surface tension is locally induced, a convection or movement of liquid will occur. Thus, the Marangoni convection contributes to the decomposition of foams. If the mass transfer rate is high, aphrons will become unstable. Therefore, the shell fluid must be designed to have certain viscosity to minimize the Marangoni effect (Bjorndalen and Kuru, 2008).

Bulk viscosity is generally controlled by the addition of polymers and clays. The IFT is usually lowered with a surfactant. In contrast to a typical bubble, an aphron is stabilized by a very high interfacial viscosity of the second phase (Brookey, 2004). Fluids with a low shear rate viscosity are helpful in controlling the invasion of a filtrate by creating an impermeable layer close to the formation openings. Since the fluid moves at a very slow rate, the viscosity becomes very high, and the depth of invasion of the fluid into the formation is kept shallow.

Specific aphron stabilizers are PVA in combination with surfactants, such as cocamidopropyl betaine or an alkyl ether sulfate (Growcock and Simon, 2006). The aphron stabilizer modifies the viscosity of the water layer to such an extent that it creates an elastomeric membrane. This elastomeric membrane allows the aphrons to display an improved stability and sealing capability.

The surfactants required to create the aphrons must be compatible with the polymers present in the fluid to create the desired low shear rate viscosity. Therefore, the surfactants will generally be of the nonionic or anionic type (Brookey, 2004).

LOW-FLUORESCENT EMULSIFIERS

Citric acid-based PA-type emulsifiers for drilling applications have been developed that exhibit a very low fluorescence. PAs are conventionally prepared by first reacting a fatty acid with diethylenetriamine in order to form the amide. Afterwards the amide is reacted with citric acid. These products exhibit a relatively high fluorescence. The reaction is shown in Fig. 13.9.

On the other hand, a PA with pendent citric acid units can be made via a cyclic intermediate, an imidazoline structure. This is followed by ring-opening to yield an isomeric amide. The reaction is shown in Fig. 13.10. The products synthesized via the reaction shown in Fig. 13.10 exhibit a much lower fluorescence. Discharged fluids having low fluorescence are less likely to impart a sheen to the ocean's surface (Cravey, 2010). In this way, discharge operations will become less obvious.

BACTERIA CONTROL

Bacterial contamination of drilling fluids contributes to a number of problems. Many of the muds contain sugar-based polymers in their formulation that provide an effective food source to bacterial populations. This can lead to direct degradation of the mud. In addition, the bacterial metabolism can generate deleterious products.

Most notable among these is hydrogen sulfide, which can lead to decomposition of mud polymers, formation of problematic solids such as iron sulfide, and corrosive action on drilling tubes and drilling hardware (Elphingstone and Woodworth, 1999). Moreover, hydrogen sulfide is a toxic gas.

Many polymers are used in drilling fluids as fluid loss control agents or viscosifiers. Because of the degradation of the polymers by bacteria in

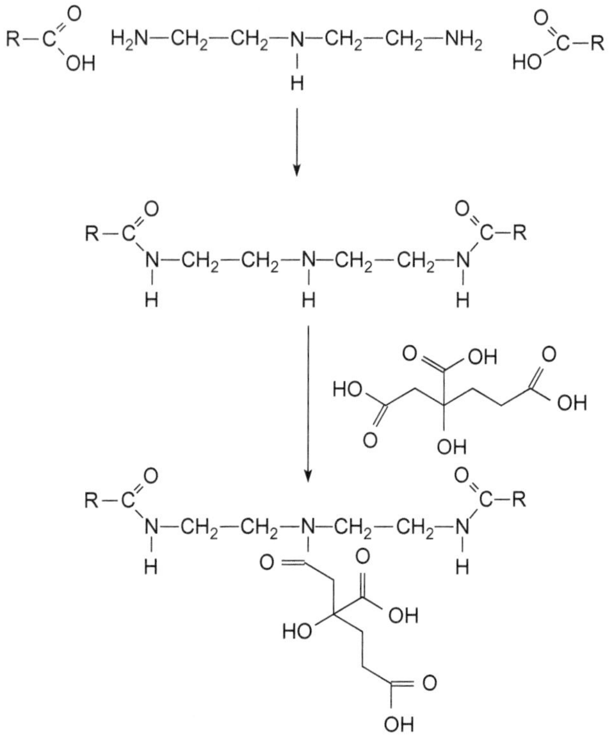

FIGURE 13.9 Conventional synthesis of poly(amide) surfactants (Cravey, 2010).

drilling fluids, an increase in fluid loss can occur. All naturally occurring polymers are capable of being degraded by bacterial action. However, some polymers are more susceptible to bacterial degradation than others. One solution, besides using bactericides, is replacing the starch with low-viscosity PAC, polyanionic lignin, or other enzyme-resistant polymers (Hodder et al., 1992). Certain additives are protected from biodegradation while drilling deep wells by quaternary ammonium salts (Rastegaev et al., 1999). This results in a considerably reduced consumption of the additives needed.

Bacteria control is important not only in drilling fluids, but also for other oil and gas operations. Some bactericides especially recommended for drilling fluids are summarized in Table 13.8 and sketched out in Fig. 13.11.

CORROSION INHIBITORS

The history of corrosion inhibitors and neutralizers and their invention, development, and application in the petroleum industry has been reviewed by Fisher (1993). Early corrosion inhibitor applications in each of the various

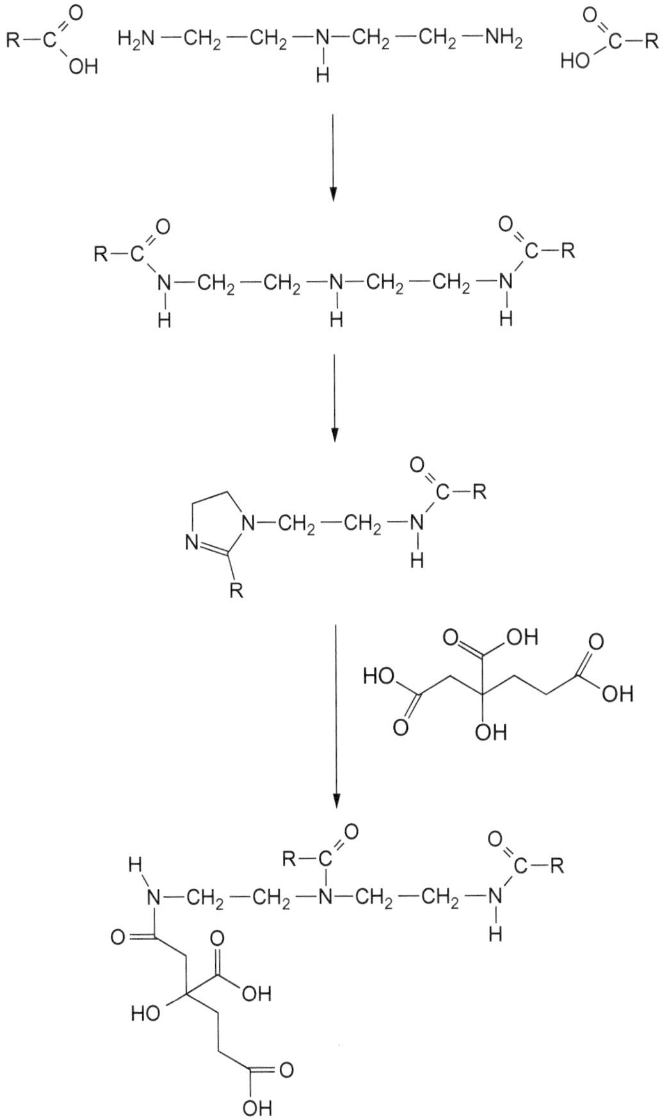

FIGURE 13.10 Synthesis of poly(amide) surfactants via imidazoline (Cravey, 2010).

segments of the industry, including oil wells, natural gas plants, refineries, and product pipelines, are reviewed.

Corrosion and scale deposition are the two most costly problems in oil industries. Corrodible surfaces are found throughout production, transport, and refining equipment. The *Corrosion and Scale Handbook* gives an overview of corrosion problems and methods of corrosion prevention (Becker, 1998).

TABLE 13.8 Bactericides for Drilling Fluids

Bactericide	References
Bis[tetrakis(hydroxymethyl)phosphonium] sulfate[a]	Elphingstone and Woodworth (1999)
Dimethyl-tetrahydro-thiadiazine-thione	Karaseva et al. (1995)
2-Bromo-4-hydroxyacetophenone[b]	Oppong and King (1995)
Thiocyanomethylthio-benzothiazole[c]	Oppong and Hollis (1995)
Dithiocarbamic acid	Austin and Morpeth (1992)
Hydroxamic acid[c]	Austin and Morpeth (1992)
1,2-Benzoisothiazolin-3-one[d]	Morpeth and Greenhalgh (1990)
3-(3,4-Dichlorophenyl)-1,1-dimethylurea	Morpeth and Greenhalgh (1990)
Di-iodomethyl-4-methylphenyl sulfone[e]	Morpeth and Greenhalgh (1990)
Isothiazolinones	Downey et al. (1995), Hsu (1990, 1995), Morpeth (1993)

[a] Absorbed on solid.
[b] Synergistically effective with organic acids.
[c] Synergistically effective with organic acids.
[d] Fungicide.
[e] Algicide.

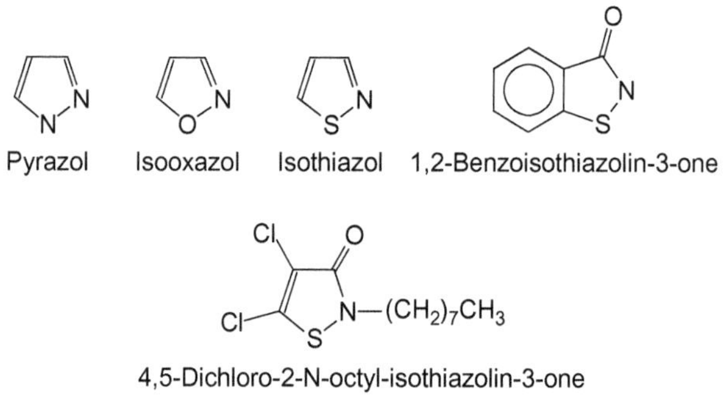

Pyrazol   Isooxazol   Isothiazol   1,2-Benzoisothiazolin-3-one

4,5-Dichloro-2-N-octyl-isothiazolin-3-one

FIGURE 13.11 Components for biozides.

Corrosion inhibitors have been divided into many groups, such as (Dietsche et al., 2007):

- Cathodic and anodic inhibitors
- Inorganic and organic corrosion inhibitors
- Filming and nonfilming inhibitors

Low molecular weight corrosion inhibitors often change the surface tension of water. Actually, these groups are acting as surfactants, as they form a protective layer on the metal surfaces (Dietsche et al., 2007). Polymeric corrosion inhibitors act in the same way as ordinary low molecular weight inhibitors. Polymeric film forming corrosion inhibitors are different from polymer coatings as they exhibit a specific interaction with the surface before the dry film is formed. Polymeric corrosion inhibitors may not form a barrier layer against oxygen and water. Instead they change the corrosion potential of the metal (Dietsche et al., 2007).

From the chemists's view, corrosion inhibitors can be classified into the following broad groupings:

- Amides and imidazolines
- Salts of nitrogenous molecules with carboxylic acids (i.e., fatty acids and naphthenic acids)
- Nitrogen quaternaries
- Polyoxylated amines, amides, imidazolines
- Nitrogen heterocyclics

A few inhibitors are shown in Fig. 13.12. Further compounds are summarized in Table 13.9. Some of these compounds are shown in Fig. 13.13.

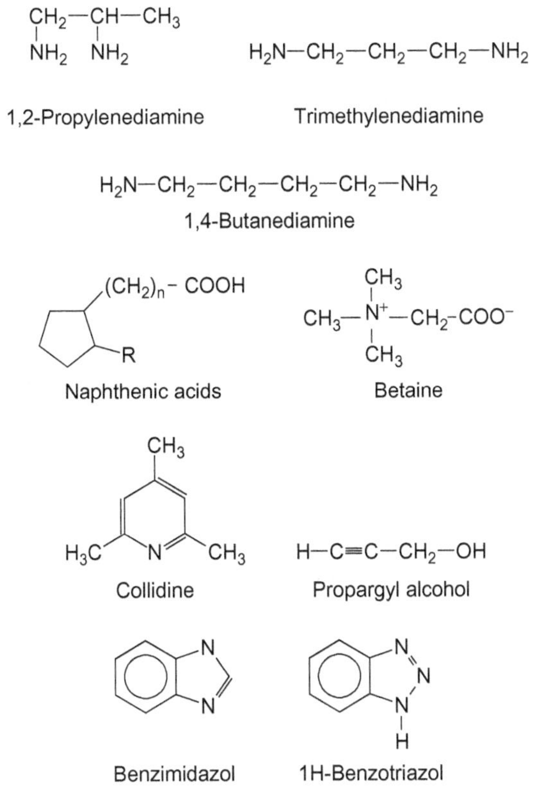

FIGURE 13.12 Diamines, acids alcohols, and azoles as corrosion inhibitors.

TABLE 13.9 Corrosion Inhibitors

Compound	References
Acetylinic alcohol[a]	Teeters (1992)
Tall oil fatty acid anhydrides	Fischer and Parker (1995, 1997)
3-Phenyl-2-propyn-1-ol[b]	Growcock and Lopp (1988)
Dicyclopentadiene dicarboxylic acid salts[c]	Darden and McEntire (1986, 1990)
Hydroxamic acid	Fong and Shambatta (1993)
Cyclohexylammonium benzoate	Johnson and Ippolito (1994, 1995)
Acyl derivatives of tris-hydroxy-ethyl- perhydro-1,3,5-triazine	Au and Hussey (1986, 1989)
2,4-Diamino-6-mercapto pyrimidine sulfate combined with oxysalts of vanadium, niobium, tantalum or titanium, zirconium, hafnium	Ramanarayanan and Vedage (1994)
Aqueous alkanol amine solution[d]	Schutt (1990), Veldman and Trahan (1999)
Quaternized fatty esters of alkoxylated alkyl-alkylene diamines	Wirtz et al. (1989)
Mercaptoalcohols	Ahn and Jovancicevic (2001)
Polysulfide[e]	Gay et al. (1993)
Polyphosphonohydroxybenzene sulfonic acid compounds[f]	Kreh (1991)
1-Hydroxyethylidene-1,1-diphosphonic acid[g]	Sekine et al. (1991)
2-Hydroxyphosphono-acetic acid[h]	Zefferi and May (1994a,b)
Water-soluble 1,2-dithiol-3-thiones[i]	Alink (1991, 1993)
Sulfonated alkyl phenol[j]	Babaian-Kibala (1993)
Polythioether	Incorvia (1988)
Thiazolidines	Alink and Outlaw (2001)
Substituted thiacrown ethers pendent on vinyl polymers	Minevski and Gaboury (1999)
Benzylsulfinylacetic acid or benzylsulfonylacetic acid	Lindstrom and Mark (1987)
Halohydroxyalkylthio-substituted and dihydroxyalkylthio-substituted polycarboxylic acids[k]	Lindstrom and Louthan (1987)
Alkyl-substituted thiourea	Tang et al. (1995)
2,5-Bis(N-pyridyl)-1,3,4-oxadiazoles	Bentiss et al. (2000)

[a] In combination with ClO_2 treatment for bacteria control.
[b] Aqueous HCl.
[c] 0.1–6% with antifreezers such as glycols.
[d] Gas stream containing H_2S or CO_2.
[e] Forms a film of iron disulfide.
[f] Relatively nontoxic, substitution of chromate-based corrosion inhibitors, conventional phosphate, and organophosphonate inhibitors and the zinc-based inhibitors.
[g] CO_2 environment.
[h] Calcium chloride brine.
[i] 10–500 ppm.
[j] 5–200 ppm to inhibit naphthenic acid corrosion.
[k] In drilling equipment.

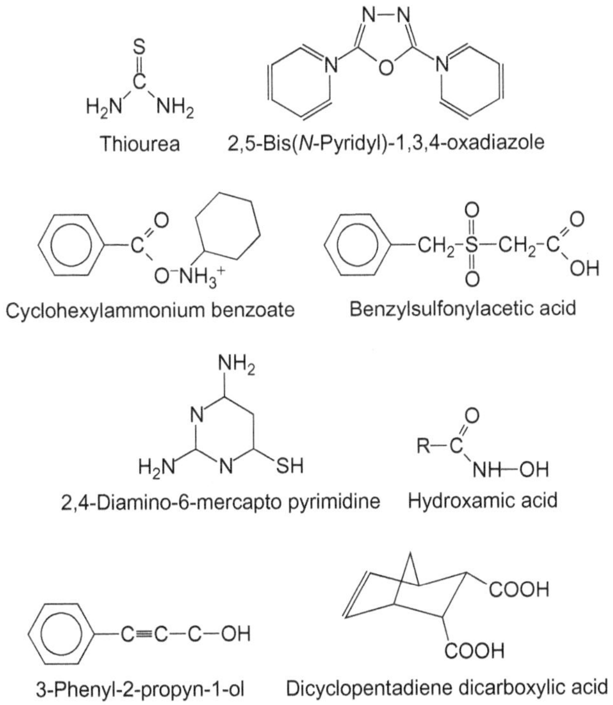

FIGURE 13.13 Miscellaneous corrosion inhibitors.

OXYGEN SCAVENGER

Oxygen corrosion is often underestimated. Studies have shown that the corrosion can be limited when proper oxygen scavengers are used. Hydrazine leads the group of chemicals that are available for oxygen removal. Because of its special properties, it is used for corrosion control in heating systems and in drilling operations, well workover, and cementing (Sikora, 1994).

HYDROGEN SULFIDE REMOVAL

Hydrogen sulfide is produced by these bacteria and is released to the protected environment, where it reacts with the dissolved iron from the corrosion process to form iron sulfide (Martin et al., 2005). So, it is sometimes necessary to remove hydrogen sulfide from a drilling mud. Techniques using iron compounds that form sparingly soluble sulfides have been developed, e.g., with iron (II) oxalate (Sunde and Olsen, 2000) and iron sulfate (Prokhorov et al., 1993). The sulfur is precipitated out as FeS. Ferrous gluconate is an organic iron-chelating agent, stable at pH levels as high as 11.5 (Davidson, 2001).

Zinc compounds have a high reactivity with regard to H_2S and therefore are suitable for the quantitative removal of even small amounts of hydrogen sulfide (Wegner and Reichert, 1990). However, at high temperatures they may negatively affect the rheology of drilling fluids.

SPECIAL ADDITIVES FOR WATER-BASED DRILLING MUDS

Improving the Thermal Stability

To avoid the problems associated with viscosity reduction in polymer-based aqueous fluids, formates, such as potassium formate and sodium formate, are commonly added to the fluids to enhance their thermal stability. However, this technology of using the formates is very expensive. The thermal stability of polymer-based aqueous fluids can be improved without the need of using formates (Maresh, 2009).

The stability of a wellbore treatment fluid may be maintained at temperatures up to 135−160°C (275−325°F). Various poly(saccharide)s may be included in the fluid. The apparent viscosities of drilling fluids containing xanthan gum and PAM before and after rolling at 120°C are shown in Table 13.10.

TABLE 13.10 Apparent Viscosity before and after Rolling

Composition	Before	After
	η [cP]	η [cP]
Brine/XC	13	3
Brin/PA	8.5	6
Brine/Filtercheck	4	4
Brine/FLC/XC	16	10.5
Brine/FLC/PA	14.6	9
Brine/XC/CLAYSEAL	12.5	3
XC/PA	30	28.5
XC/PA/FLC	38.5	16.5
XC/PA/FLC/CLAYSEAL	34	28
XC/PA/FLC/CLAYSEAL/Barite	38.5	38.5

Maresh (2009).

Dispersants

Complexes of tetravalent zirconium and ligands selected from organic acids such as citric, tartaric, malic, and lactic acid and a complex of aluminum and citric acid are suitable as dispersants (Burrafato and Carminati, 1994a,b, 1996, Burrafato et al., 1997). This type of dispersant is especially useful in dispersing bentonite suspensions. The muds can be used at pH values ranging from slightly acidic to strongly basic.

SYNTHETIC POLYMERS

Maleic Anhydride Copolymers

A mixture of sulfonated styrene−MA copolymer and polymers prepared from AA or AAm and their derivatives (Hale and Lawson, 1988) are dispersants for drilling fluids. The rheologic characteristics of aqueous well drilling fluids are enhanced by incorporating into the fluids small amounts of sulfonated styrene−itaconic acid copolymers (Hale and Rivers, 1988) and an AA or AAm polymer (Hale, 1988).

Sulfonated styrene−maleimide copolymers are similarly active (Lawson and Hale, 1989). Examples of maleimide monomers are maleimide, N-phenyl maleimide, N-ethyl maleimide, N-(2-chloropropyl) maleimide, and N-cyclohexyl maleimide (see Fig. 13.14). N-aryl and substituted aryl maleimide monomers are preferred. The polymers are obtained by free radical polymerization in solution, in bulk, or by suspension.

In copolymers containing the styrene sulfonate moiety and MA units, the MA units can be functionalized with an alkyl amine (Peiffer et al., 1991, 1992a,b,c, 1993a,b). The water-soluble polymers impart enhanced deflocculation characteristics to the mud. Typically, the deflocculants are relatively

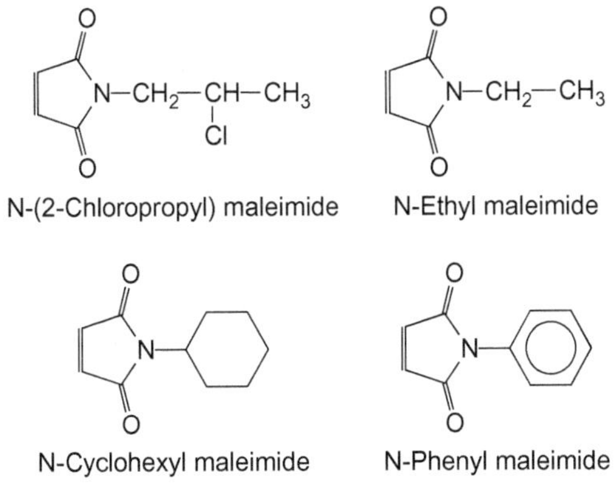

FIGURE 13.14 Imides.

low molecular weight polymers composed of styrene sodium sulfonate monomer, MA, as the anhydride or the diacid, and a zwitterionic functionalized MA. Typically, the molar ratio of styrene sulfonate units to total MA units ranges from 3:1 to 1:1. The level of alkyl amine functionalization of the MA units is 75–100 mol 45%. The molar concentrations of sulfonate and zwitterionic units are not necessarily equivalent, because the deflocculation properties of these water-soluble polymers can be controlled via changes in their ratio.

Alternating 1:1 copolymers of sodium methallylsulfonate and MA are useful as water-soluble dispersants (Grey, 1993). The copolymers are produced by free radical polymerization in acetic acid solution. Because of their high solubility in water and the high proportion of sulfonate salt functional groups, these alternating polymers are useful as dispersing agents in water-based drilling fluids.

Acrylics

Low molecular weight copolymers of AA and salts of vinyl sulfonic acid have been described as dispersants and high-temperature deflocculants for the stabilization of the rheologic properties of aqueous, clay-based drilling fluids subjected to high levels of calcium ion contamination (Portnoy, 1986, 1987). Divalent ions, such as calcium ions or magnesium ions, can cause uncontrolled thickening of the mud and large increases in filtration times of fluids from the mud into permeable formations.

A flocculation of the mud can occur in high-temperature applications. This flocculation increases the thickening effects of certain chemical contaminants and deactivates or destroys many mud thinners, which are used to stabilize the muds with respect to these effects.

Poly(acrylic acid) (PAA) or a water-soluble salt, having a molecular weight of 1.5–5 kDa, measured on the respective sodium salt, and a polydispersity of 1.05–1.45, has been described as a dispersant for a drilling or packer fluid (Farrar et al., 1992). Copolymers or terpolymers of AA, which contain from 5 to 50 mol 45% of sulfoethyl acrylamide, AAm and sulfoethyl acrylamide, ethyl acrylate and sulfoethyl acrylamide, AAm and sulfophenyl acrylamide, and AAm and sulfomethyl acrylamide, are claimed to be calcium tolerant deflocculants for drilling fluids (Giddings and Fong, 1988). In general, 0.1–2 lb of polymer per barrel of drilling fluid is sufficient to prevent the flocculation of the additives in the drilling fluid. A salt of a polymer or copolymer of acrylic or MA, in which the acid is neutralized with alkanolamines, alkyl amines, or lithium salts (Garvey et al., 1987), is suitable as a dispersing agent.

Amine Sulfide Terminal Moieties

Amine sulfide terminal moieties can be imparted into vinyl polymers by using aminethiols as chain transfer agents in aqueous radical polymerization (McCallum and Weinstein, 1994). The polymers are useful as mineral dispersants. Other uses are as water treatment additives for boiler waters, cooling towers, reverse osmosis applications, and geothermal processes in oil wells, as detergent additives acting as builders, antifilming agents, dispersants, sequestering agents, and encrustation inhibitors.

Polycarboxylates

Polycarboxylated polyalkoxylates and their sulfate derivatives may be prepared by reacting an ethoxylated or propoxylated alcohol with a water-soluble, alkali, or earth alkali metal salt of an unsaturated carboxylic acid (Chadwick and Phillips, 1995).

The reaction occurs in aqueous solution in the presence of a free radical initiator and gives products an enhanced yield and reduced impurity levels, compared with the essentially anhydrous reactions with free carboxylic acids, which have been used otherwise. The method provides products that give solutions that are clear on neutralization, remain clear and homogeneous on dilution, and are useful as cleaning agents in drilling and other oil field operations.

Allyloxybenzenesulfonate

Water-soluble polymers of allyloxybenzenesulfonate monomers can be used as dispersants in drilling fluids and in treating boiler waters in steamflooding and as plasticizers in cement slurries (Leighton and Sanders, 1988, 1990). The preferable molecular weight range is 1−500 kDa.

Sulfonated Isobutylene Maleic Anhydride Copolymer

A dispersant that can be used in drilling fluids, spacer fluids, cement slurries, completion fluids, and mixtures of drilling fluids and cement slurries controls the rheologic properties of and enhances the filtrate control in these fluids. The dispersant consists of polymers derived from monomeric residues, including low molecular weight olefins that may be sulfonated or phosphonated, unsaturated dicarboxylic acids, ethylenically unsaturated anhydrides, unsaturated aliphatic monocarboxylic acids, PVAs and diols, and sulfonated or phosphonated styrene. The sulfonic acid, phosphonic acid, and carboxylic acid groups on the polymers may be present in neutralized form as alkali metal or ammonium salts (Bloys et al., 1993, 1994).

Modified Natural Polymers

Polysaccharides

Phosphated, oxidized starch with a molecular weight of 1.5−40 kDa, with a carboxyl degree of substitution of 0.30−0.96, is useful as a dispersant for drilling fluids (Just and Nickol, 1989). Physical mixtures consist of reversibly cross-linked and uncross-linked hydrocolloid compositions and hydrocolloids. These show improved dispersion properties (Szablikowski et al., 1995).

Sulfonated Asphalt

Sulfonated asphalt can be produced as follows (Rooney et al., 1988):

1. Heating an asphaltic material
2. Mixing the asphalt with a solvent, such as hexane
3. Sulfonating the asphalt with a liquid sulfonating agent, such as liquid sulfur trioxide
4. Neutralizing the sulfonic acids with a basic neutralizing agent, such as sodium hydroxide
5. Separating solvent from the sulfonated asphalt
6. Recovering the evaporated solvent for reuse
7. Drying the separated, sulfonated asphalt by passing it through a drum dryer

This is a batch-type process in which the rates of flow of the solvent, the asphaltic material, the sulfonating agent, and the neutralizing agent and the periods of time before withdrawal of the sulfonic acids and the sulfonated asphalt are coordinated according to a predetermined time cycle.

The dried sulfonated asphalt can then be used in the preparation of drilling fluids, such as aqueous, oil-based, and emulsion types. Such drilling fluids have excellent rheologic properties, such as viscosity and gel strength, and exhibit a low rate of filtration or fluid loss.

Humic Acids

Coal with a mean particle size of less than 3 mm is slurried with water and then oxidized with oxygen or mixtures of oxygen and air at temperatures ranging from 100 to 300°C, at partial oxygen pressures ranging from 0.1 to 10 MPa and reaction periods ranging from 5−600 min (Cronje, 1989). In the absence of catalysts, such as alkaline bases, the main products of oxidation are humic acids.

These humic acids are not dissolved because the pH of this slurry is in the range of 4−9. Small amounts of fulvic acids are formed, and these are soluble in the water of the slurry. The coal-derived humic acids find applications as drilling fluid dispersants and viscosity control agents, whereas the coal-derived fulvic acids may be used to produce plasticizers and petrochemicals.

Miscellaneous Dispersants

A nonpolluting dispersing agent for drilling fluids (Bouchut et al., 1989, 1990, 1992) has been described. The agent is based on polymers or copolymers of unsaturated acids, such as AA or MA, with suitable counter ions.

SPECIAL ADDITIVES FOR NONAQUEOUS DRILLING MUDS

Poly(ether)cyclicpolyols

Poly(ether)cyclicpolyols possess enhanced molecular properties and characteristics and permit the preparation of enhanced drilling fluids that inhibit the formation of gas hydrates, prevent shale dispersion, and reduce the swelling of the formation to enhance wellbore stability, reduce fluid loss, and reduce filter cake thickness.

Drilling muds incorporating the poly(ether)cyclicpolyols are substitutes for OBMs in many applications (Blytas and Frank, 1995; Blytas et al., 1992a,b; Zuzich and Blytas, 1994, 1995). Poly(ether)cyclicpolyols are prepared by thermally condensing a polyol, e.g., glycerol to oligomers and cyclic ethers.

Emulsifier for Deep Drilling

Two major problems are encountered when using NADFs for drilling very deep wells (Dalmazzone et al., 2007). The first is a problem with the stability of the emulsions with temperature. The emulsifying agents that are stabilizing the emulsions must maintain water droplets in an emulsion up to temperatures of 400°F (200°C).

If the emulsion separates by coalescence of the water droplets, the fluid loses its rheological properties. The second is an environmental problem. The emulsification agents must not only be effective, but also as nontoxic as possible.

Fatty acid amides consisting of N-alkylated poly(ether) chains are used as emulsifiers. For those, the term polyalkoxylated superamides has been coined (Le Helloco et al., 2004). As a cosurfactant, tall oil fatty acids or their salts can be used.

Esters and Acetals

Esters of C_6 to C_{11} monocarboxylic acids (Mueller et al., 1990d,e, 1994; Müller et al., 1990a,b), acid-methyl esters (Mueller et al., 1990a), and polycarboxylic acid esters (Mueller et al., 1991b), as well as oleophilic monomeric and oligomeric diesters (Mueller et al., 1991c), have been proposed as basic materials for inverted emulsion muds. Natural oils are triglyceride ester oils

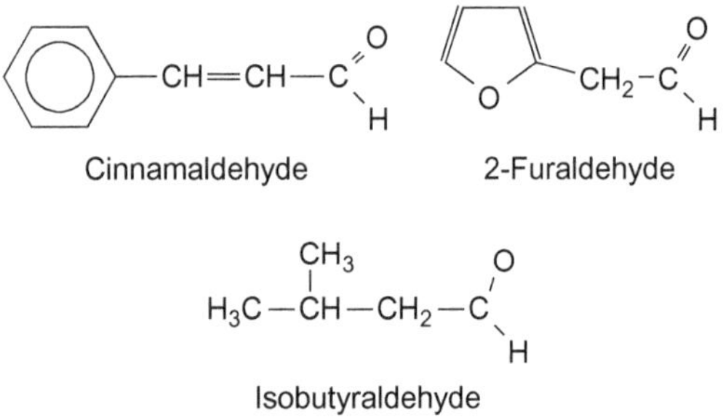

FIGURE 13.15 Aldehydes.

(Wilkinson et al., 1995) and are similar to synthetic esters. Diesters also have been proposed (Mueller et al., 1991c, 1992, 1993, 1995).

Acetals and oleophilic alcohols or oleophilic esters are suitable for the preparation of inverted emulsion drilling muds and emulsion drilling muds. They may replace the base oils, diesel oil, purified diesel oil, white oil, olefins, and alkyl benzenes (Hille et al., 1996, 1998a.). Examples are isobutyraldehyde, di-2-ethylhexyl acetal, and dihexyl formal. Also, mixtures with coconut alcohol, soya oil, and α-methyldecanol are suitable. Some aldehydes are shown in Fig. 13.15.

Inverted emulsion muds are more advantageous in stable, water-sensitive formations and in inclined boreholes. They are stable up to very high temperatures and provide excellent corrosion protection. Disadvantages are the higher price, the greater risk if gas reservoirs are bored through, the more difficult handling for the team at the tower, and greater environmental problems.

The high setting point of linear alcohols and the poor biologic degradability of branched alcohols limit their use as an environmentally friendly mineral oil substitutes. Higher alcohols, which are still just somewhat water-soluble, are eliminated for use in offshore muds because of their high toxicity to fish. Esters and acetals can be degraded anaerobically on the seabed. This possibility minimizes the environmentally damaging effect on the seabed. When such products are used, rapid recovery of the ecology of the seabed takes place after the end of drilling. Acetals, which have a relatively low viscosity and in particular a relatively low setting point, can be prepared by combining various aldehydes and alcohols (Hille et al., 1998b; Young and Young, 1994).

Antisettling Properties

Ethylene–AA copolymer neutralized with amines such as triethanol amine or N-methyl diethanol amine enhances antisettling properties (McNally et al., 1999; Santhanam and MacNally, 2001).

Glycosides

The advantage of using glycosides in the internal phase is that much of the concern for the ionic character of the internal phase is no longer required. If water is limited in the system, the hydration of the shales is greatly reduced. The reduced water activity of the internal phase of the mud and the improved efficiency of the shale is an osmotic barrier if the glycoside interacts directly with the shale. This helps to lower the water content of the shale, thus increasing rock strength, lowering effective mean stress, and stabilizing the wellbore (Hale and Loftin, 1996).

Methyl glucosides also could find applications in water-based drilling fluids and have the potential to replace OBMs (Headley et al., 1995). The use of such a drilling fluid could reduce the disposal of oil-contaminated drilling cuttings, minimize health and safety concerns, and minimize environmental effects.

Wettability

Other base materials proposed for wettability include Quaternary oleophilic esters of alkylolamines and carboxylic acids which improve the wettability of clay (Ponsati et al., 1992, 1994). Nitrates and nitrites can replace calcium chloride in inverted emulsion drilling muds (Fleming and Fleming, 1995).

The most consistently toxic bioassay phase is the suspended solid phase. This phase consists of bentonite, cuttings, and also soluble components. This toxic behavior has been explained by a certain chemical toxicity of a mud component, or by a physical toxicity by abrading or clogging epithelial tissue, i.e., respiratory or digestive body surfaces. In addition, the danger to marine animals from exposure to waste drilling muds may originate also from chemical toxicity. Further details are beyond the scope of this text and are referred to the literature (Kanz and Cravey, 1985, p. 329).

In a long-term study, the influence of increased levels of petroleum hydrocarbons upon soil and plants have been studied. Clean soil and different doses of drilling fluids and crude oil were applied. The changes in some chemical parameters of soil, plant density, and crop yields have been correlated. Drilling fluids showed a stronger impact on the chemical properties of the studied soil, while the plant density and yield were more strongly affected by crude oil. The soil levels of petroleum hydrocarbons, mineral oils, and polycyclic aromatic hydrocarbons were significantly reduced after the first trial year (Kisic et al., 2009).

For the reasons illustrated above, there is a need to take care on the waste management of drilling muds. Next, selected solutions for waste management are discussed.

Surfactants

Alkyl phenol ethoxylates are a class of surfactants that have been used widely in the drilling fluid industry. The popularity of these surfactants is based on their cost-effectiveness, availability, and range of obtainable hydrophilic–lipophilic balance values (Getliff and James, 1996). Studies have shown that alkyl phenol ethoxylates exhibit estrogenic effects and can cause sterility in some male aquatic species. This may have subsequent human consequences, and such problems have led to a banning of their use in some countries and agreements to phase out their use. Alternatives to products containing alkyl phenol ethoxylates are available, and in some cases they show an even better technical performance.

Organophilic clays treated with biodegradable cationic surfactants may be used in drilling fluids for drilling a wellbore without being concerned that the surfactant could accumulate in the environment. As such, the surfactant usually never reaches toxic levels that could harm the surrounding environment and the life supported by that environment (Miller, 2009).

MUD TO CEMENT CONVERSION

Water-based drilling fluids may be converted into cements using a hydraulic blast furnace slag (Bell, 1993; Cowan and Hale, 1995; Cowan et al., 1994; Cowan and Smith, 1993; Zhao et al., 1996). Hydraulic blast furnace slag is a unique material that has a low impact on rheologic and fluid loss properties of drilling fluids. It can be activated to set in drilling fluids that are difficult to convert to cements with other solidification technologies.

Hydraulic blast furnace slag is a more uniform and consistent quality product than are Portland well cements, and it is available in large quantities from multiple sources. Fluid and hardened solid properties of blast furnace slag and drilling fluids mixtures used for cementing operations are comparable with the properties of conventional Portland cement compositions.

REFERENCES

Ahn, Y.S., Jovancicevic, V., 2001. Mercaptoalcohol corrosion inhibitors. WO Patent 0 112 878, assigned to Baker Hughes Inc., February 22.

Alink, B.A.O., 1991. Water soluble 1,2-dithio-3-thiones. EP Patent 415 556, assigned to Petrolite Corporation, March 06.

Alink, B.A.O., 1993. Water soluble 1,2-dithio-3-thiones. US Patent 5 252 289, assigned to Petrolite Corporation, October 12.

Alink, B.A.M.O., Outlaw, B.T., 2001. Thiazolidines and use thereof for corrosion inhibition. WO Patent 0 140 205, assigned to Baker Hughes Inc., June 07.

Allan, M.L., Kukacka, L.E., 1995. Calcium phosphate cements for lost circulation control in geothermal. Geothermics 24 (2), 269–282.

Alonso-Debolt, M., Jarrett, M.A., 1994. New polymer/surfactant systems for stabilizing troublesome gumbo shale. In: Proceedings Volume, SPE International Petroleum Conference in Mexico, Veracruz, Mexico, October 10–13, 1994, pp. 699–708.

Alonso-Debolt, M., Jarrett, M., 1995a. Synergistic effects of sulfosuccinate/polymer system for clay stabilization. In: Proceedings Volume, Vol. PD-65, Asme Energy-Sources Technology Conference Drilling Technology Symposium, Houston, January 29–February 1, 1995, pp. 311–315.

Alonso-Debolt, M., Jarrett, M.A., 1995b. Drilling fluid additive for water-sensitive shales and clays, and method of drilling using the same. EP Patent 668 339, assigned to Baker Hughes Inc., August 23.

Alonso-Debolt, M.A., Bland, R.G., Chai, B.J., Eichelberger, P.B., Elphingstone, E.A., 1995. Glycol and glycol ether lubricants and spotting fluids. WO Patent 9 528 455, assigned to Baker Hughes Inc., 26 October.

Alonso-Debolt, M.A., Bland, R.G., Chai, B.J., Elchelberger, P.B., Elphingstone, E.A., 1999. Glycol and glycol ether lubricants and spotting fluids. US Patent 5 945 386, assigned to Baker Hughes Inc., 31 August.

Al-Yami, A.S.H.A.-B., 2009. Non-damaging manganese tetroxide water-based drilling fluids. US Patent 7 618 924, assigned to Saudi Arabian Oil Company, Dhahran, SA, 17 November.

Anderson, R.L., Ratcliffe, I., Greenwell, H.C., Williams, P.A., Cliffe, S., Coveney, P.V., 2010. Clay swelling: a challenge in the oilfield. Earth-Sci. Rev. 98 (3-4), 201–216.

Andreson, B.A., Abdrakhmanov, R.G., Bochkarev, G.P., Umutbaev, V.N., Fryazinov, V.V., Kudinov, V.N., et al., 1992. Lubricating additive for water-based drilling solutions–contains products of condensation of monoethanolamine and tall oils, kerosene, monoethanolamine and flotation reagent. SU Patent 1 749 226, assigned to Bashkir Oil Ind. Res. Inst. and Bashkir Oil Proc. Inst., 23 July.

API Standard, API RP 13K, 2016. Recommended Practice for Chemical Analysis of Barite. American Petroleum Institute, Washington, DC.

API Standard, API Specification 13A, 2015. Recommended Practice for Chemical Analysis of Barite. American Petroleum Institute, Washington, DC.

Arco, M.J., Blanco, J.G., Marquez, R.L., Garavito, S.M., Tovar, J.G., Farias, A.F. , et al., 2000. Field application of glass bubbles as a density-reducing agent. In: Proceedings Volume, Annual SPE Technical Conference, Dallas, TX, 1–4 October, 2000, pp. 115–126.

Argillier, J.F., Audibert, A., Marchand, P., Demoulin, A., Janssen, M., 1997. Lubricating composition including an ester-use of the composition and well fluid including the composition. US Patent 5 618 780, assigned to Inst. Francais Du Petrole, 08 April.

Argillier, J.F., Demoulin, A., Audibert-Hayet, A., Janssen, M., 1999. Borehole fluid containing a lubricating composition–method for verifying the lubrification of a borehole fluid–application with respect to fluids with a high pH. WO Patent 9 966 006, assigned to Inst. Francais Du Petrole and Fina Research SA, 23 December.

Argillier, J.-F., Demoulin, A., Audibert-Hayet, A., Janssen, M., 2004. Borehole fluid containing a lubricating composition-method for verifying the lubrification of a borehole fluid-application with respect to fluids with a high pH. US Patent 6 750 180, assigned to Institut Francais du Petrole (Rueil-Malmaison Cedex, FR) Oleon NV (Ertvelde, BE), 15 June.

Au, A.T., 1986. Acyl derivatives of tris-hydroxy-ethyl-perhydro- 1,3,5-triazine. US Patent 4 605 737, August 12.

Au, A.T., Hussey, H.F., 1989. Method of inhibiting corrosion using perhydro-s-triazine derivatives. US Patent 4 830 827, May 16.

Audebert, R., Janca, J., Maroy, P., Hendriks, H., 1994. Chemically crosslinked polyvinyl alcohol (PVA), process for synthesizing same and its applications as a fluid loss control agent in oil fluids. GB Patent 2 278 359, assigned to Sofitech NV, 30 November.

Audebert, R., Janca, J., Maroy, P., Hendriks, H., 1996. Chemically crosslinked polyvinyl alcohol (PVA), process for synthesizing same and its applications as a fluid loss control agent in oil fluids. CA Patent 2 118 070, assigned to Schlumberger Canada Ltd., 14 April.

Audebert, R., Maroy, P., Janca, J., Hendriks, H., 1998. Chemically crosslinked polyvinyl alcohol (PVA), and its applications as a fluid loss control agent in oil fluids. EP Patent 705 850, assigned to Sofitech NV, 02 September.

Audibert, A., Lecourtier, J., Bailey, L., Hall, P.L., Keall, M., 1992. The role of clay/polymer interactions in clay stabilization during drilling. In: Proceedings Volume, 6th Inst. Francais Du Petrole Exploration & Production Research Conference, Saint-Raphael, France, September 4–6, 1991, pp. 203–209.

Audibert, A., Argillier, J.F., Ladva, H.K.J., Way, P.W., Hove, A.O., 1999. Role of polymers on formation damage. In: Proceedings Volume, SPE Europe Formation Damage Conference, The Hague, Netherland, May 31—June 1, pp. 505–516.

Audibert-Hayet, A., Giard-Blanchard, C., Dalmazzone, C., 2007. Organic emulsion-breaking formula and its use in treating wellbores drilled in oil-base mud. US Patent 7 226 896, assigned to Institut Francais du Petrole, Rueil Malmaison Cedex, FR, June 5.

Austin, P.W., Morpeth, F.F., 1992. Composition and use. EP Patent 500 352, assigned to Imperial Chemical Inds Pl, August 26.

Aviles-Alcantara, C., Guzman, C.C., Rodriguez, M.A., 2000. Characterization and synthesis of synthetic drilling fluid shale stabilizer. In: Proceedings Volume, SPE International Petroleum Conference in Mexico, Villahermosa, February 1–3, 2000.

Babaian-Kibala, E., 1993. Naphthenic acid corrosion inhibitor. US Patent 5 252 254, October 12.

Ballard, T.J., Beare, S.P., Lawless, T.A., 1994. Shale inhibition with water-based muds: The influence of polymers on water transport through shales. In: Proceedings Volume, Recent Advances in Oilfield Chemistry, 5th Royal Society of Chemistry International Symposium, Ambleside, England, April 13–15, 1994, pp. 38–55.

Becker, J.R., 1998. Corrosion and Scale Handbook. Pennwell Publishing Co, Tulsa, OK.

Bel, S.L.A., Demin, V.V., Kashkarov, N.G., Konovalov, E.A., Sidorov, V.M., Bezsolitsen, V.P., et al., 1998. Lubricating composition–for treatment of clayey drilling solutions, contains additive in form of sulphonated fish fat. RU Patent 2 106 381, assigned to Shchelkovsk Agro Ent St C and Fakel Research Production Association, 10 March.

Belkin, A., Irving, M., O'Connor, R., Fosdick, M., Hoff, T., Growcock, F.B., 2005. How aphron drilling fluids work, Annual Technical Conference And Exhibition, Vol. 5. Society of Petroleum Engineers, Richardsoon, TX.

Bell, S., 1993. Mud-to-cement technology converts industry practices. Pet. Eng. Int. 65 (9), 51–52, 54–55.

Bell, S.A., Shumway, W.W., 2009. Additives for imparting fragile progressive gel structure and controlled temporary viscosity to oil based drilling fluids. US Patent 7 560 418, assigned to Halliburton Energy Services, Inc., Duncan, OK, 14 July.

Bentiss, F., Lagrenee, M., Traisnel, M., 2000. 2,5-bis(N-pyridyl)-1,3,4-oxadiazoles as corrosion inhibitors for mild steel in acidic media. Corrosion 56 (7), 733–742.

Bjorndalen, N., Kuru, E., 2008. Physico-chemical characterization of aphron based drilling fluids. J. Can. Pet. Technol. 47 (11), 15–21.

Bloys, J.B., Wilton, B.S., 1991. Control of lost circulation in wells. US Patent 5 065 820, assigned to Atlantic Richfield Company, 19 November.

Bloys, J.B., Wilson, W.N., Malachosky, E., Carpenter, R.B., Bradshaw, R.D., 1993. Dispersant compositions for subterranean well drilling and completion. EP Patent 525 037, 03 February.
Bloys, J.B., Wilson, W.N., Malachosky, E., Bradshaw, R.D., Grey, R.A., 1994. Dispersant compositions comprising sulfonated isobutylene maleic anhydride copolymer for subterranean well drilling and completion. US Patent 5 360 787, assigned to Atlantic Richfield Company, 01 November.
Blytas, G.C., Frank, H., 1995. Copolymerization of polyethercyclicpolyols with epoxy resins. US Patent 5 401 860, assigned to Shell Oil Company, March 28.
Blytas, G.C., Frank, H., Zuzich, A.H., Holloway, E.L., 1992a. Method of preparing polyethercyclicpolyols. EP Patent 505 000, assigned to Shell International Research Maatschappij BV, September 23.
Blytas, G.C., Zuzich, A.H., Holloway, E.L., Frank, H., 1992b. Method of preparing polyethercyclicpolyols. EP Patent 505 002, assigned to Shell International Research Maatschappij BV, september 23.
Bouchut, P., Rousset, J., Kensicher, Y., 1989. A non-polluting dispersing agent for drilling fluids based on freshwater or salt water. AU Patent 590 248, November 02.
Bouchut, P., Rousset, J., Kensicker, Y., 1990. Non-polluting fluidizing agents for drilling fluids having soft or salt water base (agent fluidifiant non polluant pour fluides de forage a base d'eau douce ou saline). CA Patent 1 267 777, April 17.
Bouchut, P., Kensicher, Y., Rousset, J., 1992. Non-polluting dispersing agent for drilling fluids based on freshwater or salt water. US Patent 5 099 928, March 31.
Branch, H.I., 1988. Shale-stabilizing drilling fluids and method for producing same. US Patent 4 719 021, January 12.
Brookey, T., 1998. Micro-bubbles: new aphron drill-in fluid technique reduces formation damage in horizontal wells. SPE Formation Damage Control Conference. SPE, Society of Petroleum Engineers, The Woodlands. TX.
Brookey, T.F., 2004. Aphron-containing well drilling and servicing fluids. US Patent 6 716 797, assigned to Masi Technologies, LLC, Houston, TX, 6 April.
Burba, J.L.I., Strother, G.W., 1991. Mixed metal hydroxides for thickening water or hydrophylic fluids. US Patent 4 990 268, assigned to Dow Chemical Company, 05 February.
Burba, J.L.I., Hoy, E.F., Read Jr., A.E., 1992. Adducts of clay and activated mixed metal oxides. WO Patent 9 218 238, assigned to Dow Chemical Company, 29 October.
Burrafato, G., Carminati, S., 1994a. Aqueous drilling muds fluidified by means of zirconium and aluminium complexes. EP Patent 623 663, assigned to Eniricerche SPA and Agip SPA, 09 November.
Burrafato, G., Carminati, S., 1994b. Aqueous drilling muds fluidified by means of zirconium and aluminum complexes. CA Patent 2 104 134, 08 November.
Burrafato, G., Carminati, S., 1996. Aqueous drilling muds fluidified by means of zirconium and aluminum complexes. US Patent 5 532 211, 02 July.
Burrafato, G., Gaurneri, A., Lockhart, T.P., Nicora, L., 1997. Zirconium additive improves field performance and cost of biopolymer muds. In: Proceedings Volume, SPE Oilfield Chemistry International Symposium, Houston, TX, 18–21 February 1997, pp. 707–710.
Burts Jr., B.D., 1992. Lost circulation material with rice fraction. US Patent 5 118 664, assigned to Bottom Line Industries In, June 02.
Burts Jr., B.D., 1997. Lost circulation material with rice fraction. US Patent 5 599 776, assigned to M & D Inds Louisiana Inc., February 04.
Burts Jr., B.D., 2001. Well fluid additive, well fluid made therefrom, method of treating a well fluid, method of circulating a well fluid. US Patent 6 323 158, assigned to Bottom Line Industries In, November 27.

Chadwick, R.E., Phillips, B.M., 1995. Preparation of ether carboxylates. EP Patent 633 279, assigned to Albright & Wilson Ltd., 11 January.

Clapper, D.K., Watson, S.K., 1996. Shale stabilising drilling fluid employing saccharide derivatives. EP Patent 702 073, assigned to Baker Hughes Inc., 20 March.

Coveney, P.V., Watkinson, M., Whiting, A., Boek, E.S., 1999a. Stabilising clayey formations. GB Patent 2 332 221, assigned to Sofitech NV, June 16.

Coveney, P.V., Watkinson, M., Whiting, A., Boek, E.S., 1999b. Stabilizing clayey formations. WO Patent 9 931 353, assigned to Sofitech NV, Dowell Schlumberger SA, and Schlumberger Canada Ltd., June 24.

Cowan, M.K., Hale, A.H., 1994. Restoring lost circulation. US Patent 5 325 922, assigned to Shell Oil Company, 05 July.

Cowan, K.M., Hale, A.H., 1995. High temperature well cementing with low grade blast furnace slag. US Patent 5 379 840, assigned to Shell Oil Comapny, 10 January.

Cowan, K.M., Smith, T.R., 1993. Application of drilling fluids to cement conversion with blast furnace slag in Canada. In: Proceedings Volume, no. 93-601, CADE/CAODE Spring Drilling Conference, Calgary, Canada, 14–16 April 1993 Proc.

Cowan, K.M., Hale, A.H., Nahm, J.J.W., 1994. Dilution of drilling fluid in forming cement slurries. US Patent 5 314 022, assigned to Shell Oil Comapny, 24 May.

Cravey, R.L., 2010. Citric acid based emulsifiers for oilfield applications exhibiting low fluorescence. US Patent 7 691 960, assigned to Akzo Nobel NV, Arnhem, NL, 6 April.

Crawshaw, J.P., Way, P.W., Thiercelin, M., 2002. A method of stabilizing a wellbore wall. GB Patent 2 363 810, assigned to Sofitech NV, January 09.

Cronje, I.J., 1989. Process for the oxidation of fine coal. EP Patent 298 710, assigned to Council Sci. & Ind. Research, January 11.

Dahanayake, M., Li, J., Reierson, R.L., Tracy, D.J., 1996. Amphoteric surfactants having multiple hydrophobic and hydrophilic groups. EP Patent 697 244, assigned to Rhone Poulenc Inc., February 21.

Dalmazzone, C., Audibert-Hayet, A., Langlois, B., Touzet, S., 2007. Oil-based drilling fluid comprising a temperature-stable and non-polluting emulsifying system. US Patent 7 247 604, assigned to Institut Francais du Petrole (Rueil Malmaison Cedex, FR) and Rhodia Chimie (Aubervilliers Cedex, FR), 24 July.

Darden, J.W., McEntire, E.E., 1986. Dicyclopentadiene dicarboxylic acid salts as corrosion inhibitors. EP Patent 200 850, assigned to Texaco Development Corporation, November 12.

Darden, J.W., McEntire, E.E., 1990. Dicyclopentadiene dicarboxylic acid salts as corrosion inhibitors. CA Patent 1 264 541, January 23.

Davidson, E., 2001. Method and composition for scavenging sulphide in drilling fluids. WO Patent 0 109 039, assigned to Halliburton Energy Services, 08 February.

Deem, C.K., Schmidt, D.D., Molner, R.A., 1991. Use of MMH (mixed metal hydroxide)/propylene glycol mud for minimization of formation damage in a horizontal well. In: Proceedings 4th CADE/CAODE Spring Drilling Conference, Calgary, Canada, 10–12 April 1991 Proc.

Delhommer, H.J., Walker, C.O., 1987. Method for controlling lost circulation of drilling fluids with hydrocarbon absorbent polymers. US Patent 4 633 950, 03 January.

Denton, R.M., Lockstedt, A.W., 2006. Rock bit with grease composition utilizing polarized graphite. US Patent 7 121 365, assigned to Smith International, Inc., Houston, TX, 17 October.

Dietsche, F., Essig, M., Friedrich, R., Kutschera, M., Schrepp, W., Witteler, H., et al., 2007. Organic corrosion inhibitors for interim corrosion protection. Corrosion. NACE International, Nashville, TN, p. 2007.

Dixon, J., 2009. Drilling fluids. US Patent 7 614 462, assigned to Croda International PLC, Goole, East Yorkshire, GB, 10 November.

Downey, A.B., Willingham, G.L., Frazier, V.S., 1995. Compositions comprising 4,5-dichloro-2-n- octyl-3- isothiazolone and certain commercial biocides. EP Patent 680 695, assigned to Rohm & Haas Company, November 08.

Duhon, J.J.S., 1998. Olive pulp additive in drilling operations. US Patent 5 801 127, September 01.

Durand, C., Onaisi, A., Audibert, A., Forsans, T., Ruffet, C., 1995a. Influence of clays on borehole stability: a literature survey: Pt.2: mechanical description and modelling of clays and shales drilling practices versus laboratory simulations. Rev. Inst. Franc. Pet. 50 (3), 353–369.

Durand, C., Onaisi, A., Audibert, A., Forsans, T., Ruffet, C., 1995b. Influence of clays on borehole stability: a literature survey: Pt.1: Occurrence of drilling problems physico-chemical description of clays and of their interaction with fluids. Rev. Inst. Franc. Pet. 50 (2), 187–218.

Durr Jr., A.M., Huycke, J., Jackson, H.L., Hardy, B.J., Smith, K.W., 1994. An ester base oil for lubricant compounds and process of making an ester base oil from an organic reaction by-product. EP Patent 606 553, assigned to Conoco Inc., 20 July.

Elphingstone, E.A., Woodworth, F.B., 1999. Dry biocide. US Patent 6 001 158, assigned to Baker Hughes Inc., 14 December.

Eoff, L.S., Reddy, B.R., Wilson, J.M., 2006. Compositions for and methods of stabilizing subterranean formations containing clays. US Patent 7 091 159, assigned to Halliburton Energy Services, Inc., Duncan, OK, 15 August.

Fabrichnaya, A.L., Shamraj, Y.V., Shakirzyanov, R.G., Sadriev, Z.K., Koshelev, V.N., Vakhrushev, L.P., et al., 1997. Additive for drilling solutions with high foam extinguishing properties-containing specified surfactant in hydrocarbon solvent with methyl- diethylalkoxymethyl ammonium methyl sulphate. RU Patent 2 091 420, assigned to ETN Company Ltd., 27 September.

Farrar, D., Hawe, M., Dymond, B., 1992. Use of water soluble polymers in aqueous drilling or packer fluids and as detergent builders. EP Patent 182 600, assigned to Allied Colloids Ltd., 12 August.

Felix, M.S., 1996. A surface active composition containing an acetal or ketal adduct. WO Patent 9 600 253, assigned to Dow Chemical Company, January 04.

Felixberger, J., 1996. Mixed metal hydroxides (MMH)-an inorganic thickener for water-based drilling muds. In: Proceedings DMGK Spring Conference, Celle, Germany, 25–26 April 1996, pp. 339–351.

Fischer, E.R., Parker III, J.E., 1995. Tall oil fatty acid anhydrides as corrosion inhibitor intermediates. In: Proceedings, 50th Annual NACE International Corrosion Conference (Corrosion 95), Orlando, FL, March 26–31, 1995.

Fischer, E.R., Parker III, J.E., 1997. Tall oil fatty acid anhydrides as corrosion inhibitor intermediates. Corrosion 53 (1), 62–64.

Fischer, E.R., Parker III, J.E., 1995. Tall oil fatty acid anhydrides as corrosion inhibitor intermediates. In: Proceedings, 50th Annual NACE International Corrosion Conference (Corrosion 95), Orlando, FL, March 26–31, 1995. Fischer, E.R., Parker III, J.E., 1997. Tall oil fatty acid anhydrides as corrosion inhibitor intermediates. Corrosion 53(1), 62–64.

Fisk Jr., J.V., Kerchevile, J.D., Pober, K.W., 2006. Silicic acid mud lubricants. US Patent 6 989 352, assigned to Halliburton Energy Services, Inc., Duncan, OK, 24 January.

Fleming, J.K., Fleming, H.C., 1995. Invert emulsion drilling mud. WO Patent 9 504 788, assigned to J K F Investments Ltd. and Hour Holdings Ltd., 16 February.

Fong, D.W., Shambatta, B.S., 1993. Hydroxamic acid containing polymers used as corrosion inhibitors. CA Patent 2 074 535, assigned to Nalco Chemical Company, January 25.
Fuh, G.F., Morita, N., Whitfill, D.L., Strah, D.A., 1993. Method for inhibiting the initiation and propagation of formation fractures while drilling. US Patent 5 180 020, assigned to Conoco Inc., January 19.
Garvey, C.M., Savoly, A., Weatherford, T.M., 1987. Drilling fluid dispersant. US Patent 4 711 731, December 08.
Garyan, S.A., Kuznetsova, L.P., Moisa, Y.N., 1998. Experience in using environmentally safe lubricating additive fk-1 in drilling muds during oil and gas well drilling. Stroit Neft Gaz Skvazhin Sushe More(10), 11–14.
Gay, R.J., Gay, C.C., Matthews, V.M., Gay, F.E.M., Chase, V., 1993. Dynamic polysulfide corrosion inhibitor method and system for oil field piping. US Patent 5 188 179, February 23.
Genuyt, B., Janssen, M., Reguerre, R., Cassiers, J., Breye, F., 2001. Biodegradable lubricating composition and uses thereof, in particular in a bore fluid. WO Patent 0 183 640, assigned to Total Raffinage Dist SA, 08 November.
Genuyt, B., Janssen, M., Reguerre, R., Cassiers, J., Breye, F., 2006. Biodegradable lubricating composition and uses thereof, in particular in a bore fluid. US Patent 7 071 150, assigned to Total Raffinage Distribution S.A., Puteaux, FR, 4 July.
Getliff, J.M., James, S.G., 1996. The replacement of alkyl-phenol ethoxylates to improve the environmental acceptability of drilling fluid additives. In: Proceedings Volume, Vol. 2, 3rd SPE et al. Health, Safety & Environment International Conference, New Orleans, LA, 9–12 June 1996, pp. 713–719.
Giddings, D.M., Fong, D.W., 1988. Calcium tolerant deflocculant for drilling fluids. US Patent 4 770 795, September 13.
Glowka, D.A., Loeppke, G.E., Rand, P.B., Wright, E.K., 1989. Laboratory and Field Evaluation of Polyurethane Foam for Lost Circulation Control. Vol. 13 of The Geysers—Three Decades of Achievement: A Window on the Future. Geothermal Resources Council, Davis, California, pp. 517–524.
Green, B.D., 2001. Method for creating dense drilling fluid additive and composition therefor. WO Patent 0 168 787, assigned to Grinding & Sizing Company Inc., 20 September.
Grey, R.A., 1993. Process for preparing alternating copolymers of olefinically unsaturated sulfonate salts and unsaturated dicarboxylic acid anhydrides. US Patent 5 210 163, assigned to Arco Chemical Technology Inc., 11 May.
Growcock, F.B., Lopp, V.R., 1988. The inhibition of steel corrosion in hydrochloric acid with 3-phenyl-2-propyn-1-ol. Corrosion Sci 28 (4), 397–410.
Growcock, F.B., Simon, G.A., 2006. Stabilized colloidal and colloidal-like systems. US Patent 7 037 881, 2 May.
Growcock, F.B., Belkin, A., Fosdick, M., Irving, M., O'Connor, B., Brookey, T., 2007. Recent advances in aphron drilling-fluid technology. SPE Drill. Complet. 22 (2), 74–80.
Gullett, P.D., Head, P.F., 1993. Materials incorporating cellulose fibres, methods for their production and products incorporating such materials. WO Patent 9 318 111, assigned to Stirling Design Intl Ltd., September 16.
Hale, A.H., 1988. Well drilling fluids and process for drilling wells. US Patent 4 728 445, 01 March.
Hale, A.H., Lawson, H.F., 1988. Well drilling fluids and process for drilling wells. US Patent 4 740 318, 26 April.
Hale, A.H., Loftin, R.E., 1996. Glycoside-in-oil drilling fluid system. US Patent 5 494 120, assigned to Shell Oil Company, 27 February.

Hale, A.H., Rivers, G.T., 1988. Well drilling fluids and process for drilling wells. US Patent 4 721 576, 26 January.
Hale, A.H., van Oort, E., 1997. Efficiency of ethoxylated/propoxylated polyols with other additives to remove water from shale. US Patent 5 602 082, 11 February.
Halliday, W.S., Clapper, D.K., 1998. Purified paraffins as lubricants, rate of penetration enhancers, and spotting fluid additives for water-based drilling fluids. US Patent 5 837 655, 17 November.
Halliday, W.S., Schwertner, D., 1997. Olefin isomers as lubricants, rate of penetration enhancers, and spotting fluid additives for water-based drilling fluids. US Patent 5 605 879, assigned to Baker Hughes Inc., 25 February.
Halliday, W.S., Thielen, V.M., 1987. Drilling mud additive. US Patent 4 664 818, assigned to Newpark Drilling Fluid Inc., 12 May.
Hanahan, D.J., 1997. A Guide to Phospholipid Chemistry. Oxford University Press, New York, NY.
Hatchman, K., 1999a. Concentrates for use in structured surfactant drilling fluids. GB Patent 2 329 655, assigned to Albright & Wilson Ltd., March 31.
Hatchman, K., 1999b. Drilling fluid concentrates. EP Patent 903 390, assigned to Albright & Wilson Ltd., March 24.
Headley, J.A., Walker, T.O., Jenkins, R.W., 1995. Environmentally safe water-based drilling fluid to replace oil-based muds for shale stabilization. In: Proceedings, SPE/IADC Drilling Conference, Amsterdam, Netherland, 28 February–2 March 1995, pp. 605–612.
Hille, M., Wittkus, H., Weinelt, F., 1996. Application of acetal-containing mixtures. EP Patent 702 074, assigned to Hoechst AG, March 20.
Hille, M., Wittkus, H., Weinelt, F., 1996. Application of acetal-containing mixtures. EP Patent 702 074, assigned to Hoechst AG, March 20. Hille, M., Wittkus, H., Weinelt, F., 1998a. Use of acetal-containing mixtures. US Patent 5 830 830, assigned to Clariant GmbH, 03 November.
Hille, M., Wittkus, H., Windhausen, B., Scholz, H.J., Weinelt, F., 1998b. Use of acetals. US Patent 5 759 963, assigned to Hoechst AG, 02 June.
Himes, R.E., 1992. Method for clay stabilization with quaternary amines. US Patent 5 097 904, assigned to Halliburton Company, March 24.
Hodder, M.H., Ballard, D.A., Gammack, G., 1992. Controlling drilling fluid enzyme activity. Pet. Eng. Int. 64 (11), 31,33,35.
Holinski, R., 1995. Solid lubricant composition. US Patent 5 445 748, assigned to Dow Corning GmbH, Wiesbaden, DE, August 29.
House, R.F., Wilkinson, A.H., Cowan, C., 1991. Well working compositions, method of decreasing the seepage loss from such compositions, and additive therefor. US Patent 5 004 553, assigned to Venture Innovations Inc., April 02.
Hsu, J.C., 1990. Synergistic microbicidal combinations containing 3- isothiazolone and commercial biocides. US Patent 4 906 651, assigned to Rohm & Haas Company, March 6.
Hsu, J.C., 1995. Biocidal compositions. EP Patent 685 158, assigned to Rohm & Haas Comapny, December 06.
Huber, J., Plank, J., Heidlas, J., Keilhofer, G., Lange, P., 2009. Additive for drilling fluids. US Patent 7 576 039, assigned to BASF Construction Polymers GmbH, Trostberg, DE, 18 August.
Incorvia, M.J., 1988. Polythioether corrosion inhibition system. US Patent 4 759 908, July 26.
Ivan, C.D., Blake, L.D., Quintana, J.L., 2001. Aphron-base drilling fluid: Evolving technologies for lost circulation control. In: Proceedings, Annual SPE Technical Conference, New Orleans, LA, 30 September–3 October 2001.

Jarrett, M., 1996. Nonionic alkanolamides as shale stabilizing surfactants for aqueous well fluids. WO Patent 9 632 455, assigned to Baker Hughes Inc., 17 October.

Jarrett, M., 1997a. Amphoteric acetates and glycinates as shale stabilizing surfactants for aqueous well fluids. US Patent 5 593 952, assigned to Baker Hughes Inc., January 14.

Jarrett, M., 1997b. Nonionic alkanolamides as shale stabilizing surfactants for aqueous well fluids. US Patent 5 607 904, assigned to Baker Hughes Inc., March 04.

Jarrett, M., Clapper, D., 2010. High temperature filtration control using water based drilling fluid systems comprising water soluble polymers. US Patent 7 651 980, assigned to Baker Hughes Incorporated, Houston, TX, 26 January.

Johnson, D.M., Ippolito, J.S., 1994. Corrosion inhibitor and sealable thread protector end cap for tubular goods. US Patent 5 352 383, October 04.

Johnson, D.M., Ippolito, J.S., 1995. Corrosion inhibitor and sealable thread protector end cap for tubular goods. US Patent 5 452 749, September 26.

Jones, T.A., Wentzler, T., 2008. Polymer hydration method using microemulsions. US Patent 7 407 915, assigned to Baker Hughes Incorporated, Houston, TX, 5 August.

Just, E.K., Nickol, R.G., 1989. Phosphated, oxidized starch and use of same as dispersant in aqueous solutions and coating for lithography. EP Patent 319 989, June 14.

Kanz, J.E., Cravey, M.J., 1985. Oil well drilling fluids: Their physical and chemical properties and biological impact. In: Saxena, J., Fisher, F. (Eds.), Hazard Assessment of Chemicals, Vol. 5. Elsevier, New York, pp. 291–421.

Karaseva, E.V., Dedyukhina, S.N., Dedyukhin, A.A., 1995. Treatment of water-based drilling solution to prevent microbial attack—by addition of dimethyl-tetrahydro-thiadiazine-thione bactericide. RU Patent 2 036 216, May 27.

Kashkarov, N.G., Verkhovskaya, N.N., Ryabokon, A.A., Gnoevykh, A.N., Konovalov, E.A., Vyakhirev, V.I., 1997. Lubricating reagent for drilling fluids-consists of spent sunflower oil modified with additive in form of aqueous solutions of sodium alkylsiliconate(s). RU Patent 2 076 132, assigned to Tyumen Nat Gases Research Institute, 27 March.

Kashkarov, N.G., Konovalov, E.A., Vjakhirev, V.I., Gnoevykh, A.N., Rjabokon, A.A., Verkhovskaja, N.N., 1998. Lubricant reagent for drilling muds-contains spent sunflower oil, and light tall oil and spent coolant-lubricant as modifiers. RU Patent 2 105 783, assigned to Tyumen Nat Gases Research Institute, 274 February.

Keilhofer, G., Plank, J., 2000. Solids composition based on clay minerals and use thereof. US Patent 6 025 303, assigned to SKW Trostberg AG, 15 February.

Kinchen, D., Peavy, M.A., Brookey, T., Rhodes, D., 2001. Case history: drilling techniques used in successful redevelopment of low pressure H2S gas carbonate formation. In: Proceedings Volume, Vol. 1, SPE/IADC Drilling Conference, Amsterdam, Netherlands, 27 February–1 March 2001, pp. 392–403.

Kippie, D.P., Gatlin, L.W., 2009. Shale inhibition additive for oil/gas down hole fluids and methods for making and using same. US Patent 7 566 686, assigned to Clearwater International, LLC, Houston, TX, 28 July.

Kisic, I., Mesic, S., Basic, F., Brkic, V., Mesic, M., Durn, G., et al., 2009. The effect of drilling fluids and crude oil on some chemical characteristics of soil and crops. Geoderma 149 (3-4), 209–216.

Klein, H.P., Godinich, C.E., 2006. Drilling fluids. US Patent 7 012 043, assigned to Huntsman Petrochemical Corporation, The Woodlands, TX, March 14.

Kohn, R.S., 1988. Thixotropic aqueous solutions containing a divinylsulfone-crosslinked polygalactomannan gum. US Patent 4 752 339, June 21.

Kokal, S.L., 2006. Crude oil emulsions. In: Fanchi, J.R. (Ed.), Petroleum Engineering Handbook, Vol. I. Society of Petroleum Engineers, Richardson, TX, pp. 533–570. , Ch. 12.

Kokal, S.L., Wingrove, M., 2000. Emulsion separation index: from laboratory to field case studies. Proceedings Volume, no. 63165-MS, SPE Annual Technical Conference and Exhibition, 1–4 October 2000. Society of Petroleum Engineers, Dallas, TX.

Konovalov, E.A., Ivanov, Y.A., Shumilina, T.N., Pichugin, V.F., Komarova, N.N., 1993a. Lubricating reagent for drilling solutions–contains agent based on spent sunflower oil, water, vat residue from production of oleic acid, and additionally water glass. SU Patent 1 808 861, assigned to Moscow Gubkin Oil Gas Institute, 15 April.

Konovalov, E.A., Rozov, A.L., Zakharov, A.P., Ivanov, Y.A., Pichugin, V.F., Komarova, N.N., 1993b. Lubricating reagent for drilling solutions–contains spent sunflower oil as active component, water, boric acid as emulsifier, and additionally water glass. SU Patent 1 808 862, assigned to Moscow Gubkin Oil Gas Institute, 15 April.

Kreh, R.P., 1991. Method of inhibiting corrosion and scale formation in aqueous systems. US Patent 5 073 339, December 17.

Kubena Jr., E., Whitebay, L.E., Wingrave, J.A., 1993. Method for stabilizing boreholes. US Patent 5 211 250, assigned to Conoco Inc., May 18.

Kulpa, K., Adkins, R., Walker, N.S., 1992. New testing vindicates use of barite. Am. Oil Gas Rep. 35 (4), 52–54.

Kurochkin, B.M., Tselovalnikov, V.F., 1994. Use of ellipsoidal glass granules for drilling under complicated conditions. Neft Khoz 10, 7–13.

Kurochkin, B.M., Kolesov, L.V., Biryukov, M.B., 1990. Use of ellipsoidal glass granules as an antifriction mud additive. Neft Khoz 12, 61–64.

Kurochkin, B.M., Simonyan, E.A., Simonyan, A.A., Khirazov, E.F., Ozarchuk, P.A., Voloshinivskii, V.O., et al., 1992a. New technology of drilling with the use of glass granules. Neft Khoz 7, 9–11.

Kurochkin, B.M., Kolesov, L.V., Masich, V.I., Stepanov, N.V., Tselovalnikov, V.F., Alekperov, V.T., et al., 1992b. Solution for drilling gas and oil wells–contains ellipsoidal glass beads as additive reducing friction between walls of well and casing string. SU Patent 1 740 396, assigned to Drilling Technology Research Institute, 15 June.

Lange, P., Plank, J., 1999. Mixed metal hydroxide (MMH) viscosifier for drilling fluids: Properties and mode of action. Erdöl Erdgas Kohle 115 (7–8), 349–353.

Lawson, H.F., Hale, A.H., 1989. Well drilling fluids and process for drilling wells. US Patent 4 812 244, 14 March.

Lecocumichel, N., Amalric, C., 1995. Concentrated aqueous compositions of alkylpolyglycosides, and applications thereof. WO Patent 9 504 592, assigned to Seppic SA, February 16.

Leighton, J.C., Sanders, M.J., 1988. Water soluble polymers containing allyloxybenzenesulfonate monomers. EP Patent 271 784, June 22.

Leighton, J.C., Sanders, M.J., 1990. Water soluble polymers containing allyloxybenzenesulfonate monomers. US Patent 4 892 898, January 09.

Lindstrom, M.R., Louthan, R.P., 1987. Inhibiting corrosion. US Patent 4 670 163, assigned to Phillips Petroleum Co., June 02.

Lindstrom, M.R., Mark, H.W., 1987. Inhibiting corrosion: Benzylsulfinylacetic acid or benzylsulfonylacetic acid. US Patent 4 637 833, January 20.

Lipkes, M.I., Mezhlumov, A.O., Shits, L.A., Avdeev, G.E., Fomenko, V.I., Shvetsov, A.M., 1995. Carbonate weighting material for drilling-in producing formations and well overhaul. Stroit Neft Gaz Skvazhin Sushe More 5-6, 34–41.

Maresh, J.L., 2009. Wellbore treatment fluids having improved thermal stability. US Patent 7 541 316, assigned to Halliburton Energy Services, Inc., Duncan, OK, 2 June.

Martin, R.L., Brock, G.F., Dobbs, J.B., 2005. Corrosion inhibitors and methods of use. US Patent 6 866 797, assigned to BJ Services Company, 15 March.

McCallum, T.F.I., Weinstein, B., 1994. Amine-thiol chain transfer agents. US Patent 5 298 585, assigned to Rohm & Haas Company, March 29.

McDonald, W.J., Cohen, J.H., Hightower, C.M., 1999. New lightweight fluids for underbalanced drilling, DOE/FETC Rep 99-1103, Maurer Engineering Inc.

McGlothlin, R.E., Woodworth, F.B., 1996. Well drilling process and clay stabilizing agent. US Patent 5 558 171, assigned to M I Drilling Fluids Llc., September 24.

McNally, K., Nae, H., Gambino, J., 1999. Oil well drilling fluids with improved anti-settling properties and methods of preparing them. EP Patent 906 946, assigned to Rheox Inc., April 07.

Medley Jr., G.H., Maurer, W.C., Garkasi, A.Y., 1995. Use of hollow glass spheres for underbalanced drilling fluids. In: Proceedings Volume, Annual SPE Technical Conference, Dallas, TX, 22–25 October 1995, pp. 511–520.

Medley Jr., G.H., Haston, J.E., Montgomery, R.L., Martindale, I.D., Duda, J.R., 1997a. Field application of lightweight, hollow-glass-sphere drilling fluid. J. Pet. Technol. 49 (11), 1209–1211.

Medley Jr., G.H., Haston, J.E., Montgomery, R.L., Martindale, I.D., Duda, J.R., 1997b. Field application of lightweight hollow glass sphere drilling fluid. In: Proceedings, Annual SPE Technical Conference, San Antonio, TX, 5–8 October 1997, pp. 699–707.

Messenger, J.U., 1981. Lost Circulation. PennWell Publishing Co., Tulsa, OK.

Miller, J.J., 2009. Drilling fluids containing biodegradable organophilic clay. US Patent 7 521.

Minevski, L.V., Gaboury, J.A., 1999. Thiacrown ether compound corrosion inhibitors for alkanolamine units. EP Patent 962 551, assigned to Betzdearborn Europe Inc., December 08.

Moajil, A.M.A., Nasr-El-Din, H.A., Al-Yami, A.S., Al-Aamri, A.D., Al-Agil, A.K., 2008. Removal of filter cake formed by manganese tetraoxide-based drilling fluids. SPE International Symposium and Exhibition on Formation Damage Control. Society of Petroleum Engineers, Lafayette, LA.

Morpeth, F.F., Greenhalgh, M., 1990. Composition and use. EP Patent 390 394, assigned to Imperial Chemical Inds Pl, October 03.

Mueller, H., Herold, C.P., von Tapavicza, S., 1990a. Monocarboxylic acid-methyl esters in invert- emulsion muds. EP Patent 382 071, assigned to Henkel KG Auf Aktien, August 16.

Mueller, H., Herold, C.P., von Tapavicza, S., 1990b. Oleophilic alcohols as components of invert emulsion drilling fluids. EP Patent 391 252, assigned to Henkel KG Auf Aktien, October 10.

Mueller, H., Herold, C.P., von Tapavicza, S., 1990c. Oleophilic basic amine derivatives as additives in invert emulsion muds. EP Patent 382 070, assigned to Henkel KG Auf Aktien, August 16.

Mueller, H., Herold, C.P., von Tapavicza, S., Grimes, D.J., Braun, J.M., Smith, S.P.T., 1990d. Use of selected ester oils in drilling muds, especially for offshore oil or gas recovery. EP Patent 374 672, assigned to Henkel KG Auf Aktien and Baroid Ltd., June 27.

Mueller, H., Herold, C.P., von Tapavicza, S., Grimes, D.J., Braun, J.M., Smith, S.P.T., 1990e. Use of selected ester oils in drilling muds, especially for offshore oil or gas recovery. EP Patent 374 671, assigned to Henkel KG Auf Aktien and Baroid Ltd., June 27.

Mueller, H., Herold, C.P., von Tapavicza, S., 1991a. Use of hydrated castor oil as a viscosity promoter in oil-based drilling muds. WO Patent 9 116 391, assigned to Henkel KG Auf Aktien, October 31.

Mueller, H., Herold, C.P., von Tapavicza, S., Fues, J.F., 1991b. Fluid borehole-conditioning agent based on polycarboxylic acid esters. WO Patent 9 119 771, assigned to Henkel KG Auf Aktien, December 26.

Mueller, H., Herold, C.P., Westfechtel, A., von Tapavicza, S., 1991c. Free-flowing drill hole treatment agents based on carbonic acid diesters. WO Patent 9 118 958, assigned to Henkel KG Auf Aktien, December 12.

Mueller, H., Herold, C.P., Westfechtel, A., von Tapavicza, S., 1993. Free-flowing drill hole treatment agents based on carbonic acid diesters. EP Patent 532 570, assigned to Henkel KG Auf Aktien, March 24.

Mueller, H., Herold, C.P., von Tapavicza, S., Neuss, M., Burbach, F., 1994. Use of selected ester oils of low carboxylic acids in drilling fluids. US Patent 5 318 954, assigned to Henkel KG Auf Aktien, Jun. 07.

Mueller, H., Herold, C.P., Westfechtel, A., von Tapavicza, S., 1995. Fluid-drill-hole treatment agents based on carbonic acid diesters. US Patent 5 461 028, assigned to Henkel KG Auf Aktien, October 24.

Mueller, H., Breuer, W., Herold, C.P., Kuhm, P., von Tapavicza, S., 1997. Mineral additives for setting and/or controlling the rheological properties and gel structure of aqueous liquid phases and the use of such additives. US Patent 5 663 122, assigned to Henkel KG Auf Aktien, 02 September.

Mueller, H., Herold, C.P., Bongardt, F., Herzog, N., von Tapavicza, S., 2000. Lubricants for drilling fluids (schmiermittel fuer bohrspuelungen). WO Patent 0 029 502, assigned to Cognis Deutschland GmbH, 25 May.

Nakaya, T., Li, Y.-J., 1999. Phospholipid polymers. Prog. Polym. Sci. 24 (1), 143–181.

Nguyen, P.D., Rickman, R.D., Dusterhoft, R.G., 2010. Method of stabilizing unconsolidated formation for sand control. US Patent 7 673 686, assigned to Halliburton Energy Services, Inc., Duncan, OK, 9 March.

Nicora, L.F., McGregor, W.M., 1998. Biodegradable surfactants for cosmetics find application in drilling fluids. In: Proceedings Volume, IADC/SPE Drilling Conference, Dallas, TX, 3–6 March 1998, pp. 723–730.

Oppong, D., Hollis, C.G., 1995. Synergistic antimicrobial compositions containing (thiocyanomethylthio)benzothiazole and an organic acid. WO Patent 9 508 267, assigned to Buckman Labs Internat. Inc., March 30.

Oppong, D., King, V.M., 1995. Synergistic antimicrobial compositions containing a halogenated acetophenone and an organic acid. WO Patent 9 520 319, assigned to Buckman Labs Internat. Inc., August 03.

Palumbo, S., Giacca, D., Ferrari, M., Pirovano, P., 1989. The development of potassium cellulosic polymers and their contribution to the inhibition of hydratable clays. In: Proceedings Volume, SPE Oilfield Chemistry International Symposium, Houston, February 8–10, 1989, pp. 173–182.

Patel, A.D., McLaurine, H.C., 1993. Drilling fluid additive and method for inhibiting hydration. CA Patent 2 088 344, assigned to M I Drilling Fluids Company, October 11.

Patel, A.D., McLaurine, H.C., Stamatakis, E., Thaemlitz, C.J., 1995. Drilling fluid additive and method for inhibiting hydration. EP Patent 634 468, assigned to M I Drilling Fluids Company, January 18.

Patel, A.D., Davis, E., Young, S., Stamatakis, E., 2006. Phospholipid lubricating agents in aqueous based drilling fluids. US Patent 7 094 738, assigned to M-I L.L.C., Houston, TX, 22 August.

Patel, A.D., Stamatakis, E., Davis, E., Friedheim, J., 2007. High performance water based drilling fluids and method of use. US Patent 7 250 390, assigned to M-I LLC, Houston, TX, July 31.

Peiffer, D.G., Bock, J., Elward-Berry, J., 1991. Zwitterionic functionalized polymers as deflocculants in water based drilling fluids. US Patent 5 026 490, assigned to Exxon Research & Engineering Company, June 25.

Peiffer, D.G., Bock, J., Elward-Berry, J., 1992a. Thermally stable hydrophobically associating rheological control additives for water-based drilling fluids. US Patent 5 096 603, assigned to Exxon Research & Engineering Company, March 17.

Peiffer, D.G., Bock, J., Elward-Berry, J., 1992b. Zwitterionic functionalized polymers. GB Patent 2 247 240, assigned to Exxon Research & Engineering Company, February 26.

Peiffer, D.G., Bock, J., Elward-Berry, J., 1992c. Zwitterionic functionalized polymers as deflocculants in water based drilling fluids. CA Patent 2 046 669, assigned to Exxon Research & Engineering Company, February 09.

Peiffer, D.G., Bock, J., Elward-Berry, J., 1993a. Thermally stable hydrophobically associating rheological control additives for water-based drilling fluids. CA Patent 2 055 011, assigned to Exxon Research & Engineering Company, May 07.

Peiffer, D.G., Bock, J., Elward-Berry, J., 1993b. Zwitterionic functionalized polymers as deflocculants in water based drilling fluids. AU Patent 638 917, assigned to Exxon Research & Engineering Company, July 08.

Penkov, A.I., Vakhrushev, L.P., Belenko, E.V., 1999. Characteristics of the behavior and use of polyalkylene glycols for chemical treatment of drilling muds. Stroit Neft Gaz Skvazhin Sushe More 21−24.

Ponsati, O., Trius, A., Herold, C.P., Mueller, H., Nitsch, C., von Tapavicza, S., 1992. Use of selected oleophilic compounds with quaternary nitrogen to improve the oil wettability of finely divided clay and their use as viscosity promoters. WO Patent 9 219 693, assigned to Henkel KG Auf Aktien, November 12.

Ponsati, O., Trius, A., Herold, C.P., Mueller, H., Nitsch, C., von Tapavicza, S., 1994. Use of selected oleophilic compounds with quaternary nitrogen to improve the oil wettability of finely divided clay and their use as viscosity promoters. EP Patent 583 285, assigned to Henkel KG Auf Aktien, February 23.

Portnoy, R.C., 1986. Anionic copolymers for improved control of drilling fluid rheology. GB Patent 2 174 402, assigned to Exxon Chemical Patents In, November 05.

Portnoy, R.C., 1987. Anionic copolymers for improved control of drilling fluid rheology. US Patent 4 680 128, July 14.

Prokhorov, N.M., Smirnova, L.N., Luban, V.Z., 1993. Neutralisation of hydrogen sulphide in drilling solution−by introduction of additive consisting of iron sulphate and additionally sodium aluminate, to increase hydrogen sulphide absorption. SU Patent 1 798 358, assigned to Polt. Br. Ukr. Geoprosp. Inst., 28 February.

Ramanarayanan, T.A., Vedage, H.L., 1994. Inorganic/organic inhibitor for corrosion of iron containing materials in sulfur environment. US Patent 5 279 651, January 18.

Rastegaev, B.A., Andreson, B.A., Raizberg, Y.L., 1998. Bactericidal protection of chemical agents from biodegradation while drilling deep wells. Stroit Neft Gaz Skvazhin Sushe More 7-8, 32−34.

Reid, A.L., Grichuk, H.A., 1991. Polymer composition comprising phosphorous-containing gelling agent and process thereof. US Patent 5 034 139, assigned to Nalco Chemical Company, 23 July.

Robinson, F., 1996. Polymers useful as pH responsive thickeners and monomers therefor. WO Patent 9 610 602, assigned to Rhone Poulenc Inc., 11 April.

Robinson, F., 1999. Polymers useful as pH responsive thickeners and monomers therefor. US Patent 5 874 495, assigned to Rhodia, 23 February.

Rooney, P., Russell, J.A., Brown, T.D., 1988. Production of sulfonated asphalt. US Patent 4 741 868, 03 May.

Rose, R.A., 1996. Method of drilling with fluid including nut cork and drilling fluid additive. US Patent 5 484 028, assigned to Grinding & Sizing Company Inc., January 16.

Runov, V.A., Mojsa, Y.N., Subbotina, T.V., Pak, K.S., Krezub, A.P., Pavlychev, V.N., et al., 1991. Lubricating additive for clayey drilling solution-is obtained by esterification of tall oil or tall pitch with hydroxyl group containing agent, e.g. low mol. wt. glycol or ethyl cellulose. SU Patent 1 700 044, assigned to Volgo Don Br. Sintez Pav and Burenie Sci Production Association, 23 December.

Saasen, A., Hoset, H., Rostad, E.J., Fjogstad, A., Aunan, O., Westgard, E., et al., 2001. Application of ilmenite as weight material in water based and oil based drilling fluids. In: Proceedings, Annual SPE Technical Conference, New Orleans, LA, 30 September–3 October 2001.

Sano, M., 1997. Polypropylene glycol (PPG) used as drilling fluids additive. Sekiyu Gakkaishi 40 (6), 534–538.

Santhanam, M., MacNally, K., 2001. Oil and oil invert emulsion drilling fluids with improved anti- settling properties. EP Patent 1 111 024, assigned to Rheox Inc., June 27.

Schlemmer, R.F., 2007. Membrane forming in-situ polymerization for water based drilling fluids. US Patent 7 279 445, assigned to M-I LLC, Houston, TX, 9 October.

Schutt, H.U., 1990. Reducing stress corrosion cracking in treating gases with alkanol amines. US Patent 4 959 177, Sep. 25.

Sebba, F., 1984. Preparation of biliquid foam compositions. US Patent 4 486 333, 4 December.

Sebba, F., 1987. Foams and Biliquid Foams-Aphrons. John Wiley & Sons, Chichester.

Sekine, I., Yuasa, M., Shimode, T., Takaoka, K., 1991. Inhibition of corrosion. GB Patent 2 234 501, February 06.

Senaratne, K.P.A., Lilje, K.C., 1994. Preparation of branched chain carboxylic esters. US Patent 5 322 633, assigned to Albemarle Corporation, 21 June.

Sheu, J.J., Bland, R.G., 1992a. Drilling fluid additive. GB Patent 2 245 579, assigned to Baker Hughes Inc., January 8.

Sheu, J.J., Bland, R.G., 1992b. Drilling fluid with browning reaction anionic carbohydrate. US Patent 5 106 517, assigned to Baker Hughes Inc., 21 April.

Sheu, J.J., Bland, R.G., 1992c. Drilling fluid with stabilized browning reaction anionic carbohydrate. US Patent 5 110 484, assigned to Baker Hughes Inc., 5 May.

Shuey, M.W., Custer, R.S., 1995. Quebracho-modified bitumen compositions, method of manufacture and use. US Patent 5 401 308, assigned to Saramco Inc., 28 March.

Sifferman, T.R., Muijs, H.M., Fanta, G.F., Felker, F.C., Erhan, S.M., 2003. Starch-lubricant compositions for improved lubricity and fluid loss in water-based drilling muds. Proceedings Volume, no. 80213-MS, International Symposium on Oilfield Chemistry. Society of Petroleum Engineers, Inc., Houston, TX.

Sikora, D., 1994. Hydrazine—a universal oxygen scavenger. Nafta Gaz (Pol) 50 (4), 161–168.

Smith, K.W., 2009. Well drilling fluids. US Patent 7 576 038, assigned to Clearwater International, LLC, Houston, TX, 18 August.

Smith, C.K., Balson, T.G., 2000. Shale-stabilizing additives. GB Patent 2 340 521, assigned to Sofitech NV and Dow Chemical Company, February 23.

Smith, C.K., Balson, T.G., 2004. Shale-stabilizing additives. US Patent 6 706 667, 16 March.

Smith, K.W., Thomas, T.R., 1995a. Method of treating shale and clay in hydrocarbon formation drilling. EP Patent 680 504, November 08.

Smith, K.W., Thomas, T.R., 1995b. Method of treating shale and clay in hydrocarbon formation drilling. WO Patent 9 514 066, May 26.

Smith, K.W., Thomas, T.R., 1997. Method of treating shale and clay in hydrocarbon formation drilling. US Patent 5 607 902, assigned to Clearwater Inc., March 04.

Stowe, C., Bland, R.G., Clapper, D., Xiang, T., Benaissa, S., 2002. Water-based drilling fluids using latex additives. GB Patent 2 363 622, assigned to Baker Hughes Inc., January 02.

Sunde, E., Olsen, H., 2000. Removal of H2S in drilling mud. WO Patent 0 023 538, assigned to Den Norske Stats Oljese A, 27 April.

Sydansk, R.D., 1990. Lost circulation treatment for oil field drilling operations. US Patent 4 957 166, assigned to Marathon Oil Corporation, September 18.

Szablikowski, K., Lange, W., Kiesewetter, R., Reinhardt, E., 1995. Easily dispersible blends of reversibly crosslinked and uncrosslinked hydrocolloids, with aldehydes as crosslinker. EP Patent 686 666, assigned to Wolff Walsrode AG, December 13.

Tang, Y., Han, Z., Wang, H., Chen, H., 1995. Sp-2 acid corrosion inhibitor. J. Univ. Pet. China 19 (1), 98–101.

Taylor, R.S., Funkhouser, G.P., Dusterhoft, R.G., 2009. Gelled invert emulsion compositions comprising polyvalent metal salts of an organophosphonic acid ester or an organophosphinic acid and methods of use and manufacture. US Patent 7 534 745, assigned to Halliburton Energy Services, Inc., Duncan, OK, 19 May.

Teeters, S.M., 1992. Corrosion inhibitor. US Patent 5 084 210, assigned to Chemlink Inc., January 28.

Van Oort, E., 1997. Physico-chemical stabilization of shales. In: Proceedings Volume, SPE Oilfield Chemistry International Symposium, Houston, TX, 18–21 February 1997, pp. 523–538.

Veldman, R.R., Trahan, D.O., 1999. Gas treating solution corrosion inhibitor. WO Patent 9 919 539, assigned to Coastal Fluid Technol. LLC, April 22.

Walker, C.O., 1986. Encapsulated lime as a lost circulation additive for aqueous drilling fluids. US Patent 4 614 599, September 30.

Walker, C.O., 1987. Method for controlling lost circulation of drilling fluids with water absorbent polymers. US Patent 4 635 726, 13 January.

Walker, C.D., 1989. Encapsulated lime as a lost circulation additive for aqueous drilling fluids. CA Patent 1 261 604, September 26.

Walker, C.O., 1989. Method for controlling lost circulation of drilling fluids with water absorbent polymers. CA Patent 1 259 788, assigned to Texaco Development Corporation, September 26.

Wall, K., Martin, D.W., Zard, P.W., Barclay-Miller, D.J., 1995. Temperature stable synthetic oil. WO Patent 9 532 265, assigned to Burwood Corporation Ltd., 30 November.

Watcharasing, S., Angkathunyakul, P., Chavadej, S., 2008. Diesel oil removal from water by froth flotation under low interfacial tension and colloidal gas aphron conditions. Sep. Purif. Technol. 62 (1), 118–127.

Wegner, C., Reichert, G., 1990. Hydrogen sulfide scavenger in drilling fluids (schwefelwasserstoff- scavenger in bohrspülungen). In: Proceedings, BASF et al. Chem. Prod. in Petrol. Prod. Mtg. H2S—A Hazardous Gas in Crude Oil Recovery Discuss, Clausthal-Zellerfeld, Germany, 12–13 September 1990.

Westerkamp, A., Wegner, C., Mueller, H.P., 1991. Borehole treatment fluids with clay swelling-inhibiting properties (ii). EP Patent 451 586, assigned to Bayer AG, October 16.

Whitfill, D.L., Kukena Jr., E., Cantu, T.S., Sooter, M.C., 1990. Method of controlling lost circulation in well drilling. US Patent 4 957 174, assigned to Conoco Inc., September 18.

Whitfill, D.L., Pober, K.W., Carlson, T.R., Tare, U.A., Fisk, J.V., Billingsley, J.L., 2005. Method for drilling depleted sands with minimal drilling fluid loss. US Patent 6 889 780, assigned to Halliburton.

Wilkinson, A.O., Grigson, S.J., Turnbull, R.W., 1995. Drilling mud. WO Patent 9 526 386, assigned to Heriot Watt University, October 05.

Wirtz, H., Hoffmann, H., Ritschel, W., Hofinger, M., Mitzlaff, M., Wolter, D., 1989. Optionally quaternized fatty esters of alkoxylated alkyl-alkylene diamines. EP Patent 320 769, June 21.

Wood, R.R., 2001. Improved drilling fluids. WO Patent 0 153 429, 26 July.

Xiang, T., 2010. Invert emulsion drilling fluid systems comprising an emulsified aqueous phase comprising dispersed integral latex particles. US Patent 7 749 945, assigned to Baker Hughes Incorporated, Houston, TX, 6 July.

Young, S., Young, A., 1994. Recent field experience using an acetal based invert emulsion fluid. In: Proceedings Volume, IBC Technical Services Ltd. Prevention of Oil Discharge from Drilling Opererations – The Options Conference, Aberdeen, Scotland, June 15–16, 1994.

Zaitoun, A., Berton, N., 1990. Stabilization of montmorillonite clay in porous media by high-molecular-weight polymers. In: Proceedings Volume, 9th SPE Formation Damage Control Symposium, Lafayette, LA, February 22–23, 1990, pp. 155–164.

Zaleski, P.L., Derwin, D.J., Weintritt, D.J., Russell, G.W., 1998. Drilling fluid loss prevention and lubrication additive. US Patent 5 826 669, assigned to Superior Graphite Company, 27 October.

Zefferi, S.M., May, R.C., 1994a. Corrosion inhibition of calcium chloride brine. US Patent 5 292 455, assigned to Betz Laboratories Inc., March 08.

Zefferi, S.M., May, R.C., 1994b. Corrosion inhibition of calcium chloride brine. CA Patent 2 092 207, August 26.

Zefferi, S.M., May, R.C., 1994a. Corrosion inhibition of calcium chloride brine. US Patent 5 292 455, assigned to Betz Laboratories Inc., March 08. Zefferi, S.M., May, R.C., 1994b. Corrosion inhibition of calcium chloride brine. CA Patent 2 092 207, August 26. Zhao, L., Xie, Q., Luo, Y., Sun, Z., Xu, S., Su, H., et al., 1996. Utilization of slag mix mud conversion cement in the Karamay oilfield, xinjiang. J. Jianghan Pet. Inst. 18 (3), 63–66.

Zhou, Z.J., Gunter, W.D., Jonasson, R.G., 1995. Controlling formation damage using clay stabilizers: A review. In: Proceedings Volume-2, no. CIM 95-71, 46th Annual CIM Petroleum Society Technical Meeting, Banff, Canada, 14–17 May 1995.

Zuzich, A.H., Blytas, G.C., 1994. Polyethercyclicpolyols from epihalohydrins, polyhydric alcohols and metal hydroxides or epoxy alcohol and optionally polyhydric alcohols with addition of epoxy resins. US Patent 5 286 882, assigned to Shell Oil Company, February 15.

Zuzich, A.H., Blytas, G.C., Frank, H., 1995. Polyethercyclicpolyols from epihalohydrins, polyhydric alcohols, and metal hydroxides or epoxy alcohols and optionally polyhydric alcohols with thermal condensation. US Patent 5 428 178, assigned to Shell Oil Company, June.27.

Chapter 14

Drilling and Drilling Fluids Waste Management

INTRODUCTION

Beginning in the 1970s, the drilling industry, as well as many other industrial operations, came under scrutiny by environmental organizations and government regulators with respect to the handling of oil field chemicals and waste disposal. Drilling fluids were particularly in the spotlight looking for hazardous and toxic materials (McAuliffe and Palmer, 1976; Honeycutt, 1970). Since the 1970s the drilling industry has learned that environmental technologies have a positive cost/benefit impact on drilling and drilling fluids (Brantley et al., 2013).

Although most drilling fluid materials are relatively benign, some were found to be environmentally unacceptable and therefore subject to controls or banned outright.

The years since then have produced intense discussions and negotiations and creative research and development (R&D) providing the industry with acceptable products, systems, and operations to meet the most stringent environmental regulations. The industry did not get rid of all toxic and hazardous materials, but we do manage them effectively under the current environmental climate. In general, drilling wastes that are nonhazardous are called NOW: nonhazardous oilfield wastes. Some drilling fluid chemicals, however, can be

- Hazardous: caustic soda, acids
- Toxic: diesel oil, bactericides, some cationic materials
- Ignitable: oils, alcohols

The amount of these materials used in drilling and completion fluids is extremely small. Normal inventory control can track these materials and ensure they are handled and disposed properly.

A publication by the International Association of Oil and Gas Producers (IPIECA, 2009) is a comprehensive coverage of the health risk management of drilled cuttings and drilling fluids discharges.

DRILLING WASTES

By far the largest volume of wastes produced in drilling operations is drilled cuttings. For example, the rule of thumb using traditional oil field units is to calculate the approximate volume of cuttings generated in barrels/1000 ft by doubling the hole size in inches. For a 12¼-inch hole this gives about 150 bbl/1000 ft. For a single 10,000 foot drilling operation the volume is 1500 bbl (239 m^3).

If the mud is a freshwater-based fluid disposing of these cuttings and the mud covering the cuttings, as well as excess mud from dilution, is usually not a problem, if space is available. Any hazardous chemicals, such as caustic soda, can be treated or neutralized to acceptable levels for most landfills or for spreading on the ground. The cost of handling and transporting this amount of waste solids and fluids, however, can be a significant percentage of the drilling operation. In offshore operations, the cost to haul wastes to land can be prohibitive.

If the mud is a nonaqueous drilling fluid (NADF), there are considerably more rules and regulations specifying the proper handling and disposal of cuttings and excess fluid. The base fluid for a NADF can have toxic components, e.g., diesel oil contains toxic aromatic compounds. In addition, some of the emulsifiers and other surfactants used in NADFs are toxic. Therefore, the cuttings generated when drilling with these fluids are also tightly controlled as to disposal. The current regulations in the Gulf of Mexico specify the following for use of all drilling fluids in offshore drilling.

General Discharge Limitations Offshore

- Oil-base mud (diesel, mineral oil)—prohibited
- Generic water-based muds—subject to toxicity limit of 30,000 ppm
- Additives—approved lists
- Bioassays—required, end of well
- Free oil—no discharge based on static sheen test
- Barite—1 mg/kg mercury, 3 mg/kg cadmium
- Oil-on-cuttings—no aromatic base oil discharge, no free oil
- Other—no halogenated phenols, chromates, minimum use of surfactants, dispersants, and detergents

NADF Discharge Limitations Offshore

- No discharge of whole mud
- No free oil from cuttings
- No formation oil (crude oil) on cuttings
- Base fluid on cuttings <8.0 wt % for ester base, <6.9% for olefin base
- Maximum polyaromatic content 10 ppm
- Biodegrade at least as fast as C16–C18 internal olefin (IO)
- No more toxic than olefin in 10-day sediment toxicity test

MINIMIZING WASTE PROBLEMS

Minimizing waste disposal costs starts with the drilling program. The most important factors in planning the drilling operation are

- Hole size and depth—minimize solids generated
- Mud type—inhibitive water-based, NADF
- Solids control equipment—dewatering/zero discharge systems

Of equal importance while drilling and after well completion is a plan to handle and dispose of the cuttings and excess fluids, both drilling and completion, generated. The following is from API Recommended Practice 51R (API, 2009), which is free from the API website (www.API.org).

Completion fluid selection should take into account the safety and logistics of transporting, handling, storing, and disposing of clean and contaminated fluid. For both new and existing well sites, all fuels, treatment chemicals, completion brines, and other similar liquids should be properly stored in labeled containers intended for that purpose. Containment should be constructed so spilled fuels or chemicals do not reach the ground.

Wherever practical, tanks or existing drilling pits should be used for completion and workover operations. Completion brines and other potential pollutants should be kept in lined pits, steel pits, or storage tanks. If a new earthen pit is necessary, it should be constructed in a manner that prevents contamination of soils, surface water, and groundwater, both during the construction process and during the life of the pit. Consideration should be given to the use of tanks or lined pits to protect soil and groundwater, especially for brines and oil-based fluids.

Normal operations should preclude oil in pits. However, in the event that well completion operations dictate use of pits containing oil for a brief period of time, they should be fenced, screened, netted and/or flagged, as appropriate, to protect livestock, wild game, and fowl. Refer to the Migratory Bird Treaty and Enforcement Improvement Act for additional guidance. Oil accumulated in pits should be promptly removed and recovered, recycled, or disposed.

All liquids and other materials placed in pits should be recovered, recycled, or disposed in an environmentally acceptable manner (determined by the constituents in the material and the environmental sensitivity of the location). When operations are completed, pits not required for well operation should be closed in accordance with the environmental sensitivity of the location. The surface area should be restored to a condition compatible with the uses of the adjacent land area. Any pit retained should be of minimum size commensurate with well operations. Refer to API Environmental Guidance Document: Onshore Solid Waste Management in Exploration and Production Operations for additional information and permitting requirements.

WASTE DISPOSAL OPTIONS

The following options can be considered for waste disposal of drilled cuttings and excess fluids for both water-base muds (WBMs) and NADFs:

- *Leave in place*. The cuttings and fluids are left in place, usually in a reserve pit or on the sea floor.
- *Bioremediate*. The wastes are left in place and treated with some form of bioremediation technique for oily wastes.
- *Cover*. The wastes are left in place or transported to a landfill and covered with some sort of benign material.
- *Spreading*. The wastes are land farmed by spreading in a relatively thin layer, allowing the wastes to dewater by evaporation. Oily wastes are treated with a bioremediation package.
- *Entomb*. The wastes are entombed either in place or at a separate site by some form of solidification process. This is not normally done offshore, but offshore wastes can be transported to land to be entombed.
- *Retrieve and reinject*. The cuttings and waste fluids are slurrified and injected into an appropriate waste disposal well either on land or on an offshore platform.

SLURRY FRACTURE INJECTION

The slurry fracture injection (SFI) process can be used in the petroleum industry to dispose of produced solids, oily viscous fluids/sludges, tank bottoms, contaminated soils, and drill cuttings. The SFI process advantages are:

- Permanent disposal of waste materials
- Zero surface discharge
- No adverse interaction with the environment
- Reduced risk of ground water contamination
- Relatively low cost

The following example is a land-based SFI process description from Terralog Technologies that was done in the Duri Oilfield in Indonesia. The field was discovered in 1941 by Caltex Pacific Indonesia and is now being steamflooded, producing up to 400 m^3 (2500 bbl/day) of oily viscous fluids. These fluids are sent to one of five oil production central gathering stations in the Duri oilfield.

According to the Terralog paper (Arfie et al., 2005), the waste material is screened to pass specified injection criteria, such as size.

> These wastes are then mixed with water to produce a pumpable slurry. The slurry is made with as high a waste concentration as possible, typically 10–33% by volume. Deep disposal wells are used to inject the slurry under pressure into a suitable geological "target" formation deep within the earth. Injection

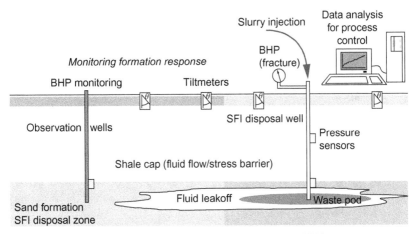

FIGURE 14.1 Terralog's SFI process in the Duri oilfield, Indonesia (2005).

pressures are in excess of the fracture gradient of the disposal formation. The target formation(s) are high-permeability, thick, unconsolidated sands...

The SFI operations at Duri are conducted on a daily basis (>8 hours/day), continuous injection cycles (>48 hours/cycle, one cycle 3 days), and long term (12–15 days/month). Under such conditions, as waste material is injected into a formation, an area around the wellbore begins to fill with the injected waste material. The water component of the slurry dissipates into the formation. This waste filling area coalesces and evolves into a "waste pod" [*Fig. 14.1*]. Relatively lower permeability, lower compressibility, and lower porosity than the surrounding formation can characterize this waste pod. The SFI process incorporates a number of unique monitoring and operating features that allow for effective process control during injection-disposal operations. Process control during SFI operations refers to: a) maintaining optimum formation injectivity; b) maintaining fracture containment; and c) maximizing formation storage capacity.

A second Terralog paper (Marika et al., 2009) describes an offshore SFI process as shown in Fig. 14.2.

The regulatory considerations for applying SFI in United States oilfields is outlined in a paper by M. Sipple-Srinivasan (1998) presented at the International Petroleum Environmental Conference in 1998. The following is the abstract of the Sipple-Srinivasan paper:

> The principal directive of the Federal Underground Injection Control (UIC) program, authorized by the Safe Drinking Water Act of 1974, is to protect underground sources of drinking water (USDWs) from contamination resulting from the injection of fluids into subsurface geologic formations. The UIC

602 Composition and Properties of Drilling and Completion Fluids

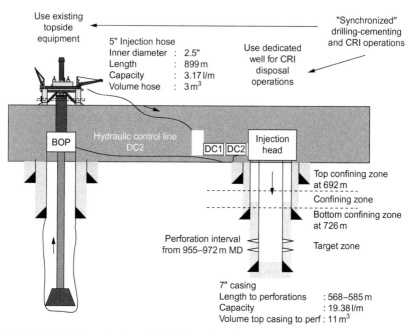

FIGURE 14.2 Schematic of a typical offshore slurry fracture injection process (2009).

program regulates injection fluids in five classes of wells; Class II wells being for injection of fluids associated with the exploration and production of oil and gas. In 1988 the Environmental Protection Agency issued a regulatory determination stating that E&P wastes, being generally lower in toxicity than other wastes regulated under RCRA, should be exempt from RCRA Subtitle C regulations. Oil field wastes are consequently designated as nonhazardous material under Federal regulations and can be injected into Class II wells. Regulatory oversight for disposal into these wells has been delegated largely to individual States (Primacy States), with the USEPA administering the UIC program in the remaining States (Direct Implementation States).

Disposal of oil field waste into Class II wells through high-pressure injection of slurried waste material into deep geologic formations has been successfully implemented in Alaska, the Gulf of Mexico, California, the North Sea, and Canada. The method of Slurry Fracture Injection provides an environmentally sound and permanent disposal solution for terminal oilfield wastes where the alternative remedial options of landfills, road spreading, thermal treatment, and separation techniques fall short. Waste injection with this method results in minimal impact to surface land use, and reduced long-term liability to the operator.

Current State regulations generally have some provision for new technologies to be approved. Injection pressures in SFI exceed the formation parting pressures and result in large volumes of waste material being deposited into disposal formations. Although individual State regulations vary, injecting above fracture pressure is often expressly prohibited. At issue is the security of proximal USDWs, and the containment of fractures, and consequently waste material, within the target formation. To minimize the potential for fracture propagation into confining zones adjacent to USDWs, an acceptable monitoring and analysis program capable of effectively tracking formation response to the SFI process must be designed.

Regulatory acceptance of this oil field waste disposal technique can be achieved through close cooperation between regulatory agencies, waste generators, and the injection project operators. The key to the success is developing sound monitoring strategies which demonstrate fracture orientation and propagation control, and reliably indicate formation response. Evaluating potential strategies to mitigate wastes of higher toxicity in the future may then be considered. Ultimately, protecting human health and the environment through implementation of intelligent remedial options will benefit regulators, operators, and society at large.

OFFSHORE WASTE DISPOSAL FOR NADFS

The International Association of Oil and Gas Producers issued a publication (IOPG, 2003) called *"Environmental aspects of the use and disposal of nonaqueous drilling fluids associated with offshore oil & gas operations*: Report No. 342, May 2003." A follow-up report was issued in 2016 called *"Environmental fates and effects of ocean discharge of drill cuttings and associated drilling fluids from offshore oil and gas operations."* These publications are free and can be ordered on their website at www.OGP.org.uk.

The IOPG has classified nonaqueous base fluids (NABFs) as to their aromatic content. Disposal options are then assigned to cuttings and whole mud disposal based on these classifications.

Group I NABFs (High Aromatic Content)

These base fluids include crude oil and oils refined from crude oil, diesel fuel, and conventional mineral oils. They are complex mixtures of liquid hydrocarbons, including paraffins, aromatic hydrocarbons, and polycyclic aromatic hydrocarbons, PAHs. These base fluids are defined as containing more than 5% by weight aromatic hydrocarbons, with PAH concentrations greater than 0.35 wt%.

These oils are toxic and will persist in the ocean. As a result, Group I drill cuttings are no longer discharged in most of the world. Local regulations, however, may allow such fluids to be used offshore. For land drilling, the restrictions aren't nearly so severe, so diesel is still used in many parts of the world.

Group II NABFs (Medium Aromatic Content)

These base fluids are usually called LTMBFs. Group II NABFs also are produced by refining crude oil, but the distillation process is refined to reduce total aromatic hydrocarbon concentrations to between 0.5 and 5 wt% and PAH concentrations to between 0.001 and 0.35 wt%.

Group III NABFs (Low to Negligible Aromatic Content)

This group of NABFs are materials that are synthesized from specific, well-defined organic nonpetroleum precursors to produce synthetic base fluids (SBFs). Also, more extensive refining of a petroleum stock can produce enhanced mineral oil-base fluids (EMOBFs). These fluids contain less than 0.5 wt% total aromatics and less than 0.001 wt% (10 mg/L) total PAH.

The following has been abstracted from the IOPG, 2009 publication on nonaqueous fluids.

> New technical challenges in offshore drilling have led to the requirement of drilling fluids with drilling properties that exceed those of water-based fluids. New concepts such as directional and extended reach drilling are required to develop many new resources economically. Such drilling requires fluids that provide high lubricity, stability at high temperatures and wellbore stability. These challenges have led to the development of more sophisticated nonaqueous drilling fluids (NADFs) that deliver high drilling performance and ensure environmentally sound operations.
>
> The introduction of NADFs into the marine environment is associated with fluid adhering to discharged cuttings following treatment, since bulk discharge of NADFs is generally not allowed. This paper does not consider bulk discharge of NADFs. Significant advances have been made to reduce the toxicity and environmental impacts of NADFs. Where NADF cuttings discharge is allowed, diesel and conventional mineral oils have largely been replaced with fluids that are less toxic and less persistent. Polyaromatic hydrocarbons, the most toxic component of drilling fluids, have been reduced from 1−4% to less than 0.001% for newer fluids. New generation drilling fluids, such as paraffins, olefins and esters are less toxic and are more biodegradable than early generation diesel and mineral oil-base fluids.
>
> The purpose of this paper is to summarize the technical knowledge about discharges of cuttings when NADFs are used. The report summarizes the results from over 75 publications and compiles the findings from all available research on the subject. It is intended to provide technical insight into this issue as regulations are considered in countries around the world. It should aid in the environmental assessment process for new projects as it provides a comprehensive synopsis of what is known about the environmental impacts resulting from discharge. A compilation of current regulations and practices from around the world is included *in Appendix C* of this report.

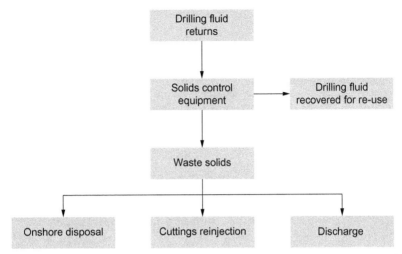

FIGURE 14.3 Disposal options for drilling and drilling fluid wastes. *Courtesy of the International Association of Oil and Gas Producers.*

As summarized in this paper, discharge is one of several options that may be considered when deciding on waste management options [*Fig. 14.3*]. Other options include injection of cuttings or hauling cuttings to shore for disposal. All waste management options have both advantages and disadvantages with regard to environmental impact. This paper shows how environmental, operational and cost considerations can be weighed to decide which options might be considered for given operational and local environmental conditions. The development of more environmentally friendly fluids has been undertaken to reduce the environmental impact associated with the discharge of drill cuttings that when NADFs are used, and make that option more broadly acceptable. When applicable, offshore discharge is the safest and most economical option.

This paper also covers the tools and methods available to predict the fate and effects of drilling discharges. These include laboratory techniques that have been used to address toxicity, biodegradation and bioaccumulation characteristics of different fluids. Numerical models that can be used to predict the distribution of cuttings that are discharged into a given environment are also described.

A compilation of field monitoring results at offshore drilling sites reveals a relatively consistent picture of the fate and effects drill cuttings associated with NADFs. The degree of impact is a function of local environmental conditions (water depth, currents, temperature), and the amount and type of waste discharged. Further, at sites where cuttings associated with early generation drilling fluids were discharge, more significant temporal and spatial impacts were observed. Cuttings discharged with newer fluids resulted in a smaller zone of impact on the sea floor, and the biological community recovered more rapidly.

It is generally thought that the largest potential impact from discharge will occur in the sediment dwelling (benthic) community. The risk of water-column impact is low due to the short residence time of cuttings as they settle to the sea floor and the low water solubility and aromatic content of the base fluid. Impacts on the benthic biota are potentially due to several factors. These include the chemical toxicity of the base fluid, oxygen depletion due to NADF biodegradation in the sediments and physical impacts from burial or changes in grain size. At sites where newer NADFs were used, field studies show that recovery is underway within one year of cessation of discharges.

The nature and degree of impacts on the benthic community tends to reflect variability between local environmental settings and differences in discharge practices. However, in sediments with substantially elevated NADF concentrations, impacts include reduced abundance and diversity of fauna. Recovery tends to follow a successional re-colonization, with initial colonization with hydrocarbon-tolerant species and/or opportunistic species that feed on bacteria that metabolize hydrocarbons. As hydrocarbon loads diminish, other species recolonize the area to more closely resemble the original state. The implications of potential sea floor impacts depend on the sensitivity and significance of the bottom-dwelling resources. In many environmental settings, the bottom sediments are already anoxic, and the addition of cuttings will have little incremental effect.

The degree and duration of impact depends on the thickness of the deposition, the original state of the sediment and the local environmental conditions. In some settings, the cuttings can be re-suspended, eliminating any substantial accumulations. Initial deposition thickness depends on a number of factors including the amount of material discharged, water depth, discharge depth, the strength of currents in the area and the rate at which cuttings fall through the water column. Greater accumulation would be expected in the case of a multiple well development when compared with a single exploration well.

The solids control system sequentially applies different technologies to remove formation solids from the drilling fluid and to recover drilling fluid so that it can be reused. The challenge faced in processing is to remove formation solids while at the same time minimizing loss of valuable components such as barite, bentonite, and NABF. Ultimately, the solids waste stream will comprise the drill cuttings (small pieces of stone, clay, shale, and sand) and solids in the drilling fluid adhering to the cuttings (barite and clays).

Some drill cuttings, particularly in water-base fluids (WBFs), disintegrate into very small particles called "fines," which can build up in the drilling fluid, increasing the drilling fluid solids content and degrading the flow properties of the drilling fluid. If drilling fluid solids cannot be controlled efficiently, dilution with fresh drilling fluids might be necessary to maintain the performance characteristics of the drilling fluid system. The increase of fluid volume resulting from dilution becomes a waste. For WBF systems, when the drilling fluid in use cannot meet the critical operational properties, then that fluid may be

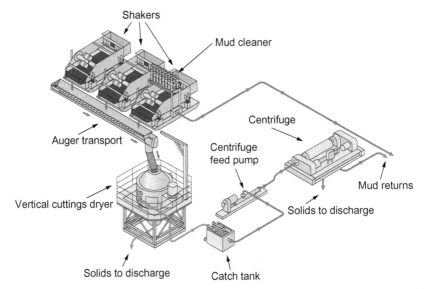

FIGURE 14.4 Solids control equipment for drying cuttings. *Courtesy of the International Association of Oil and Gas Producers.*

replaced by freshly prepared drilling fluid or a different type of fluid. Used water-based fluids are typically disposed of by discharging into the sea, in accordance with local regulations.

Unlike WBFs, used NADFs are recycled instead of being discharged because of regulatory requirements and the expense of the fluid. The components of the solids control system will depend upon the type of drilling fluid used, the formations being drilled, the available equipment on the rig, and the specific requirements of the disposal option. Solids control may involve both primary and secondary treatment steps. *Fig. 14.4* illustrates an advanced type of system in use by industry.

As part of primary treatment, cuttings are first processed through equipment designed to remove large cuttings and then through a series of shale shakers with sequentially finer mesh sizes, designed to remove progressively smaller drill cuttings. Shale shakers are the primary solids control devices. Each stage of the process produces partially dried cuttings and a liquid stream. Where no secondary treatment is employed, partially dried cuttings output will be disposed of by the selected option.

Where secondary treatment is used, the partially dried cuttings may be further processed using specialized equipment commonly called cuttings dryers followed by additional centrifugal processing. Cuttings dryers, sometimes used to process NADF cuttings, include such equipment as specialized shale shakers and centrifuges that apply higher centrifugal forces than developed by conventional shale shakers.

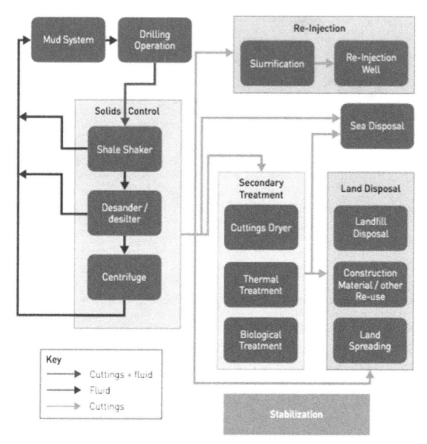

FIGURE 14.5 Updated Disposal options for drilling and drilling fluid wastes. *Courtesy of the International Association of Oil and Gas Producers.*

The IOPG, 2016 document describes an updated disposal options flow chart (Fig. 14.5), which includes thermal and biological treatments.

OFFSHORE DISCHARGE

Table 14.1 shows the relative advantages and disadvantages of offshore discharges of cuttings. Table 14.2 lists the relative advantages and disadvantages of cutting reinjection (see Fig. 14.2).

ONSHORE DISPOSAL

Fig. 14.5 shows the options for onshore disposal of cuttings generated onshore. For land drilling operations, the costs are usually less than for offshore. This is usually dependent on logistical considerations involving the

TABLE 14.1 Advantages (+) and Disadvantages (−) of Offshore Discharge

Economics	Operational	Environmental
+ Very low cost per unit volume treatment	+ Simple process with little equipment needed	+ No incremental air emissions
+ No potential liabilities at onshore facilities	+ No transportation costs involved	+ Low energy usage
− Potential future offshore liability	+ Low power requirements	+ No environmental issues at onshore sites
− Cost of analysis of discharges and potential impacts (e.g., compliance testing, discharge modeling, field monitoring programs)	+ Low personal requirements	− Potential for short-term localized impacts on seafloor biology
	+ Low safety risk	
	+ No shore-based infrastructure required	
	+ No additional space or storage requirements	
	+ No weather restrictions	
	− Management requirements of fluid constituents	

Source: IPIECA Drilling fluids and health risk management, 2009.

availability of disposal sites, the chloride content of the mud system, and local regulations.

However, in offshore operations cuttings may be processed on the drilling rig, stored, and transported to shore for disposal. Consequently, there are two components of onshore disposal that must be considered when evaluating the viability of this option. The first is marine transport (i.e., ship-to-shore, which is common to all potential onshore disposal options) and the associated advantages and disadvantages.

This option is not technically complicated, but it involves a substantial amount of equipment, effort, and cost. The option involves the following steps:

- Cuttings from the shale shakers are stored in storage containers (boxes, bags, or tanks).
- Storage containers are offloaded by crane to a workboat or other vessel or cuttings may be pumped by vacuum into tanks on a workboat.
- The vessel transports the cuttings (and containers) to shore.
- Containers are offloaded from the boat to the dock at port.

TABLE 14.2 Advantages (+) and Disadvantages (−) of Cuttings Reinjection

Economics	Operational	Environmental
+ Enables use of a less expensive drilling fluid	+ Cuttings can be injected if pretreated	+ Elimination of seafloor impact
+ No offsite transportation needed	+ Proven technology	+ Limits possibility of surface and ground water contamination
+ Ability to dispose of other wastes that would have to be taken to shore for disposal	− Extensive equipment and labor requirements	− Increase in air pollution due to large power requirements
− Expensive and labor-intensive	− Application requires receiving formations with appropriate properties	− Possible breach to seafloor if not designed correctly
− Shutdown of equipment can half drilling activities	− Casting and wellhead design limitations	
	− Overpressuring and communication between adjacent wells	
	− Variable efficiency	
	− Difficult for exploration wells due to lack of knowlege of formations	
	− Limited experience on floating drilling operations and in deep water	

Source: IPIECA Drilling fluids and health risk management, 2009.

- As trucks or other ground transport vehicles are available, cuttings (and containers) are loaded into the trucks.
- The trucks transport the cuttings to the land disposal or treatment facility.
- Equipment at the facility offloads the cuttings from the trucks, while other equipment may provide further treatment or manipulation (e.g., bulldozers, grinding, and slurrification units).
- The treated cuttings may be placed in a landfill and buried, incinerated, spread on land, or injected into a suitably isolated injection zone.
- Empty containers are transported back to the port by truck and, ultimately, back to the rig by boat.

As indicated by the description above, onshore disposal involves a substantial amount of additional equipment relative to the discharge option. On the platform itself, incremental equipment requirements are primarily limited to storage containers, such as cuttings boxes to hold the cuttings prior to and during transport. However, this option also involves increased use of existing rig equipment, such as the crane. If vacuum transfer equipment is used, less lifting will be required (Tables 14.3 and 14.4).

TABLE 14.3 Advantages (+) and Disadvantages (−) of Marine Transport to Shore

Economics	Operational	Environmental
+ Waste can be removed from drilling location, eliminating future liability at the rig site	− Safety hazards associated with loading and unloading of waste containers on workboats and at the shore base	+ No impacts on benthic community
− Transportation cost can be high for vessel rental and vary with distance of shore base from the drilling location	− Increased handling of waste is necessary at the drilling location and at shore base	+ Avoids impacts to environmentally sensitive areas offshore
− Transportation may require chartering of additional supply vessels	− Additional personnel required	− Fuel use and consequent air emissions associated with transfer of wastes to a shore base
− Additional costs associated with offshore transport equipment (vaccums, augers) cuttings boxes or bulk containers), and personnel	− Risk of exposure of personnel to aromatic hydrocarbons is greater	− Increased risk of spills in transfer (transport to shore and offloading)
− Operational shutdown due to inability to handle generated cuttings would make operations more costly	− Efficient collection and transportation of waste are necessary at the drilling location	− Disposal onshore creates new problems (e.g., potential ground water contamination)
	− May be difficult to handle logistics of cuttings generated with drilling of high rate of penetration large diameter holes	− Potential interference with shipping and fishing from increased vessel traffic and increased traffic at the port
	− Weather or logistical issues may preclude loading and transport of cuttings, resulting in a shutdown of drilling or need to discharge	

Source: IPIECA, Drilling fluids and health risk management, 2009.

TABLE 14.4 Costs for Onshore Cuttings Treatment and Disposal

Cost Parameter	Value	Units	Comments
Thermal treatment (UK)	251	$/t	UKOOA (2016)
Incineration treatment (UK)	111	$/t	UKOOA (2016)
Landfarm (USA)	37	$/t	API/NOIA (2000)
Untreated landfill (UK)	74	$/t	UKOOA (2016)
Treated landfill (median-UK, Norway, USA)	208	$/t	Calculated from Vell (1998), and Kunze and Skorve (2000)
Onshore injection (median, USA)	130	$/t	Calculated from Vell (1998)

EVALUATION OF FATE AND EFFECTS OF DRILL CUTTINGS DISCHARGE

This section is adapted from the IOPG report 342, 2003. It addresses the general environmental fate and effects of NADF cuttings once discharged into the ocean. Fig. 14.6 illustrates the deposition and fate of drill cuttings as they fall through the water column and settle on the sea floor. Initial deposition is largely dependent on water depth and currents, as well as the volume and density of the discharged cuttings. Persistence on the sea floor is related to sediment transport and resuspension as well as the biodegradation of the base fluid. Biological effects of cuttings are dependent on the toxicity of the fluid-coated cuttings and the extent of the cuttings deposition (Fig. 14.7).

Initial Seabed Deposition

The initial cuttings deposition on the seabed is the result of a number of physical processes that differ significantly from site to site. The pattern of cuttings deposition is determined by the following conditions at each site:

- Quantities and rate of cuttings discharged
- Cuttings discharge configuration (i.e., depth of discharge pipe)
- Oceanographic conditions (e.g., current velocities, water column density gradient)
- Total amount and concentration of NADFs on cuttings
- Water depth

Since the particles are covered with a NADF, the cuttings tend to aggregate once they are discharged. The aggregates fall at a greater fall velocity than the particles in the more easily dispersed WBF cuttings. Less dispersion

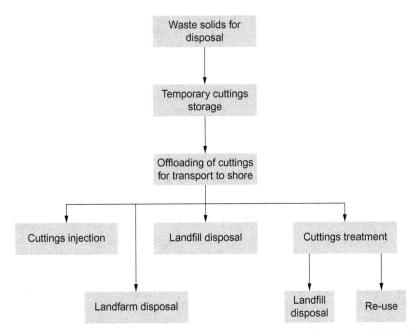

FIGURE 14.6 Onshore disposal options. *Courtesy of the International Association of Oil and Gas Producers.*

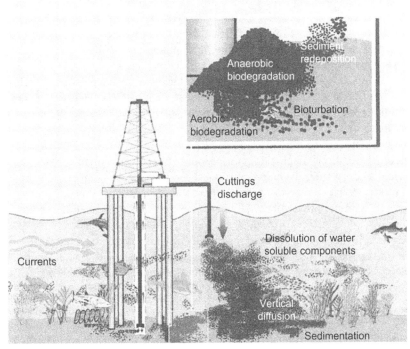

FIGURE 14.7 Fate of NADF cuttings discharged into the ocean. *Courtesy of the International Association of Oil and Gas Producers.*

and the greater fall velocity of the NADF-covered cuttings generally results in smaller, but thicker, area on the seabed, when compared to WBF cuttings discharged under the same conditions. The degree of aggregation is affected by the oil-on-cuttings content and any cleaner methods used on the rig.

Physical Persistence

Once cuttings are deposited on the seabed, the physical persistence of the cuttings and elevated NADF levels depend upon the natural energy of resuspension and transport on the sea floor, coupled with biodegradation of the base fluids. In addition, shale cuttings can weather naturally and disaggregate into silt/clay particles due to hydration upon exposure to seawater. The duration of the benthic community impact is related to the persistence of NADF cuttings accumulations and associated hydrocarbons in the sediment. Field studies indicate that for NADF cuttings discharges, the areas that recovered most rapidly were those characterized by higher energy seabed conditions. Because of the tendency for adhesion between NADF cuttings, resuspension of NADF cuttings requires higher current velocities than those required for WBF cuttings. Laboratory tests found that the critical current velocity required for erosion of NADF cuttings was 36 cm/s for cuttings with 5% oil content. Critical velocity was not found to be a strong function of oil content.

Since cuttings would be expected to be less persistent in areas with thinner deposits recovery from any impacts would be expected to be more rapid than areas with deep piles. Therefore, it is important to consider factors that govern the initial deposition thickness and the potential for erosion in assessing recovery potential. Initial deposition thickness will depend on the current profile and water depth. Stronger currents lead to wider dispersion before deposition, and greater water depth generally will lead to thinner initial deposits.

The potential for erosion is dependent on currents near the sea floor. In relatively shallow water, tidal currents and storm events often provide sufficient energy for substantial erosion. Therefore, shallow areas with high currents are not likely to experience substantial accumulations of cuttings for extended periods of time. Although it is commonly assumed that bottom currents are relatively low in deep water settings, this is not always the case. For example, in the deepwater Gulf of Mexico, currents with velocities in excess of 100 cm/s have been observed. Therefore, local environmental conditions are very important in determining the characteristics of the initial deposition and the likelihood of extended persistence of accumulations of discharged cuttings.

BENTHIC IMPACTS AND RECOVERY

The deposition of drill cuttings may result in physical smothering of benthic organisms regardless of the nature of cuttings, WBF or NADF. The initial deposition of cuttings can also have a physical impact on bottom-dwelling animals by altering the sediment particle size distribution of the substrate. Since NADFs are biodegradable organic compounds, their presence with the cuttings on the sediments increases the oxygen demand in the sediments. This organic enrichment of the sediment can lead to anoxic/anaerobic conditions as biodegradation of the organic material occurs. Anoxic conditions may also result from burial of organic matter by sediment redistribution. Most field studies following discharge of Group III NADF cuttings indicated increased anaerobic conditions in subsurface sediments.

BIODEGRADATION AND ORGANIC ENRICHMENT

NADF biodegradation rates depend upon sea floor environmental conditions (temperature, oxygen availability in sediments) as well as NADF concentrations and NADF type. Biodegradation occurs more rapidly under aerobic conditions (with oxygen) than under anaerobic conditions (in the absence of oxygen). With very few exceptions, oxygen is present in seawater at the sea floor. Therefore, aerobic conditions occur at the exposed surface of the cuttings accumulations and impacted seabed as oxygen diffuses from the water to the sediments. Laboratory studies have indicated that the activity of sediment reworker organisms in the sediments further enhances oxygen transfer into the sediments and the rate of biodegradation. Likewise, oxygen is more available to dispersed or remobilised particles containing NABF than deep inside impacted sediments.

When the rate of biodegradation in sediments is greater than the rate of diffusion of oxygen into the sediments, oxygen becomes limited and sediments become anaerobic. Anaerobic (or anoxic) conditions would be expected to occur deeper within the cuttings accumulation or impacted sediment. If anoxic conditions are generated, additional anaerobic biodegradation may occur by specific populations of microorganisms.

The term used to describe the effects of NABF biodegradation in sediments is organic enrichment. In certain environments, the subsurface is already anoxic due to natural processes, and in other cases, the anoxic zone may begin only a few centimeters from the surface. In such environments, the impacts of biodegradation following discharge may be less.

If anoxia is induced, benthic organisms, macro- and meiofauna that require oxygen for survival, may not be able to compete with bacteria for oxygen. As a consequence, the rapid biodegradation of NADF may lead indirectly to sediment toxicity. Furthermore, if the concentration of hydrogen sulphide becomes high enough in the sediments, it may impact benthic populations. As a result of these factors, benthic populations may be altered in the affected sediments until the NADF has been sufficiently removed to mitigate the organic enrichment and organisms can recolonize the sediments.

As the NABF biodegrades, the NADF cuttings aggregate becomes more hydrophilic (water soluble) and the fine particulate solids are released.

Bottom currents can then more easily disperse the cuttings. The rate and extent to which this can occur will depend upon the biodegradability of the fluid and velocity of local currents. Laboratory results have shown that the sediment characteristics may have an effect on the rate of biodegradation. Sediments with a greater fraction of clay and silt particles supported more rapid biodegradation than sediments that were sandier.

CHEMICAL TOXICITY AND BIOACCUMULATION

In addition to the potential effects from anoxia, chemical toxicity and bioaccumulation of NABF components could also lead to benthic impacts. As mentioned earlier, the impacts to the water column are minimal and the majority of impacts that are measured are in the benthic communities. The toxicity of NADF cuttings on benthic communities is a combination of fluid toxicity and anoxic conditions.

It is difficult to distinguish the effects of chemical toxicity from those of oxygen deprivation. Recent statistical analysis of North Sea monitoring data suggests that the impacts to benthic biota can be related to anoxic conditions caused by rapid biodegradation of hydrocarbons contained in the base fluid. This suggests that the biodegradation rate of hydrocarbons in the sediment may determine the extent of impacts on the benthic biota, and faster degradation rates may lead to larger initial impacts.

The potential for significant bioaccumulation of NABFs in aquatic species is believed to be low. Bioaccumulation of NABF in benthic species occurs when organisms exposed to hydrocarbons incorporate those compounds within their biomass. The extent of bioaccumulation is a function of incorporation of a compound into the tissue mass of the organisms countered by the ability of the organism to depurate or metabolize the compound. An associated condition is taint, namely alteration of the odor or taste of edible tissue resulting from the uptake of certain substances, including certain hydrocarbons. Some complications with detecting taint are that flesh that contains hydrocarbon may have no detectable taint. There is no evidence that NADFs cause taint in concentrations discharged to the environment. Davies et al. (1989) reviewed a series of taint studies on fish caught near platforms in the North Sea where cuttings containing oil-based muds were discharged. Fish testing panels were unable to determine an off taste (taint) in fish caught in the vicinity of these platforms.

RECOVERY

The recovery of the benthic communities is dependent upon the type of community affected, the thickness, area extent, and persistence of the cuttings (due to a combination of sea floor redistribution and biodegradation), and the availability of colonizing organisms. Field studies have indicated that in the short term, impacts from discharging NABFs can range from minor alterations in the biological community structure at moderate distances (i.e., hundreds of meters) from the discharge point to significant mortality of biota in the immediate vicinity of the outfall. Field studies on NABFs have shown a decrease in faunal abundance and diversity near the well sites, with a corresponding increase in opportunistic species. Typically, over the longer term, the affected areas are recolonized by biological communities in a successional manner. Initial colonization is by species that are tolerant of hydrocarbons and/or opportunistic species that feed on bacteria that metabolize hydrocarbons. As time passes and hydrocarbon loads diminish, other species return via in-migration and reproduction, and the community structure returns to something more closely resembling its former state.

As the NABF biodegrades, the NADF cuttings aggregate becomes more hydrophilic (water soluble) and the fine particulate solids are released. Bottom currents can then more easily disperse the cuttings. The rate and extent to which this can occur will depend upon the biodegradability of the fluid and velocity of local currents. Laboratory results have shown that the sediment characteristics may have an effect on the rate of biodegradation. Sediments with a greater fraction of clay and silt particles supported more rapid biodegradation than sediments that were sandier.

LABORATORY STUDIES

Laboratory tests can be used to assess the biodegradability, toxicity, and potential for bioaccumulation of NABFs (Candler et al., 1999). Test protocols have been incorporated into some national and regional regulatory frameworks. Laboratory data provide a means of differentiating products in terms of their environmental performance. It is commonly assumed that fluids that are less toxic and more biodegradable in laboratory studies have a higher likelihood of causing less impact on the sea floor. However, laboratory tests, while useful tools, are not always predictive of ecological impacts because they cannot account for the complexities and variables of the marine environment. For example, while esters have been shown to be highly biodegradable in the laboratory, they have caused low oxygen concentrations in field sediments, resulting in greater biological impact than less biodegradable NABFs.

The Harmonized Mandatory Control System (OSPAR, 2016) for the North East Atlantic, and the USEPA's effluent discharge permits in the United States (USEPA, 2001), require various laboratory tests to determine if a material is suitable for offshore discharge. Laboratory tests may be used either as a regulatory compliance tool or as a research tool to evaluate the impacts of NABF on the seabed under controlled conditions. However, caution should be used when trying to extrapolate results from laboratory tests. Laboratory data are generated under tightly controlled, constant environmental conditions, while seabed conditions are highly variable and usually much different from the conditions in the laboratory.

CHARACTERIZATION OF NADF BIODEGRADABILITY

In some parts of the world, approval for overboard discharge of cuttings drilled with nonaqueous drilling fluids (NADFs) also requires laboratory biodegradation testing of the base fluid. The concept of using biodegradation as a fluid-compliance criterion is that a material that degrades readily in laboratory tests will not persist for extensive periods of time in the environment upon discharge. Quantifying biodegradation rates of NADF base oils used in drilling fluids is complicated by a number of factors that could affect biodegradation rates both in the laboratory and in the field.

There have not been widely accepted seabed survey protocols that have been applied to test biodegradation for varying field conditions. This is a further reason why there is no mechanism to compare laboratory to field data directly. It is clear that under certain environmental conditions (e.g., low currents and static piles), evidence suggests long-term persistence of the NABF in the environment. However, under other environmental conditions, evidence suggests the environment is able to accommodate discharged cuttings, by biodegradation or other mechanisms. The biodegradability of NABFs is being studied in the laboratory using different types of experimental protocols: standard laboratory tests, solid phase tests, and simulated seabed tests. While the performance of each fluid is test specific, a few generalizations can be made from the results of laboratory biodegradation tests that have been reported in the literature to date:

- NABFs exhibit a range of degradation rates. Under comparable conditions, esters seem to degrade most quickly and other base fluids have more similar degradation rates. The extent to which the range of base fluids appears to differentiate themselves in degradation rates depends upon the testing protocol used.
- All Group II and Group III NABFs have been shown to biodegrade to some degree with at least one laboratory test.
- Laboratory tests have shown Group III fluids generally biodegrade more rapidly than Group II, although there is still disagreement or debate over how this information relates to the biodegradation occurring in a cuttings pile or on the seabed. Aspects of these tests that affect biodegradation rates include the degree of oxygen present, temperature, and the effects of inocula.
- Increased temperature increases rate of biodegradation.
- A lag phase was reported in many of the tests, and so, the reported half life used to quantitatively describe the biodegradation process should be used with caution.

- The half lives of fluids in sediments increases with concentration of base fluid in the sediments. However, the rate of biodegradation in mg/kg per day of esters increased with higher concentrations of fluid in sediment.
- Sediment type (e.g., sand versus clay/silt) affects degradation rate; degradation occurs more rapidly in silt/clay sediments than in sandier sediment.
- Degradation occurs more rapidly under aerobic conditions than under anaerobic conditions.
- Evaluation of degradation should include consideration of aerobic conditions, as might be found in the periphery of cuttings accumulation, and anaerobic conditions, as might be found in internal portions of a cuttings pile.
- Compounds that degraded in standardized freshwater tests also degraded in standardized seawater tests, but at a slower rate. However, the slower rate may be due in part to lower initial concentrations of microbial inocula used in the seawater tests.

The specific tests are discussed next.

Standard Laboratory Biodegradation Tests

The standard laboratory methods that have been used to test biodegradability of NABFs fall into two categories: methods that consider aerobic biodegradation, and methods that consider anaerobic biodegradation. A number of different biodegradation protocols are in use in different laboratories, and the reported test data show a high degree of variability. For example, test protocols using aerobic or oxygenated conditions yield different biodegradation rates than tests that reflect anaerobic conditions. Therefore, it is not appropriate to compare results when different test protocols are used. A more in-depth description of the various tests that can be used to measure biodegradation can be found elsewhere. A brief description of the standard tests is found in Table 14.5.

Generally the tests in this table are designed for the evaluation of water-soluble compounds and do not perform as well with nonaqueous fluids.

ISO 11734 Modified for NADF Biodegradation

On February 5, 1999, the EPA published the initial guidelines for the discharge of synthetic drilling fluids on cuttings for the United States, updated in 2001 (USEPA, 2001). These guidelines documented that drilling fluids must be more biodegradable than an IO that was 16 to 18 carbons long (IO1618) to be considered in compliance for discharge. Therefore, a substantial modification of ISO 11734:1995 (closed bottle test or CBT) was developed to discriminate specifically the IO1618 and more rapidly biodegrading fluids from less biodegradable fluids. The modifications of the standard

TABLE 14.5 Comparisons of Standard Biodegradation Tests

Test	A/An	F/M	Inoculum Source	Analyte Measured	Comments
OECD 301B	A	F	Sewage Sludge	CO_2 generation	
OECD 301D	A	F	Sewage Sludge	Oxygen loss	Dissolved oxygen is measured
OECD 301F	A	F	Sewage Sludge	Oxygen loss	Manometer method with greater range than the D version of the test
OECD 306	A	M	Sewage Sludge	Oxygen loss	Marine version of 301D
BODIS	A	F	Sewage Sludge	Oxygen loss	Similar to OECD 301D for insoluble substances
Marine BODIS	A	M	Sewage Sludge	Oxygen loss	
ISO 11734	An	F	Sewage Sludge	Gas formation	Anaerobic closed bottle test. Gas measured by pressure transducer.

A—aerobic; An—anaerobic; F—freshwater; M—marine water.

ISO 11734 are that natural sediment replaces freshwater and sewage sludge, seawater is used instead of a nutrient solution and the dosage of fluid added is greater than the standard 11734. The concentration of fluid is higher than normally used with ISO 11734.

The American Petroleum Institute (API) developed the modified ISO 11734 test for synthetic fluids, and this was accepted by the EPA as a method for demonstrating compliance with the biodegradation guidelines. The test was found to contain an appropriate compromise of the following characteristics while also allowing clear performance comparisons to the C1618IO standard:

- Discriminatory power between of fluids
- Reproducibility/repeatability
- Ecological relevance
- Standard performance of chemical controls
- Practicality and cost

Serum bottles are filled with a homogeneous NABF/sediment mixture with seawater and an indicator to detect oxygen. The bottles are sealed and

TABLE 14.6 Characteristics of the Modified ISO 11734 for NABF

Test	A/An	F/M	Inoculum Source	Analyte Measured	Comments
ISO 11734 (NADF)	An	M	Naturally occurring microorganisms in sediments	Gas generated by anaerobic biodegradation	This is a compliance test developed to address specific regulations by USEPA

the headspace flushed with nitrogen to remove oxygen. Gas generated by biodegradation is measured in the bottles by a pressure transducer. The amount of gas produced is compared to controls and standards. If the base fluid produces more gas after 275 days of incubation than an IO1618, then the test fluid is considered in compliance for biodegradation. A brief detail of the test is in Table 14.6.

Results of the CBT indicate that this modification of the anaerobic test is adequate for discriminating the biodegradation performance of various Group III fluids. Esters biodegraded the fastest under this test. Linear alpha olefins and IOs biodegrade next fastest with rates dependent on molecular weight. Type II fluids, paraffins, and mineral oils degrade very slowly, if at all, under the conditions of this test.

However, care must be taken so that the results of the test are not misinterpreted by being used beyond their regulatory function. The CBT test is purely a compliance test developed to discriminate between a subset of Group III fluids. It is inadequate for predicting the relative biodegradation rates of fluids that degrade more slowly than IOs and is not an appropriate tool for predicting rates of NADF removal in sediments. To the degree laboratory tests can mimic subsea conditions, those issues are better evaluated with simulated seabed studies.

The SOAEFD Solid Phase Test

In the North Sea, it was recognized that under some environmental conditions, large piles of cuttings were likely to persist. It was also recognized that base fluids in such a situation were unlikely to degrade at rates represented by aerobic test protocols of aqueous suspensions or extracts. Therefore, work was undertaken to determine relative degradation rates of NADFs should they be present in large static cuttings piles.

The SOAEFD test (also referred to as the solid phase test) was originally developed at the Scottish Office Agriculture, Environment, and Fisheries

Department. The basic approach of the SOAEFD test is to mix clean marine sediments with base fluids used in NADFs and to fill glass jars with the homogeneous NABF/sediment mixtures. The glass jars are placed in troughs through which a continuous laminar flow of natural ambient temperature seawater is passed. At various time intervals sets of three jars are removed and analyzed for NABF. The entire sediment volume is chemically analyzed to determine total losses of the base fluid. The rate of NABF removal is reported as the half life of NABF loss and with first order rate constants. Like the closed bottle test, no microbial inoculum was used. Biodegradation was catalyzed by the naturally occurring sediment bacteria. The original test was conducted at seawater temperatures of 10−15°C, but has been modified to accommodate other environmental conditions (25°C) to mimic conditions in Nigeria. A short description of the test is given in Table 14.7.

The SOAEFD solid phase sediment test is neither a purely aerobic or anaerobic marine biodegradation test for nonwater-soluble drilling chemicals. Although oxygenated water flows across the top of the jar, oxygen diffusion limitations into the sediment can limit the bulk of the sediment/fluid in the sample jars from being exposed to oxygen.

Tests using the SOAEFD approach were conducted with a suite of NABFs for three different concentrations (100, 500, 50,000 mg/kg) of NADF in sterilized sediment. Seven different base fluids were studied: acetal, IO, n-paraffin, poly alpha olefin (PAO), linear alpha olefin (LAO), mineral oil, and ester. Ester mud products were found to degrade significantly more rapidly than all the other base fluids at all concentrations. The results indicate that the IO, LAO, n-paraffin, mineral oil, and PAO fluids degrade to a substantial extent at 100 mg/kg. At 5000 mg/kg, the rates of degradation of the synthetic and paraffin base fluids were similar to the mineral oil except for the ester, which biodegraded faster than the other fluids. However, under the conditions of the test, the recovery of spiked fluid in the time zero

TABLE 14.7 Characteristics of the SOAEFD Test

Test	A/An	F/M	Inoculum Source	Analyte Measured	Comments
SOAEFD	Both	M	Naturally occurring microorganisms in sediments	Concentration of NADF base fluid in the sediments	Rates of degradation are measured from the time-dependent loss of NABF measured by TPH analysis

A—aerobic; An—anaerobic; F—freshwater; M—marine water.

analyses of the fluid in sediments was about 85% of the theoretical, and none of the fluids besides esters biodegraded more than 20% over the length of the test. Care should be taken on the interpretation of these results. The lack of discrimination between fluids may be a function of the testing conditions rather than a lack of different biodegradation rates. Candler and Friedheim (2006) reported comparable results in tests similar to the SOAEFD protocol.

A separate set of SOAEFD tests was conducted to simulate conditions offshore Nigeria (estuarine sediments at 25°C) as opposed to North Sea conditions (marine sediments at 10−15°C) simulated by earlier tests. Tests were conducted using a nonsterilized sediment and 120 mg/kg NADF. Biodegradation rates were higher at the higher temperature. The greater biodegradation rate has been attributed in part to enhanced oxygen transport into the seabed due to activity of small animal life burrowing in the sediment (sediment reworkers).

The water in the SOAEFD solid phase sediment test is neither a purely aerobic or anaerobic marine biodegradation test for nonwater-soluble drilling chemicals. Although oxygenated water flows across the top of the jar, oxygen diffusion limitations into the sediment can limit the bulk of the sediment/fluid in the sample jars from being exposed to oxygen.

Simulated Seabed Studies

Other laboratory methods have been developed to study biodegradability of nonaqueous fluids under simulated North Sea seabed conditions. The primary method of this group, the NIVA simulated seabed approach, was originally developed at the Norwegian Institute for Water Research (NIVA), and has been through numerous modifications since it was first introduced in 1991. Most simulated seabed studies are now based on variants of the NIVA test. The objective of the simulated seabed study is to determine the fate of the test compound in the environment by simulating the conditions of the seabed as closely as possible.

The test setup consists mainly of a series of replicate experimental systems that were maintained in easily accessible indoor basins called benthic chambers. The benthic chambers were approximately 50 cm × 50 cm × 35 cm deep. A cuttings and NADF mixture was suspended in seawater and added to the overlying water in the benthic chambers. The suspensions settled onto a 25 cm deep bed of natural sediment. Once the cuttings settled on the sediments, the chambers were flushed with seawater drawn from a depth of 40−60 m from the Oslofjord. The water in the chamber was being replaced at a rate of one or twice per day. The loss of NABFs deposited on cuttings was measured over a period of 150−187 days by TPH analysis of the top layer of sediments. Environmental characteristics such as Ba concentration, pH, Eh, and oxygen uptake of the chamber were also measured. Later modifications included quantification of sediment organisms throughout the test.

This test method includes many of the potential physical and biological processes affecting the behavior of cuttings on the seabed under controlled laboratory conditions. The test was designed to mimic the aerobic and anaerobic conditions, and the bioturbation processes that are present on the seabed caused by benthic organisms. The results indicate that ester degraded faster than ethers and mineral oils. Some mussel mortality was observed in all the chambers compared to the control. Eh values and oxygen uptake of the sediments were generally consistent with biodegradation of NABF. Mortality of benthic organisms was associated with the disappearance of esters.

Several problems were identified with the initial studies, however. These included the nonhomogenous distribution of cuttings on the surface of the sediments at the start of the test and the associated questions regarding initial conditions and the potential for some of the test fluids to have been washed away by water rather than biodegraded. Some of the questions have been resolved in subsequent refinements to the NIVA test method. However, Vik et al. (1996) concluded that in order to resolve all the issues with the design of the NIVA method, the cost of the experiments would increase dramatically.

Aquatic and Sediment Toxicity of Drilling Fluids

With the introduction of more highly refined mineral oils and the synthetic-based fluids, the aquatic toxicity of NABFs has been significantly reduced. Early-generation OBFs, which consisted of diesel or mineral oil, exhibited significant toxicity as a result of water-soluble aromatic and polyaromatic hydrocarbons. More recent low-toxicity mineral-base fluids (LTMBFs) possess considerably less aromatic hydrocarbons and are less toxic. New-generation enhanced mineral oils, paraffins, and synthetics have little or no aromatic content and generally are even less toxic.

Since NABFs possess low water solubility and are only present in the water column for a short time after discharge, it is becoming more widely accepted that water column toxicity testing does not fully address all the environmental risks associated with NADF discharge on cuttings. Sediment toxicity tests are probably more relevant to the discharge of NADF cuttings than are aqueous phase or water column toxicity tests because most of the fluid is anticipated to end up in the sediment.

Sediment toxicity tests are performed on sediment-dwelling organisms (e.g., *Corophium volutator* or *Leptocheirus plumulosus*). Field validation research, infaunal surveys, and bioassays of waste materials have shown that sediment-dwelling amphipods are sensitive to sediment contaminants. Furthermore, they have maximum exposure potential since they are intimately associated with sediments and have limited mobility.

Table 14.8 summarizes the toxicity data available on NABFs, including data on sediment-dwelling amphipods. Data are presented in terms of the medium lethal concentration, LC_{50}, or the effect concentration for cell reproduction EC_{50} (algae). Toxicity varies with test species and drilling fluids. However, it is clear that diesel oil is more toxic than more highly refined mineral oils and Group III fluids. Though Group III fluids have relatively low toxicity to sediment-dwelling organisms, with LC_{50}s greater than 1000 mg/L of sediment, the relative ranking of the different synthetic-base muds (SBMs) in terms of sediment toxicity is generally consistent across the different species. The esters appear to be the least toxic, followed by the IOs and LAOs. Differences in toxicity of SBMs Group III fluids may be due to differences in molecular size and polarity, which affects water solubility and bioavailability.

Aquatic Toxicity and Regulations

In some parts of the world, approval for discharge of drill cuttings into the sea requires toxicity testing to determine the potential for adverse effects on aquatic life. In the North Sea, base fluids and chemicals used in the drilling process must undergo aquatic toxicity testing for hazard evaluation. Tests are performed with three types of aquatic species representing the aquatic food chain: an alga and an herbivore (for which aqueous phase tests are conducted), and a sediment reworker (for which the sediment phase is used). The most common species tested are the marine alga, *Skeletonema costatum*, the copepod, *Acartia tonsa*, and the sediment-dwelling amphipod, *C. volutator*.

In the United States, discharge approval is based on compliance with effluent toxicity limits at the point of discharge. Present discharge permits require measurement of the aquatic toxicity of the suspended particulate phase of drilling effluents to the mysid shrimp, *Mysidopsis bahia*. Historical studies with water-based muds showed sediment toxicity tests to be considerably less sensitive than the suspended particulate phase; thus, sediment tests were dropped as a testing requirement for those types of discharges.

The USEPA published guidelines for the discharge of cuttings containing synthetic-based drilling fluids (Federal Register, Jan. 22, 2001). In these guidelines, the EPA requires all synthetic-based fluids to be discharged with drill cuttings to be no more toxic than a C16−C18 internal olefin-base fluid, as determined in a 10-day sediment toxicity test (ASTM E1367-92) with *L. plumulosus*. In addition, drilling muds used offshore must undergo a 4-day toxicity test with *L. plumulosus*, prior to drill cuttings being discharged. In this 4-day test, drilling muds must be no more toxic than a C16−C18 formulated drilling mud. Otherwise, drill cuttings discharge cannot proceed.

Overall, laboratory studies have shown that in most cases NADFs exhibit low toxicity and can generally meet toxicity requirements in countries where

TABLE 14.8 Aquatic Toxicity of Nonaqueous-Based Fluids (LC_{50} or EC_{50}, mg/L)

Test Organism	Ester	LAO	IO	Paraffin	LTMBF	EMBF	Diesel
Algae, *Skeletonema costatum*	60,000 (Vik et al., 1996)	>10,000 (McKee et al., 1995)	1000–10,000	NA		NA	NA
Mysid, *Mysidopsis bahia*	1,000,000 (Baroid)	794,450 (McKee et al., 1995)	150,000–1,000,000 (Zevallos et al., 1996)	NA	13,200	NA	NA
Copepod, *Acartia tonsa*	50,000 (Baroid)	>10,000 (McKee et al., 1995)	10,000 (M-I Drilling Fluids, 1995)	NA		NA	NA
Mussel, *Abra alba*	8000 (Vik et al., 1996)	277 (Friedheim and Conn, 1996)	303 (Friedheim and Conn, 1996)	572		NA	NA
Amphipod, *Corophium volutator*	>10,000	1028 (Friedheim and Conn 1996)	1560–7131 (Friedheim and Conn 1996)	NA	2747 (Harris, 1998)	7146 (Candler et al., 1997)	840 (Candler et al., 1997)
Amphipod, *Leptocheirus plumulosus*	13,449 (US EPA, 2001)	483 (US EPS, 2001)	2829 (US EPA, 2001)	NA		557	639 (US EPA, 200)

NA—not available.

toxicity data are required. However, in isolated instances where the NADF may fail to meet toxicity requirements due to the base fluid or chemical additives in the fluid system, alternative options (including possibly changing the base fluid) may need to be evaluated.

Characterization of NABF Bioaccumulation

Bioaccumulation is the uptake and retention in the tissues of an organism of a chemical from all possible external sources (water, food, and substrate). Bioaccumulation may be a concern when aquatic organisms accumulate chemical residues in their tissues to levels that can result in toxicity to the aquatic organism or to the consumer of that aquatic organism. Group I and Group II NADFs may contain some polyaromatic hydrocarbons (PAH) and other high molecular weight branched hydrocarbons that have the potential to bioaccumulate in benthic invertebrates. However, bioaccumulation in higher trophic levels, such as in fish or mammals, is unlikely since these animals have enzyme systems to metabolize PAH compounds. In the case of Group III NABFs, significant bioaccumulation appears unlikely due to their extremely low water solubility and consequent low bioavailability.

Two types of data are used to evaluate a chemical's bioaccumulative potential: the octanol−water partition coefficient and the bioconcentration factor. The octanol−water partition coefficient, often expressed as log P_{OW}, is a physicochemical measure of a chemical's propensity to partition into octanol relative to water. It is used as a chemical surrogate for bioaccumulative potential. Generally, it is recognized that organic chemicals with a log P_{OW} between 3 and 6 have the potential to bioaccumulate significantly. Chemicals with log P_{OW} values greater than 6 are not considered to bioaccumulate as readily because their low water solubility prevents them from being taken up by aquatic organisms and the physical size of the molecules is such that it cannot pass through membranes. Those with P_{OW} values less than 3 tend to not sorb readily into the octanol and readily desorb back into the water phase. Therefore, such compounds do not readily bioaccumulate.

Determination of a bioconcentration factor (BCF) is an in vivo measure of bioaccumulative potential. Basically, an organism is exposed to a constant concentration of the test material in water until equilibrium is reached between the concentration in the water and the concentration in the tissues of the organism. The BCF is the tissue concentration divided by the water concentration at equilibrium. BCF values greater than 1000 (or log BCF >3) can be a cause for concern. Compounds with BCFs <1000 (or log BCF <3), are less likely to accumulate in tissues due to their relatively high water solubility. However, this type of test is fraught with technical challenges when evaluating highly water insoluble substances like Group III fluids. Substances with extremely low water solubility have a tendency to precipitate out of solution

TABLE 14.9 Log P_{OW} and BCF Data on Select Synthetic-Base Muds

Base Fluid	Log POW	Log BCF
Internal olefin, C16–C18	8.6 (M-I, 1995)	*Mytilus edulis*: 10d exposure, 20d depuration – log BCF 4.18 (C16), log BCF 4.09 (C18), depuration $t_{1/2}$ <24 h (Environmental and Resource Technology, 1994)
Linear alpha olefin	7.82 (M-I, 1995)	*Mytilus edulis*: 4.84 (McKee et al., 1995)
Polyalphaolefin	11.19 (M-I, 1995)	*Fundulus grandis*: No systemic bioconcentration, though PAO present in stomach contents. (Rushing et al., 1991)
Ester	1.69 (Growcock et al., 1994)	Not available

or bind to suspended particles, which can make it difficult to determine the BCF accurately.

Table 14.9 summarizes the octanol–water partition coefficient (log P_{OW}) and BCF data for the most commonly used Group III fluids. With the exception of esters, all of the base fluids have a log P_{OW} exceeding 6. This means that bioaccumulation will be limited by their extremely low water solubility. Esters have log P_{OW} less than 3. Therefore, they are not expected to bioaccumulate.

BCF values generated in tests with the blue mussel, *Mytilus edulis*, appear high for both the IO and LAO. However, it is questionable whether these values are reflective of true systemic bioaccumulation because of the difficulty of maintaining steady concentrations of the base fluid in solution. Furthermore, at least with the IO, depuration of the SBF was very rapid after the mussels were transferred to clean water. Ninety-five percent of the IO was eliminated from the mussels within 5 days. This rapid depuration suggests that the IO was likely adhering to the surfaces of the mussel and not absorbed systemically. In her environmental assessment of drilling fluids, Meinhold (1998) concluded that SBFs are unlikely to be accumulated by shellfish. In the only study performed with fish, PAO residues were limited to the stomach, suggesting that SBFs are not readily absorbed into the tissues (Rushing et al., 1991).

In summary, the organic chemical constituents in WBMs do not bioaccumulate. Some trace components of OBMs could bioaccumulate in lower trophic levels but not in higher vertebrates such as fish and mammals, which are capably of metabolizing PAHs. NABFs are not expected to bioaccumulate significantly because of their extremely low water solubility and consequent low bioavailability. Their propensity to biodegrade further reduces the likelihood that exposures will be long enough that a significant bioaccumulative hazard will result.

The OGP document also contains information on the following topics:

- General descriptions of drilling operations
- Uses and types of drilling fluids
- Computer modeling of NADF cuttings
- Field studies of drilling fluids and cuttings discharges
- Country specific requirements for discharge of drilling fluids and cuttings (as of 2003)
- Summary of field studies
- List of NADF systems and base fluids by company trade names

OGP DOCUMENT CONCLUSIONS

This summarizes the current body of knowledge about the environmental aspects of the disposal of NADF cuttings by discharge into the marine environment. Marine discharge is one of a range of disposal methods for NADF cuttings that also includes onshore disposal (transport of cuttings onshore) and cuttings reinjection. Each method has advantages and disadvantages that should be taken into account. Considerations for choosing among the available waste management practices are presented. The following key points stem from the review of the literature on the environmental aspects of marine discharge of NADF cuttings:

- Recent advances have allowed production of a variety of NADFs with very low concentrations of toxic components.
- Initial environmental impacts on benthic organisms from the discharge of NADF cuttings are caused by physical burial. The risk of water column impact is low due to the short residence time of cuttings as they settle to the sea floor and the low water solubility and aromatic content of the base fluid.
- Potential impacts on the benthic biota can be caused by a number of mechanisms. These include chemical toxicity of the base fluid, oxygen depletion due to NADF biodegradation in the sediment, and physical impacts from burial or changes in grain size.
- Numerous field studies have been conducted to measure the initial impacts and recovery from NADF discharge. These studies have shown that benthic community disturbance is in general very localized and temporary. Biodegradation of modern fluids can be relatively rapid, particularly when NADF concentrations are low to moderate. At sites where newer NADFs were used, field studies show that recovery was underway within 1 year of cessation of discharges.

- In the North Sea, ambient conditions and discharge practices caused the formation of large cuttings piles (up to tens of meters thick near large development drill sites). The large mass of these cuttings and associated base fluids extend the amount of time needed for recovery of these sites.
- The formation of large, persistent piles will not occur under many oceanographic conditions. For example, environments with high currents tend to erode piles and decrease recovery time. Deep water also tends to increase dispersion and limit the pile heights that are initially formed.
- Enhanced treatment can reduce the organic loading on cuttings and thereby limit the effects of oxygen depletion in sediments. Enhanced treatment can include (1) methods to reduce the base fluid content of cuttings before disposal and (2) methods to increase cuttings dispersion.

Several factors will influence the acceptability of NADF cuttings discharge. These include

- Existing regulations
- Environmental sensitivity of the receiving environment
- Type of NADF used, the properties of the NADF (e.g., biodegradability, toxicity, and bioaccumulation potential)
- Total volume of drill cuttings discharged
- Methods of disposal (such as depth of discharge pipe, predischarge treatment)
- Ambient conditions of the receiving water
- Ability of the local environment to assimilate the cuttings and base fluid loads

Cuttings discharge appears to be a viable option in many environmental settings. Work continues to develop and implement new technologies for cuttings treatment to reduce fluid content on cuttings prior to discharge. Work also continues to improve and develop a full range of disposal options.

WASTE REDUCTION AND RECYCLING

Other efforts to reduce the quantity of wastes on Gulf of Mexico rigs were reported by Shell E&P (Satterlee and van Oort, 2003). Shell found 26 potential waste reduction strategies. Their strategy heiarchy was to "reduce, reuse, recycle and dispose." They used a "combination of solids control efficiency, cuttings dryer technology and new bulk mixing technology to reduce mud use by up to 20%." The paper contains detailed descriptions of all waste types and treatment methods for these waste streams.

Another project that is working on reducing environmental effects is the Environmentally Friendly Drilling Program (EFD) managed by the Houston Area Advanced Research Center (HARC) and Texas A&M's Global Petroleum Research Institute (GPRI). The EFD program combines new low-impact technologies that reduce the footprint of drilling activities, integrates light weight drilling rigs with reduced emission engine packages, addresses on-site waste management, optimizes the systems to fit the needs of a specific development sites, and provides tools to enhance the industry's stewardship of the environment. The EFD is working on several projects, outlined next.

PROTOTYPE SMALL FOOTPRINT DRILLING RIG

There are a number of emerging rig technologies found in certain areas that offer low impact. Examples include the Huisman LOC 250, LOC 400, and the National Oilwell VARCO Rapid Rig. The DOE-funded microhole drilling program identified a number of technologies to greatly reduce the cost of drilling shallow- and moderate-depth holes for exploration, field development, long-term subsurface monitoring, and, to a limited degree, actual oil and gas production. In addition, the EFD program developed the concept of alternative power solutions for unconventional natural gas activities, including rig and production activities. Further, Texas A&M and M-I SWACO have formed a partnership to integrate membrane water desalination technology into rig site waste management practices. All of these programs are at the tipping point of profitability and acceptance by the industry. If the costs of these processes can be reduced, and their benefit to low-impact drilling realized by the public and policy makers, then many unconventional natural gas developments that are uneconomic today could become economically viable in the future.

DISAPPEARING ROADS

The impact of access roads and drilling pads has been identified by the EFD as one of the major problems to be managed when conducting oil and gas operations in environmentally sensitive areas. Since 2005, the EFD program has been identifying technology and sponsoring research in reducing surface impact. Two major projects are underway specifically addressing such technology.

The Disappearing Road Competition is a yearly, nationwide scholastic competition sponsored by Halliburton to create a new concept for moving men and materials to and from well sites. The University of Wyoming (UW), in collaboration with the Bureau of Land Management and major upstream gas production companies, won first prize in the initial competition with a layered-mat, roll-out road system incorporating a modular frame design to minimize the impact of oil field access roads to well pads. The concept came from the need to minimize soil disruption and wildlife fragmentation in the Jonah Field and Pinedale Anticline Production Area (PAPA) of the upper Green River Valley. Now the UW and the EFD teams are conducting a field test of a scale model of the low-impact road concept at the Pecos Desert Research Test Center that incorporates some of the practices being planned for the project with recycled road materials (drill cuttings). The deliverables from this task will be a report documenting the development of the prototype lay down road system and documentation of a field test to be performed for sponsors. In addition, National Energy Testing Laboratories (NETL) is funding a new EFD project by TAMU to construct and demonstrate low-impact O&G lease roads designed to reduce the environmental impact of field development in sensitive new desert ecosystems. A summary of the winning projects can be found at the Low-Impact Access Roads Demonstration (Pecos Research Test Center).

NO_x AIR EMISSIONS STUDIES

Measurement and control of emissions such as oxides of nitrogen (NO_x), highly volatile organic compounds (HVOC), carbon dioxide (CO_2), and other greenhouse gases are becoming a high priority within the industry. Currently, there are no guidelines on the measurement or control approaches concerning these emissions. The project is developing guidelines concerning the measurement of oxides of nitrogen (NO_x) for a drilling/production site as well as guidelines concerning technologies that can be implemented to reduce NO_x emissions. On behalf of the State of Texas, HARC has been managing a program to develop and implement technologies to reduce diesel engine emissions. Leveraging this effort, guidelines concerning the application of selective catalytic reduction (SCR) technologies will be developed for drilling and production applications.

DRILLING WASTE MANAGEMENT WEBSITE

In a project funded by the US Department of Energy, the Argonne National Laboratory, in conjunction with Marathon and ChevronTexaco, is developing an interactive website to provide technical, unbiased information about drilling waste management options. The website will include components such as a technology description module to familiarize readers with available options, a regulatory module to summarize existing state and federal regulations governing drilling waste, a case study module to exhibit successful implementations of various technologies, and an interactive technology identification module to help users identify waste management options applicable in their region. This information system will provide operators with easy access to regulations, waste management options, technology, and cost data, thus allowing them to choose the option with the most environmental and economic benefits. Another benefit is that it will provide information to countries currently developing drilling waste regulations so that those regulations include the most economic, environmentally appropriate waste management options, not only options familiar in that country. The website can be found at http://web.ead.anl.gov/dwm/index.cfm (Veil et al., 2003).

NEW PRODUCT R&D

Candler and Friedheim (2006) describes the design of new products for the petroleum industry balancing economic issues, drilling performance, and health, safety, and environment (HSE) considerations. They discuss how greater environmental awareness and tighter regulatory controls have combined to advance fluids and drilling waste management technology, and they show specific examples of new water- and synthetic-based fluids and drilling waste management technology used in the industry.

The salt content, usually $CaCl_2$, normally found in NADFs also poses a disposal problem. Walker et al. (2016) proposed a new salt free internal phase for NADFs. Their material was unspecified, but gave similar water activities as do the chlorides. Materials that have been used in place of chlorides are formates, nitrates, and various polyols.

REFERENCES

Arfie, M., Marika, E., Purbodiningrat, E.S., Woodard, H.A., 2005. Implementation of slurry fracture injection technology for E&P wastes at duri oilfield. In: SPE Paper 96543-PP, SPE Asia Pacific Health, Safety and Environment Conference and Exhibition held in Kuala Lumpur, 19–20 September.

API, 2009. Environmental Protection for Onshore Oil and Gas Production Operations and Leases, API Recommended Practice 51R, July 2009.

Brantley, L., Kent, J., Wagner, N., 2013. Performance and Cost Benefits of Environmental Drilling Technologies: A Business Case for Environmental Solutions. Paper SPE-163559-MS, SPE/IADC Drilling Conference, 5-7 March, Amsterdam, The Netherlands. http://dx.doi.org/10.2118/163559-MS.

Candler, J., Friedheim, J., 2006. MISWACO, Designing environmental performance into new drilling fluids and waste management technology. In: 13th International Petroleum Environmental Conference, San Antonio, TX, 17–20 October.

Candler, J., Hebert, R., Leuterman, A.J., 1997. Effectiveness of a 10 day ASTM amphipod sediment test to screen drilling mud base fluids for benthic toxicity. In: SPE/EPA Exploration and Production Environmental Conference, Dallas, TX, March. Society of Petroleum Engineers, Inc. Richardson, TX.

Candler et al., 1999. Predicting the potential impact of synthetic-based muds with the use of biodegradation studies. In: SPE/EPA Exploration and Production Environmental Conference, Austin, TX, 28 February–3 March.

Davies, J.M., Bedborough, D.M., Blackman, R.A.A., Addy, J.M., Appelbee, J.F., Grogan, W.C., et al., 1989. The environmental effect of oil-based mud drilling in the North Sea. In: Engelhardt, F.R., Ray, J.P., Gillam, A.H. (Eds.), Drilling Wastes. Elsevier Applied Science, London.

Environmental and Resource Technology Co. 1994. Assessment of the Bioconcentration Factor of Iso-Teq Base Fluid in the Blue Mussel, *Mytilus edulis*. ERT report 94/061. Report to Baker Hughes INTEQ, Houston, TX.

Friedheim, J.E., Conn, H.L., 1996. Second generation synthetic fluids in the North Sea: are they better? In: IADC/SPE Drilling Conference, New Orleans, LA, 12–15 March.

Growcock, F., Andrews, S., Frederick, T., 1994. Physicochemical Properties of Synthetic Drilling Fluids. Paper SPE-27450-MS, presented at SPE/IADC Drilling Conference, 15-18 February, Dallas, Texas. http://dx.doi.org/10.2118/27450-MS.

Harris, G., 1998. Toxicity test results of five drilling muds and three base oils using benthic amphipod survival, bivalve survival, echinoid fertilisation and Microtox. Report for Sable Offshore Energy, Inc. Harris Industrial Testing Services.

Honeycutt, 1970. Environmental protection at the line level this paper was prepared for the 1970 Evangelize Section Regional Meeting of the Society of Petroleum Engineers of AIME, Lafayette, LA, 9–10 November.

IOPG, 2003. Environmental aspects of the use and disposal of non aqueous drilling fluids associated with offshore oil & gas operations: Report No. 342, May 2003.

IPIECA, 2009. Drilling fluids and health risk management: a guide for drilling personnel, managers and health professionals in the oil and gas industry, International Association of Oil and Gas Producers, Report Number 396.

IOPG, 2016. Environmental fates and effects of ocean discharge of drill cuttings and associated drilling fluids from offshore oil and gas operations: Report No. 543 2016.

Kunze, K, Skorve, H., 2000. Merits of Suspending the First Platform Well as a Cuttings Injector. Paper SPE-63124-MS presented at SPE Annual Technical Conference and Exhibition, 1-4 October, Dallas, Texas. http://dx.doi.org/10.2118/63124-MS.

Marika, E., Uriansrud, F., Bilak, R., Dusseault, B., 2009. Achieving zero discharge E&P operations using deep well disposal.

McAuliffe, C.D., Palmer, L.L., 1976. Environmental aspects of offshore disposal of drilling fluids and cuttings. In: SPE paper 5864-MS, Improved Oil Recovery Symposium, Tulsa, OK, 22–24 March 1976.

McKee, J.D.A., Dowrick, K., Astleford, S.J., 1995. A new development towards improved synthetic based mud performance. In: SPE/IADC Drilling Conference, Amsterdam, 28 February–2 March.

Meinhold, A.F., 1998. Framework for a Comparative Environmental Assessment of Drilling Fluids. Prepared for U.S. Dept. of Energy, Brookhaven National Laboratory, Upton, NY, BNL-66108.

M-I Drilling Fluids. 1995. NOVA System Technology Report.

OSPAR, 2016. Offshore Chemicals. http://www.ospar.org/work-areas/oic/chemicals.

Rushing, J.H., Churan, M.A., Jones, F.V., 1991. Bioaccumulation from mineral oil-wet and synthetic liquid-wet cuttings in an estuarine fish, fundulus grandis. First International Conference on Health. Safety and Environment, The Hague, 10–14 November.

Satterlee III, K., van Oort, E., 2003. Shell Exploration & Production Co and B. Whitlatch, Bulk Mixer Inc. Rig Waste Reduction Pilot Project. In: SPE/EPA/DOE Exploration and Production Environmental Conference, San Antonio, TX, 10–12 March.

Sipple-Srinivasan, M. Terralog Technologies USA, Inc., 1998. U.S. regulatory considerations in the application of slurry fracture injection for oil field waste disposal. In: Prepared for the International Petroleum Environmental Conference (IPEC)'98, Albuquerque, NM, 20–23 October 1998.

USEPA, 2001. 40 CFR Parts 9 and 435 [FRL-6029-8] RIN 2040AD14 Effluent Limitations Guidelines and New Source Performance Standards for the Oil and Gas Extraction Point Source. Federal Register 14. 22 January 2001.

Veil, J., Smith, K., Tomasko, D., Elcock, D., Blunt, D., Williams, G., 1998. Disposal of NORM Waste in Salt Caverns. Paper SPE-46561-MS presented at SPE International Conference on Health, Safety, and Environment in Oil and Gas Exploration and Production, 7-10 June, Caracas, Venezuela. http://dx.doi.org/10.2118/46561-MS.

Veil, J.A. SPE, Gasper, J.R., Puder, M.G., Sullivan, R.G., Richmond, P.D. et al., 2003. Innovative website for drilling waste management. Paper presented at the SPE/EPA/DOE Exploration and Production Environmental Conference in San Antonio, TX, 10–12 March.

Vik, E.A., Dempsey, S., Nesgard, B., 1996. Evaluation of Available Test Results from Environmental Studies of Synthetic Based Drilling Muds. OLF Project, Acceptance Criteria for Drilling Fluids Aquatema Report No. 96-010.

Walker, J., Miller, J., Burrows, K., Mander, T. Hoven, J., 2016. Nonaqueous, salt-free drilling fluid delivers excellent drilling performance with a smaller environmental footprint. In: SPE paper 178804-MS IADC?SPE Drilling Coference, Ft. Worth TX, 1–3 March.

Zevallos, M.A.L., Candler, J., Wood, J.H., Reuter, L.M., 1996. Synthetic-based fluids enhance environmental and drilling performance in deep water locations. In: SPE International Petroleum Conference and Exhibition of Mexico, Villahermosa, 5–7 March.

US WASTE REGULATION BIBLIOGRAPHY

Brasier, F.M., Kobelski, B.J., 1996. Injection of industrial wastes in the United States. Deep Injection Disposal of Hazardous and Industrial Waste: Scientific and Engineering Aspects. Academic Press, pp. 1–8.

CH2MHill, Inc., 1998. Tex-Tin Superfund Site, OU No. 1 Feasibility Study Report prepared for U.S. Environmental Protection Agency, Response Action Contract No. 68-W6-0036, March.

DeLeon, F., 1997. Personal communication between DeLeon, Railroad Commission of Texas, Austin, TX, and Srinivasan, M., Terralog Technologies USA, Inc., Arcadia, CA, October 8.

Dusseault, M.B., Bilak, R.A., Rodwell, G.L., 1997. Disposal of dirty liquids using slurry fracture injection. SPE 37907.

Dusseault, M.B., Danyluk, P.G., Bilak, R.A., 1998. Mitigation of heavy oil production environmental impact through slurry fracture injection. Paper Presented at the 7th UNITAR International Conference on Heavy Crude and Tar Sands, Beijing, 27 October.

Elliott, J.F., Henderson, M.A., Weinsoff, D.J., Polkabla, M.A., Demes, J.L., 1994. The complete guide to hazardous materials enforcement and liability. Prepared by Touchstone Environmental, Inc., Oakland, CA.

Hainey, B.W., Keck, R.G., Smith, M.B., Lynch, K.W., Barth, J.W., 1997. On-site fracturing disposal of oilfield waste solids in Wilmington Field, Long Beach Unit, CA. SPE #38255.

Louviere, R.J., Reddoch, J.A., 1993. Onsite disposal of rig-generated waste via slurrification and annular injection. SPE/IADC #25755.

Malachosky, E., Shannon, B.E., Jackson, J.E., 1991. Offshore disposal of oil-based drilling fluid waste: An environmentally acceptable solution. SPE #23373.

Marinello, S.A., Ballantine, W.T., Lyon, F.L., 1996. Nonhazardous oil field waste disposal into subpressured zones. Environ. Geosc 3 (4), 199–203.

Rutherford, G.J., Richardson, G.E., 1993. Disposal of naturally occurring radioactive material from operations on federal leases in the Gulf of Mexico, SPE 25940. Presented at the SPE/USEPA Exploration & Production Conference, San Antonio, TX, 7–10 March.

Schuh, P.R., Secoy, B.W., Sorrie, E., 1993. Case history: Cuttings reinjection on the Murdoch Development Project in the southern sector of the North Sea. SPE #26680.

Sipple-Srinivasan, M.M., Bruno, M.S., Hejl, K.A., Danyluk, P.G., Olmstead, S.E., 1998. Disposal of crude contaminated soil through slurry fracture injection at the West Coyote Field in California, SPE 46239, Proc. West. Reg. Mtg. SPE, Bakersfield, CA, 10–13 May.

U.S. Department of Energy Office of Fossil Energy, Interstate Oil and Gas Compact Commission, 1993. Oil and gas exploration and production waste management: A 17-state study.

U.S. Environmental Protection Agency, 1988. Regulatory determination for oil and gas and geothermal exploration, development and production wastes. 53 Federal Register, pp. 25446–25459, 6 July.

U.S. Environmental Protection Agency, 1993. Clarification of the regulatory determination for waste from the exploration, development and production of crude oil, natural gas, and geothermal energy. 58 Federal Register pp. 15284–15287, 22 March.

U.S. Environmental Protection Agency/IOCC, 1990. Study of state regulation of oil and gas exploration and production waste, December.

Veil, J.A., Smith, K.P., Tomasko, D., Elcock, D., Blunt, D.L., Williams, G.P., 1998. Disposal of NORM-Contaminated Oil Field Waste in Salt Caverns, August 1998. Argonne National Laboratory.

Appendix A

Conversion Factors

Table A.1 gives factors for converting some common US units to *coherent* (i.e., consistent) metric units (centimeter−gram−second), and to SI units (meter−kilogram−second). The advantage of using coherent units is that the US units need not be converted individually when writing equations. However, the conversion of some customary US field units to coherent SI units leads to inconveniently large or small numbers. Table A.2 gives some conversion factors that may be more convenient for use on the drilling rig.

The SI system includes two new units—the newton (N) and the pascal (Pa). The *newton* is the absolute unit of force, defined as 1 kilogram meter per second squared ($m \cdot kg/s^2$), and is obtained by multiplying kilograms by the gravitational constant, 9.806 m/s^2. The *pascal* is the absolute unit of pressure, defined as 1 newton per meter squared (N/m^2).

A complete description of the SI system and various conversion factors may be found on the SPE website (http://www.spe.org/industry/unit-conversion-factors.php) and on the PetroWiki website (http://petrowiki.org/Recommended_SI_units_and_conversion_factors?rel=1).

TABLE A.1 Factors for Converting Common US Units to Coherent Metric and SI Units

Common US Field Unit	Metric Unit (cgs)	To Convert US to Metric cgs Unit Multiply by	SI Unit (mkgs)	To Convert US Unit to SI Unit Multiply by
Barrel, 42 gal (bbl)	centimeter^3 (cm)^3	1.589×10^3	meter^3 (m^3)	1.589E−01[a]
Cubic foot (ft^3)	centimeter^3 (cm)^3	2.831×10^4	meter^3 (m^3)	2.831E−02
Foot (ft)	centimeter (cm)	30.48	meter (m)	3.48E−01
Foot/minute (ft min)	centimeter/sec (cm/s)	0.5080	meter/sec (m/s)	5.080E−03
Foot of water (at 39.2 F)[b]	dyne/cm^2	2.989×10^4	pascal (Pa)	2.989E+03
Foot second (ft/sec)	centimeter/sec (cm/s)	30.48	meter/sec (m/s)	3.048E−01
Gallon (gal)	centimeter^3 (cm^3)	3.785×10^3	meter^3 (m^3)	3.785E−03
Gallon/minute (gpm)	centimeter^3/sec (cm^3/s)	63.09	meter^3/sec (m^3/s)	6.309E−03
Inch (in)	centimeter (cm)	2.54	meter (m)	2.54E−02
Mil (in^{-3})	centimeter (cm)	2.54×10^{-3}	meter (m)	2.54E−05
Pound (lb)[c]	gram (g)	453.6	kilogram (kg)	4.536E−01
Pound (lb)	dyne	4.448×10^5	newton (N)	4.448E+00
Poundal[c]	dyne	1.382×10^4	newton (N)	1.382E−01
Pound/barrel (lb/bbl)	g/cm^3	2.854×10^{-3}	kg/m^3	2.854E+00
Pound/cubic foot (lb/ft^3)	g/cm^3	0.01602	kg/m^3	1.602E+01
Pound/gallon (lb/gal)	g/cm^3	0.1198	kg/m^3	1.198E+02
Pound/inch^2 (psi)	dyne/cm^2	6.895×10^4	pascal (Pa)	6.895E+03
Psi/foot	dyne/cm^2/cm	2.262×10^3	pascal/meter (Pa/m)	2.262E+04
Pound/100 feet^2 (lb/100 ft^2)	dyne/cm^2	4.788	pascal (Pa)	4.788E−01
Square inch (in^2)	centimeter^2 (cm^2)	6.451	meter^2 (m^2)	6.451E−04
Square foot (ft^2)	centimeter^2 (cm^2)	9290	meter^2 (m^2)	9.29E−02

[a] E = locates the decimal point.
[b] At 39.2°F (14°C), the specific gravity of water is 1 and a column of water 1 foot high exerts a pressure of 0.433 pound/inch^2 (psi).
[c] The unit of weight in the US system is the pound (avoirdupois) (lb). The unit of force is the poundal defined as pounds divided by the gravitational constant, 32,174 ft/s^2. These terms are used in this table to avoid confusion over the terms pounds (mass) and pounds (force).

TABLE A.2 Factors for Converting Common US Field Units to Convenient SI Units

Quantity Property	Previous Units	SI Unit	Symbol	Conversion Factors Multiply by
Depth	feet	meter	m	0.3048
Hole and pipe diameters, bit size	inches	millimeters	mm	25.4
Weight on bit	pounds	decanewtons	daN	0.445
Nozzle size	32nd inch	millimeters	mm	0.794
Drill rate	feet/hour	meters/hr	m/h	0.3048
Volume	barrels API	cubic meters	m^3	0.159
Pump output and flow rate	gal/stroke	cubic meters/stroke	m^3/stroke	0.00378
	gal/min	cubic meters/minute	m^3/min	0.00378
	bbl/stroke	cubic meters/stroke	m^3/stroke	0.159
	bbl/min	cubic meters/minute	m^3/min	0.159
Annular velocity Slip velocity	feet/min	meters/min	m/min	0.3048
Liner length and diameter, pressure	inches	millimeters	mm	25.4
	psi	kilopascals	kPa	6.895
		megapascals	MPa	0.006895
Bentonite yield	bbl/ton	cubic meters/tonne	m^3/t	0.175
Particle size	microns	micrometers	μm	1
Temperature	°Fahrenheit	°Celsius	C	(°F−32)/1.8
Mud density	lb/gallon	kilograms/cubic meter	kg/m^3	119.83
Mud gradient	psi/foot	kilopascals/meter	kPa/m	22.62
Funnel viscosity	sec/quart	sec/liter	s/L	1.057
Apparent and plastic viscosity	centipoise	millipascal · seconds	mPa·s	1
Yield point	$lb_f/100\ ft^2$	pascals	Pa	0.4788 (0.5 for field use)

(*Continued*)

TABLE A.2 (Continued)

Quantity Property	Previous Units	SI Unit	Symbol	Conversion Factors Multiply by
Gel strength and stress	$lb_f/100\ ft^2$	pascals	Pa	0.4788 (0.5 for field use)
Cake thickness	32nd inch	millimeters	mm	0.794
Filter loss	millimeters or cm^3	cubic centimeters	cm^3	1
MBT (bentonite equivalent)	pounds per barrel	kilograms/cubic meters	kg/m^3	2.85
Material concentration	pounds per barrel	kilograms/cubic meter	kg/m^3	2.85
Shear rate	reciprocal seconds	reciprocal seconds	s^{-1}	1
Torque	foot pounds	newton meters	N.m	1.3558
Table speed	revolutions per minute	revolutions/minute	r/min	1
Ionic mass concentrations	parts per million	milligram/liter	mg/L	1

Sand, solids, and oil content will be reported as volume fraction using the symbol θ substance, e.g., 12% by volume of solids will be reported as "θ solids = 0.12."
Source: Courtesy Baroid of Canada, Ltd.

Appendix B

Abbreviations

Used in Petroleum Operations

Abbreviated Title	Full Title
AADE	American Associate of Drilling Engineers
AAPG	American Association of Petroleum Geologists
AICHE	American Institute of Chemical Engineers
AIME	American Institute of Mining, Metallurgical, and Petroleum Engineers
API	American Petroleum Institute
ASTM	American Society for Testing Materials
AAPG Bull.	AAPG Bulletin
Am. Ceramic Soc. Bull.	American Ceramic Society Bulletin
Am. Chem. Soc.	American Chemical Society
Amer. J. Sci.	American Journal of Science
Am. Mineralogist	American Mineralogist
API Drill. Prod. Prac.	API Drilling and Production Practice
Baroid News Bull.	Baroid News Bulletin
Bull. Univ. Illinois	Bulletin of the University of Illinois
Can. Oil Gas Indust.	Canadian Oil and Gas Industry
Clays Clay Minerals	Proceedings of The National Conference on Clays and Clay Minerals
Conf. Drill. Rock Mech.	Conference on Drilling and Rock Mechanics
Disc. Faraday Soc.	Discussions of the Faraday Society
Drilling Contract.	Drilling Contractor
Eng. Min. J.	Engineering and Mining Journal
IADC	International Association of Drilling Contractors
Ind. Eng. Chem.	Industrial and Engineering Chemistry
J. Amer. Ceramic Soc.	Journal of the American Ceramic Society
J. Amer. Chem. Soc.	Journal of the American Chemical Society
J. Appl. Phys.	Journal of Applied Physics
J. Boston Soc. Civil Eng.	Journal of Boston Society of Civil Engineers
J. Can. Petrol. Technol.	Journal of Canadian Petroleum Technology
J. Chem. Ed.	Journal of Chemical Education
J. Colloid Interface Sci.	Journal of Colloid and Interface Science
J. Franklin Inst.	Journal of the Franklin Institute
J. Inst. Petrol.	Journal of the Institute of Petroleum (London)

(*Continued*)

(Continued)

Abbreviated Title	Full Title
J. Petrol. Technol.	Journal of Petroleum Technology
J. Phys. Chem.	Journal of Physical Chemistry
J. Rheology	Journal of Rheology
J. Soc. Cosmetic Chem.	Journal of the Society of Cosmetic Chemists
J. Wash. Acad. Sci.	Journal of the Washington Academy of Science
Kolloid Z.	Kolloid Zeitschrift
Mining Eng.	Mining Engineering
Nat. Sci. Foundation	National Science Foundation
Nature	Nature (London)
Oil Gas J.	Oil and Gas Journal
Petrol. Eng.	Petroleum Engineer
Proc. API	Proceedings API
Proc. ASTM	Proceedings ASTM
Proc. International Soc. Rock Mech.	Proceedings International Society of Rock Mechanics Congress
Proc. Symp. Abnormal Subsurface Pressures	Proceedings of Symposium on Abnormal Subsurface Pressures
SPE	Society of Petroleum Engineers of AIME
Soc. Petrol. Eng. J.	Society of Petroleum Engineers Journal
Soil Sci.	Soil Science
Symp. on Formation Damage Control	Symposium on Formation Damage Control
Symp. Rock Mech.	Symposium on Rock Mechanics
Trans. AIME	Transactions of the AIME
Trans. IADC Drill. Technol. Conf.	Transactions of the Drilling Technology Conference of IADC
Trans. Faraday Soc.	Transactions of the Faraday Society
Trans. Inst. Chem. Eng.	Transactions of the Institute of Chemical Engineers (London)
Trans. Mining and Geological Inst. of India	Transactions of the Mining and Geological Institute of India
U.S. Bur. Mines	United States Bureau of Mines
U.S. Geol. Survey Bull.	United States Geological Survey Bulletin
World Petrol. Cong.	World Petroleum Congress

Appendix C

The Development of Drilling Fluids Technology

INTRODUCTION

Appendix C covers a number of legacy fluid systems. Some of these systems may still be in use, in particular the lime systems and all oil asphaltic nonaqueous muds. Since silicate muds were first used in the late 1930s, they could be called a legacy system, but modern adaptions have kept them as mainstream systems for some applications. Silicate systems are covered in Chapter 9 on Wellbore Stability. The following is directly copied from Chapter 2 of the 6th Edition, most of which was taken from Walter Rogers's 3rd edition, called Composition and Properties of Oilwell Drilling Fluids, 1963.

WATER-BASED DRILLING FLUIDS TECHNOLOGY

Removal of Cuttings

If drilling fluid is defined as a material employed to aid tools in the creation of a borehole, the use of drilling fluids far antedates the petroleum industry. Water, the principal constituent of the majority of drilling fluids in use today, was the first drilling fluid. As early as the third millenium BC in Egypt, holes up to 20 feet deep were drilled in quarries by hand-driven rotary bits. J.E. Brantly, the recognized authority on the history of drilling, concludes that water probably was used to remove the cuttings in such borings (Brantly, 1961, 1971a).

According to Confucius (600 BC), wells were drilled in China for brine during the early part of the Chou dynasty (1122−250 BC). Many wells, some hundreds of feet deep, were bored near the border of Tibet for brine, gas, and water (Fig. C.1). Water was poured into these wells to soften the rock and to aid in the removal of cuttings (Brantly, 1971b; Pennington, 1949).

In making holes by the percussion method, cuttings were removed by bailing. Circulation of water was proposed in a patent application filed

FIGURE C.1   Early Chinese drilling rig. *Courtesy of NL Baroid.*

by Robert Beart in England in 1844 (Beart, 1845). He disclosed a method of boring by means of rotating hollow drill rods such that "matters cut or moved by the tools employed may be carried away by water." At about the same time in France, Fauvelle (1846) pumped water through a hollow boring rod to bring the drilled particles to the surface. In 1866, Sweeney (1866) received a United States patent on a "stone drill" that displayed many features of today's rotary rig, including the swivel head, rotary drive, and roller bit (Fig. C.2). Several U.S. patents issued between 1860 and 1880 mention the circulation of a drilling fluid to remove cuttings.

Appendix C: The Development of Drilling Fluids Technology **645**

FIGURE C.2 Sweeney's rotary drill of 1866 had some features in common with rigs of today. *Courtesy of NL Baroid.*

Hole Stabilization ("Wall Building")

An application for a U.S. patent, filed in 1887 by Chapman (1890) proposed "a stream of water and a quantity of plastic material, whereby the core formed in the casing will be washed out and an impervious wall be formed along the outside." He suggested clay, bran, grain, and cement. Here was another function of the drilling fluid: to plaster the wall of the hole and reduce caving tendencies.

During the 1890s many wells were drilled by the rotary method in Texas and Louisiana, where mud-making clays were common. Drillers became familiar with the use of mud as a means of hole stabilization ("wall building") in weak formations. Brantly's comprehensive *History of Oil Well Drilling* (1971), in reviewing the developments of this period, makes no mention of the use of plastering agents other than clay. Following the discovery of oil at Spindletop in 1901, drilling by the rotary method spread rapidly throughout the Gulf Coast and in California. The problem of unstable holes in poorly consolidated formations demanded attention.

Sufficient clay ("gumbo") usually was present in the cuttings from wells in the Gulf Coast to "make good mud" (Hayes and Kennedy, 1903). In California, however, clays from surface deposits were often mixed with water to "plaster the walls". Although most mud mixing was done with shovels by the crew, some auxiliary mixing devices were available (see Fig. C.3). Little attention was paid to the properties of the mud (Knapp, 1916). The terms "heavy" and "thick" were used interchangeably.

Pressure Control by Mud Density

A serious problem—the enormous waste of natural gas in drilling by the cable-tool method in Oklahoma—led to the general acceptance of "mud-laden fluid" as a means of controlling pressure. In May 1913, Pollard and Heggem (1913; Heggem and Pollard, 1914) of the U.S. Bureau of Mines demonstrated the practicality of adding mud to holes drilled with cable tools to "seal each gas-bearing stratum as it is encountered by drilling with the hole full of mud-laden fluid."

In a more definitive survey, Lewis and McMurray (1916) stated that mud-laden fluid is "a mixture of water with any clayey material which will remain suspended in water for a considerable time and is free from sand, lime cuttings or similar materials." The mixture's measured specific gravity was 1.05 to 1.15. Mud consistency, they said, should be such as to seal with little penetration of the sand. The distance of mud penetration into the sand was reported to depend on the mud's consistency, the pressure applied, and the sand's porosity. The barrier formed within the sand was maintained principally by the mud pressure in the hole. Advantages in using mud-laden fluids were: (1) reduction in number of casing strings; (2) protection of upper sands while drilling was continued; (3) prevention of migration between casings; (4) allowing recovery of casing; and (5) protection of the casing from corrosion. This survey related the properties of the mud to its performance and emphasized the economic importance of the control of the drilling fluid.

Most drillers continued to ignore the practical benefits to be derived from mud as a means of pressure control. In 1922, Stroud (1922), supervisor of the Mineral Division of the Louisiana Department of Conservation, wrote: "They go by the consistency of the mud fluid rather than by its actual

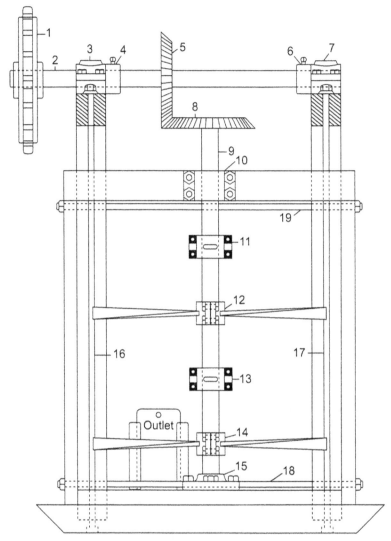

FIGURE C.3 This mud mixer used shortly after the turn of the century served the same purpose as today's mixers. *From Gray, G.R., Kellogg, W.C., 1955. The Wilcox trend—cross section of typical mud problems. World Oil, 102–117.*

weight." He pointed out the hazard of gas-cutting of thick mud and strongly recommended that "drillers should frequently weigh samples of mud." He stated that success or failure in drilling a well in the Monroe, Louisiana gas field depended upon the control of the gas pressure by heavy mud.

Stroud cited laboratory tests on cement, galena, and iron oxide (Fe_2O_3) as materials for increasing the density of mud. Iron oxide could be used to raise mud density to 16 lb/gal (specific gravity of 1.86) without an excessive increase in consistency. Iron oxide (hematite) was used successfully in several fields. In the fall of 1922, barite was used to make a heavy mud. This product was pigment-grade barite from Missouri (Stroud, 1925).

Birth of the Mud Industry

The use of barite in oil well drilling attracted the interest of Phillip E. Harth, sales manager for National Pigments and Chemical Company, a subsidiary of National Lead Company and a producer of barite for the paint industry. Mr. Stroud (1926) accepted Mr. Harth's offer of assistance in securing a patent on the addition of heavy minerals to mud in return for exclusive sales rights under any such patents. Paint-grade barite from the St. Louis, Missouri plant was sold for oil well use under the brand name Baroid. Barite mining operations were begun at El Portal, California, to supply the increasing demands for barite in drilling activities.

California Talc Company, a producer and marketer of clays, was agent for Silica Products Company of Kansas City, Missouri, in the sale of the Aguagel brand of bentonite as an admixture for cement. The success of Aquagel in stopping severe caving in a well drilling at Kettleman Hills, California, in 1928 led to widespread interest in the application of bentonite for overcoming hole problems. In 1929, to prepare a heavier mud than that made from surface clays, California Talc Company began the sale of "Plastiwate," a mixture of about 95% barite and 5% bentonite. Phillip Harth (1935), in the meantime, had noticed the problem of the settling of the heavy minerals in some muds and had filed application for a patent on bentonite as a suspending agent for the heavy minerals disclosed in the Stroud patent.

In view of the patent situation, California Talc Company agreed to discontinue the sale of Plastiwate. The announcement was made in oil industry magazines that beginning March 1, 1931, Baroid would be sold by Baroid Sales Company (a subsidiary of National Lead Company) through distributors in all areas except Louisiana, Mississippi, New Mexico, and Texas, where the Peden Company would be the agent. Aquagel would be sold by Baroid Sales Company in essentially all oil fields of the world. George L. Ratcliffe, formerly president of California Talc Company, became general manager of Baroid Sales Company, which, through the years, has become the Baroid Division of NL Industries, Inc., and is now part of Halliburton's Fluid Services.

The first proprietary thinning agent for mud, Stabilite, was introduced by T.B. Wayne in about 1930. This product, a mixture of chestnut bark extract and sodium aluminate, thinned mud without decreasing the density, released entrapped gas, and allowed further increase in mud weight (Parsons, 1931). The molecularly dehydrated phosphates became a component of Stabilite a few years later.

As a means of educating the drillers in the application of mud materials, Baroid Sales Company advertized the services of "experienced field engineers to assist at any time with any of your drilling fluid problems." Examples of field experiences and results of laboratory studies were reported in the company publication, *Drilling Mud* (Anon, 1931). The first issue compared Aquagel with several local clays, showing the weight of clay required to produce a given viscosity (measured with the Stormer viscometer). Relative performance of the clays was expressed as the "yield" in bbl/ton. Later issues of *Drilling Mud* dealt with methods and instruments for evaluating the performance of muds, and emphasized the importance of colloidal clay in mud behavior.

The George F. Mepham Corporation of St. Louis, Missouri was licensed under the Stroud patent to sell iron minerals as mudweighting material. Iron oxide was sold as Colox; bentonite was marketed as Jelox. These products, like those marketed by Baroid Sales Company, were made available in active drilling areas through local distributors. Distributors were usually suppliers of oilfield construction materials (such as lumber and cement) or of drilling equipment.

As other mud additives came into use, Baroid Sales Company and the Mepham Corporation, in contrast to the many suppliers of local surface clays, offered not only brand-name products through distributors in all major oil fields but also the advisory services of mud engineers. Through the years, numerous suppliers have entered the drilling fluids market, and direct sales to the user have replaced local distributors. However, the major suppliers of mud products continue to furnish field and laboratory services. The mud engineer continues to serve as a means of spreading information on drilling practices and on developments in mud technology.

Rapid Growth of the Mud Industry

After the California Division of Oil and Gas Operations was established in 1915, the employment of engineers in the oil industry began to increase. Usually, the engineers were concerned only with production and little attention was given to drilling (Suman, 1961). Several discussions on the use of mud were reported, however, in "Summary of Operations California Oil Fields" (Collom, 1923) and some of these were reprinted in oil industry publications (Collom, 1924).

The mechanism of wall building in the drill hole was studied to some extent (Knapp, 1923) and the observation was made by Kirwan (1924) that mud solids do not penetrate sandstone even under a pressure of more than 2000 psi (140 kg/cm^2) but that filtering action takes place, forming a sheath on the surface as shown in Fig. C.4.

The presence of this "wall" in an oil-bearing sand might either prevent oil shows while drilling or create difficulties in bringing in a well. Thus a negative requirement of the drilling fluid—that its use should not interfere

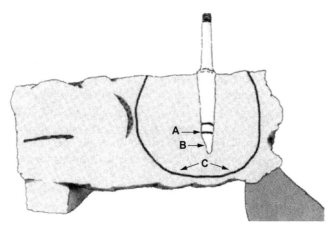

FIGURE C.4 Laboratory test showing filter cake formation on sandstone. (A) Bottom of nipple. (B) Mud sheath on wall of hole. (C) Extent to which water squeezed from mud fluid penetrated the sandstone. *From Kirwan, M.J., 1924. Mud fluid in drilling and protection of wells. Oil Weekly, 34*

with the collection of information while drilling or with final completion— was recognized.

Problems and methods of treating muds in the field began to receive some attention. Nevertheless, in most areas, "The average rotary operator did not give much serious consideration to the condition of his mud. Generally, it was either good or bad according to the individual driller's opinion and experience" (Cartwright, 1928). Based on practices of the mining industry in ore beneficiation, central mud reclamation plants were installed in some areas of concentrated drilling activity (Ockenda and Carter, 1920; Mills, 1930).

In the late 1920s, several of the major oil companies initiated research into drilling and production practices. Drilling mud was recognized as a colloidal suspension and was accepted as a subject for investigation, primarily by chemists. Sellers of clays emphasized the colloidal character of their products (Farnham, 1931). The development of gel structure when the flow of clay suspension stopped was accentuated as a desirable feature. The superior mud-making qualities of Wyoming bentonite were generally recognized.

From the ceramic industry, mud chemists adopted the methods of testing clay suspensions with such instruments as the MacMichael and the Stormer viscometers (see Fig. C.5). However, thinning agents, such as caustic soda and sodium silicate, useful in the ceramic industry, were generally not effective in muds. Alkaline solutions of various natural tannins were suitable (Doherty et al., 1931), as were the pyrophosphates and polyphosphates (Feldenheimer, 1922).

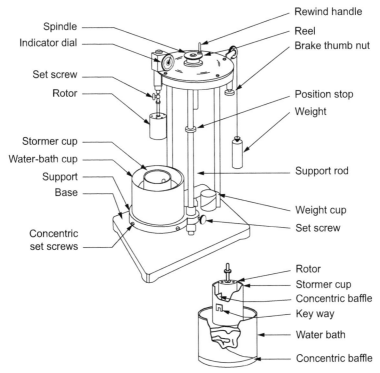

FIGURE C.5 Stormer viscometer used in measurement of mud viscosity about 1930. *Courtesy of NL Baroid.*

Marsh (1931) proposed a simple, rugged funnel as a means of measuring the apparent viscosity of muds in the field. Herrick (1932) emphasized the distinct difference between the flow behavior of muds and liquids. He pointed out that the apparent viscosity as measured with the Stormer viscometer could not be used to estimate the pressure required to pump mud through pipes. Measurements made with a pressure efflux viscometer, however, could be used.

The introduction of new and greatly improved drilling equipment in the early 1930s (Brantly) stimulated the investigation of the role mud plays in drilling performance. Through local and national meetings, both the Division of Production of the American Petroleum Institute (API) and the Petroleum Division of the American Institute of Mining and Metallurgical Engineers (AIME) afforded outlets for papers on drilling mud. The Institution of Petroleum Technologists, Trinidad Branch, provided a forum for the discussion of developments in mud practices in that area. *The Oil Weekly, The Oil and Gas Journal*, and *The Petroleum Engineer* printed many of the papers presented at API and AIME meetings as well as numerous contributed articles on muds.

At the first World Petroleum Congress in 1933, five papers were presented on mud. More papers were published on drilling mud between 1930 and 1934 than had been printed in the 40 years since the Corsicana (Texas) field was drilled by rotary tools in 1890. The first book on the subject, *Drilling Mud: Its Manufacture and Testing* (Evans and Reid, 1936), appeared in 1936. The authors not only cited the more significant papers on drilling mud, but also made numerous references to related publications in the scientific literature. In their studies of mud, they also reviewed the laboratory and field results obtained by Burmah Oil Company.

Although the interrelationship between mud problems and drilling difficulties, such as caving formations (once commonly known as "heaving shale"), high-pressure gas and saltwater flows, and bedded salt had been pointed out (Doherty et al., 1931), little attention had been given to impairment of oil production by the plugging action of the mud cake. California producers were concerned about "mudding off" oil sands, and a symposium was arranged by the Drilling Practice Committee of the American Petroleum Institute in 1932. Rubel (1932) stated that "sealing of oil sands has been responsible for the irretrievable loss of millions of barrels of oil in California operations." The laboratory tests of Farnham (1932) on the filtration of muds on sand showed that the filter cake thickness ranged from 1/16 inch with bentonite mud to over an inch with ordinary field muds. Gill (1932) cited tests on Gulf Coast sand cores that showed little penetration of mud or filter cake and concluded that "properly proportioned muds" would do little damage unless the infiltered water affected the flow of oil. Parsons (1932) reported penetration of mud up to an inch in unconsolidated sand and traces of mud to a depth of several inches.

Rubel (1932) initiated a study by Union Oil Company of mud cake formation and of filtration into sand cores. Jones and Babson (1935) reported that tests were conducted at pressures up to 4000 psi (280 kg/cm^2) and temperatures up to 275°F (135°C). Obvious differences were noted in the filtration characteristics of several field muds. The immediate application was not directed to avoidance of productivity impairment, however, but to hole stability. Tests showed that when problems existed with caving formations, mud cakes were thick, and excessive quantities of water were lost. When these muds were replaced by muds that produced thin filter cakes and allowed small losses of water, the difficulties either disappeared or were reduced greatly. The laboratory filtration equipment, however, was not suitable for use in the field. Jones (1937) later described a simple, rugged device that was practical for field use. This instrument (shown in Figs. C.6 and C.7), with minor modifications, continues to be the routine filter press for the evaluation of mud performance at the well site.

The filter press (or "wall-building tester," as it was called) provides a useful tool, helping the mud engineer relate the mud's physical properties to

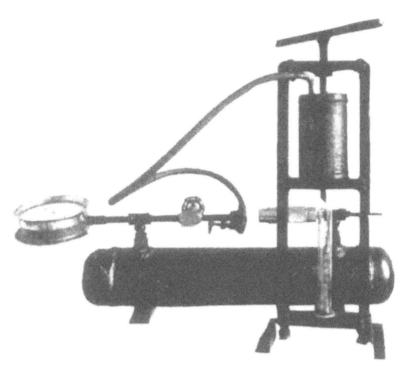

FIGURE C.6 Static filtration tester. *From Jones, P.H., 1937. Field control of drilling mud. API Dril. Prod. Prac., 24−29. Courtesty of API.*

specific hole problems. The subject of filtration is dealt with in Chapter 6, The Rheology of Drilling Fluids.

In spite of the general recognition of the role mud weight plays in the control of pressure, blowouts still occurred too frequently. In the Gulf Coast of Texas, the Conroe field (known from bottomhole pressure measurements to have had normal pressure) was the scene of several disastrous blowouts. Humble Oil and Refining Company (now Exxon) engineers investigated the problem and noted a direct association between blowouts and withdrawal of pipe from the hole, even though mud weight was more than adequate to overcome bottomhole pressure. In a field study, a subsurface pressure gauge was either attached to the bottom of the drill pipe or left in the hole while the pipe was withdrawn. Cannon (1934) reported that the pressure reduction, or "swabbing action," was sufficient to cause a blowout with muds that developed high gel strength on standing. The swabbing effect was not dependent on the viscosity of the mud (as measured in the Stormer viscometer at 600 rpm) nor on the density, but on the strength of the gel that developed on standing. The size of the pipe-hole annulus and the length of the pipe were also significant factors. The solution to the problem was to use low-gel-strength mud—not to raise mud

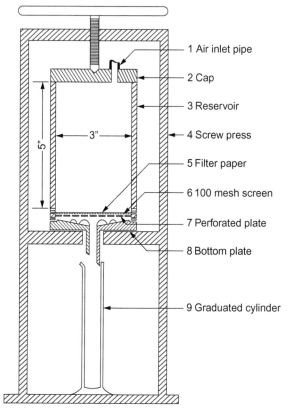

FIGURE C.7 Static-performance tester. *From Jones, P.H., 1937. Field control of drilling mud. API Dril. Prod. Prac., 24–29. Courtesty of API.*

weight. Measurement of gel strength at the well was considered necessary for safe control of mud properties.

Several investigators had emphasized the importance of gel-forming colloidal clay in mud performance (e.g., see references Lawton et al., 1932; Duckham, 1931; Ambrose and Loomis, 1931). Development of gel structure at low clay concentration was an indication of colloidity. Until the filter press became available as a field instrument, however, there was no practical method of estimating colloidity. Because filtration properties are dependent on the nature and amount of colloidal material present in the suspension, the filter press provides an overall evaluation of the colloidal fraction and, as such, has been of major importance in the development of drilling fluids technology.

With the introduction of various devices for the measurement of mud properties, and with variations in procedures as well, the need for uniformity in instruments and methods was recognized. The Houston Chapter of the

American Petroleum Institute Division of Production in 1936 began a study to formulate uniform practices (Reistle et al., 1937). This report led to the adoption of *Recommended Practice on Standard Field Procedure for Testing Drilling Fluids*, as API Code 29. Through the years, additions and modifications have been made under the Committee on Standardization of Drilling Fluid Materials. The publication has become API RP 13B (1978), which is reprinted at intervals. The number of pages has grown from 6 in 1938 to 33 in 1978.

The need for better supervision of the mud at the rig was realized as mud-related drilling problems were recognized. The American Association of Oil Well Drilling Contractors asked the Petroleum Extension Service of the University of Texas for help in crew training. Around 1945, under the supervision of John Woodruff, courses began for field men in active drilling areas. A training manual, *Principles of Drilling Mud Control*, appeared in 1946 and was revised and expanded as the program continued. The API Southwestern District Study Committee on Drilling Fluids, under the guidance of J.M. Bugbee, assumed responsibility for preparation of the manual in 1950 and edited the 8th through the 10th editions (1951–55). The 11th edition (1962) was edited by H.W. Perkins and the 12th edition (1969) by W.F. Rogers. *Principles of Drilling Mud Control* has served as the basic manual for the training of thousands of men directly concerned with the practical application of drilling fluids.

Development of Mud Types or Systems

Saltwater Muds

Early field experience, as well as laboratory studies, established that bentonite was the most practical material for improving the viscosity and wall-building properties of freshwater muds. As dissolved salt content increased, however, bentonite became progressively less effective. In saturated saltwater, bentonite did not swell, and contributed little toward reduction of filtration. To thicken salty muds, bentonite was mixed into freshwater and the resulting thick slurry was added to the saltwater mud. After a short time, however, the salty mud would become thin and additional treatment would be necessary.

Cross and Cross (1937) found that a type of fuller's earth, mined in southwestern Georgia and northwestern Florida, and consisting principally of the clay mineral attapulgite (palygorskite), could be used to thicken salty water, regardless of salt content. Although this clay greatly improved the cuttings-carrying capability of salty muds, the wall-building properties remained poor. Thick filter cakes formed on porous formations, often causing stuck pipe, and caving of shale was common. Drilling progress in the Permian Basin of West Texas was hampered by salt beds. Dome salt in the Gulf Coast created problems.

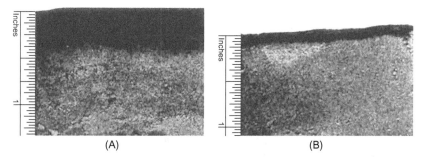

FIGURE C.8 Influence of tragacanth gum on thickness of filter cake: 20% Shafter Lake clay in salt solution filtered on sand. (A) Untreated mud. (B) Mud containing 0.5 percent tragacanth gum. *From Gray, G.R., Foster, J.L., Chapman, T.S., 1942. Control of filtration characteristics of salt water muds. Trans. AIME 146, 117−125. Copyright 1942 by AIME.*

A laboratory study on the control of filtration of saltwater muds led to field tests of tragacanth gum and gelatinized starch in 1939 (Gray et al., 1942). Field experience confirmed laboratory test results that showed great reduction in cake thickness (Fig. C.8). The major problems associated with thick filter cakes (stuck drill pipe and inability to run casings to bottom) were generally eliminated. In the United States, the ready availability and lower cost of starch excluded the imported gums from further consideration and salt-starch mud was recognized as an economical solution to problems of salt drilling. Applications of starch as a means for controlling filtration of brackish and freshwater muds spread quickly (Gray, 1944).

Muds for "Heaving Shale"

In drilling on piercement-type salt domes along the Gulf Coast in the early 1920s, the problem of heaving shale frequently led to abandonment of the hole before the objective was reached.

The term *heaving shale* was applied to any shale that sloughed into the hole in excessive quantities and thereby interfered with drilling progress. That the term described an effect, rather than a specific substance, was generally recognized (Doherty et al., 1931). Sensitivity to water was a characteristic of such shales. Some shale samples swelled on contact with water; others broke up into fragments with little indication of swelling.

The difficulties involved in obtaining reliable samples of the troublemaking shale caused much of the investigative work to be done with bentonite instead of shale. The important effect of pressure on shale behavior was ignored in these studies (see chapter: The Surface Chemistry of Drilling Fluids). Consequently, special muds were designed to prevent swelling of bentonite, either by salts dissolved in water, or by substitution of oil for water.

A patent issued in 1931 (Loomis et al., 1931) proposed the use of dissolved salts to reduce the osmotic pressure between fluid in the bore hole and that in the heaving shale. In the following years, several unsuccessful attempts were made to drill with calcium chloride−zinc chloride and sodium chloride−sodium nitrate muds (Wade, 1942). Vietti and Garrison (1939) of the Texas Company initiated a study to find an aqueous solution that would prevent the disintegration of shale. Based on such tests, sodium silicate was tried at Bryan Mound, Texas, in 1935 (Rogers, 1953). Further experimentation led to the issuance of seven patents on compositions of sodium silicate muds. These patents typically claimed "the method of preventing the heaving of shale encountered in the well which comprises circulating through the well a drilling mud comprising clay, water, sodium silicates of varying silica to sodium oxide ratios and in varying percentages and containing varying percentages of water-soluble salts" (Vietti and Garrison, 1939). Sodium silicate muds were used with some success in a number of problem wells until about 1945. W.F. Rogers devotes a chapter to sodium silicate muds in his second edition (Rogers, 1953), but silicate muds are barely mentioned in the third edition (Rogers, 1963c) because by that time other types of muds were being used successfully in drilling heaving shale.

High-pH Muds

In the 1930s the most popular thinning agent for muds was quebracho extract. This vegetable tannin, derived from a South American hardwood tree, has a deep red color when dissolved with caustic soda. High concentrations of caustic-quebracho produced *high-pH mud*, which had some desirable features in shale drilling; in particular, low gel strength and great tolerance for shale solids (Beart, 1845). From the high-pH red mud came *red lime mud*, or *lime mud*, that was consistently the most popular mud in the Gulf Coast region from 1943 to 1957. In modified compositions, it is still in use.

The origin of lime mud is obscure. As a special system, lime mud seems to have evolved from observing the improved performance of "red muds" after cement or anhydrite had been drilled. Although Rogers attributes the probable beginning of lime mud to anhydrite drilling in East Texas in 1943, Cannon (1947) cites the purposeful addition of cement to a red mud in coastal Louisiana in 1938. Whatever the introduction may have been, modifications from well to well as the lime mud found wide application throughout the Gulf Coast resulted in the development of methods for the control of properties by adjusting concentrations of lime, caustic soda, thinner, and filtration-control agents (Trout, 1948; Lancaster and Mitchell, 1949; McCray, 1949; Goins, 1950; Battle and Chaney, 1950; Van Dyke and Hermes, 1950). In time, calcium lignosulfonate (Barnes, 1949) and lignite (brown coal, leonardite) largely replaced quebracho as thinners, and sodium carboxymethylcellulose (commonly called "CMC") (Kaveler, 1946) was preferred over starch as a filtration-control agent.

Oil Emulsion Muds

At about the time that lime muds came into use in Gulf Coast drilling, oil emulsion mud was recognized as a method of "making a good mud better." Much earlier, oil had been used to loosen stuck drill pipe and to "slick up the hole" before running casing (Cartwright, 1928). Drillers in the Oklahoma City field (1934–36) added crude oil to mud to reduce sticking of pipe, and in 1937 a drilling contractor in Pottawatamie County, Oklahoma, noted a faster drilling rate after adding oil (Lummus et al., 1953).

By 1950 numerous reports of favorable field experiences with oil emulsion muds (Van Dyke, 1950; Weichert and Van Dyke, 1950) indicated such interest that the API Southwestern District Study Committee on Drilling Fluids surveyed the subject (Perkins, 1951). In brief, the conclusion was that the emulsification of either refined or crude oil improved the performance of water muds as evidenced by an increase in drilling rate and bit life, and by a reduction in hole problems. Reduction of torque, less sticking of pipe and balling of bits, and less hole enlargement were cited as major advantages of oil emulsion mud. There were no difficulties in the preparation and maintenance of the oil emulsion that were not common to the basic mud. The interpretation of electric logs and cores was not adversely affected. There was some evidence of improved productivity; none of impairment. Emulsification of oil was effected by substances already present in the mud, such as lignosulfonates, lignitic compounds, starch, CMC, or bentonite, or by the addition of surfactants, such as soaps. Similar favorable experiences with the use of oil in all types of clay-water muds was reported by the API Mid-Continent Study Committee on Drilling Fluids (Lummus et al., 1953).

Muds for Deep Holes

Muds treated with lime had definite advantages over the earlier compositions. They were less expensive to maintain while drilling thick shale sections and less affected by the common contaminants: salt, anhydrite, and cement. These benefits were attributed to the conversion of sodium clays to calcium clays by the calcium hydroxide (lime). As well depth increased, a serious problem developed in deep, hot holes, especially those requiring heavy mud. The mud in the lower portion of the hole became excessively thick when not being circulated, and actual solidification took place on prolonged heating. Studies showed that this solidification resulted from the reaction of lime with the silicious constituents of the mud (Gray et al., 1952). To minimize the problems associated with the effects of high temperatures on the strongly alkaline lime muds, compositions were proposed that were less alkaline and, consequently, less affected by high temperaturas (Watkins, 1953; Coffer and Clark, 1954).

Shale control mud was introduced by Texaco in about 1956. The objective was to gain stabilization of shale through maintenance of a high calcium ion concentration and a controlled alkalinity in the mud filtrate (Weiss et al., 1958). A premixed product composed of calcium chloride, lime, and calcium lignosulfonate furnished the desired chemical environment.

A different approach to shale stabilization was developed by Mobil research (Burdyn and Wiener, 1957). *Calcium surfactant mud* was designed to overcome the temperature limitations of lime-treated muds and to minimize the swelling and dispersion of clays by the adsorption of a nonionic surfactant (a 30-mol ethylene oxide adduct of phenol). The aggregation of the clays by the surfactant was supplemented by the addition of gypsum. Filtration was reduced by carboxymethylcellulose. If the temperature became so high as to make CMC uneconomical, the calcium ion concentration was reduced, salt was added, the system became a *sodium surfactant mud*, and polyacrylates were used to control filtration. An aqueous solution of the surfactant blended with a defoamant is sold under the trademark DMS. DMS has been found to be a distinctly useful constituent of muds for high-temperature drilling (Burdyn and Wiener, 1956; Anderson, 1961).

While lime muds were finding widespread application in drilling thick shale sections and high-pressure formations, gypsum-treated mud was introduced in Western Canada as a means of drilling anhydrite (Buffington and Horner, 1950). This *gyp mud* was prepared by adding calcium sulfate to bentonite dispersed in freshwater. Starch or CMC was added to reduce filtration rate. Properties of the gyp mud were affected only slightly by anhydrite or salt, but because of its rapid gel development, the mud was not suited to shale drilling nor to high densities. Dilution with water was the only practical means of thinning the gyp mud. Calcium lignosulfonate and the tannins required an increase in pH, and, in effect, conversion to lime mud.

The problem of controlling the flow properties of gyp muds was solved by Gray King and Carl Adolphson (King and Adolphson, 1960), who developed methods for preparing lignosulfonates of iron, chromium, aluminium, and copper from spent sulfite liquor. A ferrochrome lignosulfonate called Q-BROXIN had the unusual property of thinning gyp muds and salty muds. Roy Dawson introduced Q-BROXIN to oil field drilling in 1955. In June 1956, the first *gyp-chrome lignosulfonate mud* was used successfully in West Hackberry Field, Louisiana (King, 1976). The distinct advantages shown by the gyp-chrome lignosulfonate mud led to its rapid replacement of lime muds as the preferred drilling fluid in the Gulf Coast (Hurdle, 1957). In offshore drilling, seawater/chrome lignosulfonate muds found favor not only because seawater was readily available but also because seawater contains calcium and magnesium salts, which contribute to shale stabilization (Gill and Carnicom, 1959).

Studies showed that chrome lignosulfonate was an effective deflocculant; it afforded adequate filtration control and supplemented the action of the electrolytes in inhibiting disintegration and dispersion of shale cuttings (Simpson and Sanchez, 1961). Improvements in hole stability and reduction in break up of cuttings in shale drilling were attributed to the sealing action of the lignosulfonate. The formation on the clay surface of a multilayered adsorption film of lignosulfonate retarded penetration of water, thereby opposing the tendency of the hole walls and shale cuttings to disintegrate (Browning and Perricone, 1963).

As previously mentioned, oxidized lignite (leonardite) was applied in lime muds, particularly for counteracting high-temperature solidification. The superior heat stability of lignite was utilized also in sodium surfactant mud (Cowan, 1959). After prolonged exposure to high temperatures, the flow properties of mud containing ferrochrome lignosulfonate, lignite, and DMS were improved by the addition of sodium chromate (Newlin and Kastrop, 1960).

Alkali-solubilized lignite and sodium chromate were combined in a single package to provide a product that reduced filtration and gel development at the temperatures met in deep wells in the Gulf Coast area (Paris et al., 1961; Chisholm et al., 1961). Chrome lignite (CL) along with chrome lignosulfonate (CLS) afforded a relatively simple chemical system that was widely applicable. The *CL-CLS system* provided control of both filtration and flow properties over a wide range in pH, salinity, and solids content. Overtreatment usually produced no ill effects on properties. These features made such a program of mud treatment especially popular in foreign operations.

Low-Solids Muds

The term *low-solids muds* does not apply to any specific composition; rather, it has been applied to a number of compositions that utilize chemical and mechanical methods to maintain the minimum practical solids content. Numerous field observations, particularly in hard rock areas, showed a decrease in drilling rate when mud replaced water as the drilling fluid. A reduction in rate had been noted also as the density of mud was raised. In spite of the fact that field experience did not distinguish between density and solids effects, Wheless and Howe (1953) concluded from observations of drilling times on field development wells in the Ark-La-Tex Area that accumulation of drill solids in the mud slowed rate of penetration. The same conclusion was reached from an examination of drilling records in DeWitt County, Texas, which showed one-third less time spent in drilling from 5000 feet to 8000 feet (1500 to 2500 m) with muds averaging 13% solids by volume as compared with muds averaging 18% solids (Gray and Kellogg, 1955).

In the characteristic "clear water drilling" of West Texas, the cuttings were usually too small for satisfactory examination by the geologist. Often the hole was "mudded up" simply to obtain larger cuttings. Drilling rate then decreased (McGhee, 1956). Mallory (1957) introduced a solution to this problem with milk emulsion. Milk emulsion consisted of water to which was added approximately 5% by volume of diesel oil and about 0.03% of an emulsifier, a nonionic surfactant (a polyoxyethylene sorbitan tall-oil ester). By increasing the amount of emulsifier, oil could be emulsified in hard or salty waters. Adequate cuttings were recovered while retaining the drilling advantages of water.

Other methods in use to control the solid content were aerated mud, hydrocyclones and centrifuges, flocculants, and substitution of CMC for bentonite in the control of viscosity and filtration properties. CMC largely replaced bentonite in Phillips Petroleum Company's deep wells in Pecos County, Texas (Pope and Mesaros, 1959). One of these wells, the University E No. 1, in 1958 set a depth record of 25,340 feet (7724 m), which was not broken until 1970.

Guar gum (Mallory et al., 1960), and guar gum and starch (Collings and Griffin, 1960), provided carrying capacity for cuttings and adequate filtration control in clay-free saltwater systems.

The use of flocculants to remove small cuttings from water was studied by Pan American Petroleum Corporation in the laboratory and in the field (Gallus et al., 1958). In West Texas drilling, a 1% solution of an acrylamide-carboxylic acid copolymer was added to the discharge from the well just below the shale shaker. Circulation through earthen reserve pits allowed time for the flocculated solids to settle before the water was returned to the suction pit. While drilling permeable formations, polymer solution was injected at the pump suction to reduce water loss. Penetration rate and bit life were substantially increased by use of the polymers. Less sloughing of the "red bed" shale was observed.

Further studies (Park et al., 1960) led to the introduction of a *bentonite extender* polymer for the preparation of a low-solids, freshwater mud. This mud was composed of about 3% by weight of bentonite, 0.01% polymer, and 0.05% soda ash. A five-well test program in Wood County, Texas, showed the effect of solids content on drilling progress, as illustrated in Fig. C.9 (Lummus et al., 1961). In a later report (Lummus, 1965), the polymer was described as a vinyl acetate-maleic acid copolymer that selectively flocculated low-yield clays brought into the mud from cuttings, while increasing the thickening qualities of bentonite. The methylene blue test (Jones, 1964) was recommended as a means of measuring the effective bentonite concentration in muds (see chapter: Evaluating Drilling Fluid Performance).

Along with the progress in minimizing the solids content of the drilling fluid through modification of its composition—and vital to the success of the program—was the equally fruitful development of mechanical means of separating solids from the drilling fluid. The shale shaker (or vibrating screen) was introduced in California in 1929 and soon became a generally accepted item of rig equipment. The desander (a variation of the cone classifier), which appeared in the oil field in the early 1930s, received little attention for

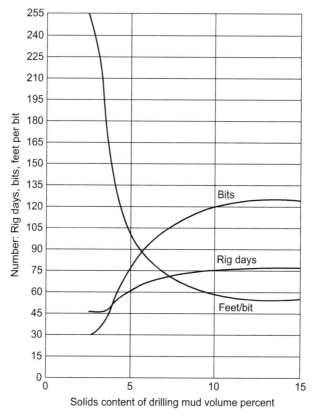

FIGURE C.9 Effect of solids on drilling performance. *From World Oil.*

many years (Kastrop, 1947). The centrifuge had found similarly limited use in California as a mud reconditioner in central mixing plants (Mills, 1930) and in the Gulf Coast at the well site (Cannon and Sullins, 1946). With deeper drilling, the requirement for prolonged use of high-density muds and the excessive cost of dilution as a method of solids control stimulated interest in mechanical methods of solids removal.

The decanting centrifuge was field tested in 1953 (Bobo and Hoch, 1954) and was accepted not only as a means of saving barite but also as an aid in penetration rate and maintenance of satisfactory mud properties (Williams and Mesaros, 1957). Interest revived in the cyclone desander and tangible savings in pump repairs and bit wear were reported (Wuth and O'Shields, 1955). Further progress in solids removal was made by passing the desanded mud through smaller hydroclones to remove silt. Removal of the silt fraction of the cuttings from the mud was credited with reduced drill pipe sticking, faster penetration rate, fewer pump repairs, and lowered cost of mud maintenance (Stone, 1964).

Appendix C: The Development of Drilling Fluids Technology

General recognition of the importance of solids control in economical drilling led to the development of more effective screening devices such as standard screens (Gill, 1966; Brandt, 1973; Cagle and Wilder, 1978), a concentric-cylinder "mud separator" (Burdyn, 1965; Burdyn et al., 1965; Burdyn and Nelson, 1968), and a "mud cleaner" (Robinson and Heilhecker, 1975; Dawson and Annis, 1977). The careful design of the entire solids-removal system as a unit operation was emphasized by Ormsby (1965, 1973), The API published a bulletin to aid in analyzing the performance of screens, hydrocyclones, and centrifuges (1974). The International Association of Oil Well Drilling Contractors, through the Rotary Drilling Committee, is preparing a series of handbooks dealing with the mechanical processing of mud at the surface, from initial mixing to final disposal.

Nondispersed Polymer Muds

Dispersion of shale cuttings had been recognized as an undesirable result of adding thinners to muds. Avoidance of thinners and substitution of polymers for bentonite seemed to be a favorable approach to faster drilling. A further advantage of polymers might come from the formation of a protective film on the surface of the cuttings and the borehole. In several publications, the general features of this approach were examined, the related factors of hole cleaning and mud properties were considered (Mondshine, 1966; Lummus and Field, 1968), and field experiences were cited (Hull, 1968; Lanman and Willingham, 1970; Anderson and Oates, 1974). The conclusion was that demonstrable savings were realized by use of nondispersed polymer muds.

Based on laboratory micro-bit tests, Eckel (1967) concluded that, rather than solids content as such, kinematic viscosity (viscosity divided by density) measured at shear rates near that in the bit nozzle was an important factor affecting drilling rate. Consequently, the shear-thinning characteristics of certain polymers offered advantages in drilling rate, and furnished adequate cuttings-carrying qualities at the same time.

The introduction of XC polymer (Deily et al., 1967) was a major contribution to the advancement of low-solids muds. XC polymer—or xanthan gum—is produced by the action of the microorganism *Xanthomonas campestris* on sugar contained in a suitable medium. XC polymer is an effective suspending agent in fresh or salty waters. At low shear rates, XC polymer has an exceptional ability to suspend solids (Carico, 1976), but its viscosity decreases markedly with increase in rate of shear. Tolerance for salts has made XC polymer an acceptable component of polymer-electrolyte drilling fluids.

Inhibited Muds: Potassium Compounds

In recent years, muds containing potassium chloride and a suitable polymer have been the subject of publications from several areas. First, however, we shall mention some earlier uses of potassium compounds.

The term *inhibited mud* (more properly, "inhibitive") was originally applied to muds (or filtrates) that suppressed the swelling of bentonite. In a study of the effect of mud filtrates on the permeability of sandstone cores from the Stevens Zone of the Paloma Field in California, Nowak and Krueger (1951) found that the relative extent of permeability impairment of water-sensitive cores could be predicted from the settled volume of Wyoming bentonite that had been standing in the mud filtrates for 24 hours. Based on this simple test, Huebotter (1954) proposed the use of a mud containing 10% by weight potassium chloride in the water and about 0.5% calcium lignosulfonate. Starch could be added to control filtration. Maintenance was similar to that of ordinary saltwater muds. This inhibited mud was used in shallow wells in South Texas and in Wyoming to give better production from dirty sands. Very little mud-making from shale drilled from 6000 to 10,000 feet (1800–3000 m) was noted in a well in Kern County, California, in which the potassium chloride mud was used. The cost of materials, the lack of solids-separation equipment at that time, and the requirement of close supervision led to the abandonment of this potassium-treated mud.

In 1960, while drilling steeply dipping shales in the Cerro Pelado area of Venezuela, Tailleur (1963) noted markedly improved hole stability when mud containing potassium ion replaced the commonly used sodium or calcium ions to inhibit clay swelling. In addition to its lubricating qualities, "Concentrate 111," which consists of potassium soaps of sulfurized tall oil acids, also served as "an emulsifier for the oil phase and as an inhibitor of clay swelling." Hole enlargement in the shale section was significantly reduced, a result attributed to the inhibitive properties of potassium ion and cited in a patent application filed in September 1963 (Tailleur, 1967).

Another possible explanation for the superior inhibiting qualities of potassium compounds was offered by Black and Hower (1965) who pointed out that potassium has an ionic diameter and a hydration number that would favor its exchange for other cations on clay surfaces. Laboratory studies of the effects of several salt solutions on the hardness of cores from water-sensitive sands showed that 2% potassium chloride was a more effective stabilizing agent than was 2% calcium chloride or 10% sodium chloride.

Shell polymer mud was introduced in the foothills area of Western Canada in 1969 (Clark et al., 1976). In this area, mechanically incompetent formations in fault zones, and water-sensitive hard shales, were successfully drilled with potassium chloride-polymer compositions. The selected polymer was a 30%-hydrolyzed polyacrylamide having a molecular weight of about 3,000,000. Subsequently, potassium chloride-polyacrylamide muds were successfully used in many other areas where troublesome shales were encountered. The results of extensive laboratory tests on the dispersive, adsorptive, and hydrational characteristics of several representative shales were reported by O'Brien and Chenevert (1973). They concluded that potassium chloride was the preferred electrolyte for shale inhibition and that XC polymer, all factors considered, was the favored polymer.

Drilling through permafrost in Arctic regions presents problems such as hole enlargement, stuck pipe, and difficulties in cementing surface casing. A drilling fluid that consisted of potassium chloride, bentonite, and XC polymer was introduced by Imperial Oil Company in 1971 for drilling permafrost in the McKenzie Delta area of northern Canada (Kljucec et al., 1974). Compared with the bentonite-water mud previously used, the potassium chloride-XC polymer addition resulted in major savings of both time and materials.

Corrosion Control

The possible effect of drilling fluid on drill string and casing corrosion received little, if any, consideration before the 1930s. With the extensive drilling activity in the West Texas and New Mexico Permian Basin, drill pipe failures became a matter of concern. Waters were usually salty—up to saturation when bedded salt was drilled—and frequently were acidic. Few drill collars were used; consequently, the drill pipe was run in compression.

In 1935, Speller (1935) stated that drill pipe failures recognized as primarily caused by corrosion were relatively rare compared with failures attributable to other causes. He believed that mud conditioning offered the most promising means of protecting drill pipe against corrosion fatigue. Colloidal bentonite would protect the pipe against oxygen corrosion, he thought, but the colloidal properties would be destroyed in the presence of salts. Therefore, sodium sulfite was suggested to remove the oxygen from salty mud.

In 1936, the American Petroleum Institute, Division of Production, Topical Committee on Materials, set up a subcommittee to investigate drill pipe corrosion fatigue. The subcommittee report (1939) contained reports by several individuals. A significant conclusion was that corrosion fatigue endurance could not be reliably predicted from loss in weight tests without stress. Also, less corrosion was observed in muds of higher pH. Sulfate-reducing bacteria were found in some field muds, although corrosion of drill pipe was not confirmed in these cases.

Grant and Texter (1941) stated that fatigue failure was the most common cause of drill pipe trouble. Notches, scars, and corrosion accelerated the failures. Use of more drill collars to keep the pipe in tension and use of more care in pipe handling were the major recommendations. In October 1945, the American Association of Oilwell Drilling Contractors (AAODC) retained Battelle Memorial Institute, Columbus, Ohio, to study drill pipe failures in Permian Basin drilling operations. The results of this extensive investigation were summarized in a series of articles in *The Drilling Contractor* from 1946 to 1948. The conclusion regarding corrosion inhibitors in the drilling fluid was that sodium chromate at a concentration of 2500 ppm was the best treatment (Jackson et al., 1947). This practice was generally adopted in drilling with salty muds in the Permian Basin.

Corrosion failure of drill pipe was not regarded as a serious problem in areas using high-pH muds containing quebracho, lignite, and lignosulfonate (the thinners serving as oxygen scavengers). Corrosion of casing through bacterial action was not found in wells drilled with high-pH mud. However, a well in which casing failure could definitely be attributed to the corrosive action of sulfate-reducing bacteria had been drilled with phosphate-treated low-pH mud (Doig and Wachter, 1951).

Corrosion began to receive more widespread attention as well depths and pipe costs increased. The National Association of Corrosion Engineers, founded in 1945, provided a medium for exchanging information on corrosion, but until the 1960s, most reports on drill stem corrosion appeared in oil industry publications.

At the annual meeting of the AAODC in 1959, King (1959) reported the results of an extensive study conducted by the Hughes Tool Company laboratory on the failures of rock-bit bearings. As had been reported in the field, the laboratory test showed rapid failure of bearings run in water that contained hydrogen sulfide. Under controlled test conditions, bearing life was reduced sixfold when 0.1% sodium sulfide was added to the 6% sodium chloride, 10% bentonite reference mud.

Various mud-treating agents were investigated in the reference mud. Tannins and lignins reduced bearing life, but caustic soda-quebracho solution did not. This study showed that corrosion affects bit life as well as that of the drill pipe life; the destructive effect of hydrogen sulfide in particular was emphasized by this investigation.

Interest in packer fluids began around 1950. The term *casing pack* is applied to the material placed in the annulus between the hole wall and the casing. The casing pack must satisfy the requirements of longtime stability under downhole conditions. It must maintain suspension and filtration properties, and serve as a noncorrosive barrier against corrosive formation fluids. When unusual local conditions required a casing pack (other than the mud used in drilling the well), a specially prepared oil mud was pumped into the annulus (see section on oil-based drilling fluids technology in this chapter).

The material placed above the packer in the annulus between the casing and the tubing to prevent pipe failure was called a *packer fluid*. Except for low filtration rate, the same requirements are placed on packer fluids as on casing packs. The packer fluid assists in maintaining the packer seal, and its density must be high enough to prevent burst or collapse of the pipe. Common practice was to leave the mud used in drilling the well in the casing-tubing annulus on completion, but severe problems developed in workover operations as well depth and temperature increased: heavy, lime-treated mud in the annulus solidified; the tubing could not be pulled; and an expensive workover became necessary. To avoid this problem, the lime mud was displaced by a packer fluid such as a freshly-prepared bentonite-barite mud (often containing soda ash for pH adjustment), an oil mud, or a salt solution.

Salt solutions had advantages in ease of preparation and long-time stability, but there were limitations in density. Because sodium nitrate was readily soluble, it was used to prepare packer fluids placed in a number of wells. Corrosion failures were prompt and severe (McGlasson et al., 1960). Examination of numerous single or mixed salt solutions led to the recommendation of sodium chloride for the density range 8.3 to 9.8 lb/gal (1.0 to 1.2 SG), calcium chloride to 11.5 lb/gal (1.4 SG), and calcium chloride-zinc chloride mixtures to 14 lb/gal (1.7 SG) (Hudgins et al., 1960; Hudgins et al., 1961). More dense solutions of the mixed calcium and zinc chlorides were considered too corrosive. Later, calcium bromide-calcium chloride solutions extended the density range to 15 lb/gal (1.8 SG) with acceptable corrosion rates (Planka, 1972), and to a maximum density of 18 lb/gal (2.15 SG) with calcium bromide. In the late 1950s, the rising cost of tubular goods and the increasing expense of workovers focused attention on corrosion mitigation (Battle, 1957; McGlasson and Greathouse, 1959). With increased emphasis on faster drilling, corrosion-induced drill pipe failures became prominent. Faster rotary speed, greater weights on bit, stronger steel, higher pressures and temperatures in deeper wells, and changes in mud composition with lower pH and clay content—all of these factors contributed to the escalation of drill pipe failures.

New methods were used to protect the drill pipe. Loss due to internal corrosion was greatly reduced by application of plastic coating to the pipe (Clark, 1965; Sheridan, 1965).

New methods of testing were introduced. As a measure of corrosion rate in a drilling well, test rings of drill pipe steel were inserted at selected intervals in the drill string; the loss in weight served to measure rate of corrosión (Behrens et al., 1962a,b). A sensitive test for conditions causing hydrogen embrittlement failure was made with prestressed roller bearings (Bush and Cowan, 1966).

Other test methods (McGlasson and Greathouse, 1959; Radd et al., 1960) and various instruments were employed in studies of corrosive attack involving the drilling fluid. No attempt is made here to list all of the publications between 1960 and 1970 that dealt with drilling fluid-related corrosion. H.E. Bush's survey of the progress made in that interval lists some pertinent references (Bush, 1973).

The serious hazards of hydrogen sulfide to the drill crew, and some spectacular failures of drill pipe, led to the development of improved analysis methods and more effective scavengers. Developments in chemical scavengers for H_2S were surveyed by Garrett et al. (1978) and current practices are reviewed in Chapter 9, Wellbore Stability.

For years, oxygen has been recognized as a major cause of drill pipe corrosion, (Bradley, 1967, 1970) and the problem continues to be a serious one (Cox, 1974). Each time the mud passes through the surface system, oxygen enters in the entrapped air. Continuous addition of an oxygen scavenger such as sodium sulfite must be made. Another approach—the use of an inert gas to sweep out the entrained oxygen—is being investigated.

Different Types of Fluid for Different Drilling Functions

In tracing the historical development of water-base muds, it would be impossible to list all the compositions that have been used. Instead, the objective here has been to emphasize ways in which technology has developed as a response to the increasing demands of ever greater well depths and the increasing complexity of drilling problems. The functional requirements placed on the mud have grown from one—cuttings removal—to many, and now include pressure control and hole stability maintenance. At the same time, the muds must not damage well productivity nor be injurious to personnel, drilling equipment, or the environment.

Certain of these functions may be performed more effectively by drilling fluids that have gas or oil as the principal component. Although the technology of these other drilling fluids developed along with that of water muds, for convenience they are reviewed separately here.

OIL-BASED DRILLING FLUIDS TECHNOLOGY

Reasons for Development

Oil-based drilling fluids have been developed to overcome certain undesirable characteristics of water-based muds. These deficiencies are primarily due to the properties of water: specifically, its abilities to dissolve salts; to interfere with the flow of oil and gas through porous rocks; to promote the disintegration and dispersion of clays; and to effect corrosion of iron. In addition to providing a means for avoiding these objectionable features of water muds, oil muds offer potential advantages: better lubricating qualities, higher boiling points, and lower freezing points. Because the cost of preparing an oil mud is always more than that of the same density water mud, the economic justification for selecting an oil mud must come from its superior performance under the particular conditions of use.

Oil for Well Completion

Oil-based drilling fluids originated with the use of crude oil in well completion, but the date of first usage is unknown. Oil was used to drill the productive zone in shallow, low-pressure wells in many early fields. A patent application filed by Swan (1923) in 1919, and granted in 1923, proposed using "a nonaqueous viscid liquid," such as coal tar, wood tar, resin, or asphalt thinned with benzene to drill wells. Although claimed for use as a drilling fluid, application of these substances to seal casing in place with an anticorrosive liquid was emphasized as a method that would facilitate recovery of casing.

Concern with the mudding off of oil sands caused California operators to use oil for completion (Jensen, 1936). Transport of cuttings up the drill pipe by reverse circulation extended the application of oil as a drilling fluid, and numerous wells were completed by this method (Beckman, 1938). Oil was used in coring shallow oil sands and coring salt in salt-dome exploration in the Gulf Coast area.

Early Oil Company Developments

Because hole collapse while drilling shale was attributed to the effect of water, an oil mud seemed likely to solve the heaving shale problem. An oil mud prepared with gas oil and spent clay from the refining of lubricating oil (Moore and Cannon, 1936) was used by Humble Oil & Refining Company (now Exxon) in an unsuccessful 1935 attempt to drill a troublesome shale interval in the Goose Creek Field in Texas. During the next two years, numerous cores were taken with oil mud in the Anahuac, Tomball, and East Texas fields in Texas to study the connate water content of reservoir sands (Schilthuis, 1938). The usual composition of this oil mud was field crude oil and spent adsorbent clay, to which about 0.5% oleic acid (Rolshausen and Bishkin, 1937) and 1% concentrated sulfuric acid were added (Moore, 1940). If higher density was needed, litharge was added (Cannon and Williams, 1941).

In 1936, Shell Oil Company began a systematic research program to develop an oil-based drilling fluid (Alexander, 1944), and in 1938, an oil mud based on this study was used in the Round Mountain Field in California (Dawson and Huisman, 1940).

Compositions consisting of stove (diesel) oil, ground oyster shells or limestone, air-blown asphalt, and lampblack—with barite for higher density if needed—were used in drilling the oil-bearing sections of wells in several California fields (Hindry, 1941). Each ingredient served a particular function: stove oil was the liquid medium; the shell, limestone, or barite furnished density and initial plastering properties; lampblack provided suspending properties; and blown asphalt supplied both suspending properties and final plastering properties. Emphasis was placed on the desirable features of the very thin filter cake and the absence of water in the filtrate (Hindry, 1941). Lampblack was later replaced as a gelling agent by alkali metal soaps of unsaturated fatty acids (Mazee, 1942), such as the reaction products of tall oil and potassium hydroxide (Dawson and Blankenhorn, 1944), and tall oil and sodium silicate (Self, 1949).

Introduction of Commercial Oil Muds

Commercial oil muds became available in 1942, when George L. Miller formed the Oil Base Drilling Fluids Company in Los Angeles, California. This company (now Oil Base, Inc.) supplied blown asphalt in the form of Black Magic, a powder which was blended with a suitable oil at the well site (Miller, 1942, 1943). Properties of the blown asphalt and of the preferred diesel oil were specified (Miller, 1943, 1949a,b). Naphthenic acid and calcium oxide were other components of the oil mud. Ready-mixed oil muds were supplied, and used muds were reconditioned.

Halliburton Oil Well Cementing Company, licensed under Shell's patents, in 1943 marketed an oil mud concentrate that contained blown asphalt, and tall oil soap formed in the oil phase by the reaction of tall oil with sodium hydroxide and sodium silicate. Salt (sodium chloride) could be added to counteract the effect of freshwater (Anderson, 1947a) on water-soluble compounds of calcium and magnesium (Anderson, 1947b). The oil mud concentrate, when added to a water mud, formed an emulsion of oil in water. After Magnet Cove Barium Corporation took over the sale of Jeloil in 1948, Jeloil E was widely used in the preparation of water-in-oil emulsions (McCray, 1949; Goins, 1950).

In September 1948, the Ken Corporation (now merged into the Imco Services Division of Halliburton Co.), of Long Beach, California, was formed to supply the oil mud compositions that had been developed by Fischer of Union Oil Co. (1951a,b,c,d). These products did not contain air-blown asphalt. Instead, suspending and sealing properties were supplied by mixed alkali and alkaline earth soaps of resin acids. Subsequently, modifications were made in the resin products (Fischer, 1952a; Hoeppel, 1956), and other water-emulsifying agents were included in the compositon (Fischer, 1952b). The liquid oil-mud concentrate was either mixed at the well site with crude oil ($10°-24°$ API gravity) or diesel oil, or supplied already mixed from central plants.

Applications of Oil Muds, 1935−50

By far the greatest use for oil muds in the period from 1935 to 1950 was for well completion, mainly in either low-pressure or low-permeability reservoirs. When oil mud was used instead of water mud, higher initial production was usually noted (Cannon and Williams, 1941; Miller, 1942; Kersten, 1946; Stuart, 1946). Coring for reservoir information was a frequent application of oil mud (Edinger, 1949), and release of stuck drill pipe was another (Miller, 1949a,b).

Although there was general agreement that hole enlargement in shale sections could usually be avoided by drilling with oil mud, there was little application solely for that purpose. Use of oil muds for drilling had its drawbacks: water was a severe contaminant; precautions were necessary to avoid fires; and the drilling rate was usually slower with oil mud than with water mud.

Oil Base—Emulsion (Invert Emulsion) Drilling Fluids

All of the commercial oil-base muds contained some water, either formed in the mud by the neutralization of organic acids, or else inadvertently introduced while in use. This water, usually not more than 5%, was emulsified in the oil. An increase in water content caused thickening, particularly with asphaltic oils. To overcome this problem of contamination by water, more effective emulsifiers for water were sought; such emulsifiers led to new formulations in which the water became a useful component instead of a contaminant. The term "oil emulsion mud" had been applied to emulsions of oil in water, so the water-in-oil emulsions were called *inverted* or *invert emulsions*. Such oil muds, containing over 10% water (sometimes as much as 60%) in the liquid phase, utilized the emulsified water as a suspending agent.

According to Wright (1954) the first field use of an oil-based emulsion mud was in August 1960, in the Los Angeles Basin. This drilling fluid was made with 40% water emulsified in a refined oil. Emulsions with more than 30% water would not support combustion, and therefore eliminated the fire hazard (Wright, 1954). Other favorable features of the emulsion system were lower *cost* and lighter color, which meant less stained clothing for the crew.

These features promoted the use of invert emulsions in low-pressure well completions. Both liquid and solid concentrates became available from mud suppliers. In 1963, Rogers listed twelve different compositions. Many others have since been developed. The composition of proprietary products usually was not revealed, and modifications were made from time to time as more effective components were found. While this review is limited to certain compositions that indicate the types of materials that have been used, some references to publications dealing with applications will be made.

In general, the invert emulsion contains both oil-soluble and water-soluble emulsifying agents. In several products, the oil-soluble emulsifier was formed in the oil phase by the addition of calcium or magnesium compounds (Dawson, 1950; Gates and Wallis, 1951). Tall oil was the source of fatty acids in several compositions.

In one composition, ethoxylation of tall oil acids produced a nonionic emulsifier that was used with lecithin, an oil-wetting agent and an emulsifier for water in oil (Lummus, 1953, 1954). Lecithin was included in another composition that involved an oxidizing agent, a fatty acid residue, and hydrogenated castor oil (Lummus, 1957).

Another composition specified the extent of oxidation of the tall oil, based on its increase in viscosity, and included a calcium compound and a quaternary amine (Watkins, 1958, 1959; Brandt et al., 1960; Walkins, 1960). Polyvalent-metal soaps of resin acids (variously modified before conversion to the soap) served as the principal emulsifier in several compositions (Hoeppel, 1956; Gates and Wallis, 1951; Dawson, 1952; Gates and Pfenning, 1952; Reddie, 1958). Organic compounds containing nitrogen, such as polyamines and polyamidoamines, were useful components (Reddie and Griffin, 1961; Hoeppel, 1961). (The basis for the selection of suitable emulsifying agents is discussed in Chapter 7, The Filtration Properties of Drilling Fluids.)

Organophilic Clays and Ammonium Humates

The development of clay compositions capable of forming gels in oil, similar to those formed by bentonite in water, was a major contribution to the technology of oil muds. Hauser (1950) discovered that hydrophilic clay could be converted to an organophilic condition by reaction with appropriate organic ammonium salts. Jordan and associates (1949), (Jordan, et al., 1950) studied the reaction of bentonite with a series of aliphatic amine salts and found that the reaction products of amines having twelve or more carbon atoms in the straight chain will swell and form gels in nitrobenzene and other organic liquids. The organo-clay complexes were formed by replacing the exchangeable cations of the bentonite with the cationic groups of the amines and by further adsorption of the hydrocarbon chain on the clay lamina surface. The oil-dispersible clays would suspend solids in oil, without requiring additional soaps and emulsifiers (Hauser, 1950a,b).

Further studies by Jordan and coworkers on reactions of the aliphatic amine salts led to the discovery of a filtration-reducing agent for oil muds (Jordan et al., 1965, 1966). This material, described as consisting of n-alkyl ammonium humates, was prepared by reactions between quaternary amines having 12 to 22 carbon atoms in one of the alkyl chains, and the alkali-soluble fraction of lignite (leonardite, brown coal). This reaction product was readily dispersed in refined or crude oils, and markedly reduced filtration without substantially increasing viscosity. The agent also aided in the emulsification of water.

The organophilic clays, the alkyl ammonium humates, the wetting agent lecithin, and a suitable emulsifier for water all made possible the relatively independent control of suspending and filtration properties in oil muds (Simpson et al., 1961). These organophilic colloidal materials that performed specific functions in a base oil could be varied in amount to furnish the properties required for satisfying the conditions of use. More extensive use of oil muds in varied applications became economically practical (Gray and Grioni, 1969).

Annulus Packs

The application of nonaqueous compositions to prevent casing corrosion was recommended in an early patent (Swan, 1923). The extensive use of thick oil muds to protect the casing exterior from corrosion by formation waters began in Kansas in the early 1950s (Bright, 1964). The practice spread to other areas where corrosive waters attacked the casings. These *casing packs* served as a stable, noncorrosive barrier against corrosive formation fluids.

In 1947, earthquakes in California caused severe casing damage to numerous producing wells. In the Long Beach area, horizontal earth movement took place along definite slippage planes at depths of 1500 to 1700 ft (460 to 510 m), usually affecting from 5 to 30 ft (2 to 9 m) of the casing. In the repair operations, the interval through which the casing had been deformed was underreamed, and a bell-hole pack was placed in the underreamed hole-casing annulus. Several oil base compositions were developed that could be pumped in to displace the mud. These slurries would develop a grease-like consistency that prevented transmission of the shock of earth movement to the casing (Miller, 1951).

As stated in the previous section, solidification of dense, lime-treated mud in the casing-tubing annulus caused very expensive workover operations. To avoid this problem, in the early 1950s, the lime-treated mud in the casing-tubing annulus of a number of deep Gulf Coast wells was displaced by a slurry of organophilic clay and barite in diesel oil. The density of the slurry was the same as that of the mud used in drilling: in some instances above 18 lb/gal (2.15 SG). Several years later, when a number of these wells were worked over, the packer was released and the tubing was pulled without difficulty.

When chromelignosulfonate-treated muds supplanted lime muds as the preferred drilling fluid for deep, hot wells in the early 1960s, a different problem became evident in later workover operations. Hydrogen sulfide, produced by thermal decomposition of the lignosulfonate, caused corrosion failure of high-strength tubing. Again, laboratory tests and field experience showed that oil mud compositions made satisfactory packer fluids (Simpson and Andrews, 1966; Simpson and Barbee, 1967; McMordie, 1968).

A casing pack especially designed for wells in Alaska's North Slope region utilized the low thermal conductivity of a gelled oil slurry (Mondshine, 1974; Remont and Nevins, 1976). A slurry consisting of organophilic clay in diesel oil displaced water mud from between the concentric strings of casing that passed through the permafrost zone. An emulsifier for water was included, and barite was added to increase the density if needed. By keeping the temperature of the slurry below 50°F (10°C), easily pumpable slurries could be prepared that contained enough organophilic clay to form a stiff grease-like gel when well production raised the temperature. Oil mud, if used in drilling, could be modified by the addition of gellant in order to make the arctic casing pack.

Borehole Stabilization by Oil Muds

As has been mentioned, oil muds were used early in drilling troublesome shales, usually with success, but with some failures. Laboratory studies by Mondshine and Kercheville (1966) showed that wet shales could be hardened by exposure to invert emulsion mud which had high-salinity water in the emulsified phase. Transfer of water from the shale to the oil mud was attributed to osmotic forces across the semipermeable membrane around the emulsified water droplets.

Not all emulsifying agents were equally effective in forming a semipermeable membrane. The greater the degree of dispersion of the emulsified water, the more rapidly water was removed from the shale. Calcium chloride, when dissolved in the emulsified water, was more effective than an equal weight of sodium chloride similarly dissolved. Successful application of the technique was reported in drilling troublesome shales in coastal Louisiana and in Algeria.

Further studies (Mondshine, 1969) showed that when the salinity of the water in the shale was higher than that of the water in the oil mud, water migrated from the oil mud into the shale. Thus, the gain or loss of water by the shale was determined by the osmotic pressure. No movement of water occurred when the surface-hydration pressure of the shale equalled the osmotic pressure of the oil mud. The salinity required to assure stability could be estimated by measuring the salinity of the water in the shale and by equating the shale's surface hydration force with the matrix stress (the overburden pressure minus pore-fluid pressure).

Chenevert (1970) used a different approach to selecting the salinity for the water phase of an oil mud. Based on the concept that shale would not adsorb water from an oil mud if the aqueous chemical potential of the mud equalled that of the shale, the water activity of shale samples was determined from adsorption isotherms. When the emulsified phase of the oil mud, in contact with a shale sample, contained enough of any salt to have the same activity (or vapor pressure) as that of the shale, no swelling of shale occurred. Thus, the activity values (vapor pressure) of shale cuttings and of the mud could be measured at the well site, as a means of field control. In general practice, a concentration of calcium chloride sufficient to balance the shale of lowest activity is maintained in the emulsified water of the oil mud.

Extreme Borehole Conditions

A review of the extreme borehole conditions encountered in record-setting United States wells before 1975 showed that oil muds had been used in wells that had met the extremes of temperature, pressure, corrosive environment, and plastic salt (Gray and Tschirley, 1975). Experience in a number of deep wells in South Texas had shown the distinct advantages of oil muds over water muds: more stability to heat and lower cost of maintenance (Weintritt, 1966). The highest measured temperature at a given depth had been recorded in Webb County. A new record was set in 1974 in the 1 Benevides well of Shell Oil Co. and El Paso Natural Gas Co. with a temperature of 555°F (291°C) measured at a depth of 23,837 ft (7,266 m) (Ives, 1974). An oil mud having a density of 18.3 lb/gal (2.20 SG) was in use at that time and no difficulty was experienced in maintaining satisfactory properties.

The extremes of pressure and corrosive environment are encountered in the deep wells of the Mississippi Salt Basin. Drilling problems in this area are compounded by highly pressured gases consisting of up to 75% hydrogen sulfide, by bottomhole temperatures approaching 400°F (204°C), and by potential losses of circulation. Record pressure gradients of 1 psi/ft (0.23 kg/cm^2/m) were measured in Shell-Murphy USA 22-7, a wildcat in Wayne County, Miss., drilled to 23,455 ft (8551 m) (Parker, 1973). In drilling a 4 1/8-inch hole from 19,904 ft (6067 m) to 23,455 ft (8551 m) with oil mud ranging in density from 19.2 lb/gal to 20.3 lb/gal (2.30 to 2.44 SG), special procedures were developed for making connections and tripping the drill string (Klementich, 1972).

Gases in this Smackover limestone of the Jurassic age vary in composition, and may contain 5% to 60% carbon dioxide and from 10% to 75% hydrogen sulfide (Kirk, 1972). Here, oil mud supplies the necessary protection against corrosion. Gas and saltwater kicks can be controlled with only minor effects on the properties of the oil mud.

Several unsuccessful attempts using water mud were made to drill through the Louann salt formation in East Texas and North Louisiana. With temperatures above 250°F (121°C) at depths below 12,000 ft (3700 m) plastic flow had to be overcome. Oil mud of density 19.2 lb/gal (2.30 SG) was used in drilling through more than 1200 ft (370 m) of salt in an East Texas well (Gray and Grioni, 1969). In Webster Parish, Louisiana, 3,590 ft (1095 m) of salt was drilled to a depth of 15,321 ft (4670 m) with oil mud having a density of 17.6 lb/gal (2.11 SG). After casing had been set through the salt, drilling continued with the oil mud to 20,395 ft (6216 m), at which depth the temperature was 405°F (206°C).

In addition to the plastic behavior of salt, another problem requires attention in drilling thick salt beds with oil muds: the tendency of the solids in the mud to become water wet. The solids in the mud become water wet because the accumulation of fine salt cuttings reduces the effectiveness of the water-in-oil emulsifying agent (Leyendecker and Murray, 1975).

Penetration Rate With Oil Mud

The effect of oil mud on drilling rate has long been of interest. Early observers agreed that rate of penetration with oil mud is slower than with water mud (Hindry, 1941; Kersten, 1946; Miller, 1951). This slower drilling rate was attributed to the high plastic viscosity of oil muds, which limits the rate of mud circulation, and to the fact that oil does not soften shales as water does (Trimble and Nelson, 1960).

Valid drilling rate comparisions could be made in 22 of approximately 200 wells drilled with oil muds in 1961 and 1962 (Tschirley, 1963). Based on these 22 comparisons, the conclusion was that drilling rate in shale and sand is somewhat faster with oil mud than with water mud of the same density and flow properties. Although not conclusive, indications were that water mud is faster in drilling limestone. By 1970 (Reid, 1970), deep drilling field experience in the Delaware Basin of West Texas and Southeast New Mexico had shown that salinity-controlled oil muds stabilize shale to the extent that wells can be drilled with oil mud of significantly lower density than is possible with water mud. The lower pressure of the oil mud column affords a faster drilling rate in shale but not in limestone (Kennedy, 1974).

A laboratory study of the effect of oil mud composition and properties on the penetration rates of microbits in limestone indicated that faster drilling rates can be expected by reducing the colloidal content (which mainly affects the filtration rate) (Fontenot and Simpson, 1974). Field experience in the Delaware Basin confirmed that higher filtration rates (less lignitic filtration-control agent) make faster drilling possible in carbonate rocks (O'Brien et al., 1977). Using the salinity-controlled invert emulsion compositions with relaxed filtration requirements, the advantages of shale stability, torque and drag reduction, quick release of trip gas, and corrosion mitigation are gained without loss of penetration rate. Major cost savings resulted (Smith, 1974). Drilling in other areas confirmed the results obtained in West Texas (Simpson, 1978).

Scope of Oil Mud Applications

In less than 50 years, the technology of oil-based drilling fluids has advanced from the use solely of crude oil as a means of improving productivity to the use of multifunctional compositions that have played parts in numerous record-setting wells. Applications have been made under conditions of extreme temperatures, high pressures, water-sensitive shales, corrosive gases, and water-soluble salts. Problems of stuck pipe, excessive torque and drag in deviated holes, and entrainment of gas in the drilling fluid have been minimized. In opposition to these favorable features, the high initial cost, the extreme precautions often required to avoid pollution while in use and on disposal, and sometimes the objections of the drilling crew must all be considered in the selection of an oil mud for a specific application.

GAS-BASED DRILLING FLUIDS TECHNOLOGY

In Chapter 1, Introduction to Drilling Fluids, principal component formed the basis for classifying three different types of drilling fluids. Of these, gas-based drilling fluids include those in which air or other gas is the continuous phase (e.g., dry gas, mist) and those in which gas is the discontinuous, or internal, phase (e.g., foam, stiff foam). The term *reduced-pressure drilling fluid* can be applied to all of these systems because they are used to reduce the pressure gradient of the drilling fluid to less than that exerted by a column of water. The original purpose for using gas-based fluids was to avoid loss of water, and the resulting damage to productive zones. A secondary benefit that became of major importance in hard rock areas was a faster drilling rate.

Dry Gas Drilling

Brantly cites a patent issued to P. Sweeney in 1866 as the earliest record he found suggesting use of compressed air to remove cuttings from a drilled hole, although air had probably been used by earlier drillers. Pressure drilling with a control head, which allows control of gas and oil flow while drilling the productive zone, was employed in the early 1920s in Mexico. The practice spread to other areas.

The first recorded injection of gas occurred in September 1932 (Foran, 1934). To keep water out of the producing zone at 8800 ft (2680 m) in the Big Lake Field in Reagan County, Texas, gas at a volume ratio of 143 to 1 was metered into the circulating water. Shortly thereafter, the closed-fluid circulating system was employed for drilling in the Fitts Pool in Pontotoc County, Oklahoma, and productivity was greatly improved compared to that of wells drilled with mud (Teis, 1936). Similar gas-injection practices were followed in California for drilling subnormal-pressure sands (Gnnsfelder and Law, 1938).

Around 1950, small rigs drilling shot-holes for seismic exploration began to use compressed air in areas where water was scarce (West Texas) or where temperatures were low (Canada) (Shallenberger, 1953). In May 1951, El Paso Natural Gas Co. in the San Juan Basin, New Mexico, began drilling with gas to avoid loss of circulation in the Mesa Verde section from 4000 to 5000 ft (1200 to 1500 m). Rate of penetration and footage per bit increased greatly. More important, well cleanup was facilitated, and productivity was much higher than when mud was used (Hollis, 1953). Economical development of the extensive San Juan Basin gas fields (1951−53) was made possible by use of gas as the drilling fluid.

The use of natural gas in successfully avoiding both loss of circulation and the accompanying damage to the producing formation led to the introduction of air drilling in Martin County, Texas, in June 1951 (Berry, 1951). Because natural gas was unavailable, compressed air was used. To supply

678 Appendix C: The Development of Drilling Fluids Technology

FIGURE C.10 Composite view of the nine two-stage and three one-stage compressors supplying air for this unusual drilling project. The compressor manifold and cooling system may be seen in the illustration. *From Berry, O.H., 1951. Air drilling the sprayberry sand. World Oil, 169–172.*

the air, nine small two-stage compressors and three single-stage boosters were assembled as shown in Fig. C.10. The volume of air supplied was not sufficient to clean the hole, however, until circulation was reversed. Air was used in drilling from 6620 to 7542 ft (2018−2300 m). Unless liquid was produced in sufficient amount to show a spray at the surface while drilling with reverse circulation, fine cuttings would adhere to the inner wall of the drill pipe. This observation led to the practice of injecting water with air whenever cuttings contained just enough moisture to be sticky.

In the next few years, attempts were made in many areas to employ air as the drilling fluid for increased penetration rate and footage per bit. Where water-bearing formations did not interfere, both air and natural gas showed outstanding advantages, wherever one or more of the following conditions existed: loss of circulation; susceptiblity of producing formation to water or water-base mud damage; and the high expense or unavailability of water or mud (Nicolson, 1953). Observation showed that significantly faster rate of penetration and more feet of hole per bit were typical. As the method was tried in different areas, both practical limitations and advantages were recognized (Cannon and Watson, 1956; Smith and Rollins, 1956; Murray and MacKay, 1957).

Angel (1957) calculated air requirements for typical hole and pipe sizes based on three assumptions. First, the annular velocity is 3000 ft/min (15 m/s). Second, a homogeneous mixture of air and cuttings with the flow properties of a perfect gas is formed. Third, the geothermal gradient is applicable as the temperature of the gas. The expanded tables (Angel, 1958) were published in 1958.

Problems With Water-Bearing Zones

Early in the use of air as a drilling fluid, water-bearing formations were found to be a major limiting factor. Often, when water-saturated formations were drilled, the wet cuttings stuck together and were not carried from the hole by the air stream. When the wet cuttings filled the annulus, a *mud ring* was formed: the air flow was shut off and the drill pipe was stuck. Yet, when water was injected with air to prevent mud ring formation, some formations became unstable.

Several methods of shutting off water were tried, including (1) forcing a liquid mixture of two polymers into the water-bearing formation, to form a stiff gel (Hower et al., 1958); (2) introducing a solution of aluminum sulfate followed by ammonia gas to form a precipitate (Goodwin, 1959); (3) injecting silicon tetraflouride gas into the water to produce a solid plug (Becker and Goodwin, 1959); and (4) injecting a liquid, a titanium ester called "Tetrakis," to form a precipitate with the water present (Stein, 1963). Certain methods had some success (Sufall, 1960); however, the problems of placement and the likelihood of drilling into other water-bearing zones rarely justified the expense. Wetting and balling of cuttings can be diminished by introducing zinc or calcium stearate into the air stream (Randall et al., 1958).

Foam

When the quantity of water entering the hole from the water-bearing formation exceeded about 2 bbl/h (0.3 m^3/h), the water could be brought out of the hole as a foam by injecting a dilute solution of a suitable foaming agent into the air stream. Foam effectively removed cuttings at lower annular velocities than was possible with air (Randall et al., 1958) and as much as 500 bbl/h (80 m^3/h) of water could be brought from the hole. With such quantities of water coming into the hole, however, the time spent unloading the hole after a trip was prolonged; the cost of foaming agent became excessive, and water disposal became a problem (Murray and MacKay, 1957). Further experience with foam led to more consistent operating practices, and the advantages and limitations in its use became more clearly defined (Murray and Eckel, 1961; Goins and Magner, 1961; Lummus and Randall, 1961).

Numerous foaming agents were on the market and several test methods had been used (Randall et al., 1958; Lummus and Randall, 1961; Dunning et al., 1959; Behrens et al., 1962a,b). The need for standardization of methods was evident. The API Mid-Continent District Committee for Air and Gas Drilling recommended test procedures that involved brine, freshwater, and fresh and saltwater containing oil (Freeze, 1964). API Recommended Practice 46 was issued in November 1966.

Aerated Mud

Another approach toward avoiding loss of circulation through reduced-pressure drilling was used by Phillips Petroleum Co. in Emory County, Utah, in May 1953 (Bobo and Barrett, 1953). In the initial test, air from a small compressor was injected into the mud stream between two mud pumps connected in series. Although circulation was maintained while drilling to 3300 ft (1000 m), the method of air injection was inefficient and, in subsequent studies in West Texas, air from a three-stage compressor was injected directly into the standpipe (Bobo et al., 1955). A special check valve placed in the drill string one joint below the kelly avoided the problem of mud spray when making connections. Drill pipe corrosion was severe in early tests of aerated mud, but by maintaining the pH of the mud above 10, corrosion was reduced. Maintaining saturation with lime minimized corrosion while drilling competent formations with water. While using the aerated mud system, water influx or mud loss often could be controlled by adjusting the volume of injected air. As the density of the mixture increased, however, the drilling rate decreased, and loss of circulation again became a problem.

Aeration of mud downhole by injecting air into the annulus between the casing and the drill pipe was the method used to avoid lost circulation in the Upper Valley field in Utah (Murray, 1968). Earlier wells had shown the static water level after loss of circulation to be about 1000 ft (300 m), while depth of the major loss was about 3000 ft (900 m). *Parasitic tubing* (tubing attached to the outside of the casing) was run to the calculated point of injection. Substantial savings in well costs resulted from this method of reduced-pressure drilling.

The introduction of dual drill strings and dual swivels supplied another method of reduced-pressure drilling (Binkley, 1968; Bobo, 1968). In this system, air is forced down either the inside or the annulus of the dual drill string to the calculated depth, where it is injected into the outer annulus. The pipe annulus that does not convey air carries mud to the bit. As with the parasitic tubing method, the air mud mixture exists only in the annulus above the injection point (see Fig. C.11).

Gel Foam or Stiff Foam

The introduction of gel foam, or stiff foam, was a notable advance in the technique of foam drilling. This form of reduced-pressure drilling contributed largely to the solution of lost circulation and hole cleaning problems at the U.S. Atomic Energy Commission's Nevada Test Site. After efforts to establish circulation by the usual methods had failed, air and foam were tried in 1962, but removal of cuttings from the large-diameter holes (64 inches, 163 cm) was extremely troublesome. In 1963, a drilling fluid was developed that, along with some changes in drilling practices, greatly reduced costs of the big holes (Crews, 1964; Quinn, 1967; Schneider, 1967). At a central mixing plant, a slurry was prepared consisting of (by weight) 98% water, 0.3% soda ash, 3.5% bentonite, and 0.17% guar gum. At the drill site, 1% by volume of foaming agent was added to the slurry. The injection rates of air and slurry

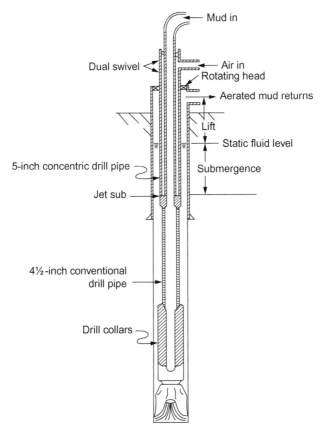

FIGURE C.11 Concentric drill pipe/air lift method for reduced-pressure drilling. *From Binkley, J.F., 1968. Concentric drill pipe/air lift — a method for combatting lost circulation. API Drill. Prod. Prac., 17–21. Courtesy of API.*

were carefully controlled to maintain returns of a foam having a consistency similar to shaving cream. With the gel foam, rising velocities as low as 100 ft per min (0.5 m/s) were adequate in drilling holes 64 inches (163 cm) in diameter. Hole stability in caving zones was improved by the gel foam. This feature of gel foam has proven especially valuable. Other polymers have been substituted for guar gum, and have also replaced bentonite in some applications.

Preformed Stable Foam

A further contribution to reduced-pressure drilling was the development of a foam generating unit by Standard Oil Co. of California, around 1965. In this device, the metered gaseous and liquid phases are mixed at the surface, and the preformed foam is introduced into the drill pipe (Anderson et al., 1966). The diagram in Fig. C.12 illustrates the equipment employed. The

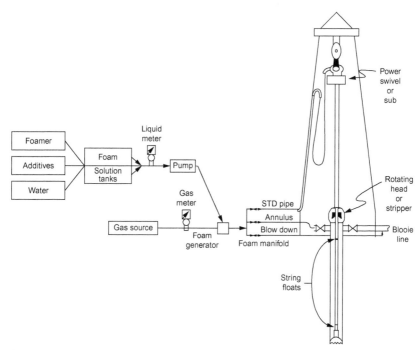

FIGURE C.12 Preformed stable foam flow diagram. *From Anderson, G.W., 1976. Foam drilling techniques and uses. In: AAODC Rotary Drilling Conf., Dallas, March 9–12, 1976. Courtesy of AAODC.*

compositions of the foaming agent and of the polymer (or polymers) can be selected to satisfy the conditions for application of the foam. For example, the composition needed for a well cleanout involving some oil and brine might differ from that employed in drilling shale. Similarly, composition of the gas might depend on availability, convenience, and cost of supply.

Optimum application of the technique requires careful planning. A mathematical model was designed and a computer program was formulated to aid in the selection of facilities for any given application (Milhone et al., 1972). In a survey of preformed foam applications (Anderson, 1976), several publications dealing with numerous uses in worldwide applications were cited.

Flow Properties of Foam

Measurement of the flow properties of foam in oil field applications began to receive attention in connection with the foam-drive process for increasing recovery of oil. In the early 1960s, viscosity of foam was measured in a modified Fann viscometer (Fried, 1961). Later, measurements were made in small diameter tubes (Marsden and Khan, 1966; Raza and Marsden, 1967; Mitchell, 1971). The major factor affecting flow behavior was found to be *foam quality*, the ratio of gas volume to total foam volume at a specified temperature and pressure. Apparent viscosity increased rapidly as foam quality increased from about 0.85 to 0.96, the limit of foam stability at the mist condition. Based on the behavior of foam as a Bingham plastic, charts were prepared for common drill pipe and hole sizes to allow estimations of air-volume and water-volume rates, and injection pressures, that minimize hydraulic horsepower (Krug and Mitchell, 1972).

Composition of foam at any temperature and pressure can be expressed also as *liquid volume fraction* (e.g., LVF = 1−quality). The particle-lifting ability of foam increases as liquid volume fraction decreases. From data obtained in pilot-scale experiments, Beyer et al. (1972) derived equations for the flow of foam in circular pipes. Two velocity components were involved: slippage at the pipe wall and fluidity based on behavior of the foam as a Bingham fluid. From the mathematical model, a computer program was prepared for effective field performance of stable foam (Hook et al., 1977).

Gas Drilling Benefits

From the controlled influx of natural gas to the injection of foam containing various functional additives, the primary objectives of reduced-pressure drilling have been to avoid loss of circulation and damage to productive formations. Other benefits have often been derived, however, such as faster drilling, improved bit performance, and ready detection of hydrocarbons. A survey of practices was made by Hook et al. (1977), and may be regarded as a summary of the state of the art at that time.

REFERENCES

Report of special sub-committee on corrosion fatigue of drill pipe. Proc. API Prod. Bull. 224, 109−135.

Recommended Practice for Testing Foaming Agents for Mist Drilling. API RP 46, first ed. API Division of Production, Dallas.

API Bulletin on Drilling Fluids Processing Equipment. API Bull. 13C, first ed. API Division of Production, Dallas.

Standard Procedure for Testing Drilling Fluids. API RP13B, seventh ed. API Division of Production, Dallas.

Alexander, W.A., 1944. Oil base drilling fluids often boost production. Oil Weekly. 36−40.

Ambrose, H.A., Loomis, A.G., 1931. Some colloidal properties of bentonite suspensions. Physics. 129−136.

Anderson, D.B., Oates, K.W., 1974. Use and maintenance of nondispersed weighted mud. Drill. Contract 59, 34−38.

Anderson, E.T., 1961. How world's hottest hole was drilled. Petrol. Eng. 47–51.

Anderson, F.M., 1947a. Oil-base drilling fluid. Oil Weekly, 43–50.

Anderson, F.M., 1947b. Oil base drilling mud. U.S. Patent No. 2,430,039 (Nov. 4).

Anderson, G.W., 1976. Foam drilling techniques and uses. In: AAODC Rotary Drilling Conf., Dallas, March 9–12, 1976.

Anderson, G.W., Harrison, T.F., Hutchison, S.O., 1966. The use of stable foam circulating fluids. Drill. Contract. 44–50.

Angel, R.R., 1957. Volume requirements for air and gas drilling. Trans. AIME 210, 325–330.

Angel, R.R., 1958. Volume Requirements for Air-Gas Drilling. Gulf Publishing Co., Houston.

Anon, 1931. Determining the comparative values of clays for use in drilling muds. Drilling Mud. 1 (1).

Barnes, W.E., 1949. Drilling mud. U.S. Patent No. 2,491,436 (Dec. 13).

Battle, J.L., 1957. Casing corrosion in the petroleum industry. Corrosion. 62–68.

Battle, J.L., Chaney, P.E., 1950. Lime-base muds. API Drill. Prod. Prac. 99–111.

Beart, R., 1845. Apparatus for Boring in the Earth and in Stone. England, Patent No. 10,258.

Becker, F.L., Goodwin, R.J., 1959. The use of silicon tetraflouride gas as a formation plugging agent. Trans. AIME 216, 168.

Beckman, F.G., 1938. Drilling with reverse oil circulation in wider use. Oil Weekly. 19–26.

Behrens, R.W., Holman, W.E., Cizek, A., 1962a. Technique for evaluation of corrosion of drilling fluids. In: API Paper 906–7-G, Southwestern Dist. Meeting, Odessa, March 21–23, 1962.

Behrens, R.W., Rice, H.L., Becker, K.W., 1962b. Laboratory method for screening foaming agents for air/gas drilling operations. In: SPE Paper 429, Annual Meeting, Los Angeles, Oct. 7–10, 1962.

Berry, O.H., 1951. Air drilling the sprayberry sand. World Oil. 169–172.

Beyer, A.H., Milhone, R.S., Foote, R.W., 1972. Flow behavior of foam as a well circulating fluid. In: SPE Paper 3986, Annual Meeting, San Antonio, Oct. 8–11, 1972.

Binkley, J.F., 1968. Concentric drill pipe/air lift – a method for combatting lost circulation. API Drill. Prod. Prac. 17–21.

Black, H.N., Hower, W.F., 1965. Advantageous use of potassium chloride water for fracturing water-sensitive formations. API Drill. Prod. Prac. 113–118.

Bobo, R.A., 1968. Aerated mud drilling innovations. Oil Gas J., Part 1: (April 1968), 64–69; Part II: (May 1968), 114–118.

Bobo, R.A., Barrett, H.M., 1953. Aeration of drilling fluids. World Oil. 145–149.

Bobo, R.A., Hoch, R.S., 1954. Mechanical treatment of weighted drilling muds. Trans. AIME 201, 93–96.

Bobo, R.A., Ormsby, G.S., Hoch, R.S., 1955. Phillips tests air-mud drilling. Oil Gas J. Part I: (Jan. 1955), 82–87; Part II: (Feb. 1955), 104–108.

Bradley, B.W., 1967. Oxygen – a major element in drill pipe corrosion. Mater. Protect. 40–43.

Bradley, B.W., 1970. Oxygen cause of drill pipe corrosion. Petrol. Eng. 50–57.

Brandt, G.W., Weintritt, D.J., Gray, G.R., 1960. An improved water-in-oil emulsion mud. J. Petrol. Technol. 14–17.

Brandt, L.K., 1973. Fine-screen shakers improve solids removal from mud systems. Oil Gas J. 76–78.

Brantly, J.E., 1961. In: Carter, D.V. (Ed.), History of Petroleum Engineering. Boyd Printing Co., Dallas, pp. 277–278.

Appendix C: The Development of Drilling Fluids Technology **685**

Brantly, J.E., 1971a. History of Oil Well Drilling. Gulf Publishing Co, Houston, pp. 3, 38, 39.
Brantly, J.E., 1971b. History of Oil Well Drilling. Gulf Publishing Co, Houston, pp. 41−47.
Bright, K., 1964. External corrosion retarded with protective coating. Drilling. 80.
Browning, W.C., Perricone, A.C., 1963. Clay chemistry and drilling fluids. In: SPE Paper 540, First University of Texas Conference on Drilling and Rock Mechanics, Austin, Jan. 23−24, 1963.
Buffington, E.C., Horner, V.V., 1950. Drilling mud problems in western Canada. World Oil 203−206.
Burdyn, R.F., 1965. A new device for field recovery of barite from drilling mud: 1. Theory and laboratory results. Soc. Petrol. Eng. J., 6−14, Trans. AIME 234.
Burdyn, R.F., Hawk, D.E., Patchen, F.D., 1965. A new device for field recovery of barite: II. Scale-up and design. Soc. Petrol. Eng. J., 100−108, Trans. AIME 234.
Burdyn, R.F., Nelson, M.D., 1968. Separator refines deep well mud control. Petrol. Eng. 68−72.
Burdyn, R.F., Wiener, L.D., 1956. That new drilling fluid for hot holes. Oil Gas J. 104−107.
Burdyn, R.F., Wiener, L.D., 1957. Calcium surfactant drilling fluids. In: Paper 719−G, Petrol. Branch AIME, Los Angeles Meeting, Oct. 14−17, 1956; World Oil (November), pp. 101−108.
Bush, H.E., 1973. Controlling corrosion in petroleum drilling and in packer fluids. In: Nathan, C.C. (Ed.), Corrosion Inhibition. National Association of Corrosion Engineers, Houston, pp. 102−113.
Bush, H.E., Cowan, J.C., 1966. Corrosion inhibitors in oilfield fluids. Mater. Protect. 25−31.
Cagle, W.S., Wilder, L.B., 1978. Layered shale shaker screens improve solids control. World Oil. 89−94.
Cannon, G.E., 1934. Changes in hydrostatic pressure due to withdrawing drillpipe from the hole. API Drill. Prod. Prac. 42−47.
Cannon, G.E., 1938. Drilling fluid for combatting heaving shale. U.S. Patent No. 2, 109, 858 (March 1).
Cannon, G.E., 1947. Developments in drilling-mud control. Oil Gas J. 101−103.
Cannon, G.E., Sullins, R.S., 1946. Problems encountered in drilling abnormal pressure formations. API Drill. Prod. Prac. 29−33.
Cannon, G.E., Watson, R.A., 1956. Review of air and gas drilling. J. Petral. Technol. 15−19.
Cannon, G.E., Williams, M., 1941. Weighted nonaqueous drilling fluid. U.S. Patent No. 2, 239, 498 (April 22).
Carico, R.D., 1976. Suspension properties of polymer fluids used in drilling, workover, and completion operations. In: SPE Paper 5870, SPE of AIME Regional Meeting, Long Beach, April 8−9, 1976.
Cartwright, R.S., 1928. Rotary drilling problems. Trans. AIME 82, 9−29.
Chapman, M.T., 1890. U.S. Patent Records, U.S. Patent No. 443, 069 (Dec. 16).
Chenevert, M.E., 1970. Shale control with balanced-activity oil-continuous muds. J. Petrol. Technol. 1309−1316.
Chisholm, F., Ruhe, R.W., Murchison, W.J., 1961. Good mud practice is not by accident. Drill. Contract. 56−64.
Clark, J.A., 1965. Experience with plastic coated drill pipe in west Texas. Drill. Contract 44−46.
Clark, R.K., Scheuerman, R.F., Rath, H., van Laar, H.G., 1976. Polyacrylamide/potassium chloride mud for drilling water-sensitive shales. J. Petrol. Technol., 719−727, Trans. AIME 261.
Coffer, H.F., Clark, R.C., 1954. An inexpensive mud for deep wells. J. Petrol. Tech. 10−14.

Collings, B.J., Griffin, R.R., 1960. Clay-free salt-water muds save rig time and bits. Oil Gas J. 115–117.

Collom, R.E., 1923. The use of mud fluid to prevent water infiltration in oil and gas wells. Summ. Oper. Calif. Oil Fields 8 (7), 26–84.

Collom, R.E., 1924. The mud problem in rotary drilling. Part 1: Oil Weekly, 37–50, Part 2: Ibid. 45–50.

Cowan, J.C., 1959. Low filtrate loss and good rheology retention at high temperatures are practical features of this new drilling mud. Oil Gas J. 83–87.

Cox, T.E., 1974. Even traces of oxygen in muds can cause corrosion. World Oil. 110–112.

Crews, S.H., 1964. Big hole drilling progress keyed to engineering. Petrol. Eng. 104–114.

Cross, R., Cross, M.F., 1937. Method of improving oil-well drilling muds. U.S. Patent No. 2, 094, 316 (Sept. 28).

Dawson, R., Annis, M.R., 1977. Total mechanical solids control. Oil Gas J. 90–100.

Dawson, R.D., 1950. Oil base drilling fluid. U.S. Patent No. 2,497,398 (Feb. 14).

Dawson, R.D., 1952. Oil base fluid for drilling wells. U.S. Patent No. 2, 588, 808 (March 11).

Dawson, R.D., Blankenhorn, C.F., 1944. Nonaqueous drilling fluid. U.S. Patent No. 2, 350, 154 (May 30).

Dawson, R.D., Huisman, P.H., 1940. Nonaqueous drilling fluid. U.S. Patent No. 2,223,027 (Nov. 26).

Deily, F.H., Holman, W.E., Lindblom, G.P., Patton, J.T., 1967. New biopolymer low-solids mud speeds drilling operation. Oil Gas J. 62–70.

Doherty, W.T., Gill, S., Parsons, C.P., 1931. Drilling fluid problems and treatment in the Gulf Coast. Proc. API. Prod. Bull 207, 100–109.

Doig, K., Wachter, A., 1951. Bacterial casing corrosion in the ventura avenue field. Corrosion. 212–214.

Duckham, A., 1931. Chemical aspect of drilling muds. J. Inst. Petrol. Tech. 17, 153–182.

Dunning, H.N., Eakin, J.L., Reinhardt, W.N., Walker, C.J., 1959. Foaming agents: cure for water-logged wells. Petrol. Eng. B28–B33.

Eckel, J.R., 1967. Microbit studies of the effect of fluid properties and hydraulics on drilling rate. J. Petrol. Technol., 541–546, Trans. AIME 240.

Edinger, W.M., 1949. Interpretation of analysis results on oil or oil-base mud cores. World Oil. 145–150.

Evans, P., Reid, A., 1936. Drilling mud: its manufacture and testing. Trans. Min. Geol. Inst. Ind. 32.

Farnham, H.H., 1931. Analytical considerations of drilling muds. Petrol Eng., 117–122.

Farnham, H.H., 1932. Discussion of sealing effect of rotary mud on productive sands. Proc. API. Prod. Bull. 209, 60–63.

Fauvelle, M., 1846. A new method of boring for artesian springs. J. Franklin Inst. 12 (3), 369–371.

Feldenheimer, W., 1922. Treatment of clay. U.S. Patent No. 1, 438, 588 (Dec.12).

Fischer, P.W., 1951a. Drilling fluids. U.S. Patent No. 2, 542, 019 (Feb. 20).

Fischer, P.W., 1951b. Drilling fluid compositions. U.S. Patent No. 2, 542, 020 (Feb. 20).

Fischer, P.W., 1951c. Drilling fluids. U.S. Patent No. 2, 573, 959 (Nov. 6).

Fischer, P.W., 1951d. Drilling fluid concentrates. U.S. Patent No. 2, 573, 960 (Nov. 6).

Fischer, P.W., 1952a. Oil base drilling fluids. U.S. Patent No. 2, 607, 731 (August 19).

Fischer, P.W., 1952b. Oil base drilling fluids. U.S. Patent No. 2, 612, 471 (Sept. 30).

Fontenot, J.E., Simpson, J.P., 1974. A microbit investigation of the potential for improving the drilling rate of oil-base muds in low-permeability rocks. J. Petrol. Technol. 507–514.

Foran, E.V., 1934. Pressure completion of wells in west Texas. API Drill. Prod Prac. 48−54.
Freeze, G.I., 1964. Testing foaming agents for air and gas drilling. In: API Paper 851-38 M. Mid- Continent Dist. Meeting, Hot Springs, May 11−13, 1964.
Fried, A.N., 1961. The foam-drive process for increasing the recovery of oil. U.S. Bur. Mines Report of Investigations 5866.
Gallus, J.P., Lummus, J.L., Fox Jr., J.E., 1958. Use of chemicals to maintain clear water for drilling. Trans. AIME 213, 70−75.
Garrett, R.L., Clark, R.K., Carney, L.L., Grantham, C.K., 1978. Chemical seavengers for sulfides in water-base drilling fluids. In: SPE Paper 7499. SPE Annual Meeting, Houston, Oct. 1−3, 1978.
Gates, J.I., Pfenning, G.E., 1952. New oil-base emulsion mud hailed as vast improvement. Oil Gas J. 166−168.
Gates, J.I., Wallis, W.M., 1951. Emulsion fluid for drilling wells. U.S. Patent No. 2, 557, 647 (June 19).
Gill, J.A., 1966. Drilling mud solids control − a new look at techniques. World Oil. 121−124.
Gill, J.A., Carnicom, W.M., 1959. Offshore Louisiana new muds and techniques improve drilling practices. World Oil. 194−206.
Gill, S., 1932. Sealing effect of rotary mud on productive sands in the southwestern district. Proc. API. Prod. Bull. 209, 42−51.
Gnnsfelder, S., Law, J., 1938. Recent pressure drilling at Dominguez. API Drill. Prod. Prac. 74−79.
Goins Jr., W.C., 1950. A study of limed mud systems. Oil Gas J., 52−54, 72.
Goins Jr., W.C., Magner, H.J., 1961. How to use foaming agents in air and gas drilling. World Oil. 59−64.
Goodwin, R.J., 1959. A water shutoff method for sand-type porosity in air drilling. Trans. AIME 216, 163−167.
Grant, R.S., Texter, H.G., 1941. Causes and prevention of drill pipe and tool joint troubles. API Drill. Prod. Prac. 9−43.
Gray, G.R., 1944. The use of modified starch in Gulf Coast drilling muds. In: Preprint Petrol. Div. AIME, Houston Meeting (May).
Gray, G.R., Foster, J.L., Chapman, T.S., 1942. Control of filtration characteristics of salt water muds. Trans. AIME 146, 117−125.
Gray, G.R., Grioni, S., 1969. Varied applications of invert emulsion muds. J. Petrol. Technol. 261−266.
Gray, G.R., Kellogg, W.C., 1955. The Wilcox trend−cross section of typical mud problems. World Oil. 102−117.
Gray, G.R., Neznayko, M., Gilkeson, P.W., 1952. Some factors affecting the solidification of lime treated muds at high temperatures. API Drill Prod. Prac. 72−81.
Gray, G.R., Tschirley, N.K., 1975. Drilling fluids programs for ultra-deep wells in the United States. Proc. Ninth World Petrol. Congress 4. Applied Science Publishers, Barking, pp. 137−149.
Harth, P.E., 1935. Application of mud-laden fluids to oil or gas wells. U.S. Patent No. 1, 991, 637 (Feb.19).
Hauser, E.A., 1950a. Application of drilling fluids. U.S. Patent No. 2, 531, 812 (Nov. 28).
Hauser, E.A., 1950b. Modified gel-forming clay and process of producing same. U.S. Patent No. 2, 531, 427 (Nov. 28).
Hayes, C.W., Kennedy, W., 1903. Oil fields of the Texas Louisiana Gulf coastal plain. U.S. Geol. Survey Bull. 212, 167.

Heggem, A.G., Pollard, J.A., 1914. Drilling wells in Oklahoma by the mud-laden fluid method. U.S. Bur. Mines Tech. Paper 68.

Herrick, H.N., 1932. Flow of drilling mud. Trans. AIME 98, 476–494.

Hindry, H.W., 1941. Characteristics and application of an oil-base mud. Trans. AIME 142, 70–75.

Hoeppel, R.W., 1956. Oil base drilling fluids. U.S. Patent No. 2, 754, 265 (July 10).

Hoeppel, R.W., 1961. Water in oil emulsion drilling and fracturing fluid. U.S. Patent No. 2, 999, 063 (Sept. 5).

Hollis, W.T., 1953. Gas drilling in San Juan Basin of New Mexico. API Drill. Prod. Prac. 310–312.

Hook, R.A., Cooper, L.W., Payne, B.R., 1977. Air, gas and foam drilling techniques. In: Trans. IAODC Drilling Technol. Conf., New Orleans, March 16–18, 1977; World Oil, Part I: (April), 95–106, Part II: Ibid. (May), 83–90.

Hower, W.F., McLaughlin, C., Ramos, J., 1958. Water shutoff in air drilling now possible with polymeric water gel. API Drill. Prod. Prac. 110–114.

Hudgins, C.M., Landers, J.E., Greathouse, W.D., 1960. Corrosion problems in the use of dense salt solutions as packer fluids. Corrosion. 91–94.

Hudgins, C.M., McGlasson, R.L., Gould, E.D., 1961. Developments in the use of dense brines as packer fluids. API Drill. Prod. Prac. 160–168.

Huebotter, E.E., 1954. Internal reports, Baroid Division, NL Industries, Inc. (Oct. 21, 1954, et seq.).

Hull, J.D., 1968. Minimum solids fluids reduce drilling costs. World Oil. 86–91.

Hurdle, J.M., 1957. Gyp muds now practical for Louisiana coastal drilling. Oil Gas J. 93–95.

Ives, G., 1974. How Shell drilled a super-hot, super-deep well. Petrol. Eng. 70–78.

Jackson, L.R., Banta, H.M., McMaster, R.C., Nordin, T.P., 1947. Use of chromate additions in drilling fluids. Drill. Contract. 77–82.

Jensen, J., 1936. Recent developments related to petroleum engineering. Trans. AIME 118, 63–68.

Jones Jr., F.O., 1964. New fast, accurate test measures bentonite in drilling mud. Oil Gas J. 76–78.

Jones, P.H., 1937. Field control of drilling mud. API Dril. Prod. Prac. 24–29.

Jones, P.H., Babson, E.C., 1935. Evaluation of rotary drilling muds. API Drill. Prod. Prac. 23–33.

Jordan, J.W., 1949. Organophilic bentonites, I. J. Phys. Chem. 53, 294–306.

Jordan, J.W., Hook, B.J., Finlayson, C.M., 1950. Organophilic bentonites, II. J. Phys. Chem. 54, 1196–1208.

Jordan, J.W., Nevins, M.J., Stearns, R.C., Cowan, J.C., Beasley Jr., A.E., 1965. Well-working fluids. U.S. Patent No. 3, 168, 475 (Feb. 2).

Jordan, J.W., Nevins, M.J., Stearns, R.O., Cowan, J.C., Beasley Jr., A.E., 1966. N-Alkyl ammonium humates. U.S. Patent No. 3, 281, 458 (Oct. 10).

Kastrop, J.E., 1947. Desanding drilling fluids. World Oil. 30–31.

Kaveler, H.H., 1946. Improved drilling muds containing carboxymethylcellulose. API Drill. Prod. Prac. 43–50.

Kennedy, J.L., 1974. Oil mud underbalancing saves money. Oil Gas J., 49–52.

Kersten, G.V., 1946. Results and use of oil base fluids in drilling and completing wells. API Drill. Prod. Prac., 61–68.

King, E.G., 1976. In personal communication to G.R. Gray (Dec. 15).

Appendix C: The Development of Drilling Fluids Technology **689**

King, E.G., Adolphson, C., 1960. Drilling fluid and process. U.S. Patent No. 2, 935, 473 (May 3).

King, G.R., 1959. Why rock-bit bearings fail. Oil Gas J. 166–182.

Kirk, W.L., 1972. Deep drilling practices in Mississippi. J. Petrol. Technol. 633–642.

Kirwan, M.J., 1924. Mud fluid in drilling and protection of wells. Oil Weekly. 34.

Klementich, E.F., 1972. Drilling a record Mississippi wildcat. In: SPE Paper 3916, SPE Annual Meeting, San Antonio, Oct, 8–11, 1972.

Kljucec, N.M., Yurkowski, K.J., Lipsett, L.R., 1974. Successful drilling of permafrost with bentonite-XC polymer–KCl mud system. J. Can. Petrol. Technol. 49–53.

Knapp, A., 1923. Action of mud-laden fluids in wells. Trans. AIME 69, 1076–1100.

Knapp, I.N., 1916. The use of mud-laden water in drilling wells. Trans. AIME 51, 571–586.

Krug, J.A., Mitchell, B.J., 1972. Charts help find volume, pressure needed for foam drilling. Oil Gas J. 61–64.

Lancaster, E.H., Mitchell, M.E., 1949. Control of conventional and lime-treated muds in southwest Texas. Trans. AIME 179, 357–371.

Lanman, D.E., Willingham, R.W., 1970. Low solids, non-dispersed muds solve hole problems. World Oil. 49–52.

Lawton, H.C., Ambrose, H.A., Loomis, A.G., 1932. Chemical treatment of rotary drilling fluids. Physics. 365–375.

Lewis, J.O., McMurray, W.F., 1916. The use of mud-laden fluid in oil and gas wells. U.S. Bur. Mines Bull. 134.

Leyendecker, E.A., Murray, S.C., 1975. Properly prepared oil muds aid massive salt drilling. World Oil. 93–95.

Loomis, A.G., Ambrose, H.A., Brown, J.S., 1931. Drilling of terrestial bores. U.S. Patent No. 1, 819, 646 (August 18).

Lummus, J.L., 1953. Water-in-oil emulsion drilling fluid. U.S. Patent No. 2, 661, 334 (Dec. 1).

Lummus, J.L., 1954. Multipurpose water-in-oil emulsion mud. Oil Gas. J. 106–108.

Lummus, J.L., 1957. Oil base drilling fluid. U.S. Patent No. 2, 793, 996 (May 28).

Lummus, J.L., 1965. Chemical removal of drilled solids. Drill. Contract. 50–54, 67.

Lummus, J.L., Barrett, H.M., Allen, H., 1953. The effects of use of oil in drilling muds. API Drill. Prod. Prac. 135–145.

Lummus, J.L., Field, L.J., 1968. Non-dispersed polymer mud – a new drilling concept. Petrol Eng. 59–65.

Lummus, J.L., Fox Jr., J.E., Anderson, D.B., 1961. New low-solids polymer mud cuts drilling costs for Pan American. Oil Gas J. 87–91.

Lummus, J.L., Randall, B.V., 1961. How new foaming agents are aiding air gas drilling. World Oil. 57–62.

Mallory, H.E., 1957. How low solid muds can cut drilling costs. Petrol. Eng. B21–B24.

Mallory, H.E., Holman, W.E., Duran, R.J., 1960. Low-solids mud resists contamination. Petrol Eng. B25–B30.

Marsden, S.S., Khan, S.A., 1966. The flow of foam through short porous media and apparent viscosity measurements. Soc. Petrol. Eng. J. 17–25, Trans. AIME, 237.

Marsh, H.N., 1931. Properties and treatment of rotary mud. Trans. AIME 92, 234–251.

Mazee, W.M., 1942. Nonaqueous drilling fluid. U.S. Patent No. 2, 297, 660 (Sept. 29).

McCray, A.W., 1949. Chemistry and control of lime base drilling muds. Petral Eng. B54–B56.

McGhee, E., 1956. New oil emulsion speeds west Texas drilling. Oil Gas J. 110–112.

McGlasson, R.L., Greathouse, W.D., 1959. Stress corrosion cracking of oil country tubular goods. Corrosion. 55–60.

McGlasson, R.L., Greathouse, W.D., Hudgins, C.M., 1960. Stress corroston cracking of carbon steels in concentrated sodium nitrate solutions. Corrosion. 113–118.

McMordie Jr., W.C., 1968. Where and why to use oil-base packer fluids. Oil Gas J. 57–59.

Milhone, R.S., Haskin, C.A., Beyer, A.H., 1972. Factors affecting foam circulation in oil wells. In: SPE Paper 4001, Annual Meeting, San Antonio, Oct. 8–11, 1972.

Miller, G., 1942. New oil base drilling fluid facilitates well completion. Petrol Eng, 104–106.

Miller, G., 1943. Oil base drilling fluid. U.S. Patent No. 2, 316, 968 (August 20).

Miller, G., 1949a. Oil base drilling fluid and mixing oil for the same. U.S. Patent No. 2, 475, 713 (July 12).

Miller, G., 1949b. Use of oil base mud to free stuck pipe. Petrol. Eng. B54–B64.

Miller, G., 1951. Oil base drilling fluids. Proc. Third World Petrol, Congress Sec. II. 2. E.J. Brill, Leiden, pp. 321–350.

Mills, B., 1930. Central mud cleaning plant proves value of elaborate reclaiming methods. Oil Weekly. 56–58, 192.

Mitchell, B.J., 1971. Test data fill theory gap on using foam as a drilling fluid. Oil Gas J. 96–100.

Mondshine, T.C., 1966. New fast-drilling muds also provide hole stability. Oil Gas J., 84–99.

Mondshine, T.C., 1969. New technique determines oil-mud salinity needs in shale drilling. Oil Gas J. 70–75.

Mondshine, T.C., 1974. Method of producing and using a gelled oil base packer fluid. U.S. Patent No. 3, 831, 678 (August 27).

Mondshine, T.C., Kercheville, J.D., 1966. Shale dehydration studies point way to successful gumbo shale drilling. Oil Gas J. 194–205.

Moore, T.V., 1940. Oil base drilling fluid and method of preparing same. U.S. Patent No. 2, 216, 955 (Oct. 8).

Moore, T.V., Cannon, G.E., 1936. Weighted oil base fluid. U.S. Patent No. 2, 055, 666 (Sept. 29).

Murray, A.S., Eckel, J.E., 1961. Foam agents and foam drilling. Oil Gas J. 125–129.

Murray, A.S., MacKay, S.P., 1957. Imperial tries air drilling. Can. Oil Gas Indust. 49–54.

Murray, J.W., 1968. Parasite tubing method of acration. API Drill. Prod. Prac. 22–28.

Newlin, F., Kastrop, J.E., 1960. World's deepest ultra-slim hole is ultra hot. Petrol. Eng. B19–B26.

Nicolson, K.M., 1953. Air drilling in California. API Drill. Prod. Prac. 300–309.

Nowak, T.J., Krueger, R.F., 1951. The effect of mud filtrates and mud particles upon the permeabilities of cores. API Drill. Prod. Prac. 164–181.

O'Brien, D.E., Chenevert, M.E., 1973. Stabilizing sensitive shales with inhibited potassium-based drilling fluids. J. Petrol. Technol. 1089–1100, Trans. AIME 255.

O'Brien, T.B., Stinson, J.P., Brownson, F., 1977. Relaxed fluid loss controls on invert muds increases ROP. World Oil. 31–34, 70.

Ockenda, M.A., Carter, A., 1920. Plant used in rotary system of drilling wells. J. Inst. Petrol 6, 249–280.

Ormsby, G.S., 1965. Desilting drilling muds with hydroclones. Drill. Contract. 55–65.

Ormsby, G.S., 1973. Proper rigging boosts efficiency of solids-removing equipment. Part 1: Oil Gas J., (March 12), 120–132; Part 2: Ibid. (March 19), 59–65.

Paris, B., Williams, C.O., Wacker, R.B., 1961. New chrome-lignite drilling mud. Oil Gas J. 86–88.

Park, A., Scott Jr., P.P., Lummus, J.L., 1960. Maintaining low-solids drilling fluids. Oil Gas J. 81−84.
Parker, C.A., 1973. Geopressures in the deep smackover of Mississippi. J. Petrol. Technol. 971−979.
Parsons, C.P., 1931. Characteristics of drilling fluids. Trans. AIME 92, 227−233.
Parsons, C.P., 1932. Sealing effect of rotary mud on productive sands in the mid-continent district. Proc. API. Prod. Bull. 209, 52−58.
Pennington, J.W., 1949. The history of drilling technology and its prospects. Proc. API. Sect. IV, Prod. Bull. 235, 481.
Perkins, H.W., 1951. Oil emulsion drilling fluids. API Drill. Prod. Prac. 349−354.
Planka, J.H., 1972. New bromide packer fluid cuts corrosion problems. World Oil. 88−89.
Pollard, J.A., Heggem, A.G., 1913. Mud-laden fluid applied to well drilling. U.S. Bur. Mines Tech. Paper 66.
Pope, P.L., Mesaros, J., 1959. Mud programs for deep wells in Pecos county, Texas. API Drill. Prod. Prac. 82−99.
Quinn, R.V., 1967. They call it Davis mix. Drilling 10 (11), 64.
Radd, F.J., Crowder, L.H., Wolfe, L.H., 1960. The effect of pH in the range 6.6−14 on the aerobic corrosion fatigue of steel. Corrosion 6, 121−124.
Randall, B.V., Lummus, J.L., Vincent, R.P., 1958. Combating wet formations while drilling with air or gas. Drill. Contract. 69−79.
Raza, S.H., Marsden, S.S., 1967. The streaming potential and rheology of foam. Soc. Petrol. Eng. J., 359−368, Trans. AIME 240.
Reddie, W.A., 1958. Water in oil emulsion. U.S. Patent No. 2, 862, 881 (Dec. 2).
Reddie, W.A., Griffin, R.N., 1961. Water in oil emulsion drilling fluid. U.S. Patent No. 2, 994, 660 (August 1).
Reid, C.A., 1970. Here are drilling fluids being used in Permian Basin's pressured formations. Oil Gas J. 80−83.
Reistle Jr., C.E., Cannon, G.E., Buchan, R.C., 1937. Standard practices for field testing of drilling fluids. Proc. API. Prod. Bull. 219, 14−22.
Remont, L.J., Nevins, M.J., 1976. Arctic casing pack. Drilling. 43−45.
Robinson, L.H., Heilhecker, J.K., 1975. Solids control in weighted drilling fluids. J. Petrol. Technol. 1141−1144.
Rogers, W.F., 1953. Composition and Properties of Oil Well Drilling Fluids, second ed. Gulf Publishing Co, Houston.
Rogers, W.F., 1963c. Composition and Properties of Oil Well Drilling Fluids, third ed. Gulf Publishing Co, Houston.
Rolshausen, F.W., Bishkin, S.L., 1937. Oil hydratable drilling fluid. U.S. Patent No. 2, 099, 825 (Nov. 23).
Rubel, A.C., 1932. The effect of drilling mud on production in California. Proc. API. Prod. Bull. 209, 33−41.
Schilthuis, R.J., 1938. Connate water in oil and gas sands. Trans. AIME 127, 199−212.
Schneider, R.P., 1967. Method of, and composition for use in, gas drilling. U.S. Patent No. 3, 313, 362 (April 11).
Self, E.S., 1949. Oil base drilling fluid. U.S. Patent No. 2, 461, 483 (Feb. 8).
Shallenberger, L.K., 1953. What about compressed air? World Oil. 155−160.
Sheridan, H., 1965. Experience with plastic coated drill pipe on Louisiana Gulf Coast. Drill. Contract. 46−50.

Simpson, J.P., 1978. A new approach to oil muds for lower cost drilling. In: SPE Paper 7500, SPE Annual Meeting, Houston, Oct. 1–3, 1978.

Simpson, J.P., Andrews, R.S., 1966. Oil mud packs for combatting casing corrosion. Mater. Protect. 21–25.

Simpson, J.P., Barbee, R.D., 1967. How corrosive are water base completion muds? Mater. Protect. 32–36.

Simpson, J.P., Cowan, J.C., Beasly Jr., A.E., 1961. The new look in oil-mud technology. J. Petrol. Technol. 1177–1183.

Simpson, J.P., Sanchez, H.V., 1961. Inhibited drilling fluids – evaluation and utilization. In: Paper 801–37C, API Pacific Coast Dist. Meeting, Los Angeles, May 11–12.

Smith, B., 1974. New oil base mud system cuts drilling costs. World Oil. 75, 76, 85.

Smith, F.W., Rollins, H.M., 1956. Air drilling practices in the permian basin. Petrol. Eng., B48–B53.

Speller, F.N., 1935. Corrosion fatigue of drill pipe. API Drill. Prod. Prac. 239–247.

Stein, N., 1963. Use of tetrakis to shut off water in wells drilled with air or gas. API Drill. Prod. Prac. 7–11.

Stone, V.D., 1964. Low-silt mud increases gulf's drilling efficiency, cuts costs. Oil Gas J. 136–142.

Stroud, B.K., 1922. Mud-laden fluids and tables on specific gravities and collapsing pressures. In: Louisiana Dept. Conservation Tech. Paper No. 1.

Stroud, B.K., 1925. Use of barytes as a mud laden fluid. Oil Weekly June 5, 29–30.

Stroud, B.K., 1926. Application of mud-laden fluids to oil or gas wells. U.S. Patent No. 1, 575, 944 and No. 1, 575, 945 (March 9).

Stuart, R.W., 1946. Use of oil base mud at Elk Hills naval petroleum reserve number one. API Drill. Prod. Prac. 69–73.

Sufall, C.K., 1960. Water shutoff techniques in air or gas drilling. API Drill. Prod. Prac. 74–77.

Suman, J.R., 1961. In: Carter, D.V. (Ed.), History of Petroleum Engineering. Boyd Printing Co., Dallas, pp. 65–132.

Swan, J.C., 1923. Method of drilling wells. U.S. Patent No. 1, 455, 010 (May 15).

Sweeney, P., 1866. U.S. Patent Records, U.S. Patent No. 51, 902 (Jan. 2).

Tailleur, R.J., 1963. Lubricating properties of drilling fluids. In: Muds, E.P. (Ed.), Proc. Sixth World Petrol. Congress 2, Förderung 6 Welt-Erdol Kongresses, Hamburg. pp. 387–404.

Tailleur, R.J., 1967. Rotary drilling process. U.S. Patent No. 3, 318, 396 (May 9).

Teis, K.R., 1936. Pressure completion of wells in the fitts pool. API Drill. Prod. Prat. pp. 23–31.

Trimble, G.A., Nelson, M.D., 1960. Use of inverted emulsion mud proves successful in zones susceptible to water damage. J. Petrol. Technol. 23–30.

Trout, K., 1948. Some notes on use of calcium-base drilling fluids. Drill. Contract. 56, 57, 80.

Tschirley, N.K., 1963. New developments in drilling fluids. In: API Paper 906-8 A. Southwestern Dist. Meeting, Fort Worth. March 13–15, 1963.

Van Dyke, O.W., 1950. Oil emulsion drilling mud. World Oil. 101–106.

Van Dyke, O.W., Hermes Jr., L.M., 1950. Chemicals used in red-lime muds. Ind. Eng. Chem. 1901–1912.

Vietti, W.V., Garrison, A.D., 1939. Method of drilling wells. U.S. Patent No. 2, 165, 824 (July 11).

Wade, G., 1942. Review of the heaving shale problem in the Gulf coast region. U.S. Bur. Mines Report of Investigations 3618 (March).

Appendix C: The Development of Drilling Fluids Technology

Walkins, T.E., 1960. New inverted emulsion mud makes good drilling and completion fluid. Oil Gas J. 176–180.

Watkins, T.E., 1953. A drilling fluid for use in drilling high-temperature formations. API Drill. Prod. Prac. 7–13.

Watkins, T.E., 1958. Emulsion fluid for wells. U.S. Patent No. 2, 861, 042 (Nov. 18).

Watkins, T.E., 1959. Component for well treating fluid. U.S. Patent No. 2, 876, 197 (March 3).

Weichert, J.P., Van Dyke, O.W., 1950. Effect of oil emulsion mud on drilling. Petrol. Eng., B16–B38.

Weintritt, D.J., 1966. Stabilizied oil mud for deep hot wells. Petrol. Eng. 68–74.

Weiss, W.J., Graves, R.H., Hall, W.L., 1958. A fundamental approach to well bore stabilization. Petrol. Eng. B43–B60.

Wheless, N.H., Howe, J.L., 1953. Low solid muds improve rate of drilling. Cut Hole Time. Drilling. 70, 75.

Williams, R.W., Mesaros, J., 1957. The centrifuge and mud technology. API Drill Prod. Prac. 185–193.

Wright, C.C., 1954. Oil-base emulsion drilling fluids. Oil Gas J. 88–90.

Wuth, D.E., O'Shields, R.L., 1955. New mud desander cuts drilling costs. Drill. Contract. 76–81.

Author Index

Note: Page numbers followed by "*f*" and "*t*" refer to figures and tables, respectively.

A

Abrams, A., 262–264, 482, 483*f*
Adams, N., 375–379, 381–382, 505
Adolphson, C., 659
Al-Ansari, A., 27
Albers, D.C., 77–78
Alexander, W.A., 669
Alford, S.E., 77, 79*f*
Ali, A., 75
Allan, M.L., 547
Allen, T.O., 485–487, 499
Allred, R.B., 353
Almon, W.R., 469–470, 469*f*
Al-Muntasheri, G., 523
Alonso-Debolt, M.A., 550
Al-Otaibi, A.M., 462
Al-Yami, A.S., 4
Al-Yami, A.S.H.A.-B., 543
Ambrose, H.A., 654
Amer, A., 372
Amott, E., 464, 468, 502
Anderson, D.B., 193, 208–209, 663
Anderson, E.T., 429, 659
Anderson, F.M., 670
Anderson, G.W., 301, 303, 681–682
Anderson, R.L., 552
Andreson, B.A., 550–551
Andrews, R.S., 673
Angel, R.R., 678
Annis, M.R., 77–78, 137, 195, 197*f*, 198*f*, 199*f*, 377, 377*f*, 378*f*, 379–381, 663
Anon., 649
Arco, M.J., 543
Arfie, M., 600–601
Argillier, J.F., 550–551
Arias, R.E., 529
Arthur, J., 526*f*
Atterbury, T.H., 485–487

Audebert, R., 550
Ayers, R.C., 22–25
Azar, J.J., 205, 218–219, 222–226

B

Babbit, H.E., 190
Babson, E.C., 167–168, 167*f*, 168*f*, 170, 652
Bagshaw, F.R., 128
Baker, C.L., 344–345
Balson, T.G., 556, 556*t*
Bannerman, J.K., 434, 436*t*
Barbee, R., 442*f*
Barbee, R.D., 673
Barclay, L.M., 102, 104*f*
Bardeen, T., 416
Bardon, C., 470–472, 471*f*, 473*f*
Barkman, J.H., 463*f*, 490–493, 492*f*
Barnes, W.E., 657
Barrett, H.M., 659
Bartko, K., 531–532
Bartlett, L.E., 195
Basan, P.B., 470
Battle, J.L., 85, 657, 667
Baumgartner, A.W., 453–454
Beart, R., 643–644, 657
Beasley, A.E., 299*f*
Becker, F.L., 679
Becker, J.R., 567
Becker, T.E., 226
Beckman, F.G., 669
Beeson, C.M., 262, 263*t*
Behrens, R.W., 80–82, 667, 679
Behrmann, L.A., 485–487
Bel, S.L.A., 550–551
Belkin, A., 562–563
Bell, S., 580
Bell, S.A., 544–545

695

Benit, B.J., 394, 394f, 395f
Benit, E.J., 14
Bennett, R.B., 10, 22–25, 203f, 204t
Bergosh, G.L., 478
Berry, L.M., 403–404
Berry, O.H., 677–678, 678f
Bertness, T.A., 479
Beyer, A.H., 300–301, 683
Bezemer, C., 68
Bikerman, J.J., 285, 288, 300–301
Binder, R.C., 190
Bingham, E.C., 155–157
Binkley, J.F., 680, 681f
Bishkin, S.L., 669
Bisson, P., 227f
Bjorndalen, N., 564
Black, A.D., 275–276, 384, 386f
Black, H.N., 664
Bland, R.G., 203f, 204t, 360–362, 427f, 554
Blankenhorn, C.F., 669–670
Blattel, S.R., 396
Bloys, J.B., 547
Bo, M.K., 253, 254f, 256–258
Bobo, R.A., 186f, 659, 662, 679–680
Bohm, B., 526f
Bol, G.M., 347–348, 397–398
Bolchover, P., 485–487
Borchardt, J.K., 128
Borst, R.L., 127
Bowie, C.P., 255–256, 258, 259f
Boyd, J.P., 360
Boyd, P.A., 22–25
Bradley, B.W., 451, 668
Bradley, W.B., 322, 325, 330f, 331f
Bradley, W.F., 106
Brandt, G.W., 672
Brandt, L.K., 663
Brannon, H., 533
Brantly, J.E., 643
Breckels, I.M., 14, 407–408
Bright, K., 673
Brindley, G.W., 100
Broms, B., 321f
Brookey, T., 562–563
Brookey, T.F., 563–565
Browning, W.C., 296–297, 370, 659–660
Bruce, G.H., 219, 220f, 221–222, 221f
Bruist, E.H., 477–478, 485
Buckingham, E., 157–158
Buffington, E.C., 659
Burba, J.L.I., 544
Burba, J.R., 235f
Burdyn, R.F., 292, 296, 429, 659, 663
Burgess, T.M., 427
Burke, L.H., 528
Burkhardt, J.A., 229, 230f, 231, 232f
Burnham, J.W., 491–493
Burrafato, G., 573
Burst, J.F., 312–313
Burton, J., 25
Bush, H.E., 82, 440–451, 442f, 445t, 453–454, 667
Busher, J.E., 190
Byck, H.T., 249–251, 255–256

C

Caenn, R., 422
Cagle, W.S., 663
Caldwell, D.H., 190
Candler, J., 22–25, 617, 622–623, 626t, 633
Cannon, G.E., 229, 255–256, 653–654, 657, 661–662, 669–670, 678
Canson, B.E., 418
Carden, R.S., 429
Carico, R.D., 128, 663
Carlton, L.A., 86, 209, 233, 234f
Carman, P., 529
Carminati, S., 573
Carney, L., 396
Carney, L.J., 432–434, 434t, 435t
Carney, L.L., 107, 448f, 449, 450f
Carnicom, W.M., 659
Carter, A., 650
Carter, T.S., 25
Cartwright, R.S., 650, 658
Chadwick, R.E., 575
Chaney, B.P., 347
Chaney, P.E., 85, 657
Chapman, M.T., 645
Chatterji, J., 128
Chauvin Jr., D.J., 503
Cheatham, J.B., 322, 325–328
Chenevert, M.E., 209, 233, 234f, 338–339, 339f, 343–344, 347, 348t, 351t, 354–356, 355f, 356f, 357f, 358, 664–665, 674
Cheng, D.G.H., 169–170
Chesser, B.G., 276, 347, 352, 393–394, 435–436, 449
Chilingar, G.V., 14, 341–342
Chisholm, F., 660
Clapper, D., 546
Clapper, D.K., 549, 555
Clark, D.E., 351

Clark, E.H., 228
Clark, J.A., 667
Clark, R.C., 658
Clark, R.K., 190, 205–206, 207f, 208f, 231, 347–348, 350t, 352, 400–401, 448f, 450f, 664–665
Clarke, R.N., 129–130
Clarke-Sturman, A.J., 511
Clarkson, B., 532
Clements, W.R., 22–25, 236–237
Cliffe, S., 28
Coberly, C.J., 263–264
Cochran, C.D., 274f
Coffer, H.F., 658
Collings, B.J., 661
Collins, R.E., 478
Collom, R.E., 649
Combs, G.D., 195, 200–201, 200f, 201f
Conn, H.L., 626t
Cook, E.L., 278–281
Courteille, J.M., 375, 375f, 376f, 377
Cowan, J.C., 87, 299f, 429, 454, 660, 667
Cowan, K.M., 580
Cowan, M.K., 547
Cox, T., 453
Cox, T.E., 451, 451f, 453, 453f, 668
Cravey, R.L., 565, 566f, 567f
Crews, J., 530
Crews, S.H., 680–681
Cromling, J., 429
Cross, M.F., 655
Cross, R., 655
Crowe, C.W., 498
Crowley, M., 208–209
Cryar Jr., H.B., 498
Cunningham, R.A., 14, 271–272, 383–389, 401, 402f
Custer, R.S., 545

D

Daemen, J.J., 322
Daines, S.R., 14, 408, 409t
Dallmus, K.F., 342–343, 343f
Dalmazzone, C., 577
Daneshy, A.A., 410
Daniel, S., 351
Darley, H.C.H., 121–122, 123f, 124f, 125f, 191–192, 194, 224, 260–262, 264, 299, 312, 325–328, 327f, 333, 334f, 335f, 336f, 339, 340f, 341f, 344f, 345f, 347, 383, 397–398, 399f, 400, 500

David, A., 300
Davidson, E., 571
Davies, J.M., 616
Davis, N., 434, 436t, 453
Dawson, D.D., 235f, 415–416
Dawson, R., 663
Dawson, R.D., 669–672
de Boisblanc, C.W., 27
Dearing, H.L., 352–353, 386f
Deem, C.K., 544
Deily, F.H., 128, 274f, 663
Delhommer, H.J., 546t, 547
Denny, J.P., 347, 352
Denton, R.M., 548
Desai, C.S., 322
Di Bona, B.G., 386f
Dickinson, G., 311–312, 313f
Dietsche, F., 568–569
Diment, W.H., 423–424, 423f
Dixon, J., 551
Dobson, M., 225
Dodd, C.G., 470
Dodge, D.W., 184–185, 185f, 187, 188f, 189f, 193f
Doherty, W.T., 650, 652, 656
Doig, K., 666
Doty, P.A., 397–398, 399t, 400–401
Downs, J., 42, 137
Duckham, A., 654
Dunn, T.H., 190
Dunning, H.N., 679
Durand, C., 551–552
Durr Jr., A.M., 550–551
Dyal, R.S., 101

E

Eaton, B.A., 310, 311f, 317, 406, 407f
Eckel, J.E., 679
Eckel, J.R., 14, 142–143, 157, 383, 401, 401f, 663
Edinger, W.M., 670
Eenik, J.E., 271–272
Eenik, J.G., 383–389
Eenink, J.G., 14
Ellis, R.C., 495
Elphingstone, E.A., 565, 568t
Engelmann, W.H., 115, 116f
Ennis, D.O., 478
Enright, D.P., 435–436
Eoff, L.S., 552–554
Evans, P., 652
Ewy, R.T., 359–361

F

Fabrichnaya, A.L., 560
Fairhurst, C., 322
Farnham, H.H., 650, 652
Farrar, D., 574
Fauvelle, M., 643–644
Feldenheimer, W., 650
Felixberger, J., 543–544
Ferguson, C.K., 68, 266, 267f, 268f, 269f, 271–272, 271f, 272t
Fertl, W.H., 14
Field, L.J., 663
Finch, W.C., 311
Fischer, P.W., 497–498, 670
Fisher, L.E., 566–567
Fisk Jr., J.V., 549–550
Fleming, C.N., 347, 352
Fleming, H.C., 579
Fleming, J.K., 579
Fleming, N., 489
Fontenot, J.E., 190, 205–206, 207f, 208f, 231, 403–404, 676
Foran, E.V., 677
Ford, T., 25
Foster, W.R., 292
Fraser, I.M., 25
Frederick, W.S., 311
Freeze, G.I., 311, 679
Freshwater, D.C., 254f
Freundlich, H., 166
Fried, A.N., 682–683
Friedheim, J., 22–25, 622–623, 633
Friedheim, J.E., 626t
Funkhouser, G.P., 529

G

Gaillard, N., 529
Galle, E.M., 322
Gallegos, T., 521
Gallus, J.P., 129–130, 396–397, 661
Garnier, A.J., 14, 383–384, 387f, 389–390, 392
Garrett, R.L., 85–86, 448–449, 448f, 450f, 667
Garrison, A.D., 121, 171, 172f, 173f, 173t, 344–345, 656–657
Garyan, S.A., 549
Gates, G.L., 255–256, 258, 259f
Gates, J.I., 671–672
Gatlin, C., 414
Gatlin, L.W., 552
Gavignet, A.A., 226–227, 227f

Geehan, T., 207
Generes, R.A., 194
Genuyt, B., 550–551
Georges, C., 227f
Gerault, R., 356
Getliff, J.M., 580
Gill, J.A., 334, 347, 659, 663
Gill, S., 652
Githens, C.J., 491–493
Glenn, E.E., 262, 264, 482
Gnirk, P.F., 322
Gnnsfelder, S., 677
Gockel, J.F., 414
Goins, W.C., 229, 233–234, 235f
Goins Jr., W.C., 401, 402f, 414–416, 657, 670, 679
Goldberg, S., 42
Golis, S.W., 28
Goode, D.L., 500
Goodeve, C.F., 166
Goodman, M.A., 25, 505–506
Goodwin, R.J., 679
Grant, R.S., 665–666
Grantham, C.K., 448f, 450f
Gravely, W., 427
Gray, D.H., 472–474
Gray, G.R., 87, 128, 141–142, 194, 347, 656, 658, 660, 672, 674–675
Gray, K.E., 262, 271–272, 383–386, 385f, 388f, 482
Greathouse, W.D., 667
Green, B.D., 546–547
Green, B.Q., 85
Green, H., 157, 166
Gregg, D.N., 334
Grey, R.A., 574
Grichuk, H.A., 547
Grieve, W.A., 25
Griffin, J.M., 347, 476–477
Griffin, R.N., 672
Griffin, R.R., 661
Griffin, W.C., 292, 296
Grim, R.E., 96, 98, 100, 107–108
Grioni, S., 672
Growcock, F.B., 562–565
Gruesbeck, C., 478
Gupta, D.V.S., 529

H

Hadden, D.M., 86
Haden, E.L., 77–78, 219f, 221f, 379–381
Haimson, B.C., 411

Hale, A.H., 547, 553t, 573, 579–580
Hall, H.N., 221–222
Halliday, W.S., 549, 553t, 557
Hamburger, C.L., 415–416
Hamby, T.W., 505
Hanahan, D.J., 549
Hanks, R.W., 159
Hanson, P.M., 236–237
Harris, G., 626t
Hartfiel, A., 191–192
Harth, P.E., 648
Hartmann, A., 258
Haskin, C.A., 500–501, 501f
Hassen, B.R., 273
Hauser, E.A., 119, 672
Hauser, J.M., 427f
Havenaar, I., 68, 190, 273, 273t
Hawk, D.E., 193–195
Hayes, C.W., 646
Hayward, J.T., 372
Headley, J.A., 579
Healey, T.W., 129–130
Hedstrom, B.O.A., 190
Heggem, A.G., 646
Heilhecker, J.K., 30–31, 663
Helmick, W.E., 21, 372
Hendricks, S.B., 101, 108, 110–111, 110f, 111f, 112f
Hermes Jr. L.M., 657
Herrick, H.N., 651
Higdon, W.T., 125f
Hight, R., 102, 122–123, 125, 125f
Hille, M., 578
Hiller, K.H., 170f, 195, 196f, 199f
Hiltabrand, R.R., 312–313
Hindry, H.W., 669–670, 675–676
Hinds, A.A., 22–25
Hlidek, B.T., 529
Hoberock, L.L., 426, 428t
Hoch, R.A., 186f
Hoch, R.S., 662
Hodder, M.H., 565–566
Hoeppel, R.W., 670, 672
Hofmann, U., 96–97
Hollis, W.T., 677
Holt, C.A., 347, 352
Holtsclaw, J., 529
Holub, R.W., 472–474
Hook, R.A., 683
Hopkin, E.A., 206f, 217, 218f, 222–225, 223f, 225f
Horner, V., 68, 271–274, 274f, 277

Horner, V.V., 659
Hottman, C.E., 325, 336, 403, 405f
Houwen, O.H., 207
Howard, G.C., 414, 414f, 415f, 419
Howe, J.L., 660
Hower, J., 312–313
Hower, W.F., 470, 664, 679
Huang, T., 530
Hubbard, J.T., 505
Hubbert, M.K., 311, 315–319, 402–403, 404f, 406
Huber, J., 555
Hudgins, C.M., 667
Hudgins Jr., C.M., 440–442
Huebotter, E.E., 664
Hughes, R.G., 172, 195
Huisman, P.H., 669
Hull, J.D., 663
Hunter, D., 377–378
Hurdle, J.M., 659
Hurst, W., 462, 481–482
Hussaini, S.M., 218–219, 222–225
Hutchison, S.O., 303

I

Ivan, C.D., 547
Ives, G., 674–675
Iyoho, A.W., 205, 226–227

J

Jackson, G.L., 505–506
Jackson, J.M., 498
Jackson, L.R., 665–666
Jackson, M.L., 107
Jackson, R.E., 304f, 305f
Jackson, S.A., 22–25
Jacobs, T., 528
Jacquin, C., 470–472, 471f, 473f
James, S.G., 580
Jamison, D.E., 236–237
Jarrett, M., 546, 561
Jefferson, M.E., 110, 110f, 111f
Jenks, L.H., 500–501
Jensen, J., 669
Jessen, F.W., 120–121
Jeu, S., 36
Jeu, S.J., 493–494
Johancsik, C.A., 25
Johnson, C.A., 120–121
Johnson, D.P., 454

Johnson, N.W., 191
Johnson, R.K., 403, 405f
Johnson, V.L., 249–251
Jones, B., 449
Jones, F.O., 73, 472–476, 477f, 491
Jones, P.H., 65–66, 167–168, 167f, 168f, 170, 424, 652
Jones, R.D., 347, 352
Jones, T.A., 562
Jones Jr. F.O., 661
Jordan, J.W., 672
Jorden, J.R., 403–404
Jordon, J.W., 10

K
Kahn, A., 101–102
Kang, K.S., 142
Kashkarov, N.G., 550–551
Kastrop, J.E., 660–662
Kaveler, H.H., 129, 147, 657
Keelan, D.K., 501–502
Keilhofer, G., 544
Kelley, J., 25
Kellogg, W.C., 660
Kelly, J., 406, 427–429
Kelly Jr., J., 193–195, 353
Kemick, J.G., 500
Kennedy, J.L., 676
Kennedy, W., 646
Kercheville, J.D., 345–346, 379–381, 674
Kersten, G.V., 499, 670, 675–676
Khan, S.A., 682–683
Kinchen, D., 563
King, E.G., 659
King, G., 45–47, 45f, 46f, 522–523
King, G.R., 666
Kippie, D.P., 552
Kirk, W.L., 675
Kirwan, M.J., 649, 650f
Kjellstrand, J.A., 444–448
Klementich, E.F., 675
Kljucec, N.M., 665
Klotz, J.A., 68, 266, 267f, 268f, 269f, 271–272, 271f, 272t, 482, 483f, 485–488, 486f, 487f, 490
Knapp, A., 649
Knapp, I.N., 646
Knight, L., 341–342
Koch, R.D., 312–313
Koch, W.M., 190
Koepf, E.H., 501–502
Kokal, S.L., 561
Konirsch, O., 227f
Konirsch, O.H., 269f, 270f
Konovalov, E.A., 550–551
Korry, D.E., 208–209
Krol, D.A., 379–381
Krueger, R.F., 275–276, 275f, 276f, 277f, 278–281, 470, 479–482, 479f, 483f, 484, 484f, 486f, 487f, 499, 664
Krueger, R.V., 262
Krug, J.A., 301, 682–683
Kruger, P., 424
Krumbein, W.C., 256, 257f, 258
Kukacka, L.E., 547
Kulpa, K., 541
Kurochkin, B.M., 549
Kuru, E., 564
Kwan, J.T., 22–25

L
La Mer, V.K., 129–130
Lachenbruch, A.H., 423f
Lal, M., 234
Lammons, R.D., 370
Lancaster, E.H., 657
Lange, P., 543–544
Lanman, D.E., 663
Larsen, D.H., 246–248, 252t
Lauzon, R.V., 128
Law, J., 677
Lawhon, C.P., 273–274
Lawson, H.F., 573
Lawton, H.C., 654
Layne, M., 526f
Lee, J., 421
Leonard, R.A., 126
Leth-Olsen, H., 37
Lewis, J.O., 646
Leyendecker, E.A., 331–333, 333f, 675
Li, Y.-J., 549
Lilje, K.C., 550–551
Lipkes, M.I., 542
Lo, S.-W., 533
Lockstedt, A.W., 548
Loftin, R.E., 579
Longley, A.J., 21, 372
Loomis, A.G., 119–120, 654, 656–657
Lord, D.L., 174, 174f, 175f
Low, P.F., 110, 126
Lu, C.F., 347
Lummus, J.L., 77–78, 82, 396–397, 420, 658, 661, 663, 671, 679
Lyons, W.C., 82

M

MacKay, S.P., 678–679
Macmillan, N.H., 304f, 305f
Macpherson, J.D., 55–56
Magalhaes, S., 55–56
Magner, H.J., 679
Mallory, H.E., 296, 397, 660–661
Maly, G.P., 490, 500
Maresh, J.L., 572
Marika, E., 601
Marsden, S.S., 300, 682–683
Marsh, H., 60
Marsh, H.N., 651
Marshall, B.V., 423f
Marshall, C.E., 96, 98–100
Martin, M., 226, 227f, 228
Martin, R.L., 571
Matthews, C.S., 462, 497–498
Matthews, W.R., 406
Maurer, W.C., 383, 391–392, 392f, 430f, 431f, 432f, 433f
Maury, V.M., 309, 322–325, 326f
Mauzy, H.L., 441–442, 443f
Mayell, M.J., 504
Mayes, T.M., 217
Mazee, W.M, 669–670
McAtee, J.L., 107–108, 120–121
McAuliffe, C.D., 22–25, 597
McCaleb, S.B., 353
McCray, A.W., 657, 670
McDonald, M., 347
McDonald, W.J., 430f, 431f, 432f, 433f, 543
McGhee, E., 660–661
McGlasson, R.L., 667
McGregor, W.M., 560
McKee, J.D.A., 626t, 628t
McLeod Jr., H.O., 490
McMordie, W.C., 195, 201, 202f, 207, 425–426, 427f
McMordie Jr., W.C., 201–203, 203f, 204t, 673
McMurray, W.F., 646
Medley Jr., G.H., 543
Meinhold, A.F., 628
Melrose, J.C., 102, 159, 190
Menzie, D.E., 174, 174f, 175f
Mering, J., 121
Mesaros, J., 661–662
Messenger, J.U., 379–381, 546
Messler, D., 511
Methven, N.E., 500
Mettath, S., 395
Metzner, A.B., 176, 178–181, 180f, 184–188, 185f, 187f, 188f, 189f, 190, 193f
M'Ewen, M.B., 126
Meyer, R.L., 107, 432–434, 434t, 435t
Milhone, R.S., 495, 505, 682
Miller, G., 10, 499, 670, 673, 675–676
Miller, J.J., 580
Miller, M., 533
Millhone, R.S., 301
Milligan, D.J., 249–251, 250f
Mills, B., 650, 661–662
Miranda, F., 14
Mitchell, B.J., 300f, 301, 302f, 682–683
Mitchell, M.E., 657
Mitchell, R.F., 234–236, 322–325, 429
Moajil, A.M.A., 542–543
Mody, F.K., 360, 360f
Monaghan, P.H., 22–25, 77–78
Monaghan, R.H., 377, 377f, 378f, 379–381
Mondshine, T.C., 77, 345–347, 352–353, 369–370, 499, 663, 673–674
Monk, G.D., 256, 257f, 258
Montgomery, C., 521, 533
Montgomery, M., 396
Mooney, M., 180–181
Moore, P.E., 208–209
Moore, P.L., 209
Moore, R.H., 25
Moore, T.F., 416, 419
Moore, T.V., 669
Moore, W.D., 332–334
Morgan, B.E., 499
Morton, E.K., 359–361
Moses, P.L., 424, 424f
Muecke, T.W., 469f, 470, 472–474, 474f, 479–481, 481f
Mueller, H., 544, 551
Mungan, N., 470, 474–475, 478
Munroe, R.J., 423f
Murray, A.S., 14, 383, 678–679
Murray, J.W., 680
Murray, S.C., 331–333, 333f, 675
Muskat, M., 463
Myers, G.M., 219f, 221f

N

Nakaya, T., 549
Nash Jr., F., 414
Nasr-El-Din, H.A., 4
Navarrete, R.C., 143
Neil, J.D., 472–474

Nelson, M.D., 74–75, 89, 499, 663, 675–676
Nemir, C.E., 414
Nesbitt, L.E., 356–358
Nevins, M.J., 73, 505–506, 673
Newlin, F., 660
Ng, F.W., 505–506
Nguyen, P.D., 560
Nicolson, K.M., 678
Nicora, L.F., 560
Nordgren, R.P., 322, 325
Norman, L., 529
Norrish, K., 111, 113f, 113t, 114f, 121, 126–127, 126f, 127f
Nowack, B., 42
Nowak, T.J., 470, 479, 479f, 499, 664

O

Oates, K.W., 663
O'Brien, D.E., 347, 348t, 351t, 664–665
O'Brien, T.B., 225, 676
Ockenda, M.A., 650
Okrajni, S.S., 226
Olivier, D.A., 490
Olsen, H., 571
Omland, T.H., 30
Oort, E., 55–56
Ormsby, G.S., 663
O'Shields, R.L., 662
Osisanya, S.O., 356, 357f, 358
Otte, C., 424
Outmans, H.D., 21, 249, 253, 255f, 265–266, 265f, 277, 373, 379–381, 382f
Ozkan, E., 462

P

Palmer, L.L., 22–25, 597
Paris, B., 660
Park, A., 77–78, 130, 661
Parker, C.A., 675
Parsons, C.P., 648, 652
Paslay, P.R., 170, 322, 325–328
Patel, A., 359–360
Patel, A.D., 549
Patroni, J.-M., 269f, 270f
Patton, C.C., 440–442
Peden, J.M., 262, 266–270
Penkov, A.I., 550
Pennington, J.W., 643
Perkins, H.W., 658
Perricone, A.C., 393–394, 437, 449, 659–660
Perry Jr., E.A., 312–313

Petit, D.J., 142
Pfenning, G.E., 672
Phillips, B.M., 575
Pickering, S.U., 296
Piggott, R.J.S., 190, 217, 222–224
Plank, J., 543–544
Planka, J.H., 667
Poletto, F., 14
Politte, M.D., 206–207
Pollard, J.A., 646
Pollard, L.D., 96, 101–104, 109t
Poole, G.L., 505
Pope, P.L., 661
Portnoy, R.C., 429–430
Powers, M.C., 312–313, 338–339, 340f
Pratt, M.I., 126
Priest, G.G., 499
Prokhorov, N.M., 571
Prokop, C.L., 68, 265–266, 266t, 277
Pruett, J.O., 352
Purcell, W.R., 464, 481–482
Pye, D.C., 483f, 486f, 487f

Q

Qiu, X., 531–532
Quinn, R.V., 680–681

R

Rabinowitsch, B., 180–181
Radd, F.J., 667
Radke, C.J., 472
Randall, B.V., 193, 208–209, 449, 679
Rastegaev, B.A., 565–566
Rath, H., 350t
Rausell-Colom, J.A., 126–127, 126f, 127f
Ray, J.D., 449
Raymond, L.R., 193–194, 194f, 425, 425f, 426f
Raza, S.H., 682–683
Reddie, W.A., 672
Reed, C.E., 119
Reed, J.C., 180–181, 185–186, 187f, 190
Reese, L.C., 322
Rehm, W.A., 430f, 431f, 432f, 433f
Reichert, G., 572
Reid, A., 652
Reid, A.L., 547
Reid, C.A., 676
Reiner, M., 160–161
Reistle Jr. C.E., 654–655
Remont, L.J., 430f, 431f, 432f, 433f, 505–506, 673

Remson, D., 322
Renner, J.L., 424
Rex, R.W., 472–474
Ribe, K.H., 465–467, 497–498
Ritter, A.J., 356
Rivers, G.T., 573
Robertson, R.E., 184
Robinson, F., 543
Robinson, L.H., 30–31, 302f, 303f, 315–316, 315f, 663
Rogers, W.F., 656–657
Rollins, H.M., 678
Rolshausen, F.W., 669
Rooney, P., 576
Roper, W.F., 61, 161, 163
Rosenberg, M., 76–77, 370
Ross, C.S., 111, 112f
Roy, R., 100
Rubel, A.C., 652
Ruehrwein, R.A., 130
Ruffin, D.R., 416
Runov, V.A., 550–551
Rupert, J.P., 396
Rushing, J.H., 628, 628t
Russell, D.A., 462, 497–498
Ryan, N.W., 191

S

Saasen, A., 236–237, 541–542
Samson, H.R., 117, 125
Samuel, M., 530
Samuels, A., 449
Sanchez, H.V., 659–660
Sano, M., 550
Santos, H., 359–360
Sargent, T.L., 347
Sartain, B.J., 381
Sass, J.H., 423f
Satterlee III, K., 630
Sauzay, J.M., 309, 322–325, 326f
Savari, S., 236–237
Savins, J.G., 61, 161, 163, 182–183, 191–193, 205
Scanley, C.S., 129–130, 258
Scarlett, B., 254f
Scharf, A.D., 396
Scheuerman, R.K., 350t
Scheuerrnan, R.F., 495, 496f, 497f
Schilthuis, R.J., 669
Schindewolf, E., 246–248, 252f, 253–254, 256, 261f
Schlemmer, R., 359
Schlemmer, R.F., 558–559, 559f
Schmidt, D.D., 505
Schmidt, G.W., 312–313, 424
Schmidt, P.W., 125f
Schneider, R.P., 680–681
Schofield, R.K., 117, 125
Schremp, F.W., 249–251
Schuh, F.J., 214–215, 231, 231f
Schwertner, D., 549
Scoppio, L., 37
Scott, P.P., 414, 414f, 415f, 419
Sebba, F., 562–564
Secor, D.T., 411, 412f, 413f
Self, E.S., 669–670
Selim, A.A., 116f
Senaratne, K.P.A., 550–551
Shah, S.N., 142–143
Shallenberger, L.K., 677
Shannon, J.L., 347, 352
Sharma, M.M., 478
Sharpe, L.H., 290–291
Sheffield, J.S., 28, 332
Shell, F.J., 127, 347
Sheridan, H., 667
Sheu, J.J., 554
Shirley, O.J., 403–404
Shuey, M.W., 545
Shumway, W.W., 544–545
Sifferman, T.R., 218–219, 219f, 221–222, 221f, 225, 551
Sikora, D., 571
Silbar, A., 170
Simms, L., 532
Simon, G.A., 565
Simpson, J.P., 10, 77–78, 81f, 249, 298–299, 299f, 347, 352–353, 379–381, 380f, 381t, 396, 442f, 505, 659–660, 672–673, 676
Sinha, B.K., 195
Sinha, V., 142–143
Sipple-Srinivasan, M., 601–603
Sitzman, J.J., 28
Skelly, W.G., 444–448
Slater, K., 372
Sloat, B., 451, 452f
Slobod, R.L., 477–478
Slusser, M.L., 262, 264, 482
Smith, B., 676
Smith, C.K., 556, 556t
Smith, F.W., 678
Smith, K.W., 554
Smith, M., 521

Smith, M.B., 389
Smith, N.R., 120–121
Smith, T.R., 580
Sobey, I.J., 226–227, 227f
Somerton, W.B., 472
Son, A.J., 436–437
Sonney, K.J., 465–467
Speer, J.W., 391, 391f
Speers, A., 179–180
Speller, F.N., 665
Sposito, G., 42
St Pierre, R., 26
Steiger, R.P., 86, 347–348, 349f, 353, 476
Stein, F.C., 504
Stein, N., 404, 679
Stiff Jr., H.A., 184
Stone, A., 42
Stone, V.D., 662
Strother, G.W., 544
Stroud, B.K., 646–648
Stuart, C.A., 14, 312
Stuart, R.W., 499, 670
Sufall, C.K., 679
Sullins, R.S., 661–662
Sullivan, P.F., 530
Suman, G.O., 498
Suman, J.R., 649
Sunde, E., 571
Swan, J.C., 668–669, 673
Swanson, B.F., 464, 465f, 466f, 467f
Sweeney, P., 645, 677

T

Tailleur, R.H., 370
Tailleur, R.J., 76–77, 664
Tare, U.A., 360, 360f
Taylor, R.S., 562
Teis, K.R., 677
ten Brink, K.C., 121
Teplitz, A.J., 235f, 416
Terichow, O., 116f
Texter, H.G., 665–666
Thielen, V.M., 553t, 557
Thomas, D.C., 505
Thomas, R.P., 221–224
Thompson, D.W., 102, 104f
Thomson, M., 427
Tjon-Joe-Pin, 533
Tomren, P.H., 226
Topping, A.D., 322, 410–411, 410t
Trapp, D.R., 159
Trimble, G.A., 499, 675–676

Trout, K., 657
Tschirley, N.K., 674–676
Tselovalnikov, V.F., 549
Tuttle, R.N., 463f, 490–493, 492f, 505

U

Underdown, D.R., 101
Urban, T.C., 423f

V

Vajargah, A.K., 55–56
Van Dyke, O.W., 657–658
Van Eekelen, H.A.M., 14, 407–408
van Everdingen, A.R., 462
van Laar, H., 350t
van Lingen, N.H., 14, 383–384, 387f,
  389–390, 390f, 392
van Olphen, H., 115, 116f, 117, 118f,
  119–120, 120f, 125, 159
Van Oort, E., 55–56, 551–552, 553t, 630
Varela, B., 521
Vaussard, A., 266, 269f, 270, 270f, 384
Veil, J.A., 632
Vidrine, D.J., 14, 394, 394f, 395f
Vietti, W.V., 656–657
Vik, E.A., 624, 626t
Vogel, L.C., 262
von Engelhardt, W., 246–248, 252f,
  253–254, 256, 261f

W

Wachter, A., 666
Wade, G., 656–657
Waite, J.M., 292
Walker, C.O., 546t, 547
Walker, J., 633
Walker, R.E., 27, 191, 208–209, 217, 226
Walker, T.O., 27, 347, 352
Wall, H.A., 219f, 221f
Wall, K., 551
Wallick, G.C., 205
Wallis, W.M., 671–672
Walton, I.C., 485–487
Ward, D.W., 130
Warpinski, N.R., 408
Warren, T.M., 389
Watcharasing, S., 563, 563f
Watkins, T.E., 89, 658, 672
Watson, R.A., 678
Watson, S.K., 555

Watts, R., 528
Watts, R.D., 396
Weaver, C.E., 96, 101–104, 109t
Webb, M.G., 500–501, 501f
Wegner, C., 572
Weibel, J., 451, 452f
Weichert, J.P., 235f, 658
Weintritt, D.J., 73, 172, 195, 249–251, 250f, 674–675
Weiss, W.J., 658–659
Welch, D., 26
Welch, G.R., 77–78, 379–381
Wentzler, T., 562
Westergaard, H.M., 322
Westwood, A.R.C., 304f, 305f
Wheless, N.H., 660
White, M.M., 274f
Whitfill, D.L., 559–560
Whitmire, L.D., 195, 200–201, 200f, 201f
Wiener, L.D., 292, 296, 429, 659
Wilder, L.B., 663
Wilhoite, J.C., 322
Willard, D.R., 77–78
Williams, C.E., 219, 220f, 221–222, 221f
Williams, F.J., 123
Williams, L.H., 101
Williams, M., 68, 255–256, 669–670
Williams, R.W., 662–663
Willis, C., 338
Willis, D.G., 311, 315–319, 402–403, 404f, 406
Willis, H.C., 208–209

Wilton, B.S., 547
Wingrove, M., 561
Wood, R.R., 547
Woodworth, F.B., 565, 568t
Wright, C.C., 671
Wright, C.W., 262, 263t
Wuth, D.E., 662
Wyant, R.E., 277–278, 377–379

X
Xiang, T., 562

Y
York, P.A., 367, 368f
Yortos, Y.C., 478
Young, S., 28, 421
Young Jr., F.S., 262, 271–272, 383–386, 385f, 388f, 482
Yuster, S.T., 465–467

Z
Zaleski, P.L., 548
Zamora, M., 222–224, 223f
Zeidler, H.U., 28, 219, 221–224
Zevallos, M.A.L., 626t
Zhao, L., 580
Zhou, Z.J., 551–552
Zilch, H.E., 86
Zoeller, W.A., 406
Zurdo, C., 375, 375f, 376f, 377

Subject Index

Note: Page numbers followed by "*f*" and "*t*" refer to figures and tables, respectively.

A

Abnormal pressures, 312–314
Abnormally pressured formations.
  See Geopressured formations
Accretion control, and shale stability, 361
Acetals, 578
  as additives for nonaqueous emulsion drilling fluids, 577–578
Acetic acid, 530
Acetone, 561*f*
Acidity, 76
Acidized fracturing fluids, 530
Acid-soluble systems, of completion and workover fluids, 498
Acrylamide (AM), 529
2-Acrylamido-2-methylpropanesulfonic acid (AMPS), 529
Acrylates, 138–139
  and acrylate copolymers, 149
Acrylic acid salts
  as dispersants, 574
Acrylics, as dispersants, 574
Additives, for drilling fluids, 537–538, 539*f*, 539*t*
Adhesion, 290–291
  and bit balling, 393
  and stuck pipe, 377–378
Adsorption/desorption isotherm, 339*f*
Aerated drilling fluid
  in drill pipe corrosion, 445*t*
Aerated mud, 679–680
Aerosol, 300
Aggregation, 121–125
  definition of, 121
  and filter cake permeability, 260
  and gel strength, 121–125

Aggregation-dispersion process, versus flocculation-deflocculation process, 121, 122*f*
Air
  entrainment, in drill pipe corrosion, 451
  entrapment, in drill pipe corrosion, 445*t*
Air drilling, 12
Air drilling fluids, temperature stability of, 429
Airfor fast drilling, 28
Alcohol(s), 549–550
Aldehydes, 578*f*
Alkalinity, 76
  analysis of, 85
Alkali-solubilized lignite and sodium chromate, 660
Alkyl phenol ethoxylates, toxicity of, and environmental protection, 580
Alkylpoly(glucoside)s (APGs), 560
Allyloxybenzenesulfonate, 575
Aluminum lignosulfonate chelate, in prevention of bit balling, 393–394
American Association of Oil Well Drilling Contractors (AAODC), 655, 665–666
American Institute of Mining and Metallurgical Engineers (AIME), 651
American Petroleum Institute (API), 86, 620, 651, 665
  drilling fluids testing procedures, 2
  filter loss test, 245, 267*f*, 268*f*, 275–276
  fluid loss, 65–70
  formate test procedures, 513–514, 514*t*
  particle size classification, 74*t*
  recommended procedure for particle size determination, 73

707

708 Subject Index

American Petroleum Institute (API) (*Continued*)
   RP 13B (Standard Procedure for Testing Drilling Fluids), 66–68, 84, 86, 245
   RP 51R, 599–600
   sand test, 86
   standard field and laboratory tests, 55
   standard filtration test, 21
   TCT procedure, 49
American Society for Testing and Materials (ASTM), recommended procedures for sieve tests, 73
Amine salts
   and corrosion control, 453
   of maleic imide, 554
Amines, and corrosion control, 453
Ammonium chloride (dry), 41
Ammonium humates, 672
Amylopectin, 141f
Amylose, 141f
Angstrom(s), 98
Aniline point, 84
Anion exchange capacity, 109
Annular shear rates, 237
Annular velocity, optimum for, 222–224
Annulus
   flow conditions in, 205
   pressure gradient in, determination of, 213–215
Annulus packs, 673
Anode, 438–439
   corrosion at, 439
Aphrons, 562–565
   stabilizers, 565
   structure of, 563, 563f
API. *See* American Petroleum Institute (API)
Apparent viscosity (AV), 17, 162
   calculation of, 61
   of drilling fluids containing PAM or xanthan gum, before and after rolling, 572t
Aquagel, 648–649
Aquatic toxicity of nonaqueous-based fluids, 626t
Arenaceous formations, 309–310
Argillaceous formations, 309–310
   plastic yielding in, 333
Ark-La-Tex Area, 660
Asphalt, sulfonated, 555, 576
Atmospheric gas, in drill pipe corrosion, 445t
Attapulgite, 106–107, 106f
   base exchange capacity (BEC) of, 109t
   oil dispersible, 299

B

Bacteria, and drill pipe corrosion, 453–454
Bacteria control, 453–454, 532, 565–566
Bactericides, for drilling fluids, 565–566, 568t
Barite, 4
   and casing/tool joint friction, 370–372
   and coefficient of friction of mud, 380
   density of, 5t
   evaluation of, 87
   in KCl muds, 351
   and mud density, 15
   specifications, 4t
   as weighting agent, 540–541
Baroid Sales Company, 648–649
Basal spacing. *See* c-spacing
Base exchange capacity (BEC), 108
   determination of, 109
Base exchange reactions, and clay blocking, 475–478
Base fluid types, 39t
Base oils, 11–12
   pneumatic drilling fluids, 12–13
Beardmore, David, 422–423
Beidellite, 101
Bentonite, 9, 27, 108, 415–416
   definition of, 107–108
   extender for, 130
   formation of, 107
   identification of, 73–74
   oil dispersible, 299
   permeability of, 312
   prehydrated, with KCl muds, 347–348
   specifications for, 86–87
Bentonite extender polymer, 661
Bentonite suspension(s), consistency curve for, in multispeed viscometer, 163, 164f
Bingham bodies. *See* Bingham plastic
Bingham plastic, 151
   consistency curve for
     in direct indicating viscometer, 161
     in rotary viscometer, 163
   definition of, 155–157
   flow model for, 155–159
   ideal, consistency curve for, 155–157, 156f
   mixed flow, in round pipe, 158f
   plug flow, in round pipe, 158f
Bingham plastic fluids, 17
   consistency curve of, 17
Bingham Plastic rheological parameters, 61
Bingham yield point, 157
   determination of, 159

Subject Index 709

Biocides, 454
Bioconcentration factor (BCF), 627–628
Biodegradability, of completion and workover fluids, 498
Biodegradable lubricants, 551
Biopolymers, 138
 fermentation, 142–146
Biotite, 102–104
Bit balling, 392–394
 inhibitors, 401
Bit nozzles, flow conditions in, 205
Bit penetration rate. *See* Penetration rate
Blowout, swab pressures and, 229
Borehole
 cleaning. *See* Hole cleaning
 hydration of
  control of, 344–345
  with water-based drilling fluids, 342–343
 plastic deformation of, 320, 321*f*
 stability. *See* Hole stability; Wellbore stability
 stresses around, 319–325
Borehole conditions, extreme, 674–675
Borehole stabilization by oil muds, 674
Borehole wall(s)
 collapse of, horizontal virgin effective stresses and equivalent depths at, 320, 324*t*
 elastic limits of, horizontal virgin effective stresses and equivalent depths at, 309, 323*t*
 failure modes at, 322*f*, 326*f*
Bottomhole balling, 391–392
Bottomhole cleaning, 393
Bottomhole density, 47
Breakers, 533
 enzyme, 533
 for invert emulsions, 562
 oxidizer, 533
Breathing seal, 415
Bridging materials
 and chip hold-down pressure, 396
 in completion and workover fluids, 498
 and drilling rate, 396
Bridging particles, 20–21, 383
 degradable, in workover fluids, 491
 in formations with no cohesive strength, 337–338
 for unconsolidated sands, 484–485
Bridging process, 260–264
Brine clarity, 49–51

Brine test procedures, 44–52
 density, 44–47
 density prediction, 46–47
 formate brines densities and PVT data, 47
 temperature and pressure effect on density, 45–46
Brine-based fluids, additives for, 538*f*
Brine(s)
 clear, as packer fluids, 504–505
 as completion fluids, 494–495, 494*t*
 for fast drilling, 28
 formate, 511–515
 properties of, 505
 saturated, for rock salt, 28
 solids-free, 494–495
 viscous, 495–497
Brittle failure, 316
Brittle-plastic yielding, 334–337
Brown coal. *See* Lignite
Brownian movement, 94, 119
Brucite, structure of, 96
Burmah Oil Company, 652

C

Cable-tool method, 646
Cabot Special Fluids, 47
Cake deposition index (CDI), 68
Calcite, density of, 5*t*
Calcium, estimation of, 85
Calcium bromide, 43–44
Calcium bromide-calcium chloride solutions, 667
Calcium carbonate
 as bridging agent, 498
 in formate fluids, test for, 513–515
 as weighting agent, 542
Calcium carbonate salt fluids, in workover wells, 488
Calcium chloride, 42–44, 674
Calcium clays, suspensions, rheology, temperature and, 197
Calcium lignosulfonate, 657
 and inhibition of bit balling, 401
Calcium oxide, 670
Calcium sulfate, estimation of, 85
Calcium surfactant mud, 659
Calcium treated fluids, 9
California Talc Company, 648
Capillary attraction, 285, 286*f*
Capillary phenomena, 464–468
Capillary pressure, 287

Subject Index

Capillary viscometer, 82, 155, 195
Carbon dioxide, in drill pipe corrosion, 444, 445t
Carbonates
  as weighting agents, 542
  in workover fluids, 491
Carboxymethylcellulose (CMC), 138, 147, 148f, 258, 659
  in completion and workover fluids, 498
  as friction reducer, 191–192, 193f
  synthesis of, 129
  unit of, 129, 129f
Casing pack(s), 503–506, 666
  functions of, 503
  oil-base, 505–506
  requirements for, 503
Cathode, 438–439
  corrosion reaction product discharge at, 439
  depolarized, 440
  polarized, 440
Cation exchange capacity (CEC), 108
Cation exchange reactions, 347–352
  and shale stabilization, 347
Cationic polymers, 139
Caustic soda, 650
Caving, 28
  in brittle shales, 343–344
  prevention of, 330
Celesite, density of, 5t
Cellulose
  polyanionic, in KCl muds, 347–348
Cellulosics, 147
Cement(s), drilling fluid conversion into, 580
Centipoise, 15–16, 154
Cesium formate (liquid), 44
Cesium formate salts, 512t
Chelate(s), 393–394
Chemical analysis, 84–86
Chemical sealant, 422
Chemical(s), in drilling fluids
  hazardous, 597
  ignitable, 597
  toxic, 597
Chemisorption, 117, 119–120
Chenevert's method, 346–347
Chip hold-down pressure (CHDP)
  bridging solids and, 397–398, 399f
  dynamic, 389–390
  static, 383–389
Chloride brines, 7
Chloride(s), analysis of, 84–85

Chlorites, 105
  base exchange capacity (BEC) of, 109t
  formula for, 105
  occurrence of, 107
  structure of, 105
Chrome lignite (CL)-chrome lignosulfonate (CLS) system, 660
Chrome lignosulfonate, 659–660
Chromelignosulfonate-treated muds, 673
Clay blocking, 472–474
  base exchange reactions and, 475–478
  pH and, 478–481
  salinity and, 469–470, 474–475, 478t
  in two-phase systems, 478–481
Clay colloids, 6–7
Clay gel, structure of, 127f
Clay mineralogy, 96–107
  knowledge of, need for, 93
Clay mineral(s), 6–7, 96. *See also* Attapulgite
  analysis of, 353
  anion exchange capacities of, 109
  base exchange capacity (BEC) of, 109t
  cation exchange capacity (CEC) of, 108
  chlorite group. *See* Chlorites
  classification of, 96
  and control of borehole hydration, 347
  crystalline, 96
  formulas for, conventional notation for, 100
  identification of, 96
  illite group. *See* Illites
  ion exchange, 108–109
  kaolinite group. *See* Kaolinite
  of mixed-layer clays, 106
  occurrence of, 107–108
  origin of, 107–108
  properties of, 96
  smectite group. *See* Smectites
Clay muds, 17, 19
Clay stabilizers, 534
Clay volume, aggregation and, 122–123
Clay(s)
  adsorption and desorption of, 338–342
  dispersible, drilling mud formulations for, 352
  equilibrium water content of, versus compacting pressure, 339, 341f
  fixation, salt types for, and shale stability, 361
  flocculation value of, 117
  indigenous
    dispersion of, prevention, 29

Subject Index 711

permeability impairment by, 469-470
mechanism of, 470-475
mica-type structure of, 96
mineral constituents. *See* Clay mineral(s)
mixed-layer, mineralogy of, 106
oil dispersible, 299
particle size, upper limit of, 96
primary, 107
secondary, 107
stabilizers, 551-559, 553*t*
swelling, 338-339. *See also* Crystalline swelling; Osmotic swelling
mechanisms of, 109-115
swelling pressures in, 338
Cleanout operations, 38
Clear brine fluid (CBF), 36, 39
chemical constituents of, 37-38
organic, 37-38
Clear brines, 35
maximum densities, 40*f*
Clear water drilling, 660-661
Closed-loop/zero-discharge drilling, 149
CMHPG (carboxymethylhydroxypropalguar), 529
Coal seams, 338
Coefficient of friction
calculated for various muds, 372
definition of, 368-369
lubricants and, 368-369
measurement of, 369*f*
Colloid chemistry, 93
Colloidal systems. *See also* Foam(s); Mist(s)
aqueous, characteristics of, 94
characteristics of, 93-96
definition of, 93-94
surface phenomena, 94
Colloid(s), 7, 93-94. *See also* Clay colloids
active, 6-7
inorganic, 6-7
organic, 7, 96
versus silt, 94
Commercial oil muds, 670
Committee on Standardization of Drilling Fluid Materials, 654-655
Compaction cell, 339, 340*f*, 341-342
Completion fluids
containing oil-soluble organic particles, 499
containing water-soluble solids, 499
formation damage by, tests for, 500-502
nonaqueous, 499-500
selection of, 493-500, 599-600

storage of, 599-600
use of, 599-600
Completion hardware, 37
Completion objectives and environment, 37
Completion operations, types of, 35
Completion workover procedures, 490-493
Composition of completion fluids, 39-44
clear brine fluids, 39
divalent inorganic brines, 42-44
monovalent inorganic brines, 39-42
monovalent organic brines, 44
Concentrate 111, 664
Concentration cells, 439
Concentric cylinder rotary viscometer, 159-162
Concentric drill pipe/air lift method for reduced-pressure drilling, 681*f*
Conroe field, 653-654
Consistency curve(s), 16-17, 151
Consistometer, 195
Contaminants, 51-52
Continuous phase, 5
Copolymer(s). *See also* Partially hydrolyzed polyacrylamide-polyacrylate copolymer (PHPA)
grafted, 556
hydrolyzed, 129-130
sulfonated isobutylene maleic anhydride, as dispersant, 575
vinyl sulfonated, 436-437
Copper salts, as hydrogen sulfide scavengers, 449
Corrosion
control, 443-454, 665-668
oil muds and, 454
troubleshooting for, 445*t*
of drill pipe, 437-454
electrochemical activity and, 439
inhibitors, 566-570, 570*t*
classification of, 569
microbiological, 453-454
packer fluids and, 505
Corrosion resistant alloy (CRA), 37
Corrosion test cell, 81*f*
Corrosion test(s), 79-82
Cost versus value, 462, 507
Counter ions, 115
Creep, 320
brittle-plastic yielding and, 334-335
in salt formations, 330
Cross-linked fracture gels, 527
characteristics of, 527

Subject Index

Cross-linked gel fluid, 530
Crude oil muds, 10
Crystal lattice, 98
Crystalline swelling, 110, 338
Crystallization of the fluid, 48–49
Crystallization temperature, 43
c-spacing, 98
and swelling, 111, 113t
Cuttings
 dispersion, clear brine polymer drilling fluids and, 400
 disposal of. See Waste disposal
 reinjection, advantages and disadvantages of, 610t
 volume of, calculation of, 598
Cuttings, removal of, 643–644

D

Darcy's law, 246, 264–265, 383–384
Data acquisition, 36
Data collection and planning, 36–37
Decanting centrifuge, 662
Deep holes, muds for, 658–660
Deepwater completion fluid selection, 493–494
Deflocculants, 119. See also Thinners
Deflocculation, 117–121
 and clay blocking, 470–472
Defoamers, 303–305
Degree of polymerization (DP), 129
Degree of substitution (DS), 129
Density. See also Mud density
 brine test, 44–47
 of common drilling fluids weighting agents, 5t
 of drilling fluids, 13–15
  evaluation, 57–60
  range of, 28
  prediction, 46–47
  temperature and pressure effect on, 45–46
  units for, 13, 58
Desanders, 398–399, 661–662
Desilters, 398–399
Desulfovibrio, and drill pipe corrosion, 453–454
Diagenesis, of sediments, 309–310
Dickite, 105
Diesel oil-bentonite (DOB) slurries, 415
 as lost circulation material, 415
Diesel oil-bentonite-cement (DOBC) slurry (ies), as lost circulation material, 415

Differential pressure sticking mechanisms, 374f
Differential sticking, 21
 of drill string, 372–382
  mechanism of, 372–379
  prevention of, 379–380
Differential-pressure sticking of drill pipe
 testing of, 379–381, 381t
Differential-pressure sticking test, 77–79, 81f
Differential-pressure test apparatus, 380f
Diffusion osmosis, 352–353
 and methyl glucoside, 352–353
Dilatant flow, 151, 176
Dioctahedral sheets, 96
Direct-indicating viscometer, 161
Disappearing roads, 631–632
Dispersed Bentonitic fluids, 9
Dispersion, 121–125
 definition of, 121
 prevention of, 123
Diutan, 143, 144f
Divalent inorganic brines, 39t, 42–44
Diverting agents, 534
Dolomite, density of, 5t
Drag
 definition of, 368
 drill string, 368–372
Dräger tube(s), 85
Drill pipe. See also Stuck drill pipe
 flow conditions in, 204–207
 pressure gradient in, determination of, 210–213
Drill piperotation, effect on velocity profiles, 204–205
Drill string
 differential sticking of, 372–379
 drag, 368–372
 torque, 368–372
Drilled solids, 7
Drill-in fluid(s), 506–510
Drilling Engineering Association, project DEA-113, 360–361
Drilling fluid selection, 22–32, 23t
 for fast drilling, 28
 for formation evaluation, 29
 for geopressured formations, 26
 for high angle holes, 29
 for high temperature, 26–27
 for hole instability, 27–28
 location and, 22–25
 for maintaining hole stability, 353–354

Subject Index   713

for mud-making shales, 25–26
for rock salt, 28
for wildcat well, 29
Drilling fluid systems, 5–9
  types, 9–13
    nonaqueous-based systems, 10–12
    water-based muds, 9–10
Drilling fluid(s)
  aerated, in drill pipe corrosion, 445t
  alkalinity of. See also Alkalinity; pH
  aquatic toxicity of, and regulation of
    discharge, 625–627
  behavior of, at low shear rates, 163–166
  characteristics of, 537
  classification of
    by base, 5–6
  composition of, 3–13
    weight materials, 4
  density of, 13–15
    evaluation, 57–60
    range of, 28
  drilling problems related to, 367
  filtration properties of, 19–22
  flow properties of, 15–19
  functions of, 2–3, 537
  health risk management of, 597
  high-temperature, 427–437
  materials
    evaluation of, 86–87
    principles of, 86–87
  negative requirements for, 3
  performance evaluation
    properties measured, 57–70
    sample preparation for, 56
  productivity impairment caused by,
    prevention of, 29–30
  properties of, 13–22
  salinity, balancing, for hole stability,
    309–310, 354–356
  and shale formations, interaction between,
    and hole instability, 338–362
  surface chemistry of, 285
  toxicity of
    assays, 579
  viscosity of
    evaluation, 60–64
    and penetration rate, 395
Drilling performance, solids and, 396
Drilling rate. See also Penetration rate
  bridging solids and, 397–399
  versus depth, 403–404
  differential pressure and, 394

drilling fluids and, 383–401
kinematic viscosity and, 395
mud properties and, 394–401
slow, 383–401
Drilling rigs
  small footprint, 627–629
Drilling wastes, 598. See also Waste disposal
Drilling well
  filtration cycle in, 270–271
  flow conditions in, 204–207
  hydraulic calculations made at, 208–210
Dry air, characteristics of, 23t
Dry gas drilling, 677–678
Drying cuttings, solids control equipment for, 607f
Du Noüy ring method, 286
Du Noüy-Padday method, 286
Du Noüy-Padday Rod Pull Tensiometer, 286
Dynamic filtration, 68–70, 264–270
  definition of, 245
Dynamic filtration rate, 68
Dynamic weight material, 236–237

E

Early oil company developments, 669–670
Effective stress, 315–316
Effective viscosity, 17–18, 157
  annular, 237–239
  calculation, from Savins-Roper viscometry, 162
  Einstein's equation for, 164–165
  of power law fluid, 178
  shear rate and, 16–18
  temperature and shear rate and, 195–197
Einstein's equation, 164–165
El Paso Natural Gas Co., 677
Electrical properties, determination of, 74–76
Electrochemical cell, 438–439, 439f
Electrochemical reactions, 437–440
Electrode(s), 438–439
Electromotive series, 438, 438t
Electrostatic double layer, 115–117, 437–438
Electroviscous effect, 129
Emulsifiers, 132, 293–295, 561
  low-fluorescent, 565
  mechanical, 296
  for oil-based drilling muds, for deep drilling, 577

Emulsion muds, 10
Emulsion(s), 10, 293–297
  classification of, 561
  continuous phase of, 293–295
  dispersed phase of, 293–295
  functions of, 561
  loose, 561
  medium, 561
  oil-in-water, 296
  and drilling rate, 396, 398f
  for gun perforating, 499
  stability of
    particle contact angle and, 296
  tight, 561
  water-in-oil, 296. See also Invert emulsion(s)
  degradable, 500
Encapsulation
  definition of, 351
  of drill cuttings, with PHPA, 351
  and shale stability, 361
Energized fracturing fluids, 528
Enhanced mineral oil-base fluids (EMOBFs), 604
Enhanced Performance Water-Based Muds, 9–10
Environmental protection, 22–25
Environmental Protection Agency, 601–603
Environmentally assisted cracking (EAC), 37
Environmentally Friendly Drilling Program (EFD), 630–631
Enzyme breakers, 533
Equivalent circulating density, 215
  calculation, 421
Equivalent spherical radii (esr), 94
Equivalent static density (ESD), 45
  comparison with hydrostatic pressure, 46f
Esters, 11
  as additives for nonaqueous emulsion drilling fluids, 577–578
  as lubricants, 550–551
Ethanolamine, 561f
Ethers, 550–551
Exchangeable cations, 98–100
Expense versus value, 461–462, 507

F

Fanning friction factor, 19, 184–185
  calculation, at well, 209
  in laminar flow, 185
  in turbulent flow, 184–185
  of Newtonian fluids, 184–185
  of non-Newtonian fluids, 185–190
Fatigue failures, 440–441
Fatty acid
  as lubricant, 370
  sulfurized, as lubricant, 370
Fault zones, drilling through, 337
Faulting
  least principal stress at, 318
  normal, tensile, 319f
  stress concentrations and, 325, 328f
  overthrust, 318–319
  stress concentrations and, 325
  stresses acting in, 317–318
Feldspars, 107
Fermentation biopolymers, 142–146
Ferrochrome lignosulfonate, 120–121, 659
Ferromagnesium minerals, 107
Fibrous materials, 413–414
Field units, conversion factors for, 637, 638t, 639t
Filter cake, 251–264
  coefficient of friction of, mud composition and, 379–381
  compaction of, and coefficient of friction, 373–374, 377
  and differential sticking of drill string, 372–373, 374f
  formation of, 19–21, 245
  permeability of, 20, 245, 252f, 255–256
    aggregation and, 260
    effect of particle size and shape, 256–258
    flocculation and, 260
    temperature and, 249–251
  thickness of, 251–253
    and differential sticking, 378–381
    equilibrium, under dynamic filtration, 266t
    measurement of, 65–66
  and pore pressure at cake/pipe interface, 375, 376f
  and pull-out force for stuck pipe, 377–378
Filter press, 66f, 68–70
Filter tester(s), 67f, 70f
Filtrate, 20
  volume
    and pressure, relationship between, 248–249
    and temperature, relationship between, 249–251
    and time, relationship between, 246–248

Filtration
  below bit, 271–274
  in borehole, 270–271
  control, 397–398
  in KCl muds, 347–348
  dynamic. *See* Dynamic filtration
  evaluation
    safety precautions with, 66–68
  static. *See* Static filtration
Filtration rate(s), 383
  downhole, evaluation of, 275–281
  and drilling rate, 401, 401*f*
First crystal to appear (FCTA), 48–49
Flaky materials, 414
Flocculants, 82, 661
Flocculation, 117–121
  definition of, 117, 121
  and effect of temperature on drilling fluid rheology, 195, 197–200, 198*f*
  and filter cake permeability, 260
  and gel strength, 121–125
  prevention of, 119
  reversal of, 119
Flocculation value, definition of, 117
Flocculation-deflocculation process, versus aggregation-dispersion process, 121, 122*f*
Flow equations, application to conditions in drilling well, 203–215
Flow model, 17*f*
Flow regimes, 15
Fluid decisions, 37
Fluid functionality, variations in, 135
Fluid loss, 532
Fluid loss control, 137–138, 141–142, 147, 530
Fluid-loss tests, 263
Foamed fracturing fluids, 528
Foaming agent(s), 82–84
  evaluation of, 82–84, 83*f*
Foam(s), 6, 12, 82–84, 300–303, 679.
    See also Aphrons; Gel foam/stiff foam
  biliquid, 562–564
  as Bingham plastic, 82, 301
  characteristics of, 23*t*
  consistency curves, 82
  effective viscosity, 82
  flow properties of, 682–683
  formation of, 300
  functions of, in drilling industry, 301
  ingredients, 23*t*
  oil-lamella, 564

plastic viscosity, 82
preformed stable, 681–682
quality, 82, 300–301, 682–683
rheological properties of, 82, 301
stability of, 300–301
stable, 23*t*
  characteristics of, 23*t*
  ingredients, 23*t*
stiff, 680–681
water-lamella, 564
yield stress, 82
Food blenders, in performance evaluation, 56, 59*f*
Formate brines, 511–515
  API tests in, 513, 514*t*
  supercooling of, 49
Formate brines densities and PVT data, 47
Formate fluids
  API retort test contraindicated with, 72, 513
  API tests in, 513, 514*t*
  calcium carbonate in, test for, 513–515
  drill solids in, determination of, 515
  solids content procedure with, 72
Formation damage
  by completion fluids, tests for, 500–502
  prevention of, 489–493, 559–560
Formation evaluation, drilling fluid selection for, 29
Formation fluids, hydrostatic pressure gradient of, 13–14
Formation pressure, 311
Formic acid, 530
Fracture initiation pressure, 522
Fracture permeability, 337
  reservoirs with, 485
Fracture propagation pressure, 522
Fracture(s)
  induced, and loss of circulation, 402–411
  natural closed, and loss of circulation, 413
  natural open, and loss of circulation, 421
  in shales, hydration of, and shale destabilization, 325
Fracturing
  hydraulic, 521
  intentional, 522
  unintentional, 521
Fracturing fluids, 521
  acidized, 530
  additives, 532–534
  composition, 523–525
    basic components and additives, 523–525

716 Subject Index

Fracturing fluids (*Continued*)
  energized, 528
  foamed, 528
  generic additives used in, 524*t*
  methanol, 531
  purpose of, 522
  selecting a fracturing fluid system, 525–532
  water-based, 525–532
  slickwater. *See* Slickwater fracturing fluids
  types, 522, 524*t*
Fragile gel(s), 544
Fresh water, characteristics of, 23*t*
Friction reducers, 191–193, 532
Friction reduction, versus shear thinning, 193

G

Galactomannans, 139
Galena
  density of, 5*t*
  as weighting agent, 542
Garrett gas train, 85–86
Gas, 12
  dry, 12. *See also* Dry gas drilling
Gas content, determination of, 71
Gas drilling, 12
Gas-based drilling fluids technology, 677–683
  aerated mud, 679–680
  dry gas drilling, 677–678
  flow properties of foam, 682–683
  foam, 679
  gas drilling benefits, 683
  gel foam/stiff foam, 680–681
  preformed stable foam, 681–682
  water-bearing zones, problems with, 678–679
Gel foam/stiff foam, 680–681
Gel rate constant(s), 171, 173*f*, 173*t*
Gel strength
  after cessation of shearing, 174
  evaluation, 64–65
  measurement of, 19
  shear stress load and, 174, 174*f*
  and swab pressure, 233, 235*f*
  temperature and, 195–197, 198*f*
  time and, 171, 171*f*, 172*f*
Gel structure, 19
Gelation. *See also* Aggregation; Flocculation
  flocculation and, 119
  mechanism of, 125–127
  salt concentrations and, 125

Gelled nonaqueous fluids, 528
Gelled water, 530
Geological conditions, and pore pressure, 13–14
Geopressured formations, 13–14
  drilling fluid selection for, 26
  and induced fractures, 403, 410–411
  and transient borehole pressures, 229
Geopressured gradients, 312–314
George F. Mepham Corporation, 649
Geostatic pressure, 310
Geostatic pressure gradient, 14, 310
Geothermal energy, drilling for, mud development for, 432
Geothermal gradient, 423–427
Gibbsite, 96
Glass beads, and drill string torque and drag reduction, 370
Glass electrode pH meter, 76
Glass granules, ellipsoidal, 549
Glass hydrometers, 45
Glass microspheres, 543
Glutaraldehyde, 533
Glycosides, 579
Goniometer/tensiometer, 286
Granular loss-circulation materials, 422
Granular materials, 414
Graphite, polarized, 548
Gravel beds, and loss of circulation, 412
Gravel pack operations, 488–489
Gravel packing, 29–30
Grease(s), 368–369
Grouting, permanent, 547
Guar gels, 523
Guar gum, 131–132, 132*f*, 138–141, 140*f*, 491–493, 661
  and starch, 661
Guar-based polymers, 523
Gumbo shales, 333–334
  and bit balling, 392–393
  definition of, 333
  drilling rates for, 333–334
  plastic yielding in, 333
Gun perforating, 29–30, 485–487
  oil-in-water emulsion for, 499
Gyp mud(s)
  for mud-making shales, 26
  for shale stabilization, 27
Gyp-chrome lignosulfonate mud, 659
Gyp-lignosulfonate muds, and control of borehole hydration, 352
Gypsum-treated mud, 659

H

Hach sulfide test, 86
Halliburton Oil Well Cementing Company, 670
Hardness, total, estimation of, 85
Harmonized Mandatory Control System, 617–618
Heaving shale, muds for, 656–657
HEC, 138
Hectorite, structure of, 101t
Hematite, 4, 5t
  density of, 5t
Hematite/Itabirite, 541
Hershel-Bulkley model, 18
High angle holes
  behavior of cuttings in, 29
  and differential sticking, 378–379
  drilling fluid selection for, 29
  water-base muds for, 29
High density brines, 531–532
High filter loss slurry(ies), 416
High Performance Water-Based Muds (HPWBM)
High-density clear brines, 38
High-pH mud, 657
High-temperature high-pressure (HTHP) filtration, 430
  and mud properties, 430, 431f, 433f
HLB number, 292
Hoffmann structure, 96–97
Hoffmeister series, 119
Hole angle, and hole stability, 325, 331f
Hole cleaning, 216–222
  optimum rheological properties for, 224–226
Hole contraction, drilling fluid selection for, 27
Hole enlargement, 337
  in brittle shales, 343–344
  drilling fluid selection for, 27–28
  in shale, 205, 206f
    clear brine polymer drilling fluids and, 400–401
Hole instability. See also Hole enlargement
  adsorption and, 343
  caused by interaction of drilling fluid and shale formations, 338–362
  desorption and, 343
  drilling fluid selection for, 27–28
  types of, 309
Hole stability, 645–646. See also Wellbore stability
  field and operational parameters affecting, 359, 359t
  in formations with no cohesive strength, 337–338
  hole angle and, 331f
  hydraulic pressure gradient and, 325–330
  maintenance of, selection of mud type for, 353–354
  mud density and, 331–332
Horneblende, 107
Houston Chapter of the American Petroleum Institute Division of Production, 654–655
HPHT fracturing, 529
HPHT polymers, synthetic, 529
Hughes Tool Company laboratory, 666
Humble Oil and Refining Company, 653–654, 669
Humic acids, 576
Hydraulic blast furnace slag, 580
Hydraulic calculations
  made at drilling well, 208–210
  metric units for use in, conversion of field units to, 210t
Hydraulic fracturing, 521
Hydraulic pressure gradient, and hole stability, 325–330
Hydraulic radius, 159
Hydrocarbons, 52
Hydrocyclones, 32
Hydrofluoric acid (HF), 530
Hydrogen embrittlement, 441–442
  test for, 82
Hydrogen ion concentration. See pH
Hydrogen sulfide, 673
  in drill pipe corrosion, 444–448, 445t
  removal, 571–572
Hydrogen sulfide scavengers, 449
Hydrolysis, 129–130
Hydrophilic clay, 672
Hydrostatic pore pressure gradients, 311
Hydroxyethylcellulose (HEC), 131–132, 491–493, 495
  in completion and workover fluids, 498
  effect on apparent viscosity of brines, 495

I

Iilmenite
  density of, 5t
Illites, 102–104
  base exchange capacity (BEC) of, 109t

718 Subject Index

Illites (*Continued*)
 and control of borehole hydration, 348*t*
 formation of, 312–313
 occurrence of, 107
Ilmenite, 396
 as weighting agent, 541–542
Imides, 573*f*
Induced fracturing, 14
Inhibited (inhibitive) muds
 and shale stabilization, 352
Inhibited muds, 663–665
Initiation pressure, 406
Injection pressure, 406
Inorganic brines, 39
 divalent, 39*t*, 42–44
 monovalent, 39–42, 39*t*
Inorganic colloids, 6–7
Instantaneous shut-in pressure (ISIP), 407–408, 410
Interfacial tension, definition of, 285
Intergranular stress, 315–316
International Organization for Standardization (ISO)
 standard 11734, modified for NADF biodegradation, 619–621, 621*t*
 standard field and laboratory tests, 55
Invert emulsion fluids, 562
Invert emulsion(s), 10, 23*t*, 293–295, 561–562, 671
 drilling fluid systems, 562
 oleophilic, rheology, effect of temperature and pressure on, 200–201, 200*f*
Ion exchange, 108–109
Iron, as contaminant, 51
Iron minerals, powdered, as hydrogen sulfide scavengers, 449
Iron oxide, 649
 density of, 5*t*
Iron oxides, 541, 647–648
ISO. *See* International Organization for Standardization (ISO)
Itabirite, 396

J

Jamin effect, 465–467
Jeffamine®, 557

K

K (power law constant), 18, 61–62, 176–178
 calculation, at well, 209
 and generalized power law, 179–184

Kaolinite, 105, 107
 base exchange capacity (BEC) of, 109*t*
 postive sites on crystal edges, 117
Ken Corporation, 670
Kinematic viscosity, 663

L

Laboratory drilling tests, 383
Laminar flow, 15–19, 152–155
 characteristics of, 151
 Fanning friction factor and Reynolds number in, relationship of, 185
 of Newtonian fluids, 152–155, 153*f*
Last crystal to dissolve (LCTD), 48–49
Lecithin, 298, 671
Legacy system, 643
Leonardite. *See* Lignite
Lignins, 666
Lignite, 657
 oil dispersible, 299
 temperature stability of, 429
Lignite muds, 545
Lignosulfonate, temperature stability of, 427–429
Lignosulfonate muds, 545
Lignosulfonates, 659
Lime content, analysis of, 85
Lime muds, 9, 657
 for mud-making shales, 26
 and shale stabilization, 27, 352
Lime-lignosulfonate muds, and control of borehole hydration, 352
Linear gels, 527
Liquid volume fraction, 683
Litholostatic pressure gradient, 14
Lithostatic pressure. *See* Geostatic pressure
Logging, drilling fluid selection for, 29
Loss of circulation, 401–421
 caused by hydrostatic pressure of mud column exceeding fracture pressure, and regaining circulation, 419–420
 curing, materials for. *See* Lost circulation material (LCM)
 diagnosis of, 416–417
 induced by marginal pressures, and regaining circulation, 418–419
 in induced fractures, 402–411
 in natural open fractures, 411
 and regaining circulation, 421

in opening with structural strength, 411–413
and regaining circulation, 416–417
into structural voids, and regaining circulation, 417–418
Lost circulation additives, 546–547, 546t
Lost circulation material (LCM), 413–416, 418–419
Low shear rate viscosity (LSRV), 142
Low shear rate yield point, 236–237
Low-solids muds, 23t, 660–663
  ingredients, 23t
  and penetration rate, 396–397
Low-solids mudscharacteristics of, 23t
Low-toxicity mineral-base fluids (LTMBFs), 624
LSR-YP, 236–237
Lubricants, 368–372, 547–550
  alcohols, 549–550
  biodegradable, 551
  and casing/tool joint friction, 370–372
  and coefficient of friction, 368–369
  comparison of, 371t
  ellipsodial glass granules, 549
  evaluation of, for torque reduction, 369–370
  extreme pressure (EP), 370
  graphite, 548
  olefins, 549
  paraffins, 549
  phospholipids, 549
Lubricity, 76–77
Lubricity evaluation monitor, 79f
Lubricity Evaluation Monitor (LEM-NT), 372
Lubricity tester(s), 77, 78f, 369f
Lyotropic series, 119

M

MacMichael viscometer, 650
  flow behavior of clay mud in, 167–168
Macroemulsions, 561
Magnesium, estimation of, 85
Magnetite, density of, 5t
Maleic anhydride, 149
Manganese tetroxide
  drilling fluid formulation with, 542–543
  as weighting agent, 542–543
L-Mannose, 143–144
Marangoni convection, 564
Marcellus shale, 523–525, 526f

Marine transport to shore, advantages and disadvantages of, 610–611, 611t
Marsh funnel, 60, 61f
Matrix stress, 315–316
Mean hydraulic radius, 155
Measured while drilling (MWD), 406
Membrane efficiency, 359–361
Membrane formation, 558–559
  by shale/mud interactions, types of, 359–360
Membrane former(s), and shale stability, 361
Mepham Corporation, 649
Methanol fracturing fluids, 531
Methyl glucosides, 579
  diffusion osmosis and, 352–353
Methyl orange end point (Mf), 85
Methylene blue test, 73–74, 354, 661
Metric units
  coherent, 637
  conversion factors for, 637, 638t
  for use in hydraulic calculations, conversion of field units to, 210t
Micas, 107
  hydrous, 102–104
Microemulsions, 561
Microfractures
  and loss of circulation, 413
  in shales, hydration of, and shale destabilization, 344, 345f
Milk emulsion, 397, 660–661
Mineral paraffin, 12
Mineral(s). See also Clay mineral(s)
  scale deposits, in drill pipe corrosion, 445t
  three-layer
    c-spacing of, 98
    unit cell of, atom arrangement in, 98, 98f
  two-layer
    c-spacing of, 98
    unit cell of, atom arrangement in, 98, 99f
Miocene drill cuttings, 333
Mist(s), 6, 12, 300–303
  characteristics of, 23t
  ingredients, 23t
Mixed metal hydroxide (MMH) fluids, in workover wells, 488
Mixed metal hydroxides, 543–544
Mixed metal silicates, 543–544
Mixer(s), 87
  in performance evaluation, 56, 57f
Modified power law fluid flow model, 18–19
Modified vegetable oils, 12
Modified-natural polymers, 137

720 Subject Index

Modifiers
  viscosity, 543–545
Mohr diagram, 318, 318f, 412f
Mohr envelope
  for plastic failure, 318f
  for unconsolidated clay, 320f
  for unconsolidated sand, 319f
Mohr failure curve, 320–322
Molecular properties, variations in, 135
Molybdenum disulfide, 548
Monomers, 129
Monovalent inorganic brines, 39–42, 39t
Monovalent organic brines, 39t, 44
Montmorillonite, 25–26
  base exchange capacity (BEC) of, 109t
  and clay blocking, 470–472
  and control of borehole hydration, 347–348
  conversion to illite, 312–313
  crystalline swelling pressure of, 338, 340f
  identification of, 73
  interlayer cations, 114, 114f
  lattice expansion of, 111, 113f
  monionic, c-spacing, in pure water, 111, 113t
  occurrence of, 107–108
  particles, dimensions of, 103t
  structure of, 100–101, 101t, 102f
  swelling, mechanisms of, 109–115, 112f
Montmorillonoids. *See* Smectites
Montney shale, 523–525, 526f
Mud balance, 57–60
  pressurized mud balance, 60f
  standard mud balance, 60f
Mud density
  barite and, 15
  and bottomhole pressure, 426, 428t
  and hole stability, 325, 331–332
  and penetration rate, 394
  and plastic flow of salt formations, 331–332, 333f
  pressure control by, 646–648
  and rate of penetration, 14
  and shale stability, 361
Mud engineer, 1
Mud gradient, conversion factors for, 58
Mud handling equipment, 30–32
Mud industry
  birth of, 648–649
  rapid growth of, 649–655
Mud overbalance pressure, 14
Mud particle damage, 481–484
  occurrence of, in field, 484–487

Mud spurt, 20, 247, 260–261, 397–398, 481–482
  filtration medium and, 262, 263t
Mud to cement conversion, 580
Mud types/systems, development of, 655–665
  deep holes, muds for, 658–660
  heaving shale, muds for, 656–657
  high-pH muds, 657
  inhibited muds, 663–665
  low-solids muds, 660–663
  nondispersed polymer muds, 663
  oil emulsion muds, 658
  saltwater muds, 655–656
Mud-laden fluid, 646
Mud(s), 1–2
  components
    and casing/tool joint friction, 370–372
    chemical degradation of, temperature and, 251
  costs of, factors affecting, 15
  particles from, permeability impairment by, 481–484
  performance evaluation
    sample preparation for, 56
    standard field and laboratory tests for, 55
  properties of, and bit penetration rate, 216
  rheological properties of
    optimum, for hole cleaning, 224–226
    for optimum performance, 215–236
  in workover wells, 487–488
Muscovite, 102–104

N

n (flow behavior index), 176, 180f
  calculation, at well, 209
  and generalized power law, 179–184
  and velocity profile, 178–179, 180f
n (power law constant), 18, 61–62
Nacrite, 105
NADF. *See also* Nonaqueous drilling fluid (NADF)
Naphthenic acid, 670
National Association of Corrosion Engineers, 666
National Energy Testing Laboratories (NETL), 631–632
Natural polymers, 136
Nephelometer, 49–50
Nephelometric turbidity units (NTU), 49–50
Nernst potential, 115, 438

Newton (N) (unit), 637
Newtonian fluid(s), 16, 151, 176, 532
  consistency curve of, 153f, 178, 179f
  laminar flow of, 152–155, 153f
    in round pipe, 154
  turbulent flow of, 184–185
  velocity profile of, in turbulent flow, 184, 185f
  viscosity of, 16
    determination of, 155
Newtonian viscous brines, 495
Nitrogen oxides (NO_x), air emissions, studies of, 632
Nonaqueous base fluids (NABFs), 603
  high aromatic content, 603
  low to negligible aromatic content, 604–608
  medium aromatic content, 604
Nonaqueous drilling fluid (NADF), 10–11, 236–237, 600
  aquatic and sediment toxicity of, 624–625
  and regulation of discharge, 625–627
  cuttings and excess fluid
    benthic impacts of, 614–615
      laboratory studies of, 617–618
    benthic recovery from, 616–617
    bioaccumulation of, 616
    characterization of, 627–629
    biodegradability of, 615–616, 618–629
    chemical toxicity of, 616
    discharge into ocean
      effect of, 603–608, 612–614, 629–630
      fate of, 612–614
    discharge limitations offshore, 598–599
    initial seabed deposition of, 612–614
    modified ISO 11734 for, 619–621, 621t
    offshore waste disposal for, 603–608
    and organic enrichment, 615–616
    physical persistence on seabed, 614
    simulated seabed studies of, 623–624
    SOAEFD solid phase test for, 621–623, 622t
    standard biodegradation tests for, 619, 620t
  solids phase, 12
  wastes associated with, 598
Nonaqueous drilling muds
  additives for, 539f, 577–580
  antisettling properties of, enhancement of, 578
  special additives for, 577–580
  wettability, materials for, 579
Nonaqueous fluid loss, 258–260
Nonaqueous fluids, 499–500
  and wellbore stability, 345–347
Nonaqueous gelling agent, 544–545
Nonaqueous-Based Drilling Fluids (NADF), 5
Nonaqueous-based fluids, aquatic toxicity of, 626t
Nonaqueous-based systems, 10–12
Non-dispersed fluids, 9
Nondispersed muds, 28
Nondispersed polymer muds, 663
Nonhazardous oilfield wastes (NOW), 597
Non-Newtonian fluids, 17
  turbulent flow of, 185–190
  viscosity of, in turbulent flow, 190
Nonproductive time (NPT), 309
  related to drilling fluids, 367, 368f
Nonpyruvylated xanthan (NPX), 143
Nontronite, structure of, 101t
Normal compactions, 310
Normally pressured formations, 13–14
Norwegian Institute for Water Research (NIVA), 623–624

O

Octahedral sheets, 96
Offshore discharge, 608
  advantages and disadvantages of, 609t
Offshore drilling
  general discharge limitations for, 598
  wastes, regulations governing, 598
Oil, distillate, grease, and pipe dope, 52
Oil Base Drilling Fluids Company, 670
Oil base–emulsion drilling fluids, 671–672.
  See also Invert emulsion(s)
Oil content, determination of, 71
Oil emulsion muds, 658, 671
  coefficient of friction for, 372
Oil mud, penetration rate with, 675–676
Oil-based drilling fluids (muds), 10
  asphaltic, rheology, effect of temperature and pressure on, 201–203, 204t
  balanced salinity, 345–346
  formulation of, 354–356
  coefficient of friction of, 379
  and control of borehole hydration, 344–345
  and corrosion control, 454
  rheology, effect of temperature and pressure on, 200

Oil-based drilling fluids (muds) (*Continued*)
  modified power law for, 201–203
  temperature stability of, 429
Oil-based drilling fluids technology, 668–676
  annulus packs, 673
  borehole stabilization by oil muds, 674
  commercial oil muds, 670
  early oil company developments, 669–670
  extreme borehole conditions, 674–675
  oil base–emulsion drilling fluids, 671–672
  oil muds, applications of (1935–50), 670–671
  organophilic clays and ammonium humates, 672
  penetration rate with oil mud, 675–676
  reasons for development, 668
  scope of oil mud applications, 676
  well completion, oil for, 668–669
Oil-based mud (OBM), 10
Oilfield, 137
Oilfield grade starches, 141–142
Oil(s), 5
  aniline point of, 84
  aromatic content of, determination of, 84
  ester-based, as lubricants, 550–551
  interstitial, *in situ* emulsification of, 468
Oil-soluble organic particles, water-base fluids containing, 497–498
Oil-wetting agents, 297–299
Oklahoma City field, drillers in, 658
Olefins. *See also* PAO, synthetic
  as lubricants, 549
Olephins, 12
Onshore disposal, 608–611
  costs for, 609–610
Onshore disposal options, 613*f*
Operating parameters, 37
Optical density, aggregation and, 122–123
Organic brines, monovalent, 39*t*, 44
Organic colloids, 7
Organo-clay complexes, 672
Organophilic clays, 672
Osmotic control, and shale stability, 361
Osmotic effectiveness, chemical, 353
Osmotic swelling, 114, 338
Ostwald-deWaele equation, 18
Overburden pressure gradient, 14
Oxides of nitrogen (NO_x), air emissions, studies of, 632
Oxidizer breakers, 533
Oxygen, in drill pipe corrosion, 440–441, 445*t*, 449–450

Oxygen corrosion cell, 441*f*
Oxygen scavenger, 453, 453*f*, 571

P

PAC (polyanionic cellulose), 138, 147
Packed hole, 379
Packer fluid(s), 503–506, 666–667
  aqueous, 503–504
  aqueous drilling muds as, 503–504
  clear brines as, 504
  and corrosion, 505
  functions of, 503
  low-solid, 504
  oil-base, 505–506
  requirements for, 503
Pad fluid, 531
Palygorskite. *See* Attapulgite
PAM-based polymer, 529
Pan American Petroleum Corporation, 661
PAO, synthetic, 550
Paraffin(s), 11, 549
Paraformaldehyde, for bacteria control, 454
Partially hydrolyzed polyacrylamide (PHPA)-AMPS-vinyl phosphonate (PAV), 529
Partially hydrolyzed polyacrylamide-polyacrylate copolymer (PHPA), and hole stability, 348
Particle association, 117–125
Particle counter, 50–51
Particle plugging apparatus (PPA), 422
Particle size
  classification of, 74*t*
  definition of, 74*t*
Pascal (Pa) (unit), 637
Penetration rate. *See also* Drilling rate
  dynamic chip hold-down pressure and, 389–390
  factors affecting, 383
  mud density and, 14, 394
  mud properties and, 216
  mud viscosity and, 395
  with oil mud, 675–676
  static chip hold-down pressure and, 383–389
Peptization, 119
Permafrost, thawing of, 25
Permeability impairment
  by indigenous clays, 469–470
  by particles from drilling mud, 481–484
pH. *See also* High-pH mud
  and clay blocking, 478–481

Subject Index 723

and clay dispersion, 478
control, for KCl muds, 351
and corrosion control, 443–451
and drill pipe corrosion, 449, 452f
of drilling fluids
  determination of, 76
  measurement of
    colorimetric method, 76
    electrometric method, 76
  thickeners responsive to, 543
Phenolphthalein end point (Pf), 84–85
Phenols, chlorinated, for bacteria control, 454
Phillips Petroleum Co., 679–680
Phosphate esters, 551
Phosphate thinned water-based muds, 545
Phosphates, 41–42
Phospholipids, as lubricants, 549
Piercement-type salt domes, drilling on, 656
Pierre shale cuttings, dispersion tests on, 355t
Pipe sticking tester, 375f
Planning, 36–37, 36f
Plastic beads, and drill string torque and drag reduction, 370
Plastic deformation, 316
  of borehole, 320, 321f
  occurrence of, in field, 330–334
Plastic failure, Mohr envelope for, 318f
Plastic viscosity (PV), 17, 155–157
  of Bingham plastic, determination of, 159
  calculation of, 61
  measurement of, 161–162
  temperature and, 195
Plastic yielding, occurrence of, in field, 330–334
Plug flow, 157–158, 158f
Pneumatic drilling, 6
  characteristics of, 23t
  ingredients, 23t
Pneumatic drilling fluids, 12–13
Poise, 15–16
Poiseulle's equation, for laminar flow of Newtonian fluid in round pipe, 154–155
Poisson's ratio, 317
  for different lithologies, 408, 409t
Poly alpha olefin. See PAO, synthetic
Polyacrylamide, 139, 149f
Poly(acrylamide), quaternized, 558f
Polyacrylamide-acrylate copolymer, 129–130, 130f
Polyacrylates, 659
Poly(acrylic acid) (PAA), 574

Polyacrylonitrile, in completion and workover fluids, 498
Poly(alkylene glycol) (PAG), 550
Poly(amide) surfactants, 566f, 567f
Polyamidoamines, 672
Polyamines, 672
Polyaromatic hydrocarbons (PAH), 627
Polycarboxylates, as dispersants, 575
Polycrylamides, 149
Polyelectrolytes, 129, 258
  as flocculation aids, 129
Poly(ether amine), 557, 557f
Poly(ether)cyclicpolyols, 577
Polyethylene glycol (PEG), in shale stabilizers, 556
Polymer fluid loss agents, 258
Polymer mud, 136
  potassium chloride, for shale stabilization, 27
Polymer systems, 9
Polymer temperature stability, 137
Polymer(s), 127–132, 135
  as additives by polymer primary function, 140t
  with amine sulfide terminal moieties, as dispersants, 575
  anionic, 557
  anionic association, 547
  brines contaminated with, 52
  in completion and workover fluids, 496–497
  degradation, by bacteria, 565–566
  and drilling rates, 398–399
  effect on apparent viscosity of brines, 495
  as friction reducers, 191–192, 398–399
  long chain anionic, and performance of KCl muds, 348, 349f
  low-molecular-weight
    as deflocculants, 435–436
    with high temperature stability, 435–436
  maleic anhydride copolymers, 573–574
  modified-natural, 137
  natural modified, 576–577
  nonionic, 131
  organic
    in drilling fluids, 132
    thermal decomposition of, 132
  and shale stabilization, 348, 350t, 351
  synthetic, 128, 137
    as dispersants, 573–577
  water swellable, 546–547

724 Subject Index

Polymer(s) (*Continued*)
water-dispersible. *See* Water-dispersible polymers
in workover fluids, 491–493
Poly(oxyyalkylene amine)s, 557
Polyphosphates, 650
Polysaccharides, 136–137, 576
  defined, 136–137
Polyvalent-metal soaps of resin acids, 672
Poly(vinyl alcohol) (PVA), 550
  quaternized etherified, 558, 558*f*
Pore pressure, 14, 311
  determination of, 403
Pore pressure transmission (PPT) test, 359–360, 360*f*
  versus traditional shale stability testing, 359–360, 361*t*
Potassium, determination of, 86
  field procedure for, 86
Potassium chloride, 40
  effect on linear swelling of shales, 347, 348*t*
Potassium chloride polymer muds, for shale stabilization, 27
Potassium chloride-polyacrylamide muds, 664–665
Potassium formate, 44, 512*t*, 554
Potassium hydrogen phosphate brine, 41–42
Potassium hydroxide, 351
Potassium lime mud (KLM), and shale stabilization, 352
Potassium muds, 28
Power law, 176
  generalized, 179–184
  for pseudo-plastic fluids, 18
Power law constants, 61–62
Power law fluids, 17–18
Preformed stable foam flow diagram, 682*f*
Preshearing time(s), 170
Pressure
  effect on density, 45–46
  and filtrate volume, 248–249
  and rheology of drilling fluids, 193–203
Pressurized crystallization temperature (PCT), 48–49
Prototype small footprint drilling rig, 631
Pseudoplastic fluids, 175–179
  consistency curve of, 175–176
  flow in round pipes, equation for, 179
Pseudo-plastic fluids, 18
Pump capacity(ies), 216
PVT data, 47
Pyrophosphates, 650
Pyrophyllite, structure of, 98–100, 101*t*

Q
Q-BROXIN, 659
Quaternary ammonium salts, 552–554
Quebracho extract, 545, 657
Quebracho muds, 545

R
Rabinowitsch-Mooney equation, 180–181
Rake, 384
Red lime mud, 657
Red muds, 545, 657
Redox potentials, 440
Reduced-pressure drilling, 679–682, 681*f*
Reduced-pressure drilling fluid, 677
References, abbreviations used in, 641–642
Reiner-Riwlin equation, 161, 179–180
Research and development (R&D), new product, 633
Reservoir drill-in fluid, 35
Reservoir drilling fluid(s) (RDF), 506–510
  density of, 508
  design of, 508–510
  and fluid loss, 508
  properties of, 507–508
  reactivity of, 508
  types of, 508
  viscosity of, 508
Residual oil saturation, 464–465
Residual water saturation, 464–465
Resin, thermoplastic, particles, in water-base fluid, 497–498
Resin acids, polyvalent-metal soaps of, 672
Resistivity meter, 75, 75*f*
Retort, 71–72
Reynolds number, 19, 184–185
  in laminar flow, 185
  in turbulent flow, 184, 186*f*
  of Newtonian fluids, 184–185
  of non-Newtonian fluids, 185–190
L-Rhamnose, 143–144
Rheogram, 16
Rheology, 151
Rheopexy, 166–167
Rig costs, and solids content, 31, 31*f*
Rigid-plug forming treatment, 422
Rittinger's formula, 217
Road(s), disappearing, 631–632
Rock salt
  drilling fluid selection for, 28
  plastic yielding in, 330

Rock(s)
  behavior of, under stress, 315–317
  brittle failure of, 316
  permeability of
    and dynamic chip hold-down pressure, 389–390
    and static chip hold-down pressure, 383–389
  plastic deformation of, 316
  strength of, average, 410t
  subsurface stress field and, 317–319
Roller oven, 88f
Rotary viscometer, 195

S

Saccharide derivatives, 555
Saccharides, defined, 136–137
Safety and the environment, 37–38
Salinity
  and clay blocking, 469–470, 474–475, 478t
  and drill pipe corrosion, 451, 451f
Salt beds
  drilling fluids for, 332
  plastic yielding in, 330
Salt solutions, saturated, water activities of, 351t
Salt water muds
  characteristics of, 23t
  ingredients, 23t
Salt(s)
  as clay/shale stabilizers, 552
  concentration, effect on linear swelling of shales, 348t
  soluble, in water-base muds, and control of borehole hydration, 347
  volume of, in mud, determination of, 71
Saltwater muds, 655–656
Saltwater systems, 10
Sand, 9
  unconsolidated, and mud particle damage, 484–485
Sand test, API, 86
Sandstone, permeability of, 312
Saponite, structure of, 101t
Sauconite, structure of, 101t
Schiff base, formation of, 558, 559f
Scleroglucan, 145–146, 147f
Seawater, characteristics of, 23t
Sedimentary basins
  formation of, 309–310
  geology and geophysics of, 309–310

Sediment(s)
  compaction of, 309–310
  consolidated, 309–310
  diagenesis of, 309–310
  unconsolidated, 309–310
Selection of completion fluid, 36–38
  corrosion resistant alloy (CRA), 37
  data collection and planning, 36–37
  flow chart for, 40f
  safety and the environment, 37–38
  wellbore cleanout, 38
Selective catalytic reduction (SCR) technologies, 632
Sepiolite, 107
  base exchange capacity (BEC) of, 109t
  as viscosifier, 429
Sepiolite muds
  for field use, 434–436
  ultra high temperature formulation, 415, 432–434, 435t
Settling efficiency, solids content and, 396–397
Shale control mud, 658–659
Shale encapsulator, 558
Shale shaker, 661–662
Shale test characteristics, 361t
Shale(s). See also Heaving shale, muds for
  abnormally pressured, 312
  adsorption and desorption of, 338–342
  adsorption/desorption isotherms of, 338–339, 339f
  brittle
    and hole instability, 343–344
    swelling pressures in, 343
  clear brine polymer drilling fluids in, 400–401
  density, effect of age and depth on, 343f
  destabilization, by hydration of fractures/microfractures, 344, 345f
  destabilizing ionic reactions within, 353
  dispersible, drilling mud formulations for, 352
  dispersion tests on, 356–358
  and drilling fluids, interaction between, and hole instability, 338–362
  geopressured, 312
    plastic yielding in, 334
  gumbo. See Gumbo shales
  laboratory analysis of, 354–361
  linear swelling of
    measurement of, 356
    salinity and, 347, 348t

Subject Index

Shale(s) (*Continued*)
  mud-making, drilling fluid selection for, 25–26
  permeability of, 312
  recovery, 556*t*
  rigsite testing of, 358
  sampling, 354
  slickensided, 337
  soft, and bit balling, 393
  soft unconsolidated, plastic yielding in, 330
  specimens, compaction of, 358
  stability testing, 359–360, 361*t*
  stability times, enhancement, mud properties for, 361–362
  stabilization, 344–345
    hydrolyzed copolymers and, 130
    muds for, 27
    stabilizers, 551–559
  swelling
    and bit balling, 392–393
    measurements of, 356
    volumetric swelling of, measurement of, 356
    water sensitive, 20–21, 27
Shear, 15–16
Shear rate, 15–16, 151
  direct-indicating viscometer, 61
  and effective viscosity, 16–18, 195–197, 198*f*
  and performance evaluation, 56
  at surface of filter cake, 68
  thickening fluid and, 415–416
Shear stress, 15–16, 18, 151, 315–316
Shear thinning, 17–18
  versus friction reduction, 193
Shell Oil Company, 669
Shell polymer mud, 664–665
SI units, conversion factors for, 637, 638*t*, 639*t*
Siderite, density of, 5*t*
Sieve tests, 73
Silica tetrahedra, 96
Silicate mud(s), and borehole stability, 344–345
Silver nitrate method, in analysis of chlorides, 84–85
Sized salt fluids, in workover wells, 488
Skin effect, 462–489
Slickensided shale, 337
Slickwater fracturing fluids, 525
  characteristics of, 525–526
  disadvantages of, 527
  low-viscosity fluids in, 526

Slurries, 414
Slurry fracture injection (SFI), 600–603
Smear effect, 422
Smectites, 98–100, 101*t*
  swelling, 338
  mechanisms of, 109–115
  SOAEFD solid phase test, 621–623, 622*t*
Sodium bentonite
  aggregation of, 123*f*
  polyvalent salts and, 123
  flocculation of, 123*f*
  polyvalent salts and, 123
Sodium bromate, 529
Sodium bromide (dry), 41
Sodium bromide (liquid), 41
Sodium chloride (dry), 39
Sodium clays, suspensions, rheology, temperature and, 197
Sodium formate (dry), 44
Sodium formate salts, 512*t*
Sodium montmorillonite
  particles, dimensions of, 103*t*
  structure of, 101–102, 104*f*
  swelling, 338
Sodium nitrate, 667
Sodium polyacrylate, 435–436
Sodium silicate, 650, 656–657
Sodium sulfite, 665
Sodium surfactant mud, 659
Sodium tetraphenylboron (NaTPB), 86
Solid(s), 51–52
  control, 31–32
  drilled, 7
  inert, 95–96
  size of, 9
Solids contamination, 49–50
Solids content
  determination of, 71
  and drilling performance, 396–397
  and settling efficiency, 396–397
Solids control equipment for drying cuttings, 607*f*
Solids removal equipment, 30–32, 398–399
Solution channels, and loss of circulation, 412
001 spacing. *See* c-spacing
Spalling, 27, 322–325, 327*f*
  brittle-plastic yielding and, 334–335
Specific gravity (SG), 13
  conversion factors for, 58
Specific surface, 6–7, 94
  versus cube size, 94

Subject Index 727

Specific surface energy, 290
Spotting oil, 381, 382f
Spud mud
  characteristics of, 23t
  ingredients, 23t
Stabilite, 648
Standard Oil Co., 681–682
Starch, 128, 131, 137–138, 141–142, 258
  in KCl muds, 347–348
Starch lubricant composites, 552t
Starch olefin copolymer
  lubricants, 551
Static filtration, 65–68, 246–251
  definition of, 245
  theory of, 246
Static filtration tester, 653f
Static-performance tester, 654f
Stern layer, 115
Sticking coefficient
  effect of set time on, 377
  measurement of, 377
Stiff foam, 680–681
Stimulation options, 510
Stirred fluid loss tester, 276
Stokes' law, 217
Stone drill, 643–644
Stormer viscometer, 650, 651f
Strain
  definition of, 315
  ultimate, 316–317
Streaming potential, 115
Strength, ultimate, 316–317
Stress cloud(s), 325, 330f
Stress concentrations
  failure caused by, 327f
  in a region of tensile faulting, 328f
Stress cracking, and corrosion, 440–443
Stress(es). *See also* Effective stress; Shear
    stress; Yield stress
  around borehole, 319–325
  behavior of rocks under, 315–317
  at Earth's crust, 317, 317f
  on subsurface rock, 317–319
Structural viscosity, 157
Stuck drill pipe. *See also* Differential
    sticking; Differential-pressure sticking
    of drill pipe
  force required to free, 79
  freeing, 381–382
  pull-out force for, factors affecting,
    377–379
  sticking point of, locating, 381–382

Styrene copolymers, 149
Subsurface stress field, 317–319
Sulfate-reducing bacteria, 665
Sulfide stress cracking, 441–442
Sulfide(s), in mud, estimation of, 86
Sulfonated styrene maleic anhydride (SSMA)
  copolymer, sodium salt of, 435–436
Surface chemistry, 285
Surface free energy, 290
Surface hydration, 110, 338
Surface potential, 94
Surface tension, 285–287
  definition of, 285
  measurement of, 285–287
  of various substances, 288t
Surfactants, 52, 77, 291–292, 560–561,
    560t
  anionic, 292
  and aphrons, 564
  cationic, 291–292
  definition of, 291
  and environmental protection, 580
  in foam formation, 300–301
  functions of, 291–292
  nonanionic, 292
Surge pressures, 28
Suspended solids content (SS), 37
Swab pressure(s), 28
  gel strength and, 233, 235f
Swabbing action, 26, 28
Swabbing effect, 653–654
Synthetic base fluids (SBFs), 604
Synthetic HPHT polymers, 529
Synthetic monomers, 147–148, 148f
Synthetic polymers, 137
Synthetic-base muds, 625
Synthetic-based muds, 11
  bioaccumulation of, characterization of,
    627–629
Synthetics, 138, 147–149

T

Talc, structure of, 98–100, 101t
Tannins, 666
Temperature
  bottomhole, versus formation temperature,
    424
  critical, 26
  downhole mud, determination of, 427–429
  and drill pipe corrosion, 451, 451f
  effect on density, 45–46

Temperature (*Continued*)
 and filtrate viscosity, 249, 250*t*
 and filtrate volume, 249–251
 high, 423–437
  aging at, 87–89
  drilling fluid selection for, 26–27
  mud degradation at, 87
  and organic polymer decomposition, 132
  and rheology of drilling fluids, 193–203
Terminal settling velocity, 216
Terralog's SFI process, 601*f*
Thermodynamic crystallization temperature (TCT), 48–49
 supercooling of formate brines, 49
Thickeners, 543
 pH responsive, 543
Thinners, 9, 20, 119
 for KCl muds, 351
Thinning agent for muds, 648, 650, 657
Thixotropy, 19
 definition of, 166
 effect on drilling muds, 166–174
 versus plasticity, 166–167
Timken lubrication tester, 76–77, 77*f*
Tragacanth gum, on thickness of filter cake, 656, 656*f*
Transport ratio, 218
Triaxial test cell, 315, 315*f*
Triglyceride/alcohol mixture, as lubricant, 370
Trioctahedral sheets, 96
True crystallization temperature (TCT), 43
Turbidity, 50–51
Turbulence, onset of, 191
Turbulent flow, 15, 19, 184–191
 characteristics of, 151
 of Newtonian fluids, 184–185
 of non-Newtonian fluids, 185–190
Two-speed viscometer, 61

U

Ultimate strain, 316–317
Ultimate strength, 316–317
Ultralow fluid loss muds, 422
Unconsolidated clay, Mohr envelope for, 320*f*
Unconsolidated sand, Mohr envelope and, 319*f*
Union Oil Company, 652
Unit layer, 96, 98

V

van der Waals forces, 98
Velocity gradient, 152

Velocity profile, 152, 154
 for laminar flow of Newtonian fluid in round pipe, 154
 n (flow behavior index) and, 178–179, 180*f*
Vermiculite
 base exchange capacity (BEC) of, 109*t*
 structure of, 101*t*
 swelling, mechanisms of, 111*f*
VES (viscoelastic surfactant), 530
Vinyl acetate-maleic acid, as bentonite extender, 130
Vinylamide/vinylsulfonate, 436–437
Viscoelastic surfactants (VES), 523
Viscometer(s), 15–16. *See also* Capillary viscometer
 direct-indicating, 61, 62*f*
 for measurement of viscosity, 61
 multispeed, 61–62
Viscosity (μ), 15–16, 152
 apparent. *See* Apparent viscosity (AV)
 effective. *See* Effective viscosity
 kinematic, 395
 plastic. *See* Plastic viscosity (PV)
 shear history and, 169–170
 structural, 157
 temperature and shear rate and, 197*f*, 198*f*
Viscosity modifiers, 543–545
Volchonskoite, structure of, 101*t*
von Karman equation, 185–186

W

Wall building, 645–646. *See also* Hole stability
Waste disposal. *See also* Offshore discharge; Onshore disposal
 costs, 598–599
 options for, 600
 problems with, minimizing, 599
Waste management, website, 632
Waste reduction and recycling, 630–631
Water, 527
 analysis, 85
 clarification, mechanism of, 130
 dispersibility, 129
 temperature stability of, 429
Water injection wells, 489
Water-base drilling fluids (muds), 5
 additives for, 538*f*
 coefficient of friction of, 379
 dispersants, 573

for high angle holes, 29
low-molecular-weight nonpolluting, 577
for maintaining hole stability, 347
as packer fluids, 503–504
resistivity of, 75
rheology, temperature and, 200
special additives for, 572–573
thermal stability, improving, additives for, 572
unweighted, coefficient of friction for, 372
weighted, coefficient of friction for, 372
for wellbore stability, characteristics of, 361–362
Water-base muds (WBMs), 600
Water-based drilling fluids, 18, 643–668
hydration of borehole with, 342–343
corrosion control, 665–668
different types of fluid for different drilling functions, 668
hole stabilization, 645–646
mud density, pressure control by, 646–648
mud industry, birth of, 648–649
mud industry, rapid growth of, 649–655
mud types/systems, development of, 655–665
high-pH muds, 657
inhibited muds, 663–665
low-solids muds, 660–663
muds for deep holes, 658–660
muds for heaving shale, 656–657
nondispersed polymer muds, 663
oil emulsion muds, 658
saltwater muds, 655–656
removal of cuttings, 643–644
Water-based fluids
properties for wellbore stability by, 361–362
Water-based muds, 5, 7–10, 236
phosphate thinned, 545
Water-bearing zones, problems with, 678–679
Waterblock, 30, 464–467
Water-dispersible polymers, 135
cellulosics, 147
fermentation biopolymers, 142–146
polymer classification, 136–137
polysaccharides, 136–137
polymers, types of, 137–139, 139t
guar gum, 139–141, 140f
starch, 141–142
synthetics, 147–149
Water-in-oil emulsifying agent, 675

Water-in-oil emulsions. *See also* Invert emulsion(s)
electrical stability of, 74–75
Water-sensitive formations, 470
Water-soluble solids, fluids with, 499
Weathering, 107
Weight materials, 4
Weighting agents, 538–542
densities of, 538–539, 540t
and drilling rate, 396
Welan, 143–144, 146f
Well completion, oil for, 668–669
Well construction, 507
Wellbore ballooning, 421
Wellbore cleanout, 38
Wellbore stability, 309. *See also* Hole stability
hole angle and, 325
mechanics of, 309–338
mud density and, 325
nonaqueous fluids and, 345–347
properties for, by water-based fluids, 361–362
Wellbore strengthening, 421–422
Wells, 643
Wettability, 287–290
capillarity and, 290
of nonaqueous drilling muds, materials for, 579
preferential, 289–290, 289f
Wetting, definition of, 287
Wetting agents, 132
While lime muds, 659
Wildcat well, drilling fluid selection for, 29
Wilhelmy Plate Tensiometer, 286
Workover fluids
nonaqueous, 499–500
nondamaging, 490
selection of, 493–500
Workover wells, 487–488
World Petroleum Congress, 652
Wyoming bentonite, 650, 664

X

Xanthan biopolymer, 128
Xanthan gum, 128, 142–143, 143f, 663
in completion and workover fluids, 497–498
with KCl muds, 347–348
Xanthan NPX monomer, 144f
XC polymer, 663–665

Y

Yield point (YP), 17, 165, 236. *See also*
  Bingham yield point
  calculation of, 61
  measurement of, 162
  temperature and, 195
Yield Power Law model, 236
Yield stress, 315–316

Z

Z stability parameter, 191
  determination of, 191, 192f
Zero-discharge drilling, 149
Zeta potential, 115, 116f
Zinc, as hydrogen sulfide scavenger, 449
Zinc bromide, 38, 43–44
Zinc chelate, as hydrogen sulfide scavenger, 449

CPSIA information can be obtained
at www.ICGtesting.com
Printed in the USA
BVHW040822060619
550280BV00004B/7/P